Group Theory With Applications in Chemical Physics

Group theory is widely used in many branches of physics and chemistry, and today it may be considered as an essential component in the training of both chemists and physicists. This book provides a thorough, self-contained introduction to the fundamentals of group theory and its applications in chemistry and molecular and solid state physics. The first half of the book, with the exception of a few marked sections, focuses on elementary topics. The second half (Chapters 11–18) deals with more advanced topics which often do not receive much attention in introductory texts. These include the rotation group, projective representations, space groups, and magnetic crystals. The book includes numerous examples, exercises, and problems, and it will appeal to advanced undergraduates and graduate students in the physical sciences. It is well suited to form the basis of a two-semester course in group theory or for private study.

PROFESSOR P. W. M. JACOBS is Emeritus Professor of Physical Chemistry at the University of Western Ontario, where he taught widely in the area of physical chemistry, particularly group theory. He has lectured extensively on his research in North America, Europe, and the former USSR. He has authored more than 315 publications, mainly in solid state chemistry and physics, and he was awarded the Solid State Medal of the Royal Society of Chemistry.

Group Theory With Applications in Chemical Physics

P. W. M. JACOBS
The University of Western Ontario

CAMBRIDGE
UNIVERSITY PRESS

CAMBRIDGE UNIVERSITY PRESS
Cambridge, New York, Melbourne, Madrid, Cape Town, Singapore, São Paulo

CAMBRIDGE UNIVERSITY PRESS
The Edinburgh Building, Cambridge, CB2 2RU, UK
Published in the United States of America by Cambridge University Press, New York

www.cambridge.org
Information on this title: www.cambridge.org/9780521642507

First published 2005

Printed in the United Kingdom at the University Press, Cambridge

Typeface Times 10/13pt. and Frutiger *System* Advent 3B2 8.07f [PND]

A catalog record for this publication is available from the British Library

ISBN-13 978-0-521-64250-7 hardback
ISBN-10 0-521-64250-7 hardback

To MFM
and to all those who love group theory

Contents

Preface

Symmetry pervades many forms of art and science, and group theory provides a systematic way of thinking about symmetry. The mathematical concept of a group was invented in 1823 by Évariste Galois. Its applications in physical science developed rapidly during the twentieth century, and today it is considered as an indispensable aid in many branches of physics and chemistry. This book provides a thorough introduction to the subject and could form the basis of two successive one-semester courses at the advanced undergraduate and graduate levels. Some features not usually found in an introductory text are detailed discussions of induced representations, the Dirac characters, the rotation group, projective representations, space groups, magnetic crystals, and spinor bases. New concepts or applications are illustrated by worked examples and there are a number of exercises. Answers to exercises are given at the end of each section. Problems appear at the end of each chapter, but solutions to problems are not included, as that would preclude their use as problem assignments. No previous knowledge of group theory is necessary, but it is assumed that readers will have an elementary knowledge of calculus and linear algebra and will have had a first course in quantum mechanics. An advanced knowledge of chemistry is not assumed; diagrams are given of all molecules that might be unfamiliar to a physicist.

The book falls naturally into two parts. Chapters 1–10 (with the exception of a few marked sections) are elementary and could form the basis of a one-semester advanced undergraduate course. This material has been used as the basis of such a course at the University of Western Ontario for many years and, though offered as a chemistry course, it was taken also by some physicists and applied mathematicians. Chapters 11–18 are at a necessarily higher level; this material is suited to a one-semester graduate course.

Throughout, explanations of new concepts and developments are detailed and, for the most part, complete. In a few instances complete proofs have been omitted and detailed references to other sources substituted. It has not been my intention to give a complete bibliography, but essential references to core work in group theory have been given. Other references supply the sources of experimental data and references where further development of a particular topic may be followed up.

I am considerably indebted to Professor Boris Zapol who not only drew all the diagrams but also read the entire manuscript and made many useful comments. I thank him also for his translation of the line from Alexander Pushkin quoted below. I am also indebted to my colleague Professor Alan Allnatt for his comments on Chapters 15 and 16 and for several fruitful discussions. I am indebted to Dr. Peter Neumann and Dr. Gabrielle Stoy of Oxford

University for their comments on the proof (in Chapter 12) that multiplication of quaternions is associative. I also thank Richard Jacobs and Professor Amy Mullin for advice on computing.

Grateful acknowledgement is made to the following for permission to make use of previously published material:

The Chemical Society of Japan, for Figure 10.3;
Taylor and Francis Ltd (http://www.tandf.co.uk/journals) for Table 10.2;
Cambridge University Press for Figure 12.5;
The American Physical Society and Dr. C. J. Bradley for Tablc 14.6.

"Сπуженье муз не терпит суеты … " — А.С. Пушкин, "19 октября"

which might be translated as:
"Who serves the muses should keep away from fuss," or, more prosaically,
"Life interferes with Art."

I am greatly indebted to my wife Mary Mullin for shielding me effectively from the daily intrusions of "Life" and thus enabling me to concentrate on this particular work of "Art."

Notation and conventions

General mathematical notation

$\equiv$	identically equal to
$\Rightarrow$	leads logically to; thus $p \Rightarrow q$ means if p is true, then q follows
$\sum$	sum of (no special notation is used when $\sum$ is applied to sets, since it will always be clear from the context when $\sum$ means a sum of sets)
$\forall$	all
iff	if and only if
$\exists$	there exists
$\overline{a}$	the negative of a (but note $\overline{\psi} = \Theta\psi$ in Chapter 13 and $\overline{R} = \overline{E}R$, an operator in the double group $\overline{\mathrm{G}}$, in Chapter 8)
C^n	n-dimensional space in which the components of vectors are complex numbers
c, s	$\cos\phi$, $\sin\phi$
c_2, s_2	$\cos 2\phi$, $\sin 2\phi$
$\mathrm{c}\,x$	$x\cos\phi$
c_n^m	$\cos(m\pi/n)$
i	imaginary unit, defined by $\mathrm{i}^2 = \sqrt{-1}$
$q_1\ q_2\ q_3$	quaternion units
$\Re^n$	n-dimensional space, in which the components of vectors are real numbers
$\Re^3$	configuration space, that is the three-dimensional space of real vectors in which symmetry operations are represented
$s\,x$	$x\sin\phi$
$T(n)$	tensor of rank n in Section 15.1

Sets and groups

$\{g_i\}$	the set of objects g_i, $i = 1, \ldots, g$, which are generally referred to as 'elements'
$\in$	belongs to, as in $g_i \in \mathrm{G}$
$\notin$	does not belong to
$\mathrm{A} \rightarrow \mathrm{B}$	map of set A onto set B
$a \rightarrow b$	map of element a (the pre-image of b) onto element b (the image of a)

A ∩ B	intersection of A and B, that is the set of all the elements that belong to both A and B
A ∪ B	the union of A and B, that is the set of all the elements that belong to A, or to B, or to both A and B
G	a group G = $\{g_i\}$, the elements g_i of which have specific properties (Section 1.1)
E, or g_1	the identity element in G
g	the order of G, that is the number of elements in G
H, A, B	groups of order h, a, and b, respectively, often subgroups of G
H ⊂ G	H is a subset of G; if $\{h_i\}$ have the group properties, H is a subgroup of G of order h
A ∼ B	the groups A and B are isomorphous
C	a cyclic group of order c
$\mathscr{C}_k$	the class of g_k in G (Section 1.2) of order c_k
C_{ij}^k	class constants in the expansion $\Omega_i\Omega_j = \sum_{k=1}^{N_c} C_{ij}^k \Omega_{\overline{k}}$ (Section A2.2)
$g_i(\mathscr{C}_k)$	ith element of the kth class
$\underline{\mathrm{G}}$	a group consisting of a unitary subgroup H and the coset AH, where A is an antiunitary operator (Section 13.2), such that $\underline{\mathrm{G}} = \{\mathrm{H}\} \oplus A\{\mathrm{H}\}$
K	the kernel of G, of order k (Section 1.6)
l_i	dimension of ith irreducible representation
l_{s}	dimension of an irreducible spinor representation
l_{v}	dimension of an irreducible vector representation
$\mathrm{N_c}$	number of classes in G
$\mathrm{N_{rc}}$	number of regular classes
$\mathrm{N_r}$	number of irreducible representations
$\mathrm{N_s}$	number of irreducible spinor representations
$\mathrm{N_v}$	number of irreducible vector representations
N(H|G)	the normalizer of H in G, of order n (Section 1.7)
t	index of a coset expansion of G on H, $\mathrm{G} = \sum_{r=1}^{t} g_r\mathrm{H}$, with $g_r \notin \mathrm{H}$ except for $g_1 = E$; $\{g_r\}$ is the set of coset representatives in the coset expansion of G, and $\{g_r\}$ is not used for G itself.
Z(h_j|G)	the centralizer of h_j in G, of order z (Section 1.7)
Z_i	an abbreviation for Z(g_i|G)
Ω_k, $\Omega(\mathscr{C}_k)$	Dirac character of $\mathscr{C}_k$, equal to $\sum_{i=1}^{c_k} g_i(\mathscr{C}_k)$
A ⊗ B	(outer) direct product of A and B, often abbreviated to DP
A ⊠ B	inner direct product of A and B
A ∧ B	semidirect product of A and B
A $\overline{\otimes}$ B	symmetric direct product of A and B (Section 5.3)
A $\underline{\otimes}$ B	antisymmetric direct product of A and B (Section 5.3)

Vectors and matrices

$\mathbf{r}$	a polar vector (often just a vector) which changes sign under inversion; $\mathbf{r}$ may be represented by the directed line segment OP, where O is the origin of the coordinate system
$x\ y\ z$	coordinates of the point P and therefore the components of a vector $\mathbf{r} = \mathrm{OP}$; independent variables in the function $f(x, y, z)$.
$\mathbf{x}\ \mathbf{y}\ \mathbf{z}$	space-fixed right-handed orthonormal axes, collinear with OX, OY, OZ
$\mathbf{e}_1\ \mathbf{e}_2\ \mathbf{e}_3$	unit vectors, initially coincident with $\mathbf{x}\ \mathbf{y}\ \mathbf{z}$, but firmly embedded in configuration space (see $R(\phi\ \mathbf{n})$ below). Note that $\{\mathbf{e}_1\ \mathbf{e}_2\ \mathbf{e}_3\}$ behave like polar vectors under rotation but are invariant under inversion and therefore they are *pseudovectors*. Since, in configuration space the vector $\mathbf{r} = \mathbf{e}_1 x + \mathbf{e}_2 y + \mathbf{e}_3 z$ changes sign on inversion, the components of $\mathbf{r}$, $\{x\ y\ z\}$, must change sign on inversion and are therefore *pseudoscalars*
$\{\mathbf{e}_i\}$	unit vectors in a space of n dimensions, $i = 1, \ldots, n$
$\{v_i\}$	components of the vector $\mathbf{v} = \sum_i \mathbf{e}_i v_i$
$\mathbb{A}$	the matrix $\mathbb{A} = [a_{rs}]$, with m rows and n columns so that $r = 1, \ldots, m$ and $s = 1, \ldots, n$. See Table A1.1 for definitions of some special matrices
$\mathbb{A}_{rs}$, a_{rs}	element of matrix $\mathbb{A}$ common to the rth row and sth column
$\mathbb{E}_n$	unit matrix of dimensions $n \times n$, in which all the elements are zero except those on the principal diagonal, which are all unity; often abbreviated to $\mathbb{E}$ when the dimensions of $\mathbb{E}$ may be understood from the context
det $\mathbb{A}$ or $\|a_{rs}\|$	determinant of the square matrix $\mathbb{A}$
$\mathbb{A} \otimes \mathbb{B}$	direct product of the matrices $\mathbb{A}$ and $\mathbb{B}$
$\mathbb{C}_{pr,qs}$	element $a_{pq}b_{rs}$ in $\mathbb{C} = \mathbb{A} \otimes \mathbb{B}$
$\mathbb{A}_{[ij]}$	ijth element (which is itself a matrix) of the supermatrix $\mathbb{A}$
$\langle a_1\ a_2 \ldots a_n\|$	a matrix of one row containing the set of elements $\{a_i\}$
$\langle a\|$	an abbreviation for $\langle a_1\ a_2 \ldots a_n\|$. The set of elements $\{a_i\}$ may be basis vectors, for example $\langle \mathbf{e}_1\ \mathbf{e}_2\ \mathbf{e}_3\|$, or basis functions $\langle \phi_1\ \phi_2 \ldots \phi_n\|$.
$\|b_1\ b_2 \ldots b_n\rangle$	a matrix of one column containing the set of elements $\{b_i\}$, often abbreviated to $\|b\rangle$; $\langle b\|$ is the transpose of $\|b\rangle$
$\langle a'\|$	the transform of $\langle a\|$ under some stated operation
$\langle \mathbf{e}\|r\rangle$	an abbreviation for the matrix representative of a vector $\mathbf{r}$; often given fully as $\langle \mathbf{e}_1\ \mathbf{e}_2\ \mathbf{e}_3 \mid x\ y\ z\rangle$

Brackets

$\langle\ \|,\ \|\ \rangle$	Dirac bra and ket, respectively; no special notation is used to distinguish the bra and ket from row and column matrices, since which objects are intended will always be clear from the context

$[A,\ B]$	commutator of A and B equal to $AB - BA$
$[a,\ A]$	complex number $a + \mathrm{i}A$
$[a\ ;\ \mathbf{A}]$	quaternion (Chapter 11)
$[g_i\ ;\ g_j]$	projective factor, or multiplier (Chapter 12); often abbreviated to $[i\ ;\ j]$
$[n_1\ n_2\ n_3]$	components of the unit vector $\mathbf{n}$, usually given without the normalization factor; for example, [1 1 1] are the components of the unit vector that makes equal angles with OX, OY, OZ, the normalization factor $3^{-1/2}$ being understood. Normalization factors will, however, be given explicitly when they enter into a calculation, as, for example, in calculations using quaternions

Angular momenta

$\mathbf{L}$, $\mathbf{S}$, $\mathbf{J}$	orbital, spin, and total angular momenta
$\hat{\mathbf{L}}$, $\hat{\mathbf{S}}$, $\hat{\mathbf{J}}$	quantum mechanical operators corresponding to $\mathbf{L}$, $\mathbf{S}$, and $\mathbf{J}$
L, S, J	quantum numbers that quantize $\mathbf{L}^2$, $\mathbf{S}^2$, and $\mathbf{J}^2$
$\hat{\mathbf{j}}$	operator that obeys the angular momentum commutation relations
$\mathbf{j} = \mathbf{j}_1 + \mathbf{j}_2$	total ($\mathbf{j}$) and individual ($\mathbf{j}_1$, $\mathbf{j}_2$, ...) angular momenta, when angular momenta are coupled

Symmetry operators and their matrix representatives

A	antiunitary operator (Section 13.1); A, B may also denote linear, Hermitian operators according to context
E	identity operator
$\overline{E}$	operator $R(2\pi\ \mathbf{n})$ introduced in the formation of the double group $\overline{G} = \{R\ \overline{R}\}$ from $\mathrm{G} = \{R\}$, where $\overline{R} = \overline{E}R$ (Section 8.1)
I	inversion operator
$I_1\ I_2\ I_3$	operators that generate infinitesimal rotations about $\mathbf{x}\ \mathbf{y}\ \mathbf{z}$, respectively (Chapter 11)
$\hat{I}_1\ \hat{I}_2\ \hat{I}_3$	function operators that correspond to $I_1\ \ I_2\ \ I_3$
$\mathtt{I}_3$	matrix representative of I_3, and similarly (note that the usual symbol $\Gamma(R)$ for the matrix representative of symmetry operator R is not used in this context, for brevity)
$\mathbf{I}$	generator of infinitesimal rotations about $\mathbf{n}$, with components I_1, I_2, I_3
$\mathtt{I}_{\mathbf{n}}$	matrix representative of $I_n = \mathbf{n} \cdot \mathbf{I}$
$\mathtt{J}_x\ \mathtt{J}_y\ \mathtt{J}_z$	matrix representatives of the angular momentum operators $\hat{J}_x$, $\hat{J}_y$, $\hat{J}_z$ for the basis $\langle m | = \langle 1/2,\ \ -1/2 |$. Without the numerical factors of ½, these are the Pauli matrices $\boldsymbol{\sigma}_1\ \boldsymbol{\sigma}_2\ \boldsymbol{\sigma}_3$

$R(\phi\ \ \mathbf{n})$	rotation through an angle ϕ about an axis which is the unit vector $\mathbf{n}$; here $\phi\ \ \mathbf{n}$ is not a product but a single symbol $\phi\mathbf{n}$ that fixes the three independent parameters necessary to describe a rotation (the three components of $\mathbf{n}$, $[n_1\ n_2\ n_3]$, being connected by the normalization condition); however, a space is inserted between ϕ and $\mathbf{n}$ in rotation operators for greater clarity, as in $R(2\pi/3\ \ \mathbf{n})$. The range of ϕ is $-\pi < \phi \leq \pi$. R acts on configuration space and on all vectors therein (including $\{\mathbf{e}_1\ \mathbf{e}_2\ \mathbf{e}_3\}$) (but not on $\{\mathbf{x}\ \mathbf{y}\ \mathbf{z}\}$, which define the space-fixed axes in the active representation)
$\hat{R}(\phi\ \ \mathbf{n})$	function operator that corresponds to the symmetry operator $R(\phi\ \ \mathbf{n})$, defined so that $\hat{R}f(\mathbf{r}) = f(R^{-1}\mathbf{r})$ (Section 3.5)
R, S, T	general symbols for point symmetry operators (point symmetry operators leave at least one point invariant)
$\hat{s}_x\ \hat{s}_y\ \hat{s}_z$	spin operators whose matrix representatives are the Pauli matrices and therefore equal to $\hat{J}_x,\ \hat{J}_y,\ \hat{J}_z$ without the common factor of 1/2
T	translation operator (the distinction between T a translation operator and T when used as a point symmetry operator will always be clear from the context)
U	a unitary operator
$\mathscr{U}$	time-evolution operator (Section 13.1)
$\Gamma(R)$	matrix representative of the symmetry operator R; sometimes just R, for brevity
$\Gamma(R)_{pq}$	pqth element of the matrix representative of the symmetry operator R
Γ	matrix representation
$\Gamma_1 \approx \Gamma_2$	the matrix representations Γ_1 and Γ_2 are equivalent, that is related by a similarity transformation (Section 4.2)
$\Gamma = \sum\limits_i c^i\Gamma_i$	the representation Γ is a direct sum of irreducible representations Γ_i, and each Γ_i occurs c^i times in the direct sum Γ; when specific representations (for example T_{1u}) are involved, this would be written $c(\mathrm{T}_{1u})$
$\Gamma \supset \Gamma_i$	the reducible representation Γ includes Γ_i
$\Gamma_i = \sum\limits_j c_{i,j}\ \Gamma_j$	the representation Γ_i is a direct sum of irreducible representations Γ_j and each Γ_j occurs $c_{i,j}$ times in the direct sum Γ_i
$\Gamma_{ij} = \sum\limits_k c_{ij,k}\ \Gamma_k$	Clebsch–Gordan decomposition of the direct product $\Gamma_{ij} = \Gamma_i \boxtimes \Gamma_j$; $c_{ij,\ k}$ are the Clebsch–Gordan coefficients
$\sigma_\mathbf{n}$	reflection in the plane normal to $\mathbf{n}$
$\sigma_1\ \sigma_2\ \sigma_3$	the Pauli matrices (Section 11.6)
Θ	time-reversal operator

Bases

$\langle \mathbf{e}_1\ \mathbf{e}_2\ \mathbf{e}_3|$	basis consisting of the three unit vectors $\{\mathbf{e}_1\ \mathbf{e}_2\ \mathbf{e}_3\}$ initially coincident with $\{\mathbf{x}\ \mathbf{y}\ \mathbf{z}\}$ but embedded in a unit sphere in configuration space so that $R\langle \mathbf{e}_1\ \mathbf{e}_2\ \mathbf{e}_3| = \langle \mathbf{e}_1{}'\ \mathbf{e}_2{}'\ \mathbf{e}_3{}'| = \langle \mathbf{e}_1\ \mathbf{e}_2\ \mathbf{e}_3|\ \Gamma(R)$. The 3×3 matrix $\Gamma(R)$ is the matrix representative of the symmetry operator R. Note that $\langle \mathbf{e}_1\ \mathbf{e}_2\ \mathbf{e}_3|$ is often abbreviated to $\langle \mathbf{e}|$. If $\mathbf{r} \in \Re^3$, $R\ \mathbf{r} = R\langle \mathbf{e}|r\rangle = \langle \mathbf{e}'|r\rangle = \langle \mathbf{e}|\Gamma(R)|r\rangle = \langle \mathbf{e}|r'\rangle$, which shows that $\langle \mathbf{e}|$ and $|r\rangle$ are *dual bases*, that is they are transformed by the same matrix $\Gamma(R)$
$\langle R_x\ R_y\ R_z|$	basis comprising the three *infinitesimal* rotations R_x, R_y, R_z about OX, OY, OZ respectively (Section 4.6)
$\langle u^j_{-j} \ldots u^j_j|$	basis consisting of the $2j+1$ functions, u^j_m, $-j \leq m \leq j$, which are eigenfunctions of the z component of the angular momentum operator $\hat{J}_z$, and of $\hat{J}^2$, with the Condon and Shortley choice of phase. The angular momentum quantum numbers j and m may be either an integer or a half-integer. For integral j the u^j_m are the spherical harmonics $Y_l^m(\theta\ \varphi)$; $y_l^m(\theta\ \varphi)$ are the spherical harmonics written without normalization factors, for brevity
$\langle u^j_m|$	an abbreviation for $\langle u^j_{-j} \ldots u^j_j|$, also abbreviated to $\langle m|$
$\langle u\ v|$	spinor basis, an abbreviation for $\langle u^{1/2}_{1/2}| = \langle |\tfrac{1}{2}\ \tfrac{1}{2}\rangle\ |\tfrac{1}{2}\ -\tfrac{1}{2}\rangle|$, or $\langle \tfrac{1}{2}\ -\tfrac{1}{2}|$ in the $\langle m|$ notation
$\langle u'\ v'|$	transform of $\langle u\ v|$ in C^2, equal to $\langle u\ v|\mathbb{A}$
$|\ u\ v\rangle$	dual of $\langle u\ v|$, such that $|u'\ v'\rangle = \mathbb{A}|\ u\ v\rangle$
$|\ U_{-1}\ U_0\ U_1\rangle$	matrix representation of the spherical vector $\mathbf{U} \in \mathrm{C}^3$ which is the dual of the basis $\langle y_1^{-1}\ y_1^0\ y_1^1|$
N	normalization factor

Crystals

$\mathbf{a}_n = \langle \mathbf{a}|\ n\rangle$	lattice translation vector; $\mathbf{a}_n = \langle \mathbf{a}_1\ \mathbf{a}_2\ \mathbf{a}_3|\ n_1\ n_2\ n_3\rangle$ (Section 16.1) (n is often used as an abbreviation for the $\mathbf{a}_n$)
$\mathbf{b}_m = \langle \mathbf{b}|\ m\rangle$	reciprocal lattice vector; $\mathbf{b}_m = \langle \mathbf{b}_1\ \mathbf{b}_2\ \mathbf{b}_3|\ m_1\ m_2\ m_3\rangle = \langle \mathbf{e}_1\ \mathbf{e}_2\ \mathbf{e}_3|\ m_x\ m_y\ m_z\rangle$ (Section 16.3); m is often used as an abbreviation for the components of $\mathbf{b}_m$

Abbreviations

1-D	one-dimensional (etc.)
AO	atomic orbital
BB	bilateral binary
bcc	body-centered cubic
CC	complex conjugate
CF	crystal field

CG	Clebsch–Gordan
CR	commutation relation
CS	Condon and Shortley
CT	charge transfer
DP	direct product
fcc	face-centered cubic
FE	free electron
FT	fundamental theorem
hcp	hexagonal close-packed
HSP	Hermitian scalar product
IR	irreducible representation
ITC	*International Tables for Crystallography* (Hahn (1983))
L, R	left and right, respectively, as in L and R cosets
LA	longitudinal acoustic
LCAO	linear combination of atomic orbitals
LI	linearly independent
LO	longitudinal optic
LS	left- side (of an equation)
LVS	linear vector space
MO	molecular orbital
MR	matrix representative
N	north, as in N pole
ORR	Onsager reciprocal relation
OT	orthogonality theorem
PBC	periodic boundary conditions
PF	projective factor
PR	projective representation
RS	right side (of an equation)
RS	Russell–Saunders, as in *RS* coupling or *RS* states
sc	simple cubic
SP	scalar product
TA	transverse acoustic
TO	transverse optic
ZOA	zero overlap approximation

Cross-references

The author (date) system is used to identify a book or article in the list of references, which precedes the index.

Equations in a different section to that in which they appear are referred to by eq. $(n_1 \cdot n_2 \cdot n_3)$, where n_1 is the chapter number, n_2 is the section number, and n_3 is the equation number within that section. Equations occurring within the same section are referred to simply by (n_3). Equations are numbered on the right, as usual, and, when appropriate, a number (or numbers) on the left, in parentheses, indicates that these equations are used in the derivation of that equation so numbered. This convention means that such phrases as "it follows from" or "substituting eq. (n_4) in eq. (n_5)" can largely be dispensed with.

Examples and Exercises are referenced, for example, as Exercise $n_1 \cdot n_2$-n_3, even within the same section. Figures and Tables are numbered $n_1 \cdot n_3$ throughout each chapter. When a Table or Figure is referenced on the left side of an equation, their titles are abbreviated to T or F respectively, as in F16.1, for example.

Problems appear at the end of each chapter, and a particular problem may be referred to as Problem $n_1 \cdot n_3$, where n_1 is the number of the chapter in which Problem n_3 is to be found.

1 The elementary properties of groups

1.1 Definitions

All crystals and most molecules possess symmetry, which can be exploited to simplify the discussion of their physical properties. Changes from one configuration to an indistinguishable configuration are brought about by sets of symmetry operators, which form particular mathematical structures called *groups*. We thus commence our study of group theory with some definitions and properties of groups of abstract elements. All such definitions and properties then automatically apply to all sets that possess the properties of a group, including symmetry groups.

Binary composition in a set of abstract elements $\{g_i\}$, whatever its nature, is always written as a multiplication and is usually referred to as "multiplication" whatever it actually may be. For example, if g_i and g_j are operators then the product $g_i\ g_j$ means "carry out the operation implied by g_j and then that implied by g_i." If g_i and g_j are both n-dimensional square matrices then $g_i\ g_j$ is the matrix product of the two matrices g_i and g_j evaluated using the usual row $\times$ column law of matrix multiplication. (The properties of matrices that are made use of in this book are reviewed in Appendix A1.) Binary composition is *unique* but is not necessarily commutative: $g_i\ g_j$ may or may not be equal to $g_j\ g_i$. In order for a set of abstract elements $\{g_i\}$ to be a G, the law of binary composition must be defined and the set must possess the following four properties.

(i) *Closure*. For all g_i, with $g_j \in \{g_j\}$,

$$g_i\ g_j = g_k \in \{g_i\}, \quad g_k \text{ a unique element of } \{g_i\}. \tag{1}$$

Because g_k is a unique element of $\{g_i\}$, if each element of $\{g_i\}$ is multiplied from the left, or from the right, by a particular element g_j of $\{g_i\}$ then the set $\{g_i\}$ is regenerated with the elements (in general) re-ordered. This result is called the *rearrangement theorem*

$$g_j\ \{g_i\} = \{g_i\} = \{g_i\}\ g_j. \tag{2}$$

Note that $\{g_i\}$ means a set of elements of which g_i is a typical member, but in no particular order. The easiest way of keeping a record of the binary products of the elements of a group is to set up a *multiplication table* in which the entry at the intersection of the g_ith row and g_jth column is the binary product $g_i\ g_j = g_k$, as in Table 1.1. It follows from the rearrangement theorem that each row and each column of the multiplication table contains each element of G once and once only.

Table 1.1. *Multiplication table for the group* $G = \{g_i\}$ *in which the product* $g_i\ g_j$ *happens to be* g_k.

G	g_i	g_j	g_k	...
g_i	g_i^2	g_k	$g_i\ g_k$	
g_j	$g_j\ g_i$	g_j^2	$g_j\ g_k$	
g_k	$g_k\ g_i$	$g_k\ g_j$	g_k^2	
$\vdots$				

(ii) Multiplication is *associative*. For all $g_i,\ g_j,\ g_k \in \{g_i\}$,

$$g_i(g_j\ g_k) = (g_i\ g_j)g_k. \tag{3}$$

(iii) The set $\{g_i\}$ contains the *identity* element E, with the property

$$E\ g_j = g_j\ E = g_j,\ \forall\ g_j \in \{g_i\}. \tag{4}$$

(iv) Each element g_i of $\{g_i\}$ has an *inverse* $g_i^{-1} \in \{g\}_i$ such that

$$g_i^{-1}\ g_i = g_i\ g_i^{-1} = E,\ g_i^{-1} \in \{g_i\},\ \forall\ g_i \in \{g_i\}. \tag{5}$$

The number of elements g in G is called the *order* of the group. Thus

$$G = \{g_i\}, \quad i = 1, 2, \ldots, g. \tag{6}$$

When this is necessary, the order of G will be displayed in parentheses G(g), as in G(4) to indicate a group of order 4.

Exercise 1.1-1 With binary composition defined to be addition: (a) Does the set of positive integers $\{p\}$ form a group? (b) Do the positive integers p, including zero (0) form a group? (c) Do the positive (p) and negative ($-p$) integers, including zero, form a group? [*Hint*: Consider the properties (i)–(iv) above that must be satisfied for $\{g_i\}$ to form a group.]

The multiplication of group elements is not necessarily commutative, but if

$$g_i\ g_j = g_j\ g_i,\ \forall\ g_i, g_j \in G \tag{7}$$

then the group G is said to be *Abelian*. Two groups that have the same multiplication table are said to be *isomorphous*. As we shall see, a number of other important properties of a group follow from its multiplication table. Consequently these properties are the same for isomorphous groups; generally it will be necessary to identify corresponding elements in the two groups that are isomorphous, in order to make use of the isomorphous property. A group G is finite if the number g of its elements is a finite number. Otherwise the group G is infinite, if the number of elements is denumerable, or it is continuous. The group of Exercise 1.1-1(c) is infinite. For finite groups, property (iv) is automatically fulfilled as a consequence of the other three.

If the sequence $g_i,\ g_i^2,\ g_i^3,\ \ldots$ starts to repeat itself at $g_i^{c+1} = g_i$, because $g_i^c = E$, then the set $\{g_i\ g_i^2\ g_i^3\ \ldots\ g_i^c = E\}$, which is the period of g_i, is a group called a *cyclic group*, C. The order of the cyclic group C is c.

Exercise 1.1-2 (a) Show that cyclic groups are Abelian. (b) Show that for a finite cyclic group the existence of the inverse of each element is guaranteed. (c) Show that $\omega = \exp(-2\pi \mathrm{i}/n)$ generates a cyclic group of order n, when binary composition is defined to be the multiplication of complex numbers.

If every element of G can be expressed as a finite product of powers of the elements in a particular subset of G, then the elements of this subset are called the *group generators*. The choice of generators is not unique: generally, a minimal set is employed and the defining relations like $g_i = (g_j)^p\ (g_k)^q$, etc., where $\{g_j\ g_k\}$ are group generators, are stated. For example, cyclic groups are generated from just one element g_i.

Example 1.1-1 A permutation group is a group in which the elements are permutation operators. A permutation operator P rearranges a set of indistinguishable objects. For example, if

$$P\{a\ b\ c\ \ldots\} = \{b\ a\ c\ \ldots\} \tag{8}$$

then P is a particular permutation operator which interchanges the objects a and b. Since $\{a\ b\ \ldots\}$ is a set of indistinguishable objects (for example, electrons), the final configuration $\{b\ a\ c\ \ldots\}$ is indistinguishable from the initial configuration $\{a\ b\ c\ \ldots\}$ and P is a particular kind of symmetry operator. The best way to evaluate products of permutation operators is to write down the original configuration, thinking of the n indistinguishable objects as allocated to n boxes, each of which contains a single object only. Then write down in successive rows the results of the successive permutations, bearing in mind that a permutation other than the identity involves the replacement of the contents of two or more boxes. Thus, if P applied to the initial configuration means "interchange the contents of boxes i and j" (which initially contain the objects i and j, respectively) then P applied to some subsequent configuration means "interchange the contents of boxes i and j, whatever they currently happen to be." A number of examples are given in Table 1.2, and these should suffice to show how the multiplication table in Table 1.3 is derived. The reader should check some of the entries in the multiplication table (see Exercise 1.1-3).

The elements of the set $\{P_0\ P_1\ \ldots P_5\}$ are the permutation operators, and binary composition of two members of the set, say $P_3\ P_5$, means "carry out the permutation specified by P_5 and then that specified by P_3." For example, P_1 states "replace the contents of box 1 by that of box 3, the contents of box 2 by that of box 1, and the contents of box 3 by that of box 2." So when applying P_1 to the configuration $\{3\ 1\ 2\}$, which resulted from P_1 (in order to find the result of applying $P_1^2 = P_1\ P_1$ to the initial configuration) the contents of box 1 (currently 3) are replaced by those of box 3 (which happens currently to be 2 – see the line labeled P_1); the contents of box 2 are replaced by those of box 1 (that is, 3); and finally the contents of box 3 (currently 2) are replaced by those of box 2 (that is, 1). The resulting configuration $\{2\ 3\ 1\}$ is the same as that derived from the original configuration $\{1\ 2\ 3\}$ by P_2, and so

Table 1.2. *Definition of the six permutation operators of the permutation group* S(3) *and some examples of the evaluation of products of permutation operators.*

In each example, the initial configuration appears on the first line and the permutation operator and the result of the operation are on successive lines. In the last example, the equivalent single operator is given on the right.

The identity $P_0 = E$

	1	2	3	original configuration (which therefore labels the "boxes")
P_0	1	2	3	final configuration (in this case identical with the initial configuration)

The two cyclic permutations

	1	2	3		1	2	3
P_1	3	1	2	P_2	2	3	1

The three binary interchanges

	1	2	3		1	2	3		1	2	3
P_3	1	3	2	P_4	3	2	1	P_5	2	1	3

Binary products with P_1

	1	2	3	
P_1	3	1	2	P_1
$P_1 P_1$	2	3	1	P_2
$P_2 P_1$	1	2	3	P_0
$P_3 P_1$	3	2	1	P_4
$P_4 P_1$	2	1	3	P_5
$P_5 P_1$	1	3	2	P_3

Table 1.3. *Multiplication table for the permutation group* S(3).

The box indicates the subgroup C(3).

S(3)	P_0	P_1	P_2	P_3	P_4	P_5
P_0	P_0	P_1	P_2	P_3	P_4	P_5
P_1	P_1	P_2	P_0	P_5	P_3	P_4
P_2	P_2	P_0	P_1	P_4	P_5	P_3
P_3	P_3	P_4	P_5	P_0	P_1	P_2
P_4	P_4	P_5	P_3	P_2	P_0	P_1
P_5	P_5	P_3	P_4	P_1	P_2	P_0

$$P_1\ P_1\{1\ 2\ 3\} = \{2\ 3\ 1\} = P_2\{1\ 2\ 3\} \tag{9}$$

so that $P_1\ P_1 = P_2$. Similarly, $P_2\ P_1 = P_0$, $P_3\ P_1 = P_4$, and so on. The equivalent single operators (products) are shown in the right-hand column in the example in the last part of Table 1.2. In this way, we build up the multiplication table of the group S(3), which is shown in Table 1.3. Notice that the rearrangement theorem (closure) is satisfied and that each element has an inverse. The set contains the identity P_0, and examples to demonstrate associativity are readily constructed (e.g. Exercise 1.1-4). Therefore this set of permutations is a group. The group of all permutations of N objects is called the symmetric group

S(N). Since the number of permutations of N objects is $N!$, the order of the symmetric group is $N!$, and so that of S(3) is $3! = 6$.

Exercise 1.1-3 Evaluate the products in the column headed P_3 in Table 1.3.

Exercise 1.1-4 (a) Using the multiplication table for S(3) in Table 1.3 show that $(P_3\ P_1)P_2 = P_3(P_1\ P_2)$. This is an example of the group property of associativity. (b) Find the inverse of P_2 and also the inverse of P_5.

Answers to Exercises 1.1

Exercise 1.1-1 (a) The set $\{p\}$ does not form a group because it does not contain the identity E. (b) The set $\{p\quad 0\}$ contains the identity 0, $p+0=p$, but the inverses $\{-p\}$ of the elements $\{p\}, p+(-p)=0$, are not members of the set $\{p\quad 0\}$. (c) The set of positive and negative integers, including zero, $\{p\ \bar{p}\ 0\}$, does form a group since it has the four group properties: it satisfies closure, and associativity, it contains the identity (0), and each element p has an inverse $\bar{p}$ or $-p$.

Exercise 1.1-2 (a) $g_i^p\, g_i^q = g_i^{p+q} = g_i^{q+p} = g_i^q\, g_i^p$. (b) *If* $p < c$, $g_i^p\, g_i^{c-p} = g_i^c = E$. Therefore, the inverse of g_i^p is g_i^{c-p}. (c) $\omega^n = \exp(-2\pi \mathrm{i}) = 1 = E$; therefore $\{\omega\ \omega^2 \ldots \omega^n = E\}$ is a cyclic group of order n.

Exercise 1.1-3

P_0	1	2	3	
P_3	1	3	2	P_3
$P_1\ P_3$	2	1	3	P_5
$P_2\ P_3$	3	2	1	P_4
$P_3\ P_3$	1	2	3	P_0
$P_4\ P_3$	2	3	1	P_2
$P_5\ P_3$	3	1	2	P_1

Exercise 1.1-4 (a) From the multiplication table, $(P_3\ P_1)\ P_2 = P_4\ P_2 = P_3$ and $P_3\ (P_1\ P_2) = P_3\ P_0 = P_3$. (b) Again from the multiplication table, $P_2\ P_1 = P_0 = E$ and so $P_2^{-1} = P_1$; $P_5\ P_5 = P_0$, $P_5^{-1} = P_5$.

1.2 Conjugate elements and classes

If $g_i, g_j, g_k \in$ G and

$$g_i\ g_j\ g_i^{-1} = g_k \tag{1}$$

then g_k is the *transform* of g_j, and g_j and g_k are *conjugate* elements. A complete set of the elements conjugate to g_i form a *class*, $\mathscr{C}_i$. The number of elements in a class is called the *order* of the class; the order of $\mathscr{C}_i$ will be denoted by c_i.

Exercise 1.2-1 Show that E is always in a class by itself.

Example 1.2-1 Determine the classes of S(3). Note that $P_0 = E$ is in a class by itself; the class of E is always named $\mathscr{C}_1$. Using the multiplication table for S(3), we find

$$\begin{aligned} P_0\,P_1\,P_0^{-1} &= P_1\,P_0 = P_1, \\ P_1\,P_1\,P_1^{-1} &= P_2\,P_2 = P_1, \\ P_2\,P_1\,P_2^{-1} &= P_0\,P_1 = P_1, \\ P_3\,P_1\,P_3^{-1} &= P_4\,P_3 = P_2, \\ P_4\,P_1\,P_4^{-1} &= P_5\,P_4 = P_2, \\ P_5\,P_1\,P_5^{-1} &= P_3\,P_5 = P_2. \end{aligned}$$

Hence $\{P_1\ P_2\}$ form a class $\mathscr{C}_2$. The determination of $\mathscr{C}_3$ is left as an exercise.

Exercise 1.2-2 Show that there is a third class of S(3), $\mathscr{C}_3 = \{P_3\ P_4\ P_5\}$.

Answers to Exercises 1.2

Exercise 1.2-1 For any group G with $g_i \in$ G,

$$g_i\,E\,g_i^{-1} = g_i\,g_i^{-1} = E.$$

Since E is transformed into itself by every element of G, E is in a class by itself.

Exercise 1.2-2 The transforms of P_3 are

$$\begin{aligned} P_0 P_3 P_0^{-1} &= P_3 P_0 = P_3, \\ P_1 P_3 P_1^{-1} &= P_5 P_2 = P_4, \\ P_2 P_3 P_2^{-1} &= P_4 P_1 = P_5, \\ P_3 P_3 P_3^{-1} &= P_0 P_3 = P_3, \\ P_4 P_3 P_4^{-1} &= P_2 P_4 = P_5, \\ P_5 P_3 P_5^{-1} &= P_1 P_5 = P_4. \end{aligned}$$

Therefore $\{P_3\ P_4\ P_5\}$ form a class, $\mathscr{C}_3$, of S(3).

1.3 Subgroups and cosets

A subset H of G, H $\subset$ G, that is itself a group with the same law of binary composition, is a *subgroup* of G. Any subset of G that satisfies closure will be a subgroup of G, since the other group properties are then automatically fulfilled. The region of the multiplication table of S(3) in Table 1.3 in a box shows that the subset $\{P_0\ P_1\ P_2\}$ is closed, so that this set is a subgroup of S(3). Moreover, since $P_1^2 = P_2$, $P_1^3 = P_1 P_2 = P_0 = E$, it is a cyclic subgroup of order 3, C(3).

Given a group G with subgroup H ⊂ G, then g_r H, where $g_r \in$ G but $g_r \notin$ H unless g_r is $g_1 = E$, is called a *left coset* of H. Similarly, H g_r is a *right coset* of H. The $\{g_r\}$, $g_r \in$ G but $g_r \notin$ H, except for $g_1 = E$, are called *coset representatives*. It follows from the uniqueness of the product of two group elements (eq. (1.1.2)) that the elements of g_r H are distinct from those of g_s H when $s \neq r$, and therefore that

$$\mathrm{G} = \sum_{r=1}^{t} g_r\,\mathrm{H}, \quad g_r \in \mathrm{G},\ g_r \notin \mathrm{H}\ (\text{except for } g_1 = E),\ t = g/h, \tag{1}$$

where t is the *index* of H in G. Similarly, G may be written as the sum of t distinct right cosets,

$$\mathrm{G} = \sum_{r=1}^{t} \mathrm{H}\,g_r, \quad g_r \in \mathrm{G},\ g_r \notin \mathrm{H}\ (\text{except for } g_1 = E),\ t = g/h. \tag{2}$$

If H $g_r = g_r$ H, so that right and left cosets are equal for all r, then

$$g_r\,\mathrm{H}\,g_r^{-1} = \mathrm{H}\,g_r\,g_r^{-1} = \mathrm{H} \tag{3}$$

and H is transformed into itself by any element $g_r \in$ G that is not in H. But for any $h_j \in$ H

$$h_j\,\mathrm{H}\,h_j^{-1} = h_j\,\mathrm{H} = \mathrm{H} \qquad (\text{closure}). \tag{4}$$

Therefore, H is transformed into itself by all the elements of G; H is then said to be an *invariant* (*or normal*) *subgroup* of G.

Exercise 1.3-1 Prove that any subgroup of index 2 is an invariant subgroup.

Example 1.3-1 Find all the subgroups of S(3); what are their indices? Show explicitly which, if any, of the subgroups of S(3) are invariant.

The subgroups of S(3) are

$$\{P_0\,P_1\,P_2\} = \mathrm{C}(3),\ \{P_0\,P_3\} = \mathrm{H}_1,\ \{P_0\,P_4\} = \mathrm{H}_2,\ \{P_0\,P_5\} = \mathrm{H}_3.$$

Inspection of the multiplication table (Table 1.3) shows that all these subsets of S(3) are closed. Since $g = 6$, their indices t are 2, 3, 3, and 3, respectively. C(3) is a subgroup of S(3) of index 2, and so we know it to be invariant. Explicitly, a right coset expansion for S(3) is

$$\{P_0\ P_1\ P_2\} + \{P_0\ P_1\ P_2\}P_4 = \{P_0\ P_1\ P_2\ P_3\ P_4\ P_5\} = \mathrm{S}(3). \tag{5}$$

The corresponding left coset expansion with the same coset representative is

$$\{P_0\ P_1\ P_2\} + P_4\{P_0\ P_1\ P_2\} = \{P_0\ P_1\ P_2\ P_4\ P_5\ P_3\} = \mathrm{S}(3). \tag{6}$$

Note that the elements of G do not have to appear in exactly the same order in the left and right coset expansions. This will only be so if the coset representatives commute with every element of H. All that is necessary is that the two lists of elements evaluated from the coset expansions both contain each element of G once only. It should be clear from eqs. (5) and (6) that H $g_r = g_r$ H, where H $= \{P_0\ P_1\ P_2\}$ and g_r is P_4. An alternative way of testing for invariance is to evaluate the transforms of H. For example,

$$P_4\{C(3)\}P_4^{-1} = P_4\{P_0\ P_1\ P_2\}P_4^{-1} = \{P_4\ P_5\ P_3\}P_4 = \{P_0\ P_2\ P_1\} = C(3). \quad (7)$$

Similarly for P_3 and P_5, showing therefore that C(3) *is* an invariant subgroup of S(3).

Exercise 1.3-2 Show that C(3) is transformed into itself by P_3 and by P_5.

$H_1 = \{P_0\ P_3\}$ is not an invariant subgroup of S(3). Although

$$\{P_0\ P_3\} + \{P_0\ P_3\}P_1 + \{P_0\ P_3\}P_2 = \{P_0\ P_3\ P_1\ P_4\ P_2\ P_5\} = S(3), \quad (8)$$

showing that H_1 is a subgroup of S(3) of index 3,

$$\{P_0\ P_3\}P_1 = \{P_1\ P_4\},\ \text{but}\ P_1\{P_0\ P_3\} = \{P_1\ P_5\}, \quad (9)$$

so that right and left cosets of the representative P_1 are not equal. Similarly,

$$\{P_0\ P_3\}P_2 = \{P_2\ P_5\},\ \text{but}\ P_2\{P_0\ P_3\} = \{P_2\ P_4\}. \quad (10)$$

Consequently, H_1 is not an invariant subgroup. For H to be an invariant subgroup of G, right and left cosets must be equal for each coset representative in the expansion of G.

Exercise 1.3-3 Show that H_2 is not an invariant subgroup of S(3).

Answers to Exercises 1.3

Exercise 1.3-1 If $t = 2$, $G = H + g_2\,H = H + H\,g_2$. Therefore, $H\,g_2 = g_2\,H$ and the right and left cosets are equal. Consequently, H is an invariant subgroup.

Exercise 1.3-2 $P_3\{P_0\,P_1\,P_2\}P_3^{-1} = \{P_3\,P_4\,P_5\}P_3 = \{P_0\,P_2\,P_1\}$ and $P_5\{P_0\,P_1\,P_2\}P_5^{-1} = \{P_5\,P_3\,P_4\}P_5 = \{P_0\,P_2\,P_1\}$, confirming that C(3) *is* an invariant subgroup of S(3).

Exercise 1.3-3 A coset expansion for H_2 is

$$\{P_0\ P_4\} + \{P_0\ P_4\}P_1 + \{P_0\ P_4\}P_2 = \{P_0\ P_4\ P_1\ P_5\ P_2\ P_3\} = S(3).$$

The right coset for P_1 is $\{P_0\ P_4\}P_1 = \{P_1\ P_5\}$, while the left coset for P_1 is $P_1\ \{P_0\ P_4\} = \{P_1\ P_3\}$, which is not equal to the right coset for the same coset representative, P_1. So H_2 is not an invariant subgroup of S(3).

1.4 The factor group

Suppose that H is an invariant subgroup of G of index t. Then the t cosets $g_r\ H$ of H (including $g_1\ H = H$) each *considered as one element*, form a group of order t called the *factor group*,

$$\mathrm{F} = \mathrm{G/H} = \sum_{r=1}^{t} (g_r\ \mathrm{H}),\ g_r \in \mathrm{G},\ g_r \notin \mathrm{H}\ (\text{except for } g_1 = E),\ t = g/h. \tag{1}$$

Each term in parentheses, g_r H, is *one* element of F. Because each element of F is a *set* of elements of G, binary composition of these sets needs to be defined. Binary composition of the elements of F is defined by

$$(g_p\ \mathrm{H})(g_q\ \mathrm{H}) = (g_p\ g_q)\ \mathrm{H},\quad g_p, g_q \in \{g_r\}, \tag{2}$$

where the complete set $\{g_r\}$ contains $g_1 =$ E as well as the $t-1$ coset representatives that $\notin$ H. It follows from closure in G that $g_p\ g_q \in$ G. Because H is an invariant subgroup

$$g_r\ \mathrm{H} = \mathrm{H}\ g_r. \tag{3}$$

(2), (3) $$g_p\ \mathrm{H}\ g_q\ \mathrm{H} = g_p\ g_p\ \mathrm{H}\ \mathrm{H} = g_p\ g_q\ \mathrm{H}. \tag{4}$$

This means that in F

(4) $$\mathrm{H}\ \mathrm{H} = \mathrm{H}, \tag{5}$$

which is the necessary and sufficient condition for H to be the identity in F.

Exercise 1.4-1 Show that $g_1\ g_1 = g_1$ is both a necessary and sufficient condition for g_1 to be E, the identity element in G. [*Hint*: Recall that the identity element E is defined by

$$E\ g_i = g_i\ E = g_i,\ \forall\ g_i \in \mathrm{G}.] \tag{1.1.5}$$

Thus, F contains the identity: that {F} is indeed a group requires the demon-stration of the validity of the other group properties. These follow from the definition of binary composition in F, eq. (2), and the invariance of H in G.

Closure: To demonstrate closure we need to show that $g_p\ g_q$ H $\in$ F for $g_p, g_q, g_r \in \{g_r\}$. Now $g_p\ g_q \in$ G and so

(1) $$g_p\ g_q \in \{g_r\ \mathrm{H}\},\ r = 1, 2, \ldots, t, \tag{6}$$

(6) $$g_p\ g_q = g_r\ h_l,\ h_l \in \mathrm{H}, \tag{7}$$

(2), (7) $$g_p\ \mathrm{H}\ g_q\ \mathrm{H} = g_p\ g_q\ \mathrm{H} = g_r\ h_l\ \mathrm{H} = g_r\ \mathrm{H} \in \mathrm{F}. \tag{8}$$

Associativity:

(2), (3), (4) $$(g_p\ \mathrm{H}\ g_q\ \mathrm{H})g_r\ \mathrm{H} = g_p\ g_q\ \mathrm{H}\ g_r\ \mathrm{H} = g_p\ g_q\ g_r\ \mathrm{H}, \tag{9}$$

(2), (3), (4) $$g_p\ \mathrm{H}(g_q\ \mathrm{H}\ g_r\ \mathrm{H}) = g_p\ H\ g_q\ g_r\ \mathrm{H} = g_p\ g_q\ g_r\ \mathrm{H}, \tag{10}$$

(9), (10) $$(g_p\ \mathrm{H}\ g_q\ \mathrm{H})g_r\ \mathrm{H} = g_p\ \mathrm{H}(g_q\ \mathrm{H} g_r\ \mathrm{H}), \tag{11}$$

and so multiplication of the elements of {F} is associative.

Table 1.4. *Multiplication table of the factor group* $F = \{E' \; P'\}$.

F	E'	P'
E'	E'	P'
P'	P'	E'

Inverse:

(2) $$(g_r^{-1} \text{ H})(g_r \text{ H}) = g_r^{-1} \, g_r \text{ H} = \text{H}, \tag{12}$$

so that the inverse of g_r H in F is g_r^{-1} H.

Example 1.4-1 The permutation group S(3) has the invariant subgroup $\text{H} = \{P_0 \; P_1 \; P_2\}$. Here $g = 6$, $h = 3$, $t = 2$, and

$$\text{G} = \text{H} + P_3 \text{ H}, \;\; \text{F} = \{\text{H} \; P_3 \text{ H}\} = \{E' \; P'\}, \tag{13}$$

where the elements of F have primes to distinguish $\text{E}' = \text{H} \in \text{F}$ from $E \in \text{G}$.

(13), (2) $$P'P' = (P_3 \text{ H})(P_3 \text{ H}) = P_3 \; P_3 \text{ H} = P_0 \text{ H} = \text{H}. \tag{14}$$

E' is the identity element in F, and so the multiplication table for the factor group of S(3), $\text{F} = \{E' \; P'\}$, is as given in Table 1.4.

Exercise 1.4-2 Using the definitions of E' and P' in eq. (13), verify explicitly that $E' \, P' = P'$, $P' \, E' = P'$. [*Hint*: Use eq. (2).]

Exercise 1.4-3 Show that, with binary composition as multiplication, the set $\{1 \;\; -1 \;\; \text{i} \;\; -\text{i}\}$, where $\text{i}^2 = -1$, form a group G. Find the factor group $\text{F} = \text{G/H}$ and write down its multiplication table. Is F isomorphous with a permutation group?

Answers to Exercises 1.4

Exercise 1.4-1

(1.1.5) $$E \, E \, g_i = E \, g_i \, E = E \, g_i, \;\; \forall \; g_i \in \text{G}, \tag{15}$$

(15) $$E \, E = E, \tag{16}$$

and so $E \, E = E$ is a *necessary* consequence of the definition of E in eq. (1.1.5). If $g_1 \, g_1 = g_1$, then multiplying each side from the left or from the right by g_1^{-1} gives $g_1 = E$, which demonstrates that $g_1 \, g_1 = g_1$ is a *sufficient* condition for g_1 to be E, the identity element in G.

Table 1.5. *Multiplication table of the group* G *of Exercise 1.4-3.*

G	1	−1	i	−i
1	1	−i	i	−i
−1	−1	1	−i	i
i	i	−i	−1	1
−i	−i	i	1	−1

Exercise 1.4-2

(13), (2) $$E'P' = (\mathrm{H})(P_3\ \mathrm{H}) = (E\ \mathrm{H})(P_3\ \mathrm{H}) = P_3\ \mathrm{H} = P',$$

(13), (2) $$P'E' = (P_3\ \mathrm{H})(\mathrm{H}) = (P_3\ \mathrm{H})(E\ \mathrm{H}) = P_3\ \mathrm{H} = P'.$$

Exercise 1.4-3 With binary composition as multiplication the set $\{1\ \ {-1}\ \ \mathrm{i}\ \ {-\mathrm{i}}\}$ is a group G because of the following.

(a) It contains the *identity* $E = 1$; $1\ g_i = g_i\ 1 = g_i$, $\forall\ g_i \in \mathrm{G}$.
(b) The set is *closed* (see Table 1.5).
(c) Since each row and each column of the multiplication table contains E once only, each $g_i \in \mathrm{G}$ has an *inverse*.
(d) *Associativity* holds; for example,

$$(-1)[(\mathrm{i})(-\mathrm{i})] = (-1)[1] = -1, \quad [(-1)(\mathrm{i})](-\mathrm{i}) = [(-\mathrm{i})](-\mathrm{i}) = -1.$$

From the multiplication table, the set $\mathrm{H} = \{1\ \ {-1}\}$ is closed and therefore it is a subgroup of G. The transforms of H for $g_i \notin \mathrm{H}$ are

$$\mathrm{i}\{1\ \ {-1}\}\mathrm{i}^{-1} = \{\mathrm{i}\ \ {-\mathrm{i}}\}(-\mathrm{i}) = \{1\ \ {-1}\} = \mathrm{H};$$
$$(-\mathrm{i})\{1\ \ {-1}\}(-\mathrm{i})^{-1} = \{-\mathrm{i}\ \mathrm{i}\}\mathrm{i} = \{1\ \ {-1}\} = \mathrm{H}.$$

Therefore H is an invariant subgroup of G. A coset expansion of G on H is $\mathrm{G} = \mathrm{H} + \mathrm{iH}$, and so $\mathrm{F} = \{\mathrm{H}\ \mathrm{iH}\}$. From binary composition in F (eq. (2)) (H) (iH) = iH, (iH) (H) = iH, (iH) (iH) = i i H = (−1) (H) = $\{-1\ \ 1\}$ = H. (Recall that H is the set of elements $\{1\ \ {-1}\}$, in no particular order.) The multiplication table of F is

F	H	iH
H	H	iH
iH	iH	H

The permutation group S(2) has just two elements $\{E\ P\}$

	1	2	
E	1	2	
	1	2	
P	2	1	
PP	1	2	E

The multiplication table of S(2)

S(2)	E	P
E	E	P
P	P	E

is the same as that of F, since both are of the form

G	E	g_2
E	E	g_2
g_2	g_2	E

F is therefore isomorphous with the permutation group S(2).

Remark Sections 1.6–1.8 are necessarily at a slightly higher level than that of the first five sections. They could be omitted at a first reading.

1.5 Minimal content of Sections 1.6, 1.7, and 1.8

1.5.1 *The direct product*

Suppose that $\mathrm{A} = \{a_i\}$, $\mathrm{B} = \{b_j\}$ are two groups of order a and b, respectively, with the same law of binary composition. If $\mathrm{A} \cap \mathrm{B} = \{E\}$ and $a_i\, b_j = b_j\, a_i$, $\forall\, a_i \in \mathrm{A}$, $\forall\, b_j \in \mathrm{B}$, then the outer *direct product* of A and B is a group G of order $g = a\, b$, written

$$\mathrm{G} = \mathrm{A} \otimes \mathrm{B}, \tag{1}$$

with elements $a_i b_j = b_j a_i$, $i = 1, \ldots, a$, $j = 1, \ldots, b$. A and B are subgroups of G, and therefore

(1.3.1) $$\mathrm{G} = \sum_{j=1}^{b} \{\mathrm{A}\}\, b_j = \sum_{j=1}^{b} b_j\, \{\mathrm{A}\},\ \ b_1 = E. \tag{2}$$

Because a_i, b_j commute for all $i = 1, \ldots, a$, $j = 1, \ldots, b$, the right and left cosets are equal, and therefore A is an invariant subgroup of G. Similarly, B is an invariant subgroup

$$\begin{array}{lllll} A & |a_1\ a_2\ a_3 & |b_1\ b_2 & |c_1\ c_2 \cdots & |\cdots \\ A' & \quad a' & \quad b' & \quad c' & \quad \cdots \end{array}$$

Figure 1.1. Diagrammatic representation of the mapping f: A = A′. Vertical bars have no significance other than to mark the fibers of a', b', c', ... , in A.

of G. It is still possible to form a direct product of A, B even when A and B are not both invariant subgroups of G.

(i) If A is an invariant subgroup of G but B is not an invariant subgroup of G, then the direct product of A and B is called the *semidirect product*, written

$$\mathrm{G} = \mathrm{A} \wedge \mathrm{B}. \tag{3}$$

Note that in semidirect products the *invariant subgroup* is always the *first* group in the product. For example,

$$\mathrm{S}(3) = \mathrm{C}(3) \wedge \mathrm{H}_1 = \{P_0\ P_1\ P_2\}\{P_0\ P_3\} = \{P_0\ P_1\ P_2\ P_3\ P_4\ P_5\}. \tag{4}$$

(ii) If neither A nor B are invariant subgroups of G, then the direct product of A with B is called the weak direct product. However, the weak direct product is not used in this book, and the term "direct product" without further qualification is taken to mean the outer direct product. (The inner direct product is explained in Section 1.6.)

1.5.2 *Mappings and homomorphisms*

A *mapping* f of the set A to the set A′, that is

$$f : \mathrm{A} \to \mathrm{A}' \tag{5}$$

involves the statement of a rule by which $a_i \in a = \{a_1\ a_2\ a_3\ \ldots\}$ in A becomes a' in A′; a' is the *image* of each $a_i \in a = \{a_i\}$ for the mapping f, and this is denoted by $a' = f(a_i)$. An example of the mapping f: A → A′ is shown in Figure 1.1. In a mapping f, every element $a_i \in$ A must have a unique image $f(a_i) = a' \in$ A′. The images of several different a_i may coincide (Figure 1.1). However, not every element in A′ is necessarily an image of some set of elements in A, and in such cases A is said to be mapped *into* A′. The set of all the elements in A′ that actually are images of some sets of elements in A is called the *range* of the mapping. The set of elements $\{a'\} = \{f(a_i)\}$, $\forall\, a_i \in$ A, is the image of the set A, and this is denoted by

$$f(\mathrm{A}) \subset \mathrm{A}',\ \forall\, a \in \mathrm{A}. \tag{6}$$

If f(A) = A′, the set A is said to be mapped *on to* A′. The set $a = \{a_i\}$ may consist of a single element, a one-to-one mapping, or $\{a_i\}$ may contain several elements, in which case the relationship of A to A′ is many-to-one. The set of elements in A that are mapped to a' is called the *fiber* of a', and the number of elements in a fiber is termed the *order of the fiber*. Thus in the example of Figure 1.1 the order of the fiber $\{a_1\ a_2\ a_3\}$ of a' is 3, while that of the fiber of $b' = \{b_1\ b_2\}$ is 2. If A, A′ are groups G, G′, and if a mapping f preserves multiplication so that

$$f(a_i\ b_j) - a'\ b' - f(a_i)\, f(b_j),\ \forall\, f(a_i) = a',\ \forall\, f(b_j) = b', \tag{7}$$

then G, G$'$ are *homomorphous*. For example, a group G and its factor group F are homomorphous. In particular, if the fibers of a', b', $\ldots$ each contain only *one* element, then G, G$'$ are *isomorphous*. In this case G and G$'$ are two different realizations of the same abstract group in which $\{g_i\}$ represents different objects, such as two different sets of symmetry operators, for example

Corollary

If multiplication is preserved in the mapping of G on to G$'$, eq. (7), then any properties of G, G$'$ that depend only on the multiplication of group elements will be the same in G, G$'$. Thus isomorphous groups have the same multiplication table and class structure.

Exercise 1.5-1 Show that in a group homomorphism the image of g_j^{-1} is the inverse of the image of g_j..

1.5.3 *More about subgroups and classes*

The *centralizer* $Z(g_j|G)$ of an element $g_j \in G$ is the subset $\{z_i\}$ of all the elements of G that commute with a particular element g_j of G, so that $z_i\, g_j = g_j\, z_i,\ g_j \in G,\ \forall\, z_i \in Z(g_j|G)$. Now $Z = Z(g_j|G)$ is a subgroup of G (of order z), and so we may write a coset expansion of G on Z as

(1.4.1) $$G = \sum_{r=1}^{t} g_r\, Z,\ \ t = g/z,\ \ g_1 = E. \tag{8}$$

It is proved in Section 1.8 that the sum of the elements $g_k\,(\mathscr{C}_i)$ that form the class $\mathscr{C}_i$ in G is given by

$$\Omega(\mathscr{C}_i) = \sum_k g_k\ (\mathscr{C}_i) = \sum_r g_r\ g_i\ g_r^{-1} \tag{9}$$

where $\Omega(\mathscr{C}_i)$ is called the *Dirac character* of the class $\mathscr{C}_i$. The distinct advantage of determining the members of $\mathscr{C}_i$ from eq. (9) instead of from the more usual procedure

$$\mathscr{C}_i = \{g_p\ g_i\ g_p^{-1}\}\quad (p = 1,\ 2,\ \ldots,\ g,\ \text{repetitions deleted}), \tag{10}$$

is that the former method requires the evaluation of only t instead of g transforms. An example of the procedure is provided in Exercise 1.8-3.

Exercise 1.5-2 Prove that $Z = Z(g_j|G)$ is a subgroup of G.

Answers to Exercises 1.5

Exercise 1.5-1 Since $E\, g_j = g_j$, $f(E)f(g_j) = f(g_j)$, and therefore $f(E) = E'$ is the identity in G$'$. Also, $g_j^{-1} g_j = E$, the identity in G. Therefore, $f(g_j^{-1} g_j) = f(g_j^{-1})\, f(g_j) = f(E) = E'$, and so the inverse of $f(g_j)$, the image of g_j, is $f(g_j^{-1})$, the image of g_j^{-1}.

Exercise 1.5-2 Since Z is the subset of the elements of G that commute with g_j, Z contains the identity E. if z_i, $z_k \in$ Z, then $(z_i \ z_k)g_j = g_j(z_i \ z_k)$, and so $\{z_i\}$ is closed. Closure, together with the inclusion of the identity, guarantee that each element of Z has an inverse which is $\in$ Z. Note that $\{z_i\} \subset$ G, and so the set of elements $\{z_i\}$ satisfy the associative property. Therefore, Z is a subgroup of G.

1.6 Product groups

If $\mathrm{A} = \{a_i\}$, $\mathrm{B} = \{b_j\}$ are two groups of order a and b, respectively, then the *outer direct product* of A and B, written $\mathrm{A} \otimes \mathrm{B}$, is a group $\mathrm{G} = \{g_k\}$, with elements

$$g_k = (a_i, \ b_j). \tag{1}$$

The product of two such elements of the new group is to be interpreted as

$$(a_i, \ b_j)(a_l, \ b_m) = (a_i \ a_l, \ b_j \ b_m) = (a_p, \ b_q) \quad \text{(closure in A and B)}. \tag{2}$$

The set $\{(a_i, b_j)\}$ therefore closes. The other necessary group properties are readily proved and so G *is* a group. "Direct product" (DP) without further qualification implies the outer direct product. Notice that binary composition is defined for each group (e.g. A and B) individually, but that, in general, a multiplication rule between elements of different groups does not necessarily exist unless it is specifically stated to do so. However, if the elements of A and B obey the same multiplication rule (as would be true, for example, if they were both groups of symmetry operators) then the product $a_i \ b_j$ *is* defined. Suppose we try to take $(a_i, \ b_j)$ as $a_i \ b_j$. This imposes some additional restrictions on the DP, namely that

$$a_l \ b_j = b_j \ a_l, \ \forall \ l, j \tag{3}$$

and

$$\mathrm{A} \cap \mathrm{B} = E. \tag{4}$$

For *if*

$$(a_i, \ b_j) = a_i, \ b_j \tag{5}$$

then

$$(a_i, \ b_j) \ (a_l, \ b_m) = (a_i \ a_l, \ b_j \ b_m) = (a_p, \ b_q) \tag{2}$$

and

$$g_k \ g_n = a_i \ b_j \ a_l \ b_m = a_i \ a_l \ b_j \ b_m = a_p \ b_q = g_s \tag{6}$$

which shows that a_l and b_j commute. The second equality in eq. (6) follows from applying eq. (5) to both sides of the first equality in eq. (2). Equation (6) demonstrates the closure of $\{\mathrm{G}\}$, provided the result $a_p \ b_q$ is unique, which it must be because A and B are groups and the products $a_i \ a_l$ and $b_j \ b_m$ are therefore unique. But, suppose the intersection of A and B contains $a_l \ (\neq E)$ which is therefore also $\in$ B. Then $a_l \ b_j \ b_m \in$ B, b_r, say, and the product $a_p \ b_q$ would also be $a_i \ b_r$, which is impossible because for eq. (6) to be a valid multiplication rule, the result must be unique. Therefore $a_l \notin$ B, $\forall\, l = 1, \ldots, a$, except when $a_l = E$. Similarly, $b_j \notin$ A, $\forall\, j = 1, 2, \ldots, b$, except when $b_j = E$. The intersection of A and B therefore contains the identity E only, which establishes eq. (4). So the multiplication rule $(a_i, b_j) = a_i\, b_j$ is only valid if the conditions in eqs. (3) and (4) also hold.

A and B are subgroups of G, and from eqs. (5) and (3) the right and left coset expansions of G are

(1.3.1) $$\mathrm{G} = \sum_{j=1}^{b} \{\mathrm{A}\} b_j, \quad b_1 = E, \tag{7}$$

$$\mathrm{G} = \sum_{j=1}^{b} b_j \{\mathrm{A}\}, \quad b_1 = E. \tag{8}$$

When eq. (5) holds, $a_i\, b_j = b_j\, a_i$, $\forall\, i = 1, 2, \ldots, a, j = 1, 2, \ldots, b$, the right and left cosets are equal

$$\{\mathrm{A}\}\; b_j = b_j\; \{\mathrm{A}\}, \;\; \forall\; b_j \in \mathrm{B}, \tag{9}$$

and therefore A is an invariant subgroup of G.

Exercise 1.6-1 Why may we not find the outer DP of the subgroups C(3) and H_1 of S(3) using the interpretation $(a_i, b_j) = a_i\, b_j$?

Exercise 1.6-2 If A $\otimes$ B = G and all binary products $a_i\, b_j$ with $a_i \in$ A, $b_j \in$ B commute, show that B is an invariant subgroup of G.

Exercise 1.6-3 Show that if the products (a_i, a_j) in the DP set A $\otimes$ A are interpreted as (a_i, a_j), as in eq. (5), then A $\otimes$ A $= a\{\mathrm{A}\}$.

To avoid redundancies introduced by the outer DP A $\otimes$ A of a group with itself (Exercise 1.6-3), the *inner direct product* A $\boxtimes$ A is defined by

$$\mathrm{A} \boxtimes \mathrm{A} = \{(a_i,\; a_i)\}, \;\; i = 1,\; 2, \ldots, a. \tag{10}$$

The *semidirect product* and the *weak direct product* have been defined in Section 1.5.

Exercise 1.6-4 (a) Show that if we attempt to use the multiplication rule $(a_i, a_i) = a_i\, a_i$ then the inner DP set does not close. (b) Show that if the inner DP is defined under the multiplication rule, eq. (2), then the inner DP set, eq. (10), is closed, and that the group A $\boxtimes$ A $\subset$ A $\otimes$ A is isomorphous with A.

Answers to Exercises 1.6

Exercise 1.6-1 In the outer DP A $\otimes$ B, the product (a_i, b_j) of elements a_i and b_j may be equated to $a_i\, b_j$ only if A $\cap$ B is E and the elements a_i, b_j commute. In C(3) $\otimes$ $\mathrm{H}_1 = \{P_0\, P_1\, P_2\}$

$\{P_0\, P_3\}$, $P_1\, P_3 = P_4$, but $P_3\, P_1 = P_5$; therefore not all pairs of elements a_i, b_j commute, and so we may not form the outer DP of C(3) and H_1 using the interpretation in eq. (5).

Exercise 1.6-2 In $G = A \otimes B$, if all binary products $a_i\, b_j$ commute then left and right cosets $a_i\{B\}$ and $\{B\}\, a_i$ are equal, for $\forall\, i = 1, 2, \ldots, a$, and so B is an invariant subgroup of G.

Exercise 1.6-3 $A \otimes A = \{(a_i, a_j)\}$; if $\{(a_i, a_j)\}$ is equated to $\{(a_i, a_j)\}$, then since $a_i\, a_j \in A$, and $i = 1, 2, \ldots, a, j = 1, 2, \ldots, a$, $A = \{a_i\}$ occurs a times in the outer DP, and so $A \otimes A = a\{A\}$.

Exercise 1.6-4 (a) The product of the ith and jth elements in the inner product $A \boxtimes A = \{(a_i, a_i)\}$, $i = 1, 2, \ldots, a$, is $(a_i, a_i)(a_j, a_j) = (a_i\, a_j, a_i\, a_j) = (a_k, a_k)$, and so the inner DP set $\{(a_i, a_i)\}$ is closed. But if we attempt to interpret (a_i, a_i) as $a_i\, a_i$, then $(a_i, a_i)(a_j, a_j) = a_i\, a_i\, a_j\, a_j$, which is not equal to $(a_i\, a_j, a_i\, a_j) = a_i\, a_j\, a_i\, a_j$, unless A is Abelian.

(b) The inner DP $A \boxtimes A = \{(a_i, a_i)\}$ is closed and is $\subset A \otimes A$, for it is a subset of $\{(a_i, a_j)\}$, which arises when $i = j$. Since the product of the ith and jth elements of A is $a_i\, a_j = a_k$, while that of $A \boxtimes A$ is $(a_i, a_i)(a_j, a_j) = (a_k, a_k)$, $A \boxtimes A$ is isomorphous with A.

1.7 Mappings, homomorphisms, and extensions

Remark If you have not yet done so, read the first part of Section 1.5.2, including eqs. (1.5.5)–(1.5.7), as this constitutes the first part of this section.

A subset $K \subset G$ that is the fiber of E' in G' is called the *kernel* of the homomorphism. If there is a homomorphism of G on to $G'(f(G) = G')$ of which K is the kernel, with $g = k\, g'$, so that all fibers of the elements of G (images in the homomorphism) have the same order, then G is called an *extension* of G' by K. An example of an extension is illustrated in Table 1.6 for the particular case of $k = 3$.

Exercise 1.7-1 (a) Show that K is an invariant subgroup of G. (b) What is the kernel of the homomorphism $f(S(3)) = F = S(3)/C(3)$. (c) If $G \to G'$ is a homomorphism, prove that $g = k\, g'$.

Table 1.6. *Example of a homomorphism* $f(G) = G'$.

G is an extension of G' by K, the kernel of the homomorphism (so that E' in G' is the image of each element in K). Similarly, g'_2 is the image of each one of g_{21}, g_{22}, g_{23}, and so on. In this example $k = 3$.

$K = \{g_{11}\, g_{12}\, g_{13}\}$
$G = \{g_{11}\, g_{12}\, g_{13};\; g_{21}\, g_{22}\, g_{23};\; \ldots\; g_{n1}\, g_{n2}\, g_{n3}\}$
$G' = \{g'_1 = E',\; g'_2,\; g'_3,\; \ldots, g'_n\}$

Answer to Exercise 1.7-1

(a) Because K is the kernel of the homomorphism $G \rightarrow G'$, $f((k_i\ k_j)) = f(k_i)\ f(k_j) = E'\ E' = E'$. Therefore, $k_i\ k_j \in K$. The set K is therefore closed and so K is a subgroup of G. Consider the mapping of $g_j\ k_i\ g_j^{-1}$, $k_i \in K$, $g_j \in G$,

$$f(g_j\ k_i\ g_j^{-1}) = f(g_j)\ f(k_i)\ f(g_j^{-1}) = f(g_i)E'f(g_j^{-1}) = E', \tag{1}$$

where we have used eq. (1.5.7) and Exercise 1.5-1. Therefore,

$$(1) \qquad g_j\ k_i\ g_j^{-1} \in K, \tag{2}$$

which shows $K = \{k_i\}$ to be an invariant subgroup of G.

(b) The subgroup C(3) is the kernel of S(3) for the homomorphism of S(3) on to its factor group F because $f(C(3)) = E'$.

(c) No two fibers in G can have a common element; otherwise this common element would have two distinct images in G′, which is contrary to the requirements for a mapping. Therefore, there are as many disjoint fibers in G as there are elements in G′, namely g'. It remains to be shown that all fibers in G have the same order, which is equal to the order k of the kernel K. Firstly, the necessary and sufficient condition for two elements g_2, g_3 that are $\in$ G to belong to the same fiber of G is that they be related by

$$g_2 = g_3\ k_i,\ \ k_i \in K. \tag{3}$$

Sufficiency:

$$f(g_2) = f(g_3\ k_i) = f(g_3)\ f(k_i) = f(g_3)E' = f(g_3). \tag{4}$$

Necessity: Suppose that $g_2 = g_3\ g_j$; then $f\,(g_2) = f(g_3)$, $f(g_j)$. But if g_2, g_3 belong to the same fiber then $f(g_j)$ must be E' and so g_j can only be $\in$ K. Secondly, if g_n is a particular element of a fiber F_n, then the other elements of F_n can all be written in the form $g_n\ k_i$, where $k_i \in K$,

$$F_n = \{g_n\ k_i\},\ \ k_i \in K. \tag{5}$$

All the distinct elements of F_n are enumerated by eq. (5) as $i = 1, 2, \ldots, k$, the order of K. Therefore, the number of elements in each one of the g' fibers in G is k, whence the order of G is

$$g = k\ g', \tag{6}$$

which establishes the required result.

1.8 More about subgroups and classes

If G and H are two groups for which a multiplication rule exists, that is to say the result $g_i\ h_j$ is defined, then the conjugate of H by an element $g_i \in G$ is

$$g_i\ H\ g_i^{-1} = \sum_j g_i\ h_j\ g_i^{-1}. \tag{1}$$

When the result is H itself, H is invariant under the element g_i,

$$g_i \, \mathrm{H} \, g_i^{-1} = \mathrm{H}. \tag{2}$$

(2) $$g_i \, \mathrm{H} = \mathrm{H} \, g_i, \tag{3}$$

which is an equivalent condition for the invariance of H under g_i. The set of elements $\{g_i\} \in \mathrm{G}$ that leave H invariant form a subgroup of G called the *normalizer* of H in G, written N(H|G). That N(H|G) does indeed form a subgroup of G follows from the fact that if $g_i, g_j \in \mathrm{N(H|G)}$

$$g_j \, \mathrm{H} \, g_j^{-1} = \mathrm{H}, \tag{4}$$

(2), (4) $$g_i \, g_j \, \mathrm{H} \, (g_i \, g_j)^{-1} = g_i \, \mathrm{H} \, g_i^{-1} = \mathrm{H}, \tag{5}$$

(5) $$g_i \, g_j \in \{g_i, g_j, \ldots\} = \mathrm{N(H|G)}, \tag{6}$$

implying closure of $\{g_i, g_j, \ldots\}$, a sufficient condition for $\{g_i, g_j, \ldots\}$ to be a subgroup of G. If the normalizer N(H|G) is G itself, so that H is invariant under all $g_i \in \mathrm{G}$, H is said to be *normal* or *invariant* under G. If H is a subgroup of G (not so far assumed) then H is an invariant subgroup of G if eqs. (2) and (3) hold.

If G, H are two groups for which a multiplication rule exists then the set of all the elements of G that commute with a particular element h_j of H form a subgroup of G called the *centralizer* of h_j in G, denoted by

$$\mathrm{Z}(h_j|\mathrm{G}) \subset \mathrm{G}. \tag{7}$$

H may be the same group as G, in which case h_j will be one element of G, say $g_j \in \mathrm{G}$. Similarly, the centralizer of H in G,

$$\mathrm{Z(H|G)} \subset \mathrm{G}, \tag{8}$$

is the set of all the elements of G that each commute with each element of H; H in eq. (8) may be a subgroup of G. If H is G itself then

$$\mathrm{Z(G|G)} \equiv \mathrm{Z(G)} \tag{9}$$

is the *center* of G, namely the set of all the elements of G that commute with *every* element of G. In general, this set is a subgroup of G, but if Z(G) = G, then G is an Abelian group.

Exercise 1.8-1 Prove that the centralizer $\mathrm{Z}(h_j|\mathrm{G})$ is a subgroup of G.

Exercise 1.8-2 (a) Find the center Z(C(3)) of C(3). (b) What is the centralizer Z(C(3)|S(3)) of C(3) in S(3)? (c) What is the centralizer $\mathrm{Z}(P_1|\mathrm{S(3)})$ of P_1 in S(3)?

A class was defined in Section 1.2 as a complete set of conjugate elements. The sum of the members $g_j(\mathscr{C}_i)$, $j = 1, 2, \ldots, c_i$ of the class $\mathscr{C}_i$ that contains the group element g_i is

$$\Omega(\mathscr{C}_i) = \sum_j g_j(\mathscr{C}_i), \tag{10}$$

$$\mathscr{C}_i = \{g_k\, g_i\, g_k^{-1}\},\ \forall\ g_k \in \mathrm{G}, \tag{11}$$

with repetitions deleted. The sum of all the elements in a class, $\Omega(\mathscr{C}_i)$, is the *Dirac character* of the class $\mathscr{C}_i$, and

(10), (11)
$$\Omega(\mathscr{C}_i) = \sum_k g_k\, g_i\, g_k^{-1}, \tag{12}$$

with repetitions deleted. It is rather a waste of effort to evaluate the transforms on the right side (RS) of eq. (12) for all $g_k \in \mathrm{G}$, since many redundancies will be found that will have to be eliminated under the "no repetitions" rule. For instance, see Example 1.2-1, where six transforms of P_1 yield a class that contains just two members, P_1 and P_2, each of which occurred three times. However, it is possible to generate the class $\mathscr{C}_i$ that contains g_i without redundancies, from the coset expansion of G that uses the centralizer of g_i as the subgroup in the expansion. Abbreviating $\mathrm{Z}(g_i|\mathrm{G})$ to Z_i, the coset expansion of G on Z_i is

$$\mathrm{G} = \sum_{r=1}^{t} g_r\ \mathrm{Z}_i,\ g_1 = E,\ t = g/z, \tag{13}$$

where z is the order of Z_i. From the definition of the coset expansion in eq. (13), the elements of $\{g_r\}$ with $r = 2, \ldots, t$, and Z are disjoint. (E is of course $\in \mathrm{Z}_i$.) We shall now prove that

$$\Omega(\mathscr{C}_i) = \sum_r g_r\, g_i\, g_r^{-1}, \tag{14}$$

where $\{g_r\}$ is the set defined by eq. (13), namely the t coset representatives.

Proof The coset expansion eq. (13) shows that $\mathrm{G} = \{g_k\}$ is the DP set of $\{z_p\}$ and $\{g_r\}$, which means that G may be generated by multiplying each of the z members of $\{z_p\}$ in turn by each of the t members of $\{g_r\}$. Therefore, g_k in eq. (12) may be written as

$$g_k = z_p\, g_r,\ g_k \in \mathrm{G},\ z_p \in \mathrm{Z}_i, \tag{15}$$

with $\{g_r\}$ defined by eq. (13). In eq. (15), p, which enumerates the z elements of Z_i, runs from 1 to z; r, which enumerates the coset representatives (including $g_1 = E$), runs from 1 to t; and k enumerates all the g elements of the group G as k runs from 1 to g.

(12), (15)
$$\sum_{r,p} g_r\, z_p\, g_i\, (g_r\, z_p)^{-1} = \sum_{r,p} g_r\, z_p\, g_i\, z_p^{-1}\, g_r^{-1} = \sum_{r,p} g_r\, g_i\, g_r^{-1} = z \sum_r g_r\, g_i\, g_r^{-1}. \tag{16}$$

The second equality in eq. (16) follows because $z_p \in \mathrm{Z}_i = \mathrm{Z}(g_i|\mathrm{G})$, which, from the definition of the centralizer, all commute with g_i. The third equality follows because the double sum consists of the same t terms repeated z times as p runs from 1 to z. It follows from the uniqueness of the binary composition of group elements that the sum over r in eq. (16) contains no repetitions. Therefore the sum over r on the RS of (16) is $\Omega(\mathscr{C}_i)$, which establishes eq. (14). Since eq. (14) gives the elements of $\mathscr{C}_i$ without repetitions, the order c_i of this class is

$$c_i \equiv t = g/z. \tag{17}$$

Equation (17) shows that the order of a class $\mathscr{C}_i$ is a divisor of the order of the group (Lagrange's theorem). It also yields the value of c_i once we determine z from $Z_i \equiv Z(g_i|G)$. The t elements g_r needed to find the Dirac character $\Omega(\mathscr{C}_i)$ of the class $\mathscr{C}_i$, and thus the members of $\mathscr{C}_i$, are the coset representatives of the centralizer $Z_i \equiv Z(g_i|G)$.

Exercise 1.8-3 Find the class of P_1 in S(3) by using the coset expansion for the centralizer $Z(P_1|S(3))$ and eq. (14).

Answers to Exercises 1.8

Exercise 1.8-1 The centralizer $Z(h_j|G)$ is the set $\{g_i\}$ of all the elements of G that commute with h_j. Let $g_k \in \{g_i\}$; then g_i, g_k each commute with h_j and

$$(g_i\, g_k)h_j = g_i\, h_j\, g_k = h_j(g_i\, g_k) \tag{18}$$

so that if g_i, $g_k \in \{g_i\}$ that commutes with h_j, then so also is $g_i\, g_k$. Equation (18) demonstrates that $\{g_i\} = Z(h_j|G)$ is closed, and that therefore it is a subgroup of G. The above argument holds for any $h_j \in H$, so that $Z(H|G)$ is a subgroup of G. It also holds if h_j is $g_j \in G$, and for any $\{g_j\}$ which is a subgroup of G, and for $\{g_j\} = G$ itself. Therefore $Z(g_j|G)$, $Z(H|G)$, where $H \subset G$, and $Z(G|G)$ are all subgroups of G, g_j being but a particular case of h_j.

Exercise 1.8-2 (a) Z(C(3)) is the set of elements of C(3) that commute with every element of C(3). From Table 1.3 we see that each element of C(3) commutes with every other element (the mutiplication table of C(3) is symmetrical about its principal diagonal from upper left to lower right) so that Z(C(3)) = C(3), and consequently C(3) is an Abelian group.

(b) The centralizer of C(3) in S(3) is the set of elements of S(3) that commute with each element of C(3). From Table 1.3 we see that none of P_3, P_4, P_5 commute with all of P_0, P_1, P_2; therefore Z(C(3)|S(3)) = C(3). Notice that here H happens to be a subgroup of G, but this is not a necessary feature of the definition of the centralizer. H needs to be a group for which binary composition with the elements of G is defined. In S(3), and therefore C(3), the product $P_i\, P_j$ means carrying out successively the permutations described by P_j first, and then P_i. Thus, Z(C(3)|S(3)) is necessarily a subgroup of S(3), in this case C(3) again (see Example 1.3-1).

(c) Again from Table 1.3, we see that only P_0, P_1, P_2 commute with P_1 so that $Z(P_1|S(3)) = C(3)$.

Exercise 1.8-3 $Z(P_1|S(3))$ is the set of elements of S(3) which commute with P_1. From Table 1.3 or Exercise 1.8-2(c), $Z(P_1|S(3)) = C(3)$. The coset expansion of S(3) on C(3) is

$$S(3) = P_0C(3) + P_3C(3) = \{P_0\ P_1\ P_2\} + \{P_3\ P_5\ P_4\}$$

so $z = 2$ and $\{g_r\} = \{P_0, P_3\}$. The Dirac character of the class of P_1 is therefore

$$\Omega(\mathscr{C}(P_1)) = \sum_r g_r\ P_1\ g_r^{-1} = P_0\ P_1\ P_0^{-1} + P_3\ P_1\ P_3^{-1} \tag{14}$$
$$= P_1 + P_3\ P_4 = P_1 + P_2.$$

Therefore, $\mathscr{C}(P_1) = \{P_1\ P_2\}$, and eq. (14) yields the class of P_1 without repetitions.

Problems

1.1 Show that the inverse of $g_i\ g_j$ is $g_j^{-1}\ g_i^{-1}$.

1.2 Prove that if each element of a group G commutes with every other element of G (so that G is an Abelian group) then each element of G is in a class by itself.

1.3 Find a generator for the group of Exercise 1.4-3.

1.4 Show that $\{P_1\ P_3\}$ is a generator for S(3).

1.5 Show that conjugation is *transitive*, that is if g_k is the transform of g_j and g_j is the transform of g_i, then g_k is the transform of g_i.

1.6 Show that conjugation is *reciprocal*, that is if g_k is the transform of g_j then g_j is the transform of g_k.

1.7 Prove that binary composition is conserved by conjugation.

1.8 There are only two groups of order 4 that are not isomorphous and so have different multiplication tables. Derive the multiplication tables of these two groups, G_4^1 and G_4^2. [*Hints*: First derive the multiplication table of the cyclic group of order 4. Call this group G_4^1. How many elements of G_4^1 are equal to their inverse? Now try to construct further groups in which a different number of elements are equal to their own inverse. Observe the rearrangement theorem.]

1.9 Arrange the elements of the two groups of order 4 into classes.

1.10 Identify the subgroups of the two groups of order 4.

1.11 Write down a coset expansion of S(3) on its subgroup $H_3 = \{P_0\ P_5\}$. Show that H_3 is not an invariant subgroup of S(3).

1.12 The *inverse class* of a class $\mathscr{C}_j = \{g_j\}$ is $\mathscr{C}_{\bar{j}} = \{g_j^{-1}\}$. Find the inverse class of the class $\{P_1\ P_2\}$ in S(3).

1.13 The classes of S(3) are $\mathscr{C}_1 = \{P_0\}$, $\mathscr{C}_2 = \{P_1\ P_2\}$, $\mathscr{C}_3 = \{P_3\ P_4\ P_5\}$. Prove that $\Omega_3\ \Omega_2 = 2\Omega_3$.

1.14 Prove that for S(3), $c_3 g^{-1} \sum_{g_i \in S(3)} g_i\, P_3\, g_i^{-1} = \Omega_3$.

2 Symmetry operators and point groups

2.1 Definitions

Symmetry operations leave a set of objects in *indistinguishable configurations* which are said to be *equivalent*. A set of symmetry operators always contains at least one element, the *identity* operator E. When operating with E the final configuration is not only indistinguishable from the initial one, it is *identical* to it. A *proper rotation*, or simply *rotation*, is effected by the operator $R(\phi\ \mathbf{n})$, which means "carry out a rotation of configuration space with respect to fixed axes through an angle ϕ about an axis along some unit vector $\mathbf{n}$." The *range* of ϕ is $-\pi < \phi \leq \pi$. Configuration space is the three-dimensional (3-D) space $\mathscr{R}^3$ of real vectors in which physical objects such as atoms, molecules, and crystals may be represented. Points in configuration space are described with respect to a system of three space-fixed right-handed orthonormal axes $\mathbf{x}$, $\mathbf{y}$, $\mathbf{z}$, which are collinear with OX, OY, OZ (Figure 2.1(a)). (A right-handed system of axes means that a right-handed screw advancing from the origin along OX would rotate OY into OZ, or advancing along OY would rotate OZ into OX, or along OZ would rotate OX into OY.) The convention in which the axes $\mathbf{x}$, $\mathbf{y}$, $\mathbf{z}$ remain fixed, while the whole of configuration space is rotated with respect to fixed axes, is called the *active representation*. Thus, the rotation of configuration space effected by $R(\phi\ \mathbf{n})$ carries with it all vectors in configuration space, including a set of unit vectors $\{\mathbf{e}_1\ \mathbf{e}_2\ \mathbf{e}_3\}$ initially coincident with $\{\mathbf{x}\ \mathbf{y}\ \mathbf{z}\}$. Figures 2.1(b) and (c) show the effect on $\{\mathbf{e}_1\ \mathbf{e}_2\ \mathbf{e}_3\}$ of $R(\pi/3\ \mathbf{x})$, expressed by

$$R(\pi/3\ \mathbf{x})\{\mathbf{e}_1\ \mathbf{e}_2\ \mathbf{e}_3\} = \{\mathbf{e}_1{}'\ \mathbf{e}_2{}'\ \mathbf{e}_3{}'\} \tag{1}$$

In the *passive representation*, symmetry operators act on the *axes*, and so on $\{\mathbf{x}\ \mathbf{y}\ \mathbf{z}\}$, but leave configuration space fixed. Clearly, one should work entirely in one representation or the other: here we shall work solely in the active representation, and we shall not use the passive representation.

An alternative notation is to use the symbol $C_n^{\pm k}$ for a rotation operator. Here n does not mean $|\mathbf{n}|$, which is 1, but is an integer that denotes the *order* of the axis, so that $C_n^{\pm k}$ means "carry out a rotation through an angle $\phi = \pm 2\pi k/n$." Here n is an integer > 1, and $k = 1, 2, \ldots, (n-1)/2$ if n is an odd integer and, if n is even, $k = 1, 2, \ldots, n/2$, with $C_n^{-n/2}$ excluded by the range of ϕ; $k = 1$ is implicit. In this notation the axis of rotation has not been specified explicitly so that it must either be considered to be self-evident (for example, to be understood from what has gone before) or to be stated separately, as in "a C_4 rotation about the $\mathbf{z}$ axis," or included as a second subscript, as in C_{4z}. (The

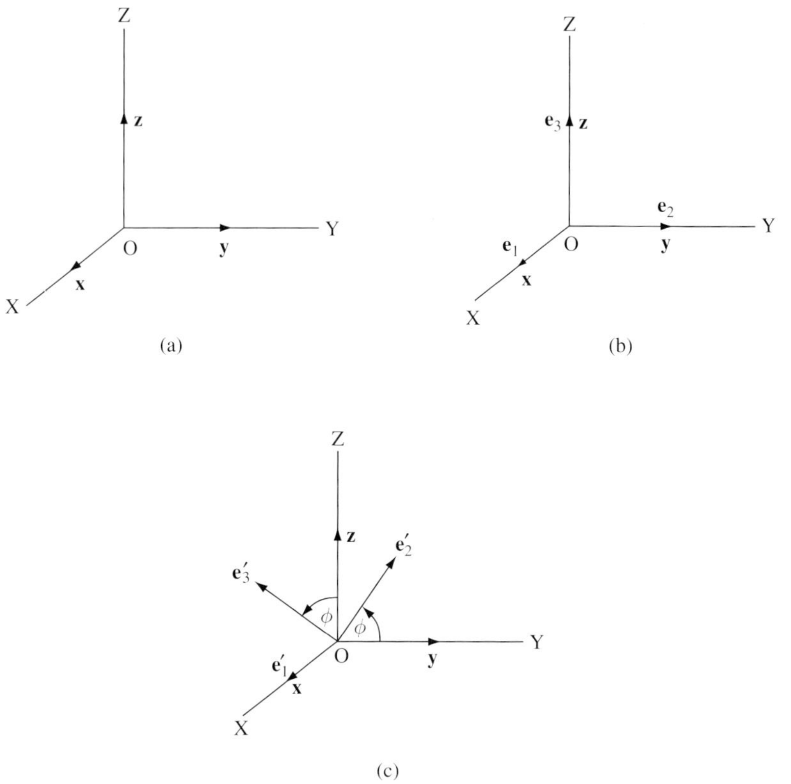

Figure 2.1. (a) Right-handed coordinate axes **x**, **y**, **z** in configuration space. A right-handed screw advancing along OX from O would rotate OY into OZ, and similarly (preserving cyclic order). (b) Initial configuration with $\{\mathbf{e}_1\ \mathbf{e}_2\ \mathbf{e}_3\}$ coincident with $\{\mathbf{x}\ \mathbf{y}\ \mathbf{z}\}$. (c) The result of a rotation of configuration space by $R(\pi/3\ \ \mathbf{x})$, expressed by eq. (1).

superscript + is often also implicit.) Thus $R(\pi/2\ \ \mathbf{z})$ and C_{4z} are equivalent notations, and we shall use either one as convenient. When the axis of rotation is not along **x** or **y** or **z**, it will be described by a unit vector **a**, **b**, ... , where **a**, for example, is defined as a unit vector parallel to the vector with components $[n_1\ \ n_2\ \ n_3]$ along **x**, **y**, and **z**, or by a verbal description, or by means of a diagram. Thus $R(\pi\ \ \mathbf{a})$ or $C_{2\mathbf{a}}$ may be used as alternative notations for the operator which specifies a rotation about a two-fold axis along the unit vector **a** which bisects the angle between **x** and **y**, or which is along the vector with components [1 1 0] (Figure 2.2(a)). A rotation is said to be *positive* $(0 < \phi \leq \pi)$ if, on looking down the axis of rotation towards the origin, the rotation appears to be *anti-clockwise* (Figure 2.2(b)). Equivalently, a positive rotation is the direction of rotation of a right-handed screw as it advances along the axis of rotation away from the origin. Similarly, a rotation that appears to be in a clockwise direction, on looking down the axis of rotation towards O, is a negative rotation with $-\pi < \phi < 0$.

Exercise 2.1-1 (a) Check the sign of the rotation shown in Figure 2.2(c) using both of the criteria given above. (b) Show the effect of $R(-\pi/2\ \ \mathbf{z})$ on $\{\mathbf{e}_1\ \mathbf{e}_2\ \mathbf{e}_3\}$.

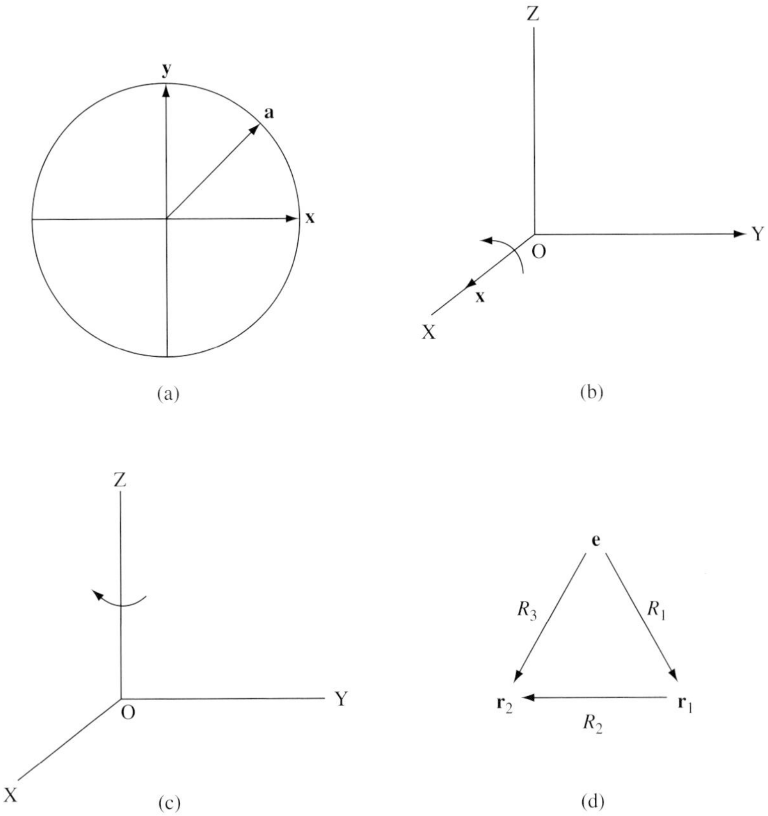

Figure 2.2. (a) The unit vector **a** bisects the angle between **x** and **y** and thus has components $2^{-1/2}$ [1 1 0]. (b) The curved arrow shows the direction of a positive rotation about **x**. (c) The curved arrow shows the direction of a negative rotation about OZ (Exercise 2.1-1(a)). (d) The product of two symmetry operators $R_2\ R_1$ is equivalent to a single operator R_3; **e**, $\mathbf{r}_1$, and $\mathbf{r}_2$ are three indistinguishable configurations of the system.

Products of symmetry operators mean "carry out the operations specified successively, beginning with the one on the right." Thus, R_2R_1 means "apply the operator R_1 *first*, and then R_2." Since the product of two symmetry operators applied to some initial configuration **e** results in an indistinguishable configuration ($\mathbf{r}_2$ in Figure 2.2(d)), it is equivalent to a single symmetry operator $R_3 = R_2\ R_1$. For example,

$$C_4\ C_4 = C_4^2 = C_2 = R(\pi\ \mathbf{n}); \tag{2}$$

$$\begin{gathered}(C_n)^{\pm k} = C_n^{\pm k} = R(\phi\ \mathbf{n}), \\ \phi = \pm 2\pi k/n\ (n > 1, k = 1, 2,\ \ldots \leq {}^{n}/_{2},\ -\pi < \phi \leq \pi).\end{gathered} \tag{3}$$

A negative sign on k in eq. (3) corresponds to a negative rotation with $-\pi < \phi < 0$. Note that $k = 1$ is implicit, as in $C_{3\mathbf{z}}^{-} = R(-2\pi/3\ \mathbf{z})$, for example. A rotation C_2 or $R(\pi\ \mathbf{n})$ is called a *binary rotation*. Symmetry operators do not necessarily commute. Thus, $R_2\ R_1$ may, or may not, be equal to $R_1\ R_2$.

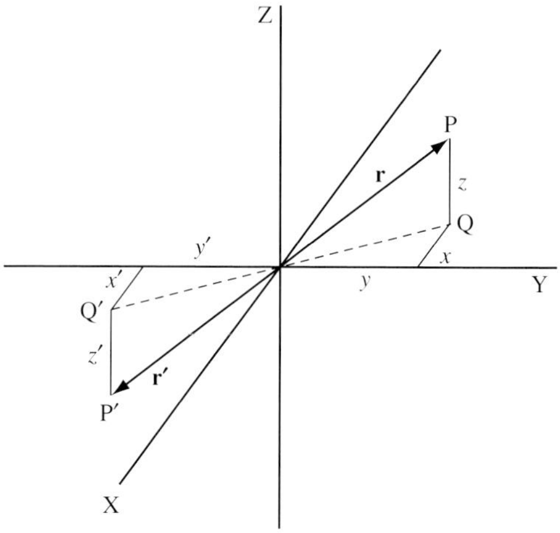

Figure 2.3. Effect of the inversion operator I on the polar vector $\mathbf{r}$. The points Q, Q′ lie in the XY plane.

Exercise 2.1-2 (a) Do successive rotations about the same axis commute? (b) Show that $R(-\phi\ \mathbf{n})$ is the inverse of $R(\phi\ \mathbf{n})$.

A *polar vector* $\mathbf{r}$ is the sum of its projections,

$$\mathbf{r} = \mathbf{e}_1 x + \mathbf{e}_2 y + \mathbf{e}_3 z. \tag{4}$$

Each projection on the RS of eq. (4) is the product of one of the set of basis vectors $\{\mathbf{e}_1\ \mathbf{e}_2\ \mathbf{e}_3\}$ and the corresponding component of $\mathbf{r}$ along that vector. The *inversion operator* I changes the vector $\mathbf{r}$ into $-\mathbf{r}$,

(4)
$$I\ \mathbf{r} = -\mathbf{r} = -\mathbf{e}_1 x - \mathbf{e}_2 y - \mathbf{e}_3 z \tag{5}$$

(see Figure 2.3). The basis vectors $\{\mathbf{e}_1\ \mathbf{e}_2\ \mathbf{e}_3\}$ are *pseudovectors*, that is they behave like ordinary polar vectors under rotation but are invariant under inversion. The components of $\mathbf{r}$, $\{x\ y\ z\}$, do change sign under inversion and are therefore *pseudoscalars* (invariant under rotation but change sign on inversion). This is made plain in Figure 2.3, which shows that under inversion $x' = -x$, $y' = -y$, $z' = -z$. A proper rotation $R(\phi\ \mathbf{n})$ followed by inversion is called an *improper rotation*, $IR(\phi\ \mathbf{n})$. Although R and IR are the only *necessary* symmetry operators that leave at least one point invariant, it is often convenient to use the *reflection* operator $\sigma_{\mathbf{m}}$ as well, where $\sigma_{\mathbf{m}}$ means "carry out the operation of reflection in a plane normal to $\mathbf{m}$." For example, the effect on $\mathbf{r}$ of reflection in the plane normal to $\mathbf{x}$ is to change x into $-x$,

$$\sigma_{\mathbf{x}}\{\mathbf{e}_1 x + \mathbf{e}_2 y + \mathbf{e}_3 z\} = \{\mathbf{e}_1 \bar{x} + \mathbf{e}_2 y + \mathbf{e}_3 z\}. \tag{6}$$

Sometimes, the plane itself rather than its normal $\mathbf{m}$ is specified. Thus $\sigma_{\mathbf{yz}}$ is equivalent to $\sigma_{\mathbf{x}}$ and means "reflect in the plane containing $\mathbf{y}$ and $\mathbf{z}$" (called the $\mathbf{yz}$ plane) which is normal to the unit vector $\mathbf{x}$. However, the notation $\sigma_{\mathbf{m}}$ will be seen to introduce simplifications in later work involving the inversion operator and is to be preferred.

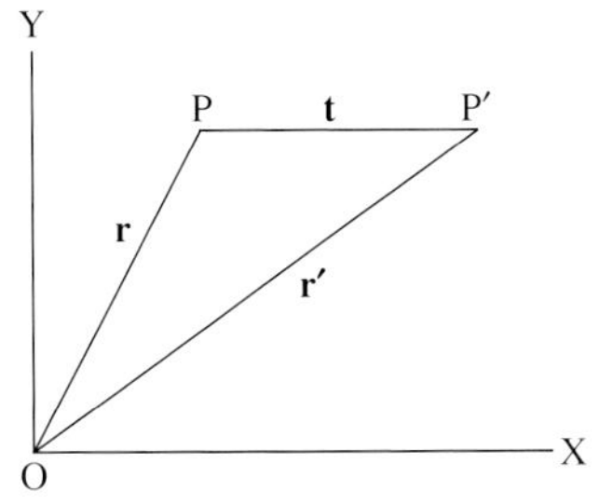

Figure 2.4. Example of a translation **t** in the active representation.

Another symmetry operator in common use is the *rotoreflection* operator

$$S_n^{\pm k} = \sigma_h R(\phi\ \mathbf{n}), \phi = \pm 2\pi k/n\ (n > 1,\ k = 1, 2, \ldots, \leq {}^n\!/_2,\ -\pi < \phi \leq \pi), \tag{7}$$

where σ_h "means reflection in a plane normal to the axis of rotation." All the symmetry operators, E, $R(\phi\ \mathbf{n}) = C_n$, $IR(\phi\ \mathbf{n})$, $\sigma_{\mathbf{m}}$, and S_n, leave at least one point invariant, and so they are called *point symmetry operators*. Contrast this with *translations*, an example of which is shown in Figure 2.4. Any point P in configuration space can be connected to the origin O by a vector **r**. In Figure 2.4, P happens to lie in the **xy** plane. Then under **t**, any point P is transformed into the point P′, which is connected to the origin by the vector **r**′, such that

$$\mathbf{r}' = \mathbf{r} + \mathbf{t}. \tag{8}$$

In Figure 2.4, **t** happens to be parallel to **x**. Translations are not point symmetry operations because *every* point in configuration space is translated with respect to the fixed axes OX, OY, OZ.

A *symmetry element* (which is not to be confused with a group element) is a point, line, or plane with respect to which a point symmetry operation is carried out. The symmetry elements, the notation used for them, the corresponding operation, and the notation used for the symmetry operators are summarized in Table 2.1. It is not necessary to use both $\tilde{n}$ and $\overline{n}$ since all configurations generated by $\tilde{n}$ can be produced by $\overline{n}'$.

Symmetry operations are conveniently represented by means of *projection diagrams*. A projection diagram is a circle which is the projection of a unit sphere in configuration space, usually on the **xy** plane, which we shall take to be the case unless otherwise stipulated. The x, y coordinates of a point on the sphere remain unchanged during the projection. A point on the hemisphere above the plane of the paper (and therefore with a positive z coordinate) will be represented in the projection by a small filled circle, and a point on the hemisphere below the plane of the paper will be represented by a larger open circle. A general point that will be transformed by point symmetry operators is marked by E. This point thus represents the initial configuration. Other points are then marked by the same symbol as the symmetry operator that produced that point from the initial one marked E. Commonly **z** is taken as normal to the plane of the paper, with **x** parallel to the top of the page, and when this is so it will not always be necessary to label the coordinate axes explicitly. An n-fold proper axis is commonly shown by an n-sided filled polygon (Figure 2.5). Improper axes are labeled by open polygons. A digon ($n = 2$) appears as

Table 2.1. *Symmetry elements and point symmetry operations.*

$\phi = 2\pi/n$, $n > 1$; **n** is a unit vector along the axis of rotation.

Symmetry element	Notation for symmetry element: Schönflies	International	Symmetry operation	Symmetry operator
None	–	–	identity	$E = R(\mathbf{0})^a$
Center	I	$\bar{1}$	inversion	I
Proper axis	C_n	n	proper rotation	$R(\phi\ \mathbf{n}) = C_n$ or $C_{n\mathbf{n}}$
Improper axis	IC_n	$\bar{n}$	rotation, then inversion	$IR(\phi\ \mathbf{n}) = IC_{n\mathbf{n}}$
Plane	$\sigma_{\mathbf{m}}$	m	reflection in a plane normal to **m**	$\sigma_{\mathbf{m}}$
Rotoreflection axis	S_n	$\tilde{n}$	rotation through $\phi = 2\pi/n$, followed by reflection in a plane normal to the axis of rotation	$S(\phi\ \mathbf{n}) = S_n$ or $S_{n\mathbf{n}}$

[a] For the identity, the rotation parameter (ϕ **n**) is zero, signifying no rotation.

$n =$	2	3	4	5	6	etc.
	digon	triangle	square	pentagon	hexagon	...
operator	C_2	C_3	C_4	C_5	C_6	...
$\phi = 2\pi/n$	π	$2\pi/3$	$\pi/2$	$2\pi/5$	$\pi/3$	...

Figure 2.5. Symbols used to show an n-fold proper axis. For improper axes the same geometrical symbols are used but they are not filled in. Also shown are the corresponding rotation operator and the angle of rotation ϕ.

though formed by two intersecting arcs. The point symmetry operations listed in Table 2.1 are illustrated in Figure 2.6.

Exercise 2.1-3 Using projection diagrams (a) prove that $IC_{2\mathbf{z}} = \sigma_{\mathbf{z}}$ and that $IC_{2\mathbf{n}} = \sigma_{\mathrm{h}}$; (b) show that I commutes with an arbitrary rotation $R(\phi\ \mathbf{n})$.

Example 2.1-1 Prove that a rotoreflection axis is an improper axis, though not necessarily of the same order.

In Figure 2.7, **n** is normal to the plane of the paper and $\phi > 0$. The open circle so marked is generated from E by $S(\phi\ \mathbf{n}) = \sigma_{\mathrm{h}} R(\phi\ \mathbf{n})$, while the second filled circle (again so marked) is generated from E by $R(\phi - \pi\ \mathbf{n})$. The diagram thus illustrates the identity

$$S(\pm|\phi|\ \mathbf{n}) = IR((\pm|\phi| \mp \pi)\ \mathbf{n}), \qquad 0 \leq |\phi| \leq \pi. \tag{9}$$

When $\phi > 0$, $R(\phi - \pi\ \mathbf{n})$ means a negative (clockwise) rotation about **n** through an angle of magnitude $\pi - \phi$. When $\phi < 0$, $R(\phi + \pi\ \mathbf{n})$ means a positive rotation through an angle $\pi + \phi$. Usually $\bar{n}$ is used in crystallography and S_n is used in molecular symmetry.

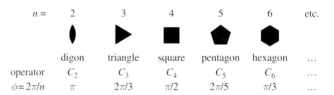

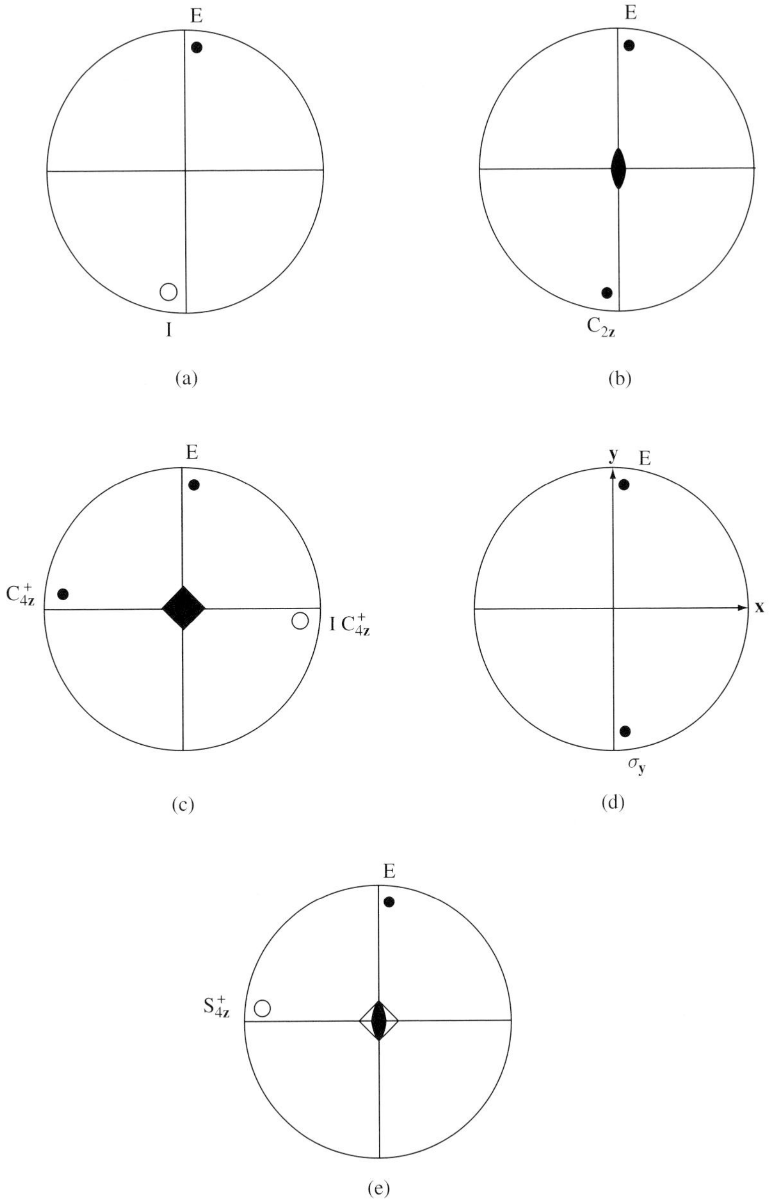

Figure 2.6. Projection diagrams showing examples of the point symmetry operators listed in Table 2.1. (a) I; (b) C_{2z}; (c) IC_{4z}^{+}; (d) σ_y; (e) S_{4z}^{+}.

It follows from Exercise 2.1-3(a) and Example 2.1-1 that the only necessary point symmetry operations are proper and improper rotations. Nevertheless, it is usually convenient to make use of reflections as well. However, if one can prove some result for R and IR, it will hold for all point symmetry operators.

As shown by Figure 2.8, $S_4^2 = C_2$. Consequently, the set of symmetry elements associated with an S_4 axis is $\{S_4\ C_2\}$, and the corresponding set of symmetry operators is

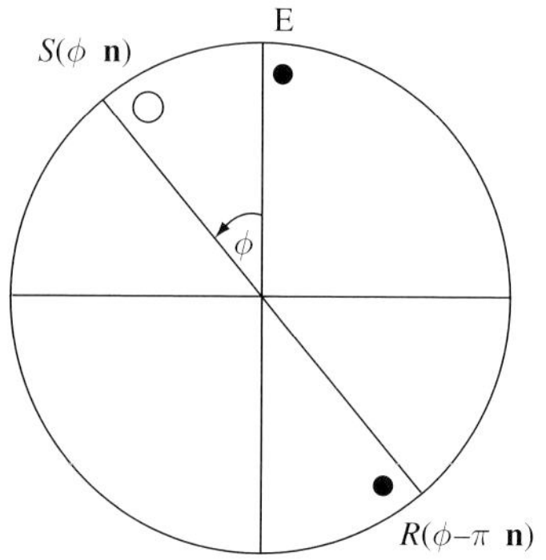

Figure 2.7. Demonstration of the equivalence of $S(\phi\ \mathbf{n})$ and $IR(\phi-\pi\ \mathbf{n})$ when $\phi > 0$ (see Example 2.1-1).

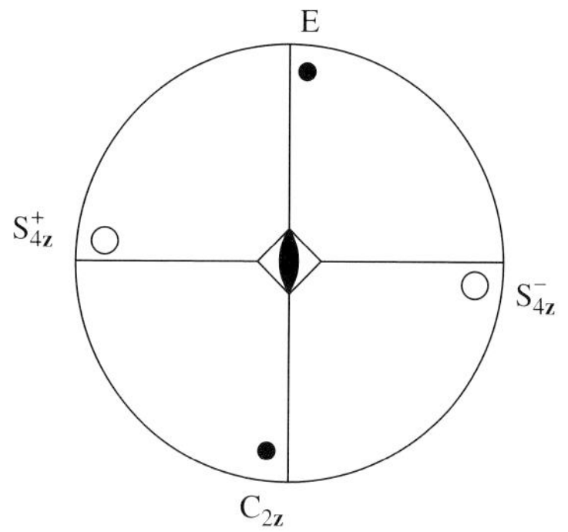

Figure 2.8. Projection diagram showing the operations connected with an S_4 axis.

$\{S_4^+\ C_2\ S_4^-\ E\}$. The identity operator E is always present (whether there is an axis of symmetry or not) and it must always be included once in any list of symmetry operators. The following convention is used in drawing up a list of symmetry operators: where the same configuration may be generated by equivalent symmetry operators we list only the "simplest form," that is the one of lowest n, with $-\pi < \phi \leq \pi$, avoiding redundancies. Thus C_2 and not S_4^2, S_4^- and not S_4^3, E and not S_4^4. The first part of this convention implies that whenever n/k in the operator $C_n^{\pm k}$ (or $S_n^{\pm k}$) is an integer p, then there is a C_p (or S_p), axis coincident with C_n (or S_n), and this should be included in the list of symmetry elements. Thus, for example, a C_6 axis implies coincident C_3 and C_2 axes, and the list of operators associated with C_6 is therefore $\{C_6^+\ C_6^-\ C_3^+\ C_3^-\ C_2\ E\}$.

The complete set of point symmetry operators that is generated from the operators $\{R_1\ R_2 \ldots\}$ that are associated with the symmetry elements (as shown, for example, in Table 2.2) by forming all possible products like $R_2\ R_1$, and including E, satisfies the necessary group properties: the set is complete (satisfies closure), it contains E, associativity is satisfied, and each element (symmetry operator) has an inverse. That this is so may be verified in any particular case: we shall see an example presently. Such groups of point symmetry operators are called *point groups*. For example, if a system has an S_4 axis and no

Table 2.2. *The multiplication table for the point group* S_4.

S_4	E	S_4^+	C_2	S_4^-
E	E	S_4^+	C_2	S_4^-
S_4^+	S_4^+	C_2	S_4^-	E
C_2	C_2	S_4^-	E	S_4^+
S_4^-	S_4^-	E	S_4^+	C_2

other symmetry elements (except the coincident C_2 axis that is necessarily associated with S_4) then the set of symmetry operators $\{E\ S_4^+\ C_2\ S_4^-\}$ satisfies all the necessary group properties and is the cyclic point group S_4.

Exercise 2.1-4 Construct the multiplication table for the set $\{E\ S_4^+\ C_2\ S_4^-\}$. Demonstrate by a sufficient number of examples that this set is a group. [*Hint*: Generally the use of projection diagrams is an excellent method of generating products of operators and of demonstrating closure.] In this instance, the projection diagram for S_4 has already been developed (see Figure 2.8).

Answers to Exercises 2.1

Exercise 2.1-1 (a) Figure 2.2(c) shows that the arrow has the opposite direction to the rotation of a right-handed screw as it moves along OZ from O. Also, on looking down the OZ axis towards O, the rotation appears to be in a clockwise direction. It is therefore a negative rotation with $-\pi < \phi < 0$.
(b) From Figure 2.9(a), $R(-\pi/2\ \mathbf{z})\{\mathbf{e}_1\ \mathbf{e}_2\ \mathbf{e}_3\} = \{\mathbf{e}_1'\ \mathbf{e}_2'\ \mathbf{e}_3'\} = \{\bar{\mathbf{e}}_2\ \mathbf{e}_1\ \mathbf{e}_3\}$.

Exercise 2.1-2 Both (a) and (b) are true from geometrical considerations. Formally, for (a) $R(\phi'\ \mathbf{n})\,R(\phi\ \mathbf{n}) = R(\phi'+\phi\ \mathbf{n}) = R(\phi+\phi'\ \mathbf{n}) = R(\phi\ \mathbf{n})\,R(\phi'\ \mathbf{n})$, and therefore rotations about the same axis commute.
(b) Following $R(\phi\ \mathbf{n})$ by $R(-\phi\ \mathbf{n})$ returns the representative point to its original position, a result which holds whether ϕ is positive or negative (see Figure 2.9(b)). Consequently, $R(-\phi\ \mathbf{n})\,R(\phi\ \mathbf{n}) = E$, so that $R(-\phi\ \mathbf{n}) = [R(\phi\ \mathbf{n})]^{-1}$.

Exercise 2.1-3 (a) Figure 2.9(c) shows that $IC_{2\mathbf{z}}$ is equivalent to $\sigma_\mathbf{z}$. Since the location of the axes is arbitrary, we may choose $\mathbf{n}$ (instead of $\mathbf{z}$) normal to the plane of the paper in Figure 2.9(c). The small filled circle would then be labeled by $C_{2\mathbf{n}}$ and the larger open circle by $I\,C_{2\mathbf{n}} = \sigma_\mathbf{n} = \sigma_\mathrm{h}$ (since σ_h means reflection in a plane normal to the axis of rotation). (b) Locate axes so that $\mathbf{n}$ is normal to the plane of the paper. Figure 2.9(d) then shows that $IR(\phi\ \mathbf{n}) = R(\phi\ \mathbf{n})I$, so that I commutes with an arbitrary rotation $R(\phi\ \mathbf{n})$.

Exercise 2.1-4 The set contains the identity E. Each column and each row of the multiplication table in Table 2.2 contains each member of the set once and once only

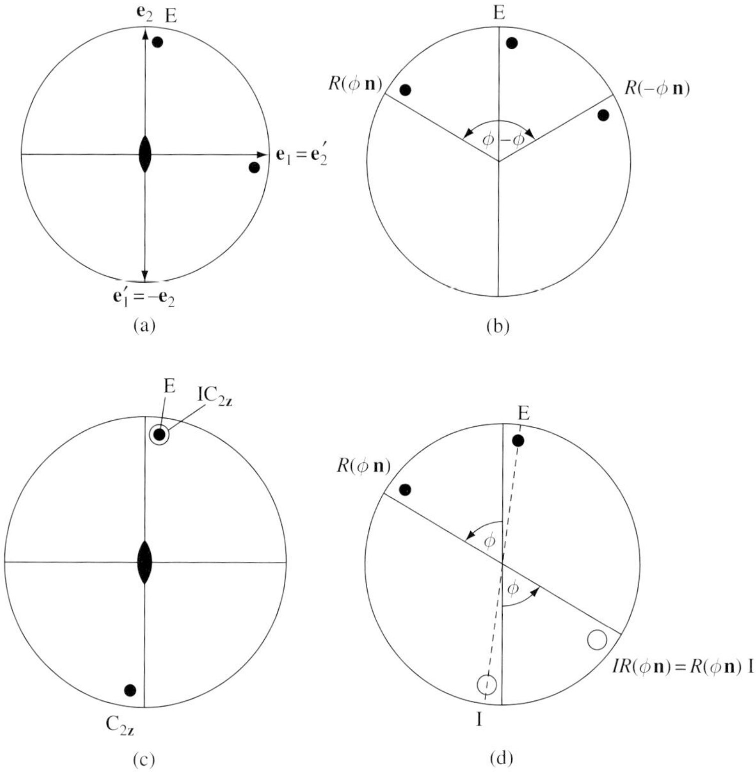

Figure 2.9. (a) The effect of $R(-\pi/2\ \mathbf{z})$ on $\{\mathbf{e}_1\ \mathbf{e}_2\ \mathbf{e}_3\}$. (b) When $\phi > 0$ the rotation $R(-\phi\ \mathbf{n})$ means a *clockwise* rotation through an angle of magnitude ϕ about $\mathbf{n}$, as illustrated. If $\phi < 0$, then $R(-\phi\ \mathbf{n})$ is an anticlockwise rotation about $\mathbf{n}$, and in either case the second rotation cancels the first. (c) This figure shows that $IC_{2\mathbf{z}} = \sigma_{\mathbf{z}}$. (d) The location of the coordinate axes is arbitrary; here the plane of the projection diagram is normal to $\mathbf{n}$.

(rearrangement theorem) so that the set is closed. Since E appears in each row or column, each element has an inverse. As a test of associativity, consider the following:

$$S_4^+(C_2\ S_4^-) = S_4^+\ S_4^+ = C_2; \quad (S_4^+\ C_2)S_4^- = S_4^-\ S_4^- = C_2,$$

which demonstrates that associativity is satisfied for this random choice of three elements from the set. Any other three elements chosen at random would also be found to demonstrate that binary combination is associative. Therefore, the group properties are satisfied. This is the cyclic group S_4.

2.2 The multiplication table – an example

Consider the set of point symmetry operators associated with a pyramid based on an equilateral triangle. Choose $\mathbf{z}$ along the C_3 axis. The set of distinct (non-equivalent) symmetry operators is $G = \{E\ C_3^+\ C_3^-\ \sigma_\mathbf{d}\ \sigma_\mathbf{e}\ \sigma_\mathbf{f}\}$ (Figure 2.10). Symmetry elements

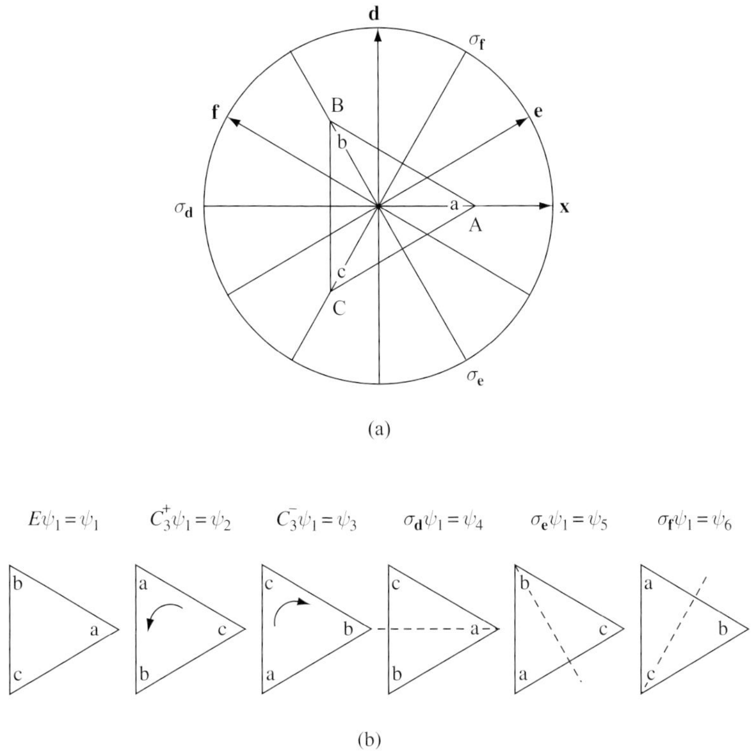

Figure 2.10. Effect of the set of symmetry operators G $= \{E \ C_3^+ \ C_3^- \ \sigma_{\mathbf{d}} \ \sigma_{\mathbf{e}} \ \sigma_{\mathbf{f}}\}$ on the triangular-based pyramid shown in (a). The C_3 principal axis is along **z**. The symmetry planes $\sigma_{\mathbf{d}}$, $\sigma_{\mathbf{e}}$, and $\sigma_{\mathbf{f}}$ contain **z** and make angles of zero, $-\pi/3$, and $+\pi/3$, respectively, with the **zx** plane. The apices of the triangle are marked a, b, and c for identification purposes only. Curved arrows in (b) show the direction of rotation under C_3^+ and C_3^-. Dashed lines show the reflecting planes.

(which here are $\{C_3 \ \sigma_{\mathbf{d}} \ \sigma_{\mathbf{e}} \ \sigma_{\mathbf{f}}\}$) are defined with respect to the Cartesian axes OX, OY, OZ, and remain *fixed*, while symmetry operators rotate or reflect the whole of configuration space including any material system – the pyramid – that exists in this space. The apices of the equilateral triangle are marked a, b, and c merely for identification purposes to enable us to keep track of the rotation or reflection of the pyramid in (otherwise) indistinguishable configurations. The three symmetry planes are *vertical* planes (σ_v) because they each contain the principal axis which is along **z**. The *reflecting plane* in the operation with $\sigma_{\mathbf{d}}$ contains the OX axis, while the reflecting planes in operations with $\sigma_{\mathbf{e}}$ and $\sigma_{\mathbf{f}}$ make angles of $-\pi/3$ and $+\pi/3$, respectively, with the **zx** plane. To help follow the configurations produced by these symmetry operators, we label the initial one ψ_1 and the other unique, indistinguishable configurations by $\psi_2, \ldots, \psi_6$. Thus, ψ_1 represents the state in which the apex marked a is adjacent to point A on the OX axis, and so on. The effect on ψ_1 of the symmetry operators that are $\in$ G is also shown in Figure 2.10, using small labeled triangles to show the configuration produced. Binary products are readily evaluated. For example,

$$C_3^+ C_3^+ \psi_1 = C_3^+ \psi_2 = \psi_3 = C_3^- \psi_1; \text{ therefore } C_3^+ C_3^+ = C_3^-; \qquad (1)$$

Table 2.3. *Multiplication table for the set* $G = \{E\ C_3^+\ C_3^-\ \sigma_\mathbf{d}\ \sigma_\mathbf{e}\ \sigma_\mathbf{f}\}$.

G	E	C_3^+	C_3^-	$\sigma_\mathbf{d}$	$\sigma_\mathbf{e}$	$\sigma_\mathbf{f}$
E	E	C_3^+	C_3^-	$\sigma_\mathbf{d}$	$\sigma_\mathbf{e}$	$\sigma_\mathbf{f}$
C_3^+	C_3^+	C_3^-	E	$\sigma_\mathbf{f}$	$\sigma_\mathbf{d}$	$\sigma_\mathbf{e}$
C_3^-	C_3^-	E	C_3^+	$\sigma_\mathbf{e}$	$\sigma_\mathbf{f}$	$\sigma_\mathbf{d}$
$\sigma_\mathbf{d}$	$\sigma_\mathbf{d}$	$\sigma_\mathbf{e}$	$\sigma_\mathbf{f}$	E	C_3^+	C_3^-
$\sigma_\mathbf{e}$	$\sigma_\mathbf{e}$	$\sigma_\mathbf{f}$	$\sigma_\mathbf{d}$	C_3^-	E	C_3^+
$\sigma_\mathbf{f}$	$\sigma_\mathbf{f}$	$\sigma_\mathbf{d}$	$\sigma_\mathbf{e}$	C_3^+	C_3^-	E

$$C_3^+ C_3^- \psi_1 = C_3^+ \psi_3 = \psi_1 = E\psi_1;\ \text{therefore } C_3^+ C_3^- = E; \tag{2}$$

$$C_3^+ \sigma_\mathbf{d} \psi_1 = C_3^+ \psi_4 = \psi_6 = \sigma_\mathbf{f}\psi_1;\ \text{therefore } C_3^+ \sigma_\mathbf{d} = \sigma_\mathbf{f}; \tag{3}$$

$$\sigma_\mathbf{d} C_3^+ \psi_1 = \sigma_\mathbf{d}\psi_2 = \psi_5 = \sigma_\mathbf{e}\psi_1;\ \text{therefore } \sigma_\mathbf{d} C_3^+ = \sigma_\mathbf{e}. \tag{4a}$$

Thus C_3^+ and $\sigma_\mathbf{d}$ do not commute. These operator equalities in eqs. (1)–(4a) are true for *any* initial configuration. For example,

$$\sigma_\mathbf{d} C_3^+ \psi_4 = \sigma_\mathbf{d}\psi_6 = \psi_3 = \sigma_\mathbf{e}\psi_4;\ \text{therefore } \sigma_\mathbf{d} C_3^+ = \sigma_\mathbf{e}. \tag{4b}$$

Exercise 2.2-1 Verify eqs. (1)–(4), using labeled triangles as in Figure 2.10.

Exercise 2.2-2 Find the products $C_3^- \sigma_\mathbf{e}$ and $\sigma_\mathbf{e} C_3^-$. The multiplication table for this set of operators $G = \{E\ C_3^+\ C_3^-\ \sigma_\mathbf{d}\ \sigma_\mathbf{e}\ \sigma_\mathbf{f}\}$ is shown in Table 2.3. The complete multiplication table has the following properties.

(a) Each column and each row contains each element of the set once and once only. This is an example of the rearrangement theorem, itself a consequence of closure and the fact that all products $g_i\ g_j$ are unique.
(b) The set contains the identity E, which occurs once in each row or column.
(c) Each element $g_i \in G$ has an inverse g_i^{-1} such that $g_i^{-1}\ g_i = E$.
(d) Associativity holds: $g_i(g_j\ g_k) = (g_i\ g_j)g_k,\ \forall\ g_i,\ g_j,\ g_k \in G$.

Exercise 2.2-3 Use the multiplication Table 2.3 to verify that $\sigma_\mathbf{d}(C_3^+\ \sigma_\mathbf{f}) = (\sigma_\mathbf{d}\ C_3^+)\sigma_\mathbf{f}$.

Any set with the four properties (a)–(d) forms a group: therefore the set G is a group for which the group elements are point symmetry operators. This point group is called C_{3v} or $3m$, because the pyramid has these symmetry elements: a three-fold principal axis and a vertical mirror plane. (If there is one vertical plane then there must be three, because of the three-fold symmetry axis.)

Exercise 2.2-4 Are the groups C_{3v} and S(3) isomorphous? [*Hint*: Compare Table 2.3 with Table 1.3.]

Answers to Exercises 2.2

Exercise 2.2-1 The orientation of the triangular base of the pyramid is shown for each of the indistinguishable configurations.

$C_3^+C_3^+\ \psi_1$		$=\ C_3^+\ \psi_2$		$=\ \psi_3$		$=\ C_3^-\ \psi_1$		
b		a		c		b		(1′)
	a		c		b		a	
c		b		a		c		

$C_3^+C_3^-\ \psi_1$		$=\ C_3^+\ \psi_3$		$=\ \psi_1$		$=\ E\ \psi_1$		
b		c		b		b		(2′)
	a		b		a		a	
c		a		c		c		

$C_3^+\sigma_{\mathbf{d}}\ \psi_1$		$=\ C_3^+\ \psi_4$		$=\ \psi_6$		$=\ \sigma_{\mathbf{f}}\ \psi_1$		
b		c		a		b		(3′)
	a		a		b		a	
c		b		c		c		

$\sigma_{\mathbf{d}}C_3^+\ \psi_1$		$=\ \sigma_{\mathbf{d}}\ \psi_2$		$=\ \psi_5$		$=\ \sigma_{\mathbf{e}}\ \psi_1$		
b		a		b		b		(4a′)
	a		c		c		a	
c		b		a		c		

$\sigma_{\mathbf{d}}C_3^+\ \psi_4$		$=\ \sigma_{\mathbf{d}}\ \psi_6$		$=\ \psi_3$		$=\ \sigma_{\mathbf{e}}\ \psi_4$		
c		a		c		c		(4b′)
	a		b		b		a	
b		c		a		b		

Exercise 2.2-2

$C_3^-\sigma_{\mathbf{e}}\psi_1$		$=\ C_3^-\psi_5$		$=\ \psi_6$		$=\ \sigma_{\mathbf{f}}\psi_1$	
b		b		a		b	
	a		c		b		a
c		a		c		c	

Exercise 2.2-3 $\sigma_{\mathbf{d}}(C_3^+\ \sigma_{\mathbf{f}}) = \sigma_{\mathbf{d}}\sigma_{\mathbf{e}} = C_3^+$ and $(\sigma_{\mathbf{d}}\ C_3^+)\sigma_{\mathbf{f}} = \sigma_{\mathbf{e}}\sigma_{\mathbf{f}} = C_3^+$.

Exercise 2.2-4 A comparison of the group multiplication tables in Table 2.3 and Table 1.3 shows that the point group C_{3v} (or $3m$) is isomorphous with the permutation group S(3). Corresponding elements in the two groups are

S(3)	P_0	P_1	P_2	P_3	P_4	P_5
C_{3v}	E	C_3^+	C_3^-	$\sigma_{\mathbf{d}}$	$\sigma_{\mathbf{e}}$	$\sigma_{\mathbf{f}}$

2.3 The symmetry point groups

We first describe the *proper point groups*, P, that is the point groups that contain the identity and proper rotations only.

(i) In the *cyclic groups*, denoted by n or C_n, with $n > 1$, there is only one axis of rotation and the group elements (symmetry operators) are E and $C_{n\mathbf{z}}^{\pm k}$, or $R(\phi\ \mathbf{n})$ with $\phi = \pm 2\pi k/n$, $-\pi < \phi \leq \pi$. Note that $C_{n\mathbf{z}}^{\pm k}$, becomes $C_{p\mathbf{z}}^{\pm}$, when $^n/_k$ is an integer p; $k = 1, 2, \ldots, (n-1)/2$, if n is an odd integer, and if n is even $k = 1, 2, \ldots, {}^n/_2$, with $C_{n\mathbf{z}}^{-n/2}$, excluded by the range of ϕ. For example, if $n = 4$, $k = 1, 2$, and $\phi = \pm\pi/2, \pi$. The symmetry elements are the C_4 axis, and a coincident C_2 axis, and the group elements (symmetry operators) are $\{E\ C_4^+\ C_4^-\ C_2\}$; $k = 1$ is implicit in $C_n^{\pm k}$. The projection diagram for C_4 is shown in Figure 2.11(a). C_1 is also a cyclic group (though not an axial group) with period $\{g_1 = E\}$ and order $c = 1$. There are no symmetry elements and the group consists solely of the identity E. The International notation used to describe the point groups is given in Table 2.4. Some International symbols are unnecessarily cumbersome, and these are abbreviated in Table 2.5.

(ii) The *dihedral groups* consist of the proper rotations that transform a regular n-sided prism into itself. The symmetry elements are C_n and $n\,C_2'$, where C_2' denotes a binary axis normal to the n-fold principal axis. (The prime is not essential but is often used to

Table 2.4. *International notation used to name the point groups comprises a minimal set of symmetry elements.*

n	n-fold proper axis ($n = 1$ means there is no axis of symmetry)
$\overline{n}$	n-fold improper axis ($\overline{n} = \overline{1}$ means an inversion center)
nm	n-fold proper axis with a vertical plane of symmetry that contains n
n/m	n-fold proper axis with a horizontal plane of symmetry normal to n
$n2$	n-fold proper axis with n binary axes normal to n

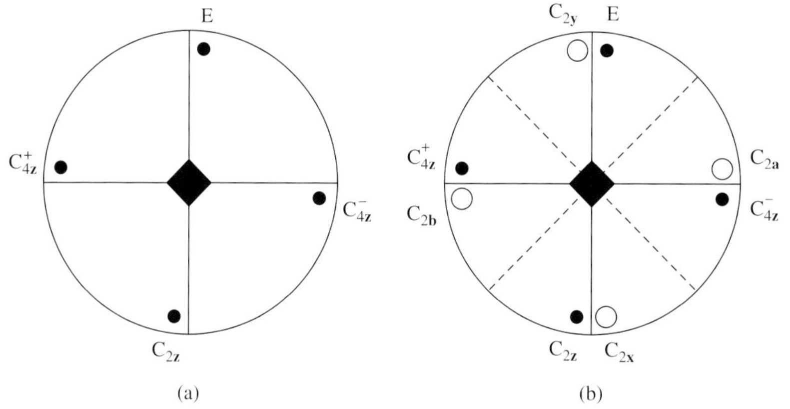

Figure 2.11. Projection diagrams (a) for the proper point group 4, or C_4, and (b) for the dihedral group 422, or D_4. The components of the unit vector **a** are $2^{-1/2}$ [1 1 0] and those of **b** are $2^{-1/2}$ [$\overline{1}$ 1 0].

Table 2.5. *Abbreviated International symbols and Schönflies notation.*

Schönflies symbol	Full International symbol	Abbreviated symbol
D_{2h}	$\frac{2}{m}\frac{2}{m}\frac{2}{m}$	mmm
D_{4h}	$\frac{4}{m}\frac{2}{m}\frac{2}{m}$	$4/mmm$
D_{6h}	$\frac{6}{m}\frac{2}{m}\frac{2}{m}$	$6/mmm$
D_{3d}	$\bar{3}\frac{2}{m}$	$\bar{3}m$
T_h	$\frac{2}{m}\bar{3}$	$m3$
O_h	$\frac{4}{m}\bar{3}\frac{2}{m}$	$m3m$

stress that a binary axis is normal to the principal axis and hence lies in the **xy** plane. In projective diagrams and descriptive text one refers to specific axes such as $C_{2\mathbf{x}}$ when greater precision is required.) The symmetry operators are $C_{nz}^{\pm k}$ or $R(\phi\ \mathbf{z})$, with ϕ and k as in (i), and $R(\pi\ \mathbf{n}_i)$, with $\mathbf{n}_i$ normal to **z** and $i = 1, \ldots, n$. In general, we shall use particular symbols for the $\mathbf{n}_i$, such as **x**, **y**, **a**, **b**, ..., with **a**, **b**, ... appropriately defined (see, for example, Figure 2.11(b)). The group symbol is D_n in Schönflies notation and in International notation it is $n2$ if n is odd and $n22$ if n is even, because there are then two sets of C_2' axes which are geometrically distinct. The projection diagram for 422 or D_4 is shown in Figure 2.11(b). The four binary axes normal to **z** lie along **x**, **y**, **a**, **b**, where **a** bisects the angle between **x** and **y** and **b** bisects that between $\bar{\mathbf{x}}$ and **y**. These axes can be readily identified in Figure 2.11(b) because each transformed point is labeled by the same symbol as that used for the operator that effected that particular transformation from the representative point E.

(iii) The *tetrahedral* point group, called 23 or T, consists of the proper rotations that transform a tetrahedron into itself. The symmetry elements are $3C_2$ and $4C_3$, and the easiest way of visualizing these is to draw a cube (Figure 2.12) in which alternate (second neighbor) points are the apices of the tetrahedron. These are marked 1, 2, 3, and 4 in Figure 2.12. The symmetry operators are

$$T = \{E\ R(\pi\ \mathbf{p})\ R(\pm 2\pi/3\ \mathbf{j})\}, \tag{1}$$

with $\mathbf{p} = \mathbf{x}, \mathbf{y}, \mathbf{z}$, and **j** a unit vector along O1, O2, O3, O4.

(iv) The *octahedral* or *cubic* group, named 432 or O, consists of the proper rotations that transform a cube or an octahedron into itself. The proper axes of the cube or octahedron are $\{3C_4\ 4C_3\ 9C_2\}$ and the symmetry operators are

$$O = \{T\} + \{R(\pi/2\ \mathbf{p})\ R(\pi\ \mathbf{n})\}, \tag{2}$$

where **n** is a unit vector along Oa, Ob, Oc, Od, Oe, Of in Figure 2.12 .

(v) The *icosahedral* group, named 532 or Y, consists of the proper rotations that transform an icosahedron or pentagonal dodecahedron into itself (Figure 2.13). The *pentagonal*

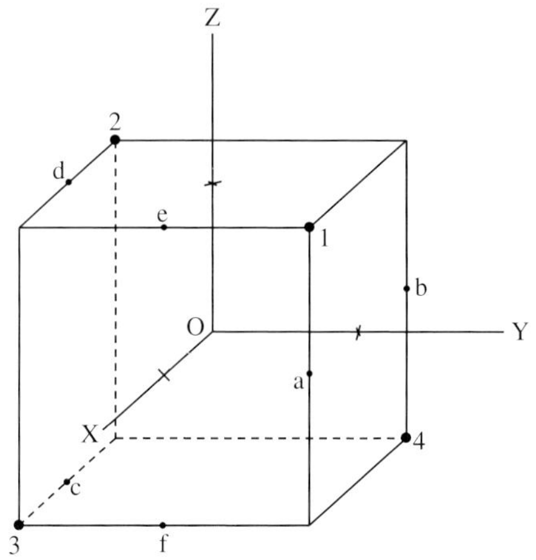

Figure 2.12. Alternate vertices of the cube (marked 1, 2, 3, and 4) are the apices of a regular tetrahedron. O1, O2, O3, and O4 are three-fold axes of symmetry. Small crosses show where the C_4 axes, OX, OY, and OZ, intersect the cube faces. Oa, Ob, Oc, Od, Oe, and Of are six binary axes.

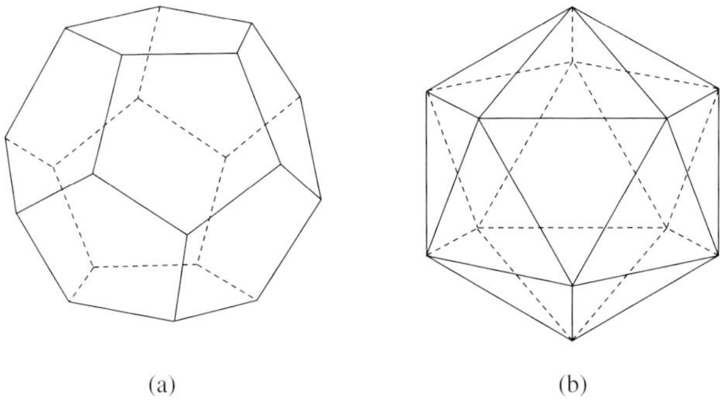

Figure 2.13. The dodecahedron and the icosahedron are two of the five Platonic solids (regular polyhedra), the others being the tetrahedron, the cube, and the octahedron. (a) The dodecahedron has twelve regular pentagonal faces with three pentagonal faces meeting at a point. (b) The icosahedron has twenty equilateral triangular faces, with five of these meeting at a point.

dodecahedron has six C_5 axes through opposite pairs of pentagonal faces, ten C_3 axes through opposite pairs of vertices, and fifteen C_2 axes that bisect opposite edges. The *icosahedron* has six C_5 axes through opposite vertices, ten C_3 axes through opposite pairs of faces, and fifteen C_2 axes that bisect opposite edges. For both these polyhedra, the symmetry elements that are proper axes are $\{6C_5 \;\; 10C_3 \;\; 15C_2\}$ and the point group of symmetry operators is therefore

$$\mathrm{Y} = \{E \;\; 6C_5^{\pm} \;\; 6C_5^{2\pm} \;\; 10C_3^{\pm} \;\; 15C_2\} \tag{3}$$

for a total $g(\mathrm{Y})$ of 60. It is isomorphous to the group of even permutations on five objects, which number $5!/2$.

This completes the list of proper point groups, P. A summary is given in the first column of Table 2.6. All the remaining axial point groups may be generated from the proper point groups P by one or other of two methods.

2.3.1 First method

This consists of taking the direct product (DP) of P with $\bar{1}$ or $C_i = \{E\ I\}$.

(i) From C_n, if n is odd,

$$C_n \otimes C_i = S_{2n},\ \ n \otimes \bar{1} = \bar{n}. \tag{4}$$

But if n is even,

$$C_n \otimes C_i = C_{nh},\ \ n \otimes \bar{1} = n/m, \tag{5}$$

where h, or $/m$, denotes a mirror plane normal to the principal axis, which arises because $IC_2 = \sigma_h$.

Example 2.3-1 (a) $C_2 \otimes C_i = \{E\ C_{2\mathbf{z}}\} \otimes \{E\ I\} = \{E\ C_{2\mathbf{z}}\ I\ \sigma_{\mathbf{z}}\} = C_{2h}$. (b) $C_3 \otimes C_i = \{E\ C_{3\mathbf{z}}^+\ C_{3\mathbf{z}}^-\} \otimes \{E\ I\} = \{E\ C_{3\mathbf{z}}^+\ C_{3\mathbf{z}}^-\ I\ S_{6\mathbf{z}}^-\ S_{6\mathbf{z}}^+\} = S_6$. Projection diagrams are illustrated in Figure 2.14.

(ii) From D_n, if n is odd,

$$D_n \otimes C_i = D_{nd},\ \ n2 \otimes \bar{1} = nm. \tag{6}$$

The subscript d denotes the presence of dihedral planes which bisect the angles between C_2' axes that are normal to the principal axis. If n is even,

$$D_n \otimes C_i = D_{nh};\ \ n22 \otimes \bar{1} = n/mmm. \tag{7}$$

If n is 2, the International symbol is abbreviated to mmm (Table 2.4).

Example 2.3-2

$$\begin{aligned} D_3 \otimes C_i &= \{E\ C_{3\mathbf{z}}^+\ C_{3\mathbf{z}}^-\ C_{2\mathbf{a}}\ C_{2\mathbf{b}}\ C_{2\mathbf{c}}\} \otimes \{E\ I\} \\ &= \{D_3\} + \{I\ S_{6\mathbf{z}}^-\ S_{6\mathbf{z}}^+\ \sigma_{\mathbf{a}}\ \sigma_{\mathbf{b}}\ \sigma_{\mathbf{c}}\} = D_{3d}; \end{aligned} \tag{8}$$

$\sigma_{\mathbf{c}}$, for example, denotes reflection in a dihedral plane **zf** that bisects the angle between **a** and **b**, which are the binary axes normal to the C_3 axis (Figure 2.10). The notation in eq. (8) is intentionally detailed, but may be compressed, as in

$$D_3 \otimes C_i = \{E\ 2C_3\ 3C_2'\} \otimes \{E\ I\} = \{E\ 2C_3\ 3C_2'\ I\ 2S_6\ 3\sigma_d\} = D_{3d}. \tag{9}$$

Exercise 2.3-1 Confirm the DP $D_3 \otimes C_i$ in eq. (9) by constructing the (labeled) projection diagram for D_{3d}. Identify the dihedral planes.

(iii) $$T \otimes C_i = T_h;\quad 23 \otimes \bar{1} = m3. \tag{10}$$

Table 2.6. *Derivation of commonly used finite point groups from proper point groups.*

If P has an invariant subgroup Q of index 2 so that $P = \{Q\} + R\{Q\}$, $R \in P$, $R \notin Q$, then $P' = \{Q\} + IR\{Q\}$ is a group isomorphous with P. In each column, the symbol for the point group is given in International notation on the left and in Schönflies notation on the right. When $n = 2$, the International symbol for D_{2h} is *mmm*. When *n* is odd, the International symbol for C_{nv} is *nm*, and when *n* is even it is *nmm*. Note that $n' = n/2$. In addition to these groups, which are either a proper point group P, or formed from P, there are the three cyclic groups: 1 or $C_1 = \{E\}$, $\bar{1}$ or $C_i = \{E\ I\}$, and *m* or $C_s = \{E\ \sigma\}$.

P		$P \otimes C_i$		$P = Q + IR\{Q\}$		Q	
n $(n = 2, 3, \ldots, 8)$	C_n	$\bar{n}$ $(n = 3, 5)$	S_{2n}	$\bar{n}$ $(n' = 3, 5)$	$C_{n'h}$	n'	$C_{n'}$
		n/m $(n = 2, 4, 6)$	C_{nh}	$\bar{n}$ $(n' = 2, 4)$	$S_{2n'}$		
$n2$ $(n = 3, 5)$	D_n	$\bar{n}m$ $(n = 3, 5)$	D_{nd}	nm, nmm $(n = 2, 3, \ldots, 6)$	C_{nv}	n	C_n
$n22$ $(n = 2, 4, 6)$	D_n	n/mmm $(n = 2, 4, 6, 8)$	D_{nh}	$\bar{n}2m$ $(n' = 3, 5)$	$D_{n'h}$	$n'2$	$D_{n'}$
				$(n' = 2, 4, 6)$	$D_{n'd}$	$n'22$	$D_{n'}$
23	T	$m3$	T_h				
432	O	$m3m$	O_h	$\bar{4}3m$	T_d	23	T
532	Y	$53m$	Y_h				

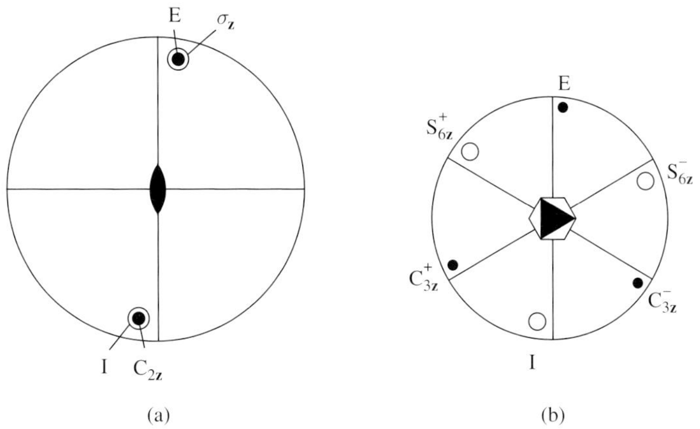

Figure 2.14. Projection diagrams for the point groups (a) C_{2h} and (b) S_6.

In abbreviated notation,

$$\begin{aligned} \mathrm{T} \otimes \mathrm{C}_i &= \{E\ 4C_3^+\ 4C_3^-\ 3C_2\} \otimes \{E\ I\} \\ &= \{\mathrm{T}\} + \{I\ 4S_6^-\ 4S_6^+\ 3\sigma_\mathrm{h}\} = \mathrm{T_h}. \end{aligned} \tag{11}$$

As shown in Figure 2.15(a), $IC_{2\mathbf{y}}$ (for example) is $\sigma_\mathbf{y}$. The plane normal to **y**, the **zx** plane, *contains* $C_{2\mathbf{z}}$ and $C_{2\mathbf{x}}$, and so this is a horizontal plane (normal to $C_{2\mathbf{y}}$) and *not* a dihedral plane, because it contains the other C_2 axes ($C_{2\mathbf{z}}$ and $C_{2\mathbf{x}}$) and does not bisect the angle between them. Note that $\mathrm{T} = C_2 \wedge C_3$ is 23 in International notation but that $D_3 = C_3 \wedge C_2$ is 32.

(iv) $$\mathrm{O} \otimes \mathrm{C}_i = \mathrm{O_h}, \quad 432 \otimes \bar{1} = m3m. \tag{12}$$

In abbreviated notation,

$$\begin{aligned} \mathrm{O} \otimes \mathrm{C}_i &= \{E\ 6C_4\ 3C_2\ 6C_2'\ 8C_3\} \otimes \{E\ I\} \\ &= \{\mathrm{O}\} + \{I\ 6S_4\ 3\sigma_\mathrm{h}\ 6\sigma_\mathrm{d}\ 8S_6\}. \end{aligned} \tag{13}$$

The three S_4 axes are coincident with the three C_4 (and coincident C_2) axes along **x**, **y**, **z**. The three horizontal planes $\sigma_\mathbf{x}$, $\sigma_\mathbf{y}$, and $\sigma_\mathbf{z}$ and two of the six dihedral planes $\sigma_\mathbf{a}$, $\sigma_\mathbf{b}$ are shown in Figures 2.15(b) and (c).

(v) $$\mathrm{Y} \otimes \mathrm{C}_i = \mathrm{Y_h}, \quad 532 \otimes \bar{1} = 53m; \tag{14}$$

$$\begin{aligned} \mathrm{Y} \otimes \mathrm{C}_i &= \{E\ 24C_5\ 20C_3\ 15C_2\} \otimes \{E\ I\} \\ &= \{\mathrm{Y}\} + \{I\ 24S_{10}\ 20S_6\ 15\sigma_\mathrm{h}\}. \end{aligned} \tag{15}$$

The six S_{10} axes are coincident with the six C_5 axes of Y, and the ten S_6 axes are coincident with the ten C_3 axes of Y. The fifteen mirror planes each contain two C_2 axes and two C_5 axes. All these DPs are given in the second column of Table 2.6.

Exercise 2.3-2 Draw a projection diagram showing that $C_5 \otimes C_i = S_{10}$.

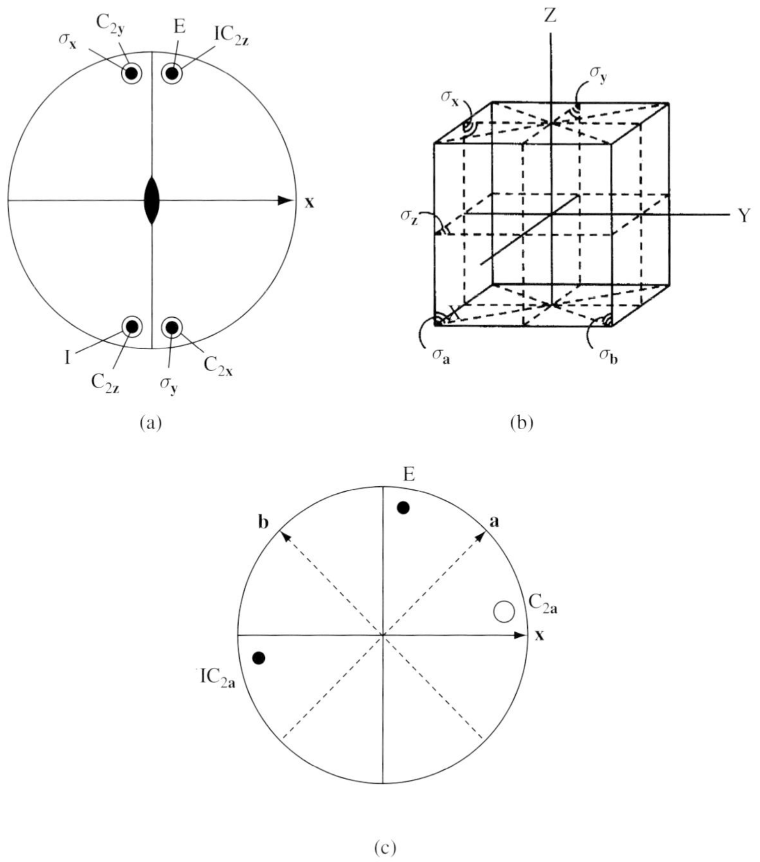

Figure 2.15. (a) In $T \otimes C_i$, $IC_{2y} = \sigma_y$, and σ_y contains the other two C_2 axes, C_{2z} and C_{2x}. Since σ_y is normal to the axis of rotation **y**, it is a horizontal plane, not a dihedral plane. (b) In $O \otimes C_i$, $IC_2 = \sigma_h$, as, for example, $IC_{2x} = \sigma_x$, which contains **y** and **z**. In (c), **a** is the unit vector along Oa in Figure 2.12, and $IC_{2a} = \sigma_a$. This dihedral plane is also shown in (b).

2.3.2 Second method

The second method is applicable to proper point groups P that have an invariant subgroup Q of index 2, so that

$$P = \{Q\} + R\{Q\}, \; R \in P, \; R \notin Q. \tag{16}$$

Then $\{Q\} + IR\{Q\}$ is a point group P′ which is isomorphous with P and therefore has the same class structure as P. The isomorphism follows from the fact that I commutes with any proper or improper rotation and therefore with any other symmetry operator. Multiplication tables for P and P′ are shown in Table 2.7; we note that these have the same structure and that the two groups have corresponding classes, the only difference being that some products X are replaced by IX in P′. Examples are given below.

Table 2.7. *Multiplication tables for* P *and* P′, *where* $P = \{Q\} + R\{Q\}$ *and* $P' = Q + IR\{Q\}$. $A, B \in Q$ and $C, D \in R\,Q$. Use has been made of the commutation property of I with any other symmetry operator.

P	{Q}	R\{Q\}	P′	{Q}	IR\{Q\}
{Q}	$\{AB\}$	$\{AD\}$	{Q}	$\{AB\}$	$I\{AD\}$
R\{Q\}	$\{CB\}$	$\{CD\}$	IR\{Q\}	$I\{CB\}$	$\{CD\}$

Exercise 2.3-3 If $X \in R\{Q\}$ and X, Y are conjugate elements in P, show that IX and IY are conjugate elements in P′.

(i) C_{2n} has the invariant subgroup C_n of index 2, because

$$C_{2n} = \{C_n\} + C_{2n}\{C_n\}. \tag{17}$$

Note that C_n means the point group C_n, but $\{C_n\}$ means the set of operators forming the point group C_n. Then

$$\{C_n\} + IC_{2n}\{C_n\} = S_{2n}\ (n \text{ even}),\ \text{or}\ = C_{nh}\ (n \text{ odd}). \tag{18}$$

In Table 2.6, n' is defined as $n/2$ to avoid any possible confusion when using International notation; $S_{2n'}$ is, of course, S_n.

Example 2.3-3

$$C_2 = E + C_2\{E\} = \{E\ C_2\}, \tag{19}$$

$$E + IC_2\{E\} = \{E\ \sigma_h\} = C_s. \tag{20}$$

The multiplication tables are

C_2	E	C_2
E	E	C_2
C_2	C_2	E

C_s	E	σ_h
E	E	σ_h
σ_h	σ_h	E

This is a rather trivial example: the classes of C_2 are E, C_2 and those of C_s are E, σ_h. Elements $X \in P$ and $IX \in P'$ are called *corresponding elements*, so here C_2 and $IC_2 = \sigma_h$ are corresponding elements.

Exercise 2.3-4 Use the second method to derive the point group P′corresponding to the proper point group C_4. Show that C_4 and P′ are isomorphous and find the classes of both groups.

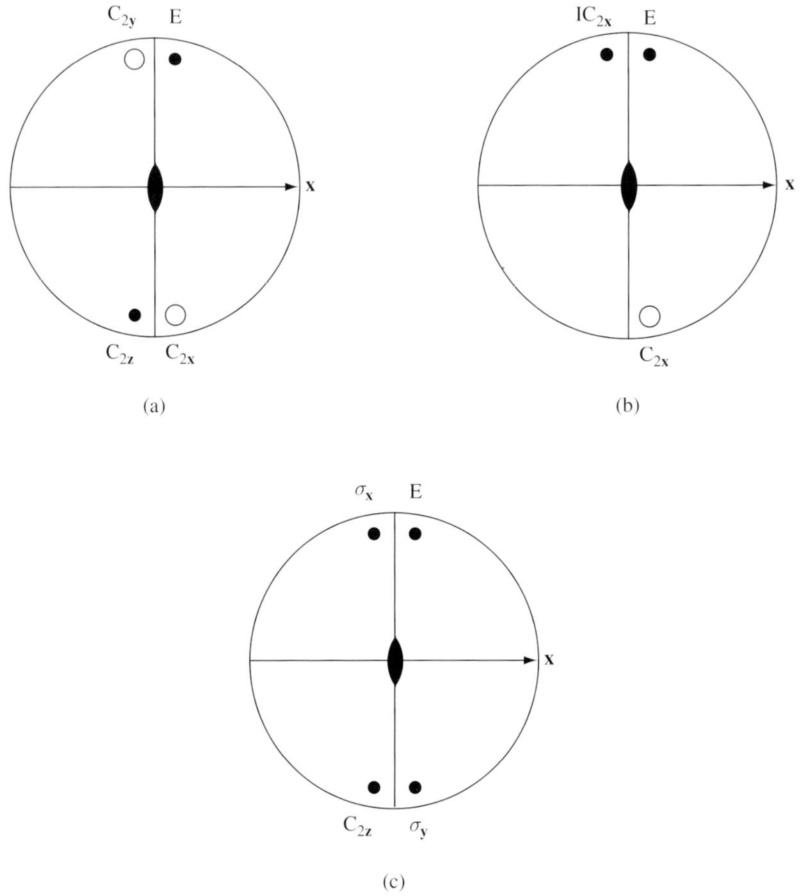

Figure 2.16. Projection diagrams (a) for D_2; (b) showing that $IC_{2x} = \sigma_{\mathbf{x}}$; and (c) for C_{2v}.

(ii) D_n has the invariant subgroup C_n. The coset expansion of D_n on C_n is

$$D_n = \{C_n\} + C_2'\{C_n\}. \tag{21}$$

(21)

$$\{C_n\} + IC_2'\{C_n\} = \{C_n\} + \sigma_v\{C_n\} = C_{nv}. \tag{22}$$

For example, for $n = 2$,

$$D_2 = \{E\ C_{2\mathbf{z}}\} + C_{2\mathbf{x}}\{E\ C_{2\mathbf{z}}\} = \{E\ C_{2\mathbf{z}}\ C_{2\mathbf{x}}\ C_{2\mathbf{y}}\}, \tag{23}$$

(23)

$$\begin{aligned}\{E\ C_{2\mathbf{z}}\} + IC_{2\mathbf{x}}\{E\ C_{2\mathbf{z}}\} &= \{E\ C_{2\mathbf{z}}\} + \sigma_{\mathbf{x}}\{E\ C_{2\mathbf{z}}\} \\ &= \{E\ C_{2\mathbf{z}}\ \sigma_{\mathbf{x}}\ \sigma_{\mathbf{y}}\} = C_{2v}.\end{aligned} \tag{24}$$

The projection diagrams illustrating D_2 and C_{2v} are in Figure 2.16.

D_{2n} has the invariant subgroup D_n of index 2, with the coset expansion

$$D_{2n} = \{D_n\} + C_{2n}\{D_n\}; \tag{25}$$

Table 2.8. *The relation of the point groups* O *and* T_d *to their invariant subgroup* T.

C_{31}^{+} means a positive rotation through $2\pi/3$ about the axis O1 and similarly (see Figure 2.12). $C_{2\mathbf{a}}$ means a rotation through π about the unit vector **a** along [1 1 0], and $\sigma_{\mathbf{a}}$ means a reflection in the mirror plane normal to **a**.

$\{T\} = \{E\ C_{2z}\ C_{2x}\ C_{2y}\ C_{31}^{+}\ C_{31}^{-}\ C_{32}^{+}\ C_{32}^{-}\ C_{33}^{+}\ C_{33}^{-}\ C_{34}^{+}\ C_{34}^{-}\}$
$C_{4z}^{+}\{T\} = \{C_{4z}^{+}\ C_{4z}^{-}\ C_{2\mathbf{a}}\ C_{2\mathbf{b}}\ C_{2\mathbf{c}}\ C_{4y}^{-}\ C_{2\mathbf{f}}\ C_{4y}^{+}\ C_{4x}^{+}\ C_{2\mathbf{d}}\ C_{4x}^{-}\ C_{2\mathbf{e}}\}$
$IC_{4z}^{+}\{T\} = \{S_{4z}^{-}\ S_{4z}^{+}\ \sigma_{\mathbf{a}}\ \sigma_{\mathbf{b}}\ \sigma_{\mathbf{c}}\ S_{4y}^{+}\ \sigma_{\mathbf{f}}\ S_{4y}^{-}\ S_{4x}^{-}\ \sigma_{\mathbf{d}}\ S_{4x}^{+}\ \sigma_{\mathbf{e}}\}$

(25) $$\{D_n\} + IC_{2n}\{D_n\} = D_{nd}\ (n \text{ even}),\ \text{or}\ D_{nh}\ (n \text{ odd}). \tag{26}$$

For example, if $n = 2$,

$$D_4 = \{D_2\} + C_{4z}^{+}\{D_2\} = \{E\ C_{2z}\ C_{2x}\ C_{2y}\ C_{4z}^{+}\ C_{4z}^{-}\ C_{2\mathbf{a}}\ C_{2\mathbf{b}}\}, \tag{27}$$

where **a** is the unit vector bisecting the angle between **x** and **y**, and **b** is that bisecting the angle between $\bar{\mathbf{x}}$ and **y**. The projection diagram for D_4 is shown in Figure 2.11(b). Applying the second method,

$$\begin{aligned} \{D_2\} + IC_{4z}^{+}\{D_2\} &= \{E\ C_{2z}\ C_{2x}\ C_{2y}\} + S_{4z}^{-}\{E\ C_{2z}\ C_{2x}\ C_{2y}\} \\ &= \{E\ C_{2z}\ C_{2x}\ C_{2y}\ S_{4z}^{-}\ S_{4z}^{+}\ \sigma_{\mathbf{a}}\ \sigma_{\mathbf{b}}\} = D_{2d}. \end{aligned} \tag{28}$$

(iv) O has the invariant subgroup T of index 2:

$$O = \{T\} + C_4^{+}\{T\} = \{E\ 3C_2\ 8C_3\ 6C_4\ 6C_2{}'\} \tag{29}$$

$$\{T\} + IC_{4z}^{+}\{T\} = \{E\ 3C_2\ 8C_3\ 6S_4\ 6\sigma_d\} = T_d. \tag{30}$$

The detailed verification of eqs. (29) and (30) is quite lengthy, but is summarized in Table 2.8.

(iii), (v) The point groups T, Y have no invariant subgroups of index 2.

This completes the derivation of the point groups that are important in molecular symmetry, with the exception of the two continuous rotation groups $C_{\infty v}$ and $D_{\infty h}$, which apply to linear molecules.

The rotation of a heteronuclear diatomic molecule like HCl through any angle ϕ about **z** (which is always chosen to lie along the molecular axis) leaves the molecule in an indistinguishable configuration. The point group therefore contains an infinite number of rotation operators $R(\phi\ \mathbf{z})$. Similarly, there are an infinite number of vertical planes of symmetry in the set of symmetry elements and the point group contains $\infty\sigma_v$. The point group is therefore called $C_{\infty v}$. For homonuclear diatomic molecules like O_2, or polyatomic linear molecules with a horizontal plane of symmetry, the point group also contains σ_h and an infinite number of C_2' axes normal to the principal axis (which is along the molecular axis). Such molecules belong to the point group $D_{\infty h}$.

For crystals, the point group must be compatible with translational symmetry, and this requirement limits n to 2, 3, 4, or 6. (This restriction applies to both proper and improper axes.) Thus the *crystallographic point groups* are restricted to ten proper point groups and a total of

Table 2.9. *The thirty-two crystallographic point groups in both International and Schönflies notation.*

In addition to the proper point groups P and the improper point groups that are either isomorpous with P or equal to P $\otimes$ C_i, there is the non-axial group 1 or $C_1 = \{E\}$.

Proper point group P		Improper group P′ isomorphous to P		P $\otimes$ C_i		Proper group isomorphous to P $\otimes$ C_i
2	C_2	m	C_s	$2/m$	C_{2h}	D_2
		$\bar{1}$	C_i			
3	C_3			$\bar{3}$	S_6	C_6
4	C_4	$\bar{4}$	S_4	$4/m$	C_{4h}	
6	C_6	$\bar{6}$	C_{3h}	$6/m$	C_{6h}	
222	D_2	$2mm$	C_{2v}	mmm	D_{2h}	
32	D_3	$3m$	C_{3v}	$\bar{3}m$	D_{3d}	D_6
422	D_4	$4mm$	C_{4v}	$4/mmm$	D_{4h}	
		$\bar{4}2m$	D_{2d}			
622	D_6	$6mm$	C_{6v}	$6/mmm$	D_{6h}	
		$\bar{6}m2$	D_{3h}			
23	T			$m3$	T_h	
432	O	$\bar{4}3m$	T_d	$m3m$	O_h	

thirty-two point groups, thirteen of which are isomorphous with at least one other crystallographic point group. The thirty-two crystallographic point groups are listed in Table 2.9.

Answers to Exercises 2.3

Exercise 2.3-1 The projection diagram is given in Figure 2.17. The dihedral planes are $\sigma_{\mathbf{x}}$, $\sigma_{\mathbf{b}}$, and $\sigma_{\mathbf{c}}$, where $\sigma_{\mathbf{x}}$ bisects the angle between $-\mathbf{b}$ and $\mathbf{c}$, $\sigma_{\mathbf{b}}$ bisects the angle between $\mathbf{x}$ and $\mathbf{c}$, and $\sigma_{\mathbf{c}}$ bisects the angle between $\mathbf{x}$ and $\mathbf{b}$.

Exercise 2.3-2 See Figure 2.18.

Exercise 2.3-3 If X, $Y \in$ P are conjugate, then for some $p_j \in$ P, $p_j\ X\ p_j^{-1} = Y$. But if $X \in R\{\mathrm{Q}\}$ in P, then $IX \in IR\{\mathrm{Q}\}$ in P′ and $p_j\ IX\ p_j^{-1} = IY$, so that IX and IY are conjugate in P′.

Exercise 2.3-4 $C_4 = \{C_2\} + C_4^+\ \{C_2\} = \{E\ C_2\} + C_4^+\ \{E\ C_2\} = \{E\ C_2\ C_4^+\ C_4^-\}$. But $\{C_2\} + IC_4^+\ \{C_2\} = \{E\ \ C_2\} + S_4^-\ \{E\ \ C_2\} = \{E\ \ C_2\ \ S_4^-\ \ S_4^+\} = S_4$. Use projection diagrams, if necessary, to verify the multiplication tables given in Tables 2.10 and 2.11. Clearly, the two multiplication tables are the same, corresponding elements being C_4^+ and $IC_4^+ = S_4^-$; C_4^- and $IC_4^- = S_4^+$. Both groups are Abelian.

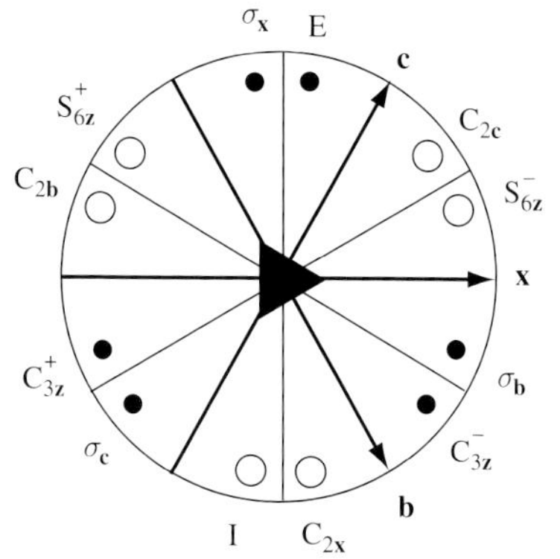

Figure 2.17. Projection diagram for the point group $D_{3d} = D_3 \otimes C_i$ (see eq. (2.3.9)). For example, $IC_{2\mathbf{b}} = \sigma_{\mathbf{b}}$, and this mirror plane normal to **b** bisects the angle between the C_2' axes $C_{2\mathbf{x}}$ and $C_{2\mathbf{c}}$ so that it is a dihedral plane. Similarly, $\sigma_{\mathbf{x}}$ and $\sigma_{\mathbf{c}}$ are dihedral planes.

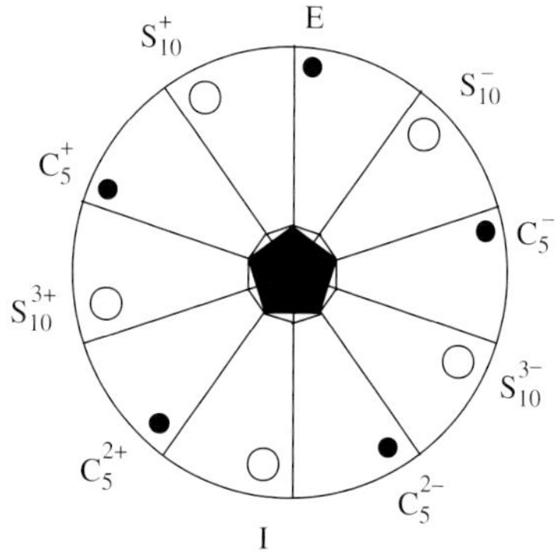

Figure 2.18. Projection diagram for the point group S_{10}.

Table 2.10. *Multiplication table for* C_4.

C_4	E	C_4^+	C_2	C_4^-
E	E	C_4^+	C_2	C_4^-
C_4^+	C_4^+	C_2	C_4^-	E
C_2	C_2	C_4^-	E	C_4^+
C_4^-	C_4^-	E	C_4^+	C_2

Table 2.11. *Multiplication table for* S_4.

S_4	E	S_4^-	C_2	S_4^+
E	E	S_4^-	C_2	S_4^+
S_4^-	S_4^-	C_2	S_4^+	E
C_2	C_2	S_4^+	E	S_4^-
S_4^+	S_4^+	E	S_4^-	C_2

2.4 Identification of molecular point groups

A systematic method for identifying the point group of any molecule is given in Figure 2.19. Some practice in the recognition of symmetry elements and in the assignment of point groups may be obtained through working through the following exercises and problems.

Exercise 2.4-1 Identify the symmetry point groups to which the following molecules belong. [*Hint*: For the two staggered configurations, imagine the view presented on looking down the C—C molecular axis.]

(a) nitrosyl chloride NOCl (non-linear),
(b) carbon dioxide O=C=O (linear),
(c) methane CH_4 (Figure 2.20),
(d) formaldehyde $H_2C{=}O$,
(e) carbonate ion CO_3^{-2} (planar),
(f) BrF_5 (pyramidal),
(g) staggered $H_3C{-}CCl_3$,
(h) $[PtCl_4]^{-2}$ (planar),
(i) staggered ethane $H_3C{-}CH_3$,
(j) $B(OH)_3$ (planar, Figure 2.20),
(k) IF_7 (pentagonal bipyramid),
(l) S_4 (non-planar).

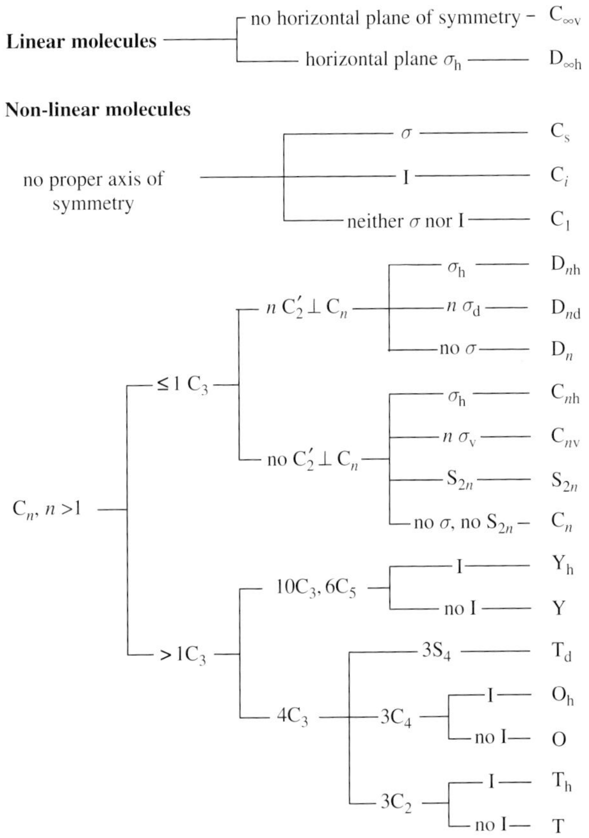

Figure 2.19. Identification of molecular point groups.

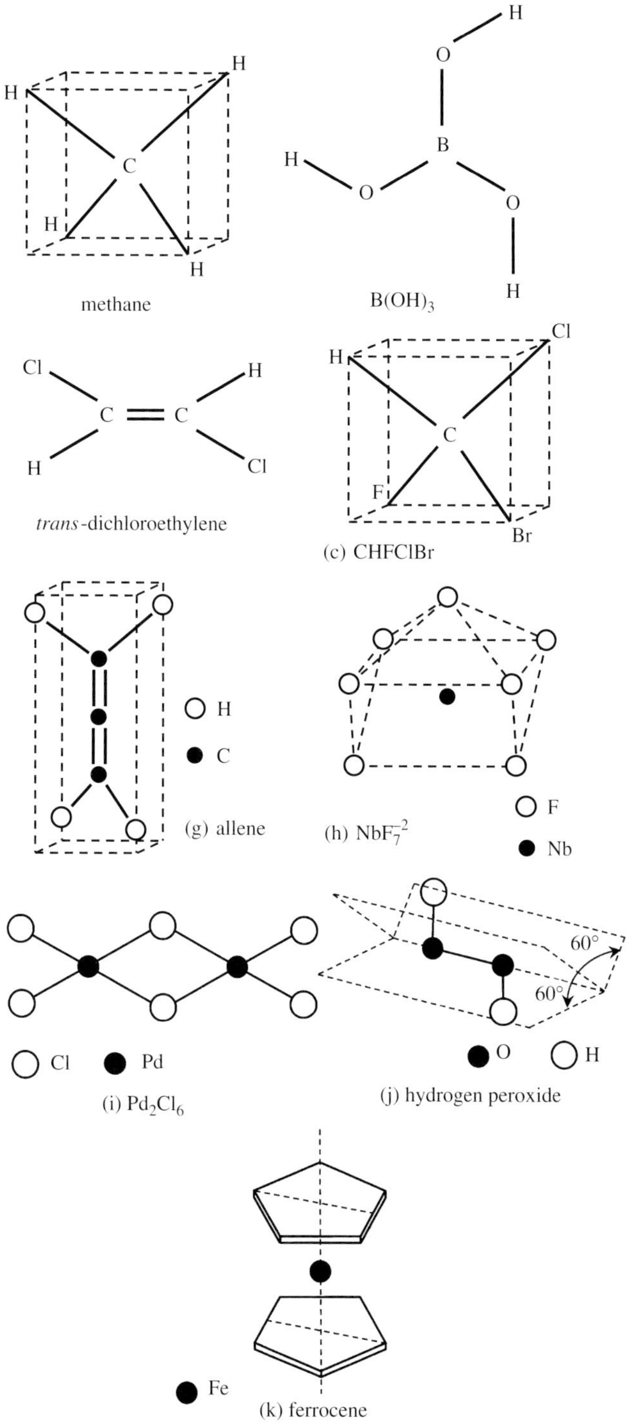

Figure 2.20. Structure of several molecules referred to in Exercise 2.4-1 and in the problems to this chapter. Lower case letters (c) and (g)–(k) refer to Problem 2.3.

Answer to Exercise 2.4-1

C_s; $D_{\infty h}$; T_d; C_{2v}; D_{3h}; C_{4v}; C_{3v}; D_{4h}; D_{3d}; C_{3h}; D_{5h}; T_d.

Problems

2.1 Prove the following results by using projection diagrams.

(a) Show that $R(\pi\ \mathbf{m})$ and $R(\pi\ \mathbf{n})$ commute when **m** is normal to **n**.

(b) Show that $\sigma_{\mathbf{y}}\sigma_{\mathbf{x}} = C_{2\mathbf{z}}$.

(c) Two planes σ_1, σ_2 intersect along **n** and make an angle $\phi/2$ with one another. Show that $\sigma_2\sigma_1 = R(\phi\ \mathbf{n})$. Do σ_1 and σ_2 commute?

(d) Show that $R(\pi\ \mathbf{x})\ R(\beta\ \mathbf{z}) = R(-\beta\ \mathbf{z})\ R(\pi\ \mathbf{x})$.

2.2 Identify the set of symmetry operators associated with the molecule *trans*-dichloroethylene (Figure 2.20). Set up the multiplication table for these operators and hence show that they form a group. Name this symmetry group. [*Hint*: Set up a right-handed system of axes with **y** along the C=C bond and **z** normal to the plane of the molecule.]

2.3 Determine the symmetry elements of the following molecules and hence identify the point group to which each one belongs. [*Hints*: Adhere to the convention stated in Section 2.1. Many of these structures are illustrated in Figure 2.20. Sketching the view presented on looking down the molecular axis will be found helpful for (k) and (l).]

(a) NH_3 (non-planar),

(b) H_3C—CCl_3 (partly rotated),

(c) CHFClBr,

(d) $C_5H_5^-$ (planar),

(e) C_6H_6 (planar),

(f) $[TiF_6]^{-3}$ (octahedral),

(g) allene,

(h) $[NbF_7]^{-2}$,

(i) Pd_2Cl_6,

(j) hydrogen peroxide,

(k) bis(cyclopentadienyl)iron or ferrocene (staggered configuration),

(l) dibenzenechromium (like ferrocene, a "sandwich compound," but the two benzene rings are in the eclipsed configuration in the crystal).

2.4 List a sufficient number of symmetry elements in the molecules sketched in Figure 2.21 to enable you to identify the point group to which each belongs. Give the point group symbol in both Schönflies and International notation.

2.5 Show that each of the following sets of symmetry operators is a generator for a point group. State the point group symbol in both Schönflies and International notation. [*Hints*: The use of projection diagrams is generally an excellent method for calculating products of symmetry operators. See Figure 2.10(a) for the location of the $C_{2\mathbf{a}}$ axis.]

(a) $\{C_{2\mathbf{y}}\ C_{2\mathbf{z}}\}$,

(b) $\{C_{4\mathbf{z}}\ I\}$,

(c) $\{S_{4\mathbf{z}}\ C_{2\mathbf{x}}\}$,

(d) $\{C_{3\mathbf{z}}\ C_{2\mathbf{a}}\ I\}$,

(e) $\{C_{4\mathbf{z}}\ \sigma_{\mathbf{x}}\}$,

(f) $\{\bar{6}\}$,

(g) $\{S_{3\mathbf{z}}\ C_{2\mathbf{a}}\}$.

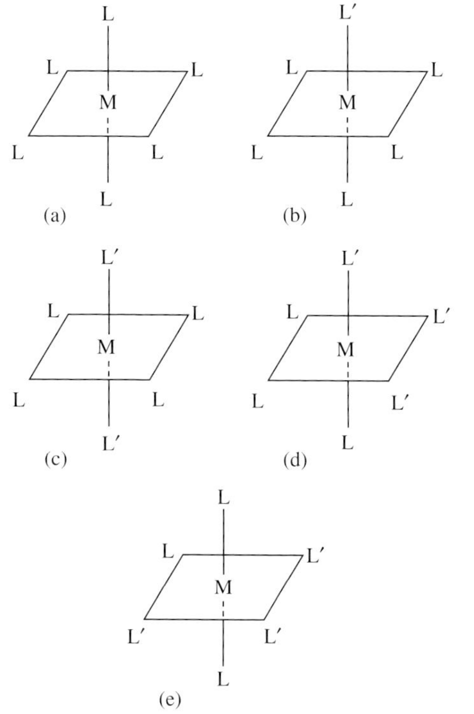

Figure 2.21. Configurations of an ML_6 complex ion and of some $ML_nL'_{6-n}$ complexes.

2.6 List a sufficient number of symmetry elements (and also significant absences) in the following *closo* $B_n H_n^{-2}$ ions that will enable you to determine the point group to which each belongs. The shapes of these molecules are shown in Figure 2.22.

(a) $B_5H_5^{-2}$,
(b) $B_6H_6^{-2}$,
(c) $B_9H_9^{-2}$,
(d) $B_{10}H_{10}^{-2}$,
(e) $B_{12}H_{12}^{-2}$.

2.7 Evaluate the following DPs showing the symmetry operators in each group. [*Hint*: For (a)–(e), evaluate products using projection diagrams. This technique is not useful for products that involve operators associated with the C_3 axes of a cube or tetrahedron, so in these cases study the transformations induced in a cube.] Explain why the DPs in (d)–(f) are semidirect products.

(a) $D_2 \otimes C_i$,
(b) $D_3 \otimes C_i$,
(c) $D_3 \otimes C_s$,
(d) $S_4 \wedge C_2$ ($C_2 = \{E\ C_{2\mathbf{x}}\}$),
(e) $D_2 \wedge C_2$ ($C_2 = \{E\ C_{2\mathbf{a}}\}$),
(f) $D_2 \wedge C_3$ ($C_3 = \{E\ C_{31}^{\pm}\}$).

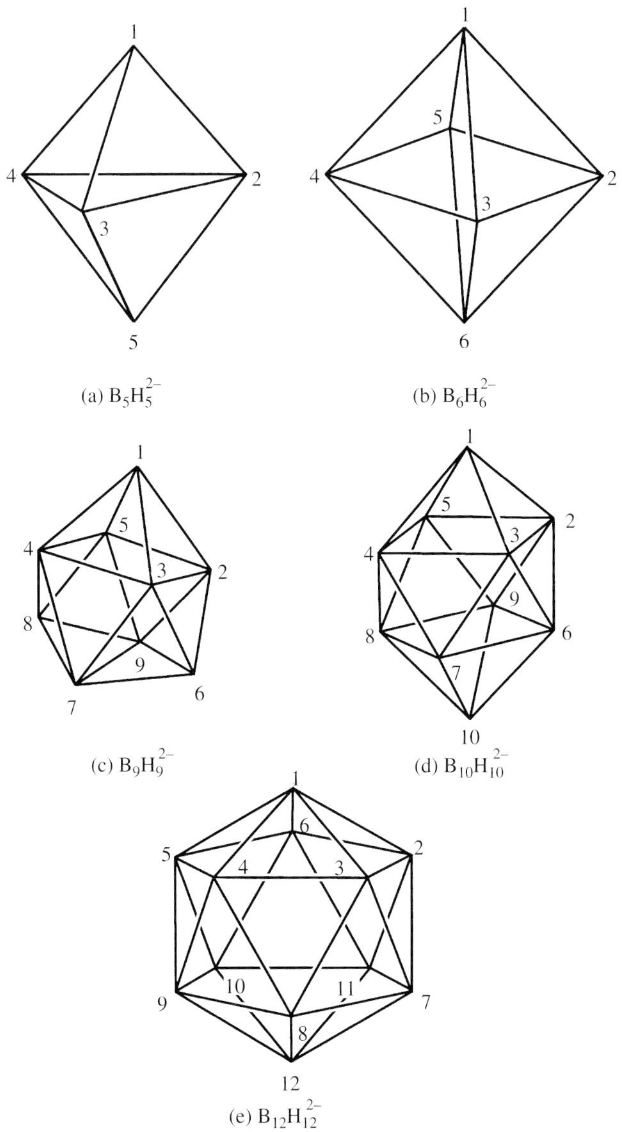

Figure 2.22. Some *closo* $B_nH_n^{-2}$ anions. The numbering scheme shown is conventional and will be an aid in identifying and describing the symmetry elements.

3 Matrix representatives

3.1 Linear vector spaces

In three-dimensional (3-D) configuration space (Figure 3.1) a position vector $\mathbf{r}$ is the *sum of its projections*,

$$\mathbf{r} = \mathbf{e}_1 x + \mathbf{e}_2 y + \mathbf{e}_3 z. \tag{1}$$

The set of three orthonormal basis vectors $\{\mathbf{e}_1\ \mathbf{e}_2\ \mathbf{e}_3\}$ in eq. (1) is the *basis* of a *linear vector space* (LVS), and the *coordinates* of the point $P(x\ y\ z)$ are the *components* of the vector $\mathbf{r}$. The *matrix representation* of $\mathbf{r}$ is

$$\mathbf{r} = \langle \mathbf{e}_1\ \mathbf{e}_2\ \mathbf{e}_3 | x\ y\ z\rangle. \tag{2}$$

$\langle \mathbf{e}_1\ \mathbf{e}_2\ \mathbf{e}_3|$ is a matrix of one row that contains the elements of the basis set, and $|\,x\ y\ z\rangle$ is a matrix of a single column containing the components of $\mathbf{r}$. The row $\times$ column law of matrix multiplication applied to the RS of eq. (2) yields eq. (1). The choice of basis vectors is arbitrary: they do not have to be mutually orthogonal but they must be linearly independent (LI) and three in number in 3-D space. Thus, $\{\mathbf{e}_1\ \mathbf{e}_2\ \mathbf{e}_3\}$ form a basis in 3-D space if it is impossible to find a set of numbers $\{v_1\ v_2\ v_3\}$ such that $\mathbf{e}_1 v_1 + \mathbf{e}_2 v_2 + \mathbf{e}_3 v_3 = 0$, except $v_j = 0, j = 1, 2, 3$. But any set of four or more vectors is linearly dependent in 3-D space. That is, the *dimensionality* of a vector space is the *maximum number of LI vectors* in that space. This is illustrated in Figure 3.2 for the example of two-dimensional (2-D) space, which is a subspace of 3-D space.

For a vector $\mathbf{v}$ in an LVS of n dimensions, eq. (1) is generalized to

$$\mathbf{v} = \sum_{i=1}^{n} \mathbf{e}_i\, v_i = \mathbf{e}_1 v_1 + \mathbf{e}_2 v_2 + \cdots + \mathbf{e}_n v_n$$
$$\neq 0,\ \text{unless}\ v_i = 0, \forall\ i = 1, \ldots, n. \tag{3}$$

In eq. (3), the vector $\mathbf{v}$ is the sum of its projections. The matrix representation of eq. (3) is

$$\mathbf{v} = \langle \mathbf{e}_1\ \mathbf{e}_2\ \ldots\ \mathbf{e}_n | v_1 v_2\ \ldots\ v_n\rangle \tag{4}$$

$$= \langle \mathbf{e}|v\rangle, \tag{5}$$

where, in eq. (5), the *row* matrix $\langle \mathbf{e}|$ implies the whole *basis* set, as given explicitly in eq. (4), and similarly v in the *column* matrix $|v\rangle$ implies the whole set of n *components*

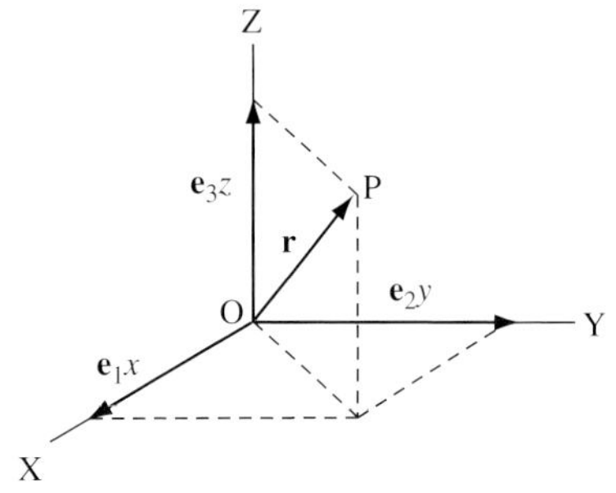

Figure 3.1. Projection of a vector OP along three orthogonal axes OX, OY, OZ.

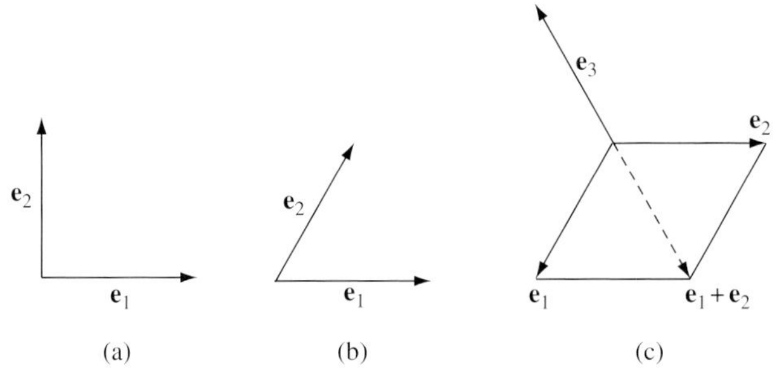

Figure 3.2. Examples, in 2-D space, of (a) an LI set of orthogonal basis vectors $\{\mathbf{e}_1\ \mathbf{e}_2\}$, (b) an LI non-orthogonal basis, and (c) a set of three basis vectors in 2-D space that are not LI because $\mathbf{e}_1 + \mathbf{e}_2 + \mathbf{e}_3 = 0$.

$\{v_1\ v_2\ \dots\ v_n\}$. If the basis $\{\mathbf{e}_i\}$ and/or the components $\{v_j\}$ are complex, the definition of the scalar product has to be generalized. The Hermitian scalar product of two vectors **u** and **v** is defined by

$$\mathbf{u}^* \cdot \mathbf{v} = \langle \mathbf{e}|u\rangle^\dagger \cdot \langle \mathbf{e}|v\rangle, \tag{6}$$

the superscript † denoting the *adjoint* or transposed complex conjugate:

(5), (6)

$$\mathbf{u}^* \cdot \mathbf{v} = \langle u^*|\mathbf{e}^*\rangle \cdot \langle \mathbf{e}|v\rangle \tag{7a}$$

$$= \langle u^*|\mathtt{M}|v\rangle \tag{7b}$$

$$= \sum_{i,j} u_i{}^* \mathtt{M}_{ij} v_j. \tag{7c}$$

The square matrix

$$\mathtt{M} = |\mathbf{e}^*\rangle \cdot \langle \mathbf{e}| \tag{8}$$

is called the *metric* of the LVS:

$$\mathrm{M} = |\mathbf{e}_1{}^* \ \mathbf{e}_2{}^* \ \ldots \ \mathbf{e}_n{}^*\rangle \cdot \langle \mathbf{e}_1 \ \mathbf{e}_2 \ \ldots \ \mathbf{e}_n| = \begin{bmatrix} \mathbf{e}_1{}^* \cdot \mathbf{e}_1 & \mathbf{e}_1{}^* \cdot \mathbf{e}_2 & \ldots \\ \mathbf{e}_2{}^* \cdot \mathbf{e}_1 & \mathbf{e}_2{}^* \cdot \mathbf{e}_2 & \ldots \\ \vdots & \vdots & \vdots \end{bmatrix}. \tag{9}$$

Note that (i)

$$\mathrm{M}_{ij} = \mathbf{e}_i{}^* \cdot \mathbf{e}_j = \mathbf{e}_j \cdot \mathbf{e}_i{}^* = (\mathbf{e}_j{}^* \cdot \mathbf{e}_i)^* = \mathrm{M}_{ji}{}^* \tag{10}$$

so that M is a Hermitian matrix ($\mathrm{M} = \mathrm{M}^\dagger$). (ii) If the basis is *orthonormal* (or *unitary*)

$$\mathrm{M}_{ij} = \mathbf{e}_i{}^* \cdot \mathbf{e}_j = \delta_{ij} \tag{11}$$

and M is just the unit matrix with n rows and columns,

$$\mathrm{M} = \mathrm{E}_n. \tag{12}$$

In this case,

$$\mathbf{u}^* \cdot \mathbf{v} = \langle u^* | v \rangle = \sum_i u_i{}^* v_i. \tag{13}$$

In eqs. (7a) and (7b) $|v\rangle$ is a matrix of one column containing the components of **v**, and $\langle u^*|$ is a matrix of one row, which is the transpose of $|u^*\rangle$, the matrix of one column containing the components of **u**, complex conjugated. In eq. (6), transposition is necessary to conform with the matrix representation of the scalar product so that the row × column law of matrix multiplication may be applied. Complex conjugation is necessary to ensure that the length of a vector **v**

$$v = |\mathbf{v}| = (\mathbf{v}^* \cdot \mathbf{v})^{1/2} \tag{14}$$

is real. A vector of unit length is said to be *normalized*, and any vector **v** can be normalized by dividing **v** by its length v.

3.2 Matrix representatives of operators

Suppose a basis $\langle \mathbf{e}|$ is transformed into a new basis $\langle \mathbf{e}'|$ under the proper rotation R, so that

$$R\langle \mathbf{e}| = \mathbf{e}'|, \tag{1}$$

or, in more detail,

$$R\langle \mathbf{e}_1 \ \mathbf{e}_2 \ \mathbf{e}_3| = \langle \mathbf{e}_1{}' \ \mathbf{e}_2{}' \ \mathbf{e}_3{}'|. \tag{2}$$

Then the new basis vectors $\{\mathbf{e}_j{}'\}$ can be expressed in terms of the old set by writing $\mathbf{e}_j{}'$ as the sum of its projections (cf. eq. (3.1.3)):

$$\mathbf{e}_j{}' = \sum_{i=1}^{3} \mathbf{e}_i \ \mathrm{r}_{ij}, \qquad j = 1, 2, 3; \tag{3}$$

$\mathbf{r}_{ij}$ is the component of $\mathbf{e}_j{}'$ along $\mathbf{e}_i$. In matrix form,

(3) $$\langle \mathbf{e}_1{}' \ \mathbf{e}_2{}' \ \mathbf{e}_3{}' | = \langle \mathbf{e}_1 \ \mathbf{e}_2 \ \mathbf{e}_3 | \Gamma(R), \tag{4}$$

where the square matrix

$$\Gamma(R) = [r_{ij}] = \begin{bmatrix} r_{11} & r_{12} & r_{13} \\ r_{21} & r_{22} & r_{23} \\ r_{31} & r_{32} & r_{33} \end{bmatrix} \tag{5}$$

and the r_{ij} in eq. (3) are seen to be the elements of the jth column of $\Gamma(R)$. In shorter notation,

(4) $$\langle \mathbf{e}' | = \langle \mathbf{e} | \Gamma(R). \tag{6}$$

Equation (6), or eq. (4), is the matrix representation of the operation of deriving the new basis $\{\mathbf{e}_j{}'\}$ from the original basis $\{\mathbf{e}_i\}$, and when we carry out the matrix multiplication on the RS of eq. (6) or eq. (4) we are using eq. (3) successively for each $\mathbf{e}_j{}'$ in turn as $j = 1, 2, 3$.

(1), (6) $$R\langle \mathbf{e} | = \langle \mathbf{e}' | = \langle \mathbf{e} \ | \Gamma(R), \tag{7}$$

which shows that $\Gamma(R)$ is the *matrix representative* (MR) of the operator R.

Example 3.2-1 When R is the identity E, $\langle \mathbf{e}' |$ is just $\langle \mathbf{e} |$ and so $\Gamma(E)$ is the 3×3 unit matrix, E_3.

Example 3.2-2 Consider a basis of three orthogonal unit vectors with $\mathbf{e}_3$ (along OZ) normal to the plane of the paper, and consider the proper rotation of this basis about OZ through an angle ϕ by the operator R (ϕ **z**) (see Figure 3.3). Any vector **v** may be expressed as the sum of its projections along the basis vectors:

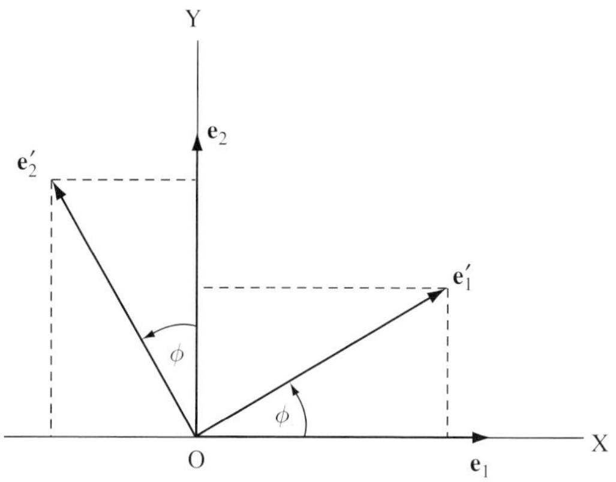

Figure 3.3. Rotation of configuration space, and therefore of all vectors in configuration space including $\{\mathbf{e}_1 \ \mathbf{e}_2 \ \mathbf{e}_3\}$, through an angle ϕ about OZ (active representation).

$$\mathbf{v} = \sum_i \mathbf{e}_i \, v_i. \tag{3.1.3}$$

To find the ith component v_i, take the scalar product of $\mathbf{e}_i$ with $\mathbf{v}$. Here the basis is real and orthonormal, so

(3.1.3)
$$\mathbf{e}_i \cdot \mathbf{v} = \mathbf{e}_i \cdot \sum_j \mathbf{e}_j \, v_j = \sum_j \delta_{ij} \, v_j = v_i. \tag{8}$$

We now represent the transformed basis vectors $\{\mathbf{e}_j'\}$ in terms of the original set $\{\mathbf{e}_i\}$ by expressing each as the sum of its projections, according to eq. (3). Writing each $\mathbf{e}_j'$ ($j = 1, 2, 3$) as the sum of their projections along $\{\mathbf{e}_i\}$ yields

$$\begin{aligned} \mathbf{e}_1' &= \mathbf{e}_1(\cos\phi) + \mathbf{e}_2(\sin\phi) + \mathbf{e}_3(0) \\ \mathbf{e}_2' &= \mathbf{e}_1(-\sin\phi) + \mathbf{e}_2(\cos\phi) + \mathbf{e}_3(0) \\ \mathbf{e}_3' &= \mathbf{e}_1(0) + \mathbf{e}_2(0) + \mathbf{e}_3(1) \end{aligned} \tag{9}$$

where we have used the fact that the scalar product of two unit vectors at an angle θ is $\cos\theta$, and that $\cos(\frac{1}{2}\pi - \phi) = \sin\phi$, $\cos(\frac{1}{2}\pi + \phi) = -\sin\{\phi\}$, and $\cos 0 = 1$. Because of the row $\times$ column law of matrix multiplication, eqs. (9) may be written as

$$\langle \mathbf{e}_1' \; \mathbf{e}_2' \; \mathbf{e}_3' \,| = \langle \mathbf{e}_1 \; \mathbf{e}_2 \; \mathbf{e}_3 | \begin{bmatrix} \cos\phi & -\sin\phi & 0 \\ \sin\phi & \cos\phi & 0 \\ 0 & 0 & 1 \end{bmatrix}. \tag{10}$$

On using eq. (7), the MR of $R(\phi\ \mathbf{z})$ is seen to be

(10)
$$\Gamma(R(\phi\ \mathbf{z})) = \begin{bmatrix} \cos\phi & -\sin\phi & 0 \\ \sin\phi & \cos\phi & 0 \\ 0 & 0 & 1 \end{bmatrix} = \begin{bmatrix} c & -s & 0 \\ s & c & 0 \\ 0 & 0 & 1 \end{bmatrix}, \tag{11}$$

where $c = \cos\phi$, $s = \sin\phi$. The proper rotation $R(\phi\ \mathbf{z})$ rotates a vector $\mathbf{r}$ in configuration space into the vector $\mathbf{r}'$ given by

(7)
$$\mathbf{r}' = R\,\mathbf{r} = R\langle \mathbf{e}|r\rangle = \langle \mathbf{e}'|r\rangle = \langle \mathbf{e}|\Gamma(R)|r\rangle = \langle \mathbf{e}|r'\rangle. \tag{12}$$

For $R = R\ (\phi\ \mathbf{z})$, the components of $\mathbf{r}'$ (which are the coordinates of the transformed point P$'$) are in

(12)
$$|r'\rangle = \Gamma(R)|r\rangle, \tag{13}$$

which provides a means of calculating the components of $|r'\rangle$ from

(13), (11)
$$\begin{bmatrix} x' \\ y' \\ z' \end{bmatrix} = \begin{bmatrix} c & -s & 0 \\ s & c & 0 \\ 0 & 0 & 1 \end{bmatrix} \begin{bmatrix} x \\ y \\ z \end{bmatrix}. \tag{14}$$

Example 3.2-3 Find the transformed components of a vector **r** when acted on by the operator $C_{4z}^{+} = R(\pi/2\ \ \mathbf{z})$.

(14) $$\begin{bmatrix} x' \\ y' \\ z' \end{bmatrix} = \begin{bmatrix} c & -s & 0 \\ s & c & 0 \\ 0 & 0 & 1 \end{bmatrix} \begin{bmatrix} x \\ y \\ z \end{bmatrix} = \begin{bmatrix} 0 & \bar{1} & 0 \\ 1 & 0 & 0 \\ 0 & 0 & 1 \end{bmatrix} \begin{bmatrix} x \\ y \\ z \end{bmatrix} = \begin{bmatrix} \bar{y} \\ x \\ z \end{bmatrix}. \qquad (15)$$

The set of components of the vector $\mathbf{r}'$ in eq. (13) is the *Jones symbol* or Jones faithful representation of the symmetry operator R, and is usually written as $(x'\ y'\ z')$ or $x'\ y'\ z'$. For example, from eq. (15) the Jones symbol of the operator $R\ (\pi/2\ \ \mathbf{z})$ is $(\bar{y}\,x\,z)$ or $\bar{y}\,x\,z$. In order to save space, particularly in tables, we will usually present Jones symbols without parentheses. A "faithful representation" is one which obeys the same multiplication table as the group elements (symmetry operators).

The inversion operator I leaves $\langle\mathbf{e}|$ invariant but changes the sign of the components of **r** (see eq. (2.1.5) and Figure 2.3):

$$I\langle\mathbf{e}|r\rangle = \langle\mathbf{e}|I|r\rangle = \langle\mathbf{e}|\Gamma(I)|r\rangle; \qquad (16)$$

(16) $$I|x\ y\ z\rangle = \Gamma(I)|x\ y\ z\rangle = |-x\ \ -y\ \ -z\rangle. \qquad (17)$$

Therefore the MR of I is

(17) $$\Gamma(I) = \begin{bmatrix} \bar{1} & 0 & 0 \\ 0 & \bar{1} & 0 \\ 0 & 0 & \bar{1} \end{bmatrix}. \qquad (18)$$

It follows that if R is a proper rotation and $R|x\ y\ z\rangle = |x'\ y'\ z'\rangle$, then

(17) $$IR|x\ y\ z\rangle = I|x'\ y'\ z'\rangle = |-x'\ -y'\ -z'\rangle. \qquad (19)$$

The improper rotation $S(\phi\ \mathbf{n}) = IR(\phi \mp \pi\ \ \mathbf{n})$, for $\phi>0$ or $\phi<0$ (see eq. (2.1.9)), so that it is sometimes convenient to have the MR of $S(\phi\ \mathbf{n})$ as well. In the improper rotation $S(\phi\ \mathbf{z}) = \sigma_{\mathbf{z}}R(\phi\ \mathbf{z})$, $\sigma_{\mathbf{z}}|x\ y\ z\rangle = |x\,y\,\bar{z}\rangle$, and so the MR of $S(\phi\ \mathbf{z})$ is

(11) $$\Gamma(S(\phi\ \mathbf{z})) = \begin{bmatrix} c & -s & 0 \\ s & c & 0 \\ 0 & 0 & \bar{1} \end{bmatrix}. \qquad (20)$$

Exercise 3.2-1 Write down the Jones symbol for the improper rotation $S_{4\mathbf{z}}^{-}$.

Exercise 3.2-2 Show that $S_n^{\pm k} = IC_n^{k \mp (n/2)}$. Find operators of the form IC_n^k that are equivalent to $S_{4\mathbf{z}}^{\pm}$ and $S_{6\mathbf{z}}^{\pm}$.

It is demonstrated in Problem 3.1 that $\Gamma(R)$ and $\Gamma(S)$ are real orthogonal matrices. An orthogonal matrix $\mathbb{A}$ has the property $\mathbb{A}^{\mathrm{T}}\mathbb{A} = \mathbb{E}$, where $\mathbb{E}$ is the unit matrix, so that $\mathbb{A}^{-1} = \mathbb{A}^{\mathrm{T}}$, which makes the calculation of $\Gamma(R)^{-1}$ and $\Gamma(S)^{-1}$ very straightforward or simple (to use space). Equations (13), (17), and (19) are of considerable importance since every point symmetry operation, apart from E and I, is equivalent to a proper or improper rotation.

Example 3.2-4 Nevertheless it is convenient to have the MR of $\sigma(\theta\ \mathbf{y})$, the operator that produces reflection in a plane whose normal **m** makes an angle θ with **y** (Figure 3.4) so that the reflecting plane makes an angle θ with the **zx** plane.

From Figure 3.4,

$$x = \cos\alpha,\ y = \sin\alpha, \tag{21}$$

$$x' = \cos(2\theta - \alpha) = x\cos(2\theta) + y\sin(2\theta), \tag{22}$$

$$y' = \sin(2\theta - \alpha) = x\sin(2\theta) - y\cos(2\theta). \tag{23}$$

(21)–(23)
$$\begin{bmatrix} x' \\ y' \\ z' \end{bmatrix} = \begin{bmatrix} \cos 2\theta & \sin 2\theta & 0 \\ \sin 2\theta & -\cos 2\theta & 0 \\ 0 & 0 & 1 \end{bmatrix} \begin{bmatrix} x \\ y \\ z \end{bmatrix} \tag{24}$$

so that the MR of $\sigma(\theta\,\mathbf{y})$ is

(24)
$$\Gamma(\sigma(\theta\ \mathbf{y})) = \begin{bmatrix} \cos 2\theta & \sin 2\theta & 0 \\ \sin 2\theta & -\cos 2\theta & 0 \\ 0 & 0 & 1 \end{bmatrix}. \tag{25}$$

Example 3.2-5 The MR of $\sigma(\pi/3\ \ \mathbf{y})$ is

(22)
$$\Gamma(\sigma(\pi/3\ \ \mathbf{y})) = \begin{bmatrix} -\frac{1}{2} & \frac{\sqrt{3}}{2} & 0 \\ \frac{\sqrt{3}}{2} & \frac{1}{2} & 0 \\ 0 & 0 & 1 \end{bmatrix}. \tag{26}$$

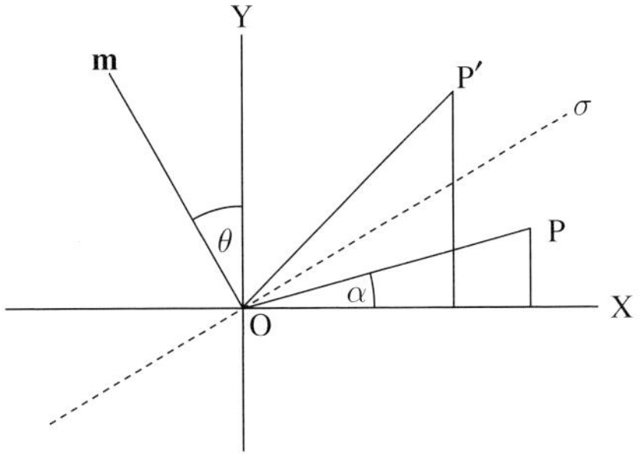

Figure 3.4. Reflection of a point P(x y) in a mirror plane σ whose normal **m** makes an angle θ with **y**, so that the angle between σ and the **zx** plane is θ. OP makes an angle α with **x**. P′(x' y') is the reflection of P in σ, and OP′ makes an angle $2\theta - \alpha$ with **x**.

Answers to Exercises 3.2

Exercise 3.2-1 From eqs. (15) and (19), the Jones symbol for S_{4z}^{-} is $y\ \bar{x}\ \bar{z}$.

Exercise 3.2-2 Let $S(\phi\ \mathbf{n}) = IC\ (\phi'\ \mathbf{n})$. Then $\phi = 2\pi k/n$ and $\phi' = 2\pi[k \mp (n/2)]/n$ so that $S_n^k = IC_n^{k \mp (n/2)}$. Therefore, $S_{4z}^{\pm} = IC_{4z}^{\mp}$ and $S_{6z}^{\pm} = IC_{3z}^{\mp}$.

3.3 Mappings

When the symmetry operator $R \in \mathrm{G}$ acts on configuration space, a vector $\mathbf{r}$ is transformed into $\mathbf{r}' = R\,\mathbf{r}$; $\mathbf{r}'$ is the *image* of $\mathbf{r}$, and the process whereby $R\{\mathbf{r}\} \rightarrow \{\mathbf{r}'\}$ is called a *mapping*. The components of $\mathbf{r}'$ are given by

(3.2.13), (3.2.19) $$|x'\ y'\ z'\rangle = \Gamma(R)|x\ y\ z\rangle, \qquad (1)$$

where $\Gamma(R)$ is the MR of the *operator* R. Equation (1) will be found to be extremely useful, for it enables us to find the effect of a symmetry operator R on the coordinates of $\mathrm{P}(x\ y\ z)$. (In eq. (1) R may be the identity, the inversion operator, or a proper or improper rotation.) The lengths of all vectors and the angles between them are invariant under symmetry operations and so, therefore, are scalar products. Consider the transformation of two vectors $\mathbf{u}$, $\mathbf{v}$ into $\mathbf{u}'$, $\mathbf{v}'$ under the symmetry operator R:

(3.2.12) $$\mathbf{u}' = R\ \mathbf{u} = R\langle\mathbf{e}|u\rangle = \langle\mathbf{e}|\Gamma(R)|u\rangle, \qquad (2)$$

(3.2.12) $$\mathbf{v}' = R\ \mathbf{v} = R\langle\mathbf{e}|v\rangle = \langle\mathbf{e}|\Gamma(\mathrm{R})|v\rangle. \qquad (3)$$

The Hermitian scalar product of $\mathbf{u}$ and $\mathbf{v}$ is

(3.1.6) $$\mathbf{u}^* \cdot \mathbf{v} = \langle\mathbf{e}|u\rangle^\dagger \cdot \langle\mathbf{e}|v\rangle$$

(3.1.7a) $$= \langle u^*|\mathtt{M}|v\rangle. \qquad (4)$$

Similarly, that of $\mathbf{u}'$ and $\mathbf{v}'$ is

(2), (3) $$\mathbf{u}'^* \cdot \mathbf{v}' = \langle\mathbf{e}|\Gamma(R)|u\rangle^\dagger \cdot \langle\mathbf{e}|\Gamma(R)|v\rangle. \qquad (5)$$

The adjoint of a product of matrices is the product of the adjoints in reverse order, so

(5) $$\mathbf{u}'^* \cdot \mathbf{v}' = \langle u^*|\Gamma(R)^\dagger|\mathbf{e}^*\rangle \cdot \langle\mathbf{e}|\Gamma(R)|v\rangle$$

(3.1.8) $$= \langle u^*|\Gamma(R)^\dagger\mathtt{M}\Gamma(R)|v\rangle. \qquad (6)$$

Because the scalar product is invariant under R, $\mathbf{u}'^* \cdot \mathbf{v}' = \mathbf{u}^* \cdot \mathbf{v}$, and

(6), (4) $$\Gamma(R)^\dagger\mathtt{M}\Gamma(R) = \mathtt{M}. \qquad (7)$$

In group theory the most important cases are those of an orthogonal or unitary basis when $\mathtt{M}$ is the 3×3 unit matrix, and consequently

(7) $$\Gamma(R)^{\dagger}\Gamma(R) = \mathrm{E}. \tag{8}$$

Equation (8) shows that $\Gamma(R)$ is a *unitary* matrix and that

$$[\Gamma(R)]^{-1} = \Gamma(R)^{\dagger} = [\Gamma(R)]^{*\mathrm{T}}, \tag{9}$$

where the superscript T denotes the transposed matrix. When the MR $\Gamma(R)$ is real,

(9) $$[\Gamma(R)]^{-1} = [\Gamma(R)]^{\mathrm{T}}. \tag{10}$$

This is a most useful result since we often need to calculate the inverse of a 3×3 MR of a symmetry operator R. Equation (10) shows that when $\Gamma(R)$ is real, $\Gamma(R)^{-1}$ is just the transpose of $\Gamma(R)$. A matrix with this property is an *orthogonal* matrix. In configuration space the basis and the components of vectors are real, so that proper and improper rotations which leave all lengths and angles invariant are therefore represented by 3×3 real orthogonal matrices. Proper and improper rotations in configuration space may be distinguished by det $\Gamma(R)$,

(10) $$\Gamma(R)\Gamma(R)^{\mathrm{T}} = \Gamma(R)^{\mathrm{T}}\Gamma(R) = \mathrm{E}. \tag{11}$$

Since

$$\det \mathbb{A}\,\mathbb{B} = \det \mathbb{A} \det \mathbb{B},$$

(11) $$\det\,\Gamma(R)^{\mathrm{T}}\Gamma(R) = \det\,\Gamma(R)^{\mathrm{T}}\,\det\,\Gamma(R) = [\det\,\Gamma(R)]^2 = 1, \tag{12}$$

(12) $$\det\,\Gamma(R) = \pm1 \quad (\Gamma(R)\text{ real}). \tag{13}$$

Real 3×3 orthogonal matrices with determinant $+1$ are called *special orthogonal* (SO) matrices and they represent proper rotations, while those with determinant -1 represent improper rotations. The set of all 3×3 real orthogonal matrices form a group called the *orthogonal group* O(3); the set of all SO matrices form a subgroup of O(3) called the *special orthogonal group* SO(3).

Exercise 3.3-1 Evaluate the matrix representative of $R(\pi/2\ \ \mathbf{z})$ by considering the rotation of the basis vectors $\{\mathbf{e}_1\ \mathbf{e}_2\ \mathbf{e}_3\}$ into $\{\mathbf{e}_1{}'\ \mathbf{e}_2{}'\ \mathbf{e}_3{}'\}$.

Exercise 3.3-2 The set of real 3×3 orthogonal matrices with determinant -1 does not form a group. Why?

Answers to Exercises 3.3

Exercise 3.3-1 As shown in Figure 3.5,

$$R(\pi/2\ \ \mathbf{z})\langle\mathbf{e}_1\ \mathbf{e}_2\ \mathbf{e}_3| = \langle\mathbf{e}_2\ -\mathbf{e}_1\ \mathbf{e}_3| = \langle\mathbf{e}_1\ \mathbf{e}_2\ \mathbf{e}_3|\begin{bmatrix} 0 & \bar{1} & 0 \\ 1 & 0 & 0 \\ 0 & 0 & 1 \end{bmatrix}.$$

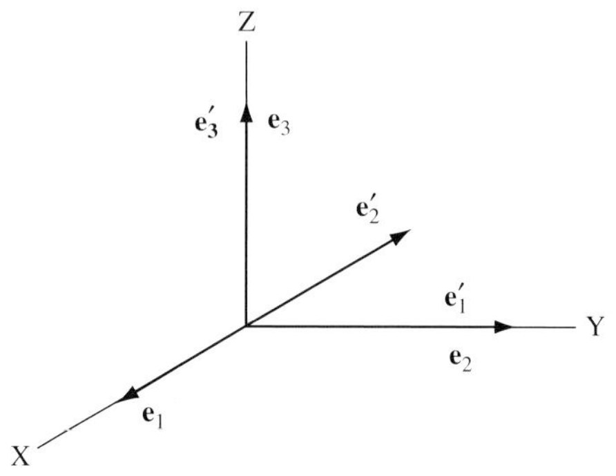

Figure 3.5. Effect of $R(\pi/2\ \mathbf{z})$ on $\{\mathbf{e}_1\ \mathbf{e}_2\ \mathbf{e}_3\}$.

Exercise 3.3-2 The identity in O(3) is $\Gamma(E) = \mathrm{E}_3$, the 3×3 unit matrix with determinant $+1$. The set of all 3×3 real orthogonal matrices with determinant -1 does not contain the identity and therefore cannot form a group.

3.4 Group representations

If $\{R, S, T, \ldots\}$ form a group G, then the set of MRs $\{\Gamma(R), \Gamma(S), \Gamma(T), \ldots\}$ forms a group that is isomorphous with G called a *group representation*. Suppose that $RS = T$; then

(3.2.12) $$T\mathbf{v} = \langle \mathbf{e}|\Gamma(T)|\mathbf{v}\rangle, \tag{1}$$

$$T\mathbf{v} = RS\mathbf{v} = R\mathbf{v}' \text{ (given)}, \tag{2}$$

$$\mathbf{v}' = S\mathbf{v} \text{ (definition of } \mathbf{v}'). \tag{3}$$

(3), (3.2.13) $$|v'\rangle = \Gamma(S)|v\rangle, \tag{4}$$

(4), (3.2.12) $$R\mathbf{v}' = \langle \mathbf{e}|\Gamma(R)|v'\rangle = \langle \mathbf{e}|\Gamma(R)\ \Gamma(S)|v\rangle, \tag{5}$$

(1), (2), (5) $$\Gamma(R)\ \Gamma(S) = \Gamma(T). \tag{6}$$

Equation (6) shows that the MRs obey the same multiplication table as the operators, and so $\{\Gamma(R), \Gamma(S), \Gamma(T), \ldots\}$ forms a group that is isomorphous with $\mathrm{G} = \{R, S, T, \ldots\}$. Such a matrix group is an example of a group representation.

3.5 Transformation of functions

We have studied the transformation of vectors induced by symmetry operators, and this led us to the concept of the MR of a symmetry operator. In order to understand how atomic orbitals transform in symmetry operations, we must now study the transformation of *functions*. To say that $f(x, y, z)$ is a *function* of the set of variables $\{x\} \equiv \{x\ y\ z\}$ means that $f(\{x\})$ has a definite value at each point $P(x, y, z)$ with coordinates $\{x, y, z\}$. Note that we will be using $\{x\}$ as an abbreviation for $\{x\ y\ z\}$ and similarly $\{x'\}$ for $\{x'\ y'\ z'\}$. Now suppose that a symmetry operator R transforms $P(x\ y\ z)$ into $P'(x'\ y'\ z')$ so that

$$R\{x\} = \{x'\}; \tag{1}$$

(3.3.3) $$|x'\rangle = \Gamma(R)|x\rangle. \tag{2}$$

$|x'\rangle$ is a matrix of one column containing the coordinates $\{x'\ y'\ z'\}$ of the transformed point P′. (Recall the correspondence between the coordinates of the point P and the components of the vector **r** that joins P to the origin O of the coordinate system, Figure 3.1.) But since a symmetry operator leaves a system in an indistinguishable configuration (for example, interchanges indistinguishable particles), the *properties* of the system are unaffected by R. Therefore R must also transform f into some new function $\hat{R}f$ in such a way that

$$\hat{R}f(\{x'\}) = f(\{x\}). \tag{3}$$

$\hat{R}$, which transforms f into a new function $f' = \hat{R}f$, is called a *function operator*. Equation (3) states that "the value of the new function $\hat{R}f$, evaluated at the transformed point $\{x'\}$, is the same as the value of the original function f evaluated at the original point $\{x\}$." Equation (3) is of great importance in applications of group theory. It is based (i) on what we understand by a function and (ii) on the invariance of physical properties under symmetry operations. The consequence of (i) and (ii) is that when a symmetry operator acts on configuration space, any function f is simultaneously transformed into a new function $\hat{R}f$. We now require a prescription for calculating $\hat{R}f$. Under the symmetry operator R, each point P is transformed into P′:

$$R\ P(x\ y\ z) = P'(x'\ y'\ z'). \tag{4}$$

(4) $$R^{-1}P'(x'\ y'\ z') = P(x\ y\ z); \tag{5}$$

(3), (5) $$\hat{R}f(\{x'\}) = f(\{x\}) = f(R^{-1}\{x'\}). \tag{6}$$

The primes in eq. (6) can be dispensed with since it is applicable at *any* point $P'(x'\ y'\ z')$:

(6) $$\hat{R}f(\{x\}) = f(R^{-1}\{x\}). \tag{7}$$

Example 3.5-1 Consider the effect of $R(\pi/2\ \mathbf{z})$ on the d orbital $d_{xy} = x\,y\,g(r)$, where $g(r)$ is a function of r only and the angular dependence is contained in the factor $x\,y$, which is therefore used as an identifying subscript on d.

(3.2.15) $$\Gamma(R) = \begin{bmatrix} 0 & \bar{1} & 0 \\ 1 & 0 & 0 \\ 0 & 0 & 1 \end{bmatrix};\tag{8}$$

(3.3.10) $$[\Gamma(R)]^{-1} = [\Gamma(R)]^{\mathrm{T}} = \begin{bmatrix} 0 & 1 & 0 \\ \bar{1} & 0 & 0 \\ 0 & 0 & 1 \end{bmatrix};\tag{9}$$

(9) $$\begin{bmatrix} 0 & 1 & 0 \\ \bar{1} & 0 & 0 \\ 0 & 0 & 1 \end{bmatrix} \begin{bmatrix} x \\ y \\ z \end{bmatrix} = \begin{bmatrix} y \\ \bar{x} \\ z \end{bmatrix};\tag{10}$$

(10) $$R^{-1}\{x\ y\ z\} = \{y\ \bar{x}\ z\}.\tag{11}$$

In other words, the Jones symbol for the operator R^{-1} is $y\ \bar{x}\ z$. Therefore $\hat{R}f(\{x\})$ is

$$\begin{aligned} \hat{R}\ d_{xy} &= d_{xy}(R^{-1}\{x\}) \\ &= d_{xy}(\{y\ \bar{x}\ z\}) \\ &= y\bar{x}\ g(r),\ \text{or}\ -xy\ g(r), \\ &= -d_{xy}\ . \end{aligned}\tag{12}$$

The second equality states that $f(\{x\,y\,z\})$ is to become $f(\{y\ \bar{x}\ z\})$ so that x is to be replaced by y, and y by $-x$ (and z by z); this is done on the third line, which shows that the function d_{xy} is transformed into the function $-d_{xy}$ under the symmetry operator $R(\pi/2\ \ \mathbf{z})$. Figure 3.6 shows that the value of $\hat{R}d_{xy} = d'_{xy} = -d_{xy}$ evaluated at the transformed point P′ has the same numerical value as d_{xy} evaluated at P. Figure 3.6 demonstrates an important result: the effect of the function operator $\hat{R}$ on d_{xy} is *just as if* the contours of the function had been rotated by $R(\pi/2\ \ \mathbf{z})$. However, eq. (7) will always supply the correct result for the transformed function, and is especially useful when it is difficult to visualize the rotation of the contours of the function.

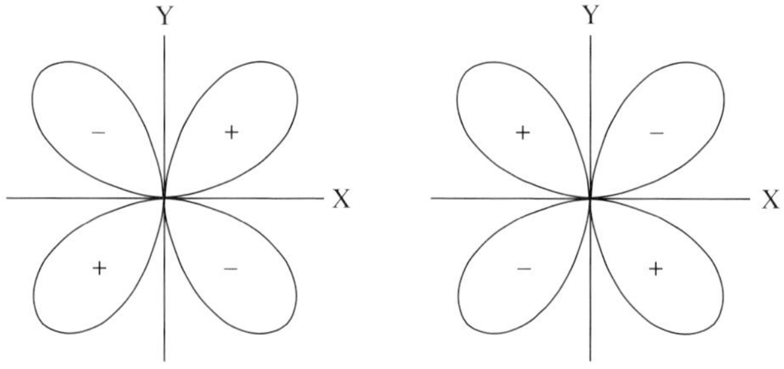

Figure 3.6. This figure shows that the effect on d_{xy} of the function operator $\hat{R}$, which corresponds to the symmetry operator $R = R(\pi/2\ \ \mathbf{z})$, is just as if the contours of the function had been rotated by R.

Exercise 3.5-1 Using $R\langle\mathbf{e}| = \langle\mathbf{e}'| = \langle\mathbf{e}|\Gamma(R)$, determine the MR $\Gamma(R)$ of the symmetry operator $R(\pi/2\ \mathbf{x})$. Hence find $R^{-1}\{x\ y\ z\}$ and then find how the three p orbitals transform under the symmetry operator $R(\pi/2\ \mathbf{x})$.

The complete set of function operators $\{\hat{R}\ \hat{S}\ \hat{T}\ldots\}$ forms a group isomorphous with the group of symmetry operators $\{R\ S\ T\ldots\}$ which transforms configuration space (and all points and vectors therein). The proof of this statement requires the inverse of the product RS. By definition, $(RS)^{-1}$ is the operator which, on multiplying RS, gives the identity E:

$$(RS)^{-1}RS = E; \tag{13}$$

$$S^{-1}R^{-1}R\ S = E \qquad (R^{-1}R = E, \forall\ R, S\ldots); \tag{14}$$

(13), (14)
$$(RS)^{-1} = S^{-1}R^{-1}. \tag{15}$$

This is the anticipated result since the MRs of symmetry operators obey the same multiplication table as the operators themselves, and it is known from the properties of matrices that

$$[\Gamma(R)\Gamma(S)]^{-1} = \Gamma(S)^{-1}\Gamma(R)^{-1}. \tag{16}$$

Suppose that $RS = T$. Then,

$$\hat{S}f(\{x\}) = f(S^{-1}\{x\}) = f'(\{x\}), \tag{17}$$

where f' denotes the transformed function $\hat{S}f$.

(17), (7)
$$\hat{R}\hat{S}f(\{x\}) = \hat{R}f'(\{x\}) = f'(R^{-1}\{x\}). \tag{18}$$

(17), (18)
$$\hat{R}\hat{S}f(\{x\}) = f(S^{-1}R^{-1}\{x\})$$

(15)
$$= f((RS)^{-1}\{x\})$$

(17)
$$= f(T^{-1}\{x\})$$

(7)
$$= \hat{T}f(\{x\}); \tag{19}$$

(18), (19)
$$\hat{R}\hat{S} = \hat{T}. \tag{20}$$

Equation (20) verifies that the set of function operators $\{\hat{R}\ \hat{S}\ \hat{T}\ldots\}$ obeys the same multiplication table as the set of symmetry operators $G = \{R\ S\ T\ldots\}$ and therefore forms a group isomorphous with G.

Answer to Exercise 3.5-1

From Figure 3.7(a),

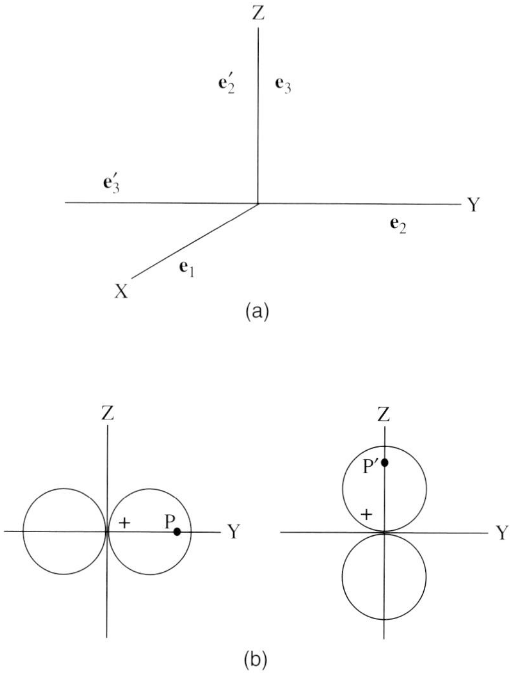

Figure 3.7. (a) Transformation of the basis set $\{\mathbf{e}_1\ \mathbf{e}_2\ \mathbf{e}_3\}$ under $R(\pi/2\ \mathbf{x})$. (b) Illustration of $\hat{R}p_y = p_y' = p_z$. The value of the original function p_y at P(0 a 0) is the same as that of the transformed function p_z at P′(0 0 a).

$$R\langle \mathbf{e}_1\,\mathbf{e}_2\,\mathbf{e}_3| = \langle \mathbf{e}_1{}'\,\mathbf{e}_2{}'\,\mathbf{e}_3{}'\,| = \langle \mathbf{e}_1\,\mathbf{e}_3\,\bar{\mathbf{e}}_2|$$
$$= \langle \mathbf{e}_1\ \mathbf{e}_2\ \mathbf{e}_3|\Gamma(R) = \langle \mathbf{e}_1\ \mathbf{e}_2\ \mathbf{e}_3| \begin{bmatrix} 1 & 0 & 0 \\ 0 & 0 & \bar{1} \\ 0 & 1 & 0 \end{bmatrix};$$

$$R^{-1}\{x\ y\ z\} = \Gamma(R^{-1}) \begin{bmatrix} x \\ y \\ z \end{bmatrix} = \begin{bmatrix} 1 & 0 & 0 \\ 0 & 0 & 1 \\ 0 & \bar{1} & 0 \end{bmatrix} \begin{bmatrix} x \\ y \\ z \end{bmatrix} = \begin{bmatrix} x \\ z \\ \bar{y} \end{bmatrix}. \tag{21}$$

$$\hat{R}\{p_x\ p_y\ p_z\} = \hat{R}\{x\ g(r)\ y\ g(r)\ z\ g(r)\}$$
$$= \{p_x(R^{-1}\{x\})\ p_y(R^{-1}\{x\})\ p_z(R^{-1}\{x\})\}$$
(21) $$= \{p_x\ p_z\ {-p_y}\}, \tag{22}$$

on replacing $\{x\ y\ z\}$ by $\{x\,y\,\bar{z}\}$ in $\{p_x\ p_y\ p_z\}$. Equation (22) states that p_z is the function which, when evaluated at the transformed point $\{x\ y\ z\}$, has the same value as the original function p_y evaluated at the original point $R^{-1}\{x\,y\,z\} = \{x\,z\,\bar{y}\}$. For example, $p_z(\{0\ 0\ a\}) = p_y(\{0\ a\ 0\})$. Note from Figure 3.7(b) that the effect of R on p_y is simply to rotate the contour of the function p_y into that of p_z.

3.6 Some quantum mechanical considerations

For a quantum mechanical state function $\psi(\{x\})$, the RS of eq. (3.5.7) requires multiplication by ω, a phase factor or complex number of modulus unity. Since the choice of phase is arbitrary and has no effect on physical properties, we generally make the most convenient choice of phase, which here is $\omega = 1$. So, for the matrix representations used in Chapters 1–11, we may use eq. (3.5.7) without modification for function operators $\hat{R}$ operating on quantum mechanical state functions, as indeed we have already done in Example 3.5-1. However, there are certain kinds of representations called *projective* or *multiplier* representations for which the conventions used result in phase factors that are not always $+1$. These representations are discussed in Chapter 12.

We already know from the invariance of the scalar product under symmetry operations that spatial symmetry operators are unitary operators, that is they obey the relation $R^{\dagger}R = R\ R^{\dagger} = E$, where E is the identity operator. It follows from eq. (3.5.7) that the set of function operators $\{\hat{R}\}$ are also unitary operators.

Exercise 3.6-1 Prove that the function operators $\{\hat{R}\}$ are unitary.

In quantum mechanics the stationary states of a system are described by the state function (or wave function) $\psi(\{x\})$, which satisfies the time-independent Schrödinger equation

$$\hat{H}\psi(\{x\}) = \mathrm{E}\psi(\{x\}). \tag{1}$$

Here $\{x\}$ stands for the positional coordinates of all the particles in the system, E is the energy of the system, and $\hat{H}$ is the Hamiltonian operator. Since a symmetry operator merely rearranges indistinguishable particles so as to leave the system in an indistinguishable configuration, the Hamiltonian is invariant under any spatial symmetry operator R. Let $\{\psi_i\}$ denote a set of eigenfunctions of $\hat{H}$ so that

$$\hat{H}\psi_i = \mathrm{E}_i\psi_i. \tag{2}$$

Suppose that a symmetry operator R acts on the physical system (atom, molecule, crystal, etc.). Then ψ_i is transformed into the function $\hat{R}\psi_i$, where $\hat{R}$ is a function operator corresponding to the symmetry operator R. Physical properties, and specifically here the energy eigenvalues $\{\mathrm{E}_i\}$, are invariant under symmetry operators that leave the system in indistinguishable configurations. Consequently, $\hat{R}\psi_i$ is also an eigenfunction of $\hat{H}$ with the same eigenvalue E_i, which therefore is degenerate:

(2) $$\hat{H}\ \hat{R}\psi_i = \mathrm{E}_i\ \hat{R}\psi_i = \hat{R}\ \mathrm{E}_i\psi_i = \hat{R}\ \hat{H}\psi_i. \tag{3}$$

Because the eigenfunctions of any linear Hermitian operator form a complete set, in the sense that any arbitrary function that satisfies appropriate boundary conditions can be expressed as a linear superposition of this set, eq. (3) holds also for such arbitrary functions. Therefore,

(3) $$[\hat{R},\ \hat{H}] = 0, \tag{4}$$

and any function operator $\hat{R}$ that corresponds to a symmetry operator R therefore commutes with the Hamiltonian. The set of all function operators $\{\hat{R}\}$ which commute with the Hamiltonian, and which form a group isomorphous with the set of symmetry operators $\{R\}$, is known as the *group of the Hamiltonian* or the *group of the Schrödinger equation*.

Answer to Exercise 3.6-1

$$\begin{aligned}\hat{R}\,\hat{R}^{\dagger}\psi(\{x\}) &= \hat{R}\,\psi'(\{x\}) = \psi'(R^{-1}\{x\}) = \hat{R}^{\dagger}\psi(R^{-1}\{x\})\\ &= \psi((R^{\dagger})^{-1}R^{-1}\{x\}) = \psi((RR^{\dagger})^{-1}\{x\})\\ &= \psi(E^{-1}\{x\}) = \hat{E}\psi(\{x\}),\end{aligned}$$

where E is the identity operator, whence it follows that the function operators $\{\hat{R}\}$ also are unitary.

Problems

3.1 Show by evaluating $[\Gamma(R)]^{\mathrm{T}}\,\Gamma(R)$, where R is the proper rotation $R(\phi\ \mathbf{z})$, that $\Gamma(R)$ is an orthogonal matrix, and hence write down $[\Gamma(R)]^{-1}$. Also write down $\Gamma(R(-\phi\ \mathbf{z}))$. Is this the same matrix as $\Gamma(R(\phi\ \mathbf{z}))^{-1}$ and, if so, is this the result you would expect? Evaluate $\det\Gamma(R(\phi\ \mathbf{z}))$ and $\det\Gamma(S(\phi\ \mathbf{z}))$.

3.2 Find the MR $\Gamma(R)$ for $R = R(2\pi/3\ \mathbf{n})$ with $\mathbf{n}$ a unit vector from O along an axis that makes equal angles with OX, OY, and OZ. What is the trace of $\Gamma(R)$? Find $|x'\ y'\ z'\rangle = \Gamma(R)|x\ y\ z\rangle$ and write down the Jones symbol for this operation. [*Hints*: Consider the effect of $R(2\pi/3\ \mathbf{n})$ by noting the action of R on $\langle\mathbf{e}_1\ \mathbf{e}_2\ \mathbf{e}_3|$ as you imagine yourself looking down $\mathbf{n}$ towards the origin. The trace of a matrix is the sum of its diagonal elements.]

3.3 (a) Find the MR $\Gamma(R)$ of R for $R(-\pi/2\ \mathbf{z})$ and hence find the matrix $\Gamma(I)\,\Gamma(R)$.
(b) Using projection diagrams, find the single operator Q that is equivalent to IR; show also that I and R commute. Give the Schönflies symbol for Q.
(c) Find the MR $\Gamma(Q)$ from $Q\langle\mathbf{e}_1\ \mathbf{e}_2\ \mathbf{e}_3| = \langle\mathbf{e}_1{}'\ \mathbf{e}_2{}'\ \mathbf{e}_3{}'| = \langle\mathbf{e}_1\ \mathbf{e}_2\ \mathbf{e}_3|\Gamma(Q)$.
(d) What can you deduce from comparing $\Gamma(Q)$ from part (c) with $\Gamma(I)\Gamma(R)$ from part (a)?

3.4 Find the MRs of the operators $\sigma_{\mathbf{a}}$, $\sigma_{\mathbf{b}}$ for the basis $\langle\mathbf{e}_1\ \mathbf{e}_2\ \mathbf{e}_3|$, where $\mathbf{a} = 2^{-1/2}[1\ 1\ 0]$, $\mathbf{b} = 2^{-1/2}[\bar{1}\ 1\ 0]$. Evaluate $\Gamma(\sigma_a)\Gamma(\sigma_b)$. Using a projection diagram find $Q = \sigma_{\mathbf{a}}\ \sigma_{\mathbf{b}}$. Find the MR of Q and compare this with $\Gamma(\sigma_{\mathbf{a}})\ \Gamma(\sigma_{\mathbf{b}})$. What can you conclude from this comparison?

3.5 Find the MRs of the operators E, C_{4z}^{+}, C_{4z}^{-}, $\sigma_{\mathbf{x}}$, $\sigma_{\mathbf{y}}$ for the basis $\langle\mathbf{e}_1\ \mathbf{e}_2\ \mathbf{e}_3|$.

3.6 Write down the Jones symbols for $R \in C_{4\mathrm{v}}$ and then the Jones symbols for $\{R^{-1}\}$. [*Hints*: You have enough information from Problems 3.4 and 3.5 to do this very easily. Remember that the MRs of $\{R\}$ are orthogonal matrices.] Write down the angular factor in the transforms of the five d orbitals under the operations of the point group C_{4v}. [*Hint*: This may be done immediately by using the substitutions provided by the Jones symbols for R^{-1}.]

3.7 Find the MR of $R(-2\pi/3\ \ [1\ \bar{1}\ 1])$ for the basis $\langle \mathbf{e}_1\ \mathbf{e}_2\ \mathbf{e}_3|$. Hence write down the Jones representations of R and of R^{-1}. Find the transformed d orbitals $\hat{R}d$, when d is d_{xy}, d_{yz}, or d_{zx}. [*Hint*: Remember that the unit vectors $\{\mathbf{e}_1\ \mathbf{e}_2\ \mathbf{e}_3\}$ are oriented initially along OX, OY, OZ, but are transformed under symmetry operations. Observe the comparative simplicity with which the transformed functions are obtained from the Jones symbol for R^{-1} instead of trying to visualize the transformation of the contours of these functions under the configuration space operator R.]

3.8 (a) List the symmetry operators of the point group D_2. Show in a projection diagram their action on a representative point E. Complete the multiplication table of D_2 and find the classes of D_2. [*Hint*: This can be done without evaluating transforms QRQ^{-1}, $Q \in D_2$.]

(b) Evaluate the direct product $D_2 \otimes C_i = G$ and name the point group G. Study the transformation of the basis $\langle \mathbf{e}_1\ \mathbf{e}_2\ \mathbf{e}_3|$ under the symmetry operators $R \in G = \{R\}$. Use the MRs of R^{-1} to find the Jones symbols for $\{R^{-1}\}$, and hence write down the transformed d orbitals when the symmetry operators of G act on configuration space.

3.9 Find the MRs of $R(\alpha\ \mathbf{x})$ and $R(\beta\ \mathbf{y})$.

4 Group representations

4.1 Matrix representations

If $\{A\ B\ C \ldots\}$ form a group G then any set of square matrices that obey the same multiplication table as that of the group elements is a *matrix representation* Γ of G. For example, we have already seen that the matrix representatives (MRs) $\Gamma(R)$ defined by

$$R\langle\mathbf{e}| = \langle\mathbf{e}'| = \langle\mathbf{e}|\Gamma(R),\ R \in \mathrm{G}, \tag{1}$$

form a representation of the group of symmetry operators. The *dimension* l of a representation is the number of rows and columns in the square matrices making up the matrix representation. In general, a matrix representation Γ is homomorphous with G, with matrix multiplication as the law of binary composition. For example, every group has a one-dimensional (1-D) representation called the *identity* representation or the *totally symmetric* representation Γ_1 for which

$$\Gamma_1(A) = 1,\ \forall A \in \mathrm{G}. \tag{2}$$

If all the matrices $\Gamma(A)$ are different, however, then Γ is isomorphous with G and it is called a *true* or *faithful* representation.

Exercise 4.1-1 Show that the MR of the inverse of A, $\Gamma(A^{-1})$, is $[\Gamma(A)]^{-1}$.

Example 4.1-1 Find a matrix representation of the symmetry group C_{3v} which consists of the symmetry operators associated with a regular triangular-based pyramid (see Section 2.2).

$C_{3v} = \{E\ C_3^+\ C_3^-\sigma_{\mathbf{d}}\ \sigma_{\mathbf{e}}\ \sigma_{\mathbf{f}}\}$. The MR for the two rotations, evaluated from eq. (1), is

$$\Gamma(R(\phi\ \mathbf{z})) = \begin{bmatrix} c & -s & 0 \\ s & c & 0 \\ 0 & 0 & 1 \end{bmatrix}, \tag{3.2.11}$$

where $c = \cos\phi$, $s = \sin\phi$. For the three reflections,

$$\Gamma(\sigma(\theta\ \mathbf{y})) = \begin{bmatrix} c_2 & s_2 & 0 \\ s_2 & -c_2 & 0 \\ 0 & 0 & 1 \end{bmatrix}, \tag{3.2.15}$$

with $c_2 = \cos 2\theta$, $s_2 = \sin 2\theta$. From Figure 2.10, the values of ϕ and θ are

$$\begin{array}{ccccc} C_3^+ & C_3^- & \sigma_{\mathbf{d}} & \sigma_{\mathbf{e}} & \sigma_{\mathbf{f}} \\ \phi = 2\pi/3 & \phi = -2\pi/3 & \theta = 0 & \theta = -\pi/3 & \theta = \pi/3 \end{array}$$

Since $\cos(2\pi/3) = -\cos(\pi/3) = {}^1\!/_2$, $\sin(2\pi/3) = \sin(\pi/3) = \sqrt{3}/2$, $\cos(-2\pi/3) = \cos(2\pi/3) = -{}^1\!/_2$, and $\sin(-2\pi/3) = -\sin(2\pi/3) = -\sqrt{3}/2$, the MRs of the elements of the symmetry group C_{3v} are as follows:

$$\begin{array}{ccc} E & C_3^+ & C_3^- \\ \begin{bmatrix} 1 & 0 & 0 \\ 0 & 1 & 0 \\ 0 & 0 & 1 \end{bmatrix} & \begin{bmatrix} -{}^1\!/_2 & -\sqrt{3}/_2 & 0 \\ \sqrt{3}/_2 & -\sqrt{1}/_2 & 0 \\ 0 & 0 & 1 \end{bmatrix} & \begin{bmatrix} -{}^1\!/_2 & \sqrt{3}/_2 & 0 \\ -\sqrt{3}/_2 & -\sqrt{1}/_2 & 0 \\ 0 & 0 & 1 \end{bmatrix} \\ \sigma_{\mathbf{d}} & \sigma_{\mathbf{e}} & \sigma_{\mathbf{f}} \\ \begin{bmatrix} 1 & 0 & 0 \\ 0 & \bar{1} & 0 \\ 0 & 0 & 1 \end{bmatrix} & \begin{bmatrix} -{}^1\!/_2 & -\sqrt{3}/_2 & 0 \\ -\sqrt{3}/_2 & \sqrt{1}/_2 & 0 \\ 0 & 0 & 1 \end{bmatrix} & \begin{bmatrix} -{}^1\!/_2 & \sqrt{3}/_2 & 0 \\ \sqrt{3}/_2 & \sqrt{1}/_2 & 0 \\ 0 & 0 & 1 \end{bmatrix}. \end{array} \qquad (3)$$

Example 4.1-2 Evaluate $\Gamma(\sigma_{\mathbf{e}})\Gamma(\sigma_{\mathbf{f}})$ and show that the result agrees with that expected from the multiplication table for the operators, Table 2.3.

$$\begin{array}{ccccc} \Gamma(\sigma_{\mathbf{e}}) & & \Gamma(\sigma_{\mathbf{f}}) & & \Gamma(C_3^+) \\ \begin{bmatrix} -{}^1\!/_2 & -\sqrt{3}/_2 & 0 \\ -\sqrt{3}/_2 & {}^1\!/_2 & 0 \\ 0 & 0 & 1 \end{bmatrix} & & \begin{bmatrix} -{}^1\!/_2 & \sqrt{3}/_2 & 0 \\ \sqrt{3}/_2 & {}^1\!/_2 & 0 \\ 0 & 0 & 1 \end{bmatrix} & = & \begin{bmatrix} -{}^1\!/_2 & -\sqrt{3}/_2 & 0 \\ \sqrt{3}/_2 & -{}^1\!/_2 & 0 \\ 0 & 0 & 1 \end{bmatrix}. \end{array}$$

From Table 2.3, we see that $\sigma_{\mathbf{e}}\sigma_{\mathbf{f}} = C_3^+$, so that multiplication of the matrix representations does indeed give the same result as binary combination of the group elements (symmetry operators) in this example.

Exercise 4.1-2 Evaluate $\Gamma(C_3^-)\Gamma(\sigma_{\mathbf{e}})$ and show that your result agrees with that expected from the multiplication table.

Answers to Exercises 4.1

Exercise 4.1-1 Since $A^{-1}A = E$, and since the matrix representations obey the same multiplication table as the group elements, $\Gamma(A^{-1})\Gamma(A) = \Gamma(E) = \mathrm{E}$, the unit matrix. Therefore, from the definition of the inverse matrix, $[\Gamma(A)]^{-1} = \Gamma(A^{-1})$. For example, $C_3^- C_3^+ = E$, and from eq. (3) $\Gamma(C_3^-) = [\Gamma(C_3^+)]^{\mathrm{T}} = [\Gamma(C_3^+)]^{-1}$.

Exercise 4.1-2 From eq. (3),

$$\Gamma(C_3^-)\Gamma(\sigma_{\mathbf{e}}) = \begin{bmatrix} -{}^1\!/_2 & \sqrt{3}/_2 & 0 \\ -\sqrt{3}/_2 & -{}^1\!/_2 & 0 \\ 0 & 0 & 1 \end{bmatrix} \begin{bmatrix} -{}^1\!/_2 & -\sqrt{3}/_2 & 0 \\ -\sqrt{3}/_2 & {}^1\!/_2 & 0 \\ 0 & 0 & 1 \end{bmatrix}$$

$$= \begin{bmatrix} -{}^1\!/_2 & \sqrt{3}/_2 & 0 \\ \sqrt{3}/_2 & {}^1\!/_2 & 0 \\ 0 & 0 & 1 \end{bmatrix} = \Gamma(\sigma_{\mathbf{f}})$$

From Table 2.3, $C_3^-\sigma_e = \sigma_f$, so for this random test the multiplication of two matrix representations again gives the same result as the group multiplication table.

4.2 Irreducible representations

Suppose that $\{\Gamma(A)\,\Gamma(B)\,\ldots\}$ forms an l-dimensional matrix representation of G and define $\Gamma'(A)$ by the *similarity transformation*

$$\Gamma'(A) = \mathrm{S}\ \Gamma(A)\ \mathrm{S}^{-1}, \tag{1}$$

where S is any non-singular $l \times l$ matrix. Then the set $\{\Gamma'(A)\ \Gamma'(B)\ \ldots\}$ also forms an l-dimensional representation of G. (Note that notation varies here, S^{-1} often being substituted for S in eq. (1).)

Proof Let AB denote the product of A and B; then

$$\begin{aligned}\Gamma'(A)\Gamma'(B) &= \mathrm{S}\Gamma(A)\mathrm{S}^{-1}\mathrm{S}\Gamma(B)\mathrm{S}^{-1} = \mathrm{S}\Gamma(A)\Gamma(B)\mathrm{S}^{-1}\\ &= \mathrm{S}\Gamma(AB)\mathrm{S}^{-1} = \Gamma'(AB),\end{aligned} \tag{2}$$

so that $\{\Gamma'(A)\,\Gamma'(B)\,\ldots\}$ is also a representation of G. Two representations that are related by a similarity transformation are said to be *equivalent*. We have seen that for an orthonormal or unitary basis, the matrix representations of point symmetry operators are unitary matrices. In fact, *any* representation of a finite group is equivalent to a unitary representation (Appendix A1.5). Hence we may consider only *unitary representations*. Suppose that Γ^1, Γ^2 are matrix representations of G of dimensions l_1 and l_2 and that for every $A \in \mathrm{G}$ an $(l_1 + l_2)$-dimensional matrix is defined by

$$\Gamma(A) = \begin{bmatrix}\Gamma^1(A) & 0\\ 0 & \Gamma^2(A)\end{bmatrix}. \tag{3}$$

Then

$$\begin{aligned}\Gamma(A)\ \Gamma(B) &= \begin{bmatrix}\Gamma^1(A) & 0\\ 0 & \Gamma^2(A)\end{bmatrix}\begin{bmatrix}\Gamma^1(B) & 0\\ 0 & \Gamma^2(B)\end{bmatrix}\\ &= \begin{bmatrix}\Gamma^1(A)\Gamma^1(B) & 0\\ 0 & \Gamma^2(A)\Gamma^2(B)\end{bmatrix}\\ &= \begin{bmatrix}\Gamma^1(AB) & 0\\ 0 & \Gamma^2(AB)\end{bmatrix} = \Gamma(AB).\end{aligned} \tag{4}$$

Therefore, $\{\Gamma(A)\,\Gamma(B)\,\ldots\}$ also forms a representation of G. This matrix representation Γ of G is called the *direct sum* of Γ^1, Γ^2 and is written as

Table 4.1.

E	C_3^+	C_3^-
$\begin{bmatrix} 1 & 0 & 0 \\ 0 & 1 & 0 \\ 0 & 0 & 1 \end{bmatrix}$	$\begin{bmatrix} -1/2 & -\sqrt{3}/2 & 0 \\ \sqrt{3}/2 & -1/2 & 0 \\ 0 & 0 & 1 \end{bmatrix}$	$\begin{bmatrix} -1/2 & \sqrt{3}/2 & 0 \\ -\sqrt{3}/2 & -1/2 & 0 \\ 0 & 0 & 1 \end{bmatrix}$
$\sigma_{\mathbf{d}}$	$\sigma_{\mathbf{e}}$	$\sigma_{\mathbf{f}}$
$\begin{bmatrix} 1 & 0 & 0 \\ 0 & \bar{1} & 0 \\ 0 & 0 & 1 \end{bmatrix}$	$\begin{bmatrix} -1/2 & -\sqrt{3}/2 & 0 \\ -\sqrt{3}/2 & 1/2 & 0 \\ 0 & 0 & 1 \end{bmatrix}$	$\begin{bmatrix} -1/2 & \sqrt{3}/2 & 0 \\ \sqrt{3}/2 & 1/2 & 0 \\ 0 & 0 & 1 \end{bmatrix}$

$$\Gamma = \Gamma^1 \oplus \Gamma^2. \tag{5}$$

Alternatively, we can regard Γ as reduced into Γ^1 and Γ^2. A representation of G is *reducible* if it can be transformed by a similarity transformation into an equivalent representation, each matrix of which has the same block-diagonal form. Then, each of the smaller representations Γ^1, Γ^2 is also a representation of G. A representation that cannot be reduced any further is called an *irreducible representation* (IR).

Example 4.2-1 Show that the matrix representation found for C_{3v} consists of the totally symmetric representation and a 2-D representation (Γ_3).

Table 4.1 shows that the MRs $\Gamma(T)$ of the symmetry operators $T \in C_{3v}$ for the basis $\langle \mathbf{e}_1\ \mathbf{e}_2\ \mathbf{e}_3|$ all have the same block-diagonal structure so that $\Gamma = \Gamma_1 \oplus \Gamma_3$. We shall soon deduce a simple rule for deciding whether or not a given representation is reducible, and we shall see then that Γ_3 is in fact irreducible.

4.3 The orthogonality theorem

Many of the properties of IRs that are used in applications of group theory in chemistry and physics follow from one fundamental theorem called the *orthogonality theorem* (OT). If Γ^i, Γ^j are two irreducible unitary representations of G which are inequivalent if $i \neq j$ and identical if $i = j$, then

$$\sum_T \sqrt{l_i/g}\ \Gamma^i(T)^*_{pq} \sqrt{l_j/g}\ \Gamma^j(T)_{rs} = \delta_{ij}\ \delta_{pr}\ \delta_{qs}. \tag{1}$$

Note that $\Gamma^i(T)^*_{pq}$ means the element common to the pth row and qth column of the MR for the group element T in the ith IR, complex conjugated. The sum is over all the elements of the group. If the matrix elements $\Gamma^i(T)_{pq}$, $\Gamma^j(T)_{rs}$ are corresponding elements, that is from the same row $p = r$ and the same column $q = s$, and from the same IR, $i = j$, then the sum is unity, but otherwise it is zero. The proof of the OT is quite lengthy, and it is therefore given in Appendix A1.5. Here we verify eq. (1) for some particular cases.

Example 4.3-1 (a) Evaluate the LS of eq. (1) for the 2-D IR Γ_3 of C_{3v} $(i=j=3)$ with $p=r=1, q=s=1$. (b) Repeat the procedure for $i=1, j=3$.

For (a), the LS $=(2/6)\times[1+¼+¼+1+¼+¼]=1$; for (b), the LS $=\sqrt{1/6}\sqrt{2/6}\times[1-½-½+1-½-½]=0$. Notice that we are multiplying together pairs of numbers as in the evaluation of the scalar product of two vectors. The Hermitian scalar product of two normalized vectors **u** and **v** in an n-dimensional linear vector space (LVS) with unitary (orthonormal) basis is

$$\begin{aligned}\mathbf{u}^*\cdot\mathbf{v} &= \sum_{i=1}^{n} u_i^* v_i = 1 \ (\mathbf{u},\ \mathbf{v}\ \text{parallel}),\\ &= 0\ (\mathbf{u},\ \mathbf{v}\ \text{orthogonal}).\end{aligned} \tag{2}$$

So we may interpret eq. (1) as a statement about the orthogonality of vectors in a g-dimensional vector space, where the components of the vectors are chosen from the elements of the l_i, l_j-dimensional matrix representations $\Gamma^i(T)$, $\Gamma^j(T)$, i.e. from the pth row and qth column of the ith IR, and from the rth row and sth column of the jth IR. If these are corresponding elements $(p=r,\ q=s)$ from the same representation $(i=j)$, then the theorem states that a vector whose components are $\Gamma^i(T)_{pq}$, $T\in G$, is of length $\sqrt{g/l_i}$. But if the components are not corresponding elements of matrices from the same representation, then these vectors are orthogonal. The maximum number of mutually orthogonal vectors in a g-dimensional space is g. Now p may be chosen in l_i ways $(p=1,\ \ldots,\ l_i)$ and similarly q may be chosen in l_i ways $(q=1,\ \ldots,\ l_i)$ so that $\Gamma^i(T)_{pq}$ may be chosen in l_i^2 from the ith IR and in $\sum_i l_i^2$ from all IRs. Therefore,

$$\sum_i l_i^2 \leq g. \tag{3}$$

In fact, we show later that the equality holds in eq. (3) so that

$$\sum_i l_i^2 = g. \tag{4}$$

4.4 The characters of a representation

The character χ^i of the MR $\Gamma^i(A)$ is the trace of the matrix $\Gamma^i(A)$, i.e. the sum of its diagonal elements $\Gamma^i(A)_{pp}$,

$$\chi^i(A) = \sum_p \Gamma^i(A)_{pp} = \mathrm{Tr}\ \Gamma^i(A). \tag{1}$$

The set of characters $\{\chi^i(A)\ \chi^i(B)\ \ldots\}$ is called the *character system* of the ith representation Γ^i.

4.4.1 Properties of the characters

(i) The character system is the same for all equivalent representations. To prove this, we need to show that $\mathrm{Tr}\ \mathsf{M}' = \mathrm{Tr}\ \mathsf{S}\ \mathsf{M}\ \mathsf{S}^{-1} = \mathrm{Tr}\ \mathsf{M}$, and to prove this result we need to show first that $\mathrm{Tr}\ \mathsf{AB} = \mathrm{Tr}\ \mathsf{BA}$:

$$\text{Tr AB} = \sum_p \sum_q a_{pq}\, b_{qp} = \sum_q \sum_p b_{qp}\, a_{pq} = \text{Tr BA}; \tag{2}$$

(2) $$\text{Tr M}' = \text{Tr (S M)S}^{-1} = \text{Tr S}^{-1}\text{S M} = \text{Tr M}. \tag{3}$$

Equation (3) shows that the character system is invariant under a similarity transformation and therefore is the same for all equivalent representations. If for some $S \in \text{G}$, $S\,R\,S^{-1} = T$, then R and T are in the same class in G. And since the MRs obey the same multiplication table as the group elements, it follows that all members of the same class have the same character. This holds too for a direct sum of IRs.

Example 4.4-1 From Table 4.1 the characters of two representations of C_{3v} are

C_{3v}	E	C_3^+	C_3^-	σ_d	σ_e	σ_f
Γ_1	1	1	1	1	1	1
Γ_3	2	−1	−1	0	0	0

(ii) The sum of the squares of the characters is equal to the order of the group. In eq. (4.3.1), set $q = p$, $s = r$, and sum over p, r, to yield

(4.3.1) $$\sum_T \sqrt{l_i/g}\; \chi^i(T)^* \sqrt{l_j/g}\; \chi^j(T)$$

$$= \delta_{ij} \sum_{p=1}^{l_i} \sum_{r=1}^{l_j} \delta_{pr} = \delta_{ij} \sum_{p=1}^{l_i} 1 = \delta_{ij} l_i;$$

$$\sum_T \chi^i(T)^* \chi^j(T) = g\sqrt{l_i/l_j}\;\delta_{ij} = g\;\delta_{ij}. \tag{4}$$

(4) $$\sum_T |\chi^i(T)|^2 = g \quad (i = j); \tag{5}$$

(4) $$\sum_T \chi^i(T)^* \chi^j(T) = 0 \quad (i \neq j). \tag{6}$$

Equation (5) provides a simple test as to whether or not a representation is reducible.

Example 4.4-2 Is the 2-D representation Γ_3 of C_{3v} reducible?

$$\chi(\Gamma_3) = \{2 \;\; -1 \;\; -1 \;\; 0 \;\; 0 \;\; 0\},$$

$$\sum_T |\chi_3(T)|^2 = 4 + 1 + 1 + 0 + 0 + 0 = 6 = g,$$

so it is irreducible. The 3 × 3 representation in Table 4.1 is clearly reducible because of its block-diagonal structure, and, as expected,

Table 4.2. *General form of the character table for a group* G.

g_k is a symbol for the type of element in the class $\mathscr{C}_k$ (e.g. C_2, σ_v); c_k is the number of elements in the kth class; g_1 is E, c_1 is 1, and Γ^1 is the totally symmetric representation.

G	$c_1\,g_1$	$c_2\,g_2$	...	$c_k g_k$	...
Γ^1	$\chi^1(\mathscr{C}_1)$	$\chi^1(\mathscr{C}_2)$		$\chi^1(\mathscr{C}_k)$	
Γ^2	$\chi^2(\mathscr{C}_1)$	$\chi^2(\mathscr{C}_2)$		$\chi^2(\mathscr{C}_k)$	
$\vdots$					
Γ^i	$\chi^i(\mathscr{C}_1)$	$\chi^i(\mathscr{C}_2)$		$\chi^i(\mathscr{C}_k)$	
Γ^j	$\chi^j(\mathscr{C}_1)$	$\chi^j(\mathscr{C}_2)$		$\chi^j(\mathscr{C}_k)$	

$$\sum_T |\chi(T)|^2 = 3^2 + 2(0)^2 + 3(1)^2 = 12 \neq g.$$

Generally, we would take advantage of the fact that all members of the same class have the same character and so perform the sums in eqs. (4), (5), and (6) over classes rather than over group elements.

(iii) First orthogonality theorem for the characters. Performing the sum over classes

(4) $$\sum_{k=1}^{N_c} \sqrt{c_k/g}\ \chi^i(\mathscr{C}_k)^* \sqrt{c_k/g}\ \chi^j(\mathscr{C}_k) = \delta_{ij}, \qquad (7)$$

where N_c is the number of classes and c_k is the number of elements in the kth class, $\mathscr{C}_k$ Equation (7) states that the vectors with components $\sqrt{c_k/g}\ \chi^i(\mathscr{C}_k)$, $\sqrt{c_k/g}\ \chi^j(\mathscr{C}_k)$ are orthonormal. If we set up a table of characters in which the columns are labeled by the elements in that class and the rows by the representations – the so-called *character table* of the group (see Table 4.2) – then we see that eq. (7) states that the rows of the character table are orthonormal. The normalization factors $\sqrt{c_k/g}$ are omitted from the character table (see Table 4.2) so that when checking for orthogonality or normalization we use eq. (7) in the form

$$g^{-1}\sum_{k=1}^{N_c} c_k\ \chi^i(\mathscr{C}_k)^* \chi^j(\mathscr{C}_k) = \delta_{ij}. \qquad (8)$$

It is customary to include c_k in the column headings along with the symbol for the elements in $\mathscr{C}_k$ (e.g. $3\sigma_v$ in Table 4.3). Since E is always in a class by itself, $E = \mathscr{C}_1$ is placed first in the list of classes and $c_1 = 1$ is omitted. The first representation is always the totally symmetric representation Γ_1.

Example 4.4-3 Using the partial character table for C_{3v} in Table 4.3, show that the character systems $\{\chi_1\}$ and $\{\chi_3\}$ satisfy the orthonormality condition for the rows.

$$g^{-1}\sum_k c_k\ |\chi_1(\mathscr{C}_k)|^2 = (1/6)[1(1)^2 + 2(1)^2 + 3(1)^2] = 1;$$

Table 4.3. *Partial character table for* C_{3v} *obtained from the matrices of the IRs* Γ_1 *and* Γ_3 *in Table 4.1.*

$\mathscr{C}_1 = \{E\}, \mathscr{C}_2 = \{C_3^+\ C_3^-\}$, and $\mathscr{C}_3 = \{\sigma_{\mathbf{d}}\ \sigma_{\mathbf{e}}\ \sigma_{\mathbf{f}}\}$, and so in C_{3v}, $c_1 = 1$, $c_2 = 2$, and $c_3 = 3$.

	E	$2C_3$	$3\sigma_v$
Γ_1	1	1	1
Γ_3	2	−1	0

$$g^{-1}\sum_k c_k\ |\chi_3(\mathscr{C}_k)|^2 = (1/6)[1(2)^2 + 2(-1)^2 + 3(0)^2] = 1;$$

$$g^{-1}\sum_k c_k\ \chi_1(\mathscr{C}_k)^*\ \chi_3(\mathscr{C}_k) = (1/6)[1(1)(2) + 2(1)(-1) + 3(1)(0)] = 0.$$

In how many ways can these vectors be chosen? We may choose the character $\chi^i(\mathscr{C}_k)$ from any of the N_r IRs. Therefore the number of mutually orthogonal vectors is the number of IRs, N_r and this must be $\leq N_c$ the dimension of the space. In fact, we shall see shortly that the number of IRs is equal to the number of classes.

(iv) Second orthogonality theorem for the characters. Set up a matrix Q and its adjoint $\mathrm{Q}^\dagger$ in which the elements of Q are the characters as in Table 4.2 but now including normalization factors, so that typical elements are

$$\mathrm{Q}_{ik} = \sqrt{c_k/g}\ \chi^i(\mathscr{C}_k),\ (\mathrm{Q}^\dagger)_{kj} = \mathrm{Q}_{jk}^* = \sqrt{c_k/g}\ \chi^j(\mathscr{C}_k)^*. \qquad (9)$$

(9) $$(\mathrm{Q}\ \mathrm{Q}^\dagger)_{ij} = \sum_k \mathrm{Q}_{ik}(\mathrm{Q}^\dagger)_{kj} = \sum_k \sqrt{c_k/g}\ \chi^i(\mathscr{C}_k)\ \sqrt{c_k/g}\ \chi^i(\mathscr{C}_k)^* = \delta_{ji}. \qquad (10)$$

(10) $$\mathrm{Q}\ \mathrm{Q}^\dagger = \mathrm{E}\quad (\mathrm{Q}\text{ a unitary matrix}); \qquad (11)$$

(11) $$\mathrm{Q}^\dagger\ \mathrm{Q} = \mathrm{E}; \qquad (12)$$

(12) $$\begin{aligned}(\mathrm{Q}^\dagger\mathrm{Q})_{kl} &= \sum_i (\mathrm{Q}^\dagger)_{ki}\mathrm{Q}_{il} = \sum_i \mathrm{Q}_{ik}^*\mathrm{Q}_{il} \\ &= \sum_{i=1}^{N_r} \sqrt{c_k/g}\ \chi^i(\mathscr{C}_k)^*\ \sqrt{c_l/g}\ \chi^i(\mathscr{C}_l) = \delta_{kl}.\end{aligned} \qquad (13)$$

Equation (13) describes the orthogonality of the columns of the character table. It states that vectors with components $\sqrt{c_k/g}\ \chi^i(\mathscr{C}_k)$ in an N_r-dimensional space are orthonormal. Since these vectors may be chosen in N_c ways (one from each of the N_c classes),

(13) $$N_c \leq N_r. \qquad (14)$$

But in eq. (7) the vectors with components $\sqrt{c_k/g}\ \chi^i(\mathscr{C}_k)$ may be chosen in N_r ways (one from each of N_r representations), and so

(7) $$N_r \leq N_c. \tag{15}$$

(14), (15) $$N_r = N_c. \tag{16}$$

The number of representations N_r is equal to N_c, the number of classes. In a more practical form for testing orthogonality

(13) $$\sum_{i=1}^{N_r} \chi^i(\mathscr{C}_k)^* \chi^i(\mathscr{C}_l) = (g/c_k)\delta_{kl}. \tag{17}$$

These orthogonality relations in eqs.(8) and (17), and also eq.(16), are very useful in setting up character tables.

Example 4.4-4 In C_{3v} there are three classes and therefore three IRs. We have established that Γ_1 and Γ_3 are both IRs, and, using $\sum_i l_i^2 = g$, we find $1 + l_2^2 + 4 = 6$, so that $l_2 = 1$. The character table for C_{3v} is therefore as given in Table 4.4(a).

From the orthogonality of the rows,

$$1(1)(1) + 2(1)\chi_2(C_3) + 3(1)\chi_2(\sigma) = 0,$$
$$1(2)(1) + 2(-1)\chi_2(C_3) + 3(0)\chi_2(\sigma) = 0,$$

so that $\chi_2(C_3) = 1$, $\chi_2(\sigma) = -1$. We check for normalization of the character system of Γ_2:

$$\sum_k c_k |\chi(\mathscr{C}_k)|^2 = 1(1)^2 + 2(1)^2 + 3(-1)^2 = 6 = g.$$

Exercise 4.4-1 Check the orthogonality of the columns in the character table for C_{3v} which was completed in Example 4.4-4.

(v) Reduction of a representation. For Γ to be a reducible representation, it must be equivalent to a representation in which each matrix $\Gamma(T)$ of T has the same block-diagonal structure. Suppose that the jth IR occurs c^j times in Γ; then

$$\chi(T) = \sum_j c^j \chi_j(T). \tag{18}$$

Multiplying by $\chi_i(T)^*$ and summing over T yields

(18), (4) $$\sum_T \chi_i(T)^* \chi(T) = \sum_j c^j \sum_T \chi_i(T)^* \chi_j(T) = \sum_j c^j\ g\ \delta_{ij} = g\ c^i; \tag{19}$$

(19) $$c^i = g^{-1} \sum_T \chi_i(T)^* \chi(T) = \sum_{k=1}^{N_c} c_k\ \chi_i(\mathscr{C}_k)^* \chi(\mathscr{C}_k). \tag{20}$$

Table 4.4(a) *Character table for* C_{3v}.

C_{3v}	E	$2C_3$	3σ
Γ_1	1	1	1
Γ_2	1	$\chi_2(C_3)$	$\chi_2(\sigma)$
Γ_3	2	-1	0

Table 4.4(b).

C_3	E	C_3^+	C_3^-
E	E	C_3^+	C_3^-
C_3^+	C_3^+	C_3^-	E
C_3^-	C_3^-	E	C_3^+

Table 4.4(c).

C_3	E	C_3^+	C_3^-
E	E	C_3^-	C_3^+
C_3^+	C_3^+	E	C_3^-
C_3^-	C_3^-	C_3^+	E

Normally we would choose to do the sum over classes rather than over group elements. Equation (20) is an extremely useful relation, and is used frequently in many practical applications of group theory.

(vi) The celebrated theorem. The number of times the ith IR occurs in a certain reducible representation called the *regular representation* Γ^r is equal to the dimension of the representation, l_i. To set up the matrices of Γ^r arrange the columns of the multiplication table so that only E appears on the diagonal. Then $\Gamma^r(T)$ is obtained by replacing T by 1 and every other element by zero (Jansen and Boon (1967)).

Example 4.4-5 Find the regular representation for the group C_3. $C_3 = \{E\ C_3^+\ C_3^-\}$. Interchanging the second and third columns of Table 4.4(b) gives Table 4.4(c).

Therefore, the matrices of the regular representation are

$$\begin{matrix} \Gamma^r(E) & \Gamma^r(C_3^+) & \Gamma^r(C_3^-) \\ \begin{bmatrix} 1 & 0 & 0 \\ 0 & 1 & 0 \\ 0 & 0 & 1 \end{bmatrix} & \begin{bmatrix} 0 & 0 & 1 \\ 1 & 0 & 0 \\ 0 & 1 & 0 \end{bmatrix} & \begin{bmatrix} 0 & 1 & 0 \\ 0 & 0 & 1 \\ 1 & 0 & 0 \end{bmatrix} \end{matrix}$$

The group C_3 is Abelian and has three classes; there are therefore three IRs and each IR occurs once in Γ^r. (But note that the matrices of Γ^r are not block-diagonal.)

Proof of the celebrated theorem

(20) $$c^i = g^{-1}\sum_T \chi_i(T)^* \, \chi_r(T)$$
$$= g^{-1}\chi_i(E)\,\chi_r(E), (\chi_r(T) = 0, \forall\, T \neq E)$$
$$= g^{-1}\, l_i\, g = l_i. \tag{21}$$

The dimension of Γ^r is g; it is also $\sum_i l_i^2$. Therefore

$$\sum_i l_i^2 = g, \tag{22}$$

as promised earlier.

Answer to Exercise 4.4-1

Normalization of the class $2C_3$: $1^2+1^2+(-1)^2=3=6/2$, and of the class 3σ: $1^2+(-1)^2+(0)^2=2=6/3$. Orthogonality of E and $2C_3$: $1(1)+1(1)+2(-1)=0$; orthogonality of E and 3σ: $1(1)+1(-1)+2(0)=0$; orthogonality of $2C_3$ and 3σ: $1(1)+1(-1)+1(-1)(0)=0$.

4.5 Character tables

Character tables are tabulations by classes of the characters of the IRs of the point groups. They are used constantly in practical applications of group theory. As an example, the character table for the point group C_{3v} (or $3m$) is given in Table 4.5. The name of the point group in either Schönflies or International notation (or both) is in the top left-hand corner. The headings to the columns are the number of elements c_k in each class $\mathscr{C}_k$ and a symbol describing the type of elements in that class. For example, the heading for the column of characters for the class $\{C_3^+ C_3^-\}$ in C_{3v} is $2C_3$. Usually Schönflies symbols are used, but some authors use other notation. Each row is labeled by the symbol for an IR; usually either Bethe or Mulliken notation is used, but sometimes one encounters other notations and examples of these will be introduced later. In Bethe's notation, the IRs are labeled

Table 4.5. *Character table for the point group* C_{3v}.

The IRs are labeled using both Bethe and Mulliken notation.

C_{3v}	E	$2C_3$	$3\sigma_v$	
Γ_1, A_1	1	1	1	z, x^2+y^2, z^2
Γ_2, A_2	1	1	-1	R_z
Γ_3, E	2	-1	0	$(x\ y), (R_x\ R_y), (x^2-y^2\ \ xy), (yz\ \ zx)$

Table 4.6. *Mulliken notation for the IRs of the point groups.*

The entry + or − signifies a positive or negative integer, respectively.

l	Notation used for IR	$\chi(C_n)^a$	$\chi(C_2')$ or $\chi(\sigma_v)^b$	$\chi(\sigma_h)$	$\chi(I)$
1	A	+1			
	B	−1			
	subscript 1		+1		
	subscript 2		−1		
2	E[c]				
3	T				
1, 2, or 3	superscript ′			+	
	superscript ″			−	
	subscript *g*				+
	subscript *u*				−

[a] Or $\chi(S_n)$ if the principal axis is an S_n axis. In D_2 the four 1-D IRs are usually designated A, B_1, B_2, B_3, because there are three equivalent C_2 axes.
[b] If no C_2' is present then subscripts 1 or 2 are used according to whether $\chi(\sigma_v)$ is +1 or −1.
[c] The symbol E for a 2-D IR is not to be confused with that used for the identity operator, E.

$\Gamma_1, \Gamma_2, \Gamma_3, \ldots$ successively; Γ_1 is always the totally symmetric representation. The remaining representations are listed in order of increasing l. Mulliken notation, which is generally used in molecular symmetry, is explained in Table 4.6. Thus, the totally symmetric representation is A_1 in C_{3v}. The second IR is labeled A_2 since $\chi(\sigma_v) = -1$, there being no C_2' axes in this group. The third IR is labeled E because $l = 2$. The dimension of any representation is given by $\chi(E)$ since the identity operator E is always represented by the unit matrix. In addition to the characters, the table includes information about how the components of a vector $\mathbf{r} = \mathbf{e}_1 x + \mathbf{e}_2 y + \mathbf{e}_3 z$ transform (or how linear functions of x, y, or z, transform) and how quadratic functions of x, y, and z transform. This information tells us to which representations p and d orbitals belong. For example, the three p orbitals and the five d orbitals are both degenerate in spherical symmetry (atoms), but in C_{3v} symmetry the maximum degeneracy is two and

$$\Gamma_p = \Gamma_1 \oplus \Gamma_3 = A_1 \oplus E,$$
$$\Gamma_d = \Gamma_1 \oplus 2\,\Gamma_3 = A_1 \oplus 2\,E.$$

We say that "z forms a *basis* for A_1," or that "z belongs to A_1," or that "z transforms according to the totally symmetric representation A_1." The s orbitals have spherical symmetry and so always belong to Γ_1. This is taken to be understood and is not stated explicitly in character tables. R_x, R_y, R_z tell us how rotations about **x**, **y**, and **z** transform (see Section 4.6). Table 4.5 is in fact only a partial character table, which includes only the *vector representations*. When we allow for the existence of electron spin, the state function $\psi(x\ y\ z)$ is replaced by $\psi(x\ y\ z)\chi(m_s)$, where $\chi(m_s)$ describes the electron spin. There are two ways of dealing with this complication. In the first one, the introduction of a new operator $\overline{E} = R(2\pi\ \mathbf{n}) \neq E$ results in additional classes and representations, and the point groups are then called *double groups*. The symbols for these new representations include information about the total angular momentum quantum number J. Double groups will be introduced in Chapter 8, and until then we shall use simplified point group character tables, like that for C_{3v} in Table 4.5, which are appropriate for discussions of the symmetry of functions of position, $f(x\ y\ z)$. The second way of arriving at the additional representations, which are called *spinor representations* (because their bases correspond to half-integral J), will be introduced in Chapter 12. This method has the advantages that the size of G is unchanged and no new classes are introduced.

Special notation is required for the complex representations of cyclic groups, and this will be explained in Section 4.7. The notation used for the IRs of the axial groups $C_{\infty v}$ and $D_{\infty h}$ is different and requires some comment. The states of diatomic molecules are classified according to the magnitude of the z component of angular momentum, L_z, using the symbols

$$\text{according to} \qquad \begin{array}{rcccc} & \Sigma & \Pi & \Delta & \Phi \\ \Lambda = |L_z| = & 0 & 1 & 2 & 3 \end{array}$$

All representations except Σ are two-dimensional. Subscripts g and u have the usual meaning, but a superscript $+$ or $-$ is used on Σ representations according to whether $\chi(\sigma_v) = \pm 1$. For $L_z > 0$, $\chi(C_2')$, and $\chi(\sigma_v)$ are zero. In double groups the spinor representations depend on the total angular momentum quantum number and are labeled accordingly.

4.6 Axial vectors

Polar vectors such as $\mathbf{r} = \mathbf{e}_1 x + \mathbf{e}_2 y + \mathbf{e}_3 z$ change sign on inversion and on reflection in a plane normal to the vector, but do not change sign on reflection in a plane that contains the vector. *Axial vectors* or *pseudovectors* do not change sign under inversion. They occur as vector products, and in symmetry operations they transform like *rotations* (hence the name axial vectors). The vector product of two polar vectors

$$\mathbf{r}_1 \times \mathbf{r}_2 = \mathbf{R} \tag{1}$$

is a pseudovector, or axial vector, of magnitude $r_1\ r_2 \sin\theta$, where θ is the included angle, $0 \leq \theta \leq \pi$ (see Figure 4.1(a)). The orientation of the axis of rotation is that it coincides with that of a unit vector $\mathbf{n}$ in a direction such that $\mathbf{r}_1$, $\mathbf{r}_2$, and $\mathbf{n}$ form a right-handed system. However, $\mathbf{R}$ is *not* a polar vector because its transformation properties under inversion and reflection are quite different to those of the polar vector $\mathbf{r}$. In Figure 4.1 the directed line segment symbols used for $\mathbf{r}_1$, $\mathbf{r}_2$ are the conventional ones for polar vectors, but the curved arrow symbol used for $\mathbf{R}$ indicates a rotation about the axis $\mathbf{n}$. The direction of rotation is that of the first-named vector $\mathbf{r}_1$ into $\mathbf{r}_2$, and the sign of $\mathbf{R}$ is positive because the direction

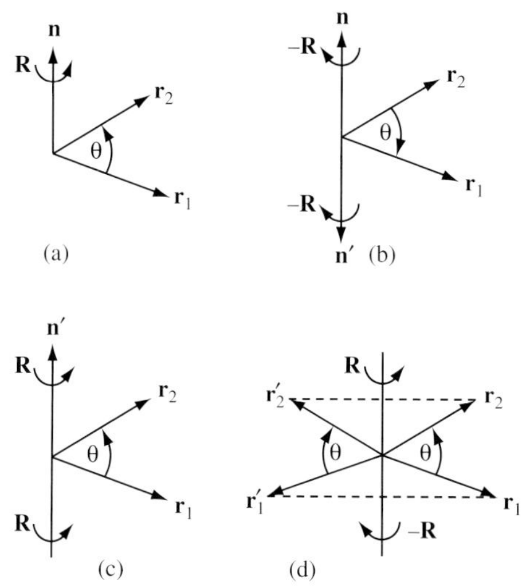

Figure 4.1. (a) The axial vector, or pseudovector, $\mathbf{r}_1 \times \mathbf{r}_2 = \mathbf{R}$. The curved arrow symbol used for **R** expresses the idea that the sense of rotation (which is that of a right-handed screw advancing along **n**, where **n**, $\mathbf{r}_1$, and $\mathbf{r}_2$ form a right-handed system) is from $\mathbf{r}_1$ into $\mathbf{r}_2$, i.e. from the first vector into the second one. (b) Reversing the order of the vectors in a vector product reverses the direction of rotation and so reverses its sign. (c) Invariance of the pseudovector $\mathbf{r}_1 \times \mathbf{r}_2 = \mathbf{R}$ under reflection in a plane normal to the axis of rotation. This figure shows why **R** must not be represented by a directed line segment normal to the plane of $\mathbf{r}_1, \mathbf{r}_2$ because such an object would change sign on reflection in the plane of $\mathbf{r}_1, \mathbf{r}_2$, whereas the sense of rotation of $\mathbf{r}_1$ into $\mathbf{r}_2$, as expressed by the curved arrow, is unchanged under this symmetry operation. (d) Reversal of the direction of rotation occurs on reflection in a plane that contains the axis of rotation.

of rotation appears anticlockwise on looking down the axis towards the origin. Reversing the order of the vectors in a vector product reverses its sign:

$$\mathbf{r}_2 \times \mathbf{r}_1 = -(\mathbf{r}_1 \times \mathbf{r}_2) \tag{2}$$

(Figure 4.1(b)). One can see in Figure 4.1(c) that reflection in a plane normal to the axis of rotation does not change the direction of rotation, but that it is reversed (Figure 4.1(d)) on reflection in a plane that contains the axis of rotation. Specification of a rotation requires a statement about both the axis of rotation and the amount of rotation. We define *infinitesimal* rotations about the axes OX, OY, and OZ by (note the cyclic order)

$$R_x = \phi(\mathbf{e}_2 \times \mathbf{e}_3), \tag{3}$$

$$R_y = \phi(\mathbf{e}_3 \times \mathbf{e}_1), \tag{4}$$

$$R_z = \phi(\mathbf{e}_1 \times \mathbf{e}_2). \tag{5}$$

Under a symmetry operator T, R_x transforms into $R_x' = \phi(\mathbf{e}_2' \times \mathbf{e}_3')$ and similarly, so that

Table 4.7. *Transformation of the basis* $\{R_x\ R_y\ R_z\}$ *under the operators in the first column.*

T	$\mathbf{e}_1{}'$	$\mathbf{e}_2{}'$	$\mathbf{e}_3{}'$	$R_x{}'$	$R_y{}'$	$R_z{}'$
E	$\mathbf{e}_1$	$\mathbf{e}_2$	$\mathbf{e}_3$	R_x	R_y	R_z
$R(\pi/2\ \mathbf{z})$	$\mathbf{e}_2$	$-\mathbf{e}_1$	$\mathbf{e}_3$	R_y	$-R_x$	R_z
$R(\pi\ \mathbf{z})$	$-\mathbf{e}_1$	$-\mathbf{e}_2$	$\mathbf{e}_3$	$-R_x$	$-R_y$	R_z
$R(\pi\ \mathbf{x})$	$\mathbf{e}_1$	$-\mathbf{e}_2$	$-\mathbf{e}_3$	R_x	$-R_y$	$-R_z$
$R(\pi\ [\bar{1}\ 1\ 0])$	$-\mathbf{e}_2$	$-\mathbf{e}_1$	$-\mathbf{e}_3$	$-R_y$	$-R_x$	$-R_z$
I	$\mathbf{e}_1$	$\mathbf{e}_2$	$\mathbf{e}_3$	R_x	R_y	R_z
$R(\phi\ \mathbf{z})$	$c\ \mathbf{e}_1 + s\ \mathbf{e}_2$	$-s\ \mathbf{e}_1 + c\ \mathbf{e}_2$	$\mathbf{e}_3$	$c\ R_x + s\ R_y$	$-s\ R_x + c\ R_y$	R_z

$$T\langle R_x\ R_y\ R_z| = \langle R_x{}'\ R_y{}'\ R_z{}'| = \langle R_x\ R_y\ R_z \mid \Gamma^{(\mathbf{R})}(T), \tag{6}$$

where

$$R_x{}' = \phi(\mathbf{e}_2{}' \times \mathbf{e}_3{}'),\ R_y{}' = \phi(\mathbf{e}_3{}' \times \mathbf{e}_1{}'),\ R_z{}' = \phi(\mathbf{e}_1{}' \times \mathbf{e}_2{}'). \tag{7}$$

$\Gamma^{(\mathbf{R})}(T)$ is not usually the same as the MR $\Gamma^{(\mathbf{r})}(T)$ for the basis $\langle\ \mathbf{e}_1\ \mathbf{e}_2\ \mathbf{e}_3|$ (previously called just $\Gamma(T)$, since there was no need then to specify the basis). With this refinement in the notation,

$$T\langle \mathbf{e}_1\ \mathbf{e}_2\ \mathbf{e}_3| \ =\ \langle \mathbf{e}_1{}'\ \mathbf{e}_2{}'\ \mathbf{e}_3{}'| \ =\ \langle \mathbf{e}_1\ \mathbf{e}_2\ \mathbf{e}_3|\Gamma^{(\mathbf{r})}(T). \tag{8}$$

The transformation properties of $\{R_x\ R_y\ R_z\}$ are then readily worked out from eq. (6) using the primed equations (7) with $\{\mathbf{e}_1{}'\ \mathbf{e}_2{}'\ \mathbf{e}_3{}'\}$ obtained from eq. (8) with the use, when necessary, of eq. (2), which simply states that reversing the order of the terms in a vector product reverses its sign.

Example 4.6-1 Find how the rotations $\{R_x\ R_y\ R_z\}$ transform under the symmetry operators: $E, R(\pi/2\ \mathbf{z}),\ R(\pi\ \mathbf{z}),\ R(\pi\ \mathbf{x}),\ R(\pi\ [\bar{1}\ 1\ 0]),\ I,\ R(\phi\ \mathbf{z})$. The solution is summarized in Table 4.7. Figure 4.2 will be found helpful in arriving at the entries in columns 2, 3, and 4.

Exercise 4.6-1 Verify in detail (from eq. (7)) the entries in columns 5, 6, and 7 of Table 4.7 for $R(\phi\ \mathbf{z})$.

The MRs of the operators in the rows 2 to 6 for the basis $\langle\ R_x\ R_y\ R_z|$ are

$$\begin{matrix} E & R(\pi/2\ \mathbf{z}) & R(\pi\ \mathbf{z}) & R(\pi\ \mathbf{x}) & R(\pi\ [\bar{1}\ 1\ 0]) \\ \begin{bmatrix} 1 & 0 & 0 \\ 0 & 1 & 0 \\ 0 & 0 & 1 \end{bmatrix} & \begin{bmatrix} 0 & \bar{1} & 0 \\ 1 & 0 & 0 \\ 0 & 0 & 1 \end{bmatrix} & \begin{bmatrix} \bar{1} & 0 & 0 \\ 0 & \bar{1} & 0 \\ 0 & 0 & 1 \end{bmatrix} & \begin{bmatrix} 1 & 0 & 0 \\ 0 & \bar{1} & 0 \\ 0 & 0 & \bar{1} \end{bmatrix} & \begin{bmatrix} 0 & \bar{1} & 0 \\ \bar{1} & 0 & 0 \\ 0 & 0 & \bar{1} \end{bmatrix} \end{matrix}.$$

This is a matrix representation of the group $D_4 = \{E\ 2C_4\ C_2\ 2C_2{}'\ 2C_2{}''\}$ and it is clearly reducible. The character systems of the two representations in the direct sum $\Gamma^{(\mathbf{R})} = \Gamma_2 \oplus \Gamma_5$ are

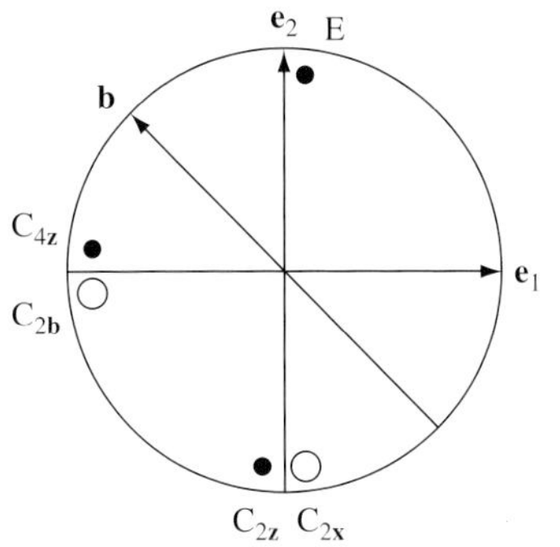

Figure 4.2. Projection in the **xy** plane of the unit sphere in configuration space, showing the initial orientation of the unit vectors $\mathbf{e}_1$, $\mathbf{e}_2$ before applying the symmetry operator T. Note that $\mathbf{e}_3$ is normal to the plane of the paper and points upwards towards the reader. Also shown are the positions of the representative point E after applying to configuration space the symmetry operators in rows 2 to 6 of Table 4.6. The unit vector **b** lies along the direction $[\bar{1}\ 1\ 0]$.

$$\begin{array}{lllllll} & & E & 2C_4 & C_2 & 2C_2' & 2C_2'' \\ \Gamma_2 & = & \{1 & 1 & 1 & -1 & -1\} \\ \Gamma_5 & = & \{2 & 0 & -2 & 0 & 0\}. \end{array}$$

Exercise 4.6-2 Show that Γ_5 is an IR of D_4. How many IRs are there in the character table of D_4? Give the names of Γ_2 and Γ_5 in Mulliken notation.

Answers to Exercises 4.6

Exercise 4.6-1 From eq. (7) and columns 2–4 of Table 4.7,

$$\mathbf{e}_2' \times \mathbf{e}_3' = (-s\ \mathbf{e}_1 + c\ \mathbf{e}_2) \times \mathbf{e}_3 = -s(\mathbf{e}_1 \times \mathbf{e}_3) + c(\mathbf{e}_2 \times \mathbf{e}_3)$$
$$= s(\mathbf{e}_3 \times \mathbf{e}_1) + c(\mathbf{e}_2 \times \mathbf{e}_3).$$

Therefore, $R_x' = c\ R_x + s\ R_y$.

$$\mathbf{e}_3' \times \mathbf{e}_1' = \mathbf{e}_3 \times (c\ \mathbf{e}_1 + s\ \mathbf{e}_2) = c(\mathbf{e}_3 \times \mathbf{e}_1) + s(\mathbf{e}_3 \times \mathbf{e}_2)$$
$$= c(\mathbf{e}_3 \times \mathbf{e}_1) - s(\mathbf{e}_2 \times \mathbf{e}_3).$$

Therefore, $R_y' = -s\ R_x + c\ R_y$.

$$\mathbf{e}_1' \times \mathbf{e}_2' = (c\ \mathbf{e}_1 + s\ \mathbf{e}_2) \times (-s\ \mathbf{e}_1 \times c\ \mathbf{e}_2) = c^2(\mathbf{e}_1 \times \mathbf{e}_2) + (-s^2)(\mathbf{e}_2 \times \mathbf{e}_1)$$
$$= (\mathbf{e}_1 \times \mathbf{e}_2).$$

Therefore $R_z' = R_z$.

Exercise 4.6-2 If Γ is an IR, the sum of the squares of the characters is equal to the order of the group. For Γ_5, $1(2)^2 + 1(-2)^2 + 2(0)^2 = 8 = g$, so Γ_5 is an IR. There are five classes and therefore five IRs. From $\sum_i l_i^2 = 8$ four are 1-D and one is 2-D. Since Γ_5 is the only IR with

$l = 2$, it is named E; Γ_2 is a 1-D IR, and in Mulliken notation it is called A_2 because $\chi(C_4) = +1$ *and* $\chi(C_2') = -1$.

4.7 Cyclic groups

If $A^n = E$, then the sequence $\{A^k\}$, with $k = 1, 2, \ldots, n$,

$$\{A \; A^2 \; A^3 \ldots A^n = E\}, \tag{1}$$

is a cyclic group of order n. All pairs of elements $A^k, A^{k'}$ commute and so $\{A^k\}$ is an Abelian group with n classes and therefore n 1-D IRs. If A is a symmetry operator then, in order to satisfy $A^n = E$, A must be either $E\,(n = 1)$, $I\,(n = 2)$, or a proper or an improper rotation, and if it is an improper rotation then n must be even. Writing the n classes in their proper order with $E = A^n$ first, a representation of

$$\{A^n = E \quad A \quad A^2 \; \ldots \; A^{n-1}\} \tag{1'}$$

is given by

$$\{\varepsilon^n = 1 \quad \varepsilon \quad \varepsilon^2 \; \ldots \; \varepsilon^{n-1}\}, \tag{2}$$

where the MRs

$$\varepsilon^k = \exp(-2\pi ik/n), \; k = 1, \ldots, n \tag{3}$$

are the n complex roots of unity. Note that

$$\varepsilon^{n-k} = \exp(-2\pi i(n-k)/n) = \exp(2\pi ik/n) = (\varepsilon^*)^k. \tag{4}$$

A second representation is

$$\{(\varepsilon^*)^n = 1 \quad \varepsilon^* \quad (\varepsilon^*)^2 \; \ldots \; (\varepsilon^*)^{n-1}\}, \tag{5}$$

so that the IRs occur in complex conjugate pairs generated from

$$\chi(A) = \exp(-2\pi ip/n), \; p = \pm 1, \pm 2, \ldots \tag{6}$$

$p = 0$ gives the totally symmetric representation

$$\Gamma_1 \text{ or } A = \{1 \; 1 \; 1 \; \ldots \; 1\}. \tag{7}$$

If n is odd, $p = 0, \pm 1, \pm 2, \ldots, \pm(n-1)/2$ generates all the representations which consist of Γ_1 and $(n-1)/2$ conjugate pairs. If n is even, $p = 0, \pm 1, \pm 2, \ldots, \pm(n-2)/2, n/2$. When $p = n/2$, $\chi(A^k) = (\varepsilon^{n/2})^k = [\exp(-i\pi)]^k = (-1)^k$, which is a representation

$$\Gamma_2 \text{ or } B = \{1 \quad -1 \quad 1 \quad -1 \quad \ldots \quad -1\} \tag{8}$$

from $k = n \; 1 \; 2 \; 3 \ldots n-1$. The character table of C_3 is given in Table 4.8.

To study the transformation of functions of $\{x \; y \; z\}$ under $R(\phi \; \mathbf{z})$ we make use of $\hat{R} f(\{x \; y \; z\}) = f(R^{-1}\{x \; y \; z\})$:

Table 4.8. *Character table for* C_3.

The form of this table with real basis functions ($E = {}^1E \oplus {}^2E$) given below the dashed line is seen in many compilations of character tables, but in practical applications the form with 1-D representations and complex basis functions should be used. If making comparisons with other compilations, note that we use the Condon and Shortley (1967) phase conventions, whereas Lax (1974) uses the Fano and Racah (1959) choice of phase (which for $j = 1$ would introduce an additional factor of i in the complex bases).

C_3	E	C_3^+	C_3^-	$\varepsilon = \exp(-\,i2\pi/3)$
A_1	1	1	1	$z,\ R_z,\ (x+iy)(x-iy),\ z^2$
${}^1E$	1	ε	ε^*	$-(x+iy),\ R_x+iR_y,\ z(x+iy),\ (x-iy)^2$
${}^2E$	1	ε^*	ε	$x-iy,\ R_x-iR_y,\ z(x-iy),\ (x+iy)^2$
A_1	1	1	1	$z,\ R_z,\ x^2+y^2,\ z^2$
E	2	-1	-1	$(x\ y),\ (R_x\ R_y),\ (yz\ zx),\ (xy\ x^2-y^2)$

$$\Gamma(R^{-1}) = \begin{bmatrix} x \\ y \\ z \end{bmatrix} = \begin{bmatrix} c & s & 0 \\ -s & c & 0 \\ 0 & 0 & 1 \end{bmatrix} \begin{bmatrix} x \\ y \\ z \end{bmatrix} = \begin{bmatrix} c\,x + s\,y \\ -s\,x + c\,y \\ z \end{bmatrix}. \tag{9}$$

Thus a proper (or improper) general rotation about **z** mixes the functions x and y. This is why $(x\ \ y)$ forms a basis for the 2-D representation E in C_{3v} while z, which transforms by itself under both $2C_3$ and $3\sigma_v$, forms a basis for the 1-D representation A_1. In C_3 there are, in addition to A, two more 1-D IRs. Since

$$R^{-1}(x \pm iy) = (c\,x + s\,y) \pm i(-s\,x + c\,y) = (c \mp is)(x \pm iy), \tag{10}$$

$-(x+iy)$ and $(x-iy)$ form 1-D bases, that is transform into themselves under $R(\phi\ \mathbf{z})$ rather than into a linear combination of functions. (The negative sign in $-(x+iy)$ comes from the Condon and Shortley phase conventions (see Chapter 11).) From eq. (10), the character for $-(x+iy)$ is $\varepsilon = \exp(-i\phi)$ for a general rotation through an angle ϕ, which becomes $\exp(-2\pi i/3)$ for a C_3^+ rotation, in agreement with eq. (6) for $p = 1$. For the basis $(x-iy)$ the character is $\exp(i\phi) = \varepsilon^*$, or $\exp(2\pi i/3)$ when $n = 3$, corresponding to $p = -1$ in eq. (6). In character tables of cyclic groups the complex conjugate (CC) representations are paired and each member of the pair is labeled by ${}^1E$, ${}^2E$ (with the addition of primes or subscripts g or u when appropriate). Because the states p and $-p$ are degenerate under time-reversal symmetry (Chapter 13), the pairs ${}^1E_p$ and ${}^2E_p$ are often bracketed together, each pair being labeled by the Mulliken symbol E, with superscripts and subscripts added when necessary. The character table for C_3 is given in Table 4.8 in both forms with complex and real representations. Complex characters should be used when reducing representations or when using projection operators (Chapter 5). However, in character tables real bases are usually given, and this practice is followed in Appendix A3.

4.8 Induced representations

Remark The material in this section is not made use of in this book until Section 16.5, in the chapter on space groups. Consequently, readers may choose to postpone their study of Section 4.8 until they reach Section 16.5.

Let $\mathrm{G} = \{g_j\}$ be a group of order g with a subgroup $\mathrm{H} = \{h_l\}$ of order h. The left coset expansion of G on H is

$$\mathrm{G} = \sum_{r=1}^{t} g_r\ \mathrm{H},\ t = g/h,\ g_1 = E, \tag{1}$$

where the coset representatives g_r for $r = 2, \ldots, t$, are $\in$ G but $\notin$ H. By closure in G, $g_j\, g_s$ ($g_s \in \{g_r\}$) is $\in$ G (g_k say) and thus a member of one of the cosets, say g_r H. Therefore, for some $h_l \in \mathrm{H}$,

$$g_j\ g_s = g_k = g_r\ h_l. \tag{2}$$

(2)
$$g_j\ g_s\ \mathrm{H} = g_r\ h_l\ \mathrm{H} = g_r\ \mathrm{H}; \tag{3}$$

(3)
$$g_j\ \langle g_s\ \mathrm{H}| = \langle g_r\ \mathrm{H}| = \langle g_s\ \mathrm{H}|\ \Gamma^{\mathrm{g}}(g_j). \tag{4}$$

In eq. (4) the cosets themselves are used as a basis for G, and from eq. (3) g_s H is transformed into g_r H by g_j. Since the operator g_j simply re-orders the basis, each matrix representation in the *ground representation* Γ^{g} is a permutation matrix (Appendix A1.2). Thus the sth column of Γ^{g} has only one non-zero element,

(4), (2)
$$\begin{aligned}[\Gamma^{\mathrm{g}}(g_j)]_{us} &= 1,\ \text{when } u = r,\ g_j\ g_s = g_r\ h_l \\ &= 0,\ \text{when } u \neq r.\end{aligned} \tag{5}$$

Because binary composition is unique (rearrangement theorem) the same restriction of only one non-zero element applies to the rows of Γ^{g}.

Exercise 4.8-1 What is the dimension of the ground representation?

Example 4.8-1 The multiplication table of the permutation group S(3), which has the cyclic subgroup $\mathrm{H} = \mathrm{C}(3)$, is given in Table 1.3. Using the coset representatives $\{g_s\} = \{P_0\ P_3\}$, write the left coset expansion of S(3) on C(3). Using eq. (2) find $g_r\ h_l$ for $\forall\, g_j \in \mathrm{G}$. [*Hint*: $g_r \in \{g_s\}$ and h_l are determined uniquely by g_i, g_s.] Hence write down the matrices of the ground representation.

The left coset expansion of S(3) on C(3) is

$$\mathrm{G} = \sum_{s=1}^{t} g_s\ \mathrm{H} = P_0\ \mathrm{H} \oplus P_3\ \mathrm{H} = \{P_0\ P_1\ P_2\} \oplus \{P_3\ P_4\ P_5\}, \tag{6}$$

with g_r and h_l, determined from $g_j\ g_s = g_k = g_r\ h_l$, given in Table 4.9. With the cosets as a basis,

$$g_j\langle P_0\ \mathrm{H}, P_3\ \mathrm{H}| = \langle P_0{}'\ \mathrm{H}, P_3{}'\ \mathrm{H}| = \langle P_0\ \mathrm{H}, P_3\ \mathrm{H}|\ \Gamma^{\mathrm{g}}(g_j). \tag{7}$$

Table 4.9. *The values of g_k and h_l determined from eq. (4.8.2) for* G = S(3) *and* H = C(3).

$g_s = P_0$					$g_s = P_3$				
g_j	g_s	g_k	g_r	h_l	g_j	g_s	g_k	g_r	h_l
P_0	P_0	P_0	P_0	P_0	P_0	P_3	P_3	P_3	P_0
P_1	P_0	P_1	P_0	P_1	P_1	P_3	P_5	P_3	P_2
P_2	P_0	P_2	P_0	P_2	P_2	P_3	P_4	P_3	P_1
P_3	P_0	P_3	P_3	P_0	P_3	P_3	P_0	P_0	P_0
P_4	P_0	P_4	P_3	P_1	P_4	P_3	P_2	P_0	P_2
P_5	P_0	P_5	P_3	P_2	P_5	P_3	P_1	P_0	P_1

Table 4.10. *The ground representation Γ^{g} determined from the cosets P_0H, P_3H by using the cosets as a basis, eq. (4.8.4).*

g_j	P_0	P_1	P_2	P_3	P_4	P_5
P_0', P_3'	P_0, P_3	P_0, P_3	P_0, P_3	P_3, P_0	P_3, P_0	P_3, P_0
$\Gamma^{\mathrm{g}}(g_j)$	$\begin{bmatrix} 1 & 0 \\ 0 & 1 \end{bmatrix}$	$\begin{bmatrix} 1 & 0 \\ 0 & 1 \end{bmatrix}$	$\begin{bmatrix} 1 & 0 \\ 0 & 1 \end{bmatrix}$	$\begin{bmatrix} 0 & 1 \\ 1 & 0 \end{bmatrix}$	$\begin{bmatrix} 0 & 1 \\ 1 & 0 \end{bmatrix}$	$\begin{bmatrix} 0 & 1 \\ 1 & 0 \end{bmatrix}$

The matrices of the ground representation are in Table 4.10. Each choice of g_j and g_s in eq. (2) leads to a particular h_l so that eq. (2) describes a mapping of G on to its subgroup H in which h_l is the image of g_j.

Example 4.8-2 Write a left coset expansion of S(3) on H = $\{P_0\ P_3\}$. Show that for $g_s = P_1$, $g_r \in \{g_s\}$ and $h_l \in$ H are determined uniquely for each choice of $g_j \in$ G.

Using Table 1.3,

$$\mathrm{S}(3) = P_0\{P_0\ P_3\} \oplus P_1\{P_0\ P_3\} \oplus P_2\{P_0\ P_3\}. \tag{8}$$

The g_r and h_l that satisfy eq. (2) are given in Table 4.11, where $\{g_r\} = \{P_0\ P_1\ P_2\}$ and $h_l \in \{P_0\ P_3\}$. Table 4.11 verifies the homomorphous mapping of G → H by $\{P_0\ P_1\ P_2\} \rightarrow P_0$ and $\{P_3\ P_4\ P_5\} \rightarrow P_3$. When necessary for greater clarity, the *subelement* h_l can be denoted by h_{sl} or by $h_{sl}(g_j)$, as in

$$g_j\ g_s = g_r\ h_{sl}(g_j). \tag{9}$$

(9), (5)
$$h_{sl}(g_j) = g_r^{-1} g_j\ g_s = \sum_u g_u^{-1} g_j\ g_s\ [\Gamma^{\mathrm{g}}(g_j)]_{us}. \tag{10}$$

The purpose of this section is to show how the representations of G may be constructed from those of its subgroup H. Let $\{\mathbf{e}_q\}$, $q = 1, \ldots, l_i$, be a subset of $\{\mathbf{e}_q\}$, $q = 1, \ldots, h$, that is an irreducible basis for H. Then

Table 4.11. *This table confirms that for $g_s = P_1$, g_r and h_l are determined by the choice of g_j, where $g_j\, g_s = g_k = g_r\, h_l$.*

g_j	g_s	g_k	g_r	h_l
P_0	P_1	P_1	P_1	P_0
P_1	P_1	P_2	P_2	P_0
P_2	P_1	P_0	P_0	P_0
P_3	P_1	P_4	P_2	P_3
P_4	P_1	P_5	P_1	P_3
P_5	P_1	P_3	P_0	P_3

$$h_l\ \mathbf{e}_q = \sum_{p=1}^{l_i} \mathbf{e}_p\ \tilde{\Gamma}_i(h_l)_{pq}, \tag{11}$$

where $\tilde{\Gamma}_i$ is the ith IR of the subgroup H. Define the set of vectors $\{\mathbf{e}_{rq}\}$ by

$$\mathbf{e}_{rq} = g_r\ \mathbf{e}_q,\ r = 1, \ldots, t;\ q = 1, \ldots, l_i. \tag{12}$$

Then $\langle \mathbf{e}_{rq}|$ is a basis for a representation of G:

(2), (5)
$$g_j\ \mathbf{e}_{sq} = g_j\ g_s\ \mathbf{e}_q = g_r\ h_l\ \mathbf{e}_q = \sum_u g_u\ [\Gamma^{\mathrm{g}}(g_j)]_{us}\ h_l\ \mathbf{e}_q. \tag{13}$$

In eq. (13) g_r has been replaced by

$$\sum_u g_u\ [\Gamma^{\mathrm{g}}(g_j)]_{us} = g_r \tag{14}$$

since the sth column of Γ^{g} consists of zeros except $u = r$.

(13), (11)
$$g_j\ \mathbf{e}_{sq} = \sum_u g_u\ [\Gamma^{\mathrm{g}}(g_j)]_{us} \sum_p \mathbf{e}_p\ \tilde{\Gamma}_i(h_{sl})_{pq} \tag{15}$$

$$= \sum_u \sum_p \mathbf{e}_{up}\ (\Gamma(g_j)_{[u\ s]})_{pq}. \tag{16}$$

In the supermatrix Γ in eq. (16) each element $[u\ s]$ is itself a matrix, in this case $\tilde{\Gamma}_i(h_{sl})$ multiplied by $\Gamma^{\mathrm{g}}(g_j)_{us}$.

(16), (15)
$$\Gamma(g_j)_{up,\ sq} = \Gamma^{\mathrm{g}}(g_j)_{us}\ \tilde{\Gamma}_i(h_{sl})_{pq}, \tag{17}$$

in which u, p label the rows and s, q label the columns; $\Gamma(g_j)$ is the matrix representation of g_j in the *induced representation* $\Gamma = \tilde{\Gamma}_i \uparrow \mathrm{G}$. Because Γ^{g} is a permutation matrix, with $\Gamma^{\mathrm{g}}(g_j)_{us} = 0$ unless $u = r$, an alternative way of describing the structure of Γ is as follows:

(15), (16), (5), (10)
$$\Gamma(g_j)_{up,\ sq} = \tilde{\Gamma}_i(g_u^{-1} g_j\ g_s)_{pq}\ \delta_{ur}. \tag{18}$$

$\tilde{\Gamma}_i(g_u^{-1} g_j\ g_s)$ is the matrix that lies at the junction of the uth row and the sth column of $\Gamma_{[u\ s]}$, and the Kronecker δ in eq. (18) ensures that $\tilde{\Gamma}_i$ is replaced by the null matrix except for $\Gamma_{[r\ s]}$.

Table 4.12. *Character table of the cyclic group* C(3) *and of the permutation group* S(3).

$\varepsilon = \exp(-2i\pi/3)$.

C(3)	P_0	P_1	P_2	S(3)	P_0	P_1, P_2	P_3, P_4, P_5
Γ_1, A	1	1	1	Γ_1, A_1	1	1	1
Γ_2, ${}^1E$	1	ε	ε^*	Γ_2, A_2	1	1	−1
Γ_3, ${}^2E$	1	ε^*	ε	Γ_3, E	2	−1	0

Table 4.13. *Subelements* $h_{sl}(g_j)$ *and MRs* $\Gamma(g_j)$ *of two representations of* S(3), $\tilde{\Gamma}\uparrow G$, *obtained by the method of induced representations.*

The third and fourth rows contain the subelements $h_{sl}(g_j)$ as determined by the values of g_s (in row 2), g_j, and g_r (in the first column). The $\Gamma^g(g_j)$ matrices were taken from Table 4.9. $\varepsilon = \exp(-i2\pi/3)$. Using Table 4.11, we see that the two representations of S(3) are $\tilde{\Gamma}_1 \uparrow G = A_1 \oplus A_2$ and $\tilde{\Gamma}_2 \uparrow G = E$.

g_j	P_0	P_1	P_2	P_3	P_4	P_5
g_r, g_s	P_0, P_3	P_0, P_3	P_0, P_3	P_0, P_3	P_0, P_3	P_0, P_3
P_0	P_0	P_1	P_2	P_0	P_2	P_1
P_3	P_0	P_2	P_1	P_0	P_1	P_2
$\tilde{\Gamma}_1 \uparrow G$	$\begin{bmatrix}1 & 0\\ 0 & 1\end{bmatrix}$	$\begin{bmatrix}1 & 0\\ 0 & 1\end{bmatrix}$	$\begin{bmatrix}1 & 0\\ 0 & 1\end{bmatrix}$	$\begin{bmatrix}0 & 1\\ 1 & 0\end{bmatrix}$	$\begin{bmatrix}0 & 1\\ 1 & 0\end{bmatrix}$	$\begin{bmatrix}0 & 1\\ 1 & 0\end{bmatrix}$
$\tilde{\Gamma}_2 \uparrow G$	$\begin{bmatrix}1 & 0\\ 0 & 1\end{bmatrix}$	$\begin{bmatrix}\varepsilon & 0\\ 0 & \varepsilon^*\end{bmatrix}$	$\begin{bmatrix}\varepsilon^* & 0\\ 0 & \varepsilon\end{bmatrix}$	$\begin{bmatrix}0 & 1\\ 1 & 0\end{bmatrix}$	$\begin{bmatrix}0 & \varepsilon^*\\ \varepsilon & 0\end{bmatrix}$	$\begin{bmatrix}0 & \varepsilon\\ \varepsilon^* & 0\end{bmatrix}$
$\chi(\tilde{\Gamma}_1 \uparrow G)$	2	2	2	0	0	0
$\chi(\tilde{\Gamma}_2 \uparrow G)$	2	−1	−1	0	0	0

Example 4.8-3 Construct the induced representations of S(3) from those of its subgroup C(3).

The cyclic subgroup C(3) has three 1-D IRs so that $\tilde{\Gamma}_i(h_{sl})$ has just one element ($p = 1$, $q = 1$). The character table of C(3) is given in Table 4.13, along with that of S(3), which will be needed to check our results. The subelements h_{sl} and coset representatives g_r depend on g_j and g_s, and our first task is to extract them from Table 4.8. They are listed in Table 4.13. Multiplying the $[\tilde{\Gamma}_i(h_{sl})]_{11} = \tilde{\chi}_i(h_{sl})$ by the elements of $\Gamma^g(g_j)$ in Table 4.12 gives the representations of S(3). An example should help clarify the procedure. In Table 4.13, when $g_j = P_4$, $g_s = P_3$, and $g_r = P_0$, the subelement $h_{sl}(g_j) = P_2$. (In rows 3 and 4 of Table 4.13 the subelements are located in positions that correspond to the non-zero elements of $\Gamma^g(g_j)$.) From Table 4.10, $[\Gamma^g(P_4)]_{12} = 1$, and in Table 4.12 $\tilde{\chi}_2(P_2) = \varepsilon^*$, so that $[\tilde{\Gamma}_2 \uparrow G]_{12} = \varepsilon^*$, as entered in the sixth row of Table 4.13.

From the character systems in Table 4.13 we see that for the IRs of S(3), $[\tilde{\Gamma}_1 \uparrow G] = A_1 \oplus A_2$ and $[\tilde{\Gamma}_2 \uparrow G] = E$. We could continue the table by finding $\tilde{\Gamma}_3 \uparrow G$, but since we already have all the representations of S(3), this could only yield an equivalent representation. Note that while this procedure $\tilde{\Gamma} \uparrow G$ does not necessarily yield IRs, it does give all the IRs of G, after reduction. A proof of this statement may be found in Altmann (1977).

4.8.1 Character system of an induced representation

We begin with

$$g_j = g_r\ h_l\ g_s^{-1}. \tag{19}$$

When $s = r$,

(19) $$g_j = g_r\ h_l\ g_r^{-1}. \tag{20}$$

Define

$$\{g_r\ h_l\ g_r^{-1}\} = \mathrm{H}^r,\ \forall\ h_l \in \mathrm{H}, \tag{21}$$

where H^r is the subgroup conjugate to H by g_r.

Exercise 4.8-2 Verify closure in H^r. Is this sufficient reason to say that H^r is a group?

The character of the matrix representation of g_j in the representation Γ induced from $\tilde{\Gamma}_i$ is

(20) $$\chi(g_j) = \sum_r \chi_r(g_j), \tag{22}$$

where the trace of the rth diagonal block $(s = r)$ of Γ is

(5), (17) $$\chi_r(g_j) = \tilde{\chi}_i\ (h_l),\ g_j \in \mathrm{H}^r$$
$$= 0,\ g_j \notin \mathrm{H}^r. \tag{23}$$

A representation Γ of $G = \{g_j\}$ is irreducible if

(4.4.5) $$\sum_j \chi(g_j)^* \chi(g_j) = g. \tag{24}$$

(24), (22) $$\sum_r \sum_j |\chi_r(g_j)|^2 + \sum_s \sum_{r \neq s} \sum_j \chi_r(g_j)^*\ \chi_s(g_j) = g. \tag{25}$$

The first term in eq. (25) is

(25), (23) $$\sum_r \sum_l |\tilde{\chi}_r(h_l)|^2 = \sum_{r=1}^{t} h = t\,h = g, \tag{26}$$

and so the second term in eq. (25) must be zero if Γ is irreducible. The irreducibility criterion eq. (25) thus becomes

(25), (23) $$\sum_{\{g_k\}} \chi_r(g_k)^*\ \chi_s(g_k) = 0,\ \forall\ r \neq s,\ \{g_k\} = \mathrm{H}^r \cap \mathrm{H}^s. \tag{27}$$

Equation (27) is known as *Johnston's irreducibility criterion* (Johnston (1960)).

The number of times c^i that the IR Γ_i occurs in the reducible representation $\Gamma = \sum_i c^i\ \Gamma_i$ of a group $G = \{g_k\}$, or *frequency* of Γ_i in Γ, is

(4.4.20) $$c^i = g^{-1} \sum_k \chi_i(g_k)^* \; \chi(g_k), \tag{28}$$

where $\chi(g_k)$ is the character of the matrix representation of g_k in the reducible representation Γ. If Γ_i is a reducible representation, we may still calculate the RS of eq. (28), in which case it is called the *intertwining number* I of Γ_i and Γ,

$$\mathrm{I}(\Gamma_i, \Gamma) = g^{-1} \sum_k \chi_i(g_k)^* \; \chi(g_k), \;\; \Gamma_i, \Gamma \text{ not IRs.} \tag{29}$$

Since I (Γ_i, Γ) is real, eq. (29) is often used in the equivalent form

$$\mathrm{I}(\Gamma_i, \Gamma) = g^{-1} \sum_k \chi_i(g_k) \; \chi(g_k)^*, \;\; \Gamma_i, \Gamma \text{ not IRs.} \tag{30}$$

If Γ_i, Γ have no IRs in common, it follows from the OT for the characters that I (Γ_i, Γ) = 0.

4.8.2 Frobenius reciprocity theorem

The frequency c^m of an IR Γ_m of G in the induced representation $\tilde{\Gamma}_i \uparrow$ G with characters $\chi_m(g_j)$ is equal to the frequency $\tilde{c}^i$ of $\tilde{\Gamma}_i$ in the *subduced* $\Gamma_m \downarrow$ H. The tilde is used to emphasize that the $\tilde{\Gamma}_i$ are representations of H. It will not generally be necessary in practical applications when the Mulliken symbols are usually sufficient identification. For example the IRs of S(3) are A_1, A_2, and E, but those of its subgroup C(3) are A, ^1E, and ^2E. *Subduction* means the restriction of the elements of G to those of H (as occurs, for example, in a lowering of symmetry). Normally this will mean that an IR Γ_m of G becomes a direct sum of IRs in H,

$$\Gamma_m = \sum_p \tilde{c}^p \; \tilde{\Gamma}_p, \quad \chi_m = \sum_p \tilde{c}^p \; \tilde{\chi}_p\,, \tag{31}$$

although if this sum contains a single term, only re-labeling to the IR of the subgroup is necessary. For example, in the subduction of the IRs of the point group T to D_2, the IR T becomes the direct sum of three 1-D IRs $B_1 \oplus B_2 \oplus B_3$ in D_2, while A_1 is re-labeled as A.

Proof

$$c^m = g^{-1} \sum_k \chi_m(g_j)^* \chi(g_j) \tag{28$'$}$$

(22) $$= g^{-1} \sum_r \sum_j \chi_m(g_j)^* \; \chi_r(g_j)$$

(20), (23) $$= g^{-1} \sum_r \sum_l \chi_m(g_r \; h_l \; g_r^{-1})^* \; \tilde{\chi}_i(h_l) \tag{32}$$

$$(31) \qquad = h^{-1}\sum_l \sum_p \tilde{c}^p \tilde{\chi}_p(h_l)^* \ \tilde{\chi}_i(h_l) \tag{33}$$

$$= \tilde{c}^i. \tag{34}$$

The tildes are not standard notation and are not generally needed in applications, but are used in this proof to identify IRs of the subgroup. In writing eq. (32), the sum over j is restricted to a sum over l (subduction) because the elements $g_r \ h_l \ g_r^{-1}$ belong to the class of h_l. In substituting eq. (31) in eq. (32) we use the fact that $\{c^p\}$ is a set of real numbers. Equation (34) follows from eq. (33) because of the OT for the characters. When H is an invariant subgroup of G, $\mathrm{H}^r = \mathrm{H}^s = \mathrm{H}$, $\forall \ r, s$. Then

$$(27), (22) \qquad \sum_l \chi_r(h_l)^* \ \chi_s(h_l) = 0, \ \forall \ r, \ s, r \neq s, \tag{35}$$

where Γ^r, Γ^s are representations of H but are not necessarily IRs.

$$(29) \qquad \sum_l \chi_r(h_l)^* \ \chi_s(h_l) = h \ \mathrm{I}(\Gamma_r, \Gamma_s). \tag{36}$$

Therefore, when H is an invariant subgroup of G,

$$(35), (36) \qquad \mathrm{I}(\Gamma_r, \Gamma_s) = 0; \tag{37}$$

that is, the representations Γ_r, Γ_s of H have, when reduced, no IRs in common.

Exercise 4.8-3 Test eq. (27) using the representations $\tilde{\Gamma}_1 \uparrow \mathrm{G}$ and $\tilde{\Gamma}_2 \uparrow \mathrm{G}$ of S(3), induced from C(3).

Answers to Exercises 4.8

Exercise 4.8-1 The dimension of the ground representation is equal to the number of cosets, $t = g/h$.

Exercise 4.8-2 Since $\{h_l\} = \mathrm{H}$ is closed, $h_l \ h_m \in \mathrm{H}$, say h_n. Then

$$g_r \ h_l \ g_r^{-1} \ g_r \ h_m \ g_r^{-1} = g_r \ h_l \ h_m \ g_r^{-1} = g_r \ h_n \ g_r^{-1} \in \mathrm{H}^r,$$

verifying that H^r is closed; h_l, h_m, h_n are $\in$ G, and therefore $\{g_r \ h_l \ g_r^{-1}\}$ satisfies the group properties of associativity and each element having an inverse. Moreover, $g_r \ E \ g_r^{-1} = E$, so that H^r does have all the necessary group properties.

Exercise 4.8-3 $\mathrm{H}^r = g_r \ \mathrm{H} \ g_r^{-1} = P_0\{P_0 \ P_1 \ P_2\}P_0^{-1} = \{P_0 \ P_1 \ P_2\} = \mathrm{H}$. $\mathrm{H}^s = P_3\{P_0 \ P_1 \ P_2\}P_3^{-1} = \{P_0 \ P_2 \ P_1\} = \mathrm{H}$. Therefore H is invariant and $\{g_k\} = \mathrm{H}^r \cap \mathrm{H}^s = \mathrm{H} = \{P_0 \ P_1 \ P_2\}$. Remember that r, s refer to different diagonal blocks. For $\tilde{\Gamma}_1 \uparrow \mathrm{G}$, $\sum_{\{g_k\}} \chi_r(g_k)^* \ \chi_s(g_k) = 1 + 1 + 1 = 3 \neq 0$, and therefore it is reducible. For $\tilde{\Gamma}_2 \uparrow \mathrm{G}$, $\sum_{\{g_k\}} \chi_r(g_k)^* \ \chi_s(g_k) = 1 + (\varepsilon^*)^2 + \varepsilon^2 = 0$, and therefore it is irreducible. This confirms the character test made in Table 4.12.

Problems

4.1 The point group of allene is $D_{2d} = \{E\ 2S_4\ C_2\ 2C_2'\ 2\sigma_d\}$ (see Problem 2.3). Choose a right-handed system of axes so that the vertical OZ axis points along the principal axis of symmetry.

(a) With the basis $\langle \mathbf{e}_1\ \mathbf{e}_2\ \mathbf{e}_3 |$, determine MRs of all eight symmetry operators of this group. Write down the character system of this matrix representation. This representation is reducible and is the direct sum of two IRs. Write down the character systems of these two IRs and check for normalization of the characters. Name these IRs using Mulliken notation.

(b) Determine how R_z transforms under the group operations. You now have sufficient information to arrange the elements of D_{2d} into classes.

(c) How many IRs are there? What are the dimensions of the IRs not yet found? From orthogonality relations find the character systems of these IRs and name them according to the Mulliken conventions. Summarize your results in a character table for D_{2d}.

(d) Find the character system of the DP representation $\Gamma_5 \otimes \Gamma_5$, where Γ_5 is the 2-D representation found in (a). Decompose this DP representation into a direct sum of IRs. [*Hint*: The characters of the DP representation are the products of the characters of the representations in the DP. Here, then, the character system for the DP representation is $\{\chi_5(T)\ \chi_5(T)\}$.]

4.2 Show that (a) $(x - iy)^2$, (b) $R_x + iR_y$, and (c) $R_x - iR_y$ form bases for the IRs of C_3, as stated in Table 4.7.

4.3 Find the character table of the improper cyclic group S_4.

4.4 Explain why the point group $D_2 = \{E\ C_{2z}\ C_{2x}\ C_{2y}\}$ is an Abelian group. How many IRs are there in D_2? Find the matrix representation based on $\langle \mathbf{e}_1\ \mathbf{e}_2\ \mathbf{e}_3 |$ for each of the four symmetry operators $R \in D_2$. The Jones symbols for R^{-1} were determined in Problem 3.8. Use this information to write down the characters of the IRs and their bases from the set of functions $\{z\,x\,y\}$. Because there are three equivalent C_2 axes, the IRs are designated A, B_1, B_2, B_3. Assign the bases R_x, R_y, R_z to these IRs. Using the result given in Problem 4.1 for the characters of a DP representation, find the IRs based on the quadratic functions x^2, y^2, z^2, xy, yz, zx.

4.5 Show that

$$\sum_k c_k\ \chi_k^{\,j} = g\ \delta_{j1}, \tag{1}$$

where j labels the IRs of G. (Since eq. (1) is based on the orthogonality of the rows, it is not an independent relation.) Verify eq. (1) for the group C_{3v}. (b) Use eq. (1) to deduce the character table of C_{2v}. [*Hint*: Is C_{2v} an Abelian group?]

4.6 (a) Show that the induction of $\tilde{\Gamma}_3 \uparrow G$, where H is C(3) and G is S(3), yields a representation equivalent to $\tilde{\Gamma}_1 \uparrow G$ in Table 4.12. (b) Show that the reducible representation $\tilde{\Gamma}_1 \uparrow G$ in Table 4.12 can be reduced into a direct sum $\Gamma_1 \oplus \Gamma_2$ by a similarity transformation using the matrix

$$\mathsf{S} = 2^{-1/2}\begin{bmatrix} 1 & 1 \\ 1 & -1 \end{bmatrix}.$$

5 Bases of representations

5.1 Basis functions

The group of the Hamiltonian, or the group of the Schrödinger equation, is the set of function operators $\{\hat{A}\ \hat{B}\ \ldots\ \hat{T}\ \ldots\}$ isomorphous with the symmetry group $\{A\ B\ \ldots\ T\ \ldots\}$ (Section 3.5). The function operators commute with the Hamiltonian operator $\hat{H}$ (Section 3.6). We will now show that the eigenfunctions of $\hat{H}$ form a basis for the group of the Hamiltonian. We make use of the fact that if $\{\phi_s\}$ is a set of degenerate eigenfunctions then a linear combination of these eigenfunctions is also an eigenfunction with the same eigenvalue. (A familiar example is the construction of the real eigenfunctions of $\hat{H}$ for the one-electron atom with $l = 1, p_x$, and p_y, from the complex eigenfunctions p_{+1}, p_{-1}; p_0, which corresponds to $m = 0$, is the real eigenfunction p_z.) The property of a basis that we wish to exploit is this. If we have a set of operators that form a group, then a basis is a set of objects, each one of which, when operated on by one of the operators, is converted into a linear combination of the same set of objects. In our work, these objects are usually a set of vectors, or a set of functions, or a set of quantum mechanical operators. For example, for the basis vectors of an n-dimensional linear vector space (LVS)

$$T\langle\mathbf{e}| = \langle\mathbf{e}'| = \langle\mathbf{e}|\Gamma(T), \tag{1}$$

or, in greater detail,

$$T\langle\mathbf{e}_1\ldots\mathbf{e}_i\ldots| = \langle\mathbf{e}'_1\ldots\mathbf{e}'_j\ldots| = \langle\mathbf{e}_1\ldots\mathbf{e}_i\ldots|\Gamma(T), \tag{2}$$

where

$$\mathbf{e}'_j = \sum_{i=1}^{l} \mathbf{e}_i\ \Gamma(T)_{ij}, \quad j = 1, \ldots, l. \tag{3}$$

The $\Gamma(T)_{ij}$ in eq. (3) are the elements of the jth column of the matrix representative $\Gamma(T)$ of the symmetry operator T. A realization of eq. (3) in 3-D space was achieved when the matrix representative (MR) of $R(\phi\ \mathbf{z})$ was calculated in Section 3.2. The MRs form a group representation, which is either an irreducible representation (IR) or a direct sum of IRs. Let $\{\phi_s\}$ be a set of degenerate eigenfunctions of $\hat{H}$ that corresponds to a particular eigenvalue E, so that

$$\hat{H}\ \phi_s = \mathrm{E}\ \phi_s, \quad s = 1, \ldots, l. \tag{4}$$

Because $\hat{H}$ and its eigenvalues are invariant when a symmetry operator T acts on the physical system, $\hat{T}\,\phi_s$ is also an eigenfunction of $\hat{H}$ with the same eigenvalue E, and therefore it is a linear combination of the $\{\phi_s\}$,

$$\hat{T}\phi_s = \sum_{r=1}^{l} \phi_r\,\Gamma(T)_{rs}, \quad s = 1, \ldots, l. \tag{5}$$

In matrix form,

$$\hat{T}\langle\phi_1 \ldots \phi_s \ldots| = \langle\phi_1' \ldots\ \phi_s' \ldots| = \langle\phi_1 \ldots \phi_r \ldots\ |\Gamma(T). \tag{6}$$

Equation (6) can be written more compactly as

$$\hat{T}\langle\phi| = \langle\phi'| = \langle\phi|\Gamma(T), \tag{7}$$

where $\langle\phi|$ implies the whole set $\langle\phi_1\ \ldots\ \phi_s\ \ldots\ |$. Equations (7) and (1) show that the $\{\phi_s\}$ are a set of basis functions in an l-dimensional LVS, called a *function space*, which justifies the use of the alternative, equivalent, terms "eigenfunction" and "eigenvector." Because of eqs. (5)–(7), every set of eigenfunctions $\{\phi_s\}$ that corresponds to the eigenvalue E forms a basis for one of the IRs of the symmetry group $G = \{T\}$. Consequently, every energy level and its associated eigenfunctions may be labeled according to one of the IRs of $\{T\}$. The notation $\{\phi_s^k\}$, E^k means that the eigenfunctions $\{\phi_s^k\}$ that correspond to the eigenvalue E^k form a basis for the kth IR. Although the converse is not true – a set of basis functions is not necessarily a set of energy eigenfunctions – there are still advantages in working with sets of basis functions. Therefore we shall now learn how to construct sets of basis functions which form bases for particular IRs.

5.2 Construction of basis functions

Just as any arbitrary vector is the sum of its projections,

$$\mathbf{v} = \sum_i \mathbf{e}_i\,\mathbf{v}_i, \tag{1}$$

where $\mathbf{e}_i\,\mathbf{v}_i$ is the projection of $\mathbf{v}$ along $\mathbf{e}_i$, so any arbitrary function

$$\phi = \sum_k \sum_{s=1}^{l_k} \phi_s^k\, b_s^k, \tag{2}$$

where $\sum_k$ is over the IRs, and $\sum_{s=1}^{l_k}$ is a sum of projections within the subspace of the kth IR. The problem is this: how can we generate $\{\phi_p^j\}$, $p = 1, \ldots, l_j$, the set of l_j orthonormal functions which form a basis for the jth IR of the group of the Schrödinger equation? We start with any *arbitrary* function ϕ defined in the space in which the set of function operators $\{\hat{T}\}$ operate. Then

(2) $$\phi = \sum_k \sum_{s=1}^{l_k} \phi_s^k\, b_s^k = \sum_k \phi^k, \tag{3}$$

where $\phi_s^k\, b_s^k$ is the projection of ϕ along ϕ_s^k. Because ϕ_s^k is a basis function,

$$\text{(3)} \qquad \hat{T}\phi_s^k = \sum_{r=1}^{l_k} \phi_r^k\ \Gamma^k(T)_{rs}. \qquad (4)$$

$$\text{(3), (4)} \qquad \begin{aligned}&(l_j/g)\sum_T \Gamma^j(T)^*_{pp}\ \hat{T}(\phi)\\ &= (l_j/g)\sum_T \Gamma^j(T)^*_{pp} \sum_k\sum_s\sum_r \phi_r^k\ \Gamma^k(T)_{rs}\ b_s^k\\ &= (l_j/g)\sum_k\sum_s\sum_r \left[\sum_T \Gamma^j(T)^*_{pp}\ \Gamma^k(T)_{rs}\right]\phi_r^k\ b_s^k.\end{aligned} \qquad (5)$$

By the orthogonality theorem, the sum in brackets in eq. (5) is $(g/l_j)\delta_{jk}\delta_{pr}\delta_{ps}$, and consequently the triple sum yields unity if $k=j$, $r=p$, and $s=p$; otherwise it is zero. Therefore,

$$\text{(5)} \qquad (l_j/g)\sum_T \Gamma^j(T)^*_{pp}\ \hat{T}(\phi) = \phi_p^j\ b_p^j, \qquad (6)$$

and so we have ϕ_p^j apart from a constant which can always be fixed by normalization. The operator

$$(l_j/g)\sum_T \Gamma^j(T)^*_{pp}\ \hat{T} = \hat{P}^j_{pp} \qquad (7)$$

is a *projection operator* because it projects out of ϕ that part which transforms as the pth column of the jth IR,

$$\text{(7), (6)} \qquad \hat{P}^j_{pp}\ \phi = \phi_p^j\ b_p^j. \qquad (8)$$

By using all the $\hat{P}^j_{pp}$, $p = 1, \ldots, l_j$, in turn, that is all the diagonal elements of $\Gamma^j(T)$, we can find all the l_j functions $\{\phi_p^j\}$ that form a basis for Γ^j.

$$\text{(8), (2)} \qquad \hat{P}^j\ \phi = \sum_p \hat{P}^j_{pp}\ \phi = \sum_{p=1}^{l_j} \phi_p^j\ b_p^j = \phi^j. \qquad (9)$$

The RS side of eq. (9) is a linear combination of the l_j functions that forms a basis for Γ^j. The operator in eq. (9) is

$$\text{(9), (7)} \qquad \hat{P}^j = \sum_p (l_j/g)\sum_T \Gamma^j(T)^*_{pp}\ \hat{T} = (l_j/g)\sum_T \chi^j(T)^*\ \hat{T}. \qquad (10)$$

It projects out from ϕ in one operation the sum of all the parts of ϕ that transform according to Γ^j. Being a linear combination of the l_j linearly independent (LI) basis functions $\{\phi_p^j\}$, ϕ^j is itself a basis function for Γ^j. Equation (9) is preferable to eq. (8), that is $\hat{P}^j$ is preferred to $\hat{P}^j_{pp}$ because it requires only the characters of $\Gamma^j(T)$ and not all its diagonal elements $\sum_T \Gamma^j(T)_{pp}$. If Γ^j is 1-D, then ϕ^j is the basis function for Γ^j. But if Γ^j is not 1-D (i.e. l_j is not equal to unity) the procedure is repeated with a new ϕ to obtain a second ϕ^j, and so on, until l_j LI functions have been obtained.

5.3 Direct product representations

The *direct product* (DP) of two matrices $\mathbb{A} \otimes \mathbb{B}$ is defined in Section A1.7. If Γ is the DP of two representations Γ^i, Γ^j, then

$$\Gamma(T) = \Gamma^i(T) \otimes \Gamma^j(T), \quad \forall\, T \in G. \tag{1}$$

But is $\Gamma(T)$ also a representation?

(1), (A1.7.7)

$$\begin{aligned} \Gamma(T_1)\,\Gamma(T_2) &= (\Gamma^i(T_1) \otimes \Gamma^j(T_1))(\Gamma^i(T_2) \otimes \Gamma^j(T_2)) \\ &= (\Gamma^i(T_1)\,\Gamma^i(T_2)) \otimes (\Gamma^j(T_1)\,\Gamma^j(T_2)) \\ &= \Gamma^i(T_1 T_2) \otimes \Gamma^j(T_1 T_2) \\ &= \Gamma(T_1 T_2), \text{ or } \Gamma^{ij}(T_1 T_2), \end{aligned} \tag{2}$$

which shows that the DP of the two representations Γ^i and Γ^j is also a representation. The second notation in eq. (2) stresses that the representation Γ^{ij} is derived from the DP of Γ^i and Γ^j. So we conclude that *the direct product of two representations is itself a representation.*

If $\{\phi_q^i\}, q = 1, \ldots, m$, is a set of functions that form a basis for Γ^i, and $\{\phi_s^j\}, s = 1, \ldots, n$, is a set of functions that form a basis for Γ^j, then the *direct product set* $\{\phi_q^i\,\phi_s^j\}$, which contains mn functions, forms a basis for the DP representation Γ^{ij}.

$$\begin{aligned} \hat{T}\,\phi_q^i\,\phi_s^j &= \phi_q^i(T^{-1}\{x\})\phi_s^j(T^{-1}\{x\}) \\ &= (\hat{T}\,\phi_q^i)(\hat{T}\,\phi_s^j) \\ &= \sum_{p=1}^{m} \phi_p^i\,\Gamma^i(T)_{pq} \sum_{r=1}^{n} \phi_r^j\,\Gamma^j(T)_{rs} \\ &= \sum_p \sum_r \phi_p^i\,\phi_r^j\,\Gamma^i(T)_{pq}\,\Gamma^j(T)_{rs} \\ &= \sum_p \sum_r \phi_p^i\,\phi_r^j\,\Gamma^{ij}(T)_{pr,\,qs}, \end{aligned} \tag{3}$$

since the product of the pqth element from the MR $\Gamma^i(T)$, and the rsth element of the MR $\Gamma^j(T)$, is the pr,qsth element of the DP matrix $\Gamma^{ij}(T)$. Therefore, *the direct product set* $\{\phi_q^i\,\phi_s^j\}$ *is a basis for the direct product representation* $\Gamma^i \otimes \Gamma^j$. The characters of the MRs in the DP representation

$$\begin{aligned} \chi_{ij}(T) &= \sum_p \sum_r \Gamma^{ij}(T)_{pr,\,pr} \\ &= \sum_p \sum_r \Gamma^i(T)_{pp}\,\Gamma^j(T)_{rr} \\ &= \chi_i(T)\,\chi_j(T). \end{aligned} \tag{4}$$

Therefore, the character of an MR in the DP representation is the product of the characters of the MRs that make up the DP. Direct product representations may be reducible or irreducible.

Table 5.1. *Some direct product representations in the point group* C_{3v}.

C_{3v}	E	$2C_3$	$3\sigma_v$
A_1	1	1	1
A_2	1	1	−1
E	2	−1	0
$E \otimes A_1$	2	−1	0
$E \otimes A_2$	2	−1	0
$E \otimes E$	4	1	0

Example 5.3-1 Find the DPs of E with all three IRs of the point group C_{3v}. The characters of the IRs of C_{3v} and their DPs with E are given in Table 5.1.

By inspection, or by using

$$c^j = g^{-1} \sum_T \chi_j(T)^* \; \chi(T),$$

we find $E \otimes A_1 = E$, $E \otimes A_2 = E$, and $E \otimes E = A_1 \oplus A_2 \oplus E$.

5.3.1 *Symmetric and antisymmetric direct products*

With $j = i$, we introduce the symbols ϕ^i_q, ψ^i_s $(q, s = 1, \ldots, m)$ to designate basis functions from two bases γ, γ' of the ith IR. (The possibility that γ and γ' might be the same basis is not excluded.) Since there is only one representation under consideration, the superscript i may be suppressed. The DP of the two bases is

$$\langle\phi_q| \otimes \langle\psi_s| = \langle\phi_q\psi_s| = \tfrac{1}{2}\langle\phi_q\psi_s + \phi_s\psi_q| \oplus \tfrac{1}{2}\langle\phi_q\psi_s - \phi_s\psi_q|. \tag{5}$$

The first term on the RS of eq. (5) is symmetric and the second term is anti-symmetric, with respect to the exchange of subscripts q and s. These two terms are called the symmetrical ($\overline{\otimes}$) and antisymmetrical ($\underline{\otimes}$) DP, respectively, and eq. (5) shows that the DP of the two bases is the direct sum of the symmetrical and antisymmetrical DPs,

$$\langle\phi_q| \otimes \langle\psi_s| = (\langle\phi_q|\overline{\otimes}\langle\psi_s|) \oplus (\langle\phi_q|\underline{\otimes}\langle\psi_s|). \tag{6}$$

If the two bases are identical, then the antisymmetrical DP vanishes and the only DP is the symmetrical one.

(3) $$\hat{T}\ \phi_q\ \psi_s = \sum_p \sum_r \phi_p\ \psi_r\ \Gamma(T)_{pq}\ \Gamma(T)_{rs}; \tag{7}$$

(3) $$\hat{T}\ \phi_s\ \psi_q = \sum_p \sum_r \phi_p\ \psi_r\ \Gamma(T)_{ps}\ \Gamma(T)_{rq}; \tag{8}$$

$$(5), (7), (8) \qquad \hat{T}\ \phi_q\ \psi_s = \tfrac{1}{2}\sum_p \sum_r \phi_p\ \psi_r[\Gamma(T)_{pq}\ \Gamma(T)_{rs} + \Gamma(T)_{ps}\ \Gamma(T)_{rq}]$$

$$+ \tfrac{1}{2}\left[\sum_p \sum_r \phi_p\ \psi_r[\Gamma(T)_{pq}\ \Gamma(T)_{rs} - \Gamma(T)_{ps}\ \Gamma(T)_{rq}]\right]. \qquad (9)$$

Restoring the index i on Γ for greater clarity,

$$(9) \qquad \hat{T}\ \phi_q\ \psi_s = \sum_p \sum_r \phi_p\ \psi_r[\Gamma^{i\overline{\otimes}i}(T)_{pr,\,qs} + \Gamma^{i\underline{\otimes}i}(T)_{pr,\,sq}]. \qquad (10)$$

$$(9), (10) \qquad \Gamma^{i\overline{\underline{\otimes}}i}(T)_{pr,\,qs} = \tfrac{1}{2}[\Gamma^i(T)_{pq}\ \Gamma^i(T)_{rs} \pm \Gamma^i(T)_{ps}\ \Gamma^i(T)_{rq}], \qquad (11)$$

where $i\overline{\underline{\otimes}}i$ means either the symmetrical or antisymmetrical DP according to whether the positive sign or the negative sign is taken on the RS of eq. (11). To find the characters, set $q = p$, $s = r$, and sum over p and r:

$$(11) \qquad \chi^{i\overline{\underline{\otimes}}i}(T) = \tfrac{1}{2}\left[\sum_p \sum_r \Gamma^i(T)_{pp}\ \Gamma^i(T)_{rr} \pm \Gamma^i(T)_{pr}\ \Gamma^i(T)_{rp}\right]$$

$$= \tfrac{1}{2}\left[\sum_p \sum_r \Gamma^i(T)_{pp}\ \Gamma^i(T)_{rr} \pm \sum_p \Gamma^i(T^2)_{pp}\right]$$

$$= \tfrac{1}{2}[(\chi^i(T))^2 \pm \chi^i(T^2)]. \qquad (12)$$

Example 5.3-2 Show that for the point group C_{3v}, $E\overline{\otimes}E = A_1 \oplus E$ and $E\underline{\otimes}E = A_2$. Using the character table for C_{3v} in Example 5.3-1, eq. (12) yields

C_{3v}	E	$2C_3$	3σ
$\chi^E(T)$	2	−1	0
$\chi^E(T^2)$	2	−1	2
$\chi^{E\overline{\otimes}E}(T)$	3	0	1
$\chi^{E\underline{\otimes}E}(T)$	1	1	−1

Therefore, $E\overline{\otimes}E = A_1 \oplus E$, $E\underline{\otimes}E = A_2$. The sum of the symmetrical and antisymmetrical DPs is $E \otimes E$, as expected from eq. (11). (See Example 5.3-1.)

5.4 Matrix elements

5.4.1 Dirac notation

In quantum mechanics, an integral of the form

$$\int \psi_u^*\ \hat{Q}\ \psi_q\ \mathrm{d}\tau = \int (\hat{Q}^\dagger\ \psi_u)^*\ \psi_q\ \mathrm{d}\tau \qquad (1)$$

is called a *matrix element*. $\hat{Q}^{\dagger}$ is the adjoint of the operator $\hat{Q}$, and the definition of $\hat{Q}^{\dagger}$ is that it is the operator which satisfies eq. (1). In Dirac notation this matrix element is written as

$$\langle\psi_u|\hat{Q}|\psi_q\rangle = \langle\psi_u|\hat{Q}\,\psi_q\rangle = \langle\hat{Q}^{\dagger}\,\psi_u|\psi_q\rangle. \tag{2}$$

In matrix notation $\langle u^*|v\rangle$ describes the matrix representation of the Hermitian scalar product of the two vectors **u, v**, in an LVS with unitary basis ($\mathrm{M} = |\mathbf{e}^*\rangle\langle\mathbf{e}| = \mathrm{E}$). The second and third expressions in eq. (2) are *scalar products* in an LVS in which the basis vectors are the functions $\{\psi_q\}$ and the scalar product is defined to be an integral over the full range of the variables. Thus, the second equality in eq. (2) conveys precisely the same information as eq. (1). The first part of the complete bracket in eq. (2), $\langle\psi_u|$, is the bra-vector or *bra*, and the last part, $|\psi_q\rangle$, is the ket-vector or *ket*, and the complete matrix element is a *bra*(c)*ket* expression. Notice that in Dirac notation, complex conjugation of the function within the bra is part of the definition of the scalar product. The ket $|\psi_q\rangle$ represents the function ψ_q, in the matrix element integral. When $\hat{Q}$ operates on the function ψ_q, it produces the new function $\hat{Q}\psi_q$ so that when $\hat{Q}$ operates to the right in eq. (2) it gives the new ket $|\hat{Q}\psi_q\rangle$. But because eqs. (2) and (1) state the same thing in different notation, when $\hat{Q}$ operates to the left it becomes the adjoint operator, $\langle\psi_u|\hat{Q} = \langle\hat{Q}^{\dagger}\psi_u|$. Some operators are self-adjoint, notably the Hamiltonian $\hat{H} = \hat{H}^{\dagger}$.

5.4.2 Transformation of operators

Suppose that $\hat{Q}\,f = g$ and that when a symmetry operator T acts on the physical system $\hat{T}\,f = f'$, $\hat{T}\,g = g'$. Now,

$$g' = \hat{T}\,g = \hat{T}\,\hat{Q}\,f = \hat{T}\,\hat{Q}\,T^{-1}\,Tf = \hat{T}\,\hat{Q}\,T^{-1}\,f'. \tag{3}$$

Comparing this with $g = \hat{Q}f$, we see that the effect of T has been to transform the operator from $\hat{Q}$ into a new operator $\hat{Q}'$, where

(3) $$\hat{Q}' = \hat{T}\,\hat{Q}\,\hat{T}^{-1}. \tag{4}$$

Operators may also form bases for the IRs of the group of the Hamiltonian, for if $\hat{Q}$ is one of the set of operators $\{\hat{Q}_s^j\}$, and if

$$\hat{T}\,\hat{Q}_s^j\,T^{-1} = \sum_r \hat{Q}_r^j\,\Gamma^j(T)_{rs} \tag{5}$$

then the $\{\hat{Q}_s^j\}$ form a basis for the jth IR.

5.4.3 Invariance of matrix elements under symmetry operations

In quantum mechanics, matrix elements (or scalar products) represent physical quantities and they are therefore invariant when a symmetry operator acts on the physical system. For example, the expectation value of the dynamical variable Q when the system is in the state described by the state function f is

$$\langle Q\rangle = \langle f|\hat{Q}|f\rangle = \langle f|g\rangle. \tag{6}$$

It follows that the function operators $\hat{T}$ are unitary operators. For

(6) $$\langle f|g\rangle = \langle\hat{T} f|\hat{T} g\rangle = \langle\hat{T}^\dagger \hat{T} f|g\rangle \tag{7}$$

(7) $$\hat{T}^\dagger \hat{T} = \hat{E}, \tag{8}$$

(8) $$\Gamma(T)^\dagger \Gamma(T) = \mathrm{E}, \tag{9}$$

so that the MRs of the function operators are unitary matrices. An important question which can be answered using group theory is: "Under what conditions is a matrix element zero?" Provided we neglect spin–orbit coupling, a quantum, mechanical state function (spin orbital) can be written as a product of a spatial part, called an orbital, and a spinor, $\Psi(\mathbf{r}, m_s) = \psi(\mathbf{r})\chi(m_s)$. Since $\hat{Q}^j_s$ acts on space and not spin variables, the matrix element $\langle\Psi^k_u|\hat{Q}^j_s|\Psi^i_q\rangle$ factorizes as

$$\langle\Psi^k_u|\hat{Q}^j_s|\Psi^i_q\rangle = \langle\psi^k_u|\hat{Q}^j_s|\psi^i_q\rangle\langle\chi_u|\chi_q\rangle. \tag{10}$$

It follows from the orthogonality of the spin functions that $\langle\chi_u|\chi_q\rangle = 0$ unless χ_u, χ_q have the same spin quantum number. Hence the matrix element in eq. (10) is zero unless $\Delta S = 0$. When the matrix element describes a transition probability, this gives *the spin selection rule*. Spin–orbit coupling, although often weak, is not zero, and so the spin selection rule is not absolutely rigid. Nevertheless it is a good guide since transitions between states with $\Delta S \neq 0$ will be weaker than those for which the spin selection rule is obeyed. Now consider what happens to a matrix element under symmetry operator T. Its value is unchanged, so

$$\langle\psi^k_u|\hat{Q}^j_s|\psi^i_q\rangle = \langle\hat{T}\,\psi^k_u|\hat{T}\,\hat{Q}^j_s\,\hat{T}^{-1}|\hat{T}\,\psi^i_q\rangle. \tag{11}$$

The LS of eq. (11) is invariant under $\{T\}$ and so it belongs to the totally symmetric representation Γ_1. The function $\hat{Q}^j_s\,\psi^i_q$ transforms according to the DP representation $\Gamma^i \otimes \Gamma^j$. To see this, consider what happens when a symmetry operator T acts on configuration space: $\hat{Q}^j_s|\psi^i_q\rangle$ becomes

$$\begin{aligned}\hat{T}\,\hat{Q}^j_s\,\hat{T}^{-1}|\hat{T}\,\psi^i_q\rangle &= \sum_p\sum_r \Gamma^i(T)_{pq}\,\Gamma^j(T)_{rs}\,\hat{Q}^j_r|\psi^i_p\rangle \\ &= \sum_p\sum_r [\Gamma^i(T)\otimes\Gamma^j(T)]_{pr,qs}\,\hat{Q}^j_r|\psi^i_p\rangle.\end{aligned} \tag{12}$$

Therefore under T, $\hat{Q}^j_s|\psi^i_q\rangle$ transforms according to the DP representation $\Gamma^i(T)\otimes\Gamma^j(T)$. The integrand in eq. (11) is the product of two functions, $(\psi^k_u)^*$ and $\hat{Q}^j_s|\psi^i_q\rangle$, and it therefore transforms as the DP $\Gamma^{k*}\otimes\Gamma^i\otimes\Gamma^j$ or $\Gamma^{k*}\otimes\Gamma^{ij}$. What is the condition that $\Gamma^{a*}\otimes\Gamma^b \supset \Gamma^1$? This DP contains Γ^1 if

$$c^1 = g^{-1} \sum_T \chi_1(T)\, \chi^{a*b}(T) = g^{-1} \sum_T \chi^a(T)^*\, \chi^b(T) \neq 0, \qquad (13)$$

which will be so if and only if $a = b$ (from the orthogonality theorem for the characters). Therefore the matrix element $\langle \psi_u^k | \hat{Q}_s^j | \psi_q^i \rangle$ is zero unless the DP $\Gamma^i \otimes \Gamma^j \supset \Gamma^k$. But $\Gamma^k \otimes \Gamma^k \supset \Gamma^1$, and so the matrix element is zero unless $\Gamma^i \otimes \Gamma^j \otimes \Gamma^k \supset \Gamma^1$. *Therefore, the matrix element is zero unless the DP of any two of the representations contains the third one.*

5.4.4 Transition probabilities

The probability of a transition being induced by interaction with electromagnetic radiation is proportional to the square of the modulus of a matrix element of the form $\langle \psi^k | \hat{Q}^j | \psi^i \rangle$, where the state function that describes the initial state transforms as Γ^i, that describing the final state transforms as Γ^k, and the operator (which depends on the type of transition being considered) transforms as Γ^j. The strongest transitions are the E1 transitions, which occur when $\hat{Q}$ is the electric dipole moment operator, $-e\mathbf{r}$. These transitions are therefore often called "electric dipole transitions." The components of the electric dipole operator transform like x, y, and z. Next in importance are the M1 transitions, for which $\hat{Q}$ is the magnetic dipole operator, which transforms like R_x, R_y, R_z. The weakest transitions are the E2 transitions, which occur when $\hat{Q}$ is the electric quadrupole operator which, transforms like binary products of x, y, and z.

Example 5.4-1 The absorption spectrum of benzene shows a strong band at 1800 Å, two weaker bands at 2000 Å and 2600 Å, and a very weak band at 3500 Å. As we shall see in Chapter 6, the ground state of benzene is ${}^1A_{1g}$, and there are singlet and triplet excited states of B_{1u}, B_{2u}, and E_{1u} symmetry. Given that in D_{6h}, (x, y) form a basis for E_{1u} and z transforms as A_{2u}, find which transitions are allowed.

To find which transitions are allowed, form the DPs between the ground state and the three excited states and check whether these contain the representations for which the dipole moment operator forms a basis:

$$A_{1g} \otimes B_{1u} = B_{1u},$$
$$A_{1g} \otimes B_{2u} = B_{2u},$$
$$A_{1g} \otimes E_{1u} = E_{1u}.$$

Only one of these (E_{1u}) contains a representation to which the electric dipole moment operator belongs. Therefore only one of the three possible transitions is symmetry allowed, and for this one the radiation must be polarized in the (x, y) plane (see Table 5.2).

The strong band at 1800 Å is due to the ${}^1A_{1g} \rightarrow {}^1E_{1u}$ transition. The two weaker bands at 2000 Å and 2600 Å are due to the ${}^1A_{1g} \rightarrow {}^1B_{1u}$ and ${}^1A_{1g} \rightarrow {}^1B_{2u}$ transitions becoming allowed through *vibronic coupling*. (We shall analyze vibronic coupling later.) The very weak transition at 3500 Å is due to ${}^1A_{1g} \rightarrow {}^3E_{1u}$ becoming partly allowed through spin–orbit coupling.

Table 5.2. *Possible transitions from* $^1A_{1g}$ *electronic ground state in benzene.*

	Symmetry-allowed	Symmetry-forbidden
Spin-allowed	$^1E_{1u}$	$^1B_{1u}$, $^1B_{2u}$
Spin-forbidden	$^3E_{1u}$	$^3B_{1u}$, $^3B_{2u}$

Problems

5.1 Find Jones symbols for $\{R^{-1}\}$, $R \in D_4$. Project the function

$$\phi = x^2 + y^2 + z^2 + xy + yz + zx$$

into the Γ_1, Γ_2, and Γ_5 subspaces and hence find bases for these IRs. [*Hints*: Do not refer to published character tables. You will need character sets for the IRs Γ_2 and Γ_5 of D_4, which were found in Exercise 4.6-1.]

5.2 Find the IRs of the point group D_{4h} for which the following Cartesian tensors form bases:

$$1,\ xyz,\ z(x^2 - y^2),\ xy(x^2 - y^2),\ xyz(x^2 - y^2).$$

[*Hint*: Use the character table for D_{4h} in Appendix A3, in which the principal axis has been chosen to lie along **z**.]

5.3 Determine correlation relations between the IRs of (a) T_d and C_{3v}, and (b) O_h and D_{3d}. [*Hints*: Use character tables from Appendix A3. For (a), choose the C_3 axis along [1 1 1] and select the three dihedral planes in T_d that are vertical planes in C_{3v}. For (b), choose one of the C_3 axes (for example, that along [1 1 1]) and identify the three C_2' axes normal to the C_3 axis.]

5.4 In the groups C_{4v}, D_{3h}, and D_{3d} which E1, M1, and E2 transitions are allowed from a Γ_1 ground state? In each of the three groups, identify the ground state in Mulliken notation. For the E1 transitions, state any polarization restrictions on the radiation.

5.5 Evaluate for the representations $i = E$, T_1, and T_2 of the group O, the DP $\Gamma^{i\otimes i}$, the symmetric DP $\Gamma^{i\overline{\otimes}i}$, and the antisymmetric DP $\Gamma^{i\underline{\otimes}i}$. Show that your results satisfy the relation $\Gamma^{i\otimes i} = \Gamma^{i\overline{\otimes}i} \oplus \Gamma^{i\underline{\otimes}i}$.

6 Molecular orbitals

6.1 Hybridization

In descriptions of chemical bonding, one distinguishes between bonds which do not have a nodal plane in the charge density along the bond and those which do have such a nodal plane. The former are called σ bonds and they are formed from the overlap of s atomic orbitals on each of the two atoms involved in the bond (ss σ bonds) or they are sp or pp σ bonds, where here p implies a p_z atomic orbital with its lobes directed along the axis of the bond, which is conventionally chosen to be the z axis. The overlap of p_x or p_y atomic orbitals on the two atoms gives rise to a π bond with zero charge density in a nodal plane which contains the bond axis. Since it is accumulation of charge density between two atoms that gives rise to the formation of a chemical bond, σ or π molecular orbitals are referred to as *bonding* orbitals if there is no nodal plane normal to the bond axis, but if there is such a nodal plane they are *antibonding* orbitals. Carbon has the electron configuration $1s^2\ 2s^2\ 2p^2$, and yet in methane the four CH bonds are equivalent. This tells us that the carbon $2s$ and $2p$ orbitals are combined in a linear combination that yields four equivalent bonds. The physical process involved in this "mixing" of s and p orbitals, which we represent as a linear combination, is described as *hybridization*. A useful application of group theory is that it enables us to determine very easily which atomic orbitals are involved in hybridization. Sometimes there is more than one possibility, but even a rough knowledge of the atomic energy levels is usually all that is required to resolve the issue.

Example 6.1-1 This example describes σ bonding in tetrahedral AB_4 molecules. The numbering of the B atoms is shown in Figure 6.1. Denote by σ_r a unit vector oriented from A along the bond between A and B_r. With $\langle \sigma_1\ \sigma_2\ \sigma_3\ \sigma_4|$ as a basis, determine the characters of the representation Γ_σ. It is not necessary to determine the matrix representatives (MRs) $\Gamma(T)$ from $T\langle \sigma| = \langle \sigma|\Gamma(T)$ since we only need the character system χ_σ of the representation Γ_σ. Every σ_r that transforms into itself under a symmetry operator T contributes $+1$ to the character of that MR $\Gamma(T)$, while every σ_r that transforms into σ_s, with $s \neq r$, makes no contribution to $\chi_\sigma(T)$. Of course, we only need to determine $\chi_\sigma(T)$ for one member of each class in the point group. The values of $\chi_\sigma(T)$ for the point group T_d are given in Table 6.1. This is a reducible representation, and to reduce it we use the prescription

$$c^j = g^{-1} \sum_T \chi_j(T)^*\ \chi_\sigma(T) = g^{-1} \sum_k c_k\ \chi_j(\mathscr{C}_k)^*\ \chi_\sigma(\mathscr{C}_k). \tag{1}$$

Table 6.1. *The character system χ_σ for the representation Γ_σ in the point group* T_d.

T_d	E	$3C_2$	$8C_3$	$6S_4$	$6\sigma_d$
χ_σ	4	0	1	0	2

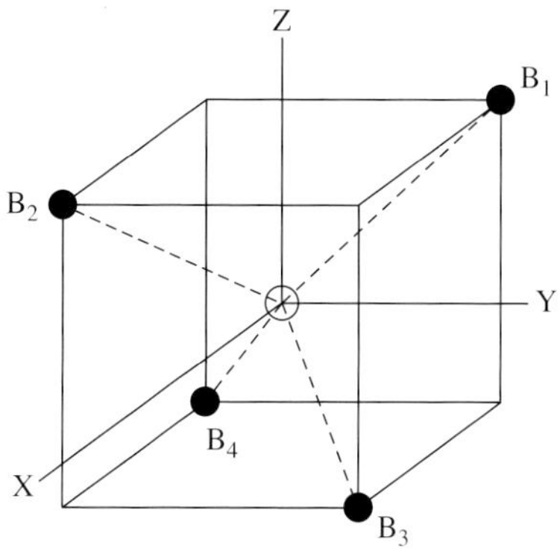

Figure 6.1. Numbering of the B atoms in a tetrahedral AB_4 molecule; σ_r is a unit vector pointing from A to atom B_r.

Using the character table of T_d in Appendix A3 we find for A_1 that

$$c(A_1) = g^{-1}[1(1)(4) + 8(1)(1) + 6(1)(2)] = 1.$$

We could proceed in a similar fashion for the remaining IRs, A_2, E, T_1, and T_2, but instead we attempt a short-cut by subtracting the character system for A_1 from that of Γ_σ:

$$\chi_\sigma - \chi^{A_1} = \{3 \;\; -1 \;\; 0 \;\; -1 \;\; 1\} = T_2. \tag{2}$$

Note that in a character system it is implied that the characters of the classes are given in the same order as in the character table. Also, when a character system is equated to the symbol for a representation, as in eq. (2), it means that it is the *character system of that representation*. Here then

$$\Gamma_\sigma = A_1 \oplus T_2. \tag{3}$$

We know that s forms a basis for A_1, and from the character table we see that (x, y, z) and also (xy, yz, zx) form bases for T_2. Therefore, σ bonds in tetrahedral AB_4 molecules are formed by sp^3 and/or sd^3 hybridization.

In general, an expression for a molecular orbital (MO) would involve linear combinations of s, and p_x, p_y, p_z and d_{xy}, d_{yz}, d_{zx} atomic orbitals (AOs), but some coefficients might be small or even negligibly small. There are two principles that control the formation of a chemical bond between two atoms: (i) the contributing AOs must be of comparable energy; and (ii) for a bonding MO, the bond should provide maximum overlap of charge density in the region between the atoms. In carbon the $3d$ orbitals lie about 10 eV above $2p$ and therefore

Table 6.2. *The character system* χ_σ *for the* AB_5 *molecule shown in Figure 6.2.*

D_{3h}	E	$2C_3$	$3C_2'$	σ_h	$2S_3$	$3\sigma_v$
χ_σ	5	2	1	3	0	3

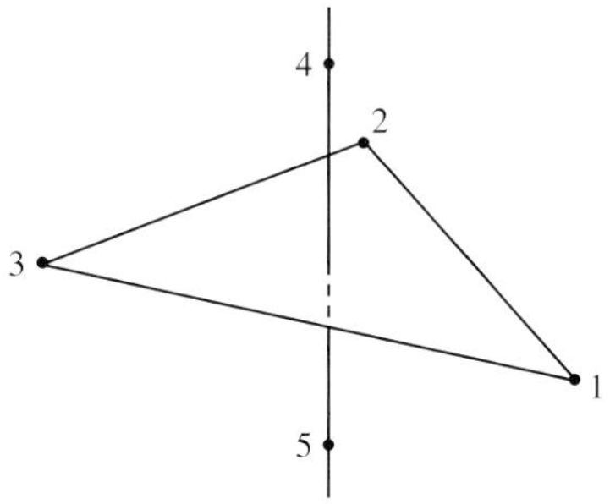

Figure 6.2. Numbering of the B atoms in the AB_5 trigonal bipyramid.

sp^3 hybridization predominates. But in manganese and chromium the $3d$ are much closer to the $4s$ than are the $4p$ orbitals, and it is likely that sd^3 hybridization predominates.

Example 6.1-2 This example describes σ bonding in the AB_5 trigonal bipyramid (e.g. PF_5). As is evident from Figure 6.2, the point group is D_{3h}. The character system for Γ_σ is given in Table 6.2. From eq. (1), with the help of the character table for D_{3h} in Appendix A3,

$$\Gamma_\sigma = 2A_1' \oplus A_2'' \oplus E'. \tag{4}$$

Exercise 6.1-1 Verify the reduction of Γ_σ into the direct sum given in eq. (4).

From the character table for D_{3h} we find that z forms a basis for A_2'' while (x, y) form a basis for E'. Similarly, $3z^2 - r^2$, as well as s, form bases for A_1' and $(xy, x^2 - y^2)$ form a basis for E'. The large difference in energy between $(n+1)s$ and ns, or between $(n+1)d_{3z^2-r^2}$ and $nd_{3z^2-r^2}$, atomic energy levels makes the contribution of two orbitals with different principal quantum numbers to hybrid MOs in AB_5 very unlikely. We conclude that one s and one $d_{3z^2-r^2}$ are involved, together with p_z, and $(p_x\ p_y)$, and/or $(d_{xy}\ d_{x^2-y^2})$. In PF_5, it is likely that $(p_x\ p_y)$ predominate, giving dsp^3 hybridization, while in molecules in which the central atom has a high atomic number Z, the p and d orbitals will both contribute, giving a mixture of dsp^3 and d^3sp hybridization. For example, in the $MoCl_5$ molecule, the molybdenum $4d$ AOs are of comparable energy to the $5p$ orbitals, so that a hybrid scheme $dsp^3 + d^3sp$ can be expected. It should be remarked that in the abbreviations used for hybridization schemes, specific d orbitals are implied; these may be found very easily by determining the character system for Γ_σ and using the character table to determine the IRs and their basis functions. The same method may be used to determine the AOs used in

π bonding, but we shall not give an example here since π bonding in the ML_6 octahedral complex will be analyzed later (in Section 6.4).

Answer to Exercise 6.1-1

For the AB_5 trigonal bipyramid, $\Gamma_\sigma = \{5\ \ 2\ \ 1\ \ 3\ \ 0\ \ 3\}$. Using the character table for D_{3h},

$$\begin{aligned} c(\mathrm{A_1}') &= (^1\!/_{12})[1(1)(5) + 2(1)(2) + 3(1)(1) + 1(1)(3) + 2(1)(0) + 3(1)(3)] \\ &= (^1\!/_{12})[5 + 4 + 3 + 3 + 0 + 9] = 2, \\ c(\mathrm{A_2}') &= (^1\!/_{12})[1(1)(5) + 2(1)(2) + 3(-1)(1) + 1(1)(3) + 2(1)(0) + 3(-1)(3)] \\ &= (^1\!/_{12})[5 + 4 - 3 + 3 + 0 - 9] = 0, \\ c(\mathrm{E}') &= (^1\!/_{12})[1(2)(5) + 2(-1)(2) + 3(0)(1) + 1(2)(3) + 2(-1)(0) + 3(0)(3)] \\ &= (^1\!/_{12})[10 - 4 + 0 + 6 + 0 + 0] = 1, \end{aligned}$$

$$\begin{aligned} 2\mathrm{A_1}' + \mathrm{E}' &= \{4\ \ 1\ \ 2\ \ 4\ \ 1\ \ 2\} \\ \Gamma_\sigma - (2\mathrm{A_1}' + \mathrm{E}') &= \{1\ \ 1\ \ -1\ \ -1\ \ -1\ \ 1\} = \mathrm{A_2}''. \end{aligned}$$

Therefore

$$\Gamma_\sigma = 2\mathrm{A_1}' \oplus \mathrm{A_2}'' \oplus \mathrm{E}'.$$

6.2 π Electron systems

The electronic charge density in an MO extends over the whole molecule, or at least over a volume containing two or more atoms, and therefore the MOs must form bases for the symmetry point group of the molecule. Useful deductions about bonding can often be made without doing any quantum chemical calculations at all by finding these symmetry-adapted MOs expressed as linear combinations of AOs (the LCAO approximation). So we seek the LCAO MOs

$$\psi^j = \sum_r \phi_r\, c_{rj}, \quad \text{or} \quad |\psi^j\rangle = \sum_r |\phi_r\rangle c_{rj} \tag{1}$$

where the AOs $\{\phi_r\}$ form an orthonormal basis set. It is common practice in Dirac notation to omit the symbol for the basis (e.g. ϕ) when this is not in doubt. For example, normalization of the ϕ basis may be expressed by

$$\int \mathrm{d}\tau\ \phi_r^*\ \phi_r \quad \text{or} \quad \langle\phi_r|\phi_r\rangle \quad \text{or} \quad \langle r|r\rangle = 1. \tag{2}$$

Example 6.2-1 This example discusses the molecular orbitals of benzene.The numbering system used for the atoms is shown in Figure 6.3. The point group of benzene is

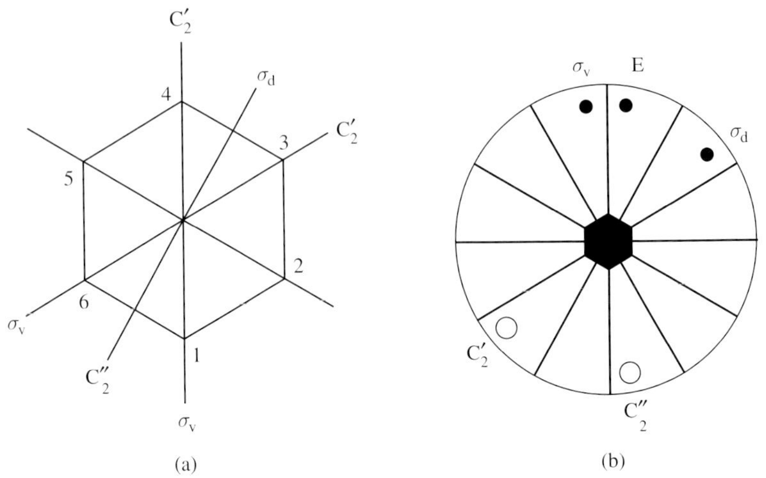

Figure 6.3. (a) Numbering scheme used for the six C atoms in the carbon skeleton of benzene. Also shown are examples of the locations of the C_2' and C_2'' axes, and of the σ_d and σ_v planes of symmetry. The C_2' axes are given precedence in naming the planes. (b) Partial projection diagram for D_{6h} showing that $IC_2' = \sigma_d$ and $IC_2'' = \sigma_v$.

$$D_{6h} = D_6 \otimes C_i = \{D_6\} \oplus I\{D_6\}, \tag{3}$$

where

$$D_6 = \{E\ 2C_6\ 2C_3\ C_2\ 3C_2'\ 3C_2''\}. \tag{4}$$

Refer to Appendix A3 for the character table of D_{6h}. Some conventions used for benzene are illustrated in Figure 6.3. The C_2 axes normal to the principal axis fall into two geometrically distinct sets. Those passing through pairs of opposite atoms are given precedence and are called C_2', and those that bisect pairs of opposite bonds are named C_2''. Consequently, the set of three vertical planes that bisect pairs of opposite bonds are designated $3\sigma_d$ (because they bisect the angles between the C_2' axes), while those that contain the C_2'' axes are called $3\sigma_v$. Note the sequence of classes in the character table: the classes in the second half of the table are derived from $I(\mathscr{C}_k)$, where $\mathscr{C}_k$ is the corresponding class in the first half. Thus $2S_3$ precedes $2S_6$ because $IC_6^+ = S_3^-$, and $3\sigma_d$ precedes $3\sigma_v$ because $IC_2' = \sigma_d$ but $IC_2'' = \sigma_v$. Notice that the characters of the u representations in the first set of classes (those of D_6) repeat those of the g representations in the top left corner of the table, but those of the u representations in the bottom right quarter have the same magnitude as those for the corresponding g representations in the top right quarter, but have the opposite sign. Some authors only give the character table for D_6, which is all that is strictly necessary, since the characters for D_{6h} can be deduced from those for D_6 using the properties explained above. The systematic presentation of character tables of direct product groups in this way can often be exploited to reduce the amount of arithmetic involved, particularly in the reduction of representations.

To find the MOs for benzene, we choose a basis comprising a $2p_z$ AO on each carbon atom and determine the characters of Γ_ϕ, the reducible representation generated by

$$\hat{T}\langle\phi| = \langle\phi'| = \langle\phi|\Gamma_\phi, \quad (5)$$

where T $\in$ D_{6h} and $\langle\phi|$ stands for the basis set $\langle\phi_1\ \phi_2\ \phi_3\ \phi_4\ \phi_5\ \phi_6|$; ϕ_r is a $2p_z$ AO on the rth carbon atom. Although we could determine $\langle\phi'|$ from the effect of the function operator $\hat{T}$ on each ϕ_r in turn, it is not necessary to do this. A much quicker method is to use the rotation of the contour of the function $\phi_r = 2p_z$ on atom r under the symmetry operator T to determine ϕ'_r, and then recognize that ϕ'_r only contributes to the character of $\Gamma_\phi(T)$ when it transforms into $\pm\phi_r$. A positive sign contributes $+1$ to the character of $\Gamma_\phi(T)$; a negative sign contributes -1; the contribution is zero if $\phi'_r = \phi_s$, $s \neq r$. The character system of Γ_ϕ may thus be written down by inspection, without doing any calculations at all. In this way we find that

$$\chi(\Gamma_\phi) = \{6\ \ 0\ \ 0\ \ 0\ \ -2\ \ 0\ \ 0\ \ 0\ \ 0\ \ -6\ \ 0\ \ 2\}. \quad (6)$$

In benzene, $T \in \{C_6\ C_3\ C_2\ I\ S_3\ S_6\ C''_2\ \sigma_d\}$ sends each ϕ_r into $\phi_{s\neq r}$ so that there are no non-zero diagonal entries in $\Gamma\phi$ for these operators and consequently $\chi(T) = 0$. For the C'_2 operators, the $2p_z$ orbitals on one pair of carbon atoms transform into their negatives, so that $\chi(C'_2) = -2$. For σ_h, each of the six atomic orbitals ϕ_r transforms into $-\phi_r$, so that $\chi(\sigma_h) = -6$. For the σ_v operators, the pair of $2p_z$ orbitals in the symmetry plane are unaffected, while the other four become $2p_z$ orbitals on different atoms, so $\chi(\sigma_v) = +2$. Finally, for the identity operator each ϕ_r remains unaffected, so $\Gamma_\phi(E)$ is the 6×6 unit matrix and $\chi(E) = 6$. Note that Γ_ϕ is a reducible representation,

$$\Gamma_\phi = \sum_j c^j\,\Gamma_j, \qquad \chi \equiv \chi(\Gamma_\phi) = \sum_j c^j\,\chi_j, \quad (7)$$

where

$$c^j = g^{-1}\sum_k c_k\,\chi_j(\mathscr{C}_k)^*\,\chi(\mathscr{C}_k), \quad (8)$$

and $\chi(\mathscr{C}_k)$ is the character for the kth class in the reducible representation.

(6), (8) $$\Gamma_\phi = A_{2u} \oplus B_{2g} \oplus E_{1g} \oplus E_{2u}. \quad (9)$$

For example,

$$c(A_{2u}) = (1/24)[1(1)(6) + 3(-1)(-2) + 1(-1)(-6) + 3(1)(2)] = 1.$$

In $\chi(\Gamma_\phi)$, the characters in the second half of the character system do not reproduce those in the first half (or reproduce their magnitudes with a change in sign). If this had been so, Γ_ϕ would have been a direct sum of g IRs (or u IRs). Here we expect the direct sum to contain both g and u representations, which turns out to be the case. The basis functions for these IRs may now be obtained by using the projection operator $\hat{P}^j$ (eq. (5.2.10)),

$$\psi^j = N_j\sum_T \chi_j(T)^*\,\hat{T}\phi; \quad (10)$$

ϕ can be any arbitrary function defined in the appropriate subspace, which here is a subspace of functions for which the six AOs $\{\phi_r\}$ form a basis. Chemical intuition tells us that a sensible choice would be $\phi = \phi_1$.

(10)
$$\begin{aligned}\psi(A_{2u}) &= N(A_{2u})[\phi_1 + (\phi_2 + \phi_6) + (\phi_3 + \phi_5) \\ &\quad + (\phi_4) - 1(-\phi_1 - \phi_3 - \phi_5) - 1(-\phi_2 - \phi_4 - \phi_6) \\ &\quad - 1(-\phi_4) - 1(-\phi_3 - \phi_5) - 1(-\phi_2 - \phi_6) - 1(-\phi_1) \\ &\quad + (\phi_2 + \phi_4 + \phi_6) + (\phi_1 + \phi_3 + \phi_5)] \\ &= N(A_{2u})[\phi_1 + \phi_2 + \phi_3 + \phi_4 + \phi_5 + \phi_6]. \end{aligned} \tag{11}$$

Normalization

In general, for $\psi^j = N_j \sum_r \phi_r \, c_{rj}$,

$$\begin{aligned}\langle\psi^j|\psi^j\rangle &= |N_j|^2 \langle \sum_r \phi_r \, c_{rj} \,|\, \sum_s \phi_s \, c_{sj}\rangle \\ &= |N_j|^2 \left[\sum_r |c_{rj}|^2 + \sum_{s \neq r}\sum c_{rj}^* \, c_{sj} \, S_{rs}\right].\end{aligned} \tag{12}$$

S_{rs} is called the *overlap integral* because the integrand is only significant in regions of space where the charge distributions described by the AOs ϕ_r and ϕ_s overlap. When either ϕ_r or ϕ_s is very small, the contribution to the integral from that volume element is small and so there are only substantial contributions from those regions of space where ϕ_r and ϕ_s overlap. A useful and speedy approximation is to invoke the *zero overlap approximation* (ZOA) which sets

$$S_{rs} = 0, \; r \neq s. \tag{13}$$

The ZOA is based more on expediency than on it being a good approximation; in fact, the value of S_{rs} is about 0.2–0.3 (rather than zero) for carbon $2p_z$ orbitals on adjacent atoms. When s is not joined to r, it is much more reasonable. Nevertheless, it is customary to use the ZOA at this level of approximation since it yields normalization constants without performing any calculations. One should remark that it affects only the N^j, the ratio of the coefficients being given by the group theoretical analysis. Using the ZOA,

$$a_{2u} = {}^1\!/\!\sqrt{6}\; [\phi_1 + \phi_2 + \phi_3 + \phi_4 + \phi_5 + \phi_6]. \tag{14}$$

In eq. (14) we have followed the usual practice of labeling the MO by the IR (here A_{2u}) for which it forms a basis, but using the corresponding lower-case letter instead of the capital letter used for the IR in Mulliken notation. It is left as a problem to find the MOs that form bases for the other IRs in the direct sum, eq. (9). In the event of l_j-fold degeneracy, there are l_j linearly independent (LI) basis functions, which we choose to make mutually orthogonal. So for $l_j = 2$, we use the projection operator $\hat{P}^j$ again, but with a different function $\phi = \phi_2$. For E_{1g}, for example, ϕ_1 and ϕ_2 give $\psi_1(E_{1g})$ and $\psi_2(E_{1g})$, which are LI but are not orthogonal. Therefore we combine them in a linear combination to ensure orthogonality while preserving normalization. Usually this can be done by inspection, although the systematic method of Schmidt orthogonalization (see, for example, Margenau and Murphy

(1943)) is available, if required. Remember that ψ can always be multiplied by an arbitrary phase factor without changing the charge density, or any other physical property, so that it is common practice to multiply ψ by -1 when this is necessary to ensure that the linear combination of atomic orbitals (LCAO) does not start with a negative sign.

6.2.1 Energy of the MOs

$$\begin{aligned} E^j &= \langle \psi^j | \hat{H} | \psi^j \rangle = |N_j|^2 \langle \sum_r \phi_r \, c_{rj} \, | \hat{H} | \, \sum_s \phi_s \, c_{sj} \rangle \\ &= |N_j|^2 \left[\sum_r |c_{rj}|^2 \; H_{rr} + \sum_{s \neq r} \sum c_{rj}{}^* \; c_{sj} \; H_{rs} \right], \end{aligned} \tag{15}$$

where

$$H_{rs} = \langle \phi_r | \hat{H} | \phi_s \rangle = H_{sr}{}^*, \tag{16}$$

the second equality following from the fact that $\hat{H}$ is an Hermitian operator. For π electron systems there are useful approximations due to Hückel. If

$$\left. \begin{array}{ll} s = r, & H_{rr} = \alpha, \\ s \leftrightarrow r, & H_{rs} = \beta \text{ (a negative quantity)}, \\ s \not\leftrightarrow r, & H_{rs} = 0 \end{array} \right\} \tag{17}$$

($s \leftrightarrow r$ means "s joined to r"). The effective energy of a bound electron in a carbon $2p_z$ atomic orbital is given by α; the delocalization energy comes from β.

(17), (15)
$$E^j = |N_j|^2 \left[\sum_r |c_{rj}|^2 \; \alpha + \sum_{s \neq r} \sum c_{rj}{}^* \; c_{sj} \; \beta \right]. \tag{18}$$

Substituting for the coefficients (see eq. (14) and Problem 6.2) and evaluating E^j from eq. (18) gives the energy-level diagram shown in Figure 6.4. Only the energies depend on

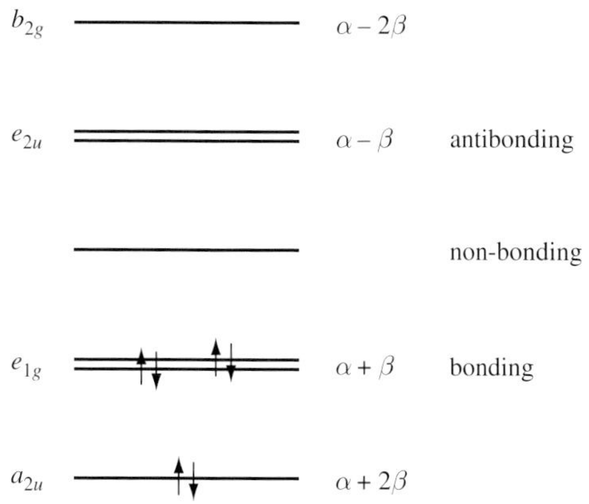

Figure 6.4. Energy-level diagram for the molecular orbitals of benzene evaluated in the Hückel approximation.

the Hückel approximations. The orbitals are correctly given within the ZOA, which only affects N_j, the ratios of the coefficients being completely determined by the symmetry of the molecule.

6.3 Equivalent bond orbitals

Example 6.3-1 In this example, we find the MOs for the nitrate ion. The numbering system employed to label the oxygen atoms is shown in Figure 6.5. Let $\{\sigma_1, \sigma_2, \sigma_3\}$ denote a set of oxygen $2p$ atomic orbitals each pointing towards the central nitrogen atom. The point group of NO_3^- is D_{3h} and the character system of the representation Γ_σ is given in Table 6.3. This is obtained in the now familiar way by studying the transformation of the basis $\langle \sigma_1 \ \sigma_2 \ \sigma_3 |$ under the symmetry operators $\hat{T}$, where $T \in D_{3h}$, and determining the characters from those orbitals σ_r which transform into $\pm \sigma_r$. The reduction of Γ_σ in the usual way (eq. (6.1.1)) gives

$$\Gamma_\sigma = A_1{}' \oplus E'. \tag{1}$$

The character table for D_{3h} tells us that the nitrogen atom orbitals involved in bonding are s, p_x, p_y. We now use the projection operator technique to find the linear combinations of oxygen ligand orbitals ψ^j that combine with s, p_x, p_y,

Table 6.3. *Character system for* Γ_σ *for the* NO_3^- *ion.*

D_{3h}	E	$2C_3$	$3C_2{}'$	σ_h	$2S_3$	$3\sigma_v$
Γ_σ	3	0	1	3	0	1

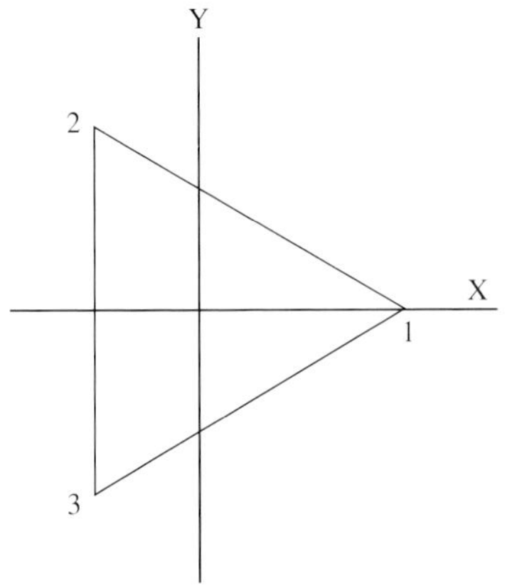

Figure 6.5. Numbering system used for the three oxygen atoms in the NO_3^- ion.

(5.2.10) $$\psi^j = N_j \sum_T \chi_j(T)^* \, \hat{T}\sigma_1, \tag{2}$$

(2) $$\psi(A_1{}') = N(A_1{}')[(\sigma_1) + (\sigma_2 + \sigma_3) + (\sigma_1 + \sigma_3 + \sigma_2) + (\sigma_1) + (\sigma_2 + \sigma_3) + (\sigma_1 + \sigma_3 + \sigma_2)], \tag{3}$$

(3) $$a_1{}' = {}^1\!/\!\sqrt{3}[\sigma_1 + \sigma_2 + \sigma_3], \tag{4}$$

on normalizing in the ZOA. An MO ψ_1 for NO_3^- is obtained by forming a linear combination of $a_1{}'$ with a central-atom AO of the same symmetry, namely s:

$$\psi_1 = s + b_1 \, a_1{}'; \tag{5}$$

ψ_1 forms a basis for $A_1{}'$. Group theory tells us which central-atom orbital (s) and which linear combination of ligand orbitals ($a_1{}'$) are involved, but to determine the mixing coefficient b_1 would require a quantum chemical calculation. Molecular orbitals like ψ_1 occur in pairs, one of which is bonding and the other antibonding, according to the sign of b_1: positive for a bonding orbital and negative for an antibonding orbital.

(2) $$\begin{aligned}\psi_1(E') &= N_1(E')[2(\sigma_1) - 1(\sigma_2 + \sigma_3) + 2(\sigma_1) - 1(\sigma_2 + \sigma_3)] \\ &= {}^1\!/\!\sqrt{6}[2\sigma_1 - (\sigma_2 + \sigma_3)]\end{aligned} \tag{6}$$

on normalizing using the ZOA. Similarly, from eq. (2), but using σ_2 and σ_3 in place of σ_1,

$$\psi_2(E') = {}^1\!/\!\sqrt{6}\ [2\sigma_2 - (\sigma_3 + \sigma_1)], \tag{7}$$

$$\psi_3(E') = {}^1\!/\!\sqrt{6}\ [2\sigma_3 - (\sigma_1 + \sigma_2)]. \tag{8}$$

But how can we have generated three basis functions for a doubly degenerate representation? The answer is that eqs. (6), (7), and (8) are not LI. So we look for two linear combinations that are LI and will overlap with the nitrogen atom orbitals p_x and p_y.

$$e_1{}' = {}^1\!/\!\sqrt{6}\,[2\sigma_1 - (\sigma_2 + \sigma_3)] \tag{9}$$

has a concentration of charge along the OX axis and overlaps satisfactorily with p_x, so

$$\psi_2 = p_x + b_2 \, e_1{}'. \tag{10}$$

Eliminating σ_1 from eqs. (7) and (8) gives

$$e_2{}' = {}^1\!/\!\sqrt{2}\ [(\sigma_2 - \sigma_3)], \tag{11}$$

$$\psi_3 = p_y + b_3 \, e_2{}'. \tag{12}$$

Subscripts 1 and 2 in $e_1{}'$, $e_2{}'$ denote the two partners that form a basis for the 2-D IR E'. Also, ψ_1, ψ_2, and ψ_3 are properly symmetrized MOs, but ψ_3, in particular, does not look much like a classical chemical bond (see Figure 6.6(a)). In order to achieve maximum overlap with the three ligand p orbitals (and hence the most stable

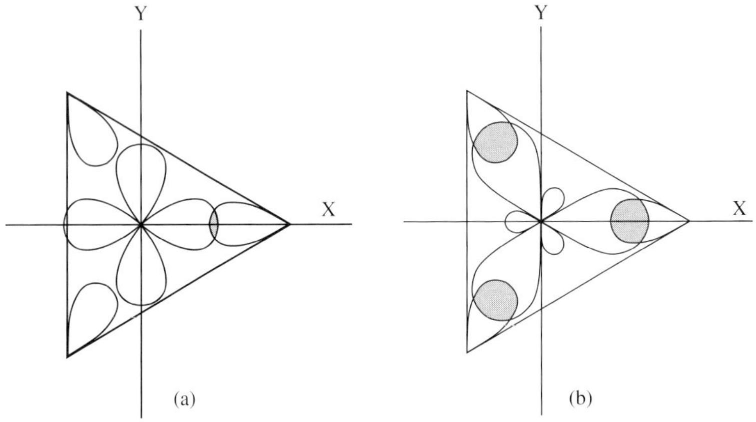

Figure 6.6. (a) Molecular orbitals for the nitrate ion that form bases for the representation E′ of the D_{3h} point group (see eqs. (6.3.9) and (6.3.11)). (b) The sp^2 hybrid orbitals in NO_3^-.

molecule), the central-atom orbitals undergo hybridization. To find the three equivalent MOs – which we call bond orbitals – we express the set of MOs given by eqs. (9), (10), and (12) in matrix form:

$$\langle \psi_1\ \psi_2\ \psi_3| = \langle s\ p_x\ p_y| + \langle a_1{}'\ e_1{}'\ e_2{}'|\mathrm{B}, \tag{13}$$

where B is the diagonal matrix

$$\mathrm{B} = \begin{bmatrix} b_1 & 0 & 0 \\ 0 & b_2 & 0 \\ 0 & 0 & b_3 \end{bmatrix}. \tag{14}$$

The ligand LCAOs are

$$\langle a_1{}'\ e_1{}'\ e_2{}'| = \langle \sigma_1\ \sigma_2\ \sigma_3|\mathtt{M}; \tag{15}$$

(4), (9), (11)
$$\mathtt{M} = \begin{bmatrix} 1/\sqrt{3} & 2/\sqrt{6} & 0 \\ 1/\sqrt{3} & -1/\sqrt{6} & 1/\sqrt{2} \\ 1/\sqrt{3} & -1/\sqrt{6} & -1/\sqrt{2} \end{bmatrix}. \tag{16}$$

(13), (15)
$$\langle \psi_1\ \psi_2\ \psi_3| = \langle s\ p_x\ p_y| + \langle \sigma_1\ \sigma_2\ \sigma_3|\mathrm{B}\ \mathtt{M}, \tag{17}$$

where we take advantage of the fact that the diagonal matrix B commutes with M.

(17)
$$\langle \psi_1{}'\ \psi_2{}'\ \psi_3{}'| = \langle \psi_1\, \psi_2\, \psi_3|\,\mathtt{M}^{-1} = \langle s\ p_x\ p_y|\,\mathtt{M}^{-1} + \langle \sigma_1\, \sigma_2\, \sigma_3|\,\mathrm{B}$$
$$= \langle h_1\, h_2\, h_3| + \langle \sigma_1\, \sigma_2\, \sigma_3|\mathrm{B}. \tag{18}$$

The sp^2 hybrid nitrogen AOs are

$$\langle h_1\ h_2\ h_3| = \langle s\ p_x\ p_y|\mathrm{M}^{\mathrm{T}} \quad (18) \qquad = \langle s\ p_x\ p_y| \begin{bmatrix} 1/\sqrt{3} & 1/\sqrt{3} & 1/\sqrt{3} \\ 2/\sqrt{6} & -1/\sqrt{6} & -1/\sqrt{6} \\ 0 & 1/\sqrt{3} & -1/\sqrt{2} \end{bmatrix}. \quad (19)$$

Exercise 6.3-1 Write down, from eqs. (18) and (19), three separate expressions for the bond orbitals ψ_1', ψ_2', and ψ_3'.

The equivalent molecular "bond orbitals" are shown diagrammatically in Figure 6.6(b). This method of finding the central-atom hybrid AOs that overlap with ligand AOs is quite general and may be applied to other situations (for example, tetrahedral AB_4) where the ligand geometry does not correspond to that of the *p* and *d* orbitals on the central atom. For square planar AB_4 and octahedral AB_6, the linear transformation to equivalent orbitals is not necessary since the disposition of the ligands corresponds to the orientation of the *p* and *d* orbitals on the central atom.

Answer to Exercise 6.3-1

$$\psi_1' = h_1 + b_1\ \sigma_1 = (1/\sqrt{3})s + (2/\sqrt{6})p_x + b_1\ \sigma_1;$$
$$\psi_2' = h_2 + b_2\ \sigma_2 = (1/\sqrt{3})s - (1/\sqrt{6})p_x + (1/\sqrt{2})p_y + b_2\ \sigma_2;$$
$$\psi_3' = h_3 + b_3\ \sigma_3 = (1/\sqrt{3})s - (1/\sqrt{6})p_x - (1/\sqrt{2})p_y + b_3\ \sigma_3.$$

6.4 Transition metal complexes

Example 6.4-1 In this example we consider the ML_6 octahedral complex. Atomic orbitals that could contribute to the MOs are the nine *nd*, $(n+1)s$, and $(n+1)p$ on M, and the eighteen *p* orbitals on the six ligands. The latter may be classified into six that point towards the central atom M, which we call σ, and twelve that are oriented at right angles to the σ *p* orbitals, which we call π and π'. We set these up so that unit vectors along σ, π, and π' (also called σ, π, and π') form a right-handed system. Figure 6.7 shows the numbering system for the ligands and the orientation of the σ, π, and π' vectors. Now M orbitals can only transform into M orbitals and similarly, so that the (27×27)-dimensional AO representation is reduced to a direct sum of representations of one, three, five, six, and twelve dimensions. The character table of $O_h = O \otimes C_i$ is given in Appendix A3. The first five classes are those of O; the second set of five classes are those of $I\{O\}$. The characters of the *u* representations are the same as those of the *g* representations for the classes of O, and the same in magnitude but of opposite sign, for the classes of $I\{O\}$. Table 6.4 shows the characters of the representations based on *p* and *d* orbitals on M. We know that an *s* orbital,

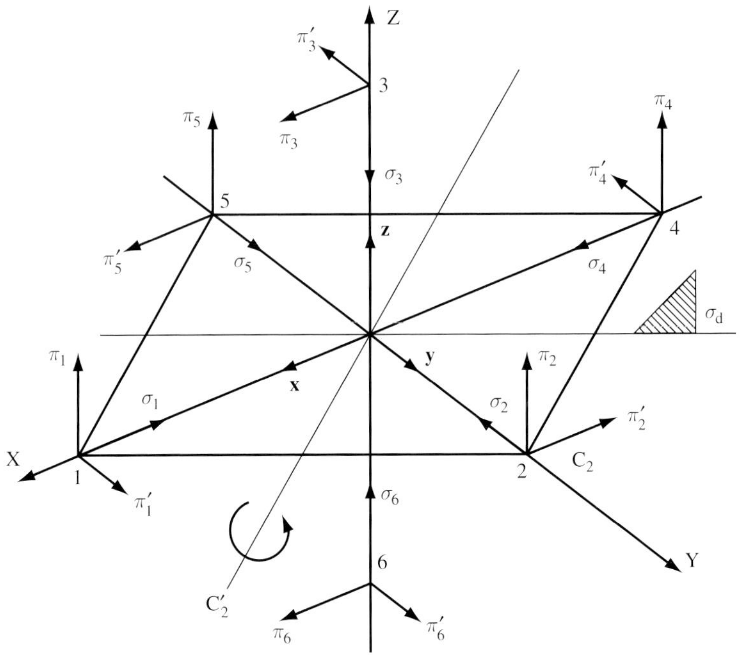

Figure 6.7. The ML_6 octahedral complex. Unit vectors σ, π, π' are oriented parallel to the orthonormal axes **x**, **y**, **z**, which have their origin at M and lie along OX, OY, OZ. The three C_2 axes that are collinear with the C_4 axes are along **x**, **y**, **z**, and the second set of six C_2 axes that bisect the angles between **x** and **y**, **y** and **z**, and **z** and **x** are designated C_2'. The symmetry planes that contain these C_2' axes are σ_d planes because they bisect the angles between C_2 axes that are normal to one of the three C_4 principal axes.

being spherically symmetrical, forms a basis for the totally symmetric representation A_{1g}. The systematic way to find the characters for Γ_p and Γ_d is as follows.

(i) First study the transformation of the basis $\langle \mathbf{e}_1\, \mathbf{e}_2\, \mathbf{e}_3|$ under one symmetry operator from each of the five classes of O. (The one actually used in this step is shown in the first row of Table 6.4.)

(ii) Write down the MRs of $\Gamma(T)$ for these symmetry operators. This is easily done by inspection.

(iii) Write down the matrices $\Gamma(T)^{-1}$, taking advantage of the fact that the MRs are orthogonal matrices so that $\Gamma(T)^{-1}$ is just the transpose of $\Gamma(T)$.

(iv) Write down the Jones symbols for the operators T^{-1}, which again can be done by inspection by just multiplying $\Gamma(T)^{-1}$ into the column matrix $|x\, y\, z\rangle$. So far, we have neglected the other five classes of O_h because the variables x, y, and z all change sign under inversion so that the Jones symbols for the operators $I(T)$ may be obtained from those of T simply by changing the sign of x, y, and z, as is done in Table 6.4.

(v) Since the three p functions are just x, or y, or z, multiplied by $f(r)$, the characters of Γ_p can be written down from the Jones symbols for T^{-1}.

(vi) The angle-dependent factors in the d orbitals can now be written down using the Jones symbols for T^{-1}, which tell us how the variables x, y, and z transform and thus how *functions* of x, y and z transform under the symmetry operators of O_h. Since the largest degeneracy in O_h is three, we expect the five-fold degeneracy of the d orbitals in spherical symmetry to be split in a cubic field, but we have no knowledge *a priori* whether the 5×5 MRs based on the five d functions will be block-diagonal. (All we know in advance is that they will form a representation equivalent to a block-diagonal representation.) In fact, as soon as we have written down the functions into which xy, yz, and zx transform, we notice that these three d orbitals only transform between themselves and never into the two remaining d orbitals. This means that the d orbital MRs are block-diagonal and the five-fold degeneracy of the d orbitals is split into at least three-fold and doubly degenerate subsets. Calling the first set $d\varepsilon$, we now write down the characters of the MRs based on xy, yz, and zx. It should be emphasized that the d orbitals transform in this way because of the cubic (= octahedral) symmetry, and that they will behave differently in different symmetries. In D_{4h} symmetry, for example, the maximum degeneracy is two, so the five d orbitals will transform in a different fashion. In general, one simply studies the transformation of the five d orbitals, and, if subsets emerge, then one can take advantage of this to reduce the arithmetic involved.

(vii) The effect of the function operators on the remaining two d orbitals is given in the next two lines of Table 6.4. When a function is a member of a basis set, in general it will transform into a linear combination of the set. In practice, this linear combination often consists of only one term (and then the entries in the corresponding column of the MR are all zero, with the exception of one that is unity). Under some operators, the basis functions transform into linear combinations, and an example of this is the class of $8C_3$, where, under the chosen operator $R(2\pi/3\ [111])$, $x^2 - y^2$ transforms into $y^2 - z^2$ and $3z^2 - r^2$ transforms into $3x^2 - r^2$. These are not d orbitals but they are linear combinations of d orbitals, for

$$y^2 - z^2 = -\tfrac{1}{2}(x^2 - y^2) - \tfrac{1}{2}(3z^2 - r^2), \tag{1}$$

$$3x^2 - r^2 = \tfrac{3}{2}(x^2 - y^2) - \tfrac{1}{2}(3z^2 - r^2). \tag{2}$$

When the function operator $\hat{R}(2\pi/3\ [111])$ acts on the basis $\langle x^2 - y^2 \quad 3z^2 - r^2|$,

(1), (2)
$$\hat{R}(2\pi/3\ [111])\langle x^2 - y^2 \quad 3z^2 - r^2| = \langle x^2 - y^2 \quad 3z^2 - r^2| \begin{bmatrix} -\tfrac{1}{2} & \tfrac{3}{2} \\ -\tfrac{1}{2} & -\tfrac{1}{2} \end{bmatrix}. \tag{3}$$

The characters of the MRs for the basis $d\gamma$ can now be written down using the transformation of the second subset of d orbitals given in Table 6.4 and eq. (3). Note that the characters for Γ_p simply change sign in the second half of the table (for the classes $I\{T\}$); this tells us that it is either a u IR, or a direct sum of u IRs. The characters for both $d\varepsilon$ and $d\gamma$ simply repeat in the second half of the table, so they are either g IRs, or direct sums of g IRs. This is because the p functions have odd parity and the d functions have even parity.

Table 6.4. *Jones symbols and character systems for AOs and MOs in the octahedral* ML_6 *complex ion.*

$\{T\} =$	E	$R(\pi\ \mathbf{z})$	$R(\phi\ \mathbf{n})$	$R(\beta\ \mathbf{z})$	$R(\pi\ \mathbf{a})$	$\{I(T)\}$				
	E	$3C_2$	$8C_3$	$6C_4$	$6C_2'$	I	$3\sigma_h$	$8S_6$	$6S_4$	$6\sigma_d$
$\chi(\Gamma_s)$	1	1	1	1	1	1	1	1	1	1
$T\{\mathbf{e}_1\mathbf{e}_2\mathbf{e}_3\}$	$\mathbf{e}_1\mathbf{e}_2\mathbf{e}_3$	$\bar{\mathbf{e}}_1\bar{\mathbf{e}}_2\mathbf{e}_3$	$\mathbf{e}_2\mathbf{e}_3\mathbf{e}_1$	$\mathbf{e}_2\bar{\mathbf{e}}_1\mathbf{e}_3$	$\mathbf{e}_2\mathbf{e}_1\bar{\mathbf{e}}_3$					
$\Gamma(T)$	$\begin{bmatrix}1&0&0\\0&1&0\\0&0&1\end{bmatrix}$	$\begin{bmatrix}\bar{1}&0&0\\0&\bar{1}&0\\0&0&1\end{bmatrix}$	$\begin{bmatrix}0&0&1\\1&0&0\\0&1&0\end{bmatrix}$	$\begin{bmatrix}0&\bar{1}&0\\1&0&0\\0&0&1\end{bmatrix}$	$\begin{bmatrix}0&1&0\\1&0&0\\0&0&\bar{1}\end{bmatrix}$					
$\Gamma(T^{-1})$	$\begin{bmatrix}1&0&0\\0&1&0\\0&0&1\end{bmatrix}$	$\begin{bmatrix}\bar{1}&0&0\\0&\bar{1}&0\\0&0&1\end{bmatrix}$	$\begin{bmatrix}0&1&0\\0&0&1\\1&0&0\end{bmatrix}$	$\begin{bmatrix}0&1&0\\\bar{1}&0&0\\0&0&1\end{bmatrix}$	$\begin{bmatrix}0&1&0\\1&0&0\\0&0&\bar{1}\end{bmatrix}$					
$T^{-1}\{xyz\}$	xyz	$\bar{x}\bar{y}z$	yzx	$y\bar{x}z$	$yx\bar{z}$	$\bar{x}\bar{y}\bar{z}$	$xy\bar{z}$	$\bar{y}\bar{z}\bar{x}$	$\bar{y}x\bar{z}$	$\bar{y}\bar{x}z$
$\chi(\Gamma_p)$	3	−1	0	1	−1	−3	1	0	−1	1
xy	xy	xy	yz	$-xy$	xy	xy	xy	yz	$-xy$	xy
yz	yz	$-yz$	zx	$-zx$	$-zx$	yz	$-yz$	zx	$-zx$	$-zx$
zx	zx	$-zx$	xy	yz	$-yz$	zx	$-zx$	xy	yz	$-yz$
$\chi(\Gamma_{d\varepsilon})$	3	−1	0	−1	1	3	−1	0	−1	1

$x^2 - y^2$	$x^2 - y^2$	$x^2 - y^2$	$y^2 - z^2$	$-(x^2 - y^2)$	$-(x^2 - y^2)$	$x^2 - y^2$	$x^2 - y^2$	$y^2 - z^2$	$-(x^2 - y^2)$	$-(x^2 - y^2)$
$3z^2 - r^2$	$3z^2 - r^2$	$3z^2 - r^2$	$3x^2 - r^2$	$3z^2 - r^2$	$3z^2 - r^2$	$3z^2 - r^2$	$3z^2 - r^2$	$3x^2 - r^2$	$3z^2 - r^2$	$3z^2 - r^2$
$\chi(\Gamma_{d\gamma})$	2	2	−1	0	0	2	2	−1	0	0
$\chi\Gamma_{\sigma})$	6	2	0	2	0	0	4	0	0	2
$\chi(\Gamma_{\pi})$	12	−4	0	0	0	0	0	0	0	0

$\phi = 2\pi/3$, $\mathbf{n} = 3^{-1/2}[1\ 1\ 1]$, $\beta = \pi/2$, $\mathbf{a} = 2^{-1/2}[1\ 1\ 0]$.

(viii) The last step in the construction of Table 6.4 is to write down the characters of the representations based on the ligand p orbitals labeled σ, which point towards M, and the ligand p orbitals labeled π or π', which are normal to unit vectors along the lines joining the ligands to M (Figure 6.7). This can be done by the "quick" method of noting how the contours of the basis functions transform under the symmetry operators: those which are invariant, or simply change sign, contribute ± 1, respectively, to the character, and the others contribute zero.

Reduction of the representations

From the characters in Table 6.4 we observe that

$$\Gamma_s = \mathrm{A}_{1g}; \tag{4}$$

$$\Gamma_p = \mathrm{T}_{1u}; \tag{5}$$

$$\Gamma_d = \Gamma_{d\varepsilon} \oplus \Gamma_{d\gamma} = \mathrm{T}_{2g} \oplus \mathrm{E}_g. \tag{6}$$

The classes for the non-zero characters of Γ_σ, its character system, and reduction, are

$$\begin{array}{rccccc} & E & 3C_2 & C_4 & 3\sigma_\mathrm{h} & 6\sigma_\mathrm{d} \\ \Gamma_\sigma = \{ & 6 & 2 & 2 & 4 & 2\}; \end{array}$$

$$\begin{aligned} c(\mathrm{A}_{1g}) &= {}^1\!/_{48}[1(1)(6) + 3(1)(2) + 6(1)(2) + 3(1)(4) + 6(1)(2)] = 1, \\ c(\mathrm{A}_{2g}) &= {}^1\!/_{48}[1(1)(6) + 3(1)(2) + 6(-1)(2) + 3(1)(4) + 6(-1)(2)] = 0, \\ c(\mathrm{E}_g) &= {}^1\!/_{48}[1(2)(6) + 3(2)(2) + 6(0)(2) + 3(2)(4) + 6(0)(2)] = 1. \end{aligned}$$

Now

$$\Gamma_\sigma = \{6\ 2\ 0\ 2\ 0\ 0\ 4\ 0\ 0\ 2\}$$

and

$$\mathrm{A}_{1g} \oplus \mathrm{E}_g = \{3\ 3\ 0\ 1\ 1\ 3\ 3\ 0\ 1\ 1\};$$

$$\Gamma_\sigma - (\mathrm{A}_{1g} \oplus \mathrm{E}_g) = \{3\ \ -1\ 0\ 1\ \ -1\ \ -3\ 1\ 0\ \ -1\ 1\} = \mathrm{T}_{1u}.$$

Therefore

$$\Gamma_\sigma = \mathrm{A}_{1g} \oplus \mathrm{E}_g \oplus \mathrm{T}_{1u}. \tag{7}$$

The non-zero characters for Γ_π are

$$\begin{array}{rcc} & E & 3C_2 \\ \Gamma_\pi = \{ & 12 & -4\}. \end{array} \tag{8}$$

The characters for E and $3C_2$ have opposite signs, and so to reach a sum of 48 in the reduction test will be unlikely except for IRs with a negative character for the class of $3C_2$. Therefore we try first those IRs for which $\chi(3C_2)$ is negative. T_{1g}, T_{2g}, T_{1u}, and T_{2u} all have $\chi(3C_2) = -1$, and

$$c(T_{1g}) = {}^1\!/_{48}[1(3)(12) + 3(-1)(-4)] = 1.$$

Since T_{2g}, T_{1u}, and T_{2u} have the same characters as T_{1g} for these classes, they must also occur once in the direct sum, which therefore is

$$\Gamma_\pi = T_g \oplus T_{2g} \oplus T_{1u} \oplus T_{2u}. \tag{9}$$

σ bonding

We need to find the linear combinations of ligand σ orbitals of symmetry A_{1g}, E_g, and T_{1u}. Omitting normalization factors, these are

$$\begin{aligned}\psi(A_{1g}) = 1(\sigma_1) &+ 1(\sigma_4 + \sigma_4 + \sigma_1)\\ &+ 1(\sigma_2 + \sigma_3 + \sigma_5 + \sigma_6 + \sigma_3 + \sigma_5 + \sigma_2 + \sigma_6)\\ &+ 1(\sigma_2 + \sigma_5 + \sigma_6 + \sigma_3 + \sigma_1 + \sigma_1)\\ &+ 1(\sigma_2 + \sigma_5 + \sigma_4 + \sigma_4 + \sigma_3 + \sigma_6)\\ &+ 1(\sigma_4) + 1(\sigma_1 + \sigma_1 + \sigma_4)\\ &+ 1(\sigma_5 + \sigma_6 + \sigma_2 + \sigma_3 + \sigma_6 + \sigma_2 + \sigma_3 + \sigma_5)\\ &+ 1(\sigma_2 + \sigma_5 + \sigma_6 + \sigma_3 + \sigma_4 + \sigma_4)\\ &+ 1(\sigma_5 + \sigma_2 + \sigma_1 + \sigma_1 + \sigma_6 + \sigma_3),\end{aligned}$$

$$\psi(A_{1g}) = \sigma_1 + \sigma_2 + \sigma_3 + \sigma_4 + \sigma_5 + \sigma_6; \tag{10}$$

$$\begin{aligned}\psi_1(E_g) = {} & 2\sigma_1 + 2(\sigma_1 + 2\sigma_4) - 2(\sigma_2 + \sigma_3 + \sigma_5 + \sigma_6) + 2\sigma_4\\ & + 2(2\sigma_1 + \sigma_4)\\ & - 2(\sigma_2 + \sigma_3 + \sigma_5 + \sigma_6),\end{aligned}$$

$$\psi_1(E_g) = 2\sigma_1 - \sigma_2 - \sigma_3 + 2\sigma_4 - \sigma_5 - \sigma_6. \tag{11}$$

Starting with σ_2, and then σ_3, as our arbitrary functions in the subspace with basis vectors (functions) $\{\sigma_1\ \sigma_2\ \sigma_3\ \sigma_4\ \sigma_5\ \sigma_6\}$ and projecting as before will simply give the cyclic permutations

$$\psi_2(E_g) = 2\sigma_2 - \sigma_3 - \sigma_4 + 2\sigma_5 - \sigma_6 - \sigma_1 \tag{12}$$

and

$$\psi_3(E_g) = 2\sigma_3 - \sigma_4 - \sigma_5 + 2\sigma_6 - \sigma_1 - \sigma_2\,. \tag{13}$$

There cannot be *three* LI basis functions of E_g symmetry, so we must choose, from eqs. (11), (12), and (13), two LI combinations that are orthogonal in the ZOA and which will overlap satisfactorily with the M atom orbitals of E_g symmetry. A suitable choice that meets these three requirements is

$$\psi_1(E_g) - \psi_2(E_g) = \sigma_1 - \sigma_2 + \sigma_4 - \sigma_5, \tag{14}$$

$$\psi_3(E_g) = 2\sigma_3 - \sigma_4 - \sigma_5 + 2\sigma_6 - \sigma_1 - \sigma_2. \tag{15}$$

Exercise 6.4-1 Verify that eqs. (11), (12), and (13) are linearly dependent and that eqs. (14) and (15) are orthogonal in the ZOA.

Continuing with the T_{1u} representation,

$$\begin{aligned}\psi_1(T_{1u}) = {} & 3\sigma_1 - 1(2\sigma_4 + \sigma_1) + 1(\sigma_2 + \sigma_5 + \sigma_6 + \sigma_3 + 2\sigma_1) \\ & - 1(\sigma_2 + \sigma_5 + \sigma_3 + \sigma_6 + 2\sigma_4) \\ & - 3\sigma_4 + 1(2\sigma_1 + \sigma_4) - 1(\sigma_2 + \sigma_5 + \sigma_6 + \sigma_3 + 2\sigma_4) \\ & + 1(\sigma_5 + \sigma_2 + 2\sigma_1 + \sigma_6 + \sigma_3),\end{aligned}$$

$$\psi_1(T_{1u}) = \sigma_1 - \sigma_4. \tag{16}$$

Since $\hat{P}(T_{1u})\sigma_1 = \sigma_1 - \sigma_4$, the other two LI linear combinations of ligand orbitals that form bases for T_{1u} are, by cyclic permutation,

$$\psi_2(T_{1u}) = \sigma_2 - \sigma_5 \tag{17}$$

and

$$\psi_3(T_{1u}) = \sigma_3 - \sigma_6. \tag{18}$$

We now have the σ bonded MOs

(10) $$a_{1g} = a_1[(n+1)s] + b_1[\sigma_1 + \sigma_2 + \sigma_3 + \sigma_4 + \sigma_5 + \sigma_6], \tag{19}$$

(14) $$e_g = a_2[nd_{x^2-y^2}] + b_2[\sigma_1 - \sigma_2 + \sigma_4 - \sigma_5], \tag{20}$$

(15) $$e_g{}' = a_3[nd_{3z^2-r^2}] + b_3[2\sigma_3 + 2\sigma_6 - \sigma_1 - \sigma_2 - \sigma_4 - \sigma_5], \tag{21}$$

(16) $$t_{1u} = a_4[(n+1)p_x] + b_4[\sigma_1 - \sigma_4], \tag{22}$$

(17) $$t_{1u}{}' = a_5[(n+1)p_y] + b_5[\sigma_2 - \sigma_5], \tag{23}$$

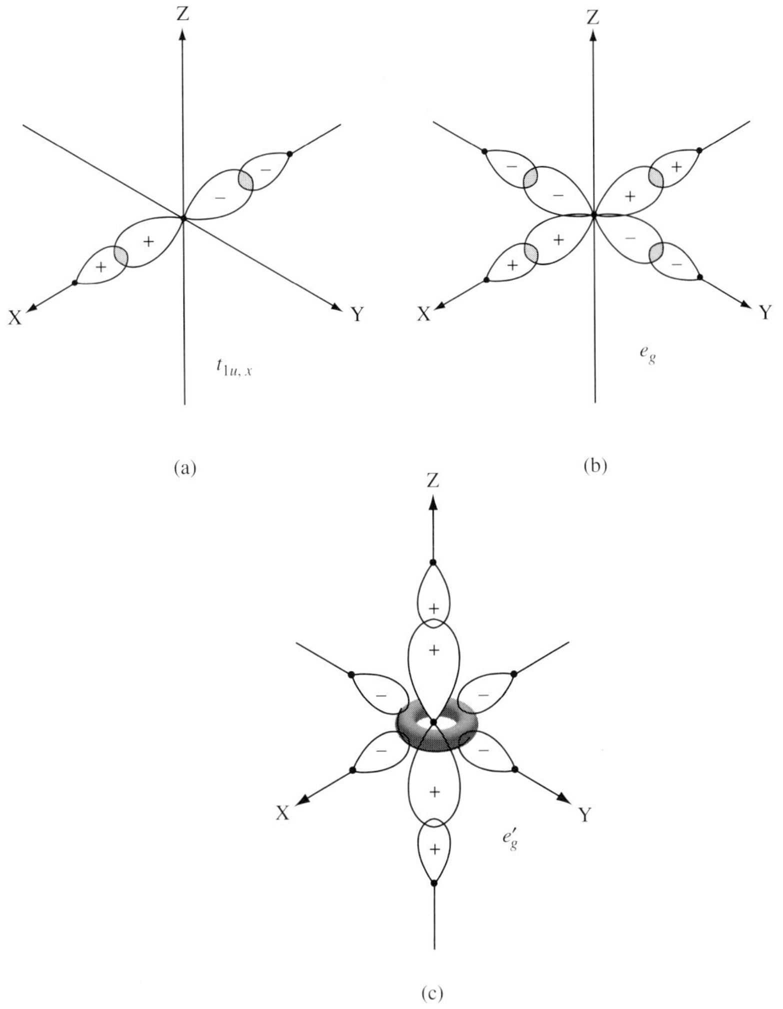

Figure 6.8. Approximate charge density prior to bonding in overlapping atomic orbitals that form σ-type molecular orbitals in ML_6: (a) $t_{1u,x}$; (b) e_g; (c) e'_g. The actual charge density in the molecule would require a quantum-chemical calculation. Only the relevant halves of the ligand p orbitals are shown in some figures. Atom centers may be marked by small filled circles for greater clarity. As usual, positive or negative signs show the sign of ψ, like signs leading to an accumulation of charge density and therefore chemical bonding. The ring depicting the region in which the $d_{3z^2-r^2}$ orbital has a negative sign has been shaded for greater clarity, but this has no other chemical significance apart from the sign.

(18)
$$t_{1u}'' = a_6[(n+1)p_z] + b_6[\sigma_3 - \sigma_6]. \tag{24}$$

The orbitals occur in bonding and antibonding pairs, according to whether a_i, b_i have the same sign or opposite sign. Rough sketches of contours of $|\psi|^2$ in the bonding AOs are shown in Figure 6.8.

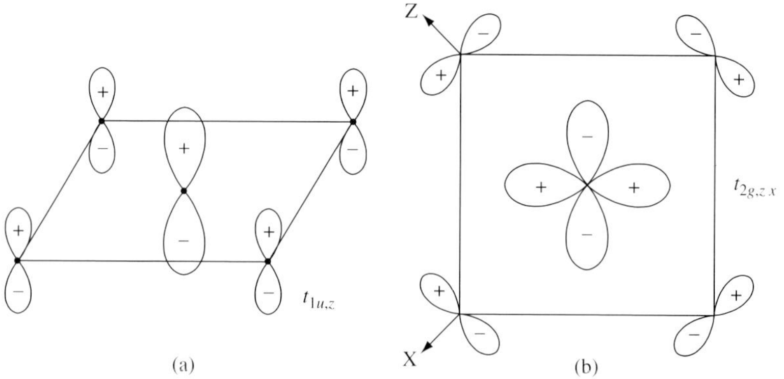

Figure 6.9. Atomic orbitals that form the π molecular orbitals: (a) $t_{1u,z}$; (b) $t_{2g,zx}$.

π bonding

We now seek linear combinations of ligand orbitals π and π' that form bases for the IRs T_{1u}, T_{2g}, T_{2u}, and T_{1g} (eq. (9)). The characters for T_{1u} in O are $\{3 \ -1 \ 0 \ 1 \ -1\}$, so with π_1 as our arbitrary function in the π, π' subspace,

$$\begin{aligned}\psi_1(T_{1u}) = {} & 3(\pi_1) - 1(\pi_4 - \pi_1 - \pi_4) + 1(\pi_2 + \pi_5 + \pi_1{}' - \pi_1{}' + \pi_6 - \pi_3) \\ & - 1(-\pi_2 - \pi_5 + \pi_4{}' - \pi_4{}' + \pi_3 - \pi_6) - 3(-\pi_4) + 1(-\pi_1 + \pi_4 + \pi_1) \\ & - 1(-\pi_5 - \pi_2 - \pi_4{}' + \pi_4{}' - \pi_3 + \pi_6) \\ & + 1(\pi_5 + \pi_2 - \pi_1{}' + \pi_1{}' - \pi_6 + \pi_3),\end{aligned}$$

giving

$$\psi_1(T_{1u}) = \pi_1 + \pi_2 + \pi_4 + \pi_5 = \pi_z. \tag{25}$$

These ligand p orbitals are symmetrically disposed to point along the OZ axis. Since the OX and OY axes are equivalent to OZ in O_h symmetry, we may write down by inspection

$$\psi_2(T_{1u}) = -\pi_2{}' + \pi_3 + \pi_5{}' + \pi_6 = \pi_x \tag{26}$$

and

$$\psi_3(T_{1u}) = \pi_1{}' + \pi_6{}' - \pi_4{}' - \pi_3{}' = \pi_y. \tag{27}$$

The MOs that form bases for T_{1u} are therefore

$$t_{1u,x} = a_7[(n+1)p_x] + b_7\ \pi_x, \tag{28}$$

$$t_{1u,y} = a_8[(n+1)p_y] + b_8\ \pi_y, \tag{29}$$

$$t_{1u,z} = a_9[(n+1)p_z] + b_9\ \pi_z. \tag{30}$$

As an example, the MO $t_{1u,z}$ is shown in Figure 6.9(a). The character system for $\chi(T_{2g})$ is $\{3 \ -1 \ 0 \ -1 \ 1\}$, and so

$$\psi_1(\mathrm{T}_{2g}) = 3(\pi_1) - 1(-\pi_1) - 1(\pi_2 + \pi_5 + \pi_6 - \pi_3) + 1(-\pi_2 - \pi_5 + \pi_3 - \pi_6) \\ + 3(-\pi_4) - 1(\pi_4) - 1(-\pi_5 - \pi_2 - \pi_3 + \pi_6) + 1(\pi_5 + \pi_2 - \pi_6 + \pi_3),$$

giving

$$\psi_1(\mathrm{T}_{2g}) = \pi_1 - \pi_4 + \pi_3 - \pi_6 = \pi_{zx}. \tag{31}$$

By inspection, the linear combinations of ligand AOs in the yz and xy planes are

$$\psi_2(\mathrm{T}_{2g}) = \pi_2 - \pi_3{}' - \pi_5 - \pi_6{}' = \pi_{yz} \tag{32}$$

and

$$\psi_3(\mathrm{T}_{2g}) = \pi_1{}' - \pi_2{}' + \pi_4{}' - \pi_5{}' = \pi_{xy}. \tag{33}$$

The MOs of T_{2g} symmetry are therefore

$$t_{2g,xy} = a_{10}[nd_{xy}] + b_{10}\ \pi_{xy}\,, \tag{34}$$

$$t_{2g,yz} = a_{11}[nd_{yz}] + b_{11}\ \pi_{yz}\,, \tag{35}$$

$$t_{2g,zx} = a_{12}[nd_{zx}] + b_{12}\ \pi_{zx}\,. \tag{36}$$

As an example, $t_{2g,zx}$ is shown in Figure 6.9(b). The character system $\chi(\mathrm{T}_{1g})$ is $\{3\ \ -1\ \ 0\ \ 1\ \ -1\}$ and so

$$\psi_1(\mathrm{T}_{1g}) = 3(\pi_1) - 1(-\pi_1) + 1(\pi_2 + \pi_5 + \pi_6 - \pi_3) - 1(-\pi_2 - \pi_5 + \pi_3 - \pi_6) \\ + 3(-\pi_4) - 1(\pi_4) + 1(-\pi_5 - \pi_2 - \pi_3 + \pi_6) - 1(\pi_5 + \pi_2 - \pi_6 + \pi_3),$$

giving

$$\psi_1(\mathrm{T}_{1g}) = \pi_1 - \pi_3 - \pi_4 + \pi_6. \tag{37}$$

There is no metal orbital of T_{1g} symmetry, so eq. (37) represents a non-bonding MO, $t_{1g,y}$. The three degenerate MOs of T_{1g} symmetry are therefore

$$t_{1g,x} = b_{13}[\pi_2 + \pi_3{}' - \pi_5 + \pi_6{}'], \tag{38}$$

$$t_{1g,y} = b_{14}[\pi_1 - \pi_3 - \pi_4 + \pi_6], \tag{39}$$

$$t_{1g,z} = b_{15}[\pi_1{}' + \pi_2{}' + \pi_4{}' + \pi_5{}']. \tag{40}$$

Finally, the character system $\chi(\mathrm{T}_{2u})$ is $\{3\ \ -1\ \ 0\ \ -1\ \ 1\}$ so that

$$\psi_1(\mathrm{T}_{2u}) = 3(\pi_1) - 1(-\pi_1) - 1(\pi_2 + \pi_5 + \pi_6 - \pi_3) + 1(-\pi_2 - \pi_5 + \pi_3 - \pi_6) \\ - 3(-\pi_4) + 1(\pi_4) + 1(-\pi_5 - \pi_2 - \pi_3 + \pi_6) - 1(\pi_5 + \pi_2 - \pi_6 + \pi_3),$$

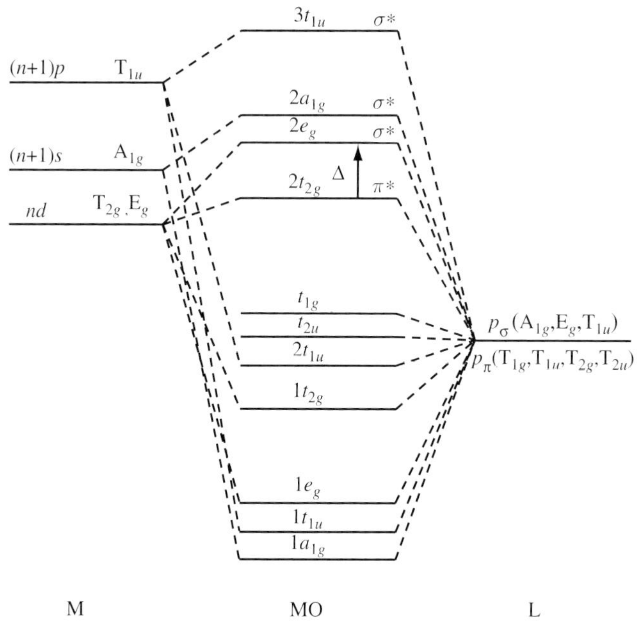

Figure 6.10. Schematic energy-level diagram for ML_6 complexes.

giving

$$\psi_1(T_{2u}) = \pi_1 - \pi_2 + \pi_4 - \pi_5. \tag{41}$$

There is no metal orbital of T_{2u} symmetry, so eq. (41) represents a non-bonding MO, $t_{2u,z}$. By inspection, the three degenerate MOs of T_{2u} symmetry are

$$t_{2u,x} = b_{16}[\pi_2{}' + \pi_3 - \pi_5{}' + \pi_6], \tag{42}$$

$$t_{2u,y} = b_{17}[\pi_1{}' - \pi_6{}' - \pi_4{}' - \pi_3{}'], \tag{43}$$

$$t_{2u,z} = b_{18}[\pi_1 - \pi_2 + \pi_4 - \pi_5]. \tag{44}$$

A schematic energy-level diagram is shown in Figure 6.10. To draw an accurate energy-level diagram for any specific molecule would require an actual quantum chemical calculation. The energy levels are labeled by the appropriate group theoretical symbols for the corresponding IRs. When the same IR occurs more than once, the convention is used that energy levels belonging to the same IR are labeled 1, 2, 3, beginning at the lowest level. In summary, in ascending order, there are

(i) the σ orbitals: $1a_{1g}$, $1t_{1u}$, $1e_g$, fully occupied by twelve electrons;
(ii) the mainly ligand π orbitals: $1t_{2g}$, $2t_{1u}$, t_{2u}, t_{1g}, which hold twenty-four electrons;
(iii) the metal d orbitals $2t_{2g}$ (or $d\varepsilon$) (with a small mixture of mainly non-bonding ligand π) and $2e_g$ (or $d\gamma$) plus ligand σ;
(iv) the anti-bonding $2a_{1g}$, $3t_{1u}$, σ* orbitals.

For example, $[MoCl_6]^{3-}$ has a total of thirty-nine valence electrons from the molybdenum $4d^5$ $5s$ and six chlorine $3p^5$ configurations, and the -3 charge on the ion. Its electron configuration is therefore σ^{12} π^{24} $(2t_{2g})^3$. Thus the MO theory of these ML_6 complex ions confirms that the electrons of prime importance are those occupying the t_{2g} and e_g levels on the metal, as predicted by crystal-field theory. However, the MO theory points the way to the more accurate calculation of electronic structure and properties.

Answer to Exercise 6.4-1

The three MOs for the E_g representation in eqs. (11), (12), and (13) are not LI since $\psi_1(E_g) + \psi_2(E_g) + \psi_3(E_g) = 0$. In the ZOA,

$$\text{(14), (15)} \qquad \langle \sigma_1 - \sigma_2 + \sigma_4 - \sigma_5 | 2\sigma_3 - \sigma_4 - \sigma_5 + 2\sigma_6 - \sigma_1 - \sigma_2 \rangle$$

$$= -1 + 1 - 1 + 1 = 0,$$

so these two basis functions are indeed orthogonal.

Problems

6.1 The point group of dodecahedral $Mo(CN)_8^{4-}$ is D_{2d}. List the symmetry operators of this point group and determine which atomic orbitals of Mo^{4+} form hybrid σ bonds in $Mo(CN)_8^{4-}$.

6.2 Show that the LCAO MOs which form bases for the MOs of benzene (in addition to a_{2u}) are

$$\psi(B_{2g}) = {}^1/_{\sqrt{6}}\, [\phi_1 - \phi_2 + \phi_3 - \phi_4 + \phi_5 - \phi_6],$$

$$\psi_1(E_{1g}) = {}^1/_{\sqrt{12}}\, [2\phi_1 + \phi_2 - \phi_3 - 2\phi_4 - \phi_5 + \phi_6],$$

$$\psi_2(E_{1g}) = {}^1/_{\sqrt{12}}\, [\phi_1 + 2\phi_2 + \phi_3 - \phi_4 - 2\phi_5 - \phi_6],$$

$$\psi_1(E_{2u}) = {}^1/_{\sqrt{12}}\, [2\phi_1 - \phi_2 - \phi_3 + 2\phi_4 - \phi_5 - \phi_6],$$

$$\psi_2(E_{2u}) = {}^1/_{\sqrt{12}}\, [\phi_1 - 2\phi_2 + \phi_3 + \phi_4 - 2\phi_5 + \phi_6].$$

Show also that the MOs

$$e_{1g} = {}^1/_{\sqrt{3}}\, [\psi_1(E_{1g}) + \psi_2(E_{1g})],$$

$$e_{1g}{}' = [\psi_1(E_{1g}) - \psi_2(E_{1g})]$$

are normalized and orthogonal. Find similarly the pair of orthonormal MOs for the representation E_{2u}. Indicate in sketches how the signs of the MOs vary around the benzene ring and mark the nodal planes. Which orbitals would you expect to be bonding and which antibonding? Confirm your conclusions by working out the energies of these MOs using the Hückel approximations. [*Hint*: Use the ZOA when determining normalization factors and when checking for orthogonality.]

6.3 Determine the MOs for the square planar molecule ML_4 of D_{4h} symmetry. [*Hint*: Set up right-handed axes σ, $\pi^{\perp}$, π'' on each ligand.]

6.4 Determine the symmetry of the σ bonded MOs in square-pyramidal ML_5. Use projection operators to find the LCAO Molecular Orbitals for ML_5, assuming d^2sp^2 hybridization to predominate.

6.5 sd^3 hybridization predominates in the tetrahedral MnO_4^- permanganate ion. Find orthonormal linear combinations t_2, t_2', t_2'' of oxygen p orbitals that are involved in σ bonding. [*Hint*: The symmetry of the molecule requires that σ_1, σ_2, σ_3, σ_4 all occur with equal weighting in the triply degenerate t_2 orbitals. This suggests that we try $\phi_1 + \phi_2$, $\phi_1 + \phi_3$, $\phi_1 + \phi_4$, where ϕ_i is the linear combination projected from σ_i.] Write down the bonding σ orbitals in matrix form,

$$[a\ t_2, t_2', t_2''] = [\sigma_1\ \sigma_2\ \sigma_3\ \sigma_4]\mathtt{M},$$

and hence determine the sd^3 hybrid orbitals. Finally, write down equations for the linear combination of σ_1, σ_2, σ_3, σ_4 with these hybrids and make sketches of the bond orbitals. [*Hint*: This requires that you decide (by inspection) which of the hybrids overlaps with which σ_r.]

6.6 (a) Assuming cyclobutadiene (C_4H_4) to be square planar, determine the symmetry of the MOs formed from a linear combination of carbon $2p_z$ AOs, one for each of the four carbon atoms. Use projection operators to determine these MOs and normalize in the ZOA. Show in sketches how the sign of each MO varies around the square and mark in the nodal planes. Hence determine the order of the stability of the MOs and show this in an energy-level diagram which shows which orbitals are occupied in the ground state. Calculate the energies of the MOs using the Hückel approximations and add these energies to your energy-level diagram, marking bonding, non-bonding and antibonding orbitals.

(b) In fact, cyclobutadiene undergoes a Jahn–Teller distortion so that its shape is rectangular rather than square. Show the energy-level splittings and re-labeling of MOs in the reduced symmetry.

(c) Removing two electrons from the highest occupied level in cyclobutadiene gives the dication $[C_4H_4]^{2+}$, which has a non-planar configuration of D_{2d} symmetry. Show the re-labeling of energy levels that occurs in the dication. Find the symmetries and spin degeneracies of the ground and first excited electronic states of $[C_4H_4]^{2+}$. Determine if an E1 transition is allowed between these two states, and, if it is, state the polarization of the allowed transition. [*Hint*: Assume that the energy gain from unpairing spins in the dication is $< 2\beta$.]

7 Crystal-field theory

7.1 Electron spin

In the early years of quantum mechanics, certain experiments, notably the anomalous Zeeman effect and the Stern–Gerlach experiment, made necessary the introduction of the idea that an electron possessed an intrinsic angular momentum in addition to the ordinary angular momentum **L**. The additional degrees of freedom were accounted for by postulating that an electron was spinning about an axis in space. According to the spin postulate of quantum mechanics, an electron possesses an intrinsic angular momentum described by the spin vector **S** and a magnetic moment

$$\boldsymbol{\mu} = -g_e \mathbf{S}(m_B/\hbar) = -g_e \mathbf{S}(e/2m_e), \tag{1}$$

where m_e is the mass of the electron and m_B is the Bohr magneton, the atomic unit of magnetic moment, with the numerical value

$$m_B = e\hbar/2m_e = 0.9274 \times 10^{-23}\ \mathrm{J\,T^{-1}}. \tag{2}$$

The only allowed value of the spin quantum number, which quantizes the square of the spin angular momentum, is $s = {}^1\!/_2$. For free electrons g_e is 2.00232. It follows from the commutation relations (CRs) obeyed by the angular momentum operators that the angular momentum quantum numbers may have integer or half-integer values (Chapter 11). The Stern–Gerlach experiment had shown that $s = {}^1\!/_2$. That g_e is equal to 2 rather than 1 comes from Dirac's theory of the electron (the precise value of g_e comes from quantum electrodynamics). The components of **S** are S_x, S_y, S_z, and the associated self-adjoint spin operators $\hat{S}_x$, $\hat{S}_y$, $\hat{S}_z$, $\hat{S}^2$ obey similar CRs to the angular momentum operators $\hat{L}_x$, $\hat{L}_y$, $\hat{L}_z$, $\hat{L}^2$. Since $\hat{S}_x$, $\hat{S}_y$, $\hat{S}_z$ all commute with $\hat{S}^2$, but not with one another, only one component of $\hat{\mathbf{S}}$, taken to be $\hat{S}_z$, can have a common set of eigenvectors with $\hat{S}^2$. The linear vector space in which the spin vectors operate (*spin space*) is separate from configuration space. Consequently, the spin operators do not act on space variables x, y, z, and therefore they commute with $\hat{L}^2$ and with the components of $\hat{\mathbf{L}}$. Because of the existence of **S**, electrons have a total angular momentum

$$\mathbf{J} = \mathbf{L} + \mathbf{S}. \tag{3}$$

$\hat{J}^2$ and the components of $\hat{\mathbf{J}}$ obey similar CRs to those of $\hat{L}^2$ and the components of $\hat{\mathbf{L}}$, so that results which follow from the CRs for $\hat{\mathbf{L}}$ therefore also hold for $\hat{\mathbf{S}}$ and for $\hat{\mathbf{J}}$. In particular, for any operator $\hat{\mathbf{j}}$ that obeys these CRs,

$$\hat{j}_+|j\ m\rangle = c_+|j\ m+1\rangle, \tag{4}$$

$$\hat{j}_-|j\ m\rangle = c_-|j\ m-1\rangle, \tag{5}$$

where $\hat{j}_+$, $\hat{j}_-$ are the raising and lowering operators

$$\hat{j}_+ = \hat{j}_x + \mathrm{i}\,\hat{j}_y, \quad \hat{j}_- = \hat{j}_x - \mathrm{i}\,\hat{j}_y, \tag{6}$$

and $|j\ m\rangle$ are the common set of eigenvectors of $\hat{j}^2$ and $\hat{j}_z$, with j and m the corresponding quantum numbers.

$$c_\pm = [j(j+1) - m(m\pm 1)]^{1/2}. \tag{7}$$

The quantum number j is an integer or half-integer. The eigenvalues of j^2 are $j(j+1)$ and the eigenvalues of j_z are $m = -j,\ -j+1,\ \ldots,\ j$ (in atomic units). These results, which are proved in Chapter 11 and in most books on quantum mechanics (for example, Atkins (1983)), follow from the CRs and therefore hold for **L**, **S**, and **J**. For **L**, l is an integer, but for **S** the only value of s is $^1\!/_2$, so that the eigenvalues of S_z are $m_s = -^1\!/_2, +^1\!/_2$ and the eigenvalue of S^2 is $s(s+1) = ^3\!/_4$. Since there are only two allowed values for the eigenvalues of S_z, there are only two spin eigenvectors $|s\ m_s\rangle$, namely $|^1\!/_2\ ^1\!/_2\rangle$ and $|^1\!/_2\ -^1\!/_2\rangle$. In function notation, the first of these is called α and the second one is called β. The eigenvectors $|s\ m_s\rangle$ have two components which describe their projections along the two basis vectors. Since one-particle spin space contains only the two vectors $|^1\!/_2\ ^1\!/_2\rangle$ and $|^1\!/_2\ \bar{1}\!/_2\rangle$, their matrix representatives (MRs) are $|1\ \ 0\rangle$ and $|0\ \ 1\rangle$, respectively, which satisfy the orthonormal conditions for the spin eigenvectors. For example, the matrix representation of the orthogonality relation

$$\left\langle \frac{1}{2}\ \frac{1}{2} \middle| \frac{1}{2}\ \frac{\bar{1}}{2} \right\rangle = 0 \tag{8}$$

is

$$\langle 1\ \ 0|0\ \ 1\rangle = 0. \tag{9}$$

Warning: Dirac notation is used in eq. (8); matrix notation is used in eq. (9).

The MRs of the spin operators are readily obtained using their known properties (as given above) and the MRs of the spin eigenvectors.

7.2 Spherical symmetry

Let $\mathbf{j}_1, \mathbf{j}_2$ denote any two angular momenta (**S** or **L** or **J**) that obey the CRs and let $\mathbf{j} = \mathbf{j}_1 + \mathbf{j}_2$. Then (in atomic units)

$$|\mathbf{j}^2| = j(j+1), \quad j_z = m, \tag{1}$$

where

$$j = j_1 + j_2,\ j_1 + j_2 - 1,\ \ldots,\ |j_1 - j_2|, \tag{2}$$

$$m = j,\ j - 1,\ \ldots,\ -j. \tag{3}$$

The quantum number j must be an integer or half-integer. These results, which are proved in Chapter 11, hold for $\mathbf{j} = \mathbf{L}$ or $\mathbf{S}$ or $\mathbf{J}$.

In a many-electron atom or ion, the Hamiltonian

$$\hat{H} = \hat{H}_0 + \hat{H}_{ee} + \hat{H}_{\mathbf{S.L}} \tag{4}$$

consists of three principal terms. $\hat{H}_0$ comprises the kinetic energy of the electrons and the electron–nucleus interactions. Approximating $\hat{H}$ by $\hat{H}_0$ alone leads to the orbital approximation

$$\Psi(1, 2, \ldots, N) = \Psi_1(1)\Psi_2(2)\ldots|\Psi_N(N), \tag{5}$$

with one-electron states characterized by the four quantum numbers n, l, m_l, and m_s. The argument (1) in eq. (5) stands for the position variables and spin of electron 1, and similarly. $\hat{H}_{ee}$ is the electron–electron interaction, and $\hat{H}_{\mathbf{S.L}}$ is the spin–orbit interaction; it is proportional to $\alpha^2 Z^2$, where α is the fine structure constant 7.29735×10^{-3} ($\simeq 1/137$), so that in atoms of low Z, $\hat{H}_{ee} \gg \hat{H}_{\mathbf{S.L}}$. The interaction $\hat{H}_{ee}$ introduces a coupling of the angular momenta of the individual electrons such that $\sum\limits_i \mathbf{L}_i = \mathbf{L}$, the total orbital angular momentum, and $\sum\limits_i \mathbf{S}_i = \mathbf{S}$, the total spin angular momentum. This is called Russell–Saunders (*RS*) coupling. The quantum numbers L and S are given by the rules in eqs. (2) and (3). These rules, together with the Pauli exclusion principle, that the state function Ψ must be antisymmetric with respect to the interchange of any two electrons, allow the determination of the quantum numbers L, S for any electron configuration. To fulfil the antisymmetry requirement, the product state function in eq. (5) must be antisymmetrized, giving

$$\Psi(1, 2, \ldots, N) = \frac{1}{\sqrt{N!}}\begin{vmatrix} \Psi_1(1) & \Psi_2(1) & \cdots & \Psi_N(1) \\ \Psi_1(2) & \Psi_2(2) & & \Psi_N(2) \\ \vdots & & & \\ \Psi_1(N) & \Psi_2(N) & & \Psi_N(N) \end{vmatrix}. \tag{6}$$

The coupled energy states in *RS* coupling are called *multiplets* and are described by *spectral terms* of the form ${}^{2S+1}\mathrm{X}$, where $2S + 1$ is the spin multiplicity and S is the total spin quantum number.

$$\mathrm{X} = \mathrm{S},\ \mathrm{P},\ \mathrm{D},\ \mathrm{F},\ \mathrm{G},\ \ldots \tag{7a}$$

when the total orbital angular momentum quantum number

$$\mathrm{L} = 0,\ 1,\ 2,\ 3,\ 4,\ \ldots. \tag{7b}$$

The spin–orbit interaction, which couples $\mathbf{L}$ and $\mathbf{S}$ to give a total angular momentum $\mathbf{J}$, splits the multiplets into their components labeled ${}^{2S+1}\mathrm{X}_J$, where J is the total angular momentum quantum number. The spin–orbit splitting is given by (Bethe (1964))

$$\Delta E_{\mathbf{S.L}} = {}^1\!/_2\zeta[J(J+1) - L(L+1) - S(S+1)], \tag{8}$$

where $\zeta(L, S)$, the spin–orbit coupling constant, is a constant for a given multiplet. For shells that are less than half filled, $\zeta > 0$, and so the state of *smallest* J lies *lowest*, but if a shell is more than half filled then $\zeta < 0$ and the state of *largest* J lies *lowest*. This is called Hund's third rule. While this rule is a good guide, there are some exceptions; for example, $\zeta < 0$ for $d^3 : {}^2F$. For a half-filled shell of maximum multiplicity, $\zeta = 0$ in first order. For *regular* multiplets (those that are less than half filled) the lowest energy state is that of highest S. For states of equal S, that with the largest L lies lowest, and in this multiplet that with the smallest J lies lowest. Straightforward procedures exist for finding all the multiplets for a particular electron configuration. For example, for d^2 the terms are ${}^3F$, ${}^1D$, ${}^1G$, ${}^3P$, and ${}^1S$. This ordering illustrates that the ${}^3F$ ground state is given correctly by the above rules and that they do not apply to excited states since ${}^1D$ lies below ${}^1G$.

7.3 Intermediate crystal field

For atoms of "low" Z the Hamiltonian $\hat{H}$, with terms in increasing order of smallness, is

$$\hat{H} = \hat{H}_0 + \hat{H}_{ee} + \hat{H}_{\mathbf{S.L}} \tag{1}$$

$\hat{H}_{CF}$ is the term to be added to the Hamiltonian which describes the electrostatic interactions of the central ion with the surrounding ions or ligands. If this term is larger than the electron–electron interactions, the electric field due to the surroundings is termed a *strong* crystal field; if it is smaller than $\hat{H}_{ee}$ but larger than the spin–orbit coupling it is called an *intermediate* crystal field; and if it is smaller than $\hat{H}_{\mathbf{S.L}}$ it is called a *weak* field. We consider first the case of an intermediate crystal field, which can be regarded as a perturbation on the Russell–Saunders multiplets defined by the values of L, S. Consider a one-electron atomic term with angular momentum l. A representation Γ_l for any group of proper rotations may be found by using the angular momentum eigenfunctions, i.e. the spherical harmonics $\{Y_l^m\}$, as a $(2l+1)$-fold degenerate basis set.

$$\begin{aligned}\hat{R}(\phi\ \mathbf{z})Y_l^m(\theta,\varphi) &= Y_l^m(R^{-1}\{\theta,\varphi\}) = Y_l^m(\theta,\varphi-\phi)\\ &= \exp(-im\phi)Y_l^m(\theta,\varphi)\end{aligned} \tag{2}$$

(see Figure 7.1) so that each member of the set is transformed into itself multiplied by the numerical coefficient $\exp(-im\phi)$. Therefore

$$\hat{R}(\phi\ \mathbf{z})\langle\{Y_l^m(\theta,\varphi)\}| = \langle\{Y_l^m(\theta,\varphi)\}|\Gamma_l(\phi) \tag{3}$$

with $\Gamma_l(\phi)$ a diagonal matrix with entries $\exp(-im\phi)$, where $m = l, l-1, \ldots, -l$:

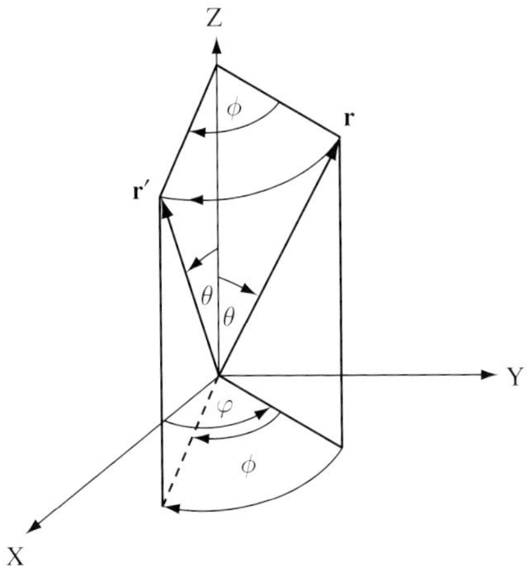

Figure 7.1. The operator $[R(\phi\ \mathbf{z})]^{-1}$ transforms the vector **r** into **r**′ and therefore the azimuthal angle φ into $\varphi - \phi$ (see eq. (7.2)).

$$\text{(2), (3)} \qquad \Gamma_l(\phi) = \begin{bmatrix} e^{-il\phi} & & & \\ & e^{-i(l-1)\phi} & & \\ & & \ddots & \\ & & & e^{il\phi} \end{bmatrix}, \qquad (4)$$

$$\text{(4)} \qquad \chi(\Gamma_l(\phi)) = e^{-il\phi} \sum_{p=0}^{2l} e^{i\phi p}. \qquad (5)$$

The sum in eq. (5) is a geometric progression, that is a series in which the ratio of any term to the preceding one has a constant value, the common ratio r. The sum to n terms is

$$S_n = a(r^n - 1)/(r - 1), \qquad (6)$$

where the first term $a = 1$, the common ratio $r = e^{i\phi}$, and the number of terms $n = 2l + 1$.

$$\text{(5), (6)} \qquad \chi(\Gamma_l(\phi)) = e^{-il\phi}[e^{i(2l+1)\phi} - 1]/[e^{i\phi} - 1]. \qquad (7)$$

On multiplying the numerator and the denominator of the RS of eq. (7) by $\exp(-i\phi/2)$, it is seen to be

$$\chi(\Gamma_l(\phi)) \equiv \chi^l(\phi) = \sin[(2l + 1)\phi/2]/\sin(\phi/2). \qquad (8)$$

It is shown in Chapter 11 that eq. (8) holds (with l replaced by j) for any operator $\hat{\mathbf{J}}$ with components $\hat{J}_x$, $\hat{J}_y$, $\hat{J}_z$ that obey the angular momentum CRs. (The quantum number j determines the eigenvalues of J^2, from eq. (11.4.40).) Consequently, eq. (8) applies also to the many-electron case with l replaced by L,

$$\chi(\Gamma_L(\phi)) = \sin\,[(2L + 1)\phi/2]/\sin(\phi/2). \qquad (9)$$

The location of the axes is arbitrary so that this result holds for any proper rotation $R(\phi\ \mathbf{n})$. (A formal proof that the MRs of all rotations through the same angle have the same

Table 7.1. *Splitting of the states of angular momentum L in an intermediate crystal field.*

See eq. (9) and Exercise 7.3-1.

$\chi(E) = 2L + 1$

$\chi(C_2) = \sin[(2L+1)\pi/2]/\sin(\pi/2) = (-1)^L$

$$\chi(C_3) = \sin[(2L+1)\pi/3]/\sin(\pi/3) = \begin{cases} 1 \text{ for } L = 0, 3, \ldots \\ 0 \text{ for } L = 1, 4, \ldots \\ -1 \text{ for } L = 2, 5, \ldots \end{cases}$$

$$\chi(C_4) = \sin[(2L+1)\pi/4]/\sin(\pi/4) = \begin{cases} 1 \text{ for } L = 0, 1, 4, 5, \ldots \\ -1 \text{ for } L = 2, 3, 6, 7, \ldots \end{cases}$$

State	Γ_L	Direct sum in cubic field
S	Γ_0	A_1
P	Γ_1	T_1
D	Γ_2	$E \oplus T_2$
F	Γ_3	$A_2 \oplus T_1 \oplus T_2$
G	Γ_4	$A_1 \oplus E \oplus T_1 \oplus T_2$

character will be given in Chapter 12.) From eq. (9) we may calculate the character system for any group of proper rotations for any L and, if this is not already irreducible, reduce this in the usual way into a direct sum of IRs.

Exercise 7.3-1 Show that for a state in which the orbital angular momentum is L,

$$\chi(E) = 2L + 1. \tag{10}$$

The characters $\chi[\Gamma_L(\phi)]$ for $\phi = \pi/2$, $2\pi/3$, and π are given in Table 7.1, which also shows the splitting of free-ion states in a cubic field when $L \geq 2$. The splitting of states in lower symmetries is given in correlation tables (see Appendix A4). Should a correlation table not be available, one can always find the direct sums using the common classes (or corresponding classes) of the two groups. An example of this procedure will be given later.

7.3.1 *Improper rotations*

The improper rotations S, I, and σ can all be expressed in the form IR, where R is a proper rotation. Let λ be the eigenvalue of the inversion operator,

$$\hat{I}\psi = \lambda\psi. \tag{11}$$

A symmetry operator leaves the physical properties of a system unchanged, and therefore $|\hat{I}\psi|^2 = |\lambda\psi|^2 = |\psi|^2$, so that $\lambda = \exp(i\gamma)$. Operating on each side of eq. (11) with I gives $\lambda^2 = \exp(2i\gamma) = 1$, so $\lambda = \pm 1$. The eigenvalue of an operator (such as I), the square of which is the unit operator, is called the *parity*. The parity of a basis function is said to be 'even' if $\lambda = +1$ and 'odd' if $\lambda = -1$. In molecular symmetry a subscript g or u is used to denote even or odd parity, but for atomic states a superscript + or − is commonly used instead. Thus, if $\lambda = +1$, $\bar{I}\psi^+ = \psi^+$, but if $\lambda = -1$, $\bar{I}\psi^- = -\psi^-$. The parity of the spherical harmonics is $(-1)^l$, and since $\hat{I}r = r$, the parity of the one-electon states is given by

	s	p	d	f
$l =$	0	1	2	3
$\lambda =$	1	−1	1	−1.

The antisymmetrized state function for N electrons is the sum of products such as $\psi_1(1)\psi_2(2)\ldots\psi_N(N)$ and similar terms with the variables permuted between the same set of one-electron eigenfunctions $\{\psi_1\ \psi_2\ \ldots\ \psi_N\}$. Thus each term contains the same product of spherical harmonics and the state therefore has parity

$$\lambda = \prod_i (-1)^{l_i} = (-1)^{\sum_i l_i}. \tag{12}$$

Notice that the parity of an atomic state is determined by its electron configuration, not by its total orbital angular momentum.

Exercise 7.3-2 Determine the parity of the atomic states derived from the electron configurations: $nsnp$, nd^3, $npn'p$.

If $I \in$ G, and the parity is even,

$$\chi[\Gamma_L^+(IR)] = \chi[\Gamma_L^+(R)], \tag{13}$$

but if the parity is odd

$$\chi[\Gamma_L^-(IR)] = -\chi[\Gamma_L^-(R)]. \tag{14}$$

We can now formulate two rules for the characters of improper groups.

(1) If $I \in$ G, then G = H ⊗ C_i = {H} + I{H} and the character table for G may be constructed from that of H. (See, for example, the character tables of D_{6h} and O_h.) O_h = O ⊗ C_i and the character table for O_h is given in Table 7.2, where g and u signify g and u IRs and $\chi\{O\}$ means the characters for the group O. (See the character table of O_h in Appendix A3.)

(2) If $I \notin$ G but G contains improper rotations then

$$G = \{Q\} + IR\{Q\}, \tag{15}$$

where Q is a subgroup of proper rotations (sometimes called a halving subgroup), and G is isomorphous with the proper point group

$$P = \{Q\} + R\{Q\} \tag{16}$$

Table 7.2.

O_h	{O}	I{O}
g	$\chi\{O\}$	$\chi\{O\}$
u	$\chi\{O\}$	$-\chi\{O\}$

Table 7.3. *Character tables for the point groups* D_2 *and* C_{2v}.

Because there is no unique principal axis in D_2, the Mulliken conventions are not used in naming the representations of D_2. These two groups are isomorphous and the character systems of the four IRs are identical, but corresponding representations are labeled differently, which tends to obscure rather than emphasize the isomorphism. Note that $C_{2\mathbf{x}}$ and $\sigma_{\mathbf{x}}$ are *corresponding* elements, and so are $C_{2\mathbf{y}}$ and $\sigma_{\mathbf{y}}$. Note that bases for the corresponding IRs are not necessarily identical (for example, z does not form a basis for the totally symmetric representation in D_2). In C_{2v}, the Mulliken designations B_1 and B_2 are arbitrary because there are two equivalent improper binary axes normal to **z**.

D_2	E	C_{2z}	$C_{2\mathbf{x}}$	$C_{2\mathbf{y}}$	
A	1	1	1	1	x^2, y^2, z^2
B_1	1	1	−1	−1	z, R_z, xy
B_2	1	−1	−1	1	y, R_y, zx
B_3	1	−1	1	−1	x, R_x, yz
C_{2v}	E	C_{2z}	$\sigma_{\mathbf{x}}$	$\sigma_{\mathbf{y}}$	
A_1	1	1	1	1	z, x^2, y^2, z^2
A_2	1	1	−1	−1	R_z, xy
B_1	1	−1	−1	1	x, R_y, zx
B_2	1	−1	1	−1	y, R_x, yz

(see Table 2.6). G has the same classes and representations as P, though we need to identify corresponding classes. If G is C_{2v}, and therefore P is D_2, the class of C_{2v} that corresponds to the class $C_{2\mathbf{y}}$ in D_2 is $IC_{2\mathbf{y}} = \sigma_{\mathbf{y}}$. Similarly, $C_{2\mathbf{x}}$ and $\sigma_{\mathbf{x}}$ are corresponding classes. Note however, that some basis functions may belong to different representations in G and in P (see Table 7.3). These rules hold also for "double groups."

Example 7.3-1 (a) Into which states does the Russell–Saunders term $d^2 : {}^3F$ split in an intermediate field of O_h symmetry? (b) Small departures from cubic symmetry often occur as a result of crystal defects, substituent ligands, and various other static and dynamic perturbations. If some of the IRs of O do not occur in the group of lower symmetry, then additional splittings of degenerate levels belonging to such IRs must occur. Consider the effect of a trigonal distortion of D_3 symmetry on the states derived in (a) above.

(a) For ${}^3F$, $L = 3$, $S = 1$. The spin quantum number is not affected by an electrostatic field, and so all the states are still triplets in a crystal field. From eq. (12), the parity $\lambda = (-1)^{2+2} = 1$. From Table 7.1, the states are ${}^3A_{2g}$, ${}^3T_{1g}$, ${}^3T_{2g}$. (b) Select the common (or corresponding) classes for the two groups and reduce the representations, where necessary, in the group of lower symmetry. The relevant characters of O and D_3, which are isomorphous with C_{3v}, are shown in Table 7.4. The representations T_1 and T_2 of O are reducible in D_3 into the direct sums shown in Table 7.5, a process called *subduction*. This is

Table 7.4. *Characters for the classes common to* O *and* D_3.

D_3	E	$2C_3$	$3C_2'$
A_1	1	1	1
A_2	1	1	-1
E	2	-1	0
O	E	$8C_3$	$6C_2'$
A_2	1	1	-1
T_1	3	0	-1
T_2	3	0	1

Table 7.5. *Correlation of the representations* A_2, T_1, *and* T_2 *of* O *with those of* D_3.

O	D_3
A_2	A_2
T_1	$A_2 \oplus E$
T_2	$A_1 \oplus E$

the way in which correlation tables are derived, and it is the method that should used in investigating the possible splitting of degeneracies when the necessary correlation table is not available.

Answers to Exercises 7.3

Exercise 7.3-1 The character of E in any representation is the degeneracy of that representation. The degeneracy of a state with orbital angular momentum L is $2L+1$ because there are $2L+1$ allowed values of the quantum number $M=L, L-1, \ldots, -L$.

Exercise 7.3-2 For *nsnp*, $\lambda=(-1)^{0+1}=-1$; for nd^3, $\lambda=(-1)^{2+2+2}=+1$; for $npn'p$, $\lambda=(-1)^{1+1}=+1$.

7.4 Strong crystal fields

A strong crystal field is one in which the electrostatic interactions due to the surroundings of an ion provide a stronger perturbation than the electron–electron interactions within the ion. One must therefore consider the effect of the field on the free-ion electron configuration and deduce which states are allowed and their degeneracies. We will then be in a position to draw a correlation diagram showing qualitatively the shift in the energy levels as the field strength varies from "intermediate" to "strong." We have seen that in O_h symmetry the five-fold degeneracy of the *d* levels in a free ion is reduced to a t_{2g} orbital

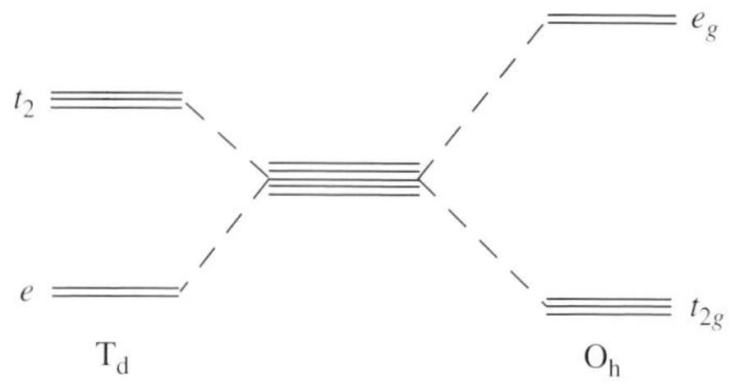

Figure 7.2. Energy-level diagram of the splitting of the five d orbitals in crystal fields of T_d and octahedral O_h symmetry.

triplet and an e_g doublet (Table 6.4). The three-fold degenerate t_{2g} levels lie below the two-fold degenerate e_g levels because of the greater Coulomb repulsion between the ligands and electrons in $d_{x^2-y^2}$ and $d_{3z^2-r^2}$ orbitals (Figure 7.2) compared with electrons in d_{xy}, d_{yz}, or d_{zx} orbitals, where the charge density lies mainly between the ligands in octahedral symmetry. This situation is in direct contrast to T_d symmetry in which the two e orbitals point between the ligands and there is greater electrostatic repulsion between electrons in t_2 orbitals and the surrounding ligands. Consequently, the e levels lie below the t_2 levels in a T_d complex (Figure 7.2).

Exercise 7.4-1 Predict qualitatively the splitting of the five-fold degenerate free-ion d states when a positive ion M is surrounded by eight negative ions which are located at the corners of a cube.

To determine the states in a strong field we shall make use of *Bethe's method of descending symmetry*. This is based on: (i) the fact that an electrostatic field does not affect the spin; and (ii) that if $\psi(1, 2) = \psi^i(1)\,\psi^j(2)$, where $\psi^i(1)$ forms a basis for Γ^i and $\psi^j(2)$ forms a basis for Γ^j, then the product $\psi^i(1)\,\psi^j(2)$ forms a basis for the DP representation $\Gamma^i \otimes \Gamma^j$. To implement Bethe's method we need to use two rules:

(1) two electrons in the *same orbital* give rise to a *singlet* state only; but
(2) two electrons in *different orbitals* give rise to a *singlet and a triplet* state.

To understand the reason for this requires a small digression on *permutation symmetry*. It is a fundamental law of nature, known as the Pauli exclusion principle, that the total state function for a system of N indistinguishable particles which are *fermions*, that is have spin $^1/_2$, $^3/_2$, ..., must be *antisymmetric* with respect to the interchange of any two particles. Let P_{ij} denote the operator that interchanges the positions and spins of indistinguishable particles i and j. Then $|P_{12}\psi(1, 2)|^2 = |\psi(1, 2)|^2$, so that $P_{12}\psi(1, 2) = \exp(i\gamma)\,\psi(1, 2)$. On repeating the interchange, $P_{12}P_{12}\psi(1, 2) = \exp(2i\gamma)\,\psi(1, 2) = \psi(1, 2)$. Therefore, $\exp(2i\gamma) = 1$, $P_{12}\psi(1, 2) = \pm\psi(1, 2)$. This means that ψ could be either symmetric or antisymmetric with respect to the interchange of indistinguishable particles, but, in fact, for fermions ψ is antisymmetric, $P_{12}\psi = -\psi$, so that the state function ψ is an eigenfunction of P_{12} with eigenvalue -1. For a two-particle system, the state function (spin orbital) $\psi \equiv \psi(1, 2)$ can be written as a product $\psi = \phi\chi$ of an orbital $\phi(1, 2)$ and a spinor $\chi(1, 2)$. For two electrons in the same orbital ϕ is symmetric, and so χ must be antisymmetric. There

Table 7.6.

	m_{s_1}	m_{s_2}	M_S (= $m_{s_1} + m_{s_2}$)	S (= largest value of M_S)
$\chi_1 = \alpha(1)\beta(2)$	$1/2$	$-1/2$	0	0
$\chi_2 = \alpha(2)\beta(1)$	$-1/2$	$1/2$	0	0

Table 7.7. *Symmetric and antisymmetric spin functions for two electrons in two different orbitals.*

$\chi(1, 2)$	M_S	S
$\alpha(1)\,\alpha(2)$	1	
$\beta(1)\,\beta(2)$	-1	1
$\mathcal{S}\alpha(1)\,\beta(2) = 2^{-1/2}[\alpha(1)\,\beta(2) + \alpha(2)\,\beta(1)]$	0	
$\mathcal{A}\alpha(1)\,\beta(2) = 2^{-1/2}[\alpha(1)\,\beta(2) - \alpha(2)\,\beta(1)]$	0	0

are two possibilities as shown in Table 7.6. Neither χ_1 nor χ_2 are eigenfunctions of P_{12} so they must be antisymmetrized by the antisymmetrizing operator

$$\mathcal{A} = 2^{-1/2}[1 - P_{12}]. \tag{1}$$

$$\mathcal{A}\alpha(1)\beta(2) = 2^{-1/2}[\alpha(1)\beta(2) - \alpha(2)\beta(1)] \equiv \chi(1, 2), \tag{2}$$

which is normalized and antisymmetric. Similarly, $\mathcal{S} = 2^{-1/2}\,[1 + P_{12}]$ is a symmetrizing operator. Note that $\mathcal{A}(\chi_2)$ gives $-\chi(1, 2)$, which differs from $\chi(1, 2)$ only by a phase factor, so there is only one independent antisymmetric spin function $\chi(1, 2)$, which describes a singlet state. It is an eigenfunction of $\hat{S}_z$ with $M_S = 0$ and of $\hat{S}^2$ with $S = 0$. For two electrons in two different orbitals, ϕ may be symmetric or antisymmetric. Consequently, χ may be antisymmetric or symmetric. There is only one antisymmetric possibility, as seen above, but there are three independent spin functions that are symmetric (see Table 7.7). The first three spin functions χ_s are symmetric and the three functions $\psi(1, 2) = \phi_{as}\chi_s$ describe the triplet state, while the state function $\psi(1, 2) = \phi_s\chi_{as}$ corresponds to the singlet state. This establishes the two rules (1) and (2) listed above.

(*Comment*: It is the action of the antisymmetrizing operator on the product state function in eq. (7.2.5) that produces the Slater determinant in eq. (7.2.6). However, the factorization into an orbital function of **r** and a spinor, that simplifies our work when $N = 2$, does not occur for $N > 2$.)

Example 7.4-1 Find all the states that arise from the configuration d^2 in a strong field of O_h symmetry. Correlate these states with those of the free ion, and of the ion in an intermediate field.

The solution using Bethe's method of descending symmetry is summarized in Table 7.8. In the strong-field limit the possible electron configurations derived from d^2 in the free ion

Table 7.8. *Application of the method of descending symmetry to the configuration d^2 in O_h symmetry.*

Point group	Configuration	Direct product representation	Irreducible representations	Allowed states	Degeneracy
O_h	e_g^2	$E_g \otimes E_g$	$A_{1g} \oplus A_{2g} \oplus E_g$		6
D_{4h}			A_{1g} B_{1g} $A_{1g} \oplus B_{1g}$		
	a_{1g}^2	$A_{1g} \otimes A_{1g}$	A_{1g}	$^1A_{1g}$	
	$a_{1g} b_{1g}$	$A_{1g} \otimes B_{1g}$	B_{1g}	$^1B_{1g}$ $^3B_{1g}$	
	b_{1g}^2	$B_{1g} \otimes B_{1g}$	A_{1g}	$^1A_{1g}$	
O_h				$^1A_{1g}$ $^3A_{2g}$ ^1E_g	6
O_h	t_{2g}^2	$T_{2g} \otimes T_{2g}$	$A_{1g} \oplus E_g \oplus T_{1g} \oplus T_{2g}$		15
C_{2h}			A_g $A_g \oplus B_g$ $A_g \oplus B_g \oplus B_g$ $A_g \oplus A_g \oplus B_g$		
	a_g^2	$A_g \otimes A_g$	A_g	^1A_g	
	$a_g a_g'$	$A_g \otimes A_g$	A_g	^1A_g ^3A_g	
	$(a_g')^2$	$A_g \otimes A_g$	A_g	^1A_g	
	$a_g b_g$	$A_g \otimes B_g$	B_g	^1B_g ^3B_g	
	$a_g' b_g$	$A_g \otimes B_g$	B_g	^1B_g ^3B_g	
	b_g^2	$B_g \otimes B_g$	A_g	^1A_g	
O_h				$^1A_{1g}$ ^1E_g $^3T_{1g}$ $^1T_{2g}$	15
O_h	$t_{2g} e_g$	$T_{2g} \otimes E_g$	$T_{1g} \oplus T_{2g}$	$^1T_{1g}$ $^3T_{1g}$ $^1T_{2g}$ $^3T_{2g}$	24

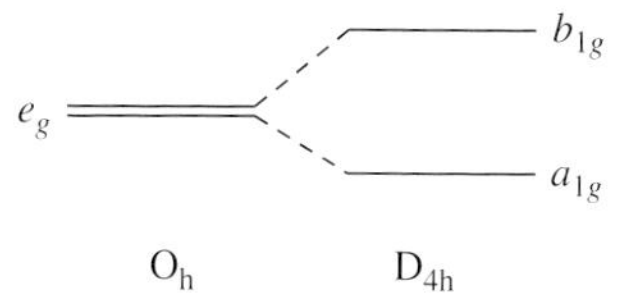

Figure 7.3. Splitting of the e_g levels when the symmetry is lowered from O_h to D_{4h}.

are: t_{2g}^2, $t_{2g}^1e_g^1$, and e_g^2. Since all the states derived from the configuration d^2 are of even parity, we may use the character table for O in reducing DPs. The result (ii) above is used continually. The configuration e_g^2 gives rise to six states in all, two singlets when the two electrons are both in the same e_g orbital (with opposed spins) and a singlet and a triplet when they are in different e_g orbitals. Since both electrons are in orbitals of e_g symmetry, these six state functions form bases for the IRs that are contained in the DP representation $E_g \otimes E_g$. From the character table for O, $E \otimes E = \{4\ 4\ 1\ 0\ 0\} = A_1 \oplus A_2 \oplus E$. At this stage we do not know which of these states are singlets and which are triplets. A reduction in symmetry has no effect on the spin, and the essence of Bethe's method is to lower the symmetry until all representations in the DP are 1-D. From a correlation table (Appendix A4) we see that if the symmetry is reduced to D_{4h}, the E_g representation becomes the direct sum $A_{1g} \oplus B_{1g}$. This means that the two-fold degeneracy of the e_g levels is lifted by the reduction in symmetry to D_{4h} and they become a_{1g}, b_{1g} (Figure 7.3). The possible configurations are a_{1g}^2, $a_{1g}\,b_{1g}$, and b_{1g}^2. The states derived from these configurations are shown in the block of Table 7.8 labeled by the D_{4h} point group. Only the $a_{1g}b_{1g}$ configuration can yield both singlet and triplet states. Therefore the B_{1g} state derived from A_{2g} in O_h is the only triplet state. The states that arise from the electron configuration e_g^2 are, therefore, ${}^1A_{1g}$, ${}^3A_{2g}$ and ${}^1E_g$, with a total degeneracy of 6. We may make use of the degeneracy to perform a final check on our deductions. Call the two e_g orbitals e_g and e_g'. There are, therefore, four one-electron spin orbitals $e_g\alpha$, $e_g\beta$, $e_g'\alpha$, and $e_g'\beta$, and the two electrons may be allotted to these four spin orbitals in a total of ${}_4C_2 = 6$ ways. The procedure for the t_{2g}^2 configuration is analogous and is summarized in Table 7.8. The DP representation $T_2 \otimes T_2 = A_1 \oplus E \oplus T_1 \oplus T_2$. In C_{2h} symmetry the T_{2g} representation becomes the direct sum $A_g \oplus A_g \oplus B_g$. Call the two orbitals of A_g symmetry a_g and a_g'. From the six possible two-electron configurations, only three can yield triplet states, and the direct sum $A_g \oplus B_g \oplus B_g$ in C_{2h} identifies the triplet state in O_h as T_{1g}. Here there are six possible one-electron spin orbitals obtained by combining each of the three orbital functions t_{2g}, t_{2g}', t_{2g}'' with either α or β for a total degeneracy of ${}_6C_2 = 15$. Finally, for the configuration $t_{2g}e_g$ there is no need to use descending symmetry. Because the two electrons are in different orbitals, one in t_{2g} and one in e_g, *both* singlets and triplets occur. The configuration $t_{2g}e_g$ requires that the twenty-four spin orbitals obtained by combining any of the six functions $t_{2g}\alpha$, $t_{2g}\beta$ with any of the four functions $e_g\alpha$, $e_g\beta$ (and antisymmetrizing where necessary) form bases for the representations $T_{2g} \otimes E_g = T_{1g} \oplus T_{2g}$. Therefore the states from configuration $t_{2g}e_g$ are ${}^1T_{1g}$, ${}^3T_{1g}$, ${}^1T_{2g}$, ${}^3T_{2g}$, with a total degeneracy of 24. Descending symmetry gives the same result. The whole procedure is summarized in Table 7.8. The correlation diagram is given in Figure 7.4. This shows qualitatively the dependence of the energy levels on the strength of the crystal field. The strong-field limit is on the right and the states

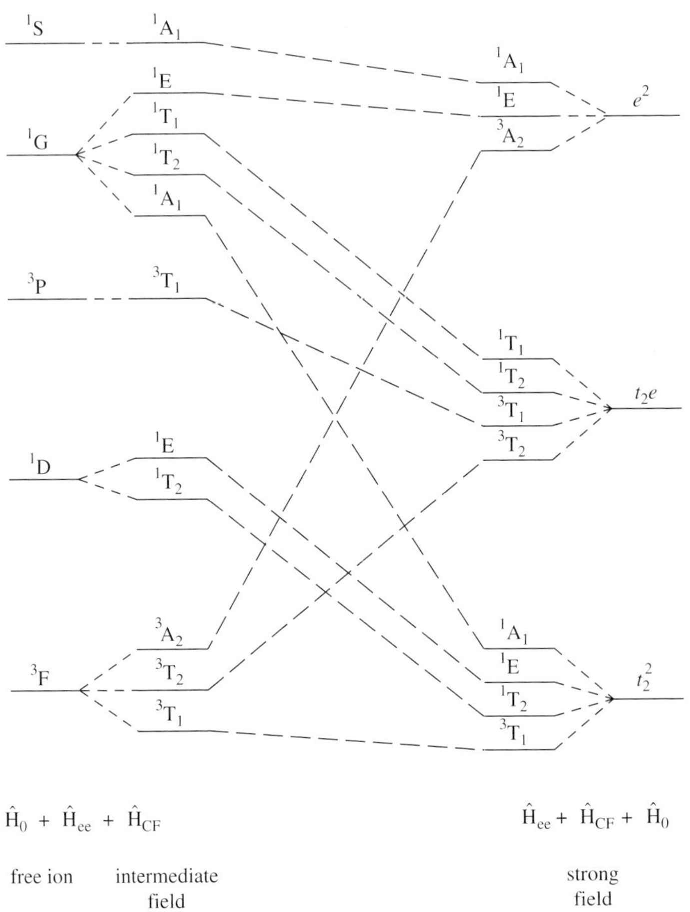

Figure 7.4. Correlation diagram for $d^2(O_h)$. All states shown have even parity.

found Table 7.8 must correlate with those of the same symmetry in the intermediate- (or medium-) field case. In drawing such diagrams we make use of some additional rules. The non-crossing rule is a strict one: this says that states of the same symmetry and spin multiplicity may not cross. When the order is not known (from either experiment or calculation), the ground state may be identified by using two rules from atomic spectroscopy (Hund's first and second rules). These are (i) that states of the highest spin degeneracy lie lowest, and (ii) that for terms with the same S, the one with the higher orbital degeneracy lies lower. Therefore in d^2 the ground state is ^3F, which lies lower than ^3P.

We now consider the configuration nd^8 in O_h. Two new principles must be observed. Firstly, we may ignore doubly occupied orbitals since they contribute A_{1g} to the DP and zero to M_s and to S. To understand this, consider a lowering of symmetry until all degeneracies have been lifted so that the electrons in doubly occupied orbitals are now paired in orbitals that form bases for 1-D representations. This is illustrated for the configuration t_{2g}^2 in Figure 7.5. Whatever the name of the representation in Mulliken

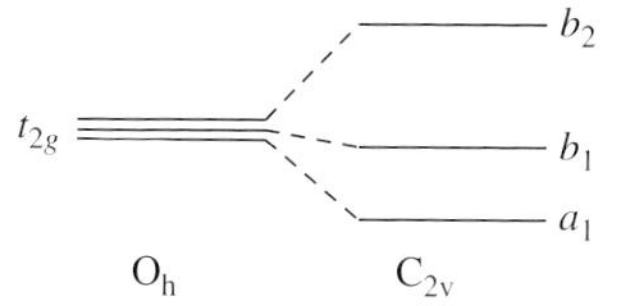

Figure 7.5. Splitting of t_{2g} levels when the symmetry is lowered to C_{2v}.

notation (it is A_1 or B_1 or B_2 in the example shown) the DP of a non-degenerate representation with itself always yields the totally symmetric representation so that the state shown is $^1A_1$ in C_{2v}, which correlates with $^1A_{1g}$ in O_h. This argument is *quite general* and applies to *any configuration* containing only doubly occupied orbitals. The second thing we must do is to take account of spin-pairing energy. Electrons in degenerate levels tend to have unpaired spins whenever possible. This is because of *Pauli repulsion*, which is a consequence of the antisymmetry requirement for a many-electron state function for fermions. If two electrons have a symmetric spin function they will tend to remain apart in space. Otherwise, since $P_{12}\phi(1, 2) = -\phi(2, 1)$, the charge density would vanish in the limit $\mathbf{r}_2 \rightarrow \mathbf{r}_1$. Since two electrons with a symmetric spin function tend to remain further apart than two electrons with an antisymmetric spin function, they shield each other from the nucleus to a smaller extent and so their Coulomb interaction with the nucleus is greater than for two electrons with antisymmetric spin functions. This is the reason why triplets lie below singlets and why it requires energy to pair up spins. But if electrons have to be promoted to higher states in order to become unpaired, then this promotion energy must be offset against the gain in energy from unpairing the spins. So the ground-state configuration will depend on the crystal-field splitting. It is an empirical fact that spin pairing requires *more energy* in e_g orbitals than in t_{2g} orbitals. Consequently, the ordering of the d^8 configurations is $t_{2g}^6e_g^2 < t_{2g}^5e_g^3 < t_{2g}^4e_g^4$, because the number of e_g pairs in the three configurations is $0 < 1 < 2$. On the left of Figure 7.6 is shown the actual configurations in a d^8 complex. Now use the first principle and re-write the configurations ignoring doubly occupied orbitals. They are, in order of increasing energy, $e_g^2 < e_gt_{2g} < t_{2g}^2$. Therefore, d^8 *behaves as if* the ordering of the t_{2g}, e_g levels had been *inverted* (Figure 7.6). Consequently, the correlation diagram for d^8 is like that for d^2 but with the ordering of the high-field states inverted. Now in T_d symmetry, the ordering of the levels is $e^2 < et_2 < t_2^2$. Therefore, the energy-level diagrams for states from d^{10-n} (O_h) are like those for d^n (T_d), and these are the inverses of d^n(O)$_h$ which are like those for d^{10-n}(T_d). Energy-level diagrams for the d^2 to d^8 configurations in octahedral symmetry have been calculated by Tanabe and Sugano (1954) (cf. Purcell and Kotz (1980), pp. 344–5).

Answer to Exercise 7.4-1

The position of the ligands is the same as that in T_d symmetry but every corner of the cube is occupied by a negative ion. Consequently, the energy-level diagram is like that for T_d symmetry (see Figure 7.2) but one should expect relatively larger crystal-field splittings for the same ligands.

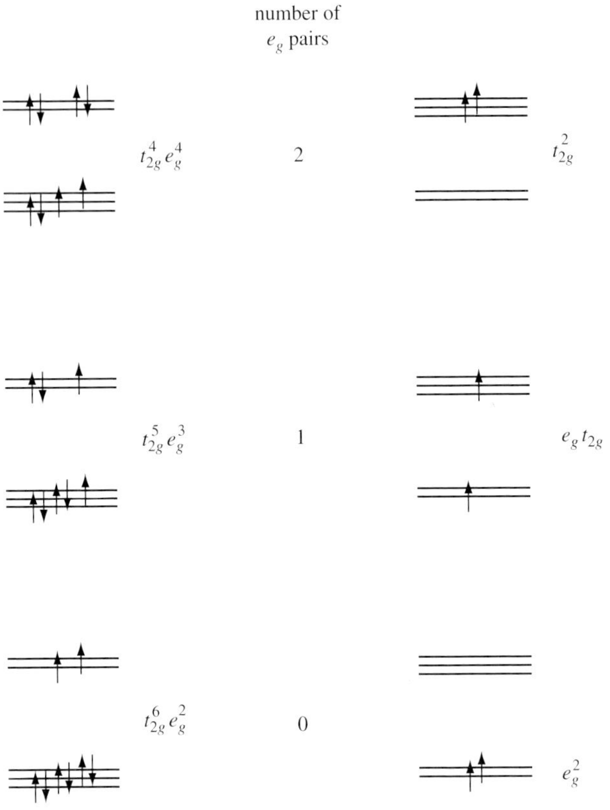

Figure 7.6. Ordering of the high-field energy levels in the d^8 configuration.

Problems

7.1 Describe the effect of an intermediate crystal field of T_d symmetry on the states of an ion with the configuration d^2. Assume *RS* coupling to hold in the free ion so that the free-ion states, in order of decreasing stability, are ^3F, ^1D, ^3P, ^1G, ^1S. (a) First work out the states, and their spin multiplicities, that arise in the strong-field limit (configurations e^2, et_2, t_2^2) using Bethe's method of descending symmetry, but employing different point groups to those used in Example 7.4-1. (b) Next work out the splittings of the Russell–Saunders states in an intermediate field. (c) Finally, draw a correlation diagram showing qualitatively the splitting of the free-ion states as a function of field strength. (*Hints*: For (a) use a correlation table (Appendix A4). For (b) use eqs. (7.3.12)–(7.3.16). As their character tables show, T_d is isomorphous with O. You will need to identify corresponding classes in these two groups in order to make use of eqs. (7.3.13) or (7.3.14).)

7.2 A Ce^{3+} ion has the configuration $4f^1$ so that its ground state is the Russell–Saunders multiplet ^4F. In a crystal of CaF_2 containing dissolved CeF_3, the Ce^{3+} ions substitute for Ca^{2+} ions at lattice sites so that each Ce^{3+} ion is at the center of a cube of F^- ions (see Exercise 7.4-1). This cubic field is of a strength intermediate between the electron–electron interactions and the spin–orbit coupling. (a) Determine the character system of the representation Γ_L and reduce this representation into a direct sum of IRs. (*Hint*: The parity of a state is determined by its electron configuration and not by its spectral term. It is sufficient, therefore, to determine the parity for $4f^1$ and then use the character table of O.) (b) Charge compensation for the extra positive charge on Ce^{3+} is provided by O^{2-} impurity ions substituting for F^-. For electrostatic reasons O^{2-} ions prefer to occupy nearest-neighbor sites to Ce^{3+} ions. What is the symmetry at these Ce^{3+} sites which have an O^{2-} ion as a nearest neighbor to the Ce^{3+} and what further splitting and re-labeling of the $4f^1$ states occur when the Ce^{3+} site symmetry is lowered by the presence of an O^{2-} in a nearest-neighbor position? (Do not use a correlation table but perform the subduction explicitly.)

7.3 (a) Use Bethe's method of descending symmetry to determine the states that arise from a d^8 configuration in a strong crystal field of O_h symmetry. (*Hint*: Correlation tables are in Appendix A4, but use different point groups to those used in Example 7.4-1.) Construct a correlation diagram showing qualitatively the shift of the free-ion energy levels with increasing field strength.

8 Double groups

8.1 Spin–orbit coupling and double groups

The spin–orbit coupling term in the Hamiltonian induces the coupling of the orbital and spin angular momenta to give a total angular momentum $\mathbf{J} = \mathbf{L} + \mathbf{S}$. This results in a splitting of the Russell–Saunders multiplets into their components, each of which is labeled by the appropriate value of the total angular momentum quantum number J. The character of the matrix representative (MR) of the operator $R(\phi\ \mathbf{n})$ in the coupled representation is

$$\text{(7.3.9)} \qquad \chi(\Gamma_L(\phi)) = \sin[(2L+1)\phi/2]/\sin(\phi/2). \qquad (1)$$

A detailed analysis (Chapter 11) shows that this result depends upon the commutation relations for the $\mathbf{L}$ operators, and, since the spin and the total angular momentum operators obey the same commutation relations (CRs), this formula holds also for $\mathbf{S}$ and for $\mathbf{J}$:

$$\text{(1)} \qquad \chi(\Gamma_J(\phi)) = \sin[(2J+1)\phi/2]/\sin(\phi/2). \qquad (2)$$

Now, L is an integer and S is an integer or half-integer; therefore the total angular momentum quantum number J is an integer or half-integer. Consider a proper rotation through an angle $\phi + 2\pi$. Then

$$\text{(2)} \qquad \begin{aligned} \chi(\Gamma_J(\phi+2\pi)) &= \frac{\sin[((2J+1)\phi/2 + (2J+1)\pi)]}{\sin[(\phi+2\pi)/2]} \\ &= \frac{\sin[(2J+1)\phi/2]\cos[(2J+1)\pi]}{-\sin(\phi/2)} \\ &= (-1)^{2J}\chi(\Gamma_J(\phi)). \end{aligned} \qquad (3)$$

If J is an integer,

$$\text{(3)} \qquad \chi(\Gamma_J(\phi+2\pi)) = \chi(\Gamma_J(\phi)), \qquad (4)$$

but if J is a half-integer, as will be the case for an atom with an odd number of electrons,

$$\text{(3)} \qquad \chi(\Gamma_J(\phi+2\pi)) = -\chi(\Gamma_J(\phi)). \qquad (5)$$

In configuration space we would expect a rotation through $\phi + 2\pi$ to be equivalent to a rotation through ϕ. The curious behavior implied by eq. (5) arises because our state functions are spin orbitals and not functions in configuration space. Equation (5) suggests the introduction of a new operator $\overline{E}$, with the property $\overline{E}R = \overline{R} = R(\phi + 2\pi\ \mathbf{n})$. Adding $\overline{E}$ to the group $G = \{R\}$ gives the *double group* $\overline{G} = \{R\} + \overline{E}\{R\}$. Note that $\overline{G}$ contains twice as many elements as G, but does not necessarily have twice the number of classes. The number of new classes in $\overline{G}$ is given by *Opechowski's rules*. (A proof of these rules is given by Altmann (1986).)

(1) $\overline{C}_{2\mathbf{n}} = \overline{E}C_{2\mathbf{n}}$ and $C_{2\mathbf{n}}$ are in the same class iff (meaning if, and only if) there is a (proper or improper) rotation about another $\mathbf{C}_2$ axis normal to $\mathbf{n}$.
(2) $\overline{C}_n = \overline{E}C_n$ and C_n are always in different classes when $n \neq 2$.
(3) For $n > 2$, $\overline{C}_n^k$ and $\overline{C}_n^{-k}$ are in the same class, as are C_n^k and C_n^{-k}.

Exercise 8.1-1 Name the classes in the double point groups $\overline{C}_4$ and $\overline{C}_{2v}$.

Rewriting eqs. (2) and (3) in a slightly more convenient notation, we have

$$\text{(2)} \qquad \chi[R(\phi\ \mathbf{n})] \equiv \chi_J(\phi) = \sin[(2J+1)\phi/2]/\sin(\phi/2), \tag{6}$$

$$\text{(3)} \qquad \chi[\overline{R}(\phi\ \mathbf{n})] = \overline{\chi}_J(\phi) = \chi_J(\phi + 2\pi) = (-1)^{2J}\chi_J(\phi). \tag{7}$$

Table 8.1 shows the characters $\chi_J(\phi)$ for $R(\phi\ \mathbf{z})$, calculated from eq. (6) for half-integral J, for the rotations of the proper point group O which occur also in $\overline{O}$. For $\overline{R}(\phi\ \mathbf{z})$, use eq. (7). For other values of ϕ use eqs. (6) and (7). For improper rotations, see Box 8.1. Equations (6) and (7) work equally well for integral J. For the classes of $\overline{G}$ (in the current example G is O), integral values of J give the same results as L (in Table 7.1) so that $J = 0, 1, 2, 3$ would generate all the standard representations of O. The characters for the new classes $c_k\overline{C}_n$ of $\overline{G}$ are the same as those of the classes C_k of G, for integral J. For half-integral J they have the same magnitude but opposite sign, in accordance with eq. (7). The new representations of $\overline{G}$ that do not occur in G, and which are generated by half-integral values of the total angular momentum quantum number J, are called the *spinor representations*.

Box 8.1. Improper rotations.

If $I \in G, I\psi^J = \pm\psi^J$, $G = \{H\} + I\{H\}$ and $\chi[\Gamma_J^{\pm}(IR)] = \pm\chi[\Gamma_J^{\pm}(R)]$.
If $I \notin G$, then $G = \{H\} + IR\{H\}$ is isomorphous with $P = \{H\} + R\{H\}$, where R is a proper rotation and $\{H\}$ is a subgroup of proper rotations.
Then

$$\chi[\Gamma_J(IR')] = \chi[\Gamma_J(R')],$$

where $R' \in R\{H\}$ and $IR' \in G$.
These rules hold for J integral or half-integral and so for L and S.

Table 8.1. *Characters of the matrix representatives* Γ_J *for half-integral* J.

For improper rotations see Table 8.2.
For $R(\phi\ \mathbf{z})$, $\chi_J(\phi) = \sin[(2J+1)\phi/2]/\sin(\phi/2)$.
For $\overline{R}(\phi\ \mathbf{z}) = \overline{E}R(\phi\ \mathbf{z})$, $\chi_J(\phi + 2\pi) = (-1)^{2J}\chi_J(\phi)$.

	E	C_2	C_3	C_4
ϕ	0	π	$2\pi/3$	$\pi/2$
$\chi(\Gamma_J)$	$2J+1$	0	$\begin{cases} 1\ (J = {}^1\!/_2, {}^7\!/_2, \ldots) \\ -1\ (J = {}^3\!/_2, {}^9\!/_2, \ldots) \\ 0\ (J = {}^5\!/_2, {}^{11}\!/_2, \ldots) \end{cases}$	$2^{1/2}\ (J = {}^1\!/_2, {}^9\!/_2, \ldots)$ $0\ (J = {}^3\!/_2, {}^7\!/_2, \ldots)$ $-2^{1/2}\ (J = {}^5\!/_2, {}^{13}\!/_2, \ldots)$
$J = {}^1\!/_2$	2	0	1	$2^{1/2}$
$J = {}^3\!/_2$	4	0	-1	0
$J = {}^5\!/_2$	6	0	0	$-2^{1/2}$

Table 8.2. $\overline{\mathrm{O}} = \{\mathrm{O}\} + \overline{\mathrm{E}}\{\mathrm{O}\}$.

		$3\overline{C}_2$			$6\overline{C}_2{}'$	$\overline{\mathrm{E}}$	$8\overline{C}_3$	$6\overline{C}_4$	
$\overline{\mathrm{O}}$	E	$3C_2$	$8C_3$	$6C_4$	$6C_2{}'$				
Γ_1 A_1	1	1	1	1	1	1	1	1	$x^2+y^2+z^2$
Γ_2 A_2	1	1	1	-1	-1	1	1	-1	
Γ_3 E	2	2	-1	0	0	2	-1	0	$(x^2-y^2\ \ 3z^2-r^2)$
Γ_4 T_1	3	-1	0	1	-1	3	0	1	$(x\ y\ z)(R_x\ R_y\ R_z)$
Γ_5 T_2	3	-1	0	-1	1	3	0	-1	$(xy\ yz\ zx)$
Γ_6 $\mathrm{E}_{1/2}$	2	0	1	$2^{1/2}$	0	-2	-1	$-2^{1/2}$	
Γ_7 $\mathrm{E}_{5/2}$	2	0	1	$-2^{1/2}$	0	-2	-1	$2^{1/2}$	
Γ_8 $\mathrm{F}_{3/2}$	4	0	-1	0	0	-4	1	0	
$\Gamma_{5/2}$	6	0	0	$-2^{1/2}$	0	-6	0	$2^{1/2}$	

Example 8.1-1 Derive the character table for the double group $\overline{\mathrm{O}}$.

This is given in Table 8.2. From Opechowski's rules there are three new classes and therefore three new representations. Since $\overline{\mathrm{O}}$ contains twice as many elements as O, the dimensions of the new representations are given by $l_6^2 + l_7^2 + l_8^2 = 24$, so that $l_6 = 2$, $l_7 = 2$, and $l_8 = 4$. The characters for $J = {}^1\!/_2, {}^3\!/_2, {}^5\!/_2$ may be written down from Table 8.1. Again, two types of notation are mainly used to label the irreducible representations (IRs). In Bethe's notation the new representations of $\overline{\mathrm{G}}$ are simply labeled by Γ_i, where i takes on as many integer values as are necessary to label all the spinor representations. Mulliken notation was extended by Herzberg to include double groups. The IRs are labeled E, F, G, H, ... according to their dimensionality 2, 4, 6, 8, ..., with a subscript that is the value of J which corresponds to the representation Γ_J in which that IR *first occurs*. First write down $\Gamma_{1/2}$ and test for irreducibility. For half-integral J, $\chi(\overline{R}) = -\chi(R)$ (see Table 8.2) so that for spinor representations we may work with the classes of O only,

$$\Gamma_{\frac{1}{2}}: \sum_T |\chi_{\frac{1}{2}}(T)|^2 = 1(4) + 8(1) + 6(2) = 24 = \overline{g}/2.$$

(Including all the classes of $\overline{\text{G}}$ in the sum would simply repeat these three terms for a total of $48 = \overline{g}$.) Therefore, $\Gamma_{\frac{1}{2}}$ is an IR, named Γ_6 or $E_{\frac{1}{2}}$. Next write down $\Gamma_{3/2}$ and test for irreducibility:

$$\Gamma_{3/2}: \sum_T |\chi_{3/2}(T)|^2 = 1(16) + 8(1) = 24 = \overline{g}/2.$$

Therefore $\Gamma_{3/2}$ is an IR, named Γ_8 or $F_{3/2}$. Similarly,

$$\Gamma_{5/2}: \sum_T |\chi_{5/2}(T)|^2 = 1(36) + 6(2) + 6(2) = 48 > \overline{g}/2,$$

so that $\Gamma_{5/2}$ is reducible. Performing the reduction in the usual way, but again just using the classes of G,

$$c(\Gamma_6) = (1/24)[1(2)(6) + 6(2^{\frac{1}{2}})(-2^{\frac{1}{2}})] = 0,$$

$$c(\Gamma_8) = (1/24)[1(4)(6)] = 1.$$

Now $\Gamma_{5/2} - \Gamma_8 = \{2\ \ 1\ \ 0\ \ -2^{\frac{1}{2}}\ \ 0\ \ -2\ \ -1\ \ 2^{\frac{1}{2}}\} = \Gamma_7 = E_{5/2}$, so the second E representation does not occur until the reduction of $\Gamma_{5/2} = \Gamma_7 + \Gamma_8$. Γ_7 is therefore $E_{5/2}$ in Mulliken–Herzberg notation. Note that all the new representations necessitated by half-integral values of J are at least doubly degenerate. This means that all energy levels corresponding to half-integral J, that is arising from a configuration with an odd number of electrons, are at least two-fold degenerate in any electrostatic field. This result is known as Kramers' theorem (Kramers (1930)). Further splittings may, however, be possible in magnetic fields.

Exercise 8.1-2 Write down the characters of $\Gamma_{7/2}$ and $\Gamma_{9/2}$ and reduce both these representations into a direct sum of IRs.

Other groups may be handled in a similar manner to O in Example 8.1-1. For improper rotations, the two rules formulated previously hold also for double groups (Box 8.1). If the group contains the inversion operator, even or odd parity is indicated by a superscript of + or − in Bethe's notation and by a subscript g or u in Mulliken–Herzberg notation.

Answers to Exercises 8.1

Exercise 8.1-1 $\overline{\text{C}}_4 = \{E\ C_4^+\ C_2\ C_4^-\ \overline{E}\ \overline{C}_4^+\ \overline{C}_2\ \overline{C}_4^-\}$; $\overline{\text{C}}_{2v} = \{E,\ C_{2z} + \overline{C}_{2z},\ \sigma_y + \overline{\sigma}_y,\ \sigma_x + \overline{\sigma}_x,\ \overline{E}\}$.

Exercise 8.1-2 $\Gamma_{7/2} = \{8\ \ 0\ \ 1\ \ 0\ \ 0\ \ -8\ \ -1\ \ 0\}$, $c(\Gamma_6) = (1/24)[1(2)(8) + 8(1)(1)] = 1$,

$$c(\Gamma_7) = (1/24)[1(2)(8) + 8(1)(1)] = 1,$$

$$\Gamma_{7/2} - (\Gamma_6 + \Gamma_7) = \{4\ 0\ \ -1\ 0\ 0\ \ -4\ 1\ 0\} = \Gamma_8,$$

$$\Gamma_{9/2} = \{10\ 0\ \ -1\ 2^{1/2}\ 0\ \ -10\ 1\ -2^{1/2}\},$$

$$c(\Gamma_6) = (1/24)[1(2)(10) + 8(1)(-1) + 6(2^{1/2})(2^{1/2})] = 1,$$

$$c(\Gamma_7) = (1/24)[1(2)(10) + 8(1)(-1) + 6(2^{1/2})(-2^{1/2})] = 0,$$

$$c(\Gamma_8) = (1/24)[1(4)(10) + 8(-1)(-1)] = 2.$$

8.2 Weak crystal fields

Since a "weak" field means one that is smaller than the spin–orbit coupling term $\hat{H}_{\mathrm{S.L}}$, a weak crystal field acts on the components of the Russell–Saunders multiplets. Depending on their degeneracy, these components may undergo further splittings in the weak crystal field. In the symmetry group O, Table 8.2 and Exercise 8.1-1 tell us thatthe following splittings occur for half-integral values of the total angular momentum quantum number J:

$$\Gamma_{3/2} = \Gamma_8,$$
$$\Gamma_{5/2} = \Gamma_7 \oplus \Gamma_8,$$
$$\Gamma_{7/2} = \Gamma_6 \oplus \Gamma_7 \oplus \Gamma_8,$$
$$\Gamma_{9/2} = \Gamma_6 \oplus 2\Gamma_8.$$

Example 8.2-1 Examine the effect of spin–orbit coupling on the states that result from an intermediate field of O symmetry on the Russell–Saunders term ^{4}F. Correlate these states with those produced by the effect of a weak crystal field of the same symmetry on the components produced by spin–orbit coupling on the ^{4}F multiplet.

The solution is summarized in Figure 8.1. The ^{4}F state has $L = 3$, and so (from Table 7.1) it is split by an intermediate field into three states which belong to the IRs $A_2 \oplus T_1 \oplus T_2 = \Gamma_2 \oplus \Gamma_4 \oplus \Gamma_5$. To examine the effect of spin–orbit coupling on these intermediate-field states, we use the fact that if $\psi = \phi^i\chi^j$, where ϕ^i forms a basis for Γ^i and χ^j forms a basis for Γ^j, then $\psi = \phi^i\chi^j$ forms a basis for the direct product (DP) representation $\Gamma^i \otimes \Gamma^j$. Here $S = 3/2$, and the representation $\Gamma_{3/2}$ is Γ_8 (Table 8.2). Take the DPs of Γ_8 with Γ_2, Γ_4, and Γ_5 to obtain

$$\Gamma_8 \otimes \Gamma_2 = \Gamma_8,$$
$$\Gamma_8 \otimes \Gamma_4 = \Gamma_6 \oplus \Gamma_7 \oplus 2\Gamma_8,$$
$$\Gamma_8 \otimes \Gamma_5 = \Gamma_6 \oplus \Gamma_7 \oplus 2\Gamma_8.$$

Exercise 8.2-1 Verify the DPs necessary to determine the spin–orbit splitting of the intermediate-field states derived from the Russell–Saunders term ^{4}F.

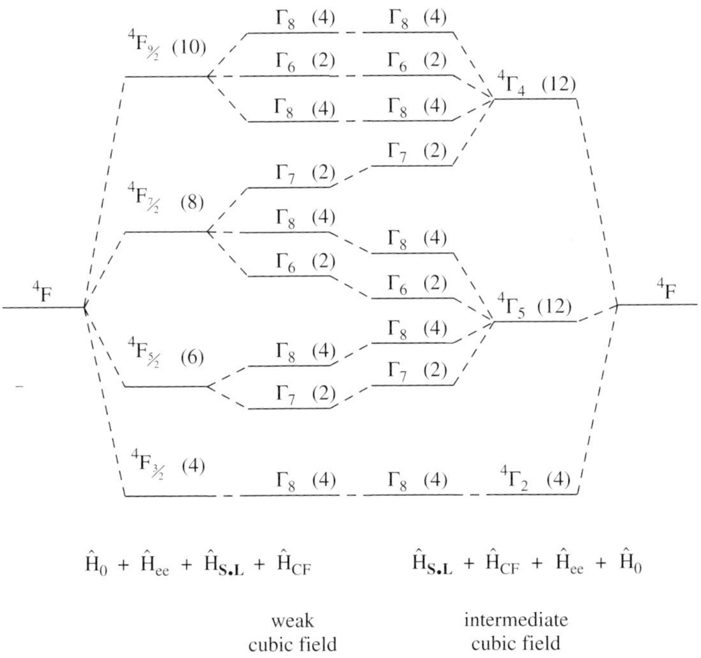

Figure 8.1. Splitting of the ^{4}F state in weak and intermediate fields of cubic symmetry.

The correlation diagram that correlates the intermediate- (or medium-) field states with the weak-field states is shown in Figure 8.1. The same states must arise independently of the order in which the crystal-field and spin–orbit coupling perturbations are applied. The numbers in parentheses are the degeneracies of the states; they provide a useful check on the accuracy of numerical calculations.

The character tables in Appendix A3 include the spinor representations of the common point groups. Double group characters are not given explicitly but, if required, these may be derived very easily. The extra classes in the double group are given by Opechowski's rules. The character of $\overline{R}$ in these new classes in vector representations is the same as that of R but in spinor representations $\chi(\overline{R}) = -\chi(R)$. The bases of spinor representations will be described in Section 12.8.

Answer to Exercise 8.2-1

$$\begin{array}{lrcl} & E & 8C_3 & \\ \Gamma_8 \otimes \Gamma_2 = & 4 & -1 & = \Gamma_8, \\ \Gamma_8 \otimes \Gamma_4 = & 12 & 0 & = \Gamma_6 \oplus \Gamma_7 \oplus 2\Gamma_8, \\ \Gamma_8 \otimes \Gamma_5 = & 12 & 0 & = \Gamma_6 \oplus \Gamma_7 \oplus 2\Gamma_8. \end{array}$$

$$c(\Gamma_6) = {}^1\!/_{24}[1(12)(2)] = 1,$$
$$c(\Gamma_7) = {}^1\!/_{24}[1(12)(2)] = 1,$$

$$c(\Gamma_8) = {}^1\!/_{24}[1(12)(4)] = 2.$$

It is easily verified that using the classes of $\overline{\mathrm{O}}$ gives the same results.

Problems

8.1. Construct the character table of the double group $\overline{\mathrm{D}}_2$.

8.2. (a) A partial character table of the double group $\overline{\mathrm{D}}_4$ is given in Table 8.3. Complete this character table by finding the missing classes and IRs. Label the IRs using both Bethe and Mulliken–Herzberg notation.
(b) An ion with an odd number of electrons has a Γ_6 ground state and a Γ_7 excited state. Are E1 (electric dipole) transitions between these two states symmetry allowed in a weak field of D_4 symmetry? State the polarization of the electromagnetic radiation involved in any allowed transitions.

8.3. A partial character table of the point group D_{3h} is given in Table 8.4. Find the missing characters of the vector and spinor representations of the double group $\overline{\mathrm{D}}_{3h}$. Determine whether E1 transitions $\mathrm{E}_{1/2} \rightarrow \mathrm{E}_{3/2}$ and $\mathrm{E}_{1/2} \rightarrow \mathrm{E}_{5/2}$ are allowed in a weak crystal field of D_{3h} symmetry. State the polarization of allowed transitions.

8.4. The electron configuration d^3 produces a number of states, one of which has the spectral term ^{2}G. Describe the splitting and/or re-labeling of this ^{2}G state under the following perturbations and summarize your discussion in the form of a correlation

Table 8.3. *Partial character table of the double group* $\overline{\mathrm{D}}_4$.

$\overline{\mathrm{D}}_4$	E	$2C_4$	C_2	$2C_2'$	$2C_2''$	
A_1	1	1	1	1	1	x^2+y^2, z^2
A_2	1	1	1	−1	−1	z, R_z
B_1	1	−1	1	1	−1	x^2-y^2
B_2	1	−1	1	−1	1	xy
E	2	0	−2	0	0	$(x\ y)(R_x\ R_y)(yz\ zx)$

Table 8.4. *Partial character table of the point group* D_{3h}.

D_{3h}	E	$2C_3$	$3C_2'$	σ_h	$2S_3$	$3\sigma_v$	
A_1'	1	1	1	1	1	1	
A_2'	1	1	−1	1	1	−1	
E'	2	−1	0	2	−1	0	$(x\ y)$
A_1''	1	1	1	−1	−1	−1	
A_2''	1	1	−1	−1	−1	1	z
E''	2	−1	0	−2	1	0	

diagram which shows the degeneracy of each level and the perturbation giving rise to each set of levels:

(i) the spin–orbit coupling perturbation $\hat{H}_{\mathbf{S.L}}$;

(ii) a crystal field of T_d symmetry that is a weaker perturbation than $\hat{H}_{\mathbf{S.L}}$;

(iii) a crystal field of T_d symmetry that is stronger than $\hat{H}_{\mathbf{S.L}}$;

(iv) the effect of spin–orbit coupling on the states derived from (iii).

Could further splittings of these states be induced by an electrostatic field of D_2 symmetry?

8.5. Describe the splitting of the multiplet ^{4}D under the conditions specified in (i)–(iv) of Problem 8.3, except that the crystal field is of O_h symmetry. [*Hint*: Since a crystal field does not affect the parity of a state, it is sufficient to work with the double group $\overline{\mathrm{O}}$.]

8.6. Investigate the effect of spin–orbit coupling on the crystal-field levels of a Ce^{3+} ion substituting for Ca^{2+} in CaF_2 with a nearest-neighbor O^{2-} ion (see Problem 7.2). Is any further splitting of these levels to be expected if the site symmetry at Ce^{3+} is lowered to C_s by a further crystal-field perturbation that is weaker than $\hat{H}_{\mathbf{S.L}}$?

9 Molecular vibrations

9.1 Classification of normal modes

The internal vibrational motion of a molecule containing N atoms may be approximated by the superposition of $3N - n$ simple harmonic vibrations called *normal modes*. Here $n = 5$ or 6 and the ambiguity arises because, although each molecule has three degrees of translational freedom, linear molecules have only two degrees of rotational freedom, whilst non-linear molecules have three degrees of rotational freedom, each corresponding to a rotational motion about one of the three Cartesian axes. The *normal coordinates* Q_k, which are linear combinations of the atomic displacements x_i, y_i, z_i from the equilibrium positions of the atoms, are *properties of the molecule* and are determined by the symmetry of the molecule, so that each transforms according to one of the IRs of the point group of the molecule.

Example 9.1-1 Classify the normal modes of vibration of the carbonate ion CO_3^{2-} according to the IRs for which the normal coordinates form bases.

To find the symmetry of the normal modes we study the transformation of the atomic displacements $\{x_i \; y_i \; z_i\}$, $i = 0, 1, 2, 3$, by setting up a local basis set $\{\mathbf{e}_{i1} \; \mathbf{e}_{i2} \; \mathbf{e}_{i3}\}$ on each of the four atoms. A sufficient number of these basis vectors are shown in Figure 9.1. The point group of this molecule is D_{3h} and the character table for D_{3h} is in Appendix A3. In Table 9.1 we give the classes of D_{3h}; a particular member R of each class; the number of atoms N_R that are invariant under any symmetry operator in that class; the 3×3 sub-matrix $\Gamma_i(R)$ for the basis $\langle \mathbf{e}_{i1} \; \mathbf{e}_{i2} \; \mathbf{e}_{i3}|$ (which is a 3×3 block of the complete matrix representative for the basis $\langle \mathbf{e}_{01} \; \ldots \; \mathbf{e}_{33}|$); the characters χ_i for the representation Γ_i; and the characters for the whole representation, which are $\chi = \chi_i N_R$. We can proceed in this way, working effectively with a basis $\langle \mathbf{e}_{i1} \; \mathbf{e}_{i2} \; \mathbf{e}_{i3}|$, because when an atom is transformed into a different atom of the same species, then the 3×3 matrix Γ_i occupies an off-diagonal block and therefore does not contribute to the character of the 12×12 matrix representative (MR). The reducible representation Γ may be reduced in the usual way to yield the direct sum

$$\Gamma = A_1' \oplus A_2' \oplus 3E' \oplus 2A_2'' \oplus E''. \tag{1}$$

This sum contains the representations for translational motion and rotational motion as well as for the vibrational motion. From the character table we see that these are

$$\Gamma_t = A_2'' \oplus E', \tag{2}$$

Table 9.1. *Deduction of the representation* Γ *based on* $\langle \mathbf{e}_{01}\, \mathbf{e}_{02}\, \ldots\, \mathbf{e}_{33}|$ *(see Figure 9.1).*

The particular R used to find $\Gamma_i(R)$ is in row 2. Only the diagonal elements of the MRs $\Gamma_i(R)$ are given. $c = \cos(2\pi/3) = -½$.

D_{3h}	E	$2C_3$	$3C_2$	σ_h	$2S_3$	$3\sigma_v$
R	E	$C_{3\mathbf{z}}^{+}$	$C_{2\mathbf{x}}$	σ_z	$S_{3\mathbf{z}}^{+}$	$\sigma_{\mathbf{y}}$
N_R	4	1	2	4	1	2
$\Gamma_i(R)$	$\begin{bmatrix} 1 & & \\ & 1 & \\ & & 1 \end{bmatrix}$	$\begin{bmatrix} c & & \\ & c & \\ & & 1 \end{bmatrix}$	$\begin{bmatrix} 1 & & \\ & \bar{1} & \\ & & \bar{1} \end{bmatrix}$	$\begin{bmatrix} 1 & & \\ & 1 & \\ & & \bar{1} \end{bmatrix}$	$\begin{bmatrix} c & & \\ & c & \\ & & \bar{1} \end{bmatrix}$	$\begin{bmatrix} 1 & & \\ & \bar{1} & \\ & & 1 \end{bmatrix}$
χ_i	3	0	-1	1	-2	1
$\chi(\Gamma)$	12	0	-2	4	-2	2

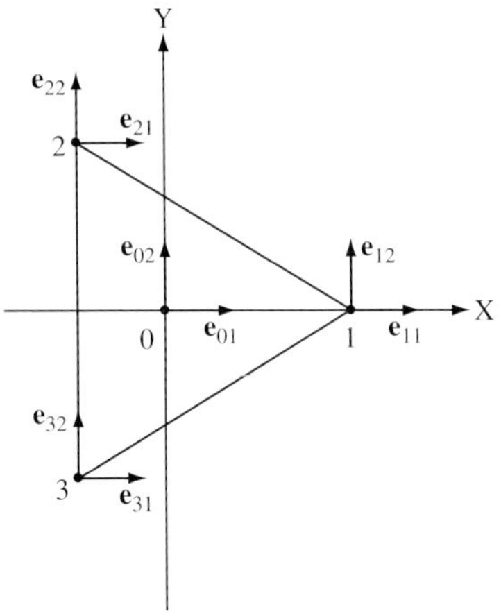

Figure 9.1. Numbering of atoms and location of basis vectors used to describe the atomic displacements in the CO_3^{2-} ion. The $\mathbf{e}_{i3}$ unit vectors are all normal to the plane of the paper, pointing towards the reader, and the displacement of the ith atom is $\mathbf{r}_i = \mathbf{e}_{i1}x_i + \mathbf{e}_{i2}y_i + \mathbf{e}_{i3}z_i$.

$$\Gamma_r = A_2{}' \oplus E'', \tag{3}$$

(1), (2), (3)

$$\Gamma_v = A_1{}' \oplus 2E' \oplus A_2{}''. \tag{4}$$

As a check, we calculate the total vibrational degeneracy from eq. (4) as 6, which is equal, as it should be, to $3N - 6$. The arithmetic involved in the reduction of the direct sum for the total motion of the atoms can be reduced by subtracting the representations for translational and rotational motion from Γ before reduction into a direct sum of IRs, but the method used above is to be preferred because it provides a useful arithmetical check on the accuracy of Γ and its reduction.

9.2 Allowed transitions

In the theory of small vibrations it is shown that, by a linear transformation of the displacements $\{x_j\ y_j\ z_j\}$ to a set of normal coordinates $\{Q_k\}$, the kinetic energy and the potential energy may be transformed simultaneously into diagonal form so that $2T = \sum\limits_k \dot{Q}_k^2$ and $2\Phi = \sum\limits_k \omega_k^2 Q_k^2$. The Hamiltonian is therefore separable into a sum of terms, each of which is the Hamiltonian for a 1-D harmonic oscillator. Consequently, the Schrödinger equation is separable and the total state function is a product of 1-D harmonic oscillator state functions

$$\Psi_{n_k}(Q_k) \equiv |n_k\rangle = N_k H_k(\gamma_k Q_k) \exp[-\tfrac{1}{2}\gamma_k{}^2 Q_k{}^2], \tag{1}$$

where

$$\gamma_k^2 = \omega_k/\hbar, \tag{2}$$

n_k is the vibrational quantum number in the kth mode, and H_k is a Hermite polynomial of order n_k. Therefore

$$\Psi_{n_1 n_2} \ldots (Q_1,\ Q_2,\ \ldots) = \prod_k^{3N-n} \Psi_{n_k}(Q_k), \tag{3}$$

or (in the occupation number representation)

$$|n_1\ n_2\ \ldots\ n_k\ \ldots\rangle = |n_1\rangle|n_2\rangle \ldots |n_k\rangle\ \ldots = \prod_{k=1}^{3N-n} |n_k\rangle. \tag{4}$$

Generally, most molecules can be assumed to be in their ground vibrational state at room temperature, in which case $n_k = 0$ for $k = 1, 2, \ldots, 3N - n$. Since H_0 is a constant, the ground state wave function is a product of exponential terms and so it is proportional to

$$\exp[-\tfrac{1}{2} \sum_k \gamma_k{}^2 Q_k{}^2], \tag{5}$$

which is invariant under any point symmetry operation.

Exercise 9.2-1 Justify the above statement about the invariance of eq. (5) under point symmetry operations.

The symmetry properties of eq. (3) in an excited state are determined by a product of Hermite polynomials. The most common vibrational transition, called a *fundamental* transition, is one in which only a single vibrational mode is excited, so that

$$\Delta n_j = 0, \quad \forall j \neq k, \quad \Delta n_k = 1, \tag{6}$$

$$|0\ 0 \ldots 0 \ldots 0\rangle \to |0\ 0 \ldots 1 \ldots 0\rangle. \tag{7}$$

The Hermite polynomial $H_1(\gamma_k Q_k) = 2\gamma_k Q_k$, so that for a fundamental transition the ground state forms a basis for Γ_1 and the first excited state transforms like Q_k. The spacing of the vibrational energy levels is such that transitions between vibrational states are induced by electromagnetic radiation in the infra-red region of the electromagnetic spectrum. The operator responsible is the dipole moment operator **D**, and so a fundamental transition $\Delta n_k = 1$ is allowed only if the matrix element

$$\langle \underset{n_k'}{0\ 0 \ldots 1 \ldots 0} | \mathbf{D} | \underset{n_k}{0\ 0 \ldots 0 \ldots 0} \rangle \neq 0. \tag{8}$$

Consequently, fundamental vibrational transitions are allowed only if the normal coordinate Q_k for that mode forms a basis for the same IR as x, y, or z.

Example 9.2-1 Find the number and degeneracy of the allowed infra-red transitions in a planar ML_3 molecule. ("Infra-red" transitions without further qualification implies a *fundamental* transition.)

From the character table for D_{3h} we observe that z forms a basis for A_2'' and x, y form a basis for E'. Since there are normal modes of symmetry $A_2'' \oplus 2E'$ we expect to see three bands in the infra-red absorption spectrum, two of which are doubly degenerate and therefore might be split in a lower symmetry. The normal mode of A_1' symmetry is inactive in the infra-red absorption spectrum.

9.2.1 *Anharmonicity*

The vibrational energy in a single mode in the harmonic approximation is

$$E_k = \hbar\omega_k(n_k + \tfrac{1}{2}). \tag{9}$$

Thus this model predicts an infinite sequence of evenly spaced levels with no allowance for dissociation. An approximation to the effective potential for a diatomic molecule $\Phi(R)$ proposed by Morse has proved to be extremely useful. The Morse potential

$$\Phi(R) = D_e\{1 - \exp[-a(R - R_0)]\}^2 \equiv D_e(1 - x)^2 \tag{10}$$

has a minimum value of zero at the equilibrium separation R_0, and $\Phi(R) \rightarrow D_e$ (the dissociation energy) as $R \rightarrow \infty$. The Schrödinger equation with a Morse potential is soluble, and yields energy eigenvalues

$$E = \hbar\omega(n + \tfrac{1}{2}) - x\hbar\omega(n + \tfrac{1}{2})^2, \tag{11}$$

where the anharmonicity constant

$$x = \hbar\omega/4D_e. \tag{12}$$

Equation (11) fits experimental data quite accurately.

9.2.2 *Overtones and combination bands*

Most infra-red spectra will show more bands than those predicted from an analysis of the fundamental transitions, although the intensity of these extra bands is usually less than that of the fundamental bands. *Combination bands* are due to the simultaneous excitation of more than one vibration. Suppose that $n_i', n_j' = 1$, $n_k' = 0$, $\forall\, k \neq i, j$, then the symmetry of the excited state is given by

$$\Gamma = \Gamma(Q_i) \otimes \Gamma(Q_j), \tag{13}$$

which may be reduced in the usual way if $\Gamma(Q_i)$, $\Gamma(Q_j)$ are degenerate. Degenerate levels may be split by anharmonic coupling.

Example 9.2-2 Acetylene HC≡CH is a linear molecule with point group $D_{\infty h}$ and has $3N - 5 = 7$ normal modes, which an analysis like that in Example 9.1-1 shows to be of symmetry

$$\Gamma_v = 2\Sigma_g^+ \oplus \Sigma_u^+ \oplus \Pi_g \oplus \Pi_u. \tag{14}$$

Suppose that a combination tone involves a single excitation of modes of Π_g and Π_u symmetry. Then, since

$$\Pi_g \otimes \Pi_u = \Sigma_u^+ \oplus \Sigma_u^- \oplus \Delta_u, \tag{15}$$

this excited state consists of one doubly degenerate and two non-degenerate levels, only the Σ_u^+ mode being infra-red active.

Overtones occur when $\Delta n_k > 1$, $\Delta n_k = 2$ being called the first overtone and so on. If only modes of one frequency are excited, and if this frequency is non-degenerate, then the excited-state wave function forms a basis for a non-degenerate representation, the characters of which are all ± 1. Therefore $R^{-1}Q_k = \pm Q_k$ and all state functions containing only even powers of Q_k (those for n_k even) belong to the totally symmetric representation, while those containing only odd powers of Q_k (those for n_k odd) belong to the same IR as Q_k. For double excitation of a degenerate mode, the characters of the symmetric direct product (DP) are given by

(5.3.22)
$$\chi(\Gamma^k \overline{\otimes} \Gamma^k) = \tfrac{1}{2}[(\chi^k(R))^2 + \chi^k(R^2)]. \tag{16}$$

In general for n quanta in the mode (that is, the $(n-1)$th overtone), eq. (16) generalizes to

$$\chi_n(\Gamma^k_{n-1} \overline{\otimes} \Gamma^k) = \tfrac{1}{2}[\chi^k_{n-1}(R)\chi^k(R) + \chi^k(R^n)]. \tag{17}$$

Example 9.2-3 If the Π_g mode of acetylene is doubly excited

(16)
$$\Pi_g \overline{\otimes} \Pi_g = \Sigma_g^+ \oplus \Delta_g \tag{18}$$

and the overtone state is split into a non-degenerate Σ_g^+ level and a doubly degenerate Δ_g level.

Answer to Exercise 9.2-1

Consider a reduction in symmetry until all representations are reduced to 1-D IRs. Then the character in any class can only be ± 1. Consequently, Q_k^2 is invariant under all the operators of the point group and so belongs to Γ_1, which correlates with the totally symmetric representation of the point symmetry group of the molecule. Therefore $\sum_k \gamma_k^2 Q_k^2$ is invariant under any of the operators of the point group of the molecule.

9.3 Inelastic Raman scattering

In inelastic Raman scattering a photon loses (or gains) one quantum of rotational or vibrational energy to (or from) the molecule. The process involves the electric field of the radiation inducing an electric dipole in the molecule and so depends on the polarizability tensor of the molecule. (A (second-order) tensor is a physical quantity with nine components.) The induced electric dipole **D** is proportional to the electric field **E**:

$$\mathbf{D} = \alpha \mathbf{E}, \tag{1}$$

the coefficient of proportionality being the polarizability α. Since both **D** and **E** are vectors, the polarizability has nine components,

$$\begin{bmatrix} D_x \\ D_y \\ D_z \end{bmatrix} = \begin{bmatrix} \alpha_{xx} & \alpha_{xy} & \alpha_{xz} \\ \alpha_{yx} & \alpha_{yy} & \alpha_{yz} \\ \alpha_{zx} & \alpha_{zy} & \alpha_{zz} \end{bmatrix} \begin{bmatrix} E_x \\ E_y \\ E_z \end{bmatrix}, \tag{2}$$

but only six are independent because the polarizability tensor is symmetric. The components of the polarizability transform like binary products of coordinates $x^2, y^2, z^2, xy, yz, zx$. Hence a fundamental transition is *Raman active* if the normal mode forms a basis for one or more components of the polarizability.

Example 9.3-1 Determine the number of bands to be expected in the Raman spectrum of a planar ML_3 molecule.

From the character table for D_{3h}, components of the polarizability form bases for the IRs A_1', E', and E''. From eq. (9.1.4) the normal coordinates form bases for the representations A_1', $2E'$, and A_2''. Therefore the Raman active modes are A_1' and $2E'$, and there are three Raman bands, with two coincidences. (A "coincidence" means that a Raman band and an infra-red band have the same frequency.)

9.3.1 General features of Raman and infra-red spectra

(1) If the symmetry of a molecule is lowered by a perturbation (for example by the substitution of a foreign ion in a crystal lattice) this may remove degeneracies and/or permit transitions that were forbidden in the more symmetric molecule.
(2) The number of Raman and infra-red bands can sometimes be used to distinguish between various possible structures.
(3) Stretching modes (for example, that of the C═O bond) can often be analyzed separately because they occur in a characteristic region of the infra-red spectrum. In such cases a full vibrational analysis is not necessary and one can simply study the transformation of unit vectors directed along the bonds in question. For example, in the molecule $ML_3(CO)_3$ shown in Figure 9.2 it is only necessary to study the transformation of the three unit vectors directed along the C═O bonds in order to determine the number of stretching modes. A stretching-mode analysis is often sufficient to distinguish between possible structures (see Problem 9.3).

9.4 Determination of the normal modes

Normal mode coordinates are linear combinations of the atomic displacements $\{x_i \ y_i \ z_i\}$, which are the components of a set of vectors $\{\mathbf{Q}\}$ in a $3N$-dimensional vector space called

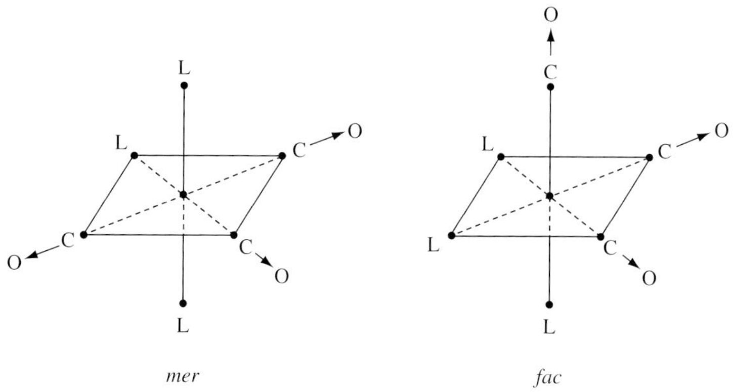

Figure 9.2. *mer* and *fac* isomers of $ML_3(CO)_3$.

displacement vector space, with basis vectors $\{\mathbf{e}_{01}\ \mathbf{e}_{02}\ \ldots\ \mathbf{e}_{N-1,\,3}\} = \{\mathbf{e}_{ij}\}$, where $i = 0, 1, \ldots, N-1$ labels the atoms and $j = 1, 2, 3$ is the orientation of each one of the subset of three orthogonal unit vectors parallel to OX, OY, OZ and centered at the equilibrium positions of the atoms (see, for example, Figure 9.1). A symmetry operation induced by the symmetry operator $R \in G$, which interchanges like particles, transforms the basis $\langle \mathbf{e}_{ij}|$ to $\langle \mathbf{e}_{ij}'|$,

$$R\langle \mathbf{e}_{ij}| = \langle \mathbf{e}_{ij}'| = \langle \mathbf{e}_{ij}|\Gamma_{\mathrm{disp}}(R) \tag{1}$$

The MR of R, $\Gamma_{\mathrm{disp}}(R)$, is a $3N{\times}3N$ matrix which consists of N $3{\times}3$ blocks labeled Γ^{lm} which are non-zero only when R transforms atom l into atom m, and then they are identical with the MR for an orthonormal basis $\{\mathbf{e}_1\ \mathbf{e}_2\ \mathbf{e}_3\}$ in 3-D space. Since a $3{\times}3$ matrix Γ^{lm} occurs on the diagonal of $\Gamma_{\mathrm{disp}}(R)$ only when $l = m$, it is a straightforward matter to determine the character system for Γ_{disp} and hence the direct sum of IRs making up Γ_{disp} and which give the symmetry of the atomic displacements in displacement vector space, in which we are describing the motion of the atomic nuclei. This basis $\{\mathbf{e}_{ij}\}$ is not a convenient one to use when solving the equations of motion since both the potential energy Φ and the kinetic energy T contain terms that involve binary products of different coordinates $\{x_i\ y_i\ z_i\}$ or their time derivatives. However, an orthogonal transformation

$$\langle \mathbf{e}(\Gamma\gamma)| = \langle \mathbf{e}_{ij}|(\mathrm{A}^{\Gamma\gamma})^{-1} \tag{2}$$

to a new basis set $\{\mathbf{e}(\Gamma\gamma)\}$ can always be found in which both T and Φ are brought to diagonal form so that

$$2T = \sum_k \dot{Q}_k^2, \quad 2\Phi = \sum_k \omega_k^2 Q_k^2. \tag{3}$$

The $\{Q_k\}$, $k = 1, 2 \ldots, 3N$, are a set of *normal coordinates*, which are the components of $\mathbf{Q}(\Gamma\gamma)$ referred to the new basis $\{\mathbf{e}(\Gamma\gamma)\}$ in which Γ denotes one of the IRs and γ denotes the component of the IR Γ when it has a dimension greater than unity. The particle masses do not appear in T and Φ because they have been absorbed into the Q_k by the definition of the normal coordinates. A displacement vector $\mathbf{Q}$ is therefore

$$\mathbf{Q} = \langle \mathbf{e}_{ij} | q_{ij} \rangle = \langle \mathbf{e}_{ij} | (\mathbb{A}^{\Gamma\gamma})^{-1} \mathbb{A}^{\Gamma\gamma} | q_{ij} \rangle = \langle \mathbf{e}(\Gamma\gamma) | Q_k \rangle, \quad (4)$$

where $\{q_{ij}\}$implies the whole set of mass-weighted displacements. Similarly $\{\mathbf{e}_{ij}\}$ in eq. (4) implies the set of Cartesian unit vectors on each of the $i = 0, 1, \ldots, N-1$ atoms. Note that $\mathbb{A}^{\Gamma\gamma}$ is the orthogonal matrix which transforms the coordinates q_{ij} into the normal coordinates $\{Q_k\}$. The normal coordinates $\{Q_k\}$ form bases for the IRs, and therefore they will now be called $\{Q(\Gamma\gamma)\}$. We do not need to evaluate $\mathbb{A}^{\Gamma\gamma}$ explicitly since the $\{Q(\Gamma\gamma)\}$ may be found by projecting an arbitrary one of the $\{q_{ij}\}$ into the appropriate Γ subspace,

$$Q(\Gamma\gamma) = N(\Gamma\gamma) \sum_R \chi_\Gamma(R)^* \hat{R} q_{ij}. \quad (5)$$

Here $q_{ij} = M_i^{1/2} x_{ij}$, M_i is the mass of atom i, and x_{ij} is the jth component of the displacement of atom i. The procedure must be repeated for each of the IRs (labeled here by Γ); $N(\Gamma\gamma)$ is a normalization factor. The projection needs to be carried out for a maximum of three times for each IR, but in practice this is often performed only once, if we are able to write down by inspection the other components $Q(\Gamma\gamma)$ of degenerate representations. It is, in fact, common practice, instead of using eq. (5), to find the transformed basis

(4) $$|\mathbf{e}(\Gamma\gamma)\rangle = \mathbb{A}^{\Gamma\gamma} |\mathbf{e}_{ij}\rangle \quad (6)$$

by projecting instead one of the $\{\mathbf{e}_i\}$ and then using the fact that $Q(\Gamma\gamma)$ is given by the same linear combination of the $\{q_{ij}\}$ as $\mathbf{e}(\Gamma\gamma)$ is of the $\{\mathbf{e}_{ij}\}$ (cf. eqs. (6) and (4)). The absolute values of the displacements are arbitrary (though they are assumed to be small in comparison with the internuclear separations) but their relative values are determined by symmetry.

There is a complication if the direct sum of IRs contains a particular representation Γ more than once, for then we must take linear combinations of the $\mathbf{e}(\Gamma\gamma)$ for this IR by making a second orthogonal transformation. This second transformation is not fully determined by symmetry, even after invoking orthogonality conditions. This is a common situation when bases for different representations of the same symmetry are combined: the linear combinations are given by symmetry, but not the numerical coefficients, the determination of which requires a separate quantum or classical mechanical calculation. (We met a similar situation when combining linear combinations of ligand orbitals with central-atom atomic orbitals (AOs) that formed bases for the same IRs in the molecular orbital (MO) theory described in Chapter 6.) Because of the assumed quadratic form for the potential energy Φ (by cutting off a Taylor expansion for Φ at the second term, valid for small-amplitude oscillations) the time dependence of the normal coordinates is simple harmonic.

Example 9.4-1 Determine the normal coordinates for the even parity modes of the ML_6 molecule or complex ion with O_h symmetry.

A diagram of the molecule showing the numbering system used for the atoms is given in Figure 9.3. Set up basis vectors $\{\mathbf{e}_{i1}\ \mathbf{e}_{i2}\ \mathbf{e}_{i3}\}$, $i = 0, 1, \ldots, 6$, on each of the seven atoms as shown in the figure. Table 9.2 shows the classes $\mathscr{C}_p$ of the point group of the molecule, the number of atoms N_R that are invariant under the symmetry operator $R \in \mathscr{C}_p$, the

Table 9.2. *Derivation of the symmetry of the normal modes of vibration for the* ML_6 *molecule or complex ion with* O_h *symmetry.*

The rows of this table contain (1) the classes of $O_h = O \otimes C_i$; (2) one element R from each class; (3) the number of atoms of type M invariant under R; (4) the number of atoms N_R of type L that are invariant under R; (5) the matrix representative of R for the basis $\{\mathbf{e}_{i1}\ \mathbf{e}_{i2}\ \mathbf{e}_{i3}\}$; (6) the character χ_r of this MR Γ_r; (7) the character $\chi = N_R \times \chi_r$ of the 18×18 MR of L_6; (8) the character of the 21×21 MR for the whole basis. $\phi = 2\pi/3$, $\mathbf{n} = 3^{-1/2}[1\ 1\ 1]$, $\mathbf{a} = 2^{-1/2}[1\ 1\ 0]$.

E	$3C_2$	$8C_3$	$6C_4$	$6C_2'$	I	$3\sigma_h$	$8S_6$	$6S_4$	$6\sigma_d$
E	$C_{2\mathbf{z}}$	$R(\phi\ \mathbf{n})$	$C_{4\mathbf{z}}^+$	$C_{2\mathbf{a}}$	I	$\sigma_{\mathbf{z}}$	$IR(\phi\ \mathbf{n})$	$S_{4\mathbf{z}}^-$	$\sigma_{\mathbf{a}}$
1	1	1	1	1	1	1	1	1	1
6	2	0	2	0	0	4	0	0	2
$\begin{bmatrix} 1 & & \\ & 1 & \\ & & 1 \end{bmatrix}$	$\begin{bmatrix} \bar{1} & & \\ & \bar{1} & \\ & & 1 \end{bmatrix}$	$\begin{bmatrix} & & 1 \\ 1 & & \\ & 1 & \end{bmatrix}$	$\begin{bmatrix} & \bar{1} & \\ 1 & & \\ & & 1 \end{bmatrix}$	$\begin{bmatrix} & 1 & \\ 1 & & \\ & & \bar{1} \end{bmatrix}$	$\begin{bmatrix} \bar{1} & & \\ & \bar{1} & \\ & & \bar{1} \end{bmatrix}$	$\begin{bmatrix} 1 & & \\ & 1 & \\ & & \bar{1} \end{bmatrix}$	$\begin{bmatrix} & & \bar{1} \\ \bar{1} & & \\ & \bar{1} & \end{bmatrix}$	$\begin{bmatrix} & 1 & \\ \bar{1} & & \\ & & \bar{1} \end{bmatrix}$	$\begin{bmatrix} & \bar{1} & \\ \bar{1} & & \\ & & 1 \end{bmatrix}$
3	−1	0	1	−1	−3	1	0	−1	1
18	−2	0	2	0	0	4	0	0	2
21	−3	0	3	−1	−3	5	0	−1	3

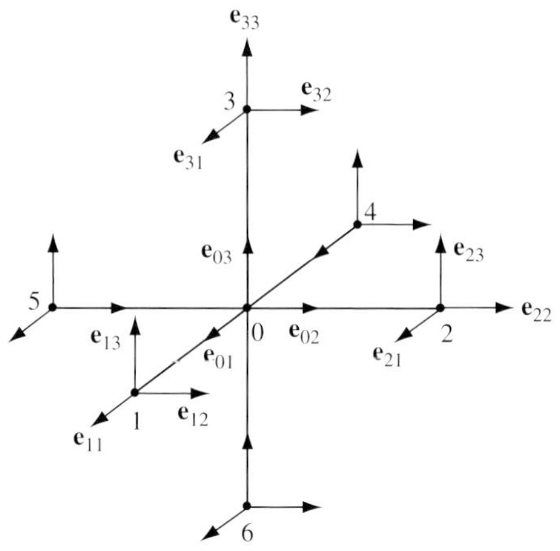

Figure 9.3. Basis vectors used to describe the atomic displacements in the ML_6 molecule, showing the numbering system used for the seven atoms. (Labeling of the unit vectors at atoms 4, 5, and 6 is not shown explicitly to avoid overcrowding the figure.)

submatrices Γ_r, the character set $\{\chi_r\}$, the characters for the 18×18 reducible matrix representation for L_6, and finally the characters for the 21×21 reducible matrix representation Γ. This representation may be reduced in the usual way to yield the direct sum

$$\Gamma = \mathrm{A}_{1g} \oplus \mathrm{E}_g \oplus \mathrm{T}_{1g} \oplus \mathrm{T}_{2g} \oplus 3\mathrm{T}_{1u} \oplus \mathrm{T}_{2u}, \tag{7}$$

with a total degeneracy of 21. The character table for O_h shows that the three rotations form a basis for T_{1g} and subtracting off T_{1g} from the direct sum in eq. (7) leaves

$$\Gamma_v \oplus \Gamma_t = \mathrm{A}_{1g} \oplus \mathrm{E}_g \oplus \mathrm{T}_{2g} \oplus 3\mathrm{T}_{1u} \oplus \mathrm{T}_{2u} \tag{8}$$

as the representations to which the $3N - 6 = 15$ normal modes and the three translations belong. We may not separate off Γ_t since there are in all three degenerate modes of T_{1u} symmetry, two vibrational modes and one translational mode. We now apply the projection operator in eq. (5) for the three even-parity representations $A_{1g} \oplus E_g \oplus T_{2g}$ to obtain

$$\mathrm{A}_{1g} = {}^{1}\!/\!\sqrt{6}[x_1 - x_4 + y_2 - y_5 + z_3 - z_6], \tag{9}$$

$$\mathrm{E}_g(\mathrm{u}) = {}^{1}\!/\!\sqrt{12}[2z_3 - 2z_6 - x_1 + x_4 - y_2 + y_5], \tag{10}$$

$$\mathrm{E}_g(\mathrm{v}) = \tfrac{1}{2}[x_1 - x_4 - y_2 + y_5], \tag{11}$$

$$\mathrm{T}_{2g}(\xi) = \tfrac{1}{2}[z_2 - z_5 + y_3 - y_6], \tag{12}$$

$$\mathrm{T}_{2g}(\eta) = \tfrac{1}{2}[z_1 - z_4 + x_3 - x_6], \tag{13}$$

$$\mathrm{T}_{2g}(\zeta) = \tfrac{1}{2}[y_1 - y_4 + x_2 - x_5]. \tag{14}$$

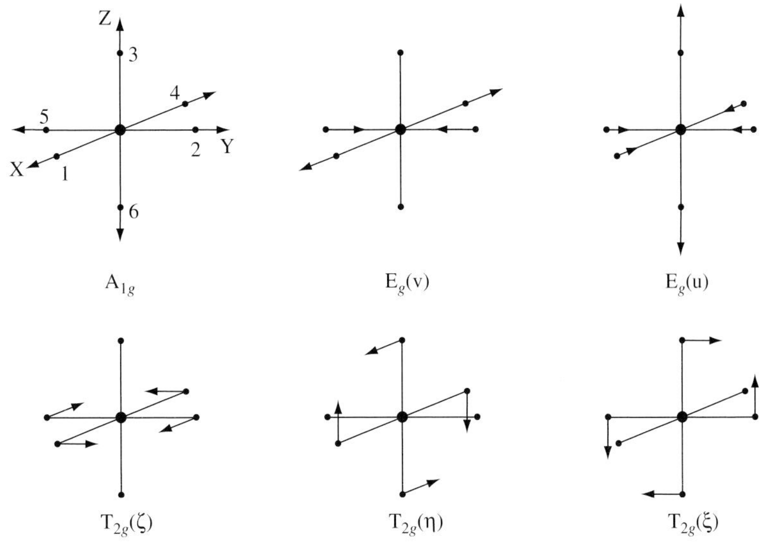

Figure 9.4. Normal modes of vibration of the even-parity modes of ML_6 in O_h symmetry. Arrows show the phases and relative magnitudes of the displacements, but the actual displacements have been enlarged for the sake of clarity.

The normal mode displacements are sketched in Figure 9.4. The notation u, v for the degenerate pair of E_g symmetry and ξ, η, ζ for the T_{2g} triplet is standard. Actually, these projections had already been done in Section 6.4, but this example has been worked in full here to illustrate the projection operator method of finding normal modes.

Similarly, for the odd-parity T_{2u} modes,

(6) or (6.4.42) $$T_{2u}(\xi) = ½[x_3 + x_6 - x_2 - x_5], \qquad (15)$$

(6) or (6.4.43) $$T_{2u}(\eta) = ½[y_1 + y_4 - y_3 - y_6], \qquad (16)$$

(6) or (6.4.44) $$T_{2u}(\zeta) = ½[z_1 + z_4 - z_2 - z_5]. \qquad (17)$$

Equation (7) tells us that there are, in all, three independent motions of T_{1u} symmetry, the three independent components of each set being designated $T_{1u}(x)$, $T_{1u}(y)$, and $T_{1u}(z)$. Omitting normalization factors, the ligand contributions to the z components of the normal coordinates of two of these modes are

(5) $$Q(T_{1u},\ z,\ 1) = q_{13} + q_{23} + q_{43} + q_{53}, \qquad (18)$$

(5) $$Q(T_{1u},\ z,\ 2) = q_{33} + q_{63}. \qquad (19)$$

The third one is just the z component for the central atom

$$Q(T_{1u},\ z,\ 3) = q_{03}. \qquad (20)$$

Superimposing the normal coordinates in eqs. (18), (19), and (20) with equal weight and phase gives

$$Q(\mathrm{T}_{1u},\ z,\ \mathrm{III}) = [q_{03} + q_{13} + q_{23} + q_{33} + q_{43} + q_{53} + q_{63}]. \tag{21}$$

Two more linear combinations of eqs. (18), (19), and (20) give the coordinates

$$Q(\mathrm{T}_{1u},\ z,\ \mathrm{I}) = [-q_{03} + a(q_{13} + q_{23} + q_{43} + q_{53}) - b(q_{33} + q_{63})], \tag{22}$$

$$Q(\mathrm{T}_{1u},\ z,\ \mathrm{II}) = [+q_{03} + a'(q_{13} + q_{23} + q_{43} + q_{53}) - b'(q_{33} + q_{63})]. \tag{23}$$

There are three orthogonality conditions between the $Q(T_{1u},\ z)$ normal coordinates but four unknown constants in eqs. (22) and (23), so this is as far as one can go without a model for the adiabatic potential.

Problems

9.1 The observed infra-red spectrum of ozone contains three fundamental bands at frequencies 705 cm^{-1}, 1043 cm^{-1}, and 1110 cm^{-1}. Use this information to decide which of I, II, and III in Figure 9.5 are possible structures for ozone. Predict what you would expect to find in the Raman spectrum of ozone.

9.2 The chromate ion CrO_4^{2-} has the shape of a tetrahedron. Deduce the symmetries of the normal modes and explain which of these are infra-red active and which are Raman active.

9.3 The following bands were found in the region of the spectrum of OsO_4N (N denotes pyridine) associated with the stretching of Os—O bonds:

$$\text{infra-red } \nu/\mathrm{cm}^{-1} = 926,\ 915,\ 908,\ 885,$$

$$\text{Raman } \nu/\mathrm{cm}^{-1} = 928(p),\ 916(p),\ 907(p),\ 886(dp),$$

where p indicates that the scattered Raman radiation is polarized and therefore can only be due to a totally symmetric vibration, and similarly dp indicates that the Raman band at 886 cm^{-1} is depolarized and therefore not associated with a totally symmetric vibration. Four possible structures of OsO_4N are shown in Figure 9.5, in each of which the four arrows indicate unit vectors along the direction of the Os—O stretching mode. State the point group symmetry of each of the four structures and determine the number of allowed infra-red and Raman bands associated with Os—O stretching in each structure, the number of coincidences, and whether the Raman bands are polarized. Hence decide on the structure of OsO_4N. [*Hint*: It is not necessary to determine the symmetries of all the normal modes.]

9.4 When XeF_4 was first prepared it was thought to be highly symmetrical, but it was not known whether it was a tetrahedral or a square-planar molecule. The infra-red absorption spectrum of XeF_4 consists of three fundamental bands and the vibrational Raman spectrum also has three bands. Determine the symmetry of the normal modes of a

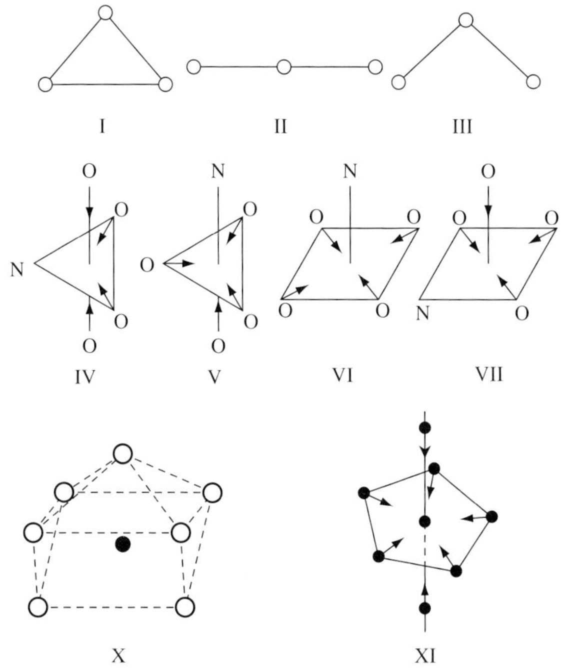

Figure 9.5. Three possible structures of ozone, I, II, and III. Four possible structures of OsO_4N (N = pyridine), IV, V, VI, and VII. Two possible structures for $Mo(CN)_7{}^{4-}$, X and XI.

square planar AB_4 molecule, and hence show that the above evidence is consistent with a square-planar configuration for XeF_4.

9.5 The infra-red spectrum of $Mo(CO)_3[P(OCH_3)_3]_3$ (VIII) shows three absorption bands at 1993, 1919, and 1890 cm^{-1} in the region in which CO stretching frequencies usually appear. But $Cr(CO)_3(CNCH)_3$ (IX) has two absorption bands in the C—O stretch region at 1942 and 1860 cm^{-1}. Octahedral $ML_3(CO)_3$ complexes can exist in either the *mer* or *fac* isomeric forms (Figure 9.2). Assign the structures of the above two molecules. How many bands would you expect to see in the vibrational Raman spectra of these two molecules, and for which of these bands would the scattered Raman radiation be polarized?

9.6 Two important geometries for seven-coordinate complex ions are the mono-capped trigonal prism (X) and the pentagonal bipyramid (XI) (Figure 9.5). Infra-red spectra have been measured for the seven-coordinate complex ion $Mo(CN)_7^{4-}$ as solid $K_4Mo(CN)_7.2H_2O$ and in aqueous solution. In the C—N stretching region the infra-red spectrum shows six bands at 2119, 2115, 2090, 2080, 2074, and 2059 cm^{-1} for the solid, and two bands at 2080 and 2040 cm^{-1} for solutions. How many Raman and infra-red bands would you expect for (X) and (XI)? What conclusions can be drawn from the experimental data given? How many Raman bands are to be expected for the solid and the solution?

9.7 The NO_3^- ion is planar (like CO_3^{2-}), but when NO_3^- is dissolved in certain crystals (called type 1 and type 2) it is observed that all four modes become both Raman active

Table 9.3.

D_{3h}	C_{3v}	C_s
A_1'	A_1	A'
A_2''	A_1	A''
E'	E	$2A'$

Table 9.4.

CO stretching frequencies/cm^{-1}	
Infra-red absorption	2028, 1994
Vibrational Raman scattering	2114, 2031, 1984

and infra-red active. In crystals of type 1 there is no splitting of degenerate modes, but in type 2 crystals the degenerate modes of NO_3^- are split. Suggest an explanation for these observations. [*Hint*: Character tables are given in Appendix A3. Table 9.3 is an extract from a correlation table for D_{3h}.]

9.8 Two likely structures for $Fe(CO)_5$ are the square pyramid and the trigonal bipyramid. Determine for both these structures the number of infra-red-active and Raman-active C—O stretching vibrations and then make use of the data given in Table 9.4 to decide on the structure of $Fe(CO)_5$.

10 Transitions between electronic states

10.1 Selection rules

As noted in Section 5.4, the transition probability between two electronic states is proportional to the square of the modulus of the matrix element

(5.4.10) $$\langle \Psi_u^k | \hat{Q}_s^j | \Psi_q^i \rangle = \langle \psi_u^k | \hat{Q}_s^j | \psi_q^i \rangle \langle \chi_u | \chi_q \rangle, \tag{1}$$

where Ψ_u^k is the uth spin orbital $\psi_u^k(\{\mathbf{r}_\mathrm{p}\})\chi_u(\{m_\mathrm{sp}\})$ belonging to the kth (possibly degenerate) (IR) of the symmetry point group (or double group) and $\hat{Q}_s^j$ is an operator that belongs to the jth IR; $\{\mathbf{r}_\mathrm{p}\}$ denotes a set of position vectors for the electrons, $\{m_\mathrm{sp}\}$ denotes the arguments of the spinor χ_u, and u is a (total) spin quantum number. In the approximation that neglects spin–orbit coupling $\hat{H}_{\mathrm{S.L}}$, because of the orthogonality of the spin functions the only allowed transitions are those for which $\Delta S = 0$. Generally, $\hat{H}_{\mathrm{S.L}}$ is not negligible and transitions in which the spin quantum number S is not conserved may occur, but with weaker intensity than those in which the spin selection rule $\Delta S = 0$ is obeyed. When the interaction of the electronic system with the radiation field is analyzed using time-dependent perturbation theory (see, for example, Flygare (1978) or Griffith (1964)) it is found that the strongest transitions are the E1 transitions for which Q is the dipole moment operator, with components x, y, z; the next strongest transitions are the M1 transitions in which Q is the magnetic dipole operator, with components R_x, R_y, R_z; while the weakest transitions are the electric quadrupole or E2 transitions in which the E2 operator transforms like binary products of the coordinates $x^2, y^2, z^2, xy, yz, zx$. In systems with a center of symmetry, the components of the dipole moment operator belong to ungerade representations. Therefore the only allowed E1 transitions are those which are accompanied by a change in parity, $g \leftrightarrow u$. This parity selection rule is known as the *Laporte rule*. As we shall see, it may be broken by vibronic interactions.

Example 10.1-1 Discuss the transitions which give rise to the absorption spectrum of benzene.

A preliminary analysis of the absorption spectrum was given in Example 5.4-1 as an illustration of the application of the direct product (DP) rule for evaluating matrix elements, but the analysis was incomplete because at that stage we were not in a position to deduce the symmetry of the electronic states from electron configurations, so these were merely stated. A more complete analysis may now be given. The molecular orbitals (MOs)

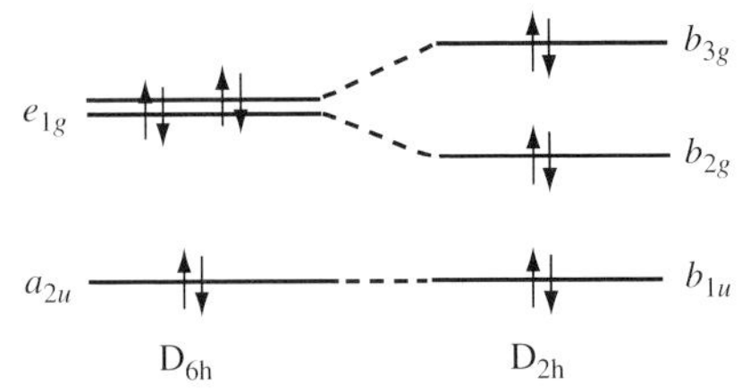

Figure 10.1. Effect of a lowering in symmetry on the occupied energy levels of benzene.

Box 10.1. Reduction of the direct product representation $E_{1g} \otimes E_{2u}$.

$$E_{1g} \otimes E_{2u} = \{4 \quad -1 \quad 1 \quad -4 \quad 0 \quad 0 \quad -4 \quad 1 \quad -1 \quad 4 \quad 0 \quad 0\}.$$

This DP representation must be a sum of u representations:

$$c(A_{1u}) = (^1\!/_{24})[4 - 2 + 2 - 4 + 4 - 2 + 2 - 4] = 0,$$
$$c(A_{2u}) = (^1\!/_{24})[4 - 2 + 2 - 4 + 4 - 2 + 2 - 4] = 0,$$
$$c(B_{1u}) = (^1\!/_{24})[4 + 2 + 2 + 4 + 4 + 2 + 2 + 4] = 1,$$
$$c(B_{2u}) = (^1\!/_{24})[4 + 2 + 2 + 4 + 4 + 2 + 2 + 4] = 1,$$
$$c(E_{1u}) = (^1\!/_{24})[8 - 2 - 2 + 8 + 8 - 2 - 2 + 8] = 1.$$

Therefore, $E_{1g} \otimes E_{2u} = B_{1u} \oplus B_{2u} \oplus E_{1u}$.

of benzene were deduced in Example 6.2-1 and Problem 6.2, and an energy-level diagram is given in Figure 6.4. This figure shows that the ground-state electron configuration of benzene is $(a_{2u})^2(e_{1g})^4$. Consider a lowering in symmetry to D_{2h}. As shown in Figure 10.1, this results in a splitting of the degenerate e_{1g} level into b_{2g} and b_{3g}. In D_{2h} symmetry the electron configuration would be $(b_{1u})^2(b_{2g})^2(b_{3g})^2$. Since the product of two functions, each of which is the basis for a particular IR, forms a basis for the DP representation of these two IRs, and the DP of any 1-D representations with itself is the totally symmetric representation, it follows that the ground-state wave function for benzene forms a basis for Γ_1, which in D_{2h} is A_g. Since each level contains two electrons with paired spins, the ground state is ^1A_g in D_{2h}, which correlates with $^1A_{1g}$ in D_{6h}. This argument is quite general and holds for *any closed shell* molecule, atom, or ion. After a lowering in symmetry until all the orbital functions form bases for 1-D representations, the only state for *any closed shell configuration* that satisfies the Pauli exclusion principle is $^1\Gamma_1$, which correlates with $^1\Gamma_1$ of the original group. The first excited state of benzene is $(a_{2u})^2(e_{1g})^3(e_{2u})^1$. The doubly occupied levels contribute $^1A_{1g}$ so only the singly occupied levels need be included in the DP. From Box 10.1, $E_{1g} \otimes E_{2u} = B_{1u} \oplus B_{2u} \oplus E_{1u}$. In D_{6h} $(x\ y)$ belong to E_{1u} and z belongs to A_{2u}, so $^1A_{1g} \rightarrow {}^1E_{1u}$ is the only symmetry- and spin-allowed E1 transition, and this will be excited by radiation polarized in the xy plane. This transition is responsible for the strongest absorption band in the absorption spectrum of benzene at 1800 Å. The very weak band at 3500 Å is due to $^1A_{1g} \rightarrow {}^3E_{1u}$, allowed (though with low intensity) through spin–orbit coupling. The spin–orbit coupling constant is proportional to $\alpha^2 Z^2$, where α is the fine-structure constant equal to 1/137.036, and so transitions allowed through spin–orbit coupling will be very weak in molecules comprising atoms of low atomic number Z. The transitions $^1A_{1g} \rightarrow {}^1B_{1u}$ and $^1A_{1g} \rightarrow {}^1B_{2u}$ are symmetry forbidden in a rigid molecule, but may become allowed through vibronic coupling.

10.2 Vibronic coupling

The electronic state function $\mathbf{\Psi}_a(\mathbf{r}, \mathbf{R})$ depends not only on electron coordinates $\{\mathbf{r}\}$ but also on the nuclear coordinates $\{\mathbf{R}\}$. The subscript a denotes a set of electronic quantum numbers. Because the mass of the electrons is much smaller than the mass of the nuclei, the electron motion follows the motion of the nuclei *adiabatically*, so it is customary to adopt the Born–Oppenheimer approximation, as a result of which the state function may be written as a product of electronic and nuclear state functions:

$$\Psi_a(\mathbf{r},\ \mathbf{R}) = \psi_{a,\mathbf{R}}(\mathbf{r})\ \phi_{a,\alpha}(\mathbf{R}), \tag{1}$$

where α denotes a set of vibrational quantum numbers. The electronic state function depends parametrically on the positions of the nuclei, and this is indicated by the subscript $\mathbf{R}$. The electronic energy $E_a(\mathbf{R})$, calculated at a series of values of the nuclear displacements $\{\mathbf{R}\}$, is the potential energy $U_a(\mathbf{R})$ for the vibrational motion. $U_a(\mathbf{R})$, which depends on the electronic state a, is called the *adiabatic potential.* Being a property of the molecule, it is invariant under any symmetry operator of the molecular point group. With $\Delta S = 0$, and making use of the orthonormal property of the spin functions, the matrix element (10.1.1) for a vibrating molecule becomes

(1) $$\langle \Psi_{a'} | \mathbf{D} | \Psi_a \rangle = \langle \psi^l \phi^k | \mathbf{D} | \psi^j \phi^i \rangle \tag{2}$$

which equals zero, unless the DP

$$\Gamma^l \otimes \Gamma^k \otimes \Gamma^j \otimes \Gamma^i \otimes \Gamma(x,y,z) \supset \Gamma_1. \tag{3}$$

For a fundamental vibrational transition $\Gamma^i = \Gamma_1$, and Γ^k is one of the representations $\Gamma(Q_k)$ to which the normal modes belong, so that the vibronic problem reduces to answering the question, does

(1) $$\Gamma^l \otimes \Gamma^j \otimes \Gamma(x,y,z) \supset \Gamma(Q_k), \tag{4}$$

where $\Gamma(Q_k)$ is one or more of the IRs for which the normal modes form a basis? The vibronic interaction is a perturbation term $\hat{H}_{\text{en}}$ in the Hamiltonian, and so transitions that are symmetry forbidden but vibronically allowed can be expected to be weaker than the symmetry-allowed transitions. Consequently, consideration of vibronic transitions is usually limited to E1 transitions. If a transition is symmetry allowed then the vibronic interaction $\hat{H}_{en}$ broadens the corresponding spectral line into a broad band, and this is the reason why absorption and emission spectra consist of broad bands in liquids and solids.

Example 10.2-1 Find if any of the symmetry-forbidden transitions in benzene can become vibronically allowed, given that in the benzene molecule there are normal modes of B_{2g} and E_{2g} symmetry.

In D_{6h} symmetry the dipole moment operator forms a basis for the representations $A_{2u} \oplus E_{1u}$. The ground-state electronic state function belongs to A_{1g} and

$$B_{1u} \otimes (A_{2u} \oplus E_{1u}) = B_{2g} \oplus E_{2g}, \tag{5}$$

$$B_{2u} \otimes (A_{2u} \oplus E_{1u}) = B_{1g} \oplus E_{2g}. \tag{6}$$

Since there are normal modes of B_{2g} and E_{2g} symmetry, both the transitions ${}^1A_{1g} \rightarrow {}^1B_{1u}$ and ${}^1A_{1g} \rightarrow {}^1B_{2u}$ (which are forbidden by symmetry in a rigid molecule) become allowed through vibronic coupling. These transitions account for the two weaker bands in the benzene spectrum at 2000 and 2600 Å.

Example 10.2–2 The ground-state configuration of an nd^1 octahedral complex is t_{2g}^1, and the first excited configuration is e_g^1 so that optical transitions between these two configurations are symmetry-forbidden by the parity selection rule. Nevertheless, $Ti(H_2O)_6^{+3}$ shows an absorption band in solution with a maximum at about 20 000 cm^{-1} and a marked "shoulder" on the low-energy side of the maximum at about 17 000 cm^{-1}. Explain the source and the structure of this absorption band.

From the character Table for O_h in Appendix A3, we find that the DP $T_{2g} \otimes E_g = T_{1g} \oplus T_{2g}$ does not contain $\Gamma(x, y, z) = T_{1u}$, so that the transition $t_{2g}^1 \rightarrow e_g^1$ is symmetry-forbidden (parity selection rule). Again using the character table for O_h,

(4) $$T_{2g} \otimes E_g \otimes T_{1u} = A_{1u} \oplus A_{2u} \oplus 2E_u \oplus 2T_{1u} \oplus 2T_{2u}. \tag{7}$$

The normal modes of ML_6 form bases for $A_{1g} \oplus E_g \oplus T_{2g} \oplus 2T_{1u} \oplus T_{2u}$. Since the DP of two g representations can give only g IRs, we may work temporarily with the group O:

$$T_{2g} \otimes E_g = \{6\ {-2}\ 0\ 0\ 0\} = T_{1g} \oplus T_{2g}.$$

The DP does not contain $\Gamma(\mathbf{r}) = T_{1u}$, so the transition $t_{2g} \rightarrow e_g$ is symmetry-forbidden. This is an example of a parity-forbidden transition. We now form the DP $T_{2g} \otimes E_g \otimes T_{1u} = \{18\ 2\ 0\ 0\ 0\}$. The direct sum must consist of u representations. Still working with O,

$$\begin{aligned}
c(T_{2u}) &= (1/24)[54 - 6] = 2,\\
c(T_{1u}) &= (1/24)[54 - 6] = 2,\\
c(E_u) &= (1/24)[36 + 12] = 2,\\
c(A_{2u}) &= (1/24)[18 + 6] = 1,\\
c(A_{1u}) &= (1/24)[18 + 6] = 1,
\end{aligned}$$

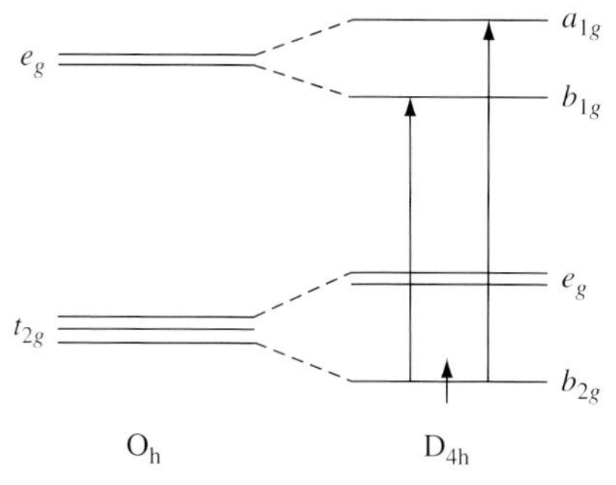

Figure 10.2. Splitting of the t_{2g} and e_g energy levels due to the Jahn–Teller effect in an octahedral d^1 complex. The short arrow indicates that in the ground state the b_{2g} level is occupied by one electron.

and so the parity-forbidden transition becomes vibronically allowed through coupling to the odd-parity vibrational modes of T_{1u} and T_{2u} symmetry. The vibronic transition $^2T_{2g} \rightarrow {}^2E_g$ accounts for the observed absorption band, but why does it show some structure? This indicates additional splitting of energy levels associated with a lowering of symmetry (the Jahn–Teller effect). The Jahn–Teller theorem (see, for example, Sugano, Tanabe, and Kamimura (1970)) states that any non-linear molecule in an orbitally degenerate state will undergo a distortion which lowers the energy of the molecule and removes the degeneracy. Consider here a lowering in symmetry from O_h to D_{4h}. The effect on the energy levels is shown in Figure 10.2. The single *d* electron is now in a b_{2g} orbital. The Jahn–Teller splitting between b_{2g} and e_g is too small for a transition between these states to appear in the visible spectrum. The relatively small splitting of the main absorption band tells us that we are looking for a relatively small perturbation of the $t_{2g} \rightarrow e_g$ transition, which is forbidden in O_h symmetry. So the observed structure of the absorption band is due to the $^2T_{2g} \rightarrow {}^2E_g$ transition in O_h symmetry being accompanied by $^2B_{2g} \rightarrow {}^2B_{1g}$ and $^2B_{2g} \rightarrow {}^2A_{1g}$ in D_{4h} symmetry due to the dynamical Jahn–Teller effect.

Example 10.2-3 Since it is the nearest-neighbor atoms in a complex that determine the local symmetry and the vibronic interactions, *trans*-dichlorobis(ethylenediamine)cobalt(III) (Figure 10.3(a)) may be regarded as having D_{4h} symmetry for the purpose of an analysis of its absorption spectrum in the visible/near-ultra-violet region (Ballhausen and Moffitt (1956)). The fundamental vibrational transitions therefore involve the $21 - 6 = 15$ normal modes of symmetry: $2A_{1g}$, B_{1g}, B_{2g}, E_g, $2A_{2u}$, B_{1u}, $3E_u$.

(1) In O_h symmetry the ground-state configuration of this low-spin complex would be t_{2g}^6. Determine the ground-state and excited-state spectral terms in O_h symmetry.
(2) Now consider a lowering of symmetry from O_h to D_{4h}. Draw an energy-level diagram in O_h symmetry showing the degeneracy and symmetry of the orbitals. Then, using a correlation table, show the splitting of these levels when the symmetry is lowered to D_{4h}. Determine the ground-state and excited-state terms, and show how these terms

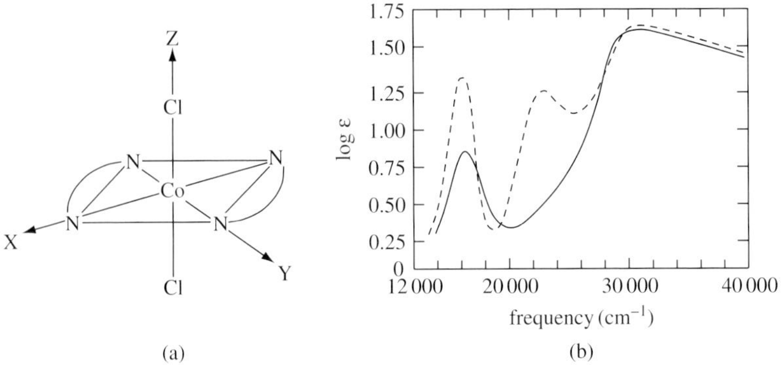

Figure 10.3. (a) The *trans*-dichlorobis(ethylenediamine)cobalt(III) ion, showing only the nearest-neighbor atoms in the ligands. (b) Absorption spectra of the *trans*-$[Co(en)_2Cl_2]^+$ ion showing the dichroism observed with light polarized nearly parallel to the Cl–Co–Cl axis OZ (——), and with light polarized in the *xy* plane perpendicular to that axis (- - - -). After Yamada and Tsuchida (1952) and Yamada *et al.* (1955).

correlate with the corresponding terms in O_h symmetry. Determine the symmetry-allowed vibronic transitions in D_{4h} symmetry between the ground state and the four excited states, noting the corresponding polarizations.

(3) Figure 10.3(b) shows the absorption spectrum: the continuous line shows the optical absorption for light polarized (nearly) parallel to OZ (the Cl–Co–Cl axis) and the dashed line indicates the absorption for light polarized perpendicular to this axis, namely in the *xy* plane. Assign transitions for the observed bands. [*Hints*: (i) The highest-energy band in the spectrum is a composite of two unresolved bands. (ii) The oscillator strengths for parallel and perpendicular transitions are not necessarily equal. (iii) The additional D_{4h} crystal-field splitting is less than the O_h splitting, named Δ or $10Dq$.]

(1) Co (atomic number 27) has the electron configuration $3d^8 4s^1$ and Co^{3+} has the configuration d^6. In O_h symmetry, the configuration is t_{2g}^6 when the crystal-field splitting is greater than the energy gain that would result from unpairing spins, as in the present case. The ground-state term is therefore ${}^1A_{1g}$. The first excited state has the configuration $t_{2g}^5 e_g^1$. Since all states for d^6 are symmetric under inversion, we may use the character table for O. As already shown in Example 10.2-2, $T_{2g} \times E_g = T_{1g} + T_{2g}$ so the excited state terms are ${}^1T_{1g}$, ${}^3T_{1g}$, ${}^1T_{2g}$, ${}^3T_{2g}$. Though parity-forbidden, the ${}^1A_{1g} \rightarrow {}^1T_{1g}$, ${}^1T_{2g}$ are vibronically allowed in O_h symmetry, it being known from calculation that the T_{1g} level lies below T_{2g}.

(2) Figure 10.4 shows the splitting of the one-electron orbital energies and states as the symmetry is lowered from O_h to D_{4h}. The ground state is $e_g{}^4 b_{2g}{}^2 : {}^1A_{1g}$. Since all states for d^6 have even parity under inversion, we may use the character table for D_4 in Appendix A3. The four excited states and their symmetries are

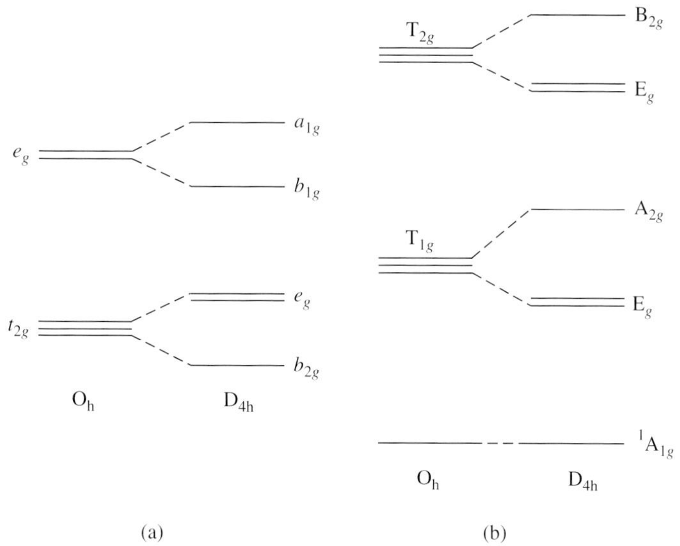

Figure 10.4. Splitting of (a) the one-electron orbital energy levels and (b) the electronic states, as the symmetry is lowered from O_h to D_{4h}.

Table 10.1. *Symmetry of the dipole moment matrix elements in trans-dichlorobis(ethylenediamine)cobalt(III) in* D_{4h} *symmetry.*

Polarization	Symmetry of operator	Final state		
		A_{2g}	B_{2g}	E_g
z	A_{2u}	A_{1u}	B_{1u}	E_u
x, y	E_u	E_u	E_u	$A_{1u} \oplus A_{2u} \oplus B_{1u} \oplus B_{2u}$

$$\begin{array}{ll} b_{2g}\ a_{1g} & B_{2g} \otimes A_{1g} = B_{2g}, \\ b_{2g}\ b_{1g} & B_{2g} \otimes B_{1g} = A_{2g}, \\ e_g\ b_{1g} & E_g \otimes B_{1g} = E_g, \\ e_g\ a_{1g} & E_g \otimes A_{1g} = E_g. \end{array}$$

There are therefore excited singlet and triplet states X_g of A_{2g}, B_{2g}, and E_g symmetry. In D_{4h} the dipole moment operator $-e\mathbf{r}$ forms a basis for $\Gamma(\mathbf{r}) = A_{2u} = E_u$. Since all states in the DP $X_g \times (A_{2u} + E_u \times {}^1A_{1g})$ are odd under inversion, we may continue to work with the D_4 character table in evaluating DPs. The symmetries of the dipole moment matrix elements for the possible transitions are shown in Table 10.1. All the transitions in Table 10.1 are forbidden without vibronic coupling. Inspection of the given list of the symmetries of the normal modes shows that there are odd-parity normal modes of A_{2u}, B_{1u}, and E_u symmetry, and consequently four allowed transitions for (x, y) polarization, namely ${}^1A_{1g} \rightarrow {}^1A_{2g}$, ${}^1B_{2g}$ and ${}^1E_g(2)$, there being two excited states of E_g symmetry, one correlating with T_{1g} in O_h and the other with T_{2g}. These transitions become allowed when there is

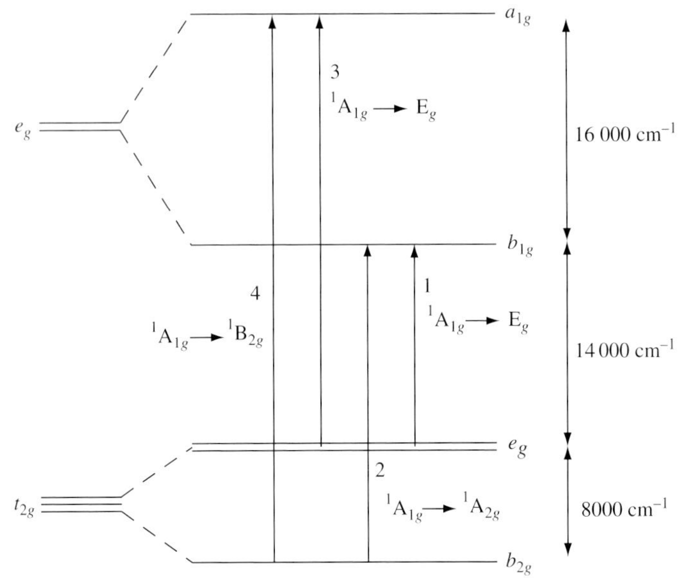

Figure 10.5. Approximate energy-level diagram showing the allowed transitions which are responsible for the observed absorption spectrum of the *trans*-$[Co(en)_2Cl_2]^+$ ion in Figure 10.4. The energies shown are optical energies, which are greater than the corresponding thermal energies because the minima in the adiabatic potential energy curves for the ground and excited states occur at different values of $\{\mathbf{R}\}$.

simultaneous excitation of a normal mode of symmetry E_u, E_u and A_{2u}, or B_{1u}, respectively. Only three bands are actually observed because the two highest energy bands overlap.

(3) In z polarization (the continuous line in the spectrum) there should be three bands due to $^1A_{1g} \rightarrow {}^1B_{2g}$, ^1E_g, and ^1E_g transitions, but only two bands are resolved. Disappearance of band 2 with z polarization identifies this with the $^1A_{1g} \rightarrow {}^1A_{2g}$ transition. Given that T_{1g} lies below T_{2g} in O_h symmetry, and since the D_{4h} splitting is less than Δ, we deduce that band 1 is due to $^1A_{1g} \rightarrow {}^1E_g$ (T_{1g}), which is allowed in both polarizations, but with somewhat different oscillator strengths. The highest-energy absorption consists of the unresolved bands 3 and 4 due to $^1A_{1g} \rightarrow {}^1E_g$ (T_{2g}) and $^1A_{1g} \rightarrow {}^1B_{2g}$, which are both allowed in both polarizations. These assignments lead to the approximate energy-level diagram shown in Figure 10.5, which also shows the observed transitions. In interpreting this diagram one must bear in mind that the energy differences given are *optical energies*, which are greater than the corresponding thermal energies because the minima in the adiabatic potential energy curves for ground and excited states do not coincide.

10.3 Charge transfer

Example 10.3-1 Table 10.2 contains a summary of observations on the optical absorption spectra of the ions $Mo(CN)_8^{3-}$ and $Mo(CN)_8^{4-}$, both of which have D_{2d} symmetry. Deduce which transitions are responsible for these absorption bands. The following additional

Table 10.2. *Absorption data for optical spectra of* $Mo(CN)_8^{3-}$ *and* $Mo(CN)_8^{4-}$.

The bracket } indicates the presence of two unresolved bands.

Ion	ν_{max}/cm^{-1}	Oscillator strength
$Mo(CN)_8^{3-}$	39 370 }	3.0×10^{-2}
	37 310 }	
	25 770	7.2×10^{-3}
$Mo(CN)_8^{4-}$	41 850	3.2×10^{-1}
	27 320	2.95×10^{-3}
	23 810	10^{-5}

After Golding and Carrington (1962).

Table 10.3. *Relative energies of metal d orbitals in* D_{2d} *symmetry.*

Orbital	Representation	Energy
d_{xy}	B_2	0.2771Δ
d_{yz}, d_{zx}	E	0.1871Δ
$d_{3z^2-r^2}$	A_1	-0.0287Δ
$d_{x^2-y^2}$	B_1	-0.6225Δ

information is available. (a) The energies of the metal *d* orbitals in a D_{2d} environment are as in Table 10.3. (b) $Mo(CN)_8^{4-}$ is diamagnetic. (c) Both complexes have an occupied non-bonding ligand orbital of a_2 symmetry. (d) Charge transfer bands, which arise from the excitation of an electron in a ligand orbital into a vacant metal orbital (or from an occupied metal orbital into a vacant ligand orbital) are rather intense bands and are usually located at the short-wavelength end of the spectrum.

The atomic number of Mo is 42: $Mo(CN)_8^{3-}$ contains Mo^{5+} and has the configuration d^1; $Mo(CN)_8^{4-}$ contains Mo^{4+} and has the configuration d^2. Inspection of the D_{2d} character table shows that the *d* orbitals transform according to the representations given in Table 10.4. Therefore the ground state of $Mo(CN)_8^{3-}$ is $b_1 : {}^2B_1$, while that of $Mo(CN)_8^{4-}$ is $b_1{}^2 : {}^1A_1$. Possible symmetry-allowed transitions are identified in Table 10.4. In the event that the transition is allowed, the necessary polarization of the radiation is given in parentheses. The column headed "Direct product" gives the DP of the ground- and excited-state representations which must contain one of the representations of the dipole moment operator for the transition to be allowed. In D_{2d}, these are $B_2(z)$ and $E(x, y)$. The DPs leading to the excited-state terms are not shown explicitly but may readily be verified with the help of the D_{2d} character table. As shown in Table 10.4 there is only one symmetry-allowed, spin-allowed, $d \rightarrow d$ transition in $Mo(CN)_8^{3-}$, from ${}^2B \rightarrow {}^2E$, which is excited by radiation polarized in the (x, y) plane. For $Mo(CN)_8^{4-}$ there is also only one symmetry-allowed, spin-allowed $d \rightarrow d$ transition, that from ${}^1A_1 \rightarrow {}^1E$, which again is (x, y)

Table 10.4. *Determination of symmetry-allowed transitions in* $Mo(CN)_8^{3-}$ *and* $Mo(CN)_8^{4-}$.

Ground state		Excited state	Direct product	Symmetry-allowed?
d → d transitions				
$Mo(CN)_8^{3-}$	$b_1 : {}^2B_1$	$a_1 : {}^2A_1$	$B_1 \otimes A_1 = B_1$	no
		$e : {}^2E$	$B_1 \otimes E = E$	yes (x, y)
		$b_2 : {}^2B_2$	$B_1 \otimes B_2 = A_2$	no
$Mo(CN)_8^{4-}$	$b_1{}^2 : {}^1A_1$	$b_1a_1 : {}^1B_1, {}^3B_1$	$A_1 \otimes B_1 = B_1$	no
		$b_1e : {}^1E, {}^3E$	$A_1 \otimes E = E$	yes (x, y)
		$b_1b_2 : {}^1A_2, {}^3A_2$	$A_1 \otimes A_2 = A_2$	no
Charge-transfer transitions				
$Mo(CN)_8^{3-}$	$a_2{}^2\, b_1 : {}^2B_1$	$a_2b_1{}^2 : {}^2A_2$	$B_1 \otimes A_2 = B_2$	yes (z)
		$a_2b_1a_1 : {}^2B_2$	$B_1 \otimes B_2 = A_2$	no
		$a_2b_1e : {}^2E$	$B_1 \otimes E = E$	yes (x, y)
		$a_2b_1b_2 : {}^2A_1$	$B_1 \otimes A_1 = B_1$	no
$Mo(CN)_8^{4-}$	$a_2{}^2\, b_1{}^2 : {}^1A_1$	$a_2b_1{}^2a_1 : {}^1A_2$	$A_1 \otimes A_2 = A_2$	no
		$a_2b_1{}^2e : {}^1E$	$A_1 \otimes E = E$	yes (x, y)
		$a_2b_1{}^2b_2 : {}^1B_1$	$A_1 \otimes B_1 = B_1$	no

Table 10.5. *Assignment of the observed bands in the absorption spectra of* $Mo(CN)_8^{3-}$ *and* $Mo(CN)_8^{4-}$.

	ν_{max}/cm^{-1}		Transition	Polarization
$Mo(CN)_8^{3-}$	39 370	CT	$a_2{}^2\, b_1 : {}^2B_1 \rightarrow a_2\, b_1{}^2 : {}^2A_2$	z
	37 310	CT	$a_2{}^2b_1 : {}^2B_1 \rightarrow a_2\, b_1e : {}^2E$	x, y
	25 770	$d \rightarrow d$	$b_1 : {}^2B_1 \rightarrow e : {}^2E$	x, y
$Mo(CN)_8^{4-}$	41 850	CT	$a_2{}^2\, b_1{}^2 : {}^1A_1 \rightarrow a_2\, b_1{}^2\, e : {}^1E$	x, y
	27 320	$d \rightarrow d$	$b_1{}^2 : {}^1A_1 \rightarrow b_1\, e : {}^1E$	x, y
	23 810	$d \rightarrow d$	$b_1{}^2 : {}^1A_1 \rightarrow b_1e : {}^3E$	x, y

polarized. These two transitions would be expected to have roughly similar intensities. The weaker band at long wavelengths in the $Mo(CN)_8^{4-}$ spectrum can be identified with the ${}^1A_1 \rightarrow {}^3E$ transition, partially allowed through spin–orbit coupling. There are still three more intense bands, and these are due to charge transfer (CT) from the non-bonding orbital of a_2 symmetry. The analysis is given in Table 10.4. Three electrons in three different orbitals give rise to doublet and quartet states, but only the former are recorded in the table because of the spin selection rule, $\Delta S = 0$. Similarly, the CT excited states for $Mo(CN)_8^{4-}$ are singlets and triplets but only the spin-allowed transitions are observed. As Table 10.4 shows, we should expect two symmetry-allowed, spin-allowed CT bands for $Mo(CN)_8^{3-}$, one z-polarized and the other with (x, y) polarization, but only one CT band for $Mo(CN)_8^{4-}$. The above analysis thus establishes the complete assignment of the observed bands in the optical absorption spectra of these two complex ions, as summarized in Table 10.5.

Problems

10.1 Determine the ground-state electron configuration and spectral term of the following octahedral complexes: low-spin $Fe(CN)_6^{3-}$; low-spin $Cr(CO)_6$; high-spin $Cr(H_2O)_6^{3+}$.

10.2 A d^2 complex ion has D_4 symmetry. It has the electronic configuration $(b_2)^2$ in the ground-state and excited-state configurations b_2e, b_2a_1, b_2b_1. Determine the electronic states that arise from these configurations. Hence decide which of the possible E1 transitions from the ground state to excited states are spin- and symmetry-allowed. If any of the possible spin-allowed E1 transitions are symmetry-forbidden, are they allowed M1 transitions?

10.3 The absorption spectrum of low-spin NiF_6^{3-} shows four absorption bands in the region below 25 000 cm^{-1}. Find the symmetry- and spin-allowed transtions in octahedral geometry and suggest a reason for any discrepancies with experiment.

10.4 The ground state of octahedral of $Co(CN)_6^{3-}$ is $t_{2g}{}^6 : {}^1A_{1g}$ and the first excited-state configuration is $t_{2g}{}^5 e_g$, which gives rise to four excited states. (a) Are transitions from the ground state to any of these excited states (i) symmetry-allowed? (ii) spin allowed? (b) Are transitions to any of these states allowed through vibronic coupling? (c) The absorption spectrum of $K_3Co(CN)_6$ in aqueous solution shows two bands at 32 050 cm^{-1} and 38 760 cm^{-1} with oscillator strengths of 5.4×10^{-3} and 3.5×10^{-3}, respectively, and a further very intense absorption band in the region of 50 000 cm^{-1}. Give an interpretation of this spectrum. [*Hint*: The energy-level diagram of this complex ion shows a vacant ligand π^* antibonding orbital of t_{1u} symmetry.] (d) $Co(CN)_6^{3-}$ ions may be dissolved in KCl crystals. The Co^{3+} ion occupies a K^+ site and the six CN^- ions occupy nearest-neighbor anion sites, thus preserving octahedral coordination. But charge compensation requires two vacant cation sites. The location of these sites is such as to lower the site symmetry at Co^{3+} sites from O_h to C_s. Explain the fact that the absorption spectrum of this crystal contains six bands in the near-ultra-violet visible spectral region (in addition to the intense band near 50 000 cm^{-1}).

10.5 (a) The tetrahedral permanganate ion MnO_4^- ion has the ground-state configuration $(1e)^4$ $(1t_2)^6$ $(2t_2)^6$ $(t_1)^6$ and the next lowest MO is an antibonding $3t_2$ orbital. Determine the symmetries of the states that correspond to the excited-state configurations $\ldots (2t_2)^6 (t_1)^5 (3t_2)^1$ and $\ldots (2t_2)^5 (t_1)^6 (3t_2)^1$. Find which E1 and M1 transitions from the ground state to these two excited states are symmetry- and spin-allowed. State the polarization of allowed E1 transitions. (b) In a crystal of $KClO_4$ containing some MnO_4^- substituting for ClO_4^-, the symmetry at an impurity anion site is reduced to C_s. Describe what splittings and relabeling of states occur in the ground state of MnO_4^- and the excited states to which E1 transitions were allowed in the free ion, when the symmetry is lowered from T_d to C_s, and state which E1 transitions are allowed now and what their polarizations are.

10.6 The absorption spectra of pink octahedral $Co(H_2O)_6^{2+}$ and of deep blue tetrahedral $Co(Cl)_4^{2-}$ show bands in the visible region at 18 500 cm^{-1} ($\varepsilon = 10$) and 15 000 cm^{-1} ($\varepsilon = 600$), respectively. Both these compounds also show infra-red absorption bands at 8350 cm^{-1} and 6300 cm^{-1}. Suggest an explanation for these observations. [*Hint:* See Griffith (1964) and Harris and Bertolucci (1978).]

11 Continuous groups

11.1 Rotations in $\Re^2$

The special orthogonal group SO(2) is the group of proper rotations in the 2-D space of real vectors, $\Re^2$, about an axis **z** normal to the plane containing **x** and **y**. Since there is only one rotation axis **z**, the notation $R(\phi\ \mathbf{z})$ for the rotation of the unit circle in $\Re^2$ will be contracted to $R(\phi)$. Then, for the orthonormal basis $\langle \mathbf{e}_1\ \mathbf{e}_2|$,

$$R(\phi)\langle \mathbf{e}_1\ \mathbf{e}_2| = \langle \mathbf{e}_1{}'\ \mathbf{e}_2{}'| = \langle \mathbf{e}_1\ \mathbf{e}_2|\Gamma(\phi), \tag{1}$$

$$\Gamma(\phi) = \begin{bmatrix} \cos\phi & -\sin\phi \\ \sin\phi & \cos\phi \end{bmatrix}. \tag{2}$$

Since det $\Gamma(\phi) = 1$ and $\Gamma(\phi)^{\mathrm{T}}\ \Gamma(\phi) = \mathrm{E}_2$, $\Gamma(\phi)$ is an orthogonal matrix with determinant $+1$, and so the group of proper rotations in $\Re^2$ is isomorphous with the group of 2×2 orthogonal matrices with determinant $+1$, which thus forms a faithful representation of the rotation group. Any function $f(x, y)$ is transformed under $R(\phi)$ into the new function

$$\hat{R}(\phi)\, f(x,y) = f(R(\phi)^{-1}\{x,y\}) = f(cx + sy,\ -sx + cy), \tag{3}$$

and so $\hat{R}(\phi)\ f(x,y)$ is $f(x',y')$ with the substitutions $x' = cx + sy$, $y' = -sx + cy$. In particular, the *functions* $\{x, y\}$ form a basis for a 2-D representation of SO(2):

$$\begin{aligned} \hat{R}(\phi)\langle x\ y| &= \langle x'\ y'| = \langle cx + sy \quad -sx + cy| \\ &= \langle x\ y|\begin{bmatrix} c & -s \\ s & c \end{bmatrix} = \langle x\ y|\Gamma(\phi). \end{aligned} \tag{4}$$

But successive rotations about the same axis commute so that the group SO(2) is Abelian with 1-D representations with bases $(x + \mathrm{i}y)^m$, $m = 0, \pm 1, \pm 2, \ldots$,

$$\begin{aligned} \hat{R}(\phi)\langle x + \mathrm{i}y| &= \langle cx + sy + \mathrm{i}(-sx + cy)| = \langle (c - \mathrm{i}s)(x + \mathrm{i}y)| \\ &= \exp(-\mathrm{i}\phi)\langle x + \mathrm{i}y|. \end{aligned} \tag{5}$$

The 1-D matrix representatives (MRs) of $R(\phi)$ for the bases $(x + \mathrm{i}y)^m$ are therefore

$$\Gamma^m(\phi) = \exp(-\mathrm{i}m\phi),\ m = 0,\ \pm 1,\ \pm 2, \ldots \tag{6}$$

Restricting m to integer values ensures that the set of functions

$$\{(2\pi)^{-1/2}(x+\mathrm{i}y)^m\} = \{(2\pi)^{-1/2}\exp(\mathrm{i}m\varphi)\} = u_m(r=1,\varphi), m=0,\ \pm 1,\ \pm 2,\ \ldots \quad (7)$$

form an orthonormal basis with

$$\langle m'|m\rangle = \delta_{mm'} \quad (8)$$

and satisfy the condition

(6) $$\Gamma^m(\phi+2\pi) = \Gamma^m(\phi). \quad (9)$$

11.2 The infinitesimal generator for SO(2)

In $\Re^2$ there is only one rotation axis, namely **z**. Rotations about this axis commute:

$$R(\phi)\ R(\phi') = R(\phi+\phi') = R(\phi')\ R(\phi), \quad (1)$$

a condition on $R(\phi)$ that is satisfied by

$$R(\phi) = \exp(-\mathrm{i}\phi I_3). \quad (2)$$

On expanding the exponential in eq. (2), we see that

$$-\mathrm{i}\ I_3 = \lim_{\phi\to 0} \frac{\mathrm{d}R(\phi)}{\mathrm{d}\phi}. \quad (3)$$

I_3 is therefore called the *infinitesimal generator* of rotations about **z** that comprise SO(2). With $r=1$,

(2) $$\begin{aligned}\hat{R}(\phi)u_m(r,\varphi) &= \exp(-\mathrm{i}\phi\hat{I}_3)u_m(r,\varphi)\\ &= u_m(R^{-1}\{r,\varphi\}) = u_m(r,\varphi-\phi),\end{aligned} \quad (4)$$

where, as usual, the carat symbol over R (or I_3) indicates the *function operator* that corresponds to the symmetry operator R (or I_3, as the case may be). For infinitesimally small rotation angles ϕ,

$$(1-\mathrm{i}\phi\hat{I}_3)\ u_m(r,\varphi) = \left(1-\frac{\phi\ \mathrm{d}}{\mathrm{d}\varphi}\right)u_m(r,\varphi) \quad (5)$$

on expanding the second and fourth members of eq. (4) to first order in ϕ.

(5) $$I_3 = -\frac{\mathrm{i}\ \mathrm{d}}{\mathrm{d}\varphi}, \quad (6)$$

which shows that the infinitesimal generator of rotations about **z** is just the angular momentum about **z**. (Atomic units, in which $\hbar=1$, are used throughout.) The MR $\Gamma(\phi)$ of the symmetry operator $R(\phi)$ is

(2) $$\Gamma(\phi) = \exp(-\mathrm{i}\phi\mathrm{I}_3). \quad (7)$$

The exponential notation in eqs. (2) and (7) means a series expansion in terms of powers of the operator I_3, or its MR I_3, respectively,

(2) $$R(\phi) = \exp(-\mathrm{i}\phi I_3) = \sum_{n=0}^{\infty} (-\mathrm{i}\phi)^n (I_3)^n / n!, \tag{8}$$

(7) $$\Gamma(\phi) = \exp(-\mathrm{i}\phi \mathrm{I}_3) = \sum_{n=0}^{\infty} (-\mathrm{i}\phi)^n (\mathrm{I}_3)^n / n! \tag{9}$$

For infinitesimally small rotation angles ϕ,

(9) $$-\mathrm{i}\mathrm{I}_3 = \lim_{\phi \to 0} \frac{\mathrm{d}\Gamma(\phi)}{\mathrm{d}\phi}. \tag{10}$$

I_3 is the MR of the infinitesimal generator I_3 of SO(2), the group of proper rotations in $\Re^2$. In the defining 2-D representation with basis $\langle \mathbf{e}_1 \ \mathbf{e}_2|$

(10), (11.1.2) $$-\mathrm{i}\mathrm{I}_3 = \begin{bmatrix} 0 & 1 \\ 1 & 0 \end{bmatrix}. \tag{11}$$

Finite rotations are generated by $R(\phi) = \exp(-\mathrm{i}\phi I_3)$. The effect of a rotation of 2-D configuration space by $R(\phi)$ on any function $f(r, \varphi)$ is given by

(4) $$\hat{R}(\phi) f(r, \varphi) = \exp(-\mathrm{i}\phi \hat{I}_3) f(r, \varphi) \tag{12}$$

$$= f(\exp(\mathrm{i}\phi I_3)\{r, \varphi\}). \tag{13}$$

In eq. (12), $\hat{R}(\phi)$ is the function operator that corresponds to the (2-D) configuration-space symmetry operator $\hat{R}(\phi)$. In eq. (13), I_3 is the infinitesimal generator of rotations about **z** (eq. (8)); $\exp(\mathrm{i}\phi I_3)$ is the operator $[R(\phi)]^{-1}$, in accordance with the general prescription eq. (3.5.7). Notice that a positive sign inside the exponential in eq. (2) would also satisfy the commutation relations (CRs), but the sign was chosen to be negative in order that I_3 could be identified with the angular momentum about **z**, eq. (6).

11.3 Rotations in $\Re^3$

The group of proper rotations in configuration space is called the special orthogonal group SO(3). There are two main complications about extending $\Re^2$ to $\Re^3$. Firstly, the group elements are the rotations $R(\phi \ \mathbf{n})$, where **n** is any unit vector in $\Re^3$, and, secondly, finite rotations about different axes do not commute. In Chapter 4 we derived the MR of $\mathrm{R}(\phi \ \mathbf{z})$ and showed this to be an orthogonal matrix of determinant +1:

$$\Gamma(\phi \ \mathbf{z}) = \begin{bmatrix} \cos\phi & -\sin\phi & 0 \\ \sin\phi & \cos\phi & 0 \\ 0 & 0 & 1 \end{bmatrix}, \qquad -\pi < \phi \leq \pi. \tag{1}$$

The $\{\Gamma(\phi\ \mathbf{z})\}$ form a group isomorphous with SO(3) and so may be regarded as merely a different realization of the same group. Since successive finite rotations about the same axis commute, the infinitesimal generator I_3 of rotations about $\mathbf{z}$ is given by

(11.2.3) $$-\mathrm{i}\, I_3 = \lim_{\phi \to 0} \frac{\mathrm{d}R(\phi\ \mathbf{z})}{\mathrm{d}\phi}, \tag{2}$$

and its MR is given by

(1), (11.2.10) $$-\mathrm{i}\, \mathtt{I}_3 = \lim_{\phi \to 0} \frac{\mathrm{d}\Gamma(\phi\ \mathbf{z})}{\mathrm{d}\phi} = \begin{bmatrix} 0 & \bar{1} & 0 \\ 1 & 0 & 0 \\ 0 & 0 & 0 \end{bmatrix}. \tag{3}$$

The matrix elements of $\mathtt{I}_3$ are thus

(3) $$(\mathtt{I}_3)_{ij} = -\mathrm{i}\, \varepsilon_{ijk},\ i,j = 1,2,3;\ k = 3, \tag{4}$$

where ε_{ijk} is the Levi–Civita antisymmetric three-index symbol: ε_{ijk} is antisymmetric under the exchange of any two indices, and ε_{123} is defined to be $+1$. Here in eq. (4), $k = 3$, so all diagonal elements are zero, as are the elements of the third row and the third column.

Exercise 11.3-1 Confirm from eq. (4) the entries $(\mathrm{I}_3)_{12}$ and $(\mathrm{I}_3)_{21}$, thus completing the verification that eq. (4) does indeed give eq. (3).

In like manner, for rotations $R(\phi\ \mathbf{x})$ about $\mathbf{x}$

$$\Gamma(\phi\ \mathbf{x}) = \begin{bmatrix} 1 & 0 & 0 \\ 0 & \cos\phi & -\sin\phi \\ 0 & \sin\phi & \cos\phi \end{bmatrix}. \tag{5}$$

(11.2.10) $$-\mathrm{i}\ \mathtt{I}_1 = \begin{bmatrix} 0 & 0 & 0 \\ 0 & 0 & \bar{1} \\ 0 & 1 & 0 \end{bmatrix}, \quad (\mathtt{I}_1)_{ij} = -\mathrm{i}\ \varepsilon_{ijk},\ i,j = 1,2,3;\ k = 1, \tag{6}$$

and for $R(\phi\ \mathbf{y})$

$$\Gamma(\phi\ \mathbf{y}) = \begin{bmatrix} \cos\phi & 0 & \sin\phi \\ 0 & 1 & 0 \\ -\sin\phi & 0 & \cos\phi \end{bmatrix}, \tag{7}$$

$$-\mathrm{i}\, \mathtt{I}_2 = \begin{bmatrix} 0 & 0 & 1 \\ 0 & 0 & 0 \\ \bar{1} & 0 & 0 \end{bmatrix}, \quad (\mathtt{I}_2)_{ij} = -\mathrm{i}\ \varepsilon_{ijk},\ i,j = 1,2,3;\ k = 2. \tag{8}$$

(4), (6), (8) $$(\mathtt{I}_k)_{ij} = -\mathrm{i}\ \varepsilon_{ijk},\ i,j,k = 1,2,3. \tag{9}$$

A general rotation $R(\phi\ \mathbf{n})$ through a *small angle* ϕ (Figure 11.1) changes a vector $\mathbf{r}$, which makes an angle θ with $\mathbf{n}$, into $\mathbf{r}' = \mathbf{r} + \delta\mathbf{r}$, where the displacement $\delta\mathbf{r}$ of R is normal to the plane of $\mathbf{n}$ and $\mathbf{r}$. Consequently,

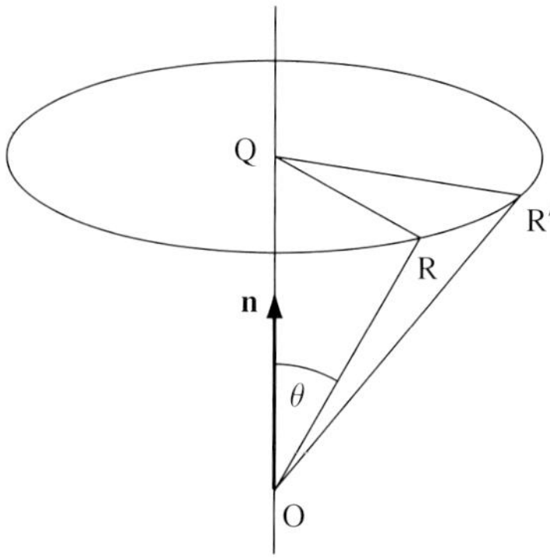

Figure 11.1. Rotation of a vector **r**, which makes an angle θ with the axis of rotation **n**, through a small angle ϕ. From the figure, $R(\phi\ \mathbf{n})\mathbf{r} = \mathbf{r}' = \mathbf{r} + \delta\mathbf{r}$, and $|\delta\mathbf{r}| = (r \sin\theta)\phi$. Note that $\delta\mathbf{r}$ is perpendicular to the plane of **n** and **r**; it is of magnitude $(r \sin\theta)\phi$; and it is of orientation such that $\delta\mathbf{r} = (\mathbf{n} \times \mathbf{r})\phi$.

$$R(\phi\ \mathbf{n})\mathbf{r} = \mathbf{r}' = \mathbf{r} + (\mathbf{n} \times \mathbf{r})\phi. \tag{10}$$

On evaluating the vector product, the components of $\mathbf{r}'$ are seen to be

$$r_i' = (\delta_{ij} - \phi\, \varepsilon_{ijk}\, n_k) r_j, \quad i = 1, 2, 3. \tag{11}$$

For notational convenience, the components of **r** are here $\{r_1, r_2, r_3\}$ rather than $\{x\ y\ z\}$. The Einstein summation convention, an implied summation over repeated indices, is used in eq. (11).

(11)
$$\Gamma(\phi\ \mathbf{n}) = \begin{bmatrix} 1 & -\phi n_3 & \phi n_2 \\ \phi n_3 & 1 & -\phi n_1 \\ -\phi n_2 & \phi n_1 & 1 \end{bmatrix}. \tag{12}$$

The MR $\mathtt{I}_n$ of the generator of infinitesimal rotations about **n** is

(12)
$$-\mathrm{i}\,\mathtt{I}_{\mathbf{n}} = \begin{bmatrix} 0 & -n_3 & n_2 \\ n_3 & 0 & -n_1 \\ -n_2 & n_1 & 0 \end{bmatrix} \tag{13}$$

with elements

$$\begin{aligned} (\mathtt{I}_{\mathrm{n}})_{ij} &= -\mathrm{i}\ \varepsilon_{ijk}\ n_k = n_k (I_k)_{ij},\ k = 1, 2, 3 \\ &= \langle n_1\ n_2\ n_3 \mid \mathtt{I}_1\ \mathtt{I}_2\ \mathtt{I}_3 \rangle_{ij}, \end{aligned} \tag{14}$$

which is the ijth element of the MR of the scalar product

$$I_{\mathbf{n}} = \mathbf{n} \cdot \mathbf{I}. \tag{15}$$

I, a vector with components I_1, I_2, I_3, is the infinitesimal generator of rotations about an arbitrary axis **n**. Successive rotations in $\Re^3$ about the same axis **n** do commute,

$$R(\phi\ \mathbf{n})R(\phi'\ \mathbf{n}) = R(\phi'\ \mathbf{n})R(\phi\ \mathbf{n}), \tag{16}$$

a condition satisfied by

$$(15) \qquad R(\phi\ \mathbf{n}) = \exp(-\mathrm{i}\phi I_{\mathbf{n}}) = \exp(-\mathrm{i}\phi\mathbf{n}\cdot\mathbf{I}). \qquad (17)$$

The MR of the symmetry operator $R(\phi\ \mathbf{n})$ is

$$(17) \qquad \Gamma(\phi\ \mathbf{n}) = \exp(-\mathrm{i}\phi\mathtt{I}_{\mathbf{n}}), \qquad (18)$$

with the matrix $\mathtt{I}_{\mathbf{n}}$ given by eq. (13). For a small angle ϕ

$$(17) \qquad R(\phi\ \mathbf{n}) = E - \mathrm{i}\phi\mathbf{n}\cdot\mathbf{I}; \qquad (19)$$

$$(18),(19) \qquad \Gamma(\phi\ \mathbf{n}) = \mathrm{E} - \mathrm{i}\phi\,\langle n_1\ n_2\ n_3|\mathtt{I}_1\ \mathtt{I}_2\ \mathtt{I}_3\rangle. \qquad (20)$$

When the symmetry operator $R(\phi\ \mathbf{n})$ acts on configuration space, a function $f(r, \theta, \varphi)$ is transformed by the function operator $\hat{R}(\phi\ \mathbf{n})$ into the new function $f'(r, \theta, \varphi)$, where

$$(17) \qquad f'(r,\theta,\varphi) = f(R(\phi\ \mathbf{n})^{-1}\{r,\theta,\varphi\}) = f(\exp(\mathrm{i}\phi\mathbf{n}\cdot\mathbf{I})\{r,\theta,\varphi\}). \qquad (21)$$

For an infinitesimal angle ϕ

$$(21) \qquad f'(r,\theta,\varphi) = f((1 + \mathrm{i}\phi\mathbf{n}\cdot\mathbf{I})\{r,\theta,\varphi\}), \qquad (22)$$

$$(22) \qquad \hat{R}(\phi\ \mathbf{n}) = \hat{E} - \mathrm{i}\phi\mathbf{n}\cdot\hat{\mathbf{I}}. \qquad (23)$$

The differences in sign in the operators in eqs. (17) and (21) and between eqs. (19) and (22) have arisen from our rules for manipulating function operators.

Answer to Exercise 11.3-1

$$(4) \qquad (\mathtt{I}_3)_{12} = -\mathrm{i}\,\varepsilon_{123} = -\mathrm{i},\ (\mathtt{I}_3)_{21} = -\mathrm{i}\,\varepsilon_{213} = \mathrm{i}.$$

Therefore, for $k = 3$,

$$-\mathrm{i}\,\mathtt{I}_3 = \begin{bmatrix} 0 & \bar{1} & 0 \\ 1 & 0 & 0 \\ 0 & 0 & 0 \end{bmatrix}.$$

11.4 The commutation relations

As a prelude to finding the irreducible representations (IRs) for SO(3) we now need to establish the CRs between the components I_1, I_2, I_3 of $\mathbf{I}$ and I^2. To do this we need, as well as eq. (11.3.9),

$$(\mathtt{I}_k)_{ij} = -\mathrm{i}\ \varepsilon_{ijk}, \quad i,j,k = 1,2,3, \qquad (1)$$

the identity

$$\varepsilon_{ilk}\ \varepsilon_{jmk} = \delta_{ij}\ \delta_{lm} - \delta_{im}\ \delta_{lj}, \tag{2}$$

$$\begin{aligned}[\mathtt{I}_i, \mathtt{I}_j]_{lm} &= (\mathtt{I}_i, \mathtt{I}_j - \mathtt{I}_j, \mathtt{I}_i)_{lm} = -\varepsilon_{lpi}\ \varepsilon_{pmj} + \varepsilon_{lqj}\ \varepsilon_{qmi} \\ &= \delta_{li}\ \delta_{jm} - \delta_{lm}\ \delta_{ji} - \delta_{lj}\ \delta_{im} + \delta_{lm}\ \delta_{ij} \\ &= \delta_{il}\ \delta_{jm} - \delta_{im}\ \delta_{jl} \\ &= \varepsilon_{ijk}\ \varepsilon_{lmk} \\ &= \mathrm{i}\ \varepsilon_{ijk}(\mathtt{I}_k)_{lm}.\end{aligned} \tag{3}$$

Equation (3) is derived using the MRs of the infinitesimal generators (symmetry operators) and therefore holds for the operators, so that

(3) $$[I_i,\ I_j] = \mathrm{i}\ \varepsilon_{ijk}\ I_k, \tag{4}$$

(4) $$[I^2, I_k] = 0. \tag{5}$$

Exercise 11.4-1 Prove eq. (2). [*Hint*: Consider $m = l$ and then $m \neq l$.]

Exercise 11.4-2 Prove eq. (5).

Exercise 11.4-3 Verify that $[\mathtt{I}_1,\ \mathtt{I}_2] = \mathrm{i}\mathtt{I}_3$ by matrix multiplication of the MRs of the infinitesimal rotation operators I_1, I_2, I_3 which are given in eqs. (11.3.3), (11.3.6), and (11.3.8).

The commutators, eqs. (4) and (5), are derived in three different ways, firstly from eq. (11.3.9) and then in Exercises (11.4-1) and (11.4-2) and Problem 11.1. Note that I_1, I_2, and I_3 are components of the symmetry operator (infinitesimal generator) **I** which acts on vectors in configuration space. Concurrently with the application of a *symmetry operator* to configuration space, all functions $f(r, \theta, \varphi)$ are transformed by the corresponding *function operator*. Therefore, the corresponding commutators for the function operators are

(4), (5) $$[\hat{I}_i,\ \hat{I}_j] = \mathrm{i}\ \varepsilon_{ijk}\ \hat{I}_k; \quad [\hat{I}^2, \hat{I}_k] = 0, \ \ k = 1, 2, 3, \tag{6}$$

where, as usual, the carat sign (^) indicates a function operator. We define the shift (or ladder) operators

$$I_+ = I_1 + \mathrm{i}\ I_2, \ \ I_- = I_1 - \mathrm{i}\ I_2. \tag{7}$$

$R(\phi\ \mathbf{n})$ is a unitary operator. Using eq. (11.3.19), which holds for small angles ϕ, and retaining only terms of first order in ϕ,

(11.3.19) $$\begin{aligned}R^\dagger R &= (E + \mathrm{i}\phi\mathbf{n}\cdot\mathbf{I}^\dagger)(E - \mathrm{i}\phi\mathbf{n}\cdot\mathbf{I}) \\ &= E - \mathrm{i}\phi\mathbf{n}\cdot(\mathbf{I} - \mathbf{I}^\dagger) = E.\end{aligned} \tag{8}$$

(8) $$\mathbf{I} = \mathbf{I}^\dagger, \tag{9}$$

which shows that the infinitesimal generator **I** of rotations about **n** is a self-adjoint operator. Therefore I^2 and the components of **I**, I_1, I_2, I_3 are also self-adjoint. Consequently,

(7) $$I_\pm{}^\dagger = I_\mp, \tag{10}$$

(7), (4) $$I_+I_- = I^2 - I_3{}^2 + I_3, \tag{11}$$

(7), (4) $$I_-I_+ = I^2 - I_3{}^2 - I_3, \tag{12}$$

(11), (12) $$[I_+, I_-] = 2\,I_3, \tag{13}$$

(5), (7) $$[I^2,\ I_\pm] = 0, \tag{14}$$

(4), (7) $$[I_3,\ I_\pm] = \pm I_\pm. \tag{15}$$

Exercise 11.4-4 Prove eqs. (11), (12), and (15).

All the above relations, eqs. (7)–(15), hold for the corresponding function operators, the presence (or absence) of the carat sign serving to indicate that the operator operates on functions (or on configuration space).

The effect on a vector **r** of a rotation through a small angle ϕ about **n** is

$$R(\phi\ \mathbf{n})\mathbf{r} = \mathbf{r}' = \mathbf{r} + (\mathbf{n} \times \mathbf{r})\phi. \tag{16}$$

(11.3.10) $$\hat{R}(\phi\ \mathbf{n})\, f(\mathbf{r}) = f(R^{-1}\mathbf{r}) = f(\mathbf{r} - (\mathbf{n} \times \mathbf{r})\phi) \tag{17}$$

$$\begin{aligned} &= f(\mathbf{r}) - \phi(\mathbf{n} \times \mathbf{r}) \cdot \nabla f(\mathbf{r}) \\ &= f(\mathbf{r}) - \phi\, \mathbf{n} \cdot \mathbf{r} \times \nabla f(\mathbf{r}), \end{aligned} \tag{18}$$

where eq. (17) has been expanded to first order because the displacement $\delta\mathbf{r} = \mathbf{r}' - \mathbf{r}$ is small. Since

$$\mathbf{r} \times \nabla = \mathrm{i}(\mathbf{r} \times \hat{\mathbf{p}}) = \mathrm{i}\ \hat{\mathbf{J}}, \tag{19}$$

where **J** is the angular momentum $\mathbf{r} \times \mathbf{p}$, the function operator corresponding to the small rotation $R(\phi\ \mathbf{n})$ is

(18), (19) $$\hat{R}(\phi\ \mathbf{n}) = \hat{E} - \mathrm{i}\,\phi\,\mathbf{n} \cdot \hat{\mathbf{J}}. \tag{20}$$

Equations (11.3.23) and (20) show that the infinitesimal generator **I** of rotations in $\Re^3$ about any axis **n** is the angular momentum about **n**. The separate symbol $\hat{\mathbf{I}}$ has now served its purpose and will henceforth be replaced by the usual symbol for the angular momentum operator, $\hat{\mathbf{J}}$, and similarly $\hat{I}_1$, $\hat{I}_2$, $\hat{I}_3$ will be replaced by $\hat{J}_x$, $\hat{J}_y$, $\hat{J}_z$.

Since $\hat{J}_x$, $\hat{J}_y$, $\hat{J}_z$ all commute with $\hat{J}^2$, but not with each other, only one of the components (taken to be $\hat{J}_z$) has a common set of eigenfunctions with $\hat{J}^2$. These eigenfunctions are called u^j_m or, in Dirac notation, $|j\ m\rangle$:

$$\hat{J}_z|j\ m\rangle = m|j\ m\rangle, \tag{21}$$

$$\hat{J}^2|j\ m\rangle = j'^2|j\ m\rangle. \tag{22}$$

At this stage, j' is just a number that determines the value of j^2 and the precise relation between j and j' is yet to be discovered.

(15), (21)
$$\begin{aligned}\hat{J}_z(\hat{J}_\pm|j\ m\rangle) &= (\pm\hat{J}_\pm + \hat{J}_\pm\,\hat{J}_z)|j\ m\rangle \\ &= (\pm\hat{J}_\pm + \hat{J}_\pm m)|j\ m\rangle \\ &= (m \pm 1)\hat{J}_\pm|j\ m\rangle.\end{aligned} \tag{23}$$

Equation (23) demonstrates the reason for the name shift operators: if $|j\ m\rangle$ is an eigenvector of $\hat{J}_z$ with eigenvalue m, then $\hat{J}_\pm|j\ m\rangle$ is also an eigenvector of $\hat{J}_z$ but with eigenvalue $m \pm 1$. However, $\hat{J}_\pm|j\ m\rangle$ is no longer normalized: let c_+ or c_- be the numerical factor that restores normalization after the application of $\hat{J}_+$ or $\hat{J}_-$, so that

(23)
$$\hat{J}_\pm|j\ m\rangle = c_\pm|j\ m \pm 1\rangle. \tag{24}$$

Then

(11), (22), (21)
$$\begin{aligned}\langle j\ m\,|\hat{J}_+\hat{J}_-|j\ m\rangle &= \langle j\ m|\hat{J}^2 - \hat{J}_z^2 + J_z|j\ m\rangle \\ &= j'^2 - m^2 + m,\end{aligned} \tag{25}$$

(10), (25)
$$\langle j\ m|\hat{J}_+\hat{J}_-|j\ m\rangle = |c_-|^2, \tag{26}$$

(25), (26)
$$c_- = [j'^2 - m(m-1)]^{1/2}\exp(\mathrm{i}\gamma). \tag{27}$$

It is customary to follow the Condon and Shortley (1967) (CS) choice of phase for the eigenfunctions $|j\ m\rangle$ by setting $\gamma = 0$, thus taking the phase factor $\exp(\mathrm{i}\gamma)$ as unity for all values of the quantum numbers j and m. The derivation of the corresponding relation

$$c_+ = [j'^2 - m(m+1)]^{1/2}\exp(\mathrm{i}\gamma) \tag{28}$$

is Problem 11.2. With the CS choice of phase,

(27), (28)
$$c_\pm = [j'^2 - m(m \pm 1)]^{1/2}, \tag{29}$$

(21), (22)
$$(\hat{J}^2 - \hat{J}_z^2)|j\ m\rangle = (j'^2 - m^2)|j\ m\rangle. \tag{30}$$

But the eigenvalues of $\hat{J}^2 - \hat{J}_z^2 = \hat{J}_x^2 + \hat{J}_y^2$ are positive; therefore

$$j'^2 - m^2 \geq 0, \quad |m| < j'. \tag{31}$$

Thus for any particular j' (defined by eq. (22)) the values of m are bounded both above and below. Define j (not so far defined) as the maximum value of m for any particular j' and let the minimum value of m (for the same j') be j''. Then the allowed values of m (for a particular j') form a ladder extending from a minimum value j'' to a maximum value j. Because the values of m are bounded,

(24) $$\hat{J}_+|j\ j\rangle = 0,\ \hat{J}_-|j\ j''\rangle = 0, \tag{32}$$

(32), (12), (22), (21) $$\hat{J}_-(\hat{J}_+|j\ j\rangle) = (j'^2 - j^2 - j)|j\ j\rangle = 0, \tag{33}$$

(33) $$j'^2 = j(j+1), \tag{34}$$

(32), (11), (22), (21) $$\hat{J}_+(\hat{J}_-|j\ j''\rangle) = (j'^2 - j''^2 + j'')|j\ j''\rangle = 0, \tag{35}$$

(35) $$j'^2 = j''^2 - j'', \tag{36}$$

(34), (36) $$j''^2 - j'' - j(j+1) = 0, \tag{37}$$

(37) $$j'' = 1 + j,\ \text{or}\ -j. \tag{38}$$

Because j is the maximum value of m, the minimum value j'' of m is $-j$. And since $\hat{J}_-$ converts $\langle|j\ j\rangle$ into $\langle|j\ -j\rangle$ in $2j$ integer steps, j is an integer or a half-integer. Therefore, for any allowed value of j (integer or half-integer) the common eigenfunctions of $\hat{J}^2$ and $\hat{J}_z$ are

$$\begin{aligned}&\{|j\ m\rangle\} = \{|j\ j\rangle, |j\ j-1\rangle, \ldots, |j\ -j\rangle\},\\ &\text{or}\ \{u_m^j\} = \{u_j^j, u_{j-1}^j, \ldots, u_{-j}^j\}.\end{aligned} \tag{39}$$

The eigenvalue equation for $\hat{J}^2$ is

(22), (34) $$\hat{J}^2|j\ m\rangle = j(j+1)|j\ m\rangle, \tag{40}$$

(24), (29), (34) $$\hat{J}_\pm|j\ m\rangle = [j(j+1) - m(m\pm 1)]^{1/2}|j\ m\pm 1\rangle \tag{41}$$

$$= [(j\mp m)(j\pm m+1)]^{1/2}|j\ m\pm 1\rangle. \tag{42}$$

Exercise 11.4-5 Prove the equivalence of the two forms for $c_\pm$ in eqs. (41) and (42).

Answers to Exercises 11.4

Exercise 11.4-1 If $m = l$, then $j = i$, and $\varepsilon_{ilk}\,\varepsilon_{jmk} = -\delta_{ij}\,\delta_{lm}$. If $m \neq l$, then $l = j$, $m = i$ and $\varepsilon_{ilk}\,\varepsilon_{jmk} = -\delta_{im}\,\delta_{lj}$. Therefore $\varepsilon_{ilk}\,\varepsilon_{jmk} = \delta_{ij}\,\delta_{lm} - \delta_{im}\,\delta_{lj}$.

Exercise 11.4-2 $[I^2,\ I_1] = [I_1{}^2 + I_2{}^2 + I_3{}^2, I_1] = [I_2{}^2 + I_3{}^2, I_1]$,

$$[I_2{}^2,\ I_1] = I_2\ [I_2,\ I_1] + [I_2,\ I_1]\ I_2 = -\mathrm{i}\ I_2 I_3 - \mathrm{i}\ I_3 I_2\,,$$

$$[I_3{}^2,\ I_1] = I_3[I_3,\ I_1] + [I_3,\ I_1]I_3 = +\mathrm{i}\ I_2 I_3 + \mathrm{i}\ I_3 I_2\,,$$

whence $[I^2, I_1] = 0$; $k = 2, 3$ follow by cyclic permutation.

Exercise 11.4-3 From eqs. (11.3.6), (11.3.8), and (11.3.3),

$$\mathrm{I}_1\,\mathrm{I}_2 - \mathrm{I}_2\,\mathrm{I}_1 = -\begin{bmatrix} 0 & 0 & 0 \\ 0 & 0 & \bar{1} \\ 0 & 1 & 0 \end{bmatrix}\begin{bmatrix} 0 & 0 & 1 \\ 0 & 0 & 0 \\ \bar{1} & 0 & 0 \end{bmatrix} + \begin{bmatrix} 0 & 0 & 1 \\ 0 & 0 & 0 \\ \bar{1} & 0 & 0 \end{bmatrix}\begin{bmatrix} 0 & 0 & 0 \\ 0 & 0 & \bar{1} \\ 0 & 1 & 0 \end{bmatrix}$$
$$= -\begin{bmatrix} 0 & 0 & 0 \\ 1 & 0 & 0 \\ 0 & 0 & 0 \end{bmatrix} + \begin{bmatrix} 0 & 1 & 0 \\ 0 & 0 & 0 \\ 0 & 0 & 0 \end{bmatrix} = -\begin{bmatrix} 0 & \bar{1} & 0 \\ 1 & 0 & 0 \\ 0 & 0 & 0 \end{bmatrix} = \mathrm{i}\,\mathrm{I}_3.$$

Exercise 11.4-4

$$\begin{aligned} I_+I_- &= (I_1 + \mathrm{i}I_2)(I_1 - \mathrm{i}I_2) = {I_1}^2 + {I_2}^2 + \mathrm{i}(I_2I_1 - I_1I_2) \\ &= I^2 - {I_3}^2 + I_3; \\ I_-I_+ &= (I_1 - \mathrm{i}I_2)(I_1 + \mathrm{i}I_2) = {I_1}^2 + {I_2}^2 + \mathrm{i}(I_1I_2 - I_2I_1) \\ &= I^2 - {I_3}^2 - I_3, \end{aligned}$$

and so

$$[I_+,\ I_-] = 2I_3. \tag{13}$$

$$[I^2, I_\pm] = [I^2, I_1 \pm \mathrm{i}I_2] = I^2(I_1 \pm \mathrm{i}I_2) - (I_1 \pm \mathrm{i}I_2)I^2 = [I^2, I_1] \pm \mathrm{i}[I^2,\ I_2] = 0. \tag{14}$$

Exercise 11.4-5

$$\begin{aligned} (j \mp m)(j \pm m + 1) &= j^2 \pm jm + j \mp jm - m^2 \mp m \\ &= j^2 + j - m^2 \mp m = j(j+1) - m(m \pm 1). \end{aligned}$$

11.5 The irreducible representations of SO(3)

From eqs. (11.3.17) and the remark following (11.4.20), any rotation $R(\phi\ \mathbf{n})$ can be expressed in terms of the angular momentum operators J_x, J_y, J_z, and therefore in terms of $J_\pm$, J_z. Since $\hat{J}_z|j\ m\rangle = m\,|j\ m\rangle$ and $\hat{J}_\pm|j\ m\rangle = c_\pm|j\ m \pm 1\rangle$, the set (11.4.39) is transformed by $R(\phi\ \mathbf{n})$ into a linear combination of the same set. Moreover, no smaller subset will suffice because $J_\pm$ always raise or lower m, while leaving j unaltered, until the ends of the ladder at $m = \pm j$ are reached. Consequently,

$$\langle u^j_m| = \langle u^j_j,\ u^j_{j-1}, \ldots,\ u^j_{-j}| \tag{1}$$

is an irreducible basis for representations of SO(3). For a given j, the notation for the basis $\langle u^j_m|$ in eq. (1) is conveniently shortened to $\langle m| = \langle\, j\ j-1\ \ \ldots\ \ -j|$.

(11.4.19)
$$\mathrm{i}\,\hat{\mathbf{J}} = \mathrm{r} \times \nabla = \begin{bmatrix} \mathbf{x} & \mathbf{y} & \mathbf{z} \\ x & y & z \\ \partial/\partial x & \partial/\partial y & \partial/\partial z \end{bmatrix}; \tag{2}$$

(2)
$$\begin{aligned} \hat{J}_x &= -\mathrm{i}\ [y\ \partial/\partial z - z\ \partial/\partial y], \\ \hat{J}_y &= -\mathrm{i}\ [z\ \partial/\partial x - x\ \partial/\partial z], \\ \hat{J}_z &= -\mathrm{i}\ [x\ \partial/\partial y - y\ \partial/\partial x]. \end{aligned} \tag{3}$$

Transformation from Cartesian coordinates $\{x\,y\,z\}$ to spherical polar coordinates $\{r\,\theta\,\varphi\}$ is a standard exercise, which yields

(3)
$$\hat{J}_x = \mathrm{i}\left[\sin\varphi\ \frac{\partial}{\partial\theta} + \cot\theta\cos\varphi\ \frac{\partial}{\partial\varphi}\right], \tag{4}$$

$$\hat{J}_y = -\mathrm{i}\left[\cos\varphi\ \frac{\partial}{\partial\theta} - \cot\theta\sin\varphi\ \frac{\partial}{\partial\varphi}\right], \tag{5}$$

$$\hat{J}_z = -\mathrm{i}\ [\partial/\partial\varphi]. \tag{6}$$

(4), (5), (11.4.7)
$$\hat{J}_\pm = \hat{J}_x \pm \mathrm{i}\,\hat{J}_y = \pm\exp(\pm\mathrm{i}\,\varphi)\left[\frac{\partial}{\partial\theta} \pm \mathrm{i}\cot\theta\ \frac{\partial}{\partial\varphi}\right]. \tag{7}$$

Integer values of j will now be distinguished by replacing j by $l\,(=0, 1, 2, \ldots)$ and u^j_m by Y^m_l. When $m = l$,

(7)
$$\hat{J}_+ Y^l_l(\theta,\varphi) = \exp(\mathrm{i}\,\varphi)\left[\frac{\partial}{\partial\theta} + \mathrm{i}\cot\theta\ \frac{\partial}{\partial\varphi}\right] Y^l_l(\theta,\varphi) = 0. \tag{8}$$

This first-order differential equation may be solved by separation of variables

$$Y^m_l(\theta,\varphi) = \Theta_{lm}(\theta)\Phi_m(\varphi), \tag{9}$$

(8), (9), (6), (11.4.21)
$$Y^l_l(\theta,\varphi) = N_l \sin^l\theta\exp(\mathrm{i}l\varphi). \tag{10}$$

Exercise 11.5-1 Fill in the steps leading from eq. (9) to eq. (10).

The remaining $2l$ eigenfunctions with $m = l-1, l-2, \ldots, -l$ follow from eq. (10) by successive use of the lowering operator $\hat{J}_-$, eq. (11.4.41). The normalization factor is

$$N_l = (-1)^l[(2l+1)!/4\pi]^{1/2}/2^l l!, \tag{11}$$

where the phase factor $(-1)^l$ is included to satisfy the CS phase convention. As anticipated by the introduction of the symbol Y^m_l for u^j_m when j is an integer, the functions $Y^m_l(\theta,\varphi)$ are the spherical harmonics

$$Y_l^m(\theta, \varphi) = \mathrm{i}^{m+|m|}\ P_l^{|m|}(\cos\theta)\exp(\mathrm{i}m\,\varphi), \tag{12}$$

where the associated Legendre functions

$$P_l^{|m|}(\cos\theta) = \left\{\frac{(2l+1)(l-|m|)!}{4\pi(l+|m|)!}\right\}^{1/2}\frac{1}{2^l l!}\sin^{|m|}\theta\,\frac{\mathrm{d}^{l+|m|}(\cos^2\theta-1)^l}{\mathrm{d}(\cos\theta)^{l+|m|}}. \tag{13}$$

These are the eigenfunctions of the Laplacian operator ∇^2 over the unit sphere. With the phase factor included in eq. (12), they satisfy the CS convention, a consequence of which is that

$$Y_l^{m\,*} = (-1)^m Y_l^{-m}. \tag{14}$$

Exercise 11.5-2 Show that eq. (14) is consistent with eqs. (12) and (13).

Exercise 11.5-3

(a) Show from eqs. (12) and (13) that the three spherical harmonics for $l = 1$ are $Y_1^m = (3/4\pi)^{1/2} y_m$, with $y_1 = -2^{-1/2}(x+\mathrm{i}y)$, $y_0 = z$, $y_{-1} = 2^{-1/2}(x-\mathrm{i}y)$.
(b) Find the MR of $R(\phi\ \mathbf{z})$ for the basis $\langle y_1\ y_0\ y^{-1}|$.
(c) Find the matrix U defined by $|\ U_1\ U_0\ U_{-1}\rangle = \mathrm{U}|x\ y\ z\rangle$, where

$$U_1 = 2^{-1/2}(x-\mathrm{i}y),\ \ U_0 = -z,\ \ U_{-1} = -2^{-1/2}(x+\mathrm{i}y). \tag{15}$$

Show explicitly that the transformation matrix U is a unitary matrix and evaluate det U.
(d) Show that the MR $\Gamma^{(U)}(R)$ defined by

$$R(\phi\ \mathbf{z})|U_1\ \ U_0\ \ U_{-1}\rangle = \Gamma^{(U)}(R)|U_1\ \ U_0\ \ U_{-1}\rangle$$

is the same matrix as the MR of $R(\phi\ \mathbf{z})$ for the basis $\langle y_1\ y_0\ y_{-1}|$.

Although this is shown in (d) only for $R(\phi\ \mathbf{z})$, $|U_1\ \ U_0\ \ U_{-1}\rangle$ and $\langle y_1\ y_0\ y_{-1}|$ are in fact transformed by the same matrix $\Gamma(R)$ under any proper or improper rotation and are thus dual bases. $|U_1\ \ U_0\ \ U_{-1}\rangle$ is called a spherical vector and the superscript $^{(U)}$ in (d) serves as a reminder of the basis.

For integer or half-integer j, the CS phase convention requires that

$$(u_m^j)^* = (-1)^m\ u_{-m}^j. \tag{16}$$

Fano and Racah (1959) employ a different convention, which results in

$$(u_m^j)^* = (-1)^{j-m} u_{-m}^j. \tag{16$'$}$$

(11.3.17)
$$\hat{R}(\phi\ \mathbf{z})u_m^j = \exp(-\mathrm{i}\phi\hat{J}_z)u_m^j = \sum_{n=0}^{\infty}(n!)^{-1}(-\mathrm{i}\phi\hat{J}_z)^n\ u_m^j$$
$$= \sum_{n=0}^{\infty}(n!)^{-1}(-\mathrm{i}\phi m)^n u_m^j = \exp(-\mathrm{i}\phi m)u_m^j; \tag{17}$$

(17) $$\hat{R}(\phi\ \mathbf{z})\langle j\ j-1\ \ldots\ -j| = \langle j\ j-1\ \ldots\ -j|$$

$$= \begin{bmatrix} e^{-i\phi j} & & & \\ & e^{-i\phi(j-1)} & & \\ & & \ldots & \\ & & & e^{i\phi j} \end{bmatrix}. \quad (18)$$

The sum of the diagonal elements of the MR $\Gamma(\phi)$ of the symmetry operator $R(\phi\ \mathbf{z})$ forms a geometric series which we have summed before in Section 7.2 with j replaced by l, an integer. As before,

$$\chi^j(\phi) = \sin[(2j+1)\phi/2]/\sin(\phi/2). \quad (19)$$

Although eq. (19) has been derived for $R(\phi\ \mathbf{z})$, all rotations $R(\phi\ \mathbf{n})$ through the same angle ϕ are in the same class (and therefore have the same character) irrespective of the orientation of the unit vector $\mathbf{n}$. Therefore, eq. (19) holds for a rotation through ϕ about *any* rotation axis. A formal proof that

$$R(\theta\ \mathbf{m})R(\phi\ \mathbf{n})R(\theta\ \mathbf{m})^{-1} = R(\phi\ \mathbf{n}'), \quad (20)$$

where $R(\theta\ \mathbf{m}) \in \{R(\phi\ \mathbf{n})\}$ and

$$\mathbf{n}' = R(\theta\ \mathbf{m})\mathbf{n}, \quad (21)$$

will be provided in Chapter 12 using the quaternion representation. Equation (21) is a particular case of the effect of a general rotation on a vector: this is called the *conical transformation* because under this transformation the vector traces out the surface of a cone. The proof of the conical transformation, namely that if $\mathbf{r}' = R(\phi\ \mathbf{n})\mathbf{r}$, then

$$\mathbf{r}' = \cos\phi\ \mathbf{r} + \sin\phi(\mathbf{n}\times\mathbf{r}) + (1-\cos\phi)(\mathbf{n}\cdot\mathbf{r})\mathbf{n}, \quad (22)$$

is set in Problem 11.4. Notice that in eq. (11.4.16) the rotation is through a small angle ϕ and that eq. (11.4.16) agrees with eq. (22) to first order in ϕ.

(19) $$\chi^j(\phi+2\pi) = (-1)^{2j}\chi^j(\phi); \quad (23)$$

(23) $$\chi^j(\phi+2\pi) = \chi^j(\phi), \text{ when } j \text{ is an integer}; \quad (24)$$

(23) $$\chi^j(\phi+2\pi) = -\chi^j(\phi) \text{ when } j \text{ is a half-integer}. \quad (25)$$

The matrices $\Gamma(\phi)$ *do not*, therefore, form a representation of $R(\phi\ \mathbf{n})$ for half-integer j. There are two ways out of this dilemma. In one, due to Bethe (1929) and described in Chapter 8, a new operator $\overline{E} = R(2\pi\ \mathbf{n})$ is introduced, thus doubling the size of the group G by replacing it by the *double group* $\overline{\mathrm{G}} = \{G\} + \overline{E}\{\mathrm{G}\}$. The other approach (Altmann (1986)) introduces no new operators but employs instead a different type of representation called a multiplier, or projective, or ray representation. This approach will be described in Chapter 12.

The usefulness of the characters $\{\chi^j(R)\}$ of a representation j stems largely from the orthogonality theorem of Section 4.4, which for a finite group of order g, is that

$$g^{-1} \sum_R \chi^j(R)^* \chi^{j'}(R) = \delta_{jj'}. \tag{26}$$

In SO(3) the characters for the different classes depend only on ϕ which varies continuously in the range $-\pi < \phi \leq \pi$. The orthogonality condition eq. (26) is therefore to be replaced by the integral

$$\int_{-\pi}^{\pi} \chi^j(R)^* \chi^{j'}(R) \, \mathrm{d}\mu(\phi) = \delta_{jj'}, \tag{27}$$

where $\mathrm{d}\mu(\phi)$ (called the measure) must satisfy the condition

$$\int_{-\pi}^{\pi} \mathrm{d}\mu(\phi) = 1 \tag{28}$$

in order that the property under the integral sign (which here is $\chi^j(R)^* \, \chi^{j'}(R)$) will form a normalized distribution. The measure $\mathrm{d}\mu(\phi)$ must also ensure that the group rearrangement theorem is satisfied, which means that the integral must be invariant when each $R(\phi \ \mathbf{n}) \in \mathrm{G}$ is multiplied by one particular element so that $R(\phi \ \mathbf{n})$ becomes $R(\phi' \ \mathbf{n}') \in \mathrm{G}$. This property of integral invariance together with eq. (28) is sufficient to determine $\mathrm{d}\mu(\phi)$. (Mathematical details are given by Wigner (1959) on p. 152, by Jones (1990) in his Appendix C, and by Kim (1999) on p. 211.) Integral invariance, eq. (28), and eq. (27), are satisfied by

$$\mathrm{d}\mu(\phi) = \mathrm{d}\phi \, (1 - \cos\phi)/2\pi. \tag{29}$$

Exercise 11.5-4 Prove that $\mathrm{d}\mu(\phi)$ given by eq. (29) satisfies eq. (28) and ensures that the integral, eq. (27), is equal to $\delta_{jj'}$.

(11.4.41), (11.4.24)
$$\hat{J}_\pm|j \ m\rangle = [\,j(j+1) - m(m \pm 1)]^{1/2}|j \ m \pm 1\rangle = c_\pm|j \ m \pm 1\rangle, \tag{30}$$

(11.4.7)
$$\hat{J}_\pm = \hat{J}_x \pm \mathrm{i}\,\hat{J}_y, \tag{31}$$

(11.4.21)
$$\hat{J}_z|j \ m\rangle = m|j \ m\rangle, \tag{32}$$

(30), (31)
$$\hat{J}_x|j \ m\rangle = \frac{1}{2}[c_+|j \ m+1\rangle + c_-|j \ m-1\rangle], \tag{33}$$

(30), (31)
$$\hat{J}_y|j \ m\rangle = \frac{-\mathrm{i}}{2}[c_+|j \ m+1\rangle - c_-|j \ m-1\rangle]. \tag{34}$$

For $j = 1$,

(33)
$$\hat{J}_x|1\ 1\rangle = 2^{-1/2}|1\ 0\rangle$$
$$\hat{J}_x|1\ 0\rangle = 2^{-1/2}(|1\ 1\rangle + |1\ \bar{1}\rangle) \tag{35}$$
$$\hat{J}_x|1\ \bar{1}\rangle = 2^{-1/2}|1\ 0\rangle.$$

Since $j = 1$, the basis functions may be abbreviated to m,

(35)
$$\hat{J}_x\langle 1\ 0\ \bar{1}| = \langle 1\ 0\ \bar{1}|2^{-1/2}\begin{bmatrix} 0 & 1 & 0 \\ 1 & 0 & 1 \\ 0 & 1 & 0 \end{bmatrix} = \langle 1\ 0\ \bar{1}|\mathrm{J}_x^{(y)}. \tag{36}$$

The MRs of J_y, J_z for the spherical harmonics basis $\langle\, m|$ derived in a similar manner are

(34), (32)
$$\mathrm{J}_y^{(y)} = 2^{-1/2}\begin{bmatrix} 0 & -\mathrm{i} & 0 \\ \mathrm{i} & 0 & -\mathrm{i} \\ 0 & \mathrm{i} & 0 \end{bmatrix}; \qquad \mathrm{J}_z^{(y)} = \begin{bmatrix} 1 & 0 & 0 \\ 0 & 0 & 0 \\ 0 & 0 & \bar{1} \end{bmatrix}. \tag{37}$$

The superscript $^{(y)}$ in eqs. (36) and (37) serves to remind us of the basis, namely the $\langle\, y_m|$, which are the spherical harmonics without the normalizing factor $(3/4\pi)^{1/2}$. From Exercise 11.5-3,

$$|U_1\ U_0\ U_{-1}\rangle = \mathrm{U}|x\ y\ z\rangle,$$

where

$$\mathrm{U} = 2^{-1/2}\begin{bmatrix} 1 & -\mathrm{i} & 0 \\ 0 & 0 & -2^{1/2} \\ -1 & -\mathrm{i} & 0 \end{bmatrix}.$$

Therefore, with $k = x, y, z$,

$$\hat{J}_k|U_1\ U_0\ U_{-1}\rangle = \mathrm{J}_k^{(\mathrm{U})}|U_1\ U_0\ U_{-1}\rangle = \mathrm{J}_k^{(\mathrm{U})}\mathrm{U}|x\ y\ z\rangle,$$

$$\hat{J}_k\ \mathrm{U}|x\ y\ z\rangle = \mathrm{U}\hat{J}_k|x\ y\ z\rangle = \mathrm{U}\mathrm{J}_k|x\ y\ z\rangle$$

$$\mathrm{J}_k = \mathrm{U}^{-1}\ \mathrm{J}_k^{(\mathrm{U})}\mathrm{U} = \mathrm{U}^\dagger\ \mathrm{J}_k^{(\mathrm{U})}\mathrm{U}, \tag{38}$$

(38)
$$\mathrm{J}_x = \begin{bmatrix} 0 & 0 & 0 \\ 0 & 0 & -\mathrm{i} \\ 0 & \mathrm{i} & 0 \end{bmatrix}, \quad \mathrm{J}_y = \begin{bmatrix} 0 & 0 & \mathrm{i} \\ 0 & 0 & 0 \\ -\mathrm{i} & 0 & 0 \end{bmatrix}, \quad \mathrm{J}_z = \begin{bmatrix} 0 & -\mathrm{i} & 0 \\ \mathrm{i} & 0 & 0 \\ 0 & 0 & 0 \end{bmatrix}, \tag{39}$$

in agreement with eqs. (11.3.6), (11.3.8) and (11.3.3). The lack of superscripts in eqs. (39) indicates that the basis is $\langle x \; y \; z \mid (= \langle r_1 \; r_2 \; r_3 |)$; however, the basis is shown by a superscript when there may be grounds for confusion.

Exercise 11.5-5 Derive the MRs in eqs. (39) from the corresponding ones for the basis $\langle u^1_m |$ (that is, $\langle m |$) in eqs. (36) and (37).

The operators and bases introduced above are not the only possible choices for SO(3). Consider the functions

$$|j \; m\rangle = [(j+m)!(j-m)!]^{-1/2} \, x^{j+m} \, y^{j-m}. \tag{40}$$

Then

$$\text{(40)} \qquad x(\partial/\partial y)|j \; m\rangle = \sqrt{\frac{(j-m)(j+m+1)}{(j+m+1)!(j-m-1)!}} \, x^{j+m+1} \, y^{j-m-1} = c_+ |j \; m+1\rangle. \tag{41}$$

Therefore, for this basis,

$$\hat{J}_+ = x \frac{\partial}{\partial y}. \tag{42}$$

$$\text{(40)} \qquad \frac{1}{2}\left[x\frac{\partial}{\partial x} - y\frac{\partial}{\partial y}\right]|j \; m\rangle = \frac{1}{2}\frac{(j+m)-(j-m)}{\sqrt{(j+m)!(j-m)!}} x^{j+m} \, y^{j-m}, \tag{43}$$

so that

$$\hat{J}_z|j \; m\rangle = \frac{1}{2}\left[x\frac{\partial}{\partial x} - y\frac{\partial}{\partial y}\right]|j \; m\rangle = m|j \; m\rangle. \tag{44}$$

(See Problem 11.5 for eqs. (45), (46), and (47).)

Answers to Exercises 11.5

Exercise 11.5-1

$$\text{(9), (8)} \qquad \frac{1}{\mathrm{i}\cot\theta \; \Theta_{ll}(\theta)} \frac{\partial \Theta_{ll}(\theta)}{\partial \theta} + \frac{1}{\Phi_l(\varphi)} \frac{\partial \Phi_l(\varphi)}{\partial \varphi} = 0.$$

Consequently, the first term is $-c$ and the second one is $+c$, where c is a constant to be determined. Equation (11.4.21) gives $c = \mathrm{i}l$, whence $\Phi_1(\varphi) = N_l \exp(\mathrm{i}l\varphi)$ (unnormalized) and $Y^l_l(\theta,\varphi) = N_l (\sin\theta)^l \exp(\mathrm{i}l\varphi)$. Normalization involves a standard integral and gives eq. (11) after including the CS phase factor.

Exercise 11.5-2

(12)
$$(-1)^m\, Y_l^{-m} = (-1)^m \mathrm{i}^{-m+|m|}\, P_l^{|m|}\,(\cos\theta)\,\exp\,(-\mathrm{i}m\varphi)$$
$$= (-\mathrm{i})^{m+|m|}\, P_l^{|m|}(\cos\theta)\,\exp(-\mathrm{i}m\varphi) = Y_l^{m*}.$$

Exercise 11.5-3

(a) (13)
$$Y_1^1 = \mathrm{i}^{1+1} P_1^1(\cos\theta)\,\exp(\mathrm{i}\varphi) = -(3/8\pi)^{1/2}\sin\theta\exp(\mathrm{i}\varphi)$$
$$= (3/4\pi)^{1/2}[-2^{-1/2}(x+\mathrm{i}y)],$$
$$Y_1^0 = (3/4\pi)^{1/2}(1/2)(\sin\theta)^0\,(2\cos\theta) = (3/4\pi)^{1/2}z,$$
$$Y_1^{-1} = (3/4\pi)^{1/2}\,2^{-1/2}\sin\theta\exp(-\mathrm{i}\varphi) = (3/4\pi)^{1/2}[2^{-1/2}(x-\mathrm{i}y)],$$
$$Y_1^m = (3/4\pi)^{1/2}\,y_m,$$

with

$$y_1 = -2^{-1/2}(x+\mathrm{i}y),\ y_0 = z,\ y_{-1} = 2^{-1/2}(x-\mathrm{i}y).$$

(b)
$$R(\phi\ \mathbf{z})\langle y_1\ y_0\ y_{-1}| = \langle y_1\ y_0\ y_{-1}|\begin{bmatrix} \mathrm{e}^{-\mathrm{i}\phi} & & \\ & 1 & \\ & & \mathrm{e}^{\mathrm{i}\phi}\end{bmatrix} = \langle y_1\ y_0\ y_{-1}|\Gamma(R)^{(y)}.$$

(c) The adjoint of $\langle y_1\ y_0\ y_{-1}|$ is

$$|-2^{-1/2}(x-\mathrm{i}y),\ z,\ 2^{-1/2}(x+\mathrm{i}y)\rangle = -2^{-1/2}\begin{bmatrix} 1 & -\mathrm{i} & 0 \\ 0 & 0 & -2^{-1/2} \\ -1 & -\mathrm{i} & 0\end{bmatrix}\begin{bmatrix} x \\ y \\ z\end{bmatrix}$$
$$= -|U_1\ U_0\ U_{-1}\rangle,$$

where $U_1 = 2^{1/2}(x-\mathrm{i}y)$, $U_0 = -z$, $U_{-1} = -2^{-1/2}(x+\mathrm{i}y)$.

$$|U_1\ U_0\ U_{-1}\rangle = \mathrm{U}|x\ y\ z\rangle, \qquad \mathrm{U} = 2^{-1/2}\begin{bmatrix} 1 & -\mathrm{i} & 0 \\ 0 & 0 & -2^{1/2} \\ -1 & -\mathrm{i} & 0\end{bmatrix}.$$

$\mathrm{U}^\dagger\,\mathrm{U} = \mathrm{E}_3$; therefore U is a unitary matrix. $|\mathrm{U}| = 1[-\mathrm{i}\ -\mathrm{i}] = -2\mathrm{i}$.

(d)
$$R(\phi\ \mathbf{z})|U_1\ U_0\ U_{-1}\rangle = R(\phi\ \mathbf{z})\mathrm{U}|x\ y\ z\rangle = \mathrm{U}R(\phi\ \mathbf{z})|x\ y\ z\rangle$$

$$= 2^{-1/2}\begin{bmatrix} 1 & -\mathrm{i} & 0 \\ 0 & 0 & -2^{1/2} \\ -1 & -\mathrm{i} & 0\end{bmatrix}\begin{bmatrix} cx-sy \\ sx+cy \\ z\end{bmatrix} = \begin{bmatrix} \mathrm{e}^{-\mathrm{i}\phi} & & \\ & 1 & \\ & & \mathrm{e}^{\mathrm{i}\phi}\end{bmatrix}\begin{bmatrix} U_1 \\ U_0 \\ U_{-1}\end{bmatrix}.$$

Since $\Gamma\,(R)^{(U)} = \Gamma(R)^{(y)}$, the bases $|U_1\ U_0\ U_{-1}\rangle$ and $\langle y_1\ y_0\ y_{-1}|$ are transformed by the same matrix under $R(\phi\ \mathbf{z})$.

Exercise 11.5-4

$$\int_{-\pi}^{\pi} \frac{(\mathrm{d}\phi/2\pi)(1-\cos\phi)\sin[(2j+1)(\phi/2)]\sin[(2j'+1)(\phi/2)]}{\sin(\phi/2)\sin(\phi/2)}$$

$$= \int_{-\pi}^{\pi} (\mathrm{d}\phi/2\pi)2\sin[(2j+1)(\phi/2)]\sin[(2j'+1)(\phi/2)]$$

$$= \int_{-\pi}^{\pi} (\mathrm{d}\phi/2\pi)[\cos(j-j')\phi - \cos(j+j'+1)\phi]$$

$$= 0,\ \text{if } j \neq j',\ \text{or} = 1,\ \text{if } j = j'.$$

Exercise 11.5-5 This exercise simply requires filling in the steps leading from eq. (38) to eqs. (39). For example, for $k=x$,

$$\mathrm{J}_x = \mathrm{U}^\dagger \mathrm{J}_x^{(\mathrm{U})} \mathrm{U} = 2^{-3/2} \begin{bmatrix} 1 & 0 & -1 \\ \mathrm{i} & 0 & \mathrm{i} \\ 0 & -2^{1/2} & 0 \end{bmatrix} \begin{bmatrix} 0 & 1 & 0 \\ 1 & 0 & 1 \\ 0 & 1 & 0 \end{bmatrix} \begin{bmatrix} 1 & -\mathrm{i} & 0 \\ 0 & 0 & -2^{1/2} \\ -1 & -\mathrm{i} & 0 \end{bmatrix}$$

$$= 2^{-3/2} \begin{bmatrix} 1 & 0 & -1 \\ \mathrm{i} & 0 & \mathrm{i} \\ 0 & -2^{1/2} & 0 \end{bmatrix} \begin{bmatrix} 0 & 0 & -2^{1/2} \\ 0 & -2\mathrm{i} & 0 \\ 0 & 0 & -2^{1/2} \end{bmatrix} = \begin{bmatrix} 0 & 0 & 0 \\ 0 & 0 & -\mathrm{i} \\ 0 & \mathrm{i} & 0 \end{bmatrix}.$$

Similarly, for $k=y$ and $k=z$.

11.6 The special unitary group SU(2)

In the same manner as in Section 11.5 (eqs. (11.5.35)–(11.5.37)), application of eqs. (11.5.32)–(11.5.34) to the basis $\langle m| = \langle \tfrac{1}{2}\ \ -\tfrac{1}{2}|$ gives the MRs

$$\mathrm{J}_x = \tfrac{1}{2}\begin{bmatrix} 0 & 1 \\ 1 & 0 \end{bmatrix}, \quad \mathrm{J}_y = \tfrac{1}{2}\begin{bmatrix} 0 & -\mathrm{i} \\ \mathrm{i} & 0 \end{bmatrix}, \quad \mathrm{J}_z = \tfrac{1}{2}\begin{bmatrix} 1 & 0 \\ 0 & \bar{1} \end{bmatrix}, \tag{1}$$

(1) $$\mathrm{J}_+ = \begin{bmatrix} 0 & 1 \\ 0 & 0 \end{bmatrix}, \quad \mathrm{J}_- = \begin{bmatrix} 0 & 0 \\ 1 & 0 \end{bmatrix}. \tag{2}$$

With the factor ½ removed, these three matrices in eqs. (1) are the Pauli matrices $\sigma_1, \sigma_2, \sigma_3$ (in the CS convention):

$$\sigma_1 = \begin{bmatrix} 0 & 1 \\ 1 & 0 \end{bmatrix}, \quad \sigma_2 = \begin{bmatrix} 0 & -\mathrm{i} \\ \mathrm{i} & 0 \end{bmatrix}, \quad \sigma_3 = \begin{bmatrix} 1 & 0 \\ 0 & \bar{1} \end{bmatrix}. \tag{3}$$

The MR of the rotation operator for the 2-D basis $\langle \tfrac{1}{2}\ \ -\tfrac{1}{2}|$ is

(11.3.18), (11.3.15) $$\Gamma^{1/2}(\phi\ \mathbf{n}) = \exp\left(-\tfrac{1}{2}\,\mathrm{i}\phi\mathbf{n}\cdot\boldsymbol{\sigma}\right), \tag{4}$$

where

$$\mathbf{n}\cdot\boldsymbol{\sigma} = n_1\sigma_1 + n_2\sigma_2 + n_3\sigma_3 = \begin{bmatrix} n_3 & n_1 - in_2 \\ n_1 + in_2 & -n_3 \end{bmatrix}, \tag{5}$$

which is a 2×2 traceless Hermitian matrix, as are the MRs of the three angular momentum operators (infinitesimal generators) in eq. (1).

Exercise 11.6-1 Prove that

$$(\mathbf{n}\cdot\boldsymbol{\sigma})^2 = {\sigma_1}^2 = {\sigma_2}^2 = {\sigma_3}^2 = \mathrm{E}_2, \tag{6}$$

where E_2 denotes the 2×2 unit matrix. Hence show that the CR $[\sigma_1,\ \sigma_2] = 2\mathrm{i}\sigma_3$ reduces to

$$\sigma_1\sigma_2 = \mathrm{i}\sigma_3 = -\sigma_2\sigma_1. \tag{7}$$

Exercise 11.6-2 Show that the 2×2 matrix $\mathbf{n}\cdot\boldsymbol{\sigma}$ is unitary.

Exercise 11.6-3 Show that

$$\Gamma^{1/2}(\phi\ \mathbf{n}) = \cos(\tfrac{1}{2}\phi)\mathrm{E}_2 - \mathrm{i}\sin(\tfrac{1}{2}\phi)\mathbf{n}\cdot\boldsymbol{\sigma}. \tag{8}$$

Exercise 11.6-4 Show that

(a)
$$\Gamma^{1/2}(\phi\ \mathbf{n}) = \begin{bmatrix} \cos(\tfrac{1}{2}\phi) - in_3\sin(\tfrac{1}{2}\phi) & -(n_2 + in_1)\sin(\tfrac{1}{2}\phi) \\ (n_2 - in_1)\sin(\tfrac{1}{2}\phi) & \cos(\tfrac{1}{2}\phi) + in_3\sin(\tfrac{1}{2}\phi) \end{bmatrix}; \tag{9}$$

(b)
$$\Gamma^{1/2}(\phi\ \mathbf{z}) = \begin{bmatrix} \exp(-\tfrac{1}{2}\phi) & 0 \\ 0 & \exp(\tfrac{1}{2}\mathrm{i}\phi) \end{bmatrix}; \tag{10}$$

(c)
$$\Gamma^{1/2}(\beta\ \mathbf{y}) = \begin{bmatrix} \cos(\tfrac{1}{2}\beta) & -\sin(\tfrac{1}{2}\beta) \\ \sin(\tfrac{1}{2}\beta) & \cos(\tfrac{1}{2}\beta) \end{bmatrix}. \tag{11}$$

Exercise 11.6-5 Show that $\det \mathbf{n}\cdot\boldsymbol{\sigma} = -1$.

Matrices which represent proper rotations are unimodular, that is they have determinant $+1$ and are unitary (orthogonal, if the space is real, as is $\Re^3$). Consider the set of all 2×2 unitary matrices with determinant $+1$. With binary composition chosen to be matrix multiplication, the set is associative and closed. It also contains the identity E and each element has an inverse. This set therefore forms a group, the special unitary group of order 2, or SU(2). The most general form for a matrix of SU(2) is

$$\mathrm{A} = \begin{bmatrix} a & b \\ -b^* & a^* \end{bmatrix}, \tag{12}$$

with

$$\det \mathrm{A} = aa^* + bb^* = |a|^2 + |b|^2 = 1, \tag{13}$$

where a and b (which are the Cayley–Klein parameters) are complex, so that there are three independent parameters which describe the rotation.

Exercise 11.6-6 The most general 2×2 matrix is $\begin{bmatrix} a & b \\ c & d \end{bmatrix}$; show that imposing the unitary and unimodular conditions results in eqs. (12) and (13).

The scalar product $\mathrm{H} = \boldsymbol{r} \cdot \boldsymbol{\sigma} = |\, \boldsymbol{r} \,|\, \boldsymbol{n} \cdot \boldsymbol{\sigma}$, where $\mathbf{r}$ is a position vector in 3-D space with (real) components $\{x\ y\ z\}$, is

(3) $$\mathrm{H} = (x\sigma_1 + y\sigma_2 + z\sigma_3) = \begin{bmatrix} z & x - \mathrm{i}y \\ x + \mathrm{i}y & -z \end{bmatrix} = \begin{bmatrix} h_{11} & h_{12} \\ h_{21} & h_{22} \end{bmatrix}. \tag{14}$$

H is also a traceless Hermitian matrix, with $\det \mathrm{H} = -(x^2 + y^2 + z^2)$.

(14) $$x = (h_{21} + h_{12})/2;\ y = (h_{21} - h_{12})/2\mathrm{i};\ z = h_{11} = -h_{22}. \tag{15}$$

Under the transformation represented by the matrix A,

(14) $$\mathrm{H} \rightarrow \mathrm{H}' = \mathrm{A\ H\ A}^{-1} = \mathbf{r}' \cdot \boldsymbol{\sigma} = \begin{bmatrix} z' & x' - \mathrm{i}y' \\ x' + \mathrm{i}y' & -z' \end{bmatrix} = \begin{bmatrix} h'_{11} & h'_{12} \\ h'_{21} & h'_{22} \end{bmatrix}, \tag{16}$$

so that $\mathbf{r}$ is transformed into the new vector $\mathbf{r}'$ with components $\{x'\ y'\ z'\}$.

16 $$\det \mathrm{H}' = -(x'^2 + y'^2 + z'^2) = \det \mathrm{H} = -(x^2 + y^2 + z^2). \tag{17}$$

Therefore, associated with each 2×2 matrix A there is a 3×3 matrix $\Gamma(\mathrm{A})$ which transforms $\mathbf{r}$ into $\mathbf{r}'$. That this transformation preserves the length of $\mathbf{r}$ follows from eq. (17). But if the lengths of all vectors are conserved, the transformation is a rigid rotation. Therefore, each matrix A of SU(2) is associated with a proper rotation in $\Re^3$. The transformation properties of A or the unimodular condition (13) are unaffected if A is replaced by $-$ A. The matrices of SU(2) therefore have an inherent double-valued nature: the replacement of A by $-$ A does not affect the values of $\{x'\ y'\ z'\}$ in eq. (16). Therefore A and $-$ A correspond to the same rotation in $\Re^3$. A unitary transformation (16) in SU(2) induces an orthogonal transformation in $\Re^3$ but because of the sign ambiguity in A the relationship of SU(2) matrices to SO(3) matrices is a 2 : 1 relationship. This relationship is in fact a homomorphous mapping, although the final proof of this must wait until Section 11.8.

(16), (12) $$|x'\ y'\ z'\rangle = \Gamma(\mathrm{A})|x\ y\ z\rangle, \tag{18}$$

$$\Gamma(\mathbb{A}) = \begin{bmatrix} \frac{1}{2}(a^2 - b^2 + a^{*^2} - b^{*^2}) & -\frac{i}{2}(a^2 + b^2 - a^{*^2} - b^{*^2}) & -(ab + a^*b^*) \\ \frac{i}{2}(a^2 - b^2 - a^{*^2} + b^{*^2}) & \frac{1}{2}(a^2 + b^2 + a^{*^2} + b^{*^2}) & -i(ab - a^*b^*) \\ ab^* + a^*b & -i(ab^* - a^*b) & aa^* - bb^* \end{bmatrix}. \quad (19)$$

The matrix in eq. (19) represents a general rotation $R(\phi\ \mathbf{n})$ so that $\Gamma(\mathbb{A}) \in SO(3)$. It is moreover clear from eq. (16) that $\mathbb{A}$ and $-\mathbb{A}$ induce the same orthogonal transformation, since H' is unaffected by the sign change. For example, when

$$a = \exp(-\tfrac{1}{2}i\phi), \quad b = 0, \quad (20)$$

(19) $$\Gamma(\phi\ \mathbf{z}) = \begin{bmatrix} \cos\phi & -\sin\phi & 0 \\ \sin\phi & \cos\phi & 0 \\ 0 & 0 & 1 \end{bmatrix}, \quad (21)$$

which is the MR of $R(\phi\ \mathbf{z})$. Again, if we choose

$$a = \cos(\beta/2) \quad b = -\sin(\beta/2), \quad (22)$$

(19) $$\Gamma(\beta\ \mathbf{y}) = \begin{bmatrix} \cos\beta & 0 & \sin\beta \\ 0 & 1 & 0 \\ -\sin\beta & 0 & \cos\beta \end{bmatrix}, \quad (23)$$

which represents a proper rotation through an angle β about $\mathbf{y}$. Any proper rotation in $\Re^3$ can be expressed as a product of three rotations: about $\mathbf{z}$, about $\mathbf{y}$, and again about $\mathbf{z}$. Therefore, the matrices in eqs. (21) and (23) assume special importance.

O(3) is the group of 3×3 orthogonal matrices, with determinant ± 1, which represent the proper and improper rotations $R(\phi\ \mathbf{n})$, $IR(\phi\ \mathbf{n})$. Removal of the inversion, and therefore all the matrices with determinant -1, gives the subgroup of proper rotations represented by 3×3 orthogonal matrices with determinant $+1$, which is called the special orthogonal group SO(3). To preserve the same kind of notation, adding the inversion to the special unitary group SU(2) of 2×2 unitary unimodular matrices would give the unitary group U(2). But the symbol U(2) is used for the group of *all* 2×2 unitary matrices, so Altmann and Herzig (1982) introduced the name SU′(2) for the group of all 2×2 unitary matrices with determinant ± 1. If $\det \mathbb{A}' = -1$, instead of $\det \mathbb{A} = \pm 1$ (as in Exercise (11.6-6) then

$$\mathbb{A}' = \begin{bmatrix} a & b \\ b^* & -a^* \end{bmatrix} \text{ with } aa^* + bb^* = 1; \quad (24)$$

$\mathbb{A}'$ is a 2×2 unitary matrix with $\det \mathbb{A}' = -1$, and therefore $\mathbb{A}' \in SU'(2)$. The fact that the transformation $\{x\ y\ z\} \rightarrow \{x'\ y'\ z'\}$ effected by eq. (16) is a rotation depends only on the unitarity of $\mathbb{A}$ (see eq. (17)), and, since $\mathbb{A}'$ is also unitary, eq. (16) also describes a rotation when $\mathbb{A}$ is replaced by $\mathbb{A}'$. The difference is that $\mathbb{A} \in SU(2)$ describes a proper rotation whereas $\mathbb{A}' \in SU'(2)$ corresponds to an improper rotation.

$$\Gamma(\mathbb{A}') = \begin{bmatrix} \frac{1}{2}(a^2 - b^2 + a^{*^2} - b^{*^2}) & -\frac{i}{2}(a^2 + b^2 - a^{*^2} - b^{*^2}) & -(ab + a^*b^*) \\ \frac{i}{2}(a^2 - b^2 - a^{*^2} + b^{*^2}) & (a^2 + b^2 + a^{*^2} + b^{*^2}) & -i(ab - a^*b^*) \\ -(ab^* + a^*b) & i(ab^* - a^*b) & -(aa^* - bb^*) \end{bmatrix}, \quad (25)$$

which represents the improper rotation $IR(\phi\ \mathbf{n})$. For $\Gamma(\mathbb{A}')$ to represent the inversion operator, the conditions $x' = -x$, $y' = -y$, $z' = -z$ are met by

$$b = 0,\ aa^* = 1. \tag{26}$$

Either

$$a = -\mathrm{i},\ a^* = +\mathrm{i} \quad \text{or} \quad a = +\mathrm{i},\ a^* = -\mathrm{i}, \tag{27}$$

(26), (27), (24)
$$\mathbb{A}' = \begin{bmatrix} -\mathrm{i} & 0 \\ 0 & -\mathrm{i} \end{bmatrix} \text{ or } \mathbb{A}' = \begin{bmatrix} \mathrm{i} & 0 \\ 0 & \mathrm{i} \end{bmatrix}. \tag{28}$$

The first choice is called the Cartan gauge; the second one is the Pauli gauge (Altmann (1986), p. 108; see also Section 11.8).

Exercise 11.6-7 Find the matrices of SU′(2) that correspond to the MRs in O(3) of the symmetry operators $\sigma_{\mathbf{x}}$, $\sigma_{\mathbf{y}}$, and $\sigma_{\mathbf{z}}$.

A general rotation $R(\phi\ \mathbf{n})$ in $\Re^3$ requires the specification of three independent parameters which can be chosen in various ways. The natural and familiar way is to specify the angle of rotation ϕ and the direction of the unit vector $\mathbf{n}$. (The normalization condition on $\mathbf{n}$ means that there are only three independent parameters.) A second parameterization $R(a\ b)$ introduced above involves the Cayley–Klein parameters a, b. A third common parameterization is in terms of the three Euler angles α, β, and γ (see Section 11.7). Yet another parameterization using the quaternion or Euler–Rodrigues parameters will be introduced in Chapter 12.

Answers to Exercises 11.6

Exercise 11.6-1

(5)
$$(\mathbf{n}\cdot\boldsymbol{\sigma})^2 = \begin{bmatrix} n_3 & n_1 - \mathrm{i}n_2 \\ n_1 + \mathrm{i}n_2 & -n_3 \end{bmatrix}\begin{bmatrix} n_3 & n_1 - \mathrm{i}n_2 \\ n_1 + \mathrm{i}n_2 & -n_3 \end{bmatrix} = \begin{bmatrix} 1 & 0 \\ 0 & 1 \end{bmatrix};$$

$$\sigma_1^2 = \begin{bmatrix} 0 & 1 \\ 1 & 0 \end{bmatrix}\begin{bmatrix} 0 & 1 \\ 1 & 0 \end{bmatrix} = \begin{bmatrix} 1 & 0 \\ 0 & 1 \end{bmatrix};$$

$$\sigma_2^2 = \begin{bmatrix} 0 & -\mathrm{i} \\ \mathrm{i} & 0 \end{bmatrix}\begin{bmatrix} 0 & -\mathrm{i} \\ \mathrm{i} & 0 \end{bmatrix} = \begin{bmatrix} 1 & 0 \\ 0 & 1 \end{bmatrix};$$

$$\sigma_3^2 = \begin{bmatrix} 1 & 0 \\ 0 & -1 \end{bmatrix}\begin{bmatrix} 1 & 0 \\ 0 & -1 \end{bmatrix} = \begin{bmatrix} 1 & 0 \\ 0 & 1 \end{bmatrix};$$

$$[\sigma_2^2\ \sigma_3] = [\mathrm{E}_2\ \sigma_3] = 0 = \sigma_2\ [\sigma_2\ \sigma_3] + [\sigma_2\ \sigma_3]\ \sigma_2 = 2\mathrm{i}(\sigma_2\sigma_1 + \sigma_1\sigma_2).$$

But $\sigma_1\ \sigma_2 = \sigma_2\ \sigma_1 + 2\mathrm{i}\sigma_3$; therefore $\sigma_1\ \sigma_2 = \mathrm{i}\sigma_3 = -\ \sigma_2\ \sigma_1$.

Exercise 11.6-2

$$(5) \qquad (\mathbf{n}\cdot\boldsymbol{\sigma})^{\dagger}(\mathbf{n}\cdot\boldsymbol{\sigma}) = (\mathbf{n}\cdot\boldsymbol{\sigma})(\mathbf{n}\cdot\boldsymbol{\sigma}) = \mathrm{E}_2.$$

Exercise 11.6-3 From eq. (4), $\Gamma^{1/2}(\phi\ \mathbf{n}) = \exp(-{}^1\!/\!_2\, \mathrm{i}\phi\, \mathbf{n}\,.\,\boldsymbol{\sigma})$; expand the exponential and collect terms of even and odd powers of $\mathbf{n}\cdot\boldsymbol{\sigma}$. The even terms give $\cos({}^1\!/\!_2\phi)\mathrm{E}_2$, while the odd powers give $-\mathrm{i}\sin({}^1\!/\!_2\phi)\ \mathbf{n}\cdot\boldsymbol{\sigma}$.

Exercise 11.6-4 Substituting from eq. (5) into eq. (8) gives eq. (9). Setting $n_3 = 1$ in eq. (9) gives eq. (10); setting $n_2 = 1$ in eq. (9) gives eq. (11).

Exercise 11.6-5 From eq. (5), $\det \mathbf{n}\cdot\boldsymbol{\sigma} = -n_3^2 - n_1^2 - n_2^2 = -1$.

Exercise 11.6-6 The unitary condition gives $aa^* + cc^* = 1$, $ab^* + cd^* = 0$, $a^*b + c^*d = 0$, $bb^* + dd^* = 1$, and the unimodular condition is $ad - bc = 1$. These five condititions are satisfied by $d = a^*$, $c = -b^*$, which gives eq. (12).

Exercise 11.6-7 The result of $\sigma_{\mathbf{x}}$ is $x' = -x$, $y' = y$, $z' = z$. Impose the first two conditions on the transformation matrix $\Gamma(\mathbb{A})$ in eq. (25); this gives $a^2 = a^{*2}$ and $b^2 = b^{*^2}$. The condition $z' = z$ gives $-aa^* + bb^* = 1$, which together with $\det \mathbb{A}' = -1$ (see eq. (24)) gives $a = 0$, and therefore (using $y' = y$ again) $b^2 + b^{*^2} = 2$. But $b^2 = b^{*^2}$, so $b^2 = 1$, $b = \pm 1$. In like manner, for $\sigma_{\mathbf{y}}$, $x' = x$, $y' = -y$, and $z' = z$, so that from eq. (24) $a = 0$ and $b = \pm\,\mathrm{i}$. Finally for $\sigma_{\mathbf{z}}$, $x' = x$, $y' = y$, $z' = -z$, so that eq. (25) yields $b = 0$, $a = \pm 1$. The three $\mathbb{A}'$ matrices are therefore

$$\mathbb{A}'(\sigma_{\mathbf{x}}) = \pm\begin{bmatrix} 0 & 1 \\ 1 & 0 \end{bmatrix}, \quad \mathbb{A}'(\sigma_{\mathbf{y}}) = \pm\begin{bmatrix} 0 & \mathrm{i} \\ -\mathrm{i} & 0 \end{bmatrix}, \quad \mathbb{A}'(\sigma_{\mathbf{z}}) = \pm\begin{bmatrix} 1 & 0 \\ 0 & -1 \end{bmatrix}.$$

Note the inherent sign ambiguity in the matrices of SU′(2); the positive and negative signs in the $\mathbb{A}'$ matrices correspond to the same improper rotation in $\Re^3$. The choice of signs $+$, $-$, and $+$ for the three $\mathbb{A}'$ matrices gives the Pauli matrices in the CS sign convention.

11.7 Euler parameterization of a rotation

A general rotation R about any axis $\mathbf{n}$ may be achieved by three successive rotations:

$$\begin{aligned} &\text{(i) } R(\gamma\ \mathbf{z}), && -\pi < \gamma \le \pi; \\ &\text{(ii) } R(\beta\ \mathbf{y}), && 0 \le \beta \le \pi; \\ &\text{(iii) } R(\alpha\ \mathbf{z}), && -\pi < \alpha \le \pi, \end{aligned} \tag{1}$$

where α, β, and γ are the three Euler angles. The total rotation, written $R(\alpha\ \beta\ \gamma)$, is therefore

$$R(\alpha\ \beta\ \gamma) = R(\alpha\ \mathbf{z})\ R(\beta\ \mathbf{y})\ R(\gamma\ \mathbf{z}). \tag{2}$$

The Euler angles have been defined in the literature in several different ways. We are (as is always the case in this book) using the active representation in which the whole of configuration space, including the three mutually perpendicular unit vectors $\{e_1\ e_2\ e_3\}$ firmly embedded in this space, is rotated with respect to fixed orthonormal axes $\{\mathbf{x}\ \mathbf{y}\ \mathbf{z}\}$, collinear with OX, OY, OZ. Here, the order of the rotations and the choice of axes in eq. (2) follow Altmann (1986) and therefore agree, *inter alia*, with the definitions of Biedenharn and Louck (1981), Fano and Racah (1959),* and Rose (1957). There are two ambiguities about the above definition of the Euler angles. Firstly, if $\beta = 0$, only $\alpha + \gamma$ is significant; the second one arises when $\beta = \pi$.

Exercise 11.7-1 Using projection diagrams, show that for an arbitrary angle ω

$$R(\alpha\ \pi\ \gamma) = R(\alpha + \omega, \pi, \gamma + \omega). \tag{3}$$

Exercise 11.7-2 Show that

$$R(-\beta\ \mathbf{y})\, R(\pi\ \mathbf{z}) = R(\pi\ \mathbf{z})\ R(\beta\ \mathbf{y}). \tag{4}$$

This relation provides a mechanism for dealing with a negative angle β. The inverse of $R(\alpha\ \beta\ \gamma)$ results from carrying out the three inverse rotations in reverse order:

(4)
$$\begin{aligned}[R(\alpha\ \beta\ \gamma)]^{-1} &= R(-\gamma\ \mathbf{z})\ R(-\beta\ \mathbf{y})\ R(-\alpha\ \mathbf{z}) \\ &= R(-\gamma\ \mathbf{z})\ R(-\beta\ \mathbf{y})\ R(\pi\ \mathbf{z})\ R(\pi\ \mathbf{z})\ R(-\alpha\ \mathbf{z}) \\ &= R(-\gamma\ \mathbf{z})\ R(\pi\ \mathbf{z})\ R(\beta\ \mathbf{y})\ R(\pi\ \mathbf{z})\ R(-\alpha\ \mathbf{z}) \\ &= R(-\gamma \pm \pi\ \mathbf{z})\ R(\beta\ \mathbf{y})\ R(-\alpha \pm \pi\ \mathbf{z}), \end{aligned} \tag{5}$$

since $R(-\pi\ \mathbf{z})$ would produce the same result as $R(\pi\ \mathbf{z})$. The sign alternatives in eq. (5) ensure that the rotations about $\mathbf{z}$ can always be kept in the stipulated range $-\pi < \alpha, \gamma \leq \pi$. The MR $\Gamma[R(\alpha\ \beta\ \gamma)]$ of $R(\alpha\ \beta\ \gamma)$ is

(2), (11.3.1), (11.3.7)
$$\Gamma[R(\alpha\ \beta\ \gamma)] = \Gamma\,[R(\alpha\ \mathbf{z})]\ \Gamma[R(\beta\ \mathbf{y})]\ \Gamma[R(\gamma\ \mathbf{z})]$$

$$\times \begin{bmatrix} \cos\alpha & -\sin\alpha & 0 \\ \sin\alpha & \cos\alpha & 0 \\ 0 & 0 & 1 \end{bmatrix} \begin{bmatrix} \cos\beta & 0 & \sin\beta \\ 0 & 1 & 0 \\ -\sin\beta & 0 & \cos\beta \end{bmatrix} \begin{bmatrix} \cos\gamma & -\sin\gamma & 0 \\ \sin\gamma & \cos\gamma & 0 \\ 0 & 0 & 1 \end{bmatrix}$$

$$= \begin{bmatrix} \cos\alpha\ \cos\beta\ \cos\gamma - \sin\alpha\ \sin\gamma & -\cos\alpha\ \cos\beta\ \sin\gamma - \sin\alpha\ \cos\gamma & \cos\alpha \sin\beta \\ \sin\alpha\ \cos\beta\ \cos\gamma + \cos\alpha\ \sin\gamma & -\sin\alpha\ \cos\beta\ \sin\gamma + \cos\alpha\ \cos\gamma & \sin\alpha\ \sin\beta \\ -\sin\beta \cos\gamma & \sin\beta \sin\gamma & \cos\beta \end{bmatrix}. \tag{6}$$

It is perhaps opportune to remind the reader that, as is always the case in this book,

$$R(\alpha\ \beta\ \gamma)\langle \mathbf{e}_1\ \mathbf{e}_2\ \mathbf{e}_3| = \langle \mathbf{e}_1{}'\ \mathbf{e}_2{}'\ \mathbf{e}_3{}'| = \langle \mathbf{e}_1\ \mathbf{e}_2\ \mathbf{e}_3|\Gamma[R(\alpha\ \beta\ \gamma)]; \tag{7}$$

$$R(\alpha\ \beta\ \gamma)\ \mathbf{r} = \mathbf{r}' = \Gamma[R(\alpha\ \beta\ \gamma)]|x\ y\ z\rangle = [x'\ y'\ z'\rangle; \tag{8}$$

$$R(\alpha\ \beta\ \gamma)\langle x\ y\ z| = \langle x\ y\ z|\Gamma[R(\alpha\ \beta\ \gamma)]. \tag{9}$$

* Fano and Racah use ψ, θ, ϕ for α, β, γ.

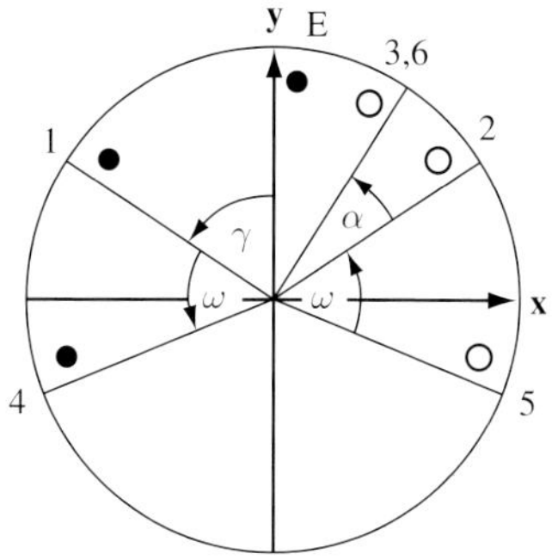

Fig 11.2. The points marked 1 through 6 show the result of acting on the representative point E with the following operators (1) $R(\gamma\ \mathbf{z})$; (2) $R(\pi\ \mathbf{y})\ R(\gamma\ \mathbf{z})$; (3) $R(\alpha\ \mathbf{z})\ R(\pi\ \mathbf{y})\ R(\gamma\ \mathbf{z})$; (4) $R(\gamma+\omega\ \mathbf{z})$; (5) $R(\pi\ \mathbf{y})\ R(\gamma+\omega\ \mathbf{z})$; (6) $R(\alpha+\omega\ \mathbf{z})\ R(\pi\ \mathbf{y})\ R(\gamma+\omega\ \mathbf{z}) = 3$.

Equation (7) describes the transformation of the set of basis vectors $\{\mathbf{e}_1\ \mathbf{e}_2\ \mathbf{e}_3\}$ that are firmly embedded in configuration space and were originally coincident with fixed orthonormal axes $\{\mathbf{x}\ \mathbf{y}\ \mathbf{z}\}$ prior to the application of the symmetry operator $R(\alpha\ \beta\ \gamma)$. In eq. (8) the column matrix $|x\ y\ z\rangle$ contains the variables $\{x\ y\ z\}$, which are the components of the vector $\mathbf{r} = \mathrm{OP}$ and the coordinates of the point P. In eq. (9) the row matrix $\langle x\ y\ z|$ contains the *functions* $\{x\ y\ z\}$ (for example, the angle-dependent factors in the three atomic p functions p_x, p_y, p_z).

Exercise 11.7-3

(a) Write down the transformation matrix $\Gamma[R(\phi\ \mathbf{n})]$ for the rotation $\phi = -2\pi/3$, $\mathbf{n} = 3^{-1/2}\,[1\ 1\ 1]$.

(b) Find the Euler angles and the rotation matrix $\Gamma[R(\alpha\ \beta\ \gamma)]$ for the rotation described in (a) and compare your result for $\Gamma(R)$ with that found in (a).

Answers to Exercises 11.7

Exercise 11.7-1 In Figure 11.2, the points marked (1–6) show the result of acting on the representative point E with the following operators: (1) $R(\gamma\ \mathbf{z})$; (2) $R(\pi\ \mathbf{y})\ R(\gamma\ \mathbf{z})$; (3) $R(\alpha\ \mathbf{z})\ R(\pi\ \mathbf{y})\ R(\gamma\ \mathbf{z})$; (4) $R(\gamma+\omega\ \mathbf{z})$; (5) $R(\pi\ \mathbf{y})\ R(\gamma+\omega\ \mathbf{z})$; and finally (6) $R(\alpha+\omega\ \mathbf{z})\ R(\pi\ \mathbf{y})\ R(\gamma+\omega\ \mathbf{z}) = 3$.

Exercise 11.7-2 In Figure 11.3 (in which the plane of the paper is the plane normal to $\mathbf{y}$), the points marked (1), (2), and (3) show the effect of acting on the representative point E with the following operators: (1) $R(\beta\ \mathbf{y})$; (2) $R(\pi\ \mathbf{z})$; (3) $R(\pi\ \mathbf{z})\ R(\beta\ \mathbf{y})$ or $R(-\beta\ \mathbf{y})\ R(\pi\ \mathbf{z})$.

Exercise 11.7-3

(a) $R(-2\pi/3\ \mathbf{n})\langle\mathbf{e}_1\ \mathbf{e}_2\ \mathbf{e}_3| = \langle\mathbf{e}_3\ \mathbf{e}_1\ \mathbf{e}_2| = \langle\mathbf{e}_1\ \mathbf{e}_2\ \mathbf{e}_3|\begin{bmatrix}0 & 1 & 0\\ 0 & 0 & 1\\ 1 & 0 & 0\end{bmatrix}.$

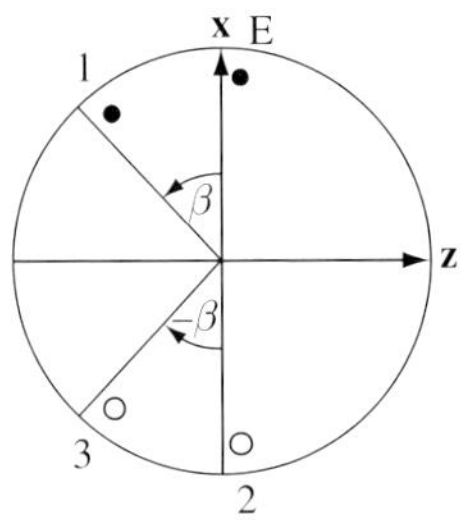

Figure 11.3. The points marked 1 through 3 show the result of acting on the representative point E with the following operators: (1) $R(\beta\ \mathbf{y})$; (2) $R(\pi\ \mathbf{z})$; (3) $R(\pi\ \mathbf{z})R(\beta\ \mathbf{y}) = R(-\beta\ \mathbf{y})R(\pi\ \mathbf{z})$.

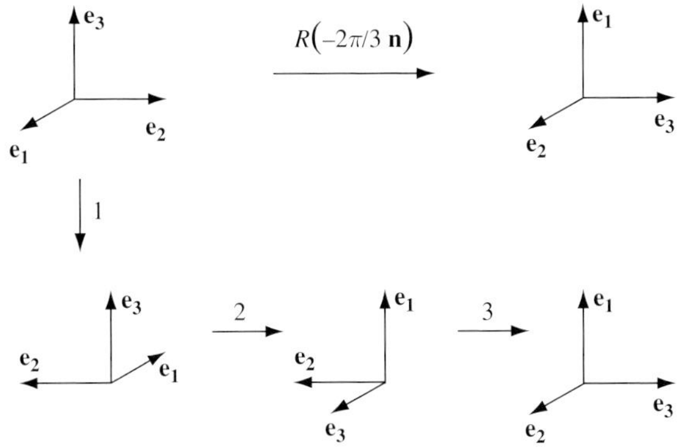

Figure 11.4. Effect on the basis vectors $\{\mathbf{e}_1\ \mathbf{e}_2\ \mathbf{e}_3\}$ of the successive operators (1) $R(\pi\ \mathbf{z})$; (2) $R(\pi/2\ \mathbf{y})$; (3) $R(\pi/2\ \mathbf{z})$. The net result is equivalent to that of the single operator $R(-2\pi/3\ \mathbf{n})$.

(b) Figure 11.4 shows the effect on the basis vectors $\{\mathbf{e}_1\ \mathbf{e}_2\ \mathbf{e}_3\}$ of the successive operators: (1) $R(\pi\ \mathbf{z})$; (2) $R(\pi/2\ \mathbf{y})$; (3) $R(\pi/2\ \mathbf{z})$. The net result is equivalent to that of the single operator $R(-2\pi/3\ \mathbf{n})$, so that for this operation the three Euler angles are $\alpha = \pi/2$, $\beta = \pi/2$, $\gamma = \pi$.

(6)
$$\Gamma[R(\alpha\ \beta\ \gamma)] = \Gamma[R(\pi/2\ \ \pi/2\ \ \pi)] = \begin{bmatrix} 0 & 1 & 0 \\ 0 & 0 & 1 \\ 1 & 0 & 0 \end{bmatrix},$$

in agreement with the transformation matrix obtained in part (a).

11.8 The homomorphism of SU(2) and SO(3)

Any basis $\langle u^j|$ which transforms under the operations of $O(3) = SO(3) \otimes C_i$, where $C_i = \{E\ I\}$, according to

$$R(\phi\ \mathbf{n})\langle u^j| = \langle u^j|\Gamma^j(\phi\ \mathbf{n}), \tag{1}$$

$$I\langle u^j| = (-1)^j\langle u^j|E_{2j+1}, \tag{2}$$

is a *tensor* of *rank* 1 and dimension $2j+1$. Note that $\Gamma^j(\phi\ \mathbf{n})$ is the MR of the rotation operator $R(\phi\ \mathbf{n})$; it is an orthogonal matrix with determinant ± 1 and of dimensions $(2j+1)\times(2j+1)$, since there are $2j+1$ functions in the basis $\langle u^j|$. Also, E_{2j+1} is the $(2j+1)\times(2j+1)$ unit matrix; $\lambda = (-1)^j$ is the *parity* of the basis $\langle u^j|$ and describes its response to the inversion operator I. When $\lambda = +1$ (or -1) the basis is gerade (even) (or ungerade (odd)). The important feature of eqs. (1) and (2) is the two MRs and not the fact that the basis is a row. For example, the spherical vector $|U_1\ U_0\ U_{-1}\rangle$ transforms under the same matrices Γ^1 and $(-1)\mathrm{E}_3$ and so it is a tensor of rank 1 and dimension 3. A tensor of rank 2 is an array that transforms under the operators of O(3) according to the DP representation $\Gamma^{j_1}\otimes\Gamma^{j_2}$, $(-1)^{j_1+j_2}\ \mathrm{E}_{(2j_1+1)(2j_2+1)}$. Such representations are generally reducible. Within SO(3),

$$\langle u^{j_1}|\otimes\langle u^{j_2}| = \langle u^{j_1+j_2}|\oplus\langle u^{j_1+j_2-1}|\oplus\ldots\oplus\langle u^{|j_1-j_2|}|, \tag{3}$$

which is the Clebsch–Gordan series. Taking $j_1\geq j_2$,

(11.5.19), (11.5.18)

$$\begin{aligned}\chi^{j_1}(\phi)\ \chi^{j_2}(\phi) &= \left[2\mathrm{i}\sin(\tfrac{1}{2}\phi)\right]^{-1}\\ &\quad\times\left\{\exp\left[\mathrm{i}(j_1+\tfrac{1}{2})\phi\right]-\exp\left[-\mathrm{i}(j_1+\tfrac{1}{2})\phi\right]\right\}\sum_{m=-j_2}^{j_2}\exp(\mathrm{i}m\phi)\\ &= \left[2\mathrm{i}\sin(\tfrac{1}{2}\phi)\right]^{-1}\sum_{m=-j_2}^{j_2}\left\{\exp\left[\mathrm{i}(j_1+m+\tfrac{1}{2})\phi\right]-\exp\left[-\mathrm{i}(j_1-m+\tfrac{1}{2})\phi\right]\right\}. \end{aligned}\tag{4}$$

But since m runs in integer steps from $-j_2$ to $+j_2$ over the same range of negative and positive values, m may be replaced by $-m$ in the second sum, giving

$$\chi^{j_1}(\phi)\ \chi^{j_2}(\phi) = \sum_{j=j_1-j_2}^{j_1+j_2}\sin\left[(j+\tfrac{1}{2})\phi\right]/\sin(\tfrac{1}{2}\phi) = \sum_{j=j_1-j_2}^{j_1+j_2}\chi^j(\phi), \tag{5}$$

which confirms eq. (3). This is an important relation since it tells us how a whole hierarchy of tensors can be constructed. When $j=0$, u_0^0 is the spherical harmonic of degree zero and the transformation matrices in eqs. (1) and (2) are both just the number 1. Such a tensor that is invariant under rotation and even under inversion is a *scalar*. For $j=½$ the basis functions $\{u_m^j\}$ (which are called spinors) are $u_{½}^{½}$, $u_{-½}^{½}$; or $|½\ \ ½\rangle$, $|½\ \ -½\rangle$; or (avoiding the awkward halves) the ordered pair ξ, η (Lax (1974)), ξ_1, ξ_2 (Tinkham (1964)); u, v (Hammermesh (1962)); or μ_1, μ_2 (Altmann (1986)). To avoid further proliferation of notation, u, v will be used here.

(11.6.10)

$$\begin{aligned}R(\phi\ \mathbf{z})\langle u\,v| &= \langle u\,v|\begin{bmatrix}\exp(-\tfrac{1}{2}\mathrm{i}\phi) & \\ & \exp(\tfrac{1}{2}\mathrm{i}\phi)\end{bmatrix}\\ &= \langle u\,v|\Gamma^{1/2}(\phi\ \mathbf{z}),\end{aligned}\tag{6}$$

where $\Gamma^{1/2}(\phi\ \mathbf{z}) \varepsilon$ SU(2). The MR of the rotation of a basis which is the DP of two such spinor bases,

$$\langle u_1\ v_1| \otimes \langle u_2\ v_2| = \langle u_1u_2\ u_1v_2\ v_1u_2\ v_1v_2|, \tag{7}$$

is

$$\begin{bmatrix} \exp(-\frac{1}{2}\mathrm{i}\phi) & \\ & \exp(\frac{1}{2}\mathrm{i}\phi) \end{bmatrix} \otimes \begin{bmatrix} \exp(-\frac{1}{2}\mathrm{i}\phi) & \\ & \exp(\frac{1}{2}\mathrm{i}\phi) \end{bmatrix}$$
$$= \begin{bmatrix} \exp(-\mathrm{i}\phi) & & & \\ & 1 & & \\ & & 1 & \\ & & & \exp(\mathrm{i}\phi) \end{bmatrix}. \tag{8}$$

The matrix on the RS of eq. (8) is reducible. The RS of eq. (7) is the uncoupled representation for the two spinors. Using a table of Clebsch–Gordan coefficients, the coupled representation is found to be

$$\langle u_1\ v_1| \otimes \langle u_2\ v_2| = \langle u_1u_2\ 2^{-1/2}(u_1v_2 + v_1u_2)\ v_1v_2| \oplus \langle 2^{-1/2}(u_1v_2 - v_1u_2)|, \tag{9}$$

which has been written as a direct sum of symmetric and antisymmetric components because they cannot be converted into one another by any of the operations of O(3).

Exercise 11.8-1 The DP of the two sets $\{u_1\ v_1\}$, $\{u_2\ v_2\}$ is $\{u_1\ v_1\} \otimes \{u_2\ v_2\} = \{u_1u_2\ u_1v_2\ v_1u_2\ v_1v_2\}$. The first and fourth components are symmetric with respect to the transposition operator P_{12}, but the second and third components are not eigenfunctions of P_{12}. Use the symmetrizing ($\mathscr{S}$) and antisymmetrizing ($\mathscr{A}$) operators to generate the symmetric and antisymmetric components in eq. (9) from u_1v_2. [*Hint*: For two objects $\mathscr{S} = 2^{-1/2}[1 + P_{12}]$, $\mathscr{A} = 2^{-1/2}[1 - P_{12}]$, where P_{12} means transpose the labels on the two spinors identified by the subscripts 1 and 2.]

The MR of $R\ (\phi\ \mathbf{n})$ for the coupled basis, eq. (9), is the direct sum

$$\begin{bmatrix} \exp(-\mathrm{i}\phi) & & \\ & 1 & \\ & & \exp(\mathrm{i}\phi) \end{bmatrix} \oplus \mathrm{E}_1. \tag{10}$$

The Clebsch–Gordan decomposition

(3), (9) $$\langle u^{1/2}| \otimes \langle u^{1/2}| = \langle u^1| \oplus \langle \overline{u}^0| \tag{11}$$

of the DP of two spinors therefore yields an object $\langle u^1|$, which transforms under rotations like a vector, and $\langle \overline{u}^0|$, which is invariant under rotation. Under inversion, the spinor basis $\langle u^{1/2}| \equiv \langle\, |{1/2}\ {1/2}\rangle\ |{1/2}\ {-1/2}\rangle\, | \equiv \langle\, u\ v|$ transforms as

(2) $$\hat{I}\langle u\,v| = (-1)^{1/2}\langle u\ v\ |\ \mathrm{E}_2 = \pm\mathrm{i}\,\langle u\ v\ |\ \mathrm{E}_2. \tag{12}$$

Clearly, eq. (2) can be satisfied by either choice of sign in eq. (12). In the *Cartan gauge* the MR of the inversion operator in eq. (12) is taken to be

$$\Gamma(I) = \begin{bmatrix} -\mathrm{i} & 0 \\ 0 & -\mathrm{i} \end{bmatrix}, \tag{13}$$

while in the *Pauli gauge*

$$\Gamma(I) = \begin{bmatrix} \mathrm{i} & 0 \\ 0 & \mathrm{i} \end{bmatrix}, \tag{14}$$

which correspond to the choices $-\mathrm{i}$ or $+\mathrm{i}$ in eq. (12). While neither choice is any more correct than the other, conventional usage favors the Pauli gauge (Altmann and Herzig (1994)). An argument sometimes used is that, since angular momentum is a pseudovector, its eigenfunctions, which include the spinors with half-integral j, must be even under inversion so that the positive sign should be taken in eq. (12). However, electron spin is not a classical object and there is a phase factor to be chosen on the RS of eq. (12), so one is free to work either in the Cartan gauge (choice of -1) or in the Pauli gauge (choice of $+1$). We shall return to this question of the choice of gauge in Section 12.7. With either choice,

(12) $$\hat{I}\ 2^{-1/2}\ \langle u_1 v_2 - v_1 u_2| = (-1)\ 2^{-1/2} \langle u_1 v_2 - v_1 u_2|\ \mathrm{E}_1, \tag{15}$$

(9), (12) $$\hat{I}\ (\langle u_1\ v_1| \otimes \langle u_2\ v_2|) = (\langle u_1\ v_1| \otimes \langle u_2\ v_2|)\ [(-1)(\mathrm{E}_3 \oplus \mathrm{E}_1)]. \tag{16}$$

Therefore $\langle\, u^1|$ is indeed a vector but $\langle\, \overline{u}^0|$ also changes sign under inversion, and so it is a *pseudoscalar*.

In order to establish the homomorphism between SU(2) and SO(3), we will consider first the dual $|u\ \ v\rangle$ of the spinor basis $|u\ \ v\rangle$. Note that no special notation, apart from the bra and ket, is used in C^2 to distinguish the *spinor basis* $\langle\, u\ \ v|$ from its dual $|u\ \ v\rangle$. A general rotation of the column spinor basis in C^2 is effected by (see Altmann (1986), Section 6.7)

$$|u'\ \ v'\rangle = \mathbb{A}\ |\ u\ \ v\rangle. \tag{17}$$

Because the column matrix $|u\ \ v\rangle$ is the dual of $\langle\, u\ \ v|$, they are transformed by the same unitary matrix $\mathbb{A} \in \mathrm{SU}(2)$, where

(11.6.12) $$\mathbb{A} = \begin{bmatrix} a & b \\ -b^* & a^* \end{bmatrix},\ \det \mathbb{A} = aa^* + bb^* = 1. \tag{18}$$

The complex conjugate (CC) spinor basis $|u^*\ \ v^*\rangle$, however, transforms not under $\mathbb{A}$ but under $\mathbb{A}^*$, as may be seen by taking the CC of eq. (17).

(17), (18) $$|u'^*\ \ v'^*\rangle = \mathbb{A}^*|u^*\ \ v^*\rangle = \begin{bmatrix} a^* & b^* \\ -b & a \end{bmatrix}\begin{bmatrix} u^* \\ v^* \end{bmatrix} = \begin{bmatrix} a^*u^* + b^*v^* \\ -bu^* + av^* \end{bmatrix}. \tag{17$'$}$$

However, the choice (Altmann and Herzig (1994))

$$u_2 = v_1^*,\ v_2 = -u_1^* \tag{19}$$

does transform correctly under $\mathbb{A}$:

$$(17),(18),(1) \qquad \langle u_2' \, v_2' \rangle = \mathbb{A} \, |u_2 \; v_2\rangle = \begin{bmatrix} a & b \\ -b^* & a^* \end{bmatrix} \begin{bmatrix} v_1^* \\ -u_1^* \end{bmatrix} = \begin{bmatrix} av_1{}^* & -bu_1{}^* \\ -b^*v_1{}^* & -a^*u_1{}^* \end{bmatrix}$$
$$= |v_1{}'^* \; - u_1{}'^*\rangle. \tag{20}$$

Now form the DP basis

$$|u \;\; v\rangle \otimes |v^* \; -u^*\rangle = |u\,v^* \; -u\,u^* \;\; v\,v^* \; -v\,u^*\rangle. \tag{21}$$

The transformation matrix for the DP basis, eq. (21), is (cf. eq. (8))

$$\mathbb{B} = \mathbb{A} \otimes \mathbb{A} = \begin{bmatrix} aa & ab & ba & b^2 \\ -ab^* & aa^* & -bb^* & ba^* \\ -ab^* & -bb^* & aa^* & a^*b \\ b^*b^* & -a^*b^* & -a^*b^* & (a^*)^2 \end{bmatrix}. \tag{22}$$

This matrix may be reduced by the same prescription as was used earlier in eq. (11), namely by forming the coupled representation

$$\mathbb{C}\; 2^{1/2}|uv^* \; -uu^* \; vv^* \; -u^*v\rangle = 2^{1/2} \begin{bmatrix} 1 & & & \\ & 2^{-1/2} & 2^{-1/2} & \\ & 2^{-1/2} & -2^{-1/2} & \\ & & & 1 \end{bmatrix} \times 2^{1/2}|uv^* \; -uu^* \; vv^* \; -u^*v\rangle$$
$$= 2^{1/2}|uv^* \;\; 2^{-1/2}(-uu^* + vv^*) \;\; -2^{-1/2}(uu^* + vv^*) \;\; -u^*v\rangle. \tag{23}$$

Equation (23) defines the transformation matrix $\mathbb{C}$, the extra factor $2^{1/2}$ having been introduced to ensure later normalization.

$$(23),(22) \qquad \mathbb{CBC}^{-1} = \begin{bmatrix} a^2 & 2^{1/2}ab & 0 & b^2 \\ -2^{1/2}ab^* & aa^* - bb^* & 0 & 2^{1/2}a^*b \\ 0 & 0 & 1 & 0 \\ (b^*)^2 & -2^{1/2}a^*b^* & 0 & (a^*)^2 \end{bmatrix}. \tag{24}$$

Equation (24) confirms that the tensor basis in eq. (23) has been reduced by the transformation $\mathbb{C}$ into the direct sum of its antisymmetric and symmetric parts,

$$2^{1/2}\,|uv^* \;\; 2^{-1/2}(-uu^* + vv^*) \;\; -u^*v\rangle \oplus |-(uu^* + vv^*)\rangle. \tag{25}$$

As eq. (24) shows, the 4×4 matrix $\mathbb{B}$ has been reduced by this basis transformation into the direct sum

$$\begin{bmatrix} a^2 & 2^{1/2}ab & b^2 \\ -2^{1/2}ab^* & aa^* - bb^* & 2^{1/2}a^*b \\ (b^*)^2 & -2^{1/2}a^*b^* & (a^*)^2 \end{bmatrix} \oplus \mathbb{E}_1 = \Gamma^1(\mathbb{A}) \oplus \mathbb{E}_1, \tag{26}$$

where the superscript 1 denotes the value of j. The matrix $\mathbb{A}$ represents a general rotation in C^2, and its symmetrized DP $\mathbb{A}\overline{\otimes}\mathbb{A}$ in eq. (26) represents a general rotation in $\Re^3$.

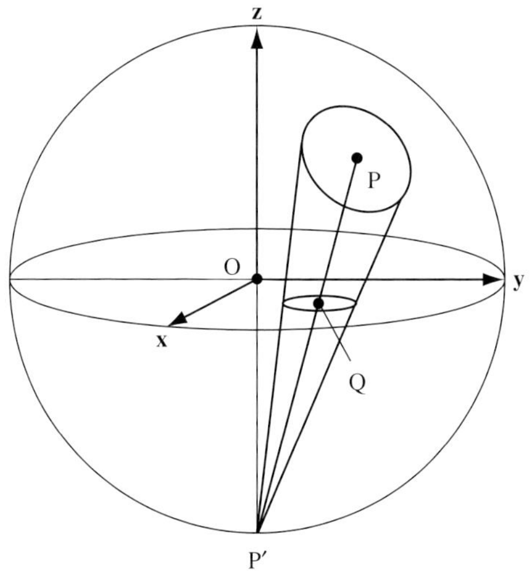

Figure 11.5. A rotation of the unit sphere describes a rotation of configuration space with respect to fixed axes $\{\mathbf{x}\ \mathbf{y}\ \mathbf{z}\}$. For example, a rotation through 2π about P′ P traces a circle on the unit sphere which projects on to a circle about Q in the **xy** plane. The figure thus shows in geometrical terms the mapping of a rotation in $\Re^3$, represented by $\Gamma^1(\mathbb{A})$, on to a rotation in 2-D, represented by $\mathbb{A}$.

This matrix $\Gamma^1(\mathbb{A})$ differs from that in eq. (11.6.19) which describes the transformation of the basis $|x\ \ y\ \ z\rangle$. The first term in the symmetrized basis in eq. (25) is the spherical vector

$$|U_1\ \ U_0\ \ U_{-1}\rangle = |2^{-1/2}(x - \mathrm{i}y)\ \ -z\ \ -2^{-1/2}(x + \mathrm{i}y)\rangle. \tag{27}$$

To see this, set

$$2^{-1/2}(x - \mathrm{i}y) = 2^{1/2}uv^*, \tag{28}$$

(28)
$$-2^{-1/2}(x + \mathrm{i}y) = -2^{1/2}u^*v; \tag{29}$$

$\{x\ \ y\ \ z\}$ are the coordinates of a point P on the surface of a sphere of unit radius, the projection of which on to the **xy** plane gives a stereographic projection of the unit sphere (Figure 11.5). Since P is a point on the surface of the unit sphere, $x^2 + y^2 + z^2 = 1$, and so

(28), (29)
$$\begin{aligned} z^2 &= 1 - (x^2 + y^2) = 1 - (x + \mathrm{i}y)(x - \mathrm{i}y) \\ &= 1 - (2\,u^*v)(2\,u\,v^*). \end{aligned} \tag{30}$$

But the spinors u, v are orthonormal so that their HSP is unity,

$$u\,u^* + v\,v^* = 1, \tag{31}$$

(30), (31)
$$\begin{aligned} z^2 &= (u\,u^* + v\,v^*)^2 - 4u\,u^*v\,v^* \\ &= (u\,u^* - v\,v^*)^2, \end{aligned} \tag{32}$$

$$(32) \qquad z = uu^* - vv^*, \tag{33}$$

on choosing the phase factor as $+1$. Equation (33) identifies the second term in the symmetrized basis, eq. (25), as $-z = U_0$. Therefore eq. (28), (29), and (33) identify the first ket in eq. (25) – that is, the symmetrized basis – with the spherical vector $|U_1 \ U_0 \ U_{-1}\rangle$,

$$|2^{1/2}uv^* \quad vv^* - uu^* \quad -2^{1/2}u^*v\rangle = |U_1 \ U_0 \ U_{-1}\rangle. \tag{34}$$

Consequently, the MR of the rotation operator Γ^1 (A) in eq. (26) will now be identified as $\Gamma^1 (\mathrm{A})^{(U)}$.

Exercise 11.8-2 Prove that $\det \Gamma^1 (\mathrm{A})^{(U)} = 1$.

Exercise 11.8-3

(a) The spherical vector $|U_1 \ U_0 \ U_{-1}\rangle$ is related to the variables $|x \ y \ z\rangle$ by $|U_1 \ U_0 \ U_{-1}\rangle = \mathrm{U} \, |x \ y \ z\rangle$. Show that the MR Γ^1 (A) of the rotation operator in the **r** basis $|x \ y \ z\rangle$ is related to that in the U basis $|U_1 \ U_0 \ U_{-1}\rangle$, namely $\Gamma^1 (\mathrm{A})^{(U)}$, by

$$\Gamma^1(\mathrm{A}) = \mathrm{U}^\dagger \, \Gamma^1(\mathrm{A})^{(U)}\mathrm{U}. \tag{35}$$

(b) Using $\Gamma^1(\mathrm{A})^{(U)}$ from eq. (26) and U from Exercise 11.5.3, find $\Gamma^1(\mathrm{A})$ from eq. (35) and show that the result agrees with that found previously in eq. (11.6.19).

At last we may prove the homomorphism of SU(2) and SO(3). That the mapping of SU(2) on to SO(3) involves a 2 : 1 correspondence has been shown in Section 11.6, but the fact that this relationship is a homomorphism could not be proved until now.

$$(35) \quad \Gamma^1(\mathrm{A}) = \mathrm{U}^{-1}\Gamma^1(\mathrm{A})^{(U)}\mathrm{U} = \mathrm{U}^{-1}\mathrm{A} \,\overline{\otimes}\, \mathrm{A}\mathrm{U}, \ \mathrm{A} \in SU(2), \ \Gamma^1(\mathrm{A}) \in SO(3); \tag{36}$$

$$(36) \qquad \Gamma^1(-\mathrm{A}) = \mathrm{U}^{-1}\mathrm{A} \,\overline{\otimes}\, \mathrm{A}\mathrm{U} = \Gamma^1(\mathrm{A}), \tag{37}$$

so that A and $-$A correspond to the same $\Gamma^1(\mathrm{A})$, and are the only matrices of SU(2) that map on to $\Gamma^1(\mathrm{A})$. Since the theorem that a product of DPs is the DP of the products holds also for symmetrized DPs,

$$(36) \qquad \begin{aligned} \Gamma^1(\mathrm{A}_1)\,\Gamma^1(\mathrm{A}_2) &= \mathrm{U}^{-1}\mathrm{A}_1 \,\overline{\otimes}\, \mathrm{A}_1\mathrm{U}\,\mathrm{U}^{-1}\mathrm{A}_2 \,\overline{\otimes}\, \mathrm{A}_2\mathrm{U} \\ &= \mathrm{U}^{-1}(\mathrm{A}_1 \,\overline{\otimes}\, \mathrm{A}_1)(\mathrm{A}_2 \,\overline{\otimes}\, \mathrm{A}_2)\mathrm{U} \\ &= \mathrm{U}^{-1}(\mathrm{A}_1\mathrm{A}_2) \,\overline{\otimes}\, (\mathrm{A}_1\mathrm{A}_2)\mathrm{U} \\ &= \Gamma^1(\mathrm{A}_1\mathrm{A}_2). \end{aligned} \tag{38}$$

Equation (38) verifies that the mutiplication rules in SU(2) are preserved in SO(3) and therefore that the mapping described by eqs. (36) and (37) is a homomorphism. This mapping is a homomorphism rather than an isomorphism because the two matrices A and $-$A of SU(2) both map on to the same matrix $\Gamma^1(\mathrm{A})$ of SO(3).

Because of the homomorphism between the groups SU(2) and SO(3), we may take eq. (11.5.40) as a basis for SU(2),

$$\text{(11.5.40)} \qquad \langle |j\ \ m\rangle| = \langle [(j+m)!(j-m)!]^{-1/2}\ u^{j+m}\ \ v^{j-m}|. \tag{39}$$

A rotation in configuration space is effected in SU(2) by

$$\text{(11.6.12)} \qquad \mathrm{R}(a\ \ b)\langle u\ \ v| = \langle u\ \ v\ |\mathbb{A} = \langle u\,v|\begin{bmatrix} a & b \\ -b^* & a^* \end{bmatrix}$$
$$= \langle au - b^* v \quad bu + a^* v|. \tag{40}$$

The transformed basis $R(a\ \ b)|j\ \ m\rangle$ is, from eqs. (40) and (39),

$$[(j+m)!(j-m)!]^{-1/2}(u')^{j+m}(v')^{j-m} = [(j+m)!(j-m)!]^{-1/2}(au - b^* v)^{j+m}(bu + a^* v)^{j-m}$$

$$= [(j+m)!(j-m)!]^{-1/2}\sum_{k=0}^{j+m}\frac{(j+m)!(au)^{j+m-k}(-b^* v)^k}{k!(j+m-k)!}\sum_{k'=0}^{j-m}\frac{(j-m)!(bu)^{j-m-k'}(a^* v)^{k'}}{k'!(j-m-k')!}$$

$$= \sum_{k=0}^{j+m}\sum_{k'=0}^{j-m}\frac{[(j+m)!(j-m)!]^{1/2}(a)^{j+m-k}(a^*)^{k'}(b)^{j+m-k'}(-b^*)^k u^{2j-k-k'}v^{k+k'}}{k!k'!(j+m-k)!(j-m-k')!}$$

$$= \sum_{m'=-j}^{j}\sum_{k=0}^{j+m}\frac{[(j+m')!(j-m')!(j+m)!(j-m)!]^{1/2}(a)^{j+m-k}(a^*)^{j-m'-k}(b)^{m'-m+k}(-b^*)^k u^{j+m'}v^{j-m'}}{k!(j+m-k)!(j-m'-k)!(m'-m+k)!} \tag{41}$$

$$= \sum_{m'=-j}^{j}[(j+m')!(j-m')!]^{-1/2}\ u^{j+m'}v^{j-m'}\ \Gamma^j_{m'm}, \tag{42}$$

where $m' = j - k - k'$, $-j \leq m' \leq j$, and

$$\Gamma^j_{m'm} = \sum_{k=0}^{j+m}\frac{[(j+m)!(j-m)!(j+m')!(j-m')!]^{1/2}(a)^{j+m-k}(a^*)^{j-m'-k}(b)^{m'-m+k}(-b^*)^k}{k!(j+m-k)!(j-m'-k)!(m'-m+k)!}. \tag{43}$$

The sum in eq. (43) runs over all values of $0 \leq k \leq j+m$, using $(-n)! = 1$ when $n = 0$ and $(-n)! = \infty$ when $n > 0$. The formula (43) for matrix elements can be expressed in an alternative form that involves binomial coefficients instead of factorials (Altmann and Herzig (1994)) and this may be rather more useful for computational purposes.

Exercise 11.8-4 Evaluate from eq. (43) the term Γ^1_{00} of the matrix Γ^1 and show that this agrees with the corresponding term in Γ^1 in eq. (28).

Answers to Exercises 11.8

Exercise 11.8-1

$$\mathscr{S} u_1 v_2 = 2^{-1/2}[1 + P_{12}]\, u_1 v_2 = 2^{-1/2}[u_1 v_2 + v_1 u_2];$$
$$\mathscr{A} u_1 v_2 = 2^{-1/2}[1 - P_{12}] u_1 v_2 = 2^{-1/2}[u_1 v_2 - v_1 u_2].$$

Exercise 11.8-2

$$\begin{aligned}|\Gamma^1(\mathbb{A})^{(\mathrm{U})}| &= a^2 a^{*^2}[(aa^* - bb^*) + 2bb^*] + 2aba^*b^*[aa^* + bb^*] + b^2 b^{*^2}[aa^* + bb^*]\\ &= a^2 a^{*^2} + 2(aa^* + bb^*) + b^2 b^{*^2} = 1.\end{aligned}$$

Exercise 11.8-3 (a)

$$R(\phi\ \mathbf{n})\ |\mathrm{U}_1\ \ \mathrm{U}_0\ \ \mathrm{U}_{-1}\rangle = R(\phi\ \mathbf{n})\ \mathrm{U}\ |x\ \ y\ \ z\rangle = \mathrm{U}R(\phi\,\mathbf{n})|x\ y\ z\rangle = \mathrm{U}\Gamma^1(\mathbb{A})|x\ \ y\ \ z\rangle.$$

But

$$\mathrm{LS} = \Gamma^1(\mathbb{A})^{(\mathrm{U})}|\mathrm{U}_1\ \ \mathrm{U}_0\ \ \mathrm{U}_{-1}\rangle = \Gamma^1(\mathbb{A})^{(\mathrm{U})}\mathrm{U}|x\ \ y\ \ z\rangle.$$

Therefore $\Gamma^1(\mathbb{A})^{(\mathrm{U})}\mathrm{U} = \mathrm{U}\Gamma^1(\mathbb{A})$, and so $\Gamma^1(\mathbb{A}) = \mathrm{U}^{-1}\Gamma^1(\mathbb{A})^{(\mathrm{U})}\mathrm{U} = \mathrm{U}^\dagger\Gamma^1(\mathbb{A})^{(\mathrm{U})}\mathrm{U}$.

(b) Take U from Exercise 11.5-3 and $\Gamma^1(\mathbb{A})^{(\mathrm{U})}$ from eq. (26), multiply the three matrices in $\mathrm{U}^\dagger\Gamma^1\ (\mathbb{A})^{(\mathrm{U})}\mathrm{U}$, and check your result with $\Gamma^1(\mathbb{A})$ in eq. (11.6.19).

Exercise 11.8-4 For $j = 1$, if $m = 0$, then $k = 0$ or 1, and with $m' = 0$, eq. (43) gives

$$\begin{aligned}\Gamma^1_{00} &= \frac{(1!1!1!1!)^{1/2}}{0!1!1!0!} \times a^1(a^*)^1 b^0(-b^*)^0 + \frac{(1!1!1!1!)^{1/2}}{1!0!0!1!} \times a^0(a^*)^0 b^1(-b^*)^1\\ &= aa^* - bb^*.\end{aligned}$$

Problems

11.1 Prove that the commutator $[R(\alpha\ \mathbf{1}), R(\beta\ \mathbf{2})] = R(\alpha\beta\ \mathbf{3}) - E$, where **1**, **2**, and **3** are unit vectors along OX, OY, and OZ, respectively. [*Hint*: Use eq. (11.3.10)]. Then use eq. (11.3.19) to show that $[I_1, I_2] = \mathrm{i}\, I_3$.

11.2 Prove eq. (11.4.28), $c_+ = [j'^2 - m(m+1)]^{1/2} \exp(\mathrm{i}\gamma)$.

11.3 Prove that eq. (11.5.16), $(u^j_m)^* = (-1)^m u^j_{-m}$, is true for $m = 1$ and $m = 2$. [*Hints*: Drop the j superscript since none of the equations involved depend on the value of j. Use $\hat{J}_\pm\ u_0$ to generate u_1, u_{-1}; substitute for $\hat{J}_+$ from eq. (11.5.7), when $u_1^* = -u_{-1}$ follows. To prove $u_2^* = u_{-2}$, proceed similarly, starting from $\hat{J}_+ u_1$ and then $\hat{J}_- u_{-1}$. You will also need $\hat{J}_-^* = -\hat{J}_+$, which follows from eq. (11.5.7).]

11.4 Prove the conical transformation, eq. (11.5.22). [*Hints*: The proof of the conical transformation is not trivial. Your starting point is a figure similar to Figure 11.1, except that the angle of rotation ϕ is now not an infinitesimal angle; **n** is the unit vector along OQ; OR = **r**; and OR′ = **r**′. In Figure 11.6(a), define vectors $\Delta\mathbf{r}$ and **s** by

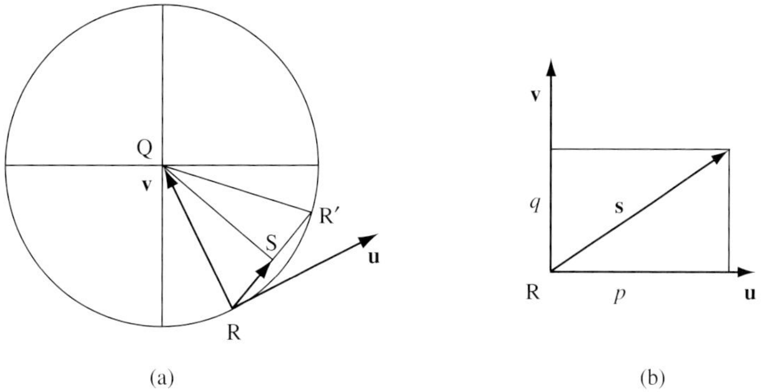

Figure 11.6. (a) Compare with Figure 11.1, in which ϕ is a small angle. RS = **s**; angle RQR′ = ϕ; $\mathbf{r}' = \mathbf{r} + \Delta\mathbf{r} = \mathbf{r} + 2\mathbf{s}$; $\mathbf{u} = \mathbf{n} \times \mathbf{r}$; RQ = $\mathbf{v} = \mathbf{n} \times \mathbf{u}$, so $|\mathbf{v}| = |\mathbf{u}|$. From (b), $\mathbf{s} = p\,(\mathbf{u}/|\mathbf{u}|) + q\,(\mathbf{v}/|\mathbf{v}|)$, where p and q are the components of **s** along **u** and **v**, respectively.

$\mathbf{r}' = \mathbf{r} + \Delta\mathbf{r} = \mathbf{r} + 2\mathbf{s}$. Write down expressions for **p** and **q**, which are defined in Figure 11.6(b), and hence evaluate **s** and $\Delta\mathbf{r}$.]

11.5 (a) Show that for the basis, eq. (11.5.40),

$$\hat{J}_- = y\frac{\partial}{\partial x},$$
$$\hat{J}_x = \frac{1}{2}\left[x\frac{\partial}{\partial y} + y\frac{\partial}{\partial x}\right], \tag{11.5.45}$$
$$\hat{J}_y = \frac{-\mathrm{i}}{2}\left[x\frac{\partial}{\partial y} - y\frac{\partial}{\partial x}\right];$$

$$\hat{\mathbf{J}}^2 = \frac{1}{4}\left[x^2\frac{\partial^2}{\partial x^2} - y^2\frac{\partial^2}{\partial y^2}\right] + \frac{1}{2}xy\left[\frac{\partial^2}{\partial x\partial y}\right] + \frac{3}{4}\left[x\frac{\partial}{\partial x} + y\frac{\partial}{\partial y}\right]. \tag{11.5.46}$$

(b) Prove that

$$\hat{\mathbf{J}}^2|j\ m\rangle = j(j+1)|j\ m\rangle. \tag{11.5.47}$$

11.6 (a) Make the assignment of a and b given in eq. (11.6.20) in the matrix $\mathbb{A}$ of eq. (11.6.12). Evaluate $\mathrm{H}' = \mathbb{A}\ \mathrm{H}\ \mathbb{A}^{-1}$ explicitly and show (using eq. (11.6.14) for $\{x'\ y'\ z'\}$ and $\{x\ y\ z\}$) that $\Gamma(\mathbb{A})$ becomes the matrix in eq. (11.6.21).
(b) Proceed similarly, starting from the assignment of a and b in eq. (11.6.22), and show that in this case the matrix $\mathbb{A}$ corresponds to eq. (11.6.23).
(c) Make the same two substitutions directly in eq. (11.6.19) and confirm that this yields eqs. (11.6.21) and (11.6.22).

11.7 Evaluate the complete matrix Γ^1 from eq. (11.8.43) and show that this agrees with the result for Γ^1 in eqs. (11.8.26).

12 Projective representations

12.1 Complex numbers

Complex numbers are numbers of the form $a + iA$, where a and A are real numbers and i is the unit imaginary number with the property $i^2 = -1$. The ordinary operations of the algebra of real numbers can be performed in exactly the same way with complex numbers by using the multiplication table for the complex number units $\{1, i\}$ shown in Table 12.1. Thus, the multiplication of two complex numbers yields

$$(a + iA)(b + iB) = ab - AB + i(aB + bA). \tag{1}$$

Complex numbers may be represented by ordered number pairs $[a, A]$ by defining $[a, 0] = a[1, 0]$ to be the real number a and $[0, 1]$ to be the pure imaginary i. Then

$$a + iA = a[1, 0] + A[0, 1] = [a, 0] + [0, A] = [a, A]; \tag{2}$$

(2), (1)
$$[a, A][b, B] = [ab - AB, \ aB + bA]. \tag{3}$$

The *complex conjugate* (CC) of $a + iA$ is defined to be $a - iA$. The product of the complex number $[a, A]$ and its CC $[a, -A]$ is

(3)
$$[a, A][a, -A] = [a^2 + A^2, 0] = a^2 + A^2, \tag{4}$$

which is non-zero except when $[a, A]$ is $[0, 0]$. Division by a complex number $[a, A]$ is defined as multiplication by its inverse

(4)
$$[a, A]^{-1} = (a^2 + A^2)^{-1}[a, -A], \tag{5}$$

whence it follows that division of a complex number by another complex number yields a complex number.

Exercise 12.1-1 Prove the associative property of the multiplication of complex numbers.

A complex number $[a, A]$ may be represented by a point P, whose Cartesian coordinates are a, A in a plane called the complex plane C (Figure 12.1). There is then a 1:1 correspondence between the complex numbers and points in this plane. Let r, θ be the polar coordinates of the point P in Figure 12.1. Then

(2)
$$[a, A] = [r\cos\theta, \ r\sin\theta] = r[\cos\theta, \ \sin\theta] = r(\cos\theta + i\sin\theta). \tag{6}$$

Table 12.1. *Multiplication table for the complex number units 1, i.*

	1	i
1	1	i
i	i	−1

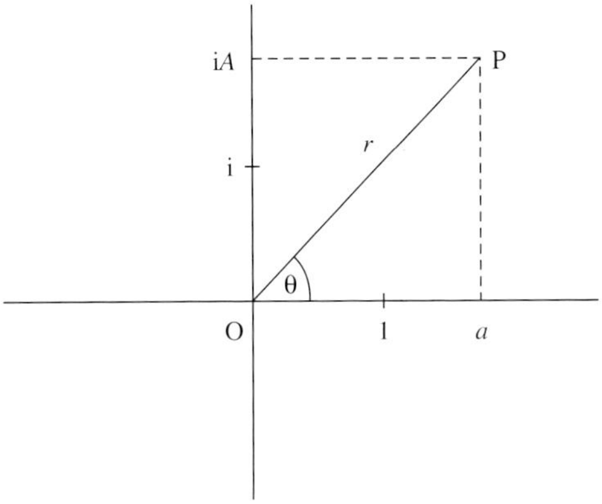

Figure 12.1. Argand diagram for the representation of complex numbers in the complex plane C.

The distance from the origin O to the point P is $r = (a^2 + A^2)^{1/2}$, which is called the *norm* or *modulus* of the complex number $[a, A]$. The product of two complex numbers is as follows:

$$[a_1, A_1][a_2, A_2] = r_1 r_2[\cos(\theta_1 + \theta_2) + \mathrm{i}\sin(\theta_1 + \theta_2)]. \quad (7)$$

(6)

Since the exponential function may be defined everywhere in the complex plane, we may expand $\exp(\mathrm{i}\theta)$ and, using the series expansions for the trigonometric functions, obtain Euler's formula

$$\exp(\mathrm{i}\theta) = \cos\theta + \mathrm{i}\sin\theta; \quad (8)$$

(6), (8)

$$[a, A] = a + \mathrm{i}A = r\exp(\mathrm{i}\theta). \quad (9)$$

Answer to Exercise 12.1-1

$$\begin{aligned}
[a, A]([b, B][c, C]) &= [a, A][bc - BC,\ bC + cB] \\
&= [abc - aBC - bCA - cAB,\ abC + caB + bcA - ABC]; \\
([a, A][b, B])[c, C] &= [ab - AB,\ aB + bA][c, C] \\
&= [abc - cAB - aBC - bCA,\ abC - ABC + caB + bcA].
\end{aligned}$$

12.2 Quaternions

A quaternion q is a hypercomplex number

$$q = a + \mathrm{q}_1 A_1 + \mathrm{q}_2 A_2 + \mathrm{q}_3 A_3 \tag{1}$$

in which a, A_1, A_2, and A_3 are scalars and $\mathrm{q}_1, \mathrm{q}_2$, and q_3 are three imaginary *quaternion units* with the properties

$$\left.\begin{array}{c} \mathrm{q}_1{}^2 = \mathrm{q}_2{}^2 = \mathrm{q}_3{}^2 = -1; \\ \mathrm{q}_1\ \mathrm{q}_2 = \mathrm{q}_3 = -\mathrm{q}_2\ \mathrm{q}_1;\ \mathrm{q}_2\ \mathrm{q}_3 = \mathrm{q}_1 = -\mathrm{q}_3\ \mathrm{q}_2;\ \mathrm{q}_3\ \mathrm{q}_1 = \mathrm{q}_2 = -\mathrm{q}_1\ \mathrm{q}_3. \end{array}\right\} \tag{2}$$

Thus the multiplication of quaternion units is non-commutative. In eq. (1) q is to be interpreted as a compound symbol that stands for two different objects: the *real quaternion*, identified with the real number a, and the *pure quaternion* $\sum_{i=1}^{3} \mathrm{q}_i\ A_i$. This is analogous to the compound symbol $a + \mathrm{i}A$ that denotes the complex number $[a, A]$ (Whittaker and Watson (1927)).

Exercise 12.2-1 Show that multiplication of quaternion units is associative. [*Hint*: The multiplication rules in eq. (2) may be summarized by $\mathrm{q}_l\ \mathrm{q}_m = -1$ if $l = m$ and $\mathrm{q}_l\ \mathrm{q}_m = \varepsilon_{lmn}\ \mathrm{q}_n$ if $l \neq m \neq n$, where l, m, and n = 1, 2, or 3 and $\varepsilon_{lmn} = +1\ (-1)$ according to whether lmn is an even (odd) permutation of 123.]

Quaternions are thus seen to form a 4-D real linear space $\Re \oplus \Re^3$, comprising the real linear space $\Re$ (basis 1) and a 3-D real linear space $\Re^3$ with basis $\{\mathrm{q}_1, \mathrm{q}_2, \mathrm{q}_3\}$. An ordered pair representation can be established for q by defining

$$a = a[1\ ;\ \mathbf{0}] = [a\ ;\ \mathbf{0}],\ \ \mathrm{q}_i = [0\ ;\ \mathbf{e}_i],\ \ i = 1, 2, 3; \tag{3}$$

(1), (3)

$$\begin{aligned} q = a + \sum_{i=1}^{3} \mathrm{q}_i A_i &= [a\ ;\ \mathbf{0}] + \sum_{i=1}^{3} [0\ ;\ \mathbf{e}_i] A_i \\ &= [a\ ;\ \mathbf{0}] + \left[0\ ;\ \sum_{i=1}^{3} \mathbf{e}_i A_i\right] \\ &= [a\ ;\ \mathbf{0}] + [0\ ;\ \mathbf{A}] = [a\ ;\ \mathbf{A}], \end{aligned} \tag{4}$$

in which the vector

$$\mathbf{A} = \sum_{i=1}^{3} \mathbf{e}_i A_i \tag{5}$$

is a pseudovector. (Recall that the basis vectors $\{\mathbf{e}_1\ \mathbf{e}_2\ \mathbf{e}_3\}$ are pseudovectors while $\{A_1\ A_2\ A_3\}$ is a set of scalars.) The pure quaternion is

$$[0\ ;\ \mathbf{A}] = [0\ ;\ A\mathbf{n}] = A[0\ ;\ \mathbf{n}] = \mathbf{q}A, \tag{6}$$

in which the unit pure quaternion $\mathbf{q}$ is related to the unit vector $\mathbf{n}$ by

(6) $$\mathbf{q} = [0 \; ; \; \mathbf{n}]. \tag{7}$$

(4), (3), (6), (7) $$[a \; ; \; \mathbf{A}] = [a \; ; \; \mathbf{0}] + [0 \; ; \; \mathbf{A}] = a[1 \; ; \; \mathbf{0}] + A[0 \; ; \; \mathbf{n}] = a + \mathbf{q}A, \tag{8}$$

which expresses a quaternion $q = [a \; ; \; \mathbf{A}]$ as the sum of a real number a and the product of another real number A with the unit pure quaternion $\mathbf{q}$, in close analogy with the complex number $[a, A] = a + \mathrm{i}A$.

The product of two quaternions

(1), (2), (5), (3), (4) $$\begin{aligned} [a \; ; \; \mathbf{A}][b \; ; \; \mathbf{B}] &= \left(a + \sum_{i=1}^{3} \mathrm{q}_i A_i\right)\left(b + \sum_{j=1}^{3} \mathrm{q}_j B_j\right) \\ &= ab + a\sum_{j=1}^{3} \mathrm{q}_j B_j + b\sum_{i=1}^{3} \mathrm{q}_i A_i - \sum_{i=1}^{3} A_i B_i + \sum_{i \neq j} \mathrm{q}_i \mathrm{q}_j A_i B_j \\ &= [ab - \mathbf{A} \cdot \mathbf{B} \; ; \; 0] + [0 \; ; \; a\mathbf{B} + b\mathbf{A} + \mathbf{A} \times \mathbf{B}] \\ &= [ab - \mathbf{A} \cdot \mathbf{B} \; ; \; a\mathbf{B} + b\mathbf{A} + \mathbf{A} \times \mathbf{B}]. \end{aligned} \tag{9}$$

Exercise 12.2-2 Show that $\mathbf{q}^2 = -1$.

The *quaternion conjugate* of the quaternion $q = [a \; ; \; \mathbf{A}]$ is

$$q^* = [a \; ; \; -\mathbf{A}]. \tag{10}$$

(4), (10), (9) $$q\,q^* = [a \; ; \; \mathbf{A}][a \; ; \; -\mathbf{A}] = a^2 + A^2, \tag{11}$$

which is a real positive number or zero, but is zero only if $q = 0$.

(11) $$(q\,q^*)^{1/2} = (a^2 + A^2)^{1/2} \tag{12}$$

is called the *norm* of q and a quaternion of unit norm is said to be *normalized*. The inverse of q is given by

(11) $$q^{-1} = q^*/(a^2 + A^2) = [a \; ; \; -\mathbf{A}]/(a^2 + A^2). \tag{13}$$

If q is normalized, $q^{-1} = q^*$. Division by q is effected by multiplying by q^{-1} so that the division of one quaternion by another results in a third quaternion,

(4), (13), (9) $$\begin{aligned} q_1/q_2 &= [a_1 \; ; \; \mathbf{A}_1][a_2 \; ; \; -\mathbf{A}_2]/({a_2}^2 + {A_2}^2) \\ &= [a_1 a_2 + \mathbf{A}_1 . \mathbf{A}_2 \; ; \; -a_1\mathbf{A}_2 + a_2\mathbf{A}_1 - \mathbf{A}_1 \times \mathbf{A}_2]/({a_2}^2 + {A_2}^2) \\ &= [a_3 \; ; \; \mathbf{A}_3] = q_3. \end{aligned} \tag{14}$$

Equation (9) shows that q_1/q_2 always exists except when $q_2 = 0$, which would require $a_2 = 0$ and $\mathbf{A}_2 = 0$. The quaternion algebra is therefore an associative, division algebra. There are in fact only three associative division algebras: the algebra of real numbers, the algebra of complex numbers, and the algebra of quaternions. (A proof of this statement may be found in Littlewood (1958), p. 251.)

Answers to Exercises 12.2

Exercise 12.2-1 There are five possibilities for the product $q_l\ q_m\ q_n$.

(i) If $l=m=n$, $(q_n\ q_n)q_n=-q_n=q_n(q_n\ q_n)$.
(ii) If $l\neq m\neq n$, $(q_l\ q_m)q_n=\varepsilon_{lmn}\ q_n\ q_n=-\varepsilon_{lmn}$; $q_l(q_m\ q_n)=q_l\,\varepsilon_{mnl}\ q_l=-\varepsilon_{mnl}=-\varepsilon_{lmn}$.
(iii) If $l=m\neq n$, $(q_m\ q_m)q_n=-q_n$, $q_m(q_m\ q_n)=q_m\varepsilon_{mnl}q_l=\varepsilon_{mnl}\varepsilon_{mln}q_n=-q_n$.
(iv) If $l\neq m$, $n=m$, $(q_l\ q_m)q_m=\varepsilon_{lmn}q_nq_m=\varepsilon_{lmn}\ \varepsilon_{nml}\ q_l=-q_l=q_l(q_m\ q_m)$.
(v) If $l\neq m$, $n=l$, $(q_l\ q_m)q_l=\varepsilon_{lmn}q_nq_l=\varepsilon_{lmn}\varepsilon_{nlm}q_m=q_m$; $q_l(q_mq_l)=q_l\ \varepsilon_{mln}\ q_n$
$=\varepsilon_{mln}\varepsilon_{nlm}q_m=q_m$.

The inclusion of one, two or three negative signs in $q_l\ q_m\ q_n$ does not change the proof of associativity so that the associative property of multiplication for the set $\{1\ \ -1\ \ q_i\ \ -q_i\}$, where $i=1, 2, 3$, is established.

Exercise 12.2-2 $\mathbf{q}^2=\mathbf{qq}=[0\ ;\ \mathbf{n}][0\ ;\ \mathbf{n}]=[0-\mathbf{n}\cdot\mathbf{n}\ ;\ \mathbf{n}\times\mathbf{n}]=-1$.

12.3 Geometry of rotations

Rotations may be studied geometrically with the aid of the unit sphere shown in Figure 12.2. The unit vector $\mathbf{OP}=\mathbf{n}$ is the axis of the rotation $R(\phi\ \ \mathbf{n})$; P is called the pole of the rotation, and is defined as the point on the sphere which is invariant under $R(\phi\ \ \mathbf{n})$ such that the rotation appears anticlockwise when viewed from outside the sphere. It follows from this definition that $R(-\phi\ \ \mathbf{n})=R(\phi\ \ \overline{\mathbf{n}})$; that is, a negative (clockwise) rotation about $\mathbf{n}$ is the same operation as a positive (anticlockwise) rotation through the same angle ϕ about $-\mathbf{n}$, the pole of this rotation being P′ in Figure 12.2. As a consequence, we may concern ourselves only with positive rotations in the range $0\leq\phi\leq\pi$. For a positive rotation, P belongs to the positive hemisphere h, whereas for negative rotations about $\mathbf{n}$ the pole is P′, which is the intersection of $-\mathbf{n}$ with

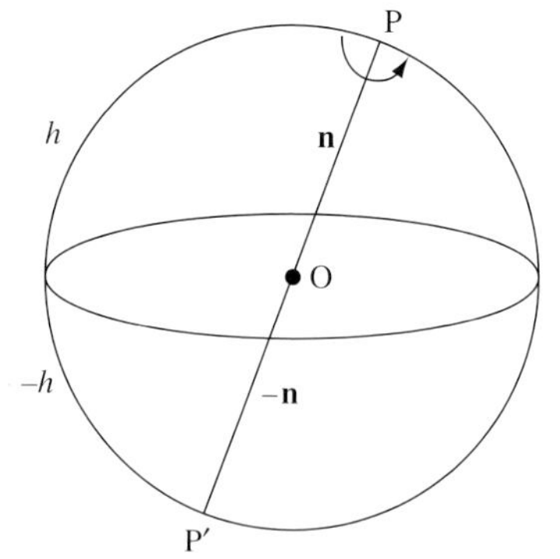

Figure 12.2. The curved arrow shows the direction of a positive rotation $R(\phi\ \ \mathbf{n})$ about $\mathbf{n}$. P is the pole of this rotation since it is invariant under the rotation and is the point from which the rotation appears anticlockwise when viewed from outside the sphere. From P′, which is the antipole of P, the same rotation appears to be clockwise.

the negative hemisphere $-h$. P′ is referred to as the antipole of P. Equivalently (Altmann and Herzig (1994)), the unit vector **n** rather than its point of intersection with the sphere could be taken as the pole of the rotation. The disjoint hemispheres h and $-h$ shown in Figure 12.2 are appropriate for O(3), but for the point groups, which are subgroups of O(3), h and $-h$ may have to be discontinuous. An example will occur later in this chapter. Since a rotation through $-\pi$ is equivalent to a rotation through π about the same axis, the antipole of the pole of a binary rotation is not required. The poles of binary rotations must therefore all be chosen within h, even if this means that h has to be discontinuous. No pole is assigned to the identity operation since for E, $\mathbf{n} = 0$ and $\phi = 0$. The rotation parameter $\phi\mathbf{n}$ for E is therefore zero and so the identity is $R(\mathbf{0})$ rather than $R(0\ \mathbf{n})$, for the latter would imply an infinite choice for $(0\ \mathbf{n})$ from the set $\{(0\ \mathbf{n})\}$. The pseudovector $\phi\mathbf{n}$ is a single rotation parameter, the specification of which requires a statement of the rotation angle ϕ and the components of **n**, n_1, n_2, n_3, only two of which are independent. The choice of a set of poles obeys some conventions, which ensure that the character χ is a class property. The following two rules have been adopted in the extensive tables compiled by Altmann and Herzig (1994).

(I) Under the operations of $G = \{g_i\}$, the pole of g_i must either be invariant or transformed into the pole (not the antipole) of an operation in the class of g_i.
(II) If G contains a subgroup H then the choice of the set of poles made for G should be such that rule I is still valid for H, otherwise the representations of G will not subduce properly to those of H. 'Subduction' means the omission of those elements of G that are not members of H and 'properly' means that the matrix representatives (MRs) of the operators in a particular class have the same characters in H as they do in G.

The product of two rotations $R(\alpha\ \mathbf{a})\ R(\beta\ \mathbf{b})$, that is, the effect of a second rotation $R(\alpha\ \mathbf{a})$ on the rotation $R(\beta\ \mathbf{b})$, may be studied by observing how the pole of a rotation is transformed by another rotation about a different axis, using a construction due to Euler. The Euler construction is shown in Figure 12.3: **a** and **b** intersect the unit sphere at A and B, respectively, which are the poles of $R(\alpha\ \mathbf{a})$ and $R(\beta\ \mathbf{b})$. (To aid visualization, A happens to be at the N pole of the unit sphere, but this is not essential.) Rotate the great circle through A and B about A to the *left* (that is, in the direction of an anticlockwise rotation) by $\alpha/2$, and again to the *right* about B (a clockwise rotation) through $\beta/2$. Let the two arcs thus generated intersect at C. Similarly, rotate this great circle to the *right* about A through $\alpha/2$ and to the *left* about B through $\beta/2$ so that the two arcs intersect at C′. Then C is the pole of the rotation $R(\alpha\ \mathbf{a})\ R(\beta\ \mathbf{b})$ because it is left invariant by this pair of successive rotations. (The first rotation $R(\beta\ \mathbf{b})$ transforms C into C′ and the second rotation $R(\alpha\ \mathbf{a})$ transforms C′ back into C.) Let the supplementary angle at C be $\gamma/2$ (see Figure 12.3). Consider now Figure 12.4 in which two planes OW and OV with the dihedral angle $\phi/2$ intersect along **n**. The reflection σ_1 in OW sends X into Y and a second reflection σ_2 in OV sends Y into Z. Then the symmetry operator that sends X into Z is the rotation about **n** through $\phi_1 + \phi_1 + \phi_2 + \phi_2 = 2(\phi_1 + \phi_2) = 2(\phi/2) = \phi$. Thus the product of the two reflections, $\sigma_2\sigma_1$, is equivalent to the rotation $R(\phi\ \mathbf{n})$. In the Euler construction (Figure 12.3) planes OCB and OAB intersect along $\mathrm{OB} = \mathbf{b}$. A reflection σ_1 in OCB followed by a reflection σ_2 in OAB is equivalent to the rotation $R(\beta\ \mathbf{b})$. The planes OAB and OAC intersect along $\mathrm{OA} = \mathbf{a}$. A reflection σ_3 in OAB followed by a reflection σ_4 in OAC is equivalent to the

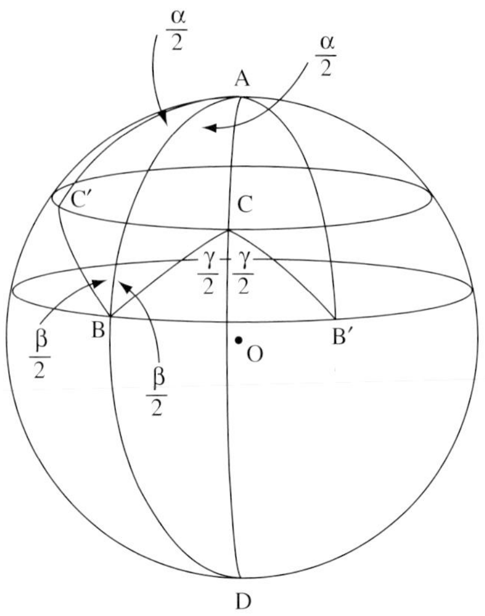

Figure 12.3. Euler construction which shows that $R(\alpha\ \mathbf{a})\ R(\beta\ \mathbf{b})=R(\gamma\ \mathbf{c})$. A, B, and C are the respective poles of the rotations $R(\alpha\ \mathbf{a})$, $R(\beta\ \mathbf{b})$, and $R(\gamma\ \mathbf{c})$; $R(\beta\ \mathbf{b})$ leaves B invariant and $R(\alpha\ \mathbf{a})$ rotates B into B′. The angle of rotation about C that sends B into B′ is γ. Angle BAC $=\alpha/2=$ angle BAC′; angle ABC $=\beta/2=$ ABC′; angle BCD $=\gamma/2=$ angle DCB′; OA $=\mathbf{a}$; OB $=\mathbf{b}$; OC $=\mathbf{c}$.

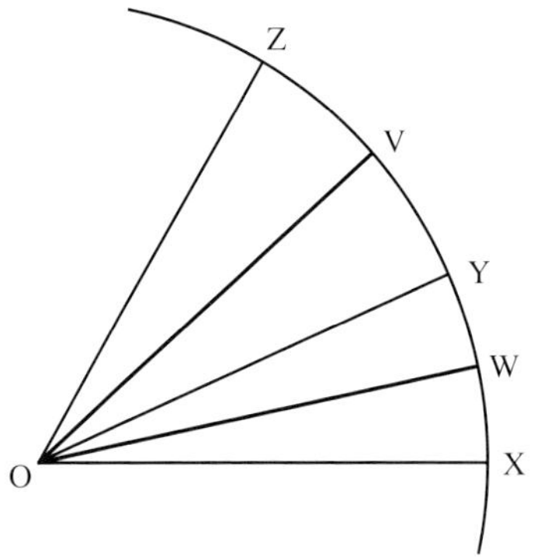

Figure 12.4. Angle XOW = angle WOY $=\phi_1$. Angle YOV = angle VOZ $=\phi_2$. Angle WOY + angle YOV $=\phi_1+\phi_2=\phi/2$. Angles ϕ_1 and ϕ_2 are arbitrary as long as their sum equals $\phi/2$, which is the dihedral angle VOW.

rotation $R(\alpha\ \mathbf{a})$. Therefore, $R(\alpha\ \mathbf{a})\,R(\beta\ \mathbf{b})=\sigma_4\,\sigma_3\,\sigma_2\,\sigma_1=\sigma_4\,\sigma_1=\sigma(\text{OAC})\sigma(\text{OCB})$, since the successive reflections $\sigma_3\,\sigma_2$ in the plane OAB cancel one another. The planes OAC and OCB intersect along OC $=\mathbf{c}$ with dihedral angle $\gamma/2$. Hence the product $\sigma(\text{OAC})\sigma(\text{OCB})$is $R(\gamma\ \mathbf{c})$. Therefore, the product of the two rotations is

$$R(\alpha\ \mathbf{a})R(\beta\ \mathbf{b}) = R(\gamma\ \mathbf{c}). \tag{1}$$

With the aid of the Euler construction, we have proved that the product of two rotations $R(\alpha\ \mathbf{a})R(\beta\ \mathbf{b})$ is a third rotation $R(\gamma\ \mathbf{c})$, but we do not yet have explicit formulae for the

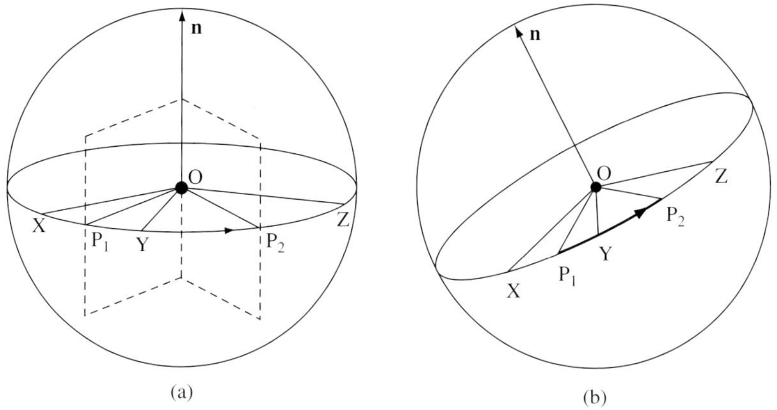

Figure 12.5. Turns P_1P_2 which are part of (a) an equatorial great circle and (b) a general great circle. Angle $XOP_1 = $ angle $P_1OY = \phi_1$; angle $YOP_2 = $ angle $P_2OZ = \phi_2$; and $\phi_1 + \phi_2 = \phi/2$. Successive reflections in planes normal to the arc P_1P_2 of the great circle through P_1P_2 (and containing OP_1 or OP_2, respectively) generate the rotation $R(\phi\ \mathbf{n})$, which traces out the arc XZ equal to twice the length of the turn $P_1P_2 = \phi/2$. The location of X is arbitrary, but the length of XZ is always equal to ϕ. After Biedenharn and Louck (1981).

rotation angle γ and the axis of rotation $\mathbf{c}$. These were first derived by Rodrigues (Altmann (1986)) though we shall obtain them by a different method in Section 12.5.

12.4 The theory of turns

The essence of this theory (Biedenharn and Louck (1981)) will be described here because of the connection it provides between rotations and the algebra of quaternions. The rotation of a unit vector OP in configuration space can be followed by the path traced out by P on the surface of the unit sphere centered on O (Figure 12.5(a)). A *turn* is defined as half the directed arc traced out on the unit sphere by the rotation, and it is parameterized by the ordered pair of points (P_1, P_2) on the surface of the unit sphere. One of these points is designated the *tail* and the other the *head.* A rotation is generated by a reflection first in the plane normal to the arc at the tail P_1 and then by reflection in the plane normal to the arc at the head P_2. If the angle between these planes is $\phi/2$, then these two reflections generate a rotation through ϕ (which is twice the arc length of the turn) about an axis $\mathbf{n}$, which is the intersection of the two planes (Figure 12.5(b)). Two turns are *equivalent* if they can be superimposed by displacing either one along the great circle through P_1P_2. (This is analogous to the superposition of two equal vectors by displacing one of the vectors parallel to itself.) We now consider the properties of turns.

(a) Binary composition in the set of turns $\{\mathbf{T}\}$ is taken to be addition, with the sum $\mathbf{T}_1 + \mathbf{T}_2$ being defined to mean "carry out $\mathbf{T}_2$ *first* and then $\mathbf{T}_1$." Choose either of the points Q where the great circles through $\mathbf{T}_1$ and $\mathbf{T}_2$ intersect (Figure 12.6). Place the head of $\mathbf{T}_2$ and the tail of $\mathbf{T}_1$ at Q. Then the turn from the tail of $\mathbf{T}_2$ to the head of $\mathbf{T}_1$ is defined to be

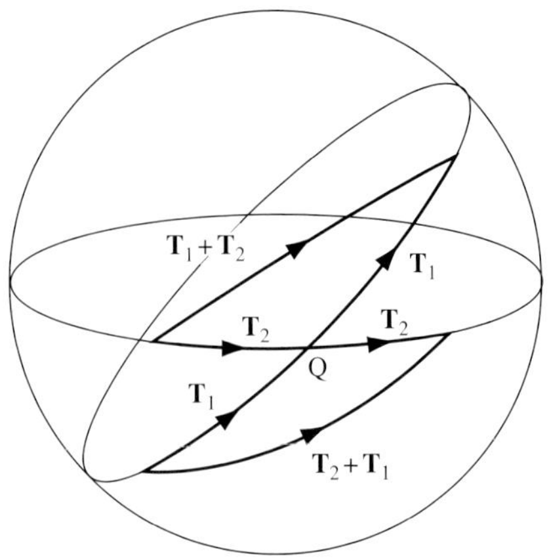

Figure 12.6. Illustration of the procedure for carrying out the addition of turns $\mathbf{T}_1 + \mathbf{T}_2$, defined to be the turn from the tail of $\mathbf{T}_2$ to the head of $\mathbf{T}_1$. The notation $\mathbf{T}_2 + \mathbf{T}_1$ means the turn from the tail of $\mathbf{T}_1$ to the head of $\mathbf{T}_2$; this result is *not* equal to $\mathbf{T}_1 + \mathbf{T}_2$.

$\mathbf{T}_1 + \mathbf{T}_2$. Figure 12.6 also shows that the addition of turns is non-commutative: $\mathbf{T}_2 + \mathbf{T}_1 \neq \mathbf{T}_1 + \mathbf{T}_2$.

(b) The set of turns $\{\mathbf{T}\}$ is *closed* since the addition of two turns always produces another turn.
(c) The addition of turns is *associative*. This may be verified by a geometrical construction in the manner of Figure 12.6.
(d) The set of turns $\{\mathbf{T}\}$ contains the *identity* $\mathbf{T}_0$, which is a turn of zero length.
(e) The *inverse* $-\mathbf{T}$ of a turn $\mathbf{T}$ is a turn of the same length as $\mathbf{T}$ on the same great circle but of opposite sense.

Properties (a)–(e) are just those necessary to ensure that $\{\mathbf{T}\}$ is a group, called by Biedenharn and Louck (1981) "Hamilton"s group of turns."

Let QP be a diameter of the unit sphere; then, since great circles defined by Q and P are not unique, all turns $\mathbf{T}_\pi$ defined by pairs of opposite points are equivalent. Since $\mathbf{T}_\pi$ can be chosen on any great circle, it commutes with any turn $\mathbf{T}$. The operation of adding $\mathbf{T}_\pi$ to a turn $\mathbf{T}$ is described as *conjugation*,

$$\mathbf{T}^{\mathrm{c}} = \mathbf{T}_\pi + \mathbf{T} = \mathbf{T} + \mathbf{T}_\pi. \tag{1}$$

Turns of length $\pi/2$ have some unique properties and are denoted by the special symbol $\mathbf{E}$.

Exercise 12.4-1 (a) Show that $\mathbf{T}_0^{\mathrm{c}} = \mathbf{T}_\pi$. (b) Show that $\mathbf{E}^{\mathrm{c}} = -\mathbf{E}$. (c) Prove that any turn $\mathbf{T}$ may be written as the sum $\mathbf{E}' + \mathbf{E}$, where $\mathbf{E}$ and $\mathbf{E}'$ are each turns of length $\pi/2$. [*Hint*: For ease of visualization take Q, the point of intersection of $\mathbf{E}$ and $\mathbf{E}'$, to be at the N pole and $\mathbf{T}$ (therefore) along the equator. (The only necessity is that Q be the intersection of the normal to the great circle of $\mathbf{T}$ with the unit sphere.) Take $\mathbf{E}$ from the tail of $\mathbf{T}$ to Q.]

Figure 12.7 shows three turns $\mathbf{E}_1$, $\mathbf{E}_2$ and $\mathbf{E}_3$ which sum to $\mathbf{T}_0$. The set of turns $\{\mathbf{T}_0\ \mathbf{T}_\pi\ \mathbf{E}_1\ \mathbf{E}_1^{\mathrm{c}}\ \mathbf{E}_2\ \mathbf{E}_2^{\mathrm{c}}\ \mathbf{E}_3\ \mathbf{E}_3^{\mathrm{c}}\}$ form a group of order eight which is isomorphous with the quaternion group $\mathrm{Q} = \{1\ \ -1\ \ q_1\ \ -q_1\ \ q_2\ \ -q_2\ \ q_3\ \ -q_3\}$ with the mapping given by the ordering of the terms in each set. Therefore

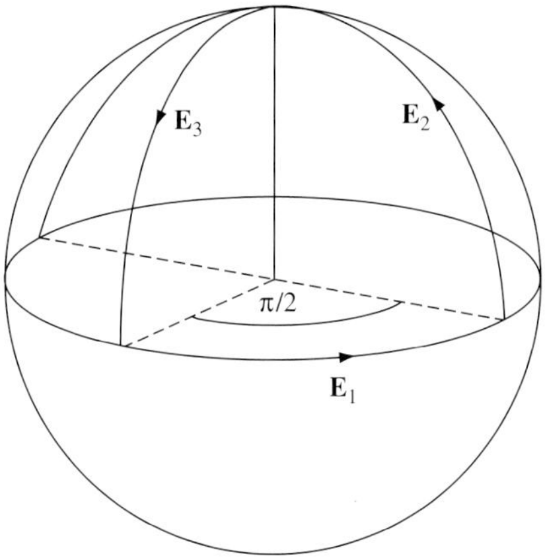

Figure 12.7. $\mathbf{E}_1$, $\mathbf{E}_2$, and $\mathbf{E}_3$ mark out a spherical triangle which is the surface of one octant of the unit sphere. This figure shows that $\mathbf{E}_3 + \mathbf{E}_2 + \mathbf{E}_1 = \mathbf{T}_0$.

$$1, -1 \rightarrow \mathbf{T}_0, \ \mathbf{T}_0^c = \mathbf{T}_\pi; \tag{2}$$

$$q_i^{\,2} = -1 \rightarrow \mathbf{E}_i + \mathbf{E}_i = \mathbf{T}_\pi \quad (i = 1, 2, 3); \tag{3}$$

$$-q_i = q_i^* \rightarrow -\mathbf{E}_i = \mathbf{E}_i^c \quad (i = 1, 2, 3); \tag{4}$$

$$q_l \, q_m = \varepsilon_{lmn} \, q_n \rightarrow \mathbf{E}_l + \mathbf{E}_m = \varepsilon_{lmn} \, \mathbf{E}_n \quad (l \neq m \neq n; \ l, m, n = 1, 2, 3). \tag{5}$$

Note that ε_{lmn}=+1 if $l \; m \; n$ is an even permutation of 1 2 3, but ε_{lmn} is the conjugation operator if $l \; m \; n$ is an odd permutation of 1 2 3. This is not inconsistent with the earlier statement that $\varepsilon_{lmn} = -1$ if $l \; m \; n$ is an odd permutation of 1 2 3 (because by eq. (12.2.4) q_i^c is $q_i^* = -q_i$). Remember that $\mathbf{E}_l + \mathbf{E}_m$ means perform $\mathbf{E}_m$ first, then $\mathbf{E}_l$ (Figure 12.6). For example (Figure 12.7),

$$q_3 \; q_2 \; q_1 = -q_1 \, q_1 = 1 \rightarrow \mathbf{E}_3 + \mathbf{E}_2 + \mathbf{E}_1 = \mathbf{T}_0. \tag{6}$$

Exercise 12.4-2 (a) From eq. (5) $q_1 \; q_2 = q_3$, $q_1 \; q_3 = -q_2$. Prove that $\mathbf{E}_1 + \mathbf{E}_2 = \mathbf{E}_3$; $\mathbf{E}_1 + \mathbf{E}_3 = \mathbf{E}_2^c$. (b) Write down the mapping that corresponds to $(q_1 \, q_2)^2$ and show that this equals $\mathbf{T}_\pi$.

The above analysis shows that the set of turns $\{\mathbf{T}_0 \; \mathbf{T}_\pi \; \mathbf{E}_i \; \mathbf{E}_i^c\}$, $i = 1, 2, 3$, provides a geometric realization of the quaternion group and thus establishes the connection between the quaternion units and turns through $\pi/2$, and hence rotations through π (binary rotations). This suggests that the whole set of turns might provide a geometric realization of the set of unit quaternions. Section 12.5 will not only prove this to be the case, but will also provide us with the correct parameterization of a rotation.

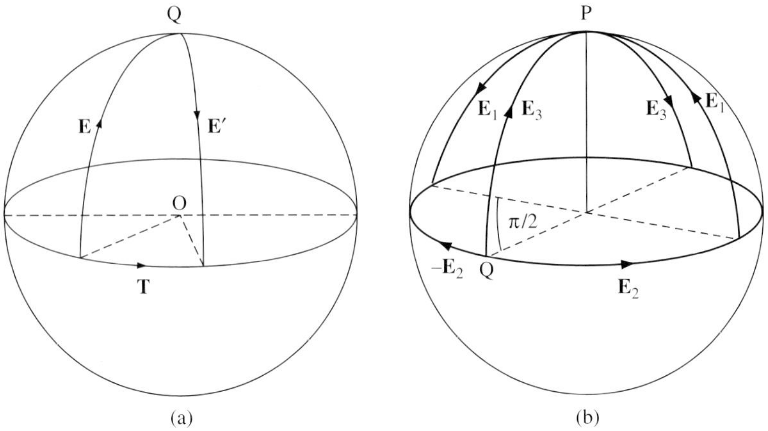

Figure 12.8. (a) Solution to Exercise 12.4-1(c). (b) Solutions to Exercise 12.4-2(a) and (b): $\mathbf{E}_1 + \mathbf{E}_2 = \mathbf{E}_3$; $\mathbf{E}_1 + \mathbf{E}_3 = -\mathbf{E}_2 = \mathbf{E}_2^{\mathrm{c}}$; $(\mathrm{q}_1\ \mathrm{q}_2)^2 \rightarrow \mathbf{E}_3 + \mathbf{E}_3 = \mathbf{T}_\pi$.

Answers to Exercises 12.4

Exercise 12.4-1 (a) From the definition of conjugation $\mathbf{T}^{\mathrm{c}} = \mathbf{T} + \mathbf{T}_\pi$; therefore $\mathbf{T}_0^{\mathrm{c}} = \mathbf{T}_0 + \mathbf{T}_\pi = \mathbf{T}_\pi$. (b) $\mathbf{E}^{\mathrm{c}} = \mathbf{E} + \mathbf{T}_\pi = -\mathbf{E}$. (c) See Figure 12.8(a).

Exercise 12.4-2 See Figure 12.8(b). (a) $\mathbf{E}_1 + \mathbf{E}_3 = -\mathbf{E}_2 = \mathbf{E}_2^{\mathrm{c}}$. (b) $(\mathrm{q}_1\ \ \mathrm{q}_2)^2 = \mathrm{q}_3^2 = -1 \rightarrow$ $\mathbf{E}_1 + \mathbf{E}_2 + \mathbf{E}_1 + \mathbf{E}_2 = \mathbf{E}_3 + \mathbf{E}_3 = \mathbf{T}_\pi$.

12.5 The algebra of turns

Let **m**, **p** be the unit vectors from O to the points M, P that define, respectively, the tail and head of the turn $\mathbf{T}_1$ of length $|\mathbf{T}_1| = {}^1\!/_2\,\phi_1$, and let $\mathbf{n}_1$ be the unit vector along the axis of rotation (Figure 12.9). Then $\mathbf{T}_1 = \mathbf{T}(a_1, \mathbf{A}_1)$ is described by the two parameters

$$a_1 = \mathbf{m} \cdot \mathbf{p} = \cos\left(\tfrac{1}{2}\phi_1\right) \tag{1}$$

and

$$\mathbf{A}_1 = \mathbf{m} \times \mathbf{p} = \sin\left(\tfrac{1}{2}\phi_1\right)\mathbf{n}_1; \tag{2}$$

(1), (2)

$$a_1{}^2 + \mathbf{A}_1 \cdot \mathbf{A}_1 = 1. \tag{3}$$

Exercise 12.5-1 Write down $\mathbf{T} = \mathbf{T}(a, \mathbf{A})$ for (a) $|\mathbf{T}| = 0$ and (b) $|\mathbf{T}| = \pi$. If $\mathbf{T} = \mathbf{T}(a, \mathbf{A})$, find (c) $-\mathbf{T}$ and (d) $-\mathbf{T}^{\mathrm{c}}$.

Figure 12.9(a) shows the addition of two turns $\mathbf{T}_2 + \mathbf{T}_1$, where $\mathbf{T}_1$ is characterized by the unit vectors **m** and **p** from O to the tail of $\mathbf{T}_1$ and the head of $\mathbf{T}_1$, respectively, and $\mathbf{T}_2$

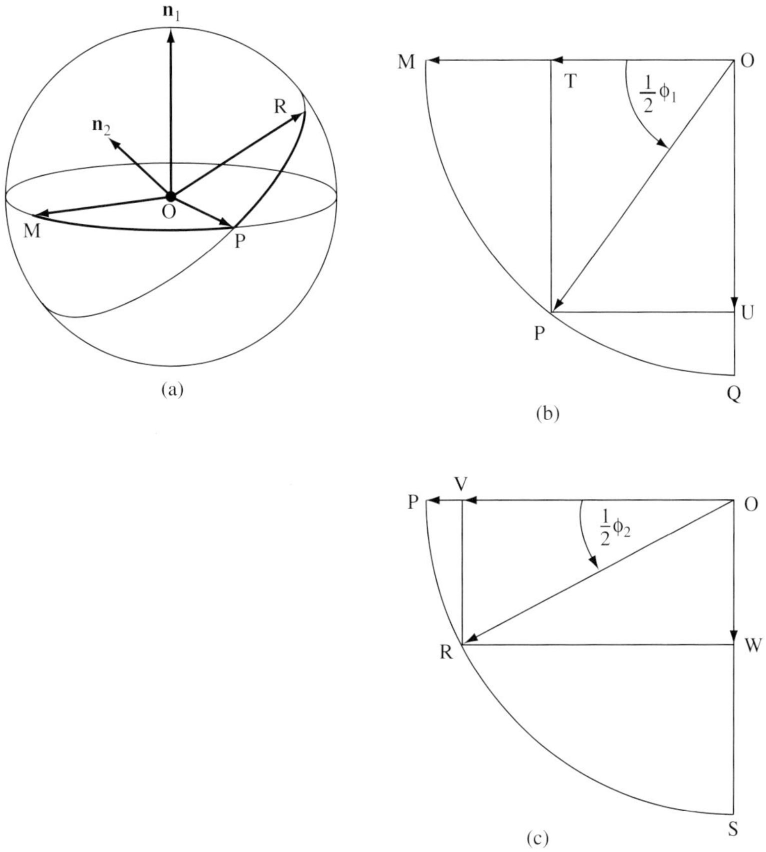

Figure 12.9. (a) Addition of two turns $\mathbf{T}_2+\mathbf{T}_1$; OM $=\mathbf{m}$, OP $=\mathbf{p}$, OR $=\mathbf{r}$, MP $=\mathbf{T}_1$, PR $=\mathbf{T}_2$, MR $=\mathbf{T}_2+\mathbf{T}_1$. (b) Part of the plane of the great circle that contains $\mathbf{T}_1$ so that $\mathbf{n}_1$ is normal to the plane of $\mathbf{m}$ and $\mathbf{p}$; OQ $=\mathbf{q}$, where $\mathbf{q}$ is normal to the plane containing $\mathbf{n}_1$ and $\mathbf{m}$, which we write as $\mathbf{q}\perp(\mathbf{n}_1,\mathbf{m})$. Note that OT $=a_1\mathbf{m}=\cos(\tfrac{1}{2}\phi_1)\mathbf{m}$; OU $=\mathbf{A}_1\times\mathbf{m}=\sin(\tfrac{1}{2}\phi_1)\mathbf{q}$. (c) Part of the plane of the great circle that contains $\mathbf{T}_2$ so that $\mathbf{n}_2$ is normal to the plane of $\mathbf{p}$ and $\mathbf{r}$; OS $=\mathbf{s}\perp(\mathbf{n}_2,\mathbf{p})$; OV $=a_2\mathbf{p}=\cos(\tfrac{1}{2}\phi_2)\mathbf{p}$; OW $=\mathbf{A}_2\times\mathbf{p}=\sin(\tfrac{1}{2}\phi_2)\mathbf{s}$. These diagrams are for angles ϕ_1 and ϕ_2 both positive.

similarly by $\mathbf{p}$ and $\mathbf{r}$. Figure 12.9(b) shows part of the plane of the great circle that contains $\mathbf{T}_1$ so that $\mathbf{n}_1$ is normal to the plane of $\mathbf{m}$ and $\mathbf{p}$, which we signify by $\mathbf{n}_1\perp(\mathbf{m},\mathbf{p})$. Note that $\mathrm{A}_1=\mathbf{m}\times\mathbf{p}$ is a pseudovector of magnitude $\sin(\tfrac{1}{2}\phi_1)$ along $\mathbf{n}_1$; then from Figure 12.9(b)

$$\mathbf{A}_1\times\mathbf{m}=\sin(\tfrac{1}{2}\phi_1)\mathbf{q},\quad \mathbf{q}\perp(\mathbf{n}_1,\ \mathbf{m}),\tag{4}$$

$$\mathbf{p}=a_1\mathbf{m}+\mathbf{A}_1\times\mathbf{m},\tag{5}$$

(5), (1), (2)
$$\mathbf{p}=(\mathbf{m}\cdot\mathbf{p})\mathbf{m}+(\mathbf{m}\times\mathbf{p})\times\mathbf{m}.\tag{6}$$

Similarly, for $\mathbf{T}_2=\mathbf{T}(a_2,\mathbf{A}_2)$,

$$a_2=\mathbf{p}\cdot\mathbf{r}=\cos(\tfrac{1}{2}\phi_2),\ \ \mathbf{A}_2=\mathbf{p}\times\mathbf{r}=\sin(\tfrac{1}{2}\phi_2)\mathbf{n}_2,\tag{7}$$

and by the construction shown in Figure 12.9(c)

$$\mathbf{A}_2 \times \mathbf{p} = \sin(\tfrac{1}{2}\phi_2)\mathbf{s}, \quad \mathbf{s} \perp (\mathbf{n}_2,\ \mathbf{p}), \tag{8}$$

$$\mathbf{r} = a_2\mathbf{p} + \mathbf{A}_2 \times \mathbf{p}, \tag{9}$$

(9), (5) $$\mathbf{r} = a_2(a_1\mathbf{m} + \mathbf{A}_1 \times \mathbf{m}) + \mathbf{A}_2 \times (a_1\mathbf{m} + \mathbf{A}_1 \times \mathbf{m}). \tag{10}$$

Now the triple vector product

$$\mathbf{A} \times (\mathbf{B} \times \mathbf{C}) = \mathbf{B}(\mathbf{A}\cdot\mathbf{C}) - \mathbf{C}(\mathbf{A}\cdot\mathbf{B}) = (\mathbf{C} \times \mathbf{B}) \times \mathbf{A}. \tag{11}$$

Remembering that $\mathbf{A}_1 \perp \mathbf{m}$ so that $\mathbf{m}.\mathbf{A}_1 = 0$,

(11) $$\begin{aligned}(\mathbf{A}_2 \times \mathbf{A}_1) \times \mathbf{m} - (\mathbf{A}_1\cdot\mathbf{A}_2)\mathbf{m} &= \mathbf{A}_1(\mathbf{m}\cdot\mathbf{A}_2) - \mathbf{A}_2(\mathbf{m}\cdot\mathbf{A}_1) - (\mathbf{A}_1\cdot\mathbf{A}_2)\mathbf{m} \\ &= \mathbf{A}_1(\mathbf{m}\cdot\mathbf{A}_2) - (\mathbf{A}_1\cdot\mathbf{A}_2)\mathbf{m} = \mathbf{A}_2 \times (\mathbf{A}_1 \times \mathbf{m}). \end{aligned} \tag{12}$$

(12), (10) $$\begin{aligned}\mathbf{r} &= a_2(a_1\mathbf{m} + \mathbf{A}_1 \times \mathbf{m}) + a_1(\mathbf{A}_2 \times \mathbf{m}) + (\mathbf{A}_2 \times \mathbf{A}_1) \times \mathbf{m} - (\mathbf{A}_1\cdot\mathbf{A}_2)\mathbf{m} \\ &= \mathbf{m}(a_1a_2 - \mathbf{A}_1\cdot\mathbf{A}_2) + (a_1\mathbf{A}_2 + a_2\mathbf{A}_1 + \mathbf{A}_2 \times \mathbf{A}_1) \times \mathbf{m} \\ &= a_3\mathbf{m} + \mathbf{A}_3 \times \mathbf{m}, \end{aligned} \tag{13}$$

where

$$a_3 = a_1a_2 - \mathbf{A}_1\cdot\mathbf{A}_2, \ \ \mathbf{A}_3 = a_1\mathbf{A}_2 + a_2\mathbf{A}_1 + \mathbf{A}_2 \times \mathbf{A}_1. \tag{14}$$

Equations (13) and (14) show that the addition of turns

$$\mathbf{T}_2 + \mathbf{T}_1 = \mathbf{T}(a_2,\ \mathbf{A}_2) + \mathbf{T}(a_1,\ \mathbf{A}_1) = \mathbf{T}(a_3,\ \mathbf{A}_3) \tag{15}$$

corresponds to the multiplication of normalized quaternions

$$[a_2\ ;\ \mathbf{A}_2][a_1\ ;\ \mathbf{A}_1] = [a_3\ ;\ \mathbf{A}_3]. \tag{16}$$

It follows that a *turn* is the geometric realization of a *normalized quaternion*, that the addition of turns is the geometric realization of the multiplication of quaternions of unit norm, and that the group of turns is isomorphous with the group of normalized quaternions. Furthermore, eqs. (1) and (2) provide the correct parameterization of a normalized quaternion as

(1), (2) $$q = [\cos(\tfrac{1}{2}\phi)\ ;\ \sin(\tfrac{1}{2}\phi)\mathbf{n}], \tag{17}$$

which corresponds to a rotation $R(\phi\ \mathbf{n})$ through an angle ϕ, since a turn $\mathbf{T}$ was defined as *half* the directed arc traced out by the rotation $R(\phi\ \mathbf{n})$ on the surface of the unit sphere. To make contact with the work of Rodrigues a rotation $R(\phi\ \mathbf{n})$ will now be re-written in terms of new parameters $[\lambda\ ;\ \mathbf{\Lambda}]$, called the *quaternion* or *Euler–Rodrigues* parameters, as

$$R(\phi\ \mathbf{n}) \to [\lambda\ ;\ \mathbf{\Lambda}],\ \lambda = \cos(\tfrac{1}{2}\phi), \quad \mathbf{\Lambda} = \sin(\tfrac{1}{2}\phi)\mathbf{n}, \tag{18}$$

where λ is a scalar and $\mathbf{\Lambda}$ is a pseudovector. The multiplication rule for two rotations $[\lambda_2\ ;\ \mathbf{\Lambda}_2]$, $[\lambda_1\ ;\ \mathbf{\Lambda}_1]$ is then the quaternion multiplication rule eq. (12.2.9) or, equivalently, eq. (14) above,

$$[\lambda_2\ ;\ \mathbf{\Lambda}_2][\lambda_1\ ;\ \mathbf{\Lambda}_1] = [\lambda_3\ ;\ \mathbf{\Lambda}_3], \tag{19}$$

with

$$\lambda_3 = \lambda_1\,\lambda_2 - \mathbf{\Lambda}_1 \cdot \mathbf{\Lambda}_2\ ,\ \ \mathbf{\Lambda}_3 = \lambda_1\,\mathbf{\Lambda}_2 + \lambda_2\,\mathbf{\Lambda}_1 + \mathbf{\Lambda}_2 \times \mathbf{\Lambda}_1. \tag{20}$$

We have now introduced four parameterizations for SO(3): $R(\phi\ \mathbf{n})$ (Section 2.1); $R(\alpha, \beta, \gamma)$, where α, β, γ are the three Euler angles (Section 11.7); $R(a, b)$, where a, b are the complex Cayley–Klein parameters (Section 11.6); and the Euler–Rodrigues parameterization $[\lambda\ ;\ \mathbf{\Lambda}]$.

Exercise 12.5-2 Deduce the relations

$$a = \lambda - \mathrm{i}\Lambda_z\ ,\ \ b = -\Lambda_y - \mathrm{i}\Lambda_x. \tag{21}$$

[*Hint*: Recall the homomorphism of SU(2) and SO(3).]

Exercise 12.5-3 Show that as $\phi \to 0$, $[\lambda\ ;\ \mathbf{\Lambda}]$ tends continuously to the identity $[1\ ;\ \mathbf{0}]$.

Exercise 12.5-4 Show that two rotations through infinitesimally small angles commute.

Exercise 12.5-5 Prove that the product of two bilateral binary (BB) rotations ($\phi_1 = \phi_2 = \pi$, $\mathbf{n}_1 \perp \mathbf{n}_2$) is a binary rotation about $\mathbf{n}_3 \perp (\mathbf{n}_1, \mathbf{n}_2)$.

One immediate application of the quaternion formulae (19) and (20) for the multiplication of rotations is the proof that all rotations through the same angle are in the same class. To find the rotations in the same class as $R(\phi\ \mathbf{n})$ we need to evaluate

$$R(\theta\ \mathbf{m})R(\phi\ \mathbf{n})R(\theta\ \mathbf{m})^{-1}, \tag{22}$$

where $\theta\ \mathbf{m} \in \{\phi\ \mathbf{n}\}$. In the quaternion representation,

$$\begin{aligned}
(22)\quad & \left[\cos\left(\tfrac{1}{2}\theta\right)\ ;\ \sin\left(\tfrac{1}{2}\theta\right)\mathbf{m}\right]\left[\cos\left(\tfrac{1}{2}\phi\right)\ ;\ \sin\left(\tfrac{1}{2}\phi\right)\mathbf{n}\right]\left[\cos\left(-\tfrac{1}{2}\theta\right)\ ;\ \sin\left(-\tfrac{1}{2}\theta\right)\mathbf{m}\right] \\
&= \left[\cos\left(\tfrac{1}{2}\theta\right)\ ;\ \sin\left(\tfrac{1}{2}\theta\right)\mathbf{m}\right]\left[\cos\left(\tfrac{1}{2}\phi\right)\cos\left(\tfrac{1}{2}\theta\right) + \sin\left(\tfrac{1}{2}\phi\right)\sin\left(\tfrac{1}{2}\theta\right)\mathbf{n}\cdot\mathbf{m}\ ;\right. \\
&\qquad \left. -\cos\left(\tfrac{1}{2}\phi\right)\sin\left(\tfrac{1}{2}\theta\right)\mathbf{m} + \sin\left(\tfrac{1}{2}\phi\right)\cos\left(\tfrac{1}{2}\theta\right)\mathbf{n} - \sin\left(\tfrac{1}{2}\phi\right)\sin\left(\tfrac{1}{2}\theta\right)\mathbf{n}\times\mathbf{m}\right] \\
&= \left[\cos\left(\tfrac{1}{2}\phi\right)\ ;\ \sin\left(\tfrac{1}{2}\phi\right)\{(\cos\theta)\mathbf{n} + (\sin\theta)\mathbf{m}\times\mathbf{n} + (1-\cos\theta)(\mathbf{m}\cdot\mathbf{n})\mathbf{m}\}\right]; \quad (23)
\end{aligned}$$

$$= \left[\cos\left(\tfrac{1}{2}\phi\right)\ ;\ \sin\left(\tfrac{1}{2}\phi\right)\mathbf{n}'\right],\ \ \mathbf{n}' = R(\theta\ \mathbf{m})\mathbf{n}, \tag{24}$$

which is the quaternion representation of $R(\phi\ \mathbf{n}')$. The expression in { } in eq. (23) will be recognized from the formula for the conical transformation (eq. (11.5.22)) as the effect of the rotation $R(\phi\ \mathbf{m})$ on the vector $\mathbf{n}$. Equation (24) shows that in SO(3) all rotations through the same angle ϕ are in the same class. In other words, we have proved that conjugation leaves the rotation angle invariant but transforms the rotation axis $\mathbf{n}$ into the new axis $\mathbf{n}'$ (and therefore the pole of the rotation P into P′).

Answers to Exercises 12.5

Exercise 12.5-1 (a) If $|\mathbf{T}| = 0$, $a = 1$, $\mathbf{A} = 0$ and $\mathbf{T} = \mathbf{T}(1, \mathbf{0}) = \mathbf{T}_0$. (b) If $|\mathbf{T}| = \pi$, $a = -1$, $\mathbf{A} = 0$ and $\mathbf{T} = \mathbf{T}(-1, \mathbf{0}) = \mathbf{T}_\pi$. (c) $-\mathbf{T}(a, \mathbf{A}) = \mathbf{T}(a, -\mathbf{A})$. (d) $-\mathbf{T}^c(a, \mathbf{A}) = -\mathbf{T}(-a, -\mathbf{A}) = \mathbf{T}(-a, \mathbf{A})$.

Exercise 12.5-2

$$\text{(11.6.9), (11.6.12), (18)} \quad \begin{aligned} a &= \cos(\tfrac{1}{2}\phi) - in_z \sin(\tfrac{1}{2}\phi) = \lambda - i\Lambda_z, \\ b &= -(n_y + in_x)\sin(\tfrac{1}{2}\phi) = -\Lambda_y - i\Lambda_x. \end{aligned}$$

Exercise 12.5-3 $[\lambda \; ; \; \mathbf{\Lambda}] = [\cos(\tfrac{1}{2}\phi) \; ; \; \sin(\tfrac{1}{2}\phi)\mathbf{n}]$. As $\phi \to 0$,

$$\cos(\tfrac{1}{2}\phi) = 1 - (\tfrac{1}{2}\phi)^2/2! \ldots \to 1, \qquad \sin(\tfrac{1}{2}\phi) = \tfrac{1}{2}\phi - (\tfrac{1}{2}\phi)^3/3! \ldots \to 0,$$

so that $[\lambda \; ; \; \mathbf{\Lambda}] \to [1 \; ; \; \mathbf{0}] = E$.

Exercise 12.5-4 In the product $[\lambda_1 \; ; \; \mathbf{\Lambda}_1][\lambda_2 \; ; \; \mathbf{\Lambda}_2] = [\lambda_3 \; ; \; \mathbf{\Lambda}_3]$, to first order in ϕ, $\lambda_1 = 1$, $\mathbf{\Lambda}_1 = \tfrac{1}{2}\phi_1\mathbf{n}_1$; $\lambda_2 = 1$, $\mathbf{\Lambda}_2 = \tfrac{1}{2}\phi_2\mathbf{n}_2$; $\lambda_3 = 1$, $\mathbf{\Lambda}_3 = \tfrac{1}{2}\phi_1\mathbf{n}_1 + \tfrac{1}{2}\phi_2\mathbf{n}_2$, whence it follows that the two rotations $R(\phi_1 \; \mathbf{n}_1)$, $R(\phi_2 \; \mathbf{n}_2)$ commute when ϕ_1, ϕ_2 are infinitesimally small angles.

Exercise 12.5-5 For BB rotations $\phi_1 = \phi_2 = \pi$, $\lambda_1 = \lambda_2 = 0$, $\mathbf{\Lambda}_1 \cdot \mathbf{\Lambda}_2 = \mathbf{n}_1 \cdot \mathbf{n}_2 = 0$, and the product of two BB rotations is $[0 \; ; \; \mathbf{n}_1 \times \mathbf{n}_2] = [0 \; ; \; \mathbf{n}_3]$, $\mathbf{n}_3 \perp (\mathbf{n}_1, \mathbf{n}_2)$.

12.6 Projective representations

The double group $\overline{\mathrm{G}}$ was introduced in Chapter 8 in order to deal with irreducible representations (IRs) that correspond to half-integral values of j. Because

$$\chi_j(R(\phi + 2\pi \; \mathbf{n})) = (-1)^{2j}\chi_j R(\phi \; \mathbf{n}), \tag{1}$$

$R(2\pi \; \mathbf{z}) \neq E$ for j equal to a half-integer. Bethe (1929) therefore introduced the new operator $\overline{E} = R(2\pi \; \mathbf{z}) \neq E$, thus doubling the size of $\mathrm{G} = \{g_i\}$ by forming the *double group*

$$\overline{\mathrm{G}} = \{g_i, \, \overline{g}_i\} = \{g_i\} \oplus \{\overline{g}_i\}, \qquad \overline{g}_i = \overline{E}g_i. \tag{2}$$

The IRs of $\overline{\mathrm{G}}$ comprise the *vector representations*, which are the IRs of G, and new representations called the *spinor* or *double group representations*, which correspond to half-integral j. The double group $\overline{\mathrm{G}}$ contains twice as many elements as G but not twice as many classes: g_i and $\overline{g}_i$ are in different classes in $\overline{\mathrm{G}}$ except when g_i is a proper or improper BB rotation (that is, a rotation about a binary axis that is normal to another binary axis), in which case g_i and $\overline{g}_i$ are in the same class and $\chi(g_i)$, $(\chi\overline{g}_i)$ are necessarily zero in spinor representations. Operations which satisfy this condition are called *irregular operations*, all other operations being termed *regular* (Altmann and Herzig (1994)). The distinction is important, since the number of spinor (or double group) representations is equal to the number of regular classes in G.

Example 12.6-1 The point group $C_{2v} = \{E\ C_{2\mathbf{z}}\ \sigma_{\mathbf{x}}\ \sigma_{\mathbf{y}}\}$, where $\sigma_{\mathbf{x}} = IC_{2\mathbf{x}}$ and $\sigma_{\mathbf{y}} = IC_{2\mathbf{y}}$. Because **x**, **y**, **z** are mutually perpendicular axes, all operations except E are irregular and there is consequently only one doubly degenerate spinor representation, $E_{1/2}$. Contrast $C_{2h} = \{E\ C_{2\mathbf{z}}\ I\ \sigma_h\}$ in which σ_h is $\sigma_z = IC_{2\mathbf{z}}$ and thus an improper binary rotation about **z**. There are therefore no irregular operations, and consequently there are four spinor representations which occur in two doubly degenerate pairs $E_{1/2,g}$ and $E_{1/2,u}$ because of time-reversal symmetry (Chapter 13). In C_{3v} there are three improper binary axes but they are not mutually perpendicular.

Example 12.6-2 The classes of $\mathbf{T}_d$ are $\{E\ 4C_3\ 3C_2\ 6S_4\ 6\sigma_d\}$. The three binary rotations are BB rotations. The six dihedral planes occur in three pairs of perpendicular improper BB rotations so both $3C_2$ and $6\sigma_d$ are irregular classes. There are therefore $N_v = 5$ vector representations and $N_s = 3$ spinor representations.

Exercise 12.6-1 Determine the number of spinor representations in the point group D_{3h}.

In an alternative approach, no new elements are introduced; so G is not altered, but instead there is a new kind of representation called a *projective representation.* In a projective representation (PR) the set of MRs only closes if a numerical factor called a *projective factor* (PF) and written $[g_i\ ;\ g_j]$ is introduced so that $\Gamma(g_i)\Gamma(g_j) = [g_i\ ;\ g_j]\ \Gamma(g_ig_j)$. The advantages of this approach are that the group, its multiplication rules, and class structure remain unchanged. Of course, we need a method of finding the PFs for any given group. Once these are determined, the spinor representations may be used in the same way as vector representations, except that when two MRs are multiplied together, the resulting matrix has to be multiplied by the appropriate PF. One potential difficulty is that the character of an MR in a PR is not *necessarily* a property of its class. But with the conventions applied in choosing the poles of rotations, the character *is* a class property for all point groups (Altmann (1979)). With the conventions to be applied in the calculation of PFs (which will be given shortly) and the pole conventions used in Section 12.3, the orthogonality relations, and their consequences, are also valid for unitary PRs except for one, and that is that the number of IRs N_r is not equal to the number of classes. However, the relation that the sum of the squares of the dimensions of the representations is equal to the order of the group (which followed from the orthogonality theorem) holds for spinor representations as well as for vector representations. Consequently,

$$\sum_s l_s^2 = g, \tag{3}$$

where $s = 1, \ldots, N_s$ enumerates the spinor representations (Altmann (1977)).

We begin by reiterating the definition of a PR and listing some conventions regarding PFs. A *projective unitary representation* of a group $G = \{g_i\}$ of dimension g is a set of matrices that satisfy the relations

$$\Gamma(g_i)^\dagger \Gamma(g_i) = \Gamma(g_i)\Gamma(g_i)^\dagger = \mathrm{E}, \tag{4}$$

$$\Gamma(g_i)\Gamma(g_j) = [g_i\ ;\ g_j]\Gamma(g_i\ g_j) \qquad \forall g_i, g_j \in \mathrm{G}. \tag{5}$$

The PFs $[g_i\ ;\ g_j]$ are a set of g^2 complex numbers, which by convention are all chosen to be square roots of unity. (For vector representations the PFs are all unity.) PFs have the following properties (Altmann (1977)):

(a) associativity

$$[g_i\ ;\ g_j][g_i\ g_j\ ;\ g_k] = [g_i\ ;\ g_j\ g_k][g_j\ ;\ g_k], \tag{6}$$

(b) standardization

$$[E\ ;\ E] = [E\ ;\ g_i] = [g_i\ ;\ E] = 1 \qquad \forall g_i \in \mathrm{G}, \tag{7}$$

(c) normalization

$$[g_i\ ;\ g_j]^*[g_i\ ;\ g_j] = 1 \qquad \forall g_i, g_j \in \mathrm{G}, \tag{8}$$

(d) symmetry

$$[g_i\ ;\ g_i^{-1}] = [g_i^{-1}\ ;\ g_i] \qquad \forall g_i \in \mathrm{G}. \tag{9}$$

The set of PFs $\{[g_i\ ;\ g_j]\}$ is called the *factor system*. Associativity (a) and the symmetry of $[g_i\ ;\ g_i^{-1}]$ (d) are true for all factor systems. The standardization (b) and normalization (c) properties are conventions chosen by Altmann and Herzig (1994) in their standard work *Point Group Theory Tables*. Associativity (a) follows from the associativity property of the multiplication of group elements. For a spinor representation Γ of G, on introducing $[i\ ;\ j]$ as an abbreviation for $[g_i\ ;\ g_j]$,

(4)
$$\begin{aligned}\Gamma(g_i)\{\Gamma(g_j)\Gamma(g_k)\} &= \Gamma(g_i)[j\ ;\ k]\Gamma(g_j\ g_k) \\ &= [i\ ;\ jk][j\ ;\ k]\Gamma(g_i\ g_j\ g_k),\end{aligned} \tag{10}$$

(4)
$$\begin{aligned}\{\Gamma(g_i)\Gamma(g_j)\}\Gamma(g_k) &= [i\ ;\ j]\Gamma(g_i\ g_j)\Gamma(g_k) \\ &= [i\ ;\ j][ij\ ;\ k]\Gamma(g_i\ g_j\ g_k),\end{aligned} \tag{11}$$

whence eq. (6) follows.

From the pole conventions in Section 11.3 it follows that $R(-\phi\ \mathbf{n}) = R(-\phi\ \bar{\mathbf{n}})$, and this restricts the rotation angle ϕ to the range $0 \le \phi \le \pi$, which, in turn, restricts the range of the quaternion or Euler–Rodrigues parameters to

$$\lambda = \cos(\tfrac{1}{2}\phi) \ge 0, \quad |\boldsymbol{\Lambda}| = \sin(\tfrac{1}{2}\phi) \ge 0. \tag{12}$$

Parameters that satisfy eq. (12) are referred to as *standard quaternion* (or Euler–Rodrigues) parameters. Note that $\mathbf{\Lambda} = \sin(\frac{1}{2}\phi)\mathbf{n}$ belongs to the positive hemisphere h for positive rotations and to $-h$ for negative rotations. For binary rotations, $\phi = \pi$, and so $\lambda = 0$, $\mathbf{\Lambda} = 1$, and $\mathbf{\Lambda}$ belongs to h because there is no rotation $R(-\pi \ \mathbf{n})$. Therefore for any point group, h must be defined so as to contain the poles of all positive rotations, including binary rotations. Due to the range of ϕ, standard quaternion parameters must satisfy either

$$\lambda > 0 \quad (0 \leq \phi < \pi) \text{ or } \lambda = 0,\ \mathbf{\Lambda} \in h\ (\phi = \pi). \tag{13}$$

The PF for the product of any two rotations may now be determined using the quaternion representation.

Example 12.6-3 Determine the PF for the multiplication of a binary rotation with itself.

For $R(\pi \ \mathbf{n})$, $\cos(\frac{1}{2}\pi) = 0$, $\sin(\frac{1}{2}\pi) = 1$ and the product $C_{2\mathbf{n}}\, C_{2\mathbf{n}}$ in the quaternion representation is

$$[0\ ;\ \mathbf{n}][0\ ;\ \mathbf{n}] = [0 - \mathbf{n}\cdot\mathbf{n}\ ;\ \mathbf{0} + \mathbf{0} + \mathbf{n}\times\mathbf{n}] = [-1\ ;\ \mathbf{0}] = -1[1\ ;\ \mathbf{0}]; \tag{14}$$

(4), (14) $$\Gamma(C_{2\mathbf{n}})\Gamma(C_{2\mathbf{n}}) = [C_{2\mathbf{n}}\ ;\ C_{2\mathbf{n}}]\Gamma(E) = -\Gamma(E), \quad [C_{2\mathbf{n}}\ ;\ C_{2\mathbf{n}}] = -1. \tag{15}$$

Example 12.6-4 Determine the PF for $C_{3\mathbf{z}}^+\, C_{3\mathbf{z}}^+$.

For $C_{3\mathbf{z}}^+$, $\phi = 2\pi/3$ and

$$\tfrac{1}{2}\phi = \pi/3, \quad \lambda_1 = \tfrac{1}{2}, \quad \mathbf{\Lambda}_1 = \tfrac{\sqrt{3}}{2}\mathbf{z} = \tfrac{\sqrt{3}}{2}[0\ 0\ 1]. \tag{16}$$

For $C_{3\mathbf{z}}^-$, $\phi = 2\pi/3$ and

$$\tfrac{1}{2}\phi = \pi/3, \quad \lambda_2 = \tfrac{1}{2}, \quad \mathbf{\Lambda}_2 = \tfrac{\sqrt{3}}{2}\bar{\mathbf{z}} = \tfrac{\sqrt{3}}{2}[0\ 0\ \bar{1}], \tag{17}$$

(16) $$\begin{aligned}[\lambda_1\ ;\ \mathbf{\Lambda}_1][\lambda_1\ ;\ \mathbf{\Lambda}_1] &= \left[\tfrac{1}{2}\ ;\ \tfrac{\sqrt{3}}{2}[0\ 0\ 1]\right]\left[\tfrac{1}{2}\ ;\ \tfrac{\sqrt{3}}{2}[0\ 0\ 1]\right] \\ &= \left[\tfrac{1}{4} - \tfrac{3}{4}\ ;\ \tfrac{\sqrt{3}}{4}[0\ 0\ 1]\right] + \tfrac{\sqrt{3}}{2}[0\ 0\ 1] = \left[-\tfrac{1}{2}\ ;\ \tfrac{\sqrt{3}}{2}[0\ 0\ 1]\right] \\ &= -\left[\tfrac{1}{2}\ ;\ \tfrac{\sqrt{3}}{2}[0\ 0\ \bar{1}]\right] = -\left[\tfrac{1}{2}\ ;\ \tfrac{\sqrt{3}}{2}\bar{\mathbf{z}}\right] = -[\lambda_2\ ;\ \mathbf{\Lambda}_2],\end{aligned} \tag{18}$$

(18), (17) $$\Gamma(C_{3\mathbf{z}}^+)\Gamma(C_{3\mathbf{z}}^+) = -\Gamma((C_{3\mathbf{z}}^-)), [C_{3\mathbf{z}}^+\,; C_{3\mathbf{z}}^+] = -1. \tag{19}$$

Example 12.6-5 Determine the PFs for $[\mathbf{C_{2z}}\ ;\ \mathbf{C_{2x}}]$ and $[\mathbf{C_{2x}}\ ;\ \mathbf{C_{2z}}]$.

For $R(\pi \ \mathbf{z})$

$$\lambda_1 = 0,\ \mathbf{\Lambda}_1 = \mathbf{z} = [0\ 0\ 1], \tag{20}$$

and for $R(\pi \ \mathbf{x})$

$$\lambda_2 = 0,\ \mathbf{\Lambda}_2 = \mathbf{x} = [1\ 0\ 0]. \tag{21}$$

Table 12.2. *Multiplication table for* $\overline{C}_2$.

The multiplication table for the subset $\{E\ C_2\}$ is not closed, showing that C_2 is not a subgroup of $\overline{C}_2$.

$\overline{C}_2$	E	C_2	$\overline{E}$	$\overline{C}_2$
E	E	C_2	$\overline{E}$	$\overline{C}_2$
C_2	C_2	$\overline{E}$	$\overline{C}_2$	E
$\overline{E}$	$\overline{E}$	$\overline{C}_2$	E	C_2
$\overline{C}_2$	$\overline{C}_2$	E	C_2	$\overline{E}$

(20), (21) $$[0\ ;\ [0\ 0\ 1]][0\ ;\ [1\ 0\ 0]] = [0\ ;\ [0\ 1\ 0]] = [0\ ;\ \mathbf{y}]; \tag{22}$$

(22) $$\Gamma(C_{2z})\Gamma(C_{2x}) = \Gamma(C_{2y}),\ [C_{2z}\ ;\ C_{2x}] = 1; \tag{23}$$

(21), (20) $$[0\ ;\ [1\ 0\ 0]][0\ ;\ [0\ 0\ 1]] = [0\ ;\ [0\ \bar{1}\ 0]] = -[0\ ;\ \mathbf{y}]; \tag{24}$$

(24) $$\Gamma(C_{2x})\Gamma(C_{2z}) = -\Gamma(C_{2y}),\ [C_{2x}\ ;\ C_{2z}] = -1. \tag{25}$$

Equation (15) applies to PRs, and the multiplication rule it obeys is not one of $G = \{E\ \ C_2\}$ in which $C_2\,C_2 = E$. The only way to maintain a 1:1 correspondence between R and $\Gamma(R)$ without introducing PFs is to enlarge the size of $G = \{g_i\}$ to $\overline{G} = \{g_i\} \oplus \{\overline{g}_i\}$; $\overline{G}$ is the *double group* of G, but G is not a subgroup of $\overline{G}$ since the multiplication rules of $\overline{G}$ are different from those of G; $\overline{G}$ is a *covering group* of G because the vector representations of $\overline{G}$ subduce to the PRs of G by omitting the MRs of $\{\overline{g}_i\}$. The quaternion representation of $\overline{E}$ is $[-1\ ;\ \mathbf{0}]$ and so if R has the parameters $[\lambda\ ;\ \mathbf{\Lambda}]$ then the parameters of $\overline{R} = \overline{E}R$ are $[-1\ ;\ 0][\lambda\ ;\ \mathbf{\Lambda}] = [-\lambda\ ;\ -\mathbf{\Lambda}]$. Thus the sign of the quaternion parameters shows whether the product of two operators is R or $\overline{R}$. This rule is exemplified in Table 12.2, which gives the multiplication table for $\overline{C}_2$. However, it is not necessary to use the covering group to find spinor representations since they may be found directly using the quaternion representation.

12.6.1 Inverse and conjugate in the quaternion parameterization

The inverse of g_i g_i^{-1} is defined by

$$g_i^{-1}\ g_i = g_i\ g_i^{-1} = E. \tag{26}$$

In the quaternion representation, using the abbreviated notation $\lambda(g_i) \rightarrow \lambda_i,\ \lambda(g_i^{-1}) \rightarrow \lambda_{i^{-1}}$, and similarly for $\mathbf{\Lambda}$,

(26) $$[\lambda_i\ ;\ \boldsymbol{\Lambda}_i][\lambda_{i^{-1}}\ ;\ \boldsymbol{\Lambda}_{i^{-1}}] = [1\ ;\ \mathbf{0}]$$
$$= [\lambda_i\lambda_{i^{-1}} - \boldsymbol{\Lambda}_i\,.\,\boldsymbol{\Lambda}_{i^{-1}}\ ;\ \lambda_i\boldsymbol{\Lambda}_{i^{-1}} + \lambda_{i^{-1}}\boldsymbol{\Lambda}_i + \boldsymbol{\Lambda}_i \times \boldsymbol{\Lambda}_{i^{-1}}]; \tag{27}$$

(27) $$\lambda(g_i{}^{-1}) = \lambda(g_i),\ \ \boldsymbol{\Lambda}(g_i{}^{-1}) = -\boldsymbol{\Lambda}(g_i). \tag{28}$$

If $\phi \neq \pi$ so that g_i is not a binary rotation and $\lambda(g_i) > 0$, then, from eq. (28), $\lambda(g_i{}^{-1}) > 0$ and its parameters are standardized (see eqs. (12) and (13)). For $\phi = \pi$, $\lambda(g_i) = 0$, and $\boldsymbol{\Lambda}(g_i) \in h$. But for $\phi = \pi$, $g_i{}^{-1} = R(-\pi\ \mathbf{n})$, and $-\pi$ is out of range. However, $g_i{}^{-1}$ is then equivalent to g_i, which has been chosen to have standard parameters by ensuring that $\boldsymbol{\Lambda}(g_i) \in h$ for all binary rotations.

From the quaternion representation of $g_k\ g_i\ g_k{}^{-1}$, when $g_k = R(\theta\ \mathbf{m})$ and $g_i = R(\phi\ \mathbf{n})$,

(12.5.23) $$[\lambda(g_k\ g_i\ g_k{}^{-1})\ ;\ \boldsymbol{\Lambda}(g_k\ g_i\ g_k{}^{-1})] \equiv [\lambda_{k\,i\,k^{-1}}\ ;\ \boldsymbol{\Lambda}_{k\,i\,k^{-1}}] \tag{29}$$

$$= [\lambda_i\ ;\ (1 - 2\Lambda_k^2)\boldsymbol{\Lambda}_i + 2\lambda_k\boldsymbol{\Lambda}_k \times \boldsymbol{\Lambda}_i + 2(\boldsymbol{\Lambda}_k\,.\,\boldsymbol{\Lambda}_i)\boldsymbol{\Lambda}_k]. \tag{30}$$

The scalar parts of eqs. (29) and (30) coincide because conjugation leaves the rotation angle invariant. We now consider three cases:

(1) If g_i is not binary, the range of $-\phi$ is $0 \leq -\phi < \pi/2$, and therefore

(13) $$\lambda_{k\,i\,k^{-1}} = \lambda_i > 0 \text{ is standard.} \tag{31}$$

(2) If g_i is binary ($\phi = \pi$) but $g_i\,g_k$ are not BB, $\lambda_{k\,i\,k^{-1}} = \lambda_i = 0$, and standardization can be ensured by choosing h such that

$$\boldsymbol{\Lambda}_{k\,i\,k^{-1}} \in h\ (\lambda_i = 0). \tag{32}$$

(3) If $g_i\,g_k$ are BB, $\lambda_i = \lambda_k = 0$, $\Lambda_i^2 = \Lambda_k^2 = 1$, $\boldsymbol{\Lambda}_i$, $\boldsymbol{\Lambda}_k \in h$, $\boldsymbol{\Lambda}_i . \boldsymbol{\Lambda}_k = 0$,

(30) $$\boldsymbol{\Lambda}_{k\,i\,k^{-1}} = (1 - 2\Lambda_k^2)\boldsymbol{\Lambda}_i = -\boldsymbol{\Lambda}_i,\ \ \boldsymbol{\Lambda}_{k\,i\,k^{-1}} \in -h. \tag{33}$$

Equation (33) shows that the conjugate pole of g_i is the antipole of g_i, a situation that arises only when g_i, g_k are BB rotations.

12.6.2 The characters

From the definition of a class $\mathscr{C}(g_i) = \{g_k\ g_i\ g_k{}^{-1}\}$, $\forall\ g_k \in \mathrm{G}$ (with repetitions deleted) it follows that for vector representations

$$\chi(g_k\ g_i\ g_k{}^{-1}) = \operatorname{Tr}\Gamma(g_k\ g_i\ g_k{}^{-1}) = \operatorname{Tr}\Gamma(g_k{}^{-1}\ g_k\ g_i) = \chi(g_i). \tag{34}$$

For spinor representations (in abbreviated notation, where k^{-1} means $g_k{}^{-1}$)

$$\begin{aligned}\chi(g_k\, g_i\, g_k{}^{-1}) &= \mathrm{Tr}\,\Gamma(g_k\, g_i\, g_k^{-1}) = [k\ ;\ i\,k^{-1}]^{-1}\ \mathrm{Tr}\,\Gamma(g_k)\Gamma(g_i\, g_k^{-1})\\ &= [k\ ;\ i\,k^{-1}]^{-1}[i\ ;\ k^{-1}]^{-1}\mathrm{Tr}\,\Gamma(g_k)\Gamma(g_i)\Gamma(g_k{}^{-1})\\ &= [k\ ;\ i\,k^{-1}]^{-1}[i\ ;\ k^{-1}]^{-1}\mathrm{Tr}\,\Gamma(g_k^{-1})\Gamma(g_k)\Gamma(g_i)\\ &= [k\ ;\ i\,k^{-1}]^{-1}[i\ ;\ k^{-1}]^{-1}[k^{-1}\ ;\ k]\ \mathrm{Tr}\,\Gamma(E)\Gamma(g_i)\\ &= [k\ ;\ ik^{-1}]^{-1}[i\ ;\ k^{-1}]^{-1}[k^{-1}\ ;\ k][E\ ;\ g_i]\mathrm{Tr}\,\Gamma(g_i)\\ &= [k\ ;\ i\,k^{-1}]^{-1}[i\ ;\ k^{-1}]^{-1}[k^{-1}\ ;\ k]\mathrm{Tr}\,\Gamma(g_i)\end{aligned} \tag{35}$$

$$= [k\,i\,k^{-1}\ ;\ k][k\ ;\ i]^{-1}\ \chi(g_i) \tag{36}$$

on using associativity twice. Reverting to full notation,

(36) $$\chi(g_k\, g_i\, g_k{}^{-1}) = [g_k\, g_i\, g_k{}^{-1}\ ;\ g_k][g_k\ ;\ g_i]^{-1}\ \chi(g_i). \tag{37}$$

Exercise 12.6-2 Complete the derivation of eq. (36) by filling in the steps between eqs. (35) and (36). [*Hint*: You will need to use associativity twice.]

In evaluating the PF in eq. (37) note that the conjugation operation $g_k\, g_i\, g_k{}^{-1}$, or $k\,i\,k^{-1}$, is to be regarded as a single operation. The quaternion parameters λ, $\mathbf{\Lambda}$ for the product of $g_k\, g_i\, g_k{}^{-1}$ with g_k are

$$\begin{aligned}\lambda_{k\,i\,k^{-1}\,k} &= \pm(\lambda_i\lambda_k - \mathbf{\Lambda}_i\,.\,\mathbf{\Lambda}_k - 2\Lambda_k^2\ \mathbf{\Lambda}_i\,.\,\mathbf{\Lambda}_k + 2\mathbf{\Lambda}_k\cdot\mathbf{\Lambda}_i\Lambda_k^2)\\ &= \pm(\lambda_i\lambda_k - \mathbf{\Lambda}_i\,.\,\mathbf{\Lambda}_k);\end{aligned} \tag{38}$$

$$\begin{aligned}\mathbf{\Lambda}_{k\,i\,k^{-1}\,k} &= \pm[\lambda_i\mathbf{\Lambda}_k + \lambda_k\{\mathbf{\Lambda}_i - 2\Lambda_k^2\mathbf{\Lambda}_i + 2\lambda_k\mathbf{\Lambda}_k\times\mathbf{\Lambda}_i + 2(\mathbf{\Lambda}_k\,.\,\mathbf{\Lambda}_i)\mathbf{\Lambda}_k\} + \mathbf{\Lambda}_i\times\mathbf{\Lambda}_k\\ &\quad - 2\Lambda_k^2\mathbf{\Lambda}_i\times\mathbf{\Lambda}_k + 2\lambda_k\mathbf{\Lambda}_k\times\mathbf{\Lambda}_i\times\mathbf{\Lambda}_k]\\ &= \pm[\lambda_i\mathbf{\Lambda}_k + \lambda_k\mathbf{\Lambda}_i + \mathbf{\Lambda}_k\times\mathbf{\Lambda}_i].\end{aligned} \tag{39}$$

For the product $g_k\, g_i$,

$$\lambda_{ki} = \lambda_k\lambda_i - \mathbf{\Lambda}_k\,.\,\mathbf{\Lambda}_i, \tag{40}$$

$$\mathbf{\Lambda}_{ki} = \lambda_k\mathbf{\Lambda}_i + \lambda_i\mathbf{\Lambda}_k + \mathbf{\Lambda}_k\times\mathbf{\Lambda}_i, \tag{41}$$

(37) – (41) $$\chi(g_k\, g_i\, g_k{}^{-1}) = \pm\chi(g_i). \tag{42}$$

Provided h has been properly chosen to contain all conjugate binary poles, the negative sign will arise only when g_k and g_i are BB, in which case $g_k\, g_i\, g_k{}^{-1} = g_i$, and eq. (42) shows that χ vanishes.

12.6.3 Direct product representations

Given M = G ⊗ H so that M = $\{m_k\}$, where $m_k = g_i\, h_j = h_j\, g_i$, and PRs Γ^1 of G and Γ^2 of H, so that

$$\Gamma^1(g_i)\Gamma^1(g_p) = [g_i\ ;\ g_p]\Gamma^1(g_i\ g_p), \qquad \Gamma^2(h_j)\Gamma^2(h_r) = [h_j\ ;\ h_r]\Gamma^2(h_jh_r), \tag{43}$$

it may be shown that

$$\Gamma^3(g_ih_j) = [g_i\ ;\ h_j]^{-1}\Gamma^1(g_i) \otimes \Gamma^2(h_j) \tag{44}$$

are PRs of M with PFs

$$[g_ih_j\ ;\ g_ph_r] = [g_i\ ;\ g_p][h_j\ ;\ h_r][g_i\ g_p\ ;\ h_jh_r][g_i\ ;\ h_j]^{-1}[g_p\ ;\ h_r]^{-1}. \tag{45}$$

Proof In the abbreviated notation for PFs

$$\begin{aligned}\Gamma^3(g_ih_j)\Gamma^3(g_ph_r) &= [i\ ;\ j]^{-1}[p\ ;\ r]^{-1}\{\Gamma^1(g_i) \otimes \Gamma^2(h_j)\}\{\Gamma^1(g_p) \otimes \Gamma^2(h_r)\}\\ &= [i\ ;\ j]^{-1}[p\ ;\ r]^{-1}\{\Gamma^1(g_i)\Gamma^1(g_p)\} \otimes \{\Gamma^2(h_j)\Gamma^2(h_r)\}\\ &= [i\ ;\ j]^{-1}[p\ ;\ r]^{-1}[i\ ;\ p][j\ ;\ r]\{\Gamma^1(g_i\ g_p)\} \otimes \{\Gamma^2(h_j\ h_r)\}\end{aligned}$$

(44) $$= [i\ ;\ j]^{-1}[p\ ;\ r]^{-1}[i\ ;\ p][j\ ;\ r][ip\ ;\ jr]\Gamma^3(g_ig_ph_jh_r). \tag{46}$$

But

$$\Gamma^3(g_ih_j)\Gamma^3(g_ph_r) = [ij\ ;\ pr]\Gamma^3(g_ig_p\ h_jh_r); \tag{47}$$

(46), (47) $$[ij\ ;\ pr] = [i\ ;\ p][j\ ;\ r][i\ ;\ j]^{-1}[p\ ;\ r]^{-1}[ip\ ;\ jr],$$

which is eq. (45) in the abbreviated notation. In the event that Γ^1, Γ^2 are two different PRs of the same group G with different factor systems a and b so that

$$\Gamma^1_a(g_i)\Gamma^2_a(g_p) = [i\ ;\ p]_a\Gamma^3_a(g_ig_p),\ \ \Gamma^1_b(g_i)\Gamma^2_b(g_p) = [i\ ;\ p]_b\Gamma^3_b(g_ig_p) \tag{48}$$

then the direct product (DP)

$$\Gamma^3 = \Gamma^1_a \otimes \Gamma^2_b \tag{49}$$

is a PR with factor system

$$[i\ ;\ p] = [i\ ;\ p]_a[i\ ;\ p]_b. \tag{50}$$

The proof of eq. (50) is similar to that of eq. (45) and is assigned as Problem 12.9.

Answers to Exercises 12.6

Exercise 12.6-1 $D_{3h} = \{E\ 2C_3\ 3C_2{}'\ \sigma_h\ 2S_3\ 3\sigma_v\}$, where $\sigma_h = IC_2$, $\sigma_v = IC_2{}''$, $C_2 \perp (C_2{}',\ C_2{}'')$. Thus the improper binary axis C_2 is normal to the $3C_2{}'$ proper binary axes and the three improper $C_2{}''$ binary axes. There are, therefore, three irregular classes σ_h, $3C_2{}'$, and $3\sigma_v$. There are six classes in all and therefore six vector representations ($N_v = 6$). There are three regular classes and therefore three spinor representations, each of which is doubly degenerate since $\sum_{s=1}^{3} l_s^2 = 2^2 + 2^2 + 2^2 = 12 = g$.

Exercise 12.6-2

$$(35)\ [k\,;\,ik^{-1}]^{-1}[i\,;\,k^{-1}]^{-1}[k^{-1}\,;\,k] = [k\,;\,ik^{-1}]^{-1}[i\,;\,k^{-1}]^{-1}[i\,;\,k^{-1}][ik^{-1}\,;\,k][i\,;\,k^{-1}k]$$
$$= [k\,;\,ik^{-1}]^{-1}[k\,;\,ik^{-1}][kik^{-1}\,;\,k][k\,;\,ik^{-1}k]^{-1}$$
$$= [kik^{-1}\,;\,k][k\,;\,i]^{-1},$$

which proves eq. (36).

12.7 Improper groups

The group O(3) comprises all proper and improper rotations in configuration space $\mathfrak{R}^3$. It is obtained from the DP

$$\mathrm{O}(3) = \mathrm{SO}(3) \otimes \mathrm{C}_i\,, \quad \mathrm{C}_i = \{E, I\}. \tag{1}$$

Since $I\,I = E$, for PRs

$$\Gamma(I)\Gamma(I) = [I\,;\,I]\Gamma(E). \tag{2}$$

With the spinor basis $\langle u\ v|$

(11.8.13) $$[I\,;\,I] = -1 \text{ (Cartan gauge)}, \tag{3}$$

(11.8.14) $$[I\,;\,I] = +1 \text{ (Pauli gauge)}. \tag{4}$$

The PFs for the other three products *EE*, *EI*, *IE* are all unity because of standardization, eq. (12.6.6). Irreducible PRs that are related by a gauge transformation may be converted one into the other by multiplying the characters for each class by a specific phase factor. Such a gauge transformation does not alter the energy eigenvalues, so, for that purpose, gauge equivalence may be ignored. However, a choice of phase factors may have other implications. Character tables for the PRs of the point groups are generally given in the *Pauli gauge* (see Altmann and Herzig (1994), equations (13.13) and (13.17)). Table 12.3 shows the multiplication table for C_i as well as the factor tables (that is, the set of PFs $\{[g_i\,;\,g_j]\}$ for each PR) and the character tables in both the Cartan gauge and the Pauli gauge. The matrices $\Gamma(I)$ in eq. (11.8.13) provide us with two inequivalent reducible PRs of C_i:

$$\Gamma_3 = \overset{E}{\begin{bmatrix} 1 & \\ & 1 \end{bmatrix}} \overset{I}{\begin{bmatrix} -\mathrm{i} & \\ & -\mathrm{i} \end{bmatrix}}; \quad \Gamma_4 = \overset{E}{\begin{bmatrix} 1 & \\ & 1 \end{bmatrix}} \overset{I}{\begin{bmatrix} \mathrm{i} & \\ & \mathrm{i} \end{bmatrix}}, \tag{5}$$

where $\Gamma_3 = \mathrm{A}_{1/2,g} \otimes \mathrm{A}_{1/2,g}$, $\Gamma_4 = \mathrm{A}_{1/2,u} \otimes \mathrm{A}_{1/2,u}$ (Table 12.3). If these two IRs are each multiplied by the phase factors 1 (for E) and i (for I) (a gauge transformation from the Cartan gauge $\Gamma(I) = -\mathrm{i}\,\mathrm{E}_2$, to the Pauli gauge $\Gamma(I) = \mathrm{E}_2$) then they are transformed into the IRs A_g, A_u as shown in Table 12.3. The apparent simplification that results from the use of the Pauli gauge has disadvantages. The fact that in PRs the inversion operator I behaves just like the identity E (see eq. (11.8.16)) is in sharp contrast with our treatment of vector representations in which I means I, a distinction in O(3) that applies to all the improper point groups that are subgroups of O(3). The choice of gauge is important when forming

Table 12.3. *Multiplication table, factor tables, and character tables for the point group* C_i.

u^0 is the spherical harmonic for $j=0$ (a scalar); $\bar{u}^0$ is the pseudoscalar $2^{-1/2}(u_1 v_2 - v_1 u_2)$ (see eqs. (11.8.11) and (11.8.15)); $u = |1/2\ 1/2\rangle, v = |1/2\ -1/2\rangle$.

Multiplication table

C_i	E	I
E	E	I
I	I	E

Cartan gauge

Factor table

$\{[g_i\ ;\ g_j]\}$	E	I
E	1	1
I	1	−1

Character table

C_i	E	I	basis
A_g	1	1	u^0
A_u	1	−1	$\bar{u}^0$
$A_{1/2,g}$	1	−i	u
$A_{1/2,u}$	1	i	u^*

Pauli gauge

Factor table

$\{[g_i\ ;\ g_j]\}$	E	I
E	1	1
I	1	1

Character table

C_i	E	I	$j=0$	$j=1/2$
A_g	1	1	u^0	u
A_u	1	−1	$\bar{u}^0$	u^*

tensor products, as the reduction of the DP basis in eq. (11.8.15) into a vector and a pseudoscalar depends on the two spinors being in the Cartan gauge (see eq. (11.8.14)).

12.7.1 Factor system for O(3)

The elements of $O(3) = SO(3) \otimes C_i$ are $m_k = g_i\, h_j$, where $g_i \in SO(3)$ and $h_j \in C_i = \{E\ I\}$. The MR of m_k is

$$\Gamma(m_k) = [g_i\ ;\ h_j]^{-1}\Gamma(g_i) \otimes \Gamma(h_j), \tag{6}$$

and the factor system for this representation is

(12.6.45)
$$\begin{aligned}[m_k\ ;\ m_s] &= [g_i h_j\ ;\ g_p h_r] \\ &= [g_i\ ;\ g_p][h_j\ ;\ h_r][g_i g_p\ ;\ h_j h_r][g_i\ ;\ h_j]^{-1}[g_p\ ;\ h_r]^{-1},\end{aligned} \tag{7}$$

where h_j and $h_r \in \{E\ I\}$. Because of standardization,

$$[g_i\ ;\ E] = [E\ ;\ g_i] = [E\ ;\ I] = [I\ ;\ E] = 1, \qquad g_i \in \mathrm{SO}(3). \tag{8}$$

In the Cartan gauge $[I\ ;\ I] = -1$ and in the Pauli gauge $[I\ ;\ I] = 1$. Consequently, the only remaining PFs to evaluate are those involving I, namely $[g_i\ ;\ I]$ and $[I\ ;\ g_i]$, assuming the factor system for SO(3) to be known. Because the pole of a rotation is invariant under the inversion operation,

$$[g_i\ ;\ I] = [I\ ;\ g_i] = 1 \qquad \forall\ g_i \in \mathrm{SO}(3). \tag{9}$$

This is because inversion transforms the point P, which is the pole of g_i, into its antipole P′, but at the same time the sense of the rotation is reversed so that P′ is in fact the antipole of $I\, g_i$ and P remains the pole of the improper rotation $I\, g_i$.

(8), (9)
$$[g_i\ ;\ h_j] = 1 \qquad \forall\ g_i \in \mathrm{SO}(3),\ h_j \in \mathrm{C}_i, \tag{10}$$

(10), (7), (8)
$$[g_i h_j\ ;\ g_p h_r] = [g_i\ ;\ g_p][h_j\ ;\ h_r] \quad \text{(Cartan gauge)}, \tag{11}$$

where $[h_j\ ;\ h_r] = \pm 1$, the minus sign applying only when h_j, h_r are both I. In the Pauli gauge, $[I\ ;\ I] = 1$ and eq. (11) becomes

$$[g_i h_j\ ;\ g_p h_r] = [g_i\ ;\ g_p] \text{ (Pauli gauge)}. \tag{12}$$

We may now consider the character theorem for the PRs of improper point groups, which are all subgroups of O(3), with factor systems defined by eqs. (11) and (12) above. Using abbreviated notation for PFs,

(12.6.36)
$$\chi(m_s\ m_k\ m_s{}^{-1}) = [s\ k\ s^{-1}\ ;\ s][s\ ;\ k]^{-1}\chi(g_k), \tag{13}$$

$$m_k = g_i\ h_j, \qquad m_s = g_p\ h_r = h_r\ g_p \qquad (h_j, h_r = E \text{ or } I), \tag{14}$$

(14)
$$\begin{aligned} m_s\ m_k\ m_s^{-1} &= g_p\ h_r\ m_k\ h_r{}^{-1}\ g_p{}^{-1} = g_p\ m_k\ g_p{}^{-1} \\ &= g_p\ g_i\ h_j\ g_p{}^{-1} = g_p\ g_i\ g_p{}^{-1}\ h_j, \end{aligned} \tag{15}$$

(15), (13)
$$\chi(m_s\ m_k\ m_s{}^{-1}) = [p\ i\ p^{-1} j\ ;\ p\ r][p\ r\ ;\ i\ j]^{-1}\chi(m_k), \tag{16}$$

(11) or (12), (16)
$$\chi(m_s\ m_k\ m_s^{-1}) = [p\ i\ p^{-1}\ ;\ p][p\ ;\ i]^{-1}\chi(m_k), \tag{17}$$

which holds in either the Cartan gauge or the Pauli gauge. For the point groups in eqs. (12.6.37) and (12.6.42)

$$[p\,i\,p^{-1}\,;\,p][p\,;\,i]^{-1} = \pm 1, \tag{18}$$

the negative sign applying only when g_i, g_p are proper or improper rotations about BB axes and therefore when m_k , m_s are both proper or improper BB rotations. Except for these *irregular* cases

(17), (18) $$\chi(m_s\; m_k\; m_s^{-1}) = \chi(m_k) \tag{19}$$

and $\mathscr{C}(m_k)$, called a *regular class*, is given by the conjugates of m_k (with repetitions deleted) as for proper point groups. When g_i, g_p are rotations about proper or improper BB axes and therefore commute,

(15) $$m_s\; m_k\; m_s{}^{-1} = g_p\; g_i\; g_p{}^{-1}\; h_j = g_i\; h_j = m_k. \tag{20}$$

In these irregular cases the negative sign applies in eq. (18) and

(17), (18), (20) $$\chi(m_s\; m_k\; m_s^{-1}) = -\chi(m_k) = 0 \quad \text{(irregular classes).} \tag{21}$$

Thus for improper point groups that are formed by the DP of a proper point group with C_i, the character is a class property which is zero for all irregular classes, namely those formed from rotations about proper or improper BB axes. All other improper point groups are isomorphous ($\sim$) with a proper point group and have the same characters and representations as that proper point group. For example: $C_{2v} \sim D_2$; $D_{2d} \sim C_{4v} \sim D_4$.

12.8 The irreducible representations

We now have all the necessary machinery for working out the matrix elements $\Gamma^j_{m'm}$ in the MRs of the proper rotations R in any point group for any required value of j. The $\Gamma^j_{m'm}$ are given in terms of the Cayley–Klein parameters a, b and their CCs by eq. (11.8.43). The parameters a, b may be evaluated from the quaternion parameters λ, $\mathbf{\Lambda}$ for R, using

(12.5.21) $$a = \lambda - \mathrm{i}\,\Lambda_z, \qquad b = -\Lambda_y - \mathrm{i}\Lambda_x,$$
$$a^* = \lambda + \mathrm{i}\,\Lambda_z, \;\; (-b^*) = \Lambda_y - \mathrm{i}\Lambda_x. \tag{1}$$

Improper rotations are expressed as *IR* and for *j* an odd integer the basis is ungerade so that the matrix $\Gamma^j(a\;b)$ must be multiplied by$(-1)^j$. For half-integral j in *the Pauli gauge* the matrix for *IR* is the same as that for R. The sum over k in eq. (11.8.43) runs over all values of $0 \le k \le j+m$ for which n in $(-n!)$ is <1 $(0! = 1)$. Certain simplifications occur. When $j=0$, $m=m'=0$, so $k=0$, the basis is $u_0^0 = |0\,\rangle$ and

(11.8.43) $$\Gamma^0(a\;b)_{00} = 1 \quad \forall\, R \in \mathrm{G}, \tag{2}$$

that is, the totally symmetric representation. Cyclic point groups involve rotations about $\mathbf{z}$ only, so that $\mathbf{\Lambda}_x$, $\mathbf{\Lambda}_y$ are zero. Therefore $b=0$, the matrix is diagonal, and all bases are 1-D.

Because $b = 0$, the only non-vanishing matrix elements $\Gamma^j{}_{m'm}$ are those for which $k = 0$ and $m' = m$, giving

$$(11.8.43) \qquad \Gamma^j_{m'm} = a^{j+m}(a^*)^{j-m}\ \delta_{m'm} \qquad (C_{n\mathbf{z}}). \qquad (3)$$

Dihedral groups D_n consist of the operators $C_{n\mathbf{z}}$ and $nC_{2\mathbf{m}}{}'$, where $\mathbf{m}$ is perpendicular to $\mathbf{z}$. For these $C_2{}'$ rotations $\lambda = \cos(\pi/2) = 0$ and $\mathbf{\Lambda}_z = 0$, so that $a = 0$, $a^* = 0$, and the exponents of a, a^* must vanish,

$$(11.8.43) \qquad j + m - k = 0\ ,\ j - m' - k = 0 \Rightarrow m' = -m, \qquad (4)$$

$$(4),\ (11.8.43) \qquad \Gamma^j_{m'm} = b^{j-m}(-b^*)^{j+m}\ \delta_{m',-m} \qquad (C_{2\mathbf{m}},\ \mathbf{m}\bot\mathbf{n}). \qquad (5)$$

Exercise 12.8-1 Justify the remark above eq. (4) that, in order for the matrix element to remain finite, the exponent of a must vanish when a is zero.

Only the cubic or icosahedral groups contain operations for which neither a nor b is zero. When $a \neq 0$ and $b \neq 0$, then when $m' = \pm j$ and $m = \pm j$, to ensure non-vanishing factorials in the denominator, $k \geq 0$ and

$$(11.8.43) \qquad \text{if } m' = j,\ j - m' - k = -k \Rightarrow k = 0; \qquad (6)$$

$$(11.8.43) \qquad \text{if } m' = -j,\ j + m - k \geq 0 \ \text{ and } \ -j - m + k \geq 0; \qquad (7)$$

$$(7) \qquad k = j + m \quad (m' = -j). \qquad (8)$$

$$(6), (8), (11.8.43) \qquad m', m = \pm j, \qquad \Gamma^j = \begin{bmatrix} a^{2j} & b^{2j} \\ (-b^*)^{2j} & (a^*)^{2j} \end{bmatrix} \begin{matrix} m' = j \\ m' = -j \end{matrix}. \qquad (9)$$
$$m = j \quad m = -j$$

The matrix Γ^j in eq. (9) is not necessarily irreducible so this must be checked. It follows from eqs. (3) and (5) that, for dihedral groups, when $m = j$, the basis $\langle\, |j\ j\rangle\, |$ can transform only into itself or into $\langle\, |j\ -j\rangle\, |$. For these groups therefore, the matrix in eq. (9) assumes a particular importance. The general case includes $j = 1/2$, in which case

$$(9) \qquad \Gamma^{\frac{1}{2}} = \begin{bmatrix} a & b \\ -b^* & a^* \end{bmatrix}. \qquad (10)$$

An alternative to determining a and b is to use the complex quaternion parameters ρ, τ defined by

$$\rho = \lambda + \mathrm{i}\ \Lambda_z, \quad \tau = \Lambda_x + \mathrm{i}\ \Lambda_y, \qquad (11)$$

$$(1), (11), (10) \qquad \Gamma^{\frac{1}{2}} = \begin{bmatrix} a & b \\ -b^* & a^* \end{bmatrix} = \begin{bmatrix} \rho^* & -\mathrm{i}\tau^* \\ -\mathrm{i}\tau & \rho \end{bmatrix}. \qquad (12)$$

Example 12.8-1 Determine the IRs and the character table for the point group D_3. Hence find the IRs of C_{3v}.

$D_3 = \{E\ 2C_3\ 3C_2\}$. There are no BB rotations so that the groups both consist of three regular classes. There are therefore three vector representations ($N_v = N_c$) and three spinor representations ($N_s = N_{rc} = N_c$). The dimensions of the N_v vector representations are $\{l_v\} = \{1\ 1\ 2\}$ (because $\sum_v l_v^2 = g = 6$) and of the N_s spinor representations also $\{l_s\} = \{1\ 1\ 2\}$ (because $\sum_s l_s^2 = g = 6$). Figure 12.10 shows the **xy** plane ($\theta = \pi/2$) in the unit sphere and the location of the three binary axes **a**, **b**, **c**. The positive half-sphere h is defined by either

$$0 \leq \theta < \pi/2 \tag{13a}$$

or

$$\theta = \pi/2 \tag{13b}$$

and

$$-\pi/6 \leq \varphi < \pi/6 \text{ or } \pi/2 \leq \varphi < 5\pi/6 \text{ or } 7\pi/6 \leq \varphi < 3\pi/2, \tag{13c}$$

a choice that ensures that the pole conventions are observed not only for D_3 but also for D_6 and for $C_{3v} = \{E\ 2C_3\ 3\sigma_v\}$, $3\sigma_v = \{\sigma_{\mathbf{d}}\ \sigma_{\mathbf{e}}\ \sigma_{\mathbf{f}}\}$. (The pole of $IC_{2\mathbf{m}}$ is the same as the pole of $C_{2\mathbf{m}}$.) Furthermore, it is a choice that ensures that during a reduction of symmetry from D_6 to D_3 the character theorem is preserved in D_3, something that is not necessarily true for other choices of h (Altmann (1986)). The quaternion parameters for the operators of D_3 are given in Table 12.4 along with the rotation parameter ϕ **n**. Remember that $0 \leq \phi \leq \pi$ so that the sign of ϕ **n** depends on whether **n** lies in h or $-h$ as defined by eqs. (13). The parameters for $\sigma_{\mathbf{d}}$, $\sigma_{\mathbf{e}}$, $\sigma_{\mathbf{f}}$, where $\sigma_{\mathbf{m}}$ means reflection in the plane normal to **m**, are also given in the table to enable later discussion of C_{3v}. Unit vectors **d**, **e**, **f** are defined in Figure 12.10. Multiplication tables for D_3 and C_{3v} are given in a compact form in Table 12.5. As we have often stressed, a diagram showing the transformation of the projection on the **xy** plane of a representative point on the surface of the unit sphere is an invaluable aid in determining the group

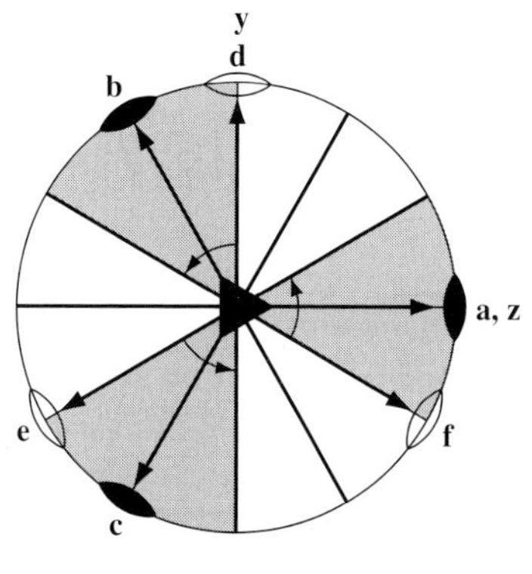

Figure 12.10. The **xy** plane ($\theta = \pi/2$) of the unit sphere. The section of the positive half-sphere defined by eq. (12.8.13) is shown by the shaded regions (which include the tails of the curved arrows but not their heads). Poles of the proper rotations $C_{2\mathbf{a}}$, $C_{2\mathbf{b}}$, and $C_{2\mathbf{c}}$ are shown by filled digons and the poles of the improper rotations $IC_{2\mathbf{m}}$, $\mathbf{m} = \mathbf{d}$, **e**, **f**, are indicated by unfilled digons.

Table 12.4. *Rotation parameters ϕ **n** or ϕ **m**, real (λ, $\mathbf{\Lambda}$), and complex (ρ, τ) quaternion parameters, and the Cayley–Klein parameters a, b for the operators $R \in D_3$.*

Also given, for future reference, are the parameters for $\{\sigma_{\mathbf{d}}\ \sigma_{\mathbf{e}}\ \sigma_{\mathbf{f}}\}$, where $\sigma_{\mathbf{m}} = IC_{2\mathbf{m}}$, **m** normal to **n**. Since the pole of a rotation is invariant under inversion, the parameters for $IC_{2\mathbf{m}}$ are those for $C_{2\mathbf{m}}$. The unit vectors **d**, **e**, **f** are defined in Figure 12.10.

D_3	ϕ	**n** or **m**	λ	$\mathbf{\Lambda}$	ρ	τ	a	b
E	0	[0 0 0]	1	**0**	1	0	1	0
C_{3z}^{+}	$2\pi/3$	[0 0 1]	$^1/_2$	$[0\ 0\ \sqrt{3}/2]$	ε	0	ε^*	0
C_{3z}^{-}	$2\pi/3$	$[0\ 0\ \bar{1}]$	$^1/_2$	$[0\ 0\ -\ \sqrt{3}/2]$	ε^*	0	ε	0
$C_{2\mathbf{a}}$	π	[1 0 0]	0	[1 0 0]	0	1	0	$-$i
$C_{2\mathbf{b}}$	π	$[-^1/_2\ \sqrt{3}/2\ 0]$	0	$[-^1/_2\ \sqrt{3}/2\ 0]$	0	$-\varepsilon^*$	0	iε
$C_{2\mathbf{c}}$	π	$[-^1/_2 - \sqrt{3}/2\ 0]$	0	$[-^1/_2 - \sqrt{3}/2\ 0]$	0	$-\varepsilon$	0	iε^*
$\sigma_{\mathbf{d}}$	π	[0 1 0]	0	[0 1 0]	0	i	0	-1
$\sigma_{\mathbf{e}}$	π	$[-\sqrt{3}/2 - ^1/_2\ 0]$	0	$[-\sqrt{3}/2 - ^1/_2\ 0]$	0	$-$iε^*	0	ε
$\sigma_{\mathbf{f}}$	π	$[\sqrt{3}/2 - ^1/_2\ 0]$	0	$[\sqrt{3}/2 - ^1/_2\ 0]$	0	$-$iε	0	ε^*

$\lambda = \cos(\frac{1}{2}\phi), \mathbf{\Lambda} = \sin(\frac{1}{2}\phi)\mathbf{n},\ \rho = \lambda + \mathrm{i}\ \Lambda_z,\ \tau = \Lambda_x + \mathrm{i}\ \Lambda_y,\ a = \lambda - \mathrm{i}\ \Lambda_z = \rho^*,\ b = -\Lambda_y - \mathrm{i}\ \Lambda_x = -\mathrm{i}\tau^*$,
$\varepsilon = \exp(\mathrm{i}\ \pi/3)$.

multiplication table. We then multiply the quaternions for $g_r, g_s \in G$ (from Table 12.4); the result is equal to the quaternion for the product $g_r\ g_s$ (from Tables 12.5 and Table 12.4) multiplied by a numerical factor which is $[g_r\ ;\ g_s]$. In this way we build up Table 12.6.

Exercise 12.8-2 Determine the PFs $[C_{3\mathbf{z}}^{+}\ ;\ C_{2\mathbf{b}}]$ and $[\sigma_{\mathbf{e}}\ ;\ \sigma_{\mathbf{f}}]$.

When $j = 0$, the basis $u_0^0 = |0\rangle$ generates the totally symmetric representation Γ_1 in the first row of the tables for D_3 and C_{3v} in Table 12.7. Next, the matrices of the standard representation $\Gamma^{1/2}(\rho\ \tau)$ for D_3 and C_{3v} (see eq. (12)) are written down using the complex quaternion parameters from Table 12.4.

Exercise 12.8-3 Write down the matrices of the standard representation for $\sigma_{\mathbf{d}}$, $\sigma_{\mathbf{e}}$, $\sigma_{\mathbf{f}}$. [*Hint*: Use the Pauli gauge.] Show, when (a) $g_r = C_{3\mathbf{z}}^{+}$, $g_s = C_{2\mathbf{b}}$, and (b) $g_r = \sigma_{\mathbf{e}}$, $g_s = \sigma_{\mathbf{f}}$, that the product $\Gamma(g_r)\Gamma(g_r) = [g_r\ ;\ g_s]\Gamma(g_r\ g_s)$, with $g_r\ g_s$ and the PF $[g_r\ ;\ g_s]$ as given in Tables 12.5 and 12.6.

Because D_3 is a dihedral group, all its representations may be generated from eqs. (9) and (10) using the standard representation in Table 12.7. These representations generated in this way are also shown in Table 12.7. For the reflections (which are also in Table 12.7), one uses $\sigma_{\mathbf{m}} = IC_{2\mathbf{m}}$ and evaluates first the matrices for $C_{2\mathbf{m}}$, where $\mathbf{m} = \mathbf{d}$, **e**, or **f**. For the vector representations, a matrix $\Gamma(C_{2\mathbf{m}})$ has to be multiplied by $(-1)^j$ to give that for $\sigma_{\mathbf{m}}$ but for the PRs no change is required in the Pauli gauge. For both vector and PRs the representation is irreducible if $\sum\limits_r |\chi_r|^2 = g$, the order of the group. This test shows that all the representations in Table 12.7 are irreducible with the exception of $\Gamma_{3/2}$. The matrices of

Table 12.5. *Multiplication tables for the point groups* D_3 *and* C_{3v}.

Read sections I and II for D_3, sections I and III for C_{3v}.

			I			II			III	
	D_3, C_{3v}	E	C_{3z}^{+}	C_{3z}^{-}	C_{2a}	C_{2b}	C_{2c}	σ_d	σ_e	σ_f
	E	E	C_{3z}^{+}	C_{3z}^{-}	C_{2a}	C_{2b}	C_{2c}	σ_d	σ_e	σ_f
I	C_{3z}^{+}	C_{3z}^{+}	C_{3z}^{-}	E	C_{2c}	C_{2a}	C_{2b}	σ_f	σ_d	σ_e
	C_{3z}^{-}	C_{3z}^{-}	E	C_{3z}^{+}	C_{2b}	C_{2c}	C_{2a}	σ_e	σ_f	σ_d
	C_{2a}	C_{2a}	C_{2b}	C_{2c}	E	C_{3z}^{+}	C_{3z}^{-}			
II	C_{2b}	C_{2b}	C_{2c}	C_{2a}	C_{3z}^{-}	E	C_{3z}^{+}			
	C_{2c}	C_{2c}	C_{2a}	C_{2b}	C_{3z}^{+}	C_{3z}^{-}	E			
	σ_d	σ_d	σ_e	σ_f				E	C_{3z}^{+}	C_{3z}^{-}
III	σ_e	σ_e	σ_f	σ_d				C_{3z}^{-}	E	C_{3z}^{+}
	σ_f	σ_f	σ_d	σ_e				C_{3z}^{+}	C_{3z}^{-}	E

Table 12.6. *Factor systems for the point groups* D_3 *and* C_{3v}.

The projective factor $[g_r\,;\,g_s]$ appears at the intersection of the row g_r with the column g_s. As in Table 12.5, read sections I and II for D_3, sections I and III for C_{3v}.

			I			II			III	
	D_3, C_{3v}	E	C_{3z}^{+}	C_{3z}^{-}	C_{2a}	C_{2b}	C_{2c}	σ_d	σ_e	σ_f
	E	1	1	1	1	1	1	1	1	1
I	C_{3z}^{+}	1	−1	1	−	−1	−1	−1	−1	−1
	C_{3z}^{-}	1	1	−1	−1	−1	−1	−1	−1	−1
	C_{2a}	1	−1	−1	−1	1	1			
II	C_{2b}	1	−1	−1	1	−1	1			
	C_{2c}	1	−1	−1	1	1	−1			
	σ_d	1	−1	−1				−1	1	1
III	σ_e	1	−1	−1				1	−1	1
	σ_f	1	−1	−1				1	1	−1

this representation may be reduced by constructing the symmetric and antisymmetric linear combinations

$$\left\langle \frac{\sqrt{3}}{2}\right|_{\pm} = \left\langle \frac{1}{\sqrt{2}}\left[\left|\frac{\sqrt{3}}{2}\quad \frac{\sqrt{3}}{2}\right> \pm \left|\frac{\sqrt{3}}{2}\quad -\frac{\sqrt{3}}{2}\right\rangle\right]\right|, \tag{14}$$

and the resulting 1-D matrix representations are shown in rows 6 and 7 of Table 12.7. In dihedral groups the spherical harmonics basis for $j = 1$ always reduces to $\langle 1\ -1| \oplus \langle 0|$, the representation based on $\langle 1\ -1|$ being easily obtained from eq. (9). For $\langle 0|$ we use

Table 12.7. *The representations* $\Gamma^j(ab)$ *for the point groups* D_3 *and* C_{3v}.

The basis for each representation is a row containing the ket(s) m shown. Thus, for example, $\langle 1 \quad -1|$ means $\langle |1\ 1\rangle\ |1 \quad -1\rangle|$. In the last two rows $\langle {}^3\!/_2|_\pm$ are abbreviations for the linear combinations $\langle 2^{-1/2}[|{}^3\!/_2 \quad {}^3\!/_2\rangle \pm |{}^3\!/_2 \quad -{}^3\!/_2\rangle|$ which diagonalize the MRs for the basis $\langle {}^3\!/_2 \quad -{}^3\!/_2|$, or $\langle |{}^3\!/_2 \quad {}^3\!/_2\rangle |{}^3\!/_2 \quad -{}^3\!/_2\rangle|$ in the $|jm\rangle$notation. $\varepsilon = \exp(\mathrm{i}\pi/3)$.

Point group D_3

j	basis	E	C_{3z}^+	C_{3z}^-	$C_{2\mathbf{a}}$	$C_{2\mathbf{b}}$	$C_{2\mathbf{c}}$
0	$\langle 0\|$	1	1	1	1	1	1
1	$\langle 0\|$	1	1	1	1	-1	-1
1	$\langle 1 \quad -1\|$	$\begin{bmatrix} 1 & \\ & 1 \end{bmatrix}$	$\begin{bmatrix} -\varepsilon & \\ & -\varepsilon^* \end{bmatrix}$	$\begin{bmatrix} -\varepsilon^* & \\ & -\varepsilon \end{bmatrix}$	$\begin{bmatrix} & -1 \\ -1 & \end{bmatrix}$	$\begin{bmatrix} & \varepsilon^* \\ \varepsilon & \end{bmatrix}$	$\begin{bmatrix} & \varepsilon \\ \varepsilon^* & \end{bmatrix}$
$^1\!/_2$	$\langle {}^1\!/_2 \quad -{}^1\!/_2\|$	$\begin{bmatrix} 1 & \\ & 1 \end{bmatrix}$	$\begin{bmatrix} \varepsilon^* & \\ & \varepsilon \end{bmatrix}$	$\begin{bmatrix} \varepsilon & \\ & \varepsilon^* \end{bmatrix}$	$\begin{bmatrix} & -\mathrm{i} \\ -\mathrm{i} & \end{bmatrix}$	$\begin{bmatrix} & \mathrm{i}\varepsilon \\ \mathrm{i}\varepsilon^* & \end{bmatrix}$	$\begin{bmatrix} & \mathrm{i}\varepsilon^* \\ \mathrm{i}\varepsilon & \end{bmatrix}$
$^3\!/_2$	$\langle {}^3\!/_2 \quad -{}^3\!/_2\|$	$\begin{bmatrix} 1 & \\ & 1 \end{bmatrix}$	$\begin{bmatrix} -1 & \\ & -1 \end{bmatrix}$	$\begin{bmatrix} -1 & \\ & -1 \end{bmatrix}$	$\begin{bmatrix} & \mathrm{i} \\ \mathrm{i} & \end{bmatrix}$	$\begin{bmatrix} & \mathrm{i} \\ \mathrm{i} & \end{bmatrix}$	$\begin{bmatrix} & \mathrm{i} \\ \mathrm{i} & \end{bmatrix}$
$^3\!/_2$	$\langle {}^3\!/_2\|_+$	1	-1	-1	i	i	i
$^3\!/_2$	$\langle {}^3\!/_2\|_-$	1	-1	-1	$-$i	$-$i	$-$i

Point group C_{3v}

j	basis	E	C_{3z}^+	C_{3z}^-	$\sigma_{\mathbf{d}}$	$\sigma_{\mathbf{e}}$	$\sigma_{\mathbf{f}}$
0	$\langle 0\|$	1	1	1	1	1	1
3	$\langle 3\ \|_+$	1	1	1	-1	-1	-1
1	$\langle 1 \quad -1\ \|$	$\begin{bmatrix} 1 & \\ & 1 \end{bmatrix}$	$\begin{bmatrix} -\varepsilon & \\ & -\varepsilon^* \end{bmatrix}$	$\begin{bmatrix} -\varepsilon^* & \\ & -\varepsilon \end{bmatrix}$	$\begin{bmatrix} & -1 \\ -1 & \end{bmatrix}$	$\begin{bmatrix} & \varepsilon^* \\ \varepsilon & \end{bmatrix}$	$\begin{bmatrix} & \varepsilon \\ \varepsilon^* & \end{bmatrix}$
$^1\!/_2$	$\langle {}^1\!/_2 \quad -{}^1\!/_2\|$	$\begin{bmatrix} 1 & \\ & 1 \end{bmatrix}$	$\begin{bmatrix} \varepsilon^* & \\ & \varepsilon \end{bmatrix}$	$\begin{bmatrix} \varepsilon & \\ & \varepsilon^* \end{bmatrix}$	$\begin{bmatrix} & -1 \\ 1 & \end{bmatrix}$	$\begin{bmatrix} & \varepsilon \\ -\varepsilon^* & \end{bmatrix}$	$\begin{bmatrix} & \varepsilon^* \\ -\varepsilon & \end{bmatrix}$
$^3\!/_2$	$\langle {}^3\!/_2 \quad -{}^3\!/_2\|$	$\begin{bmatrix} 1 & \\ & 1 \end{bmatrix}$	$\begin{bmatrix} -1 & \\ & -1 \end{bmatrix}$	$\begin{bmatrix} -1 & \\ & -1 \end{bmatrix}$	$\begin{bmatrix} & -1 \\ 1 & \end{bmatrix}$	$\begin{bmatrix} & -1 \\ 1 & \end{bmatrix}$	$\begin{bmatrix} & -1 \\ 1 & \end{bmatrix}$
$^3\!/_2$	$\langle {}^3\!/_2\|_-$	1	-1	-1	i	i	i
$^3\!/_2$	$\langle {}^3\!/_2\|_+$	1	-1	-1	$-$i	$-$i	$-$i

(11.8.43) $$\Gamma^j_{00} = aa^* - bb^* \Rightarrow \Gamma^1_{00} = aa^*(C_{nz}) \text{ or } -bb^*(C_{2\mathbf{m}}, \mathbf{m}\perp\mathbf{z}). \qquad (15)$$

The 1-D representation for $\langle 0|$ in the second row of the Table 12.7 for D_3 may therefore be written down at once using a, a^* and b, b^* from Table 12.4. For C_{3v} ,$\langle j \ \ m| = \langle 1\ 0|$ is also a basis for Γ_1 and the basis $\langle 3|_+$ must be used to generate Γ_2. We now have all the IRs for D_3 and for C_{3v}. Their characters are given in Table 12.8. Note that, although these two groups are isomorphous, the basis functions for the representations are not necessarily the same (see Γ_2, Γ_5, and Γ_6). This completes the work for Example 12.8-1. It would be straightforward to construct the characters for the double groups $\overline{D}_3$ and $\overline{C}_{3v}$ by a continuation of the same method. The quaternion parameters for $\overline{g}_r$ are the negatives of the parameters for g_r. Products involving $\overline{g}_r$, $\overline{g}_s$ are obtained from the products of the corresponding quaternions,

Table 12.8. *Character tables for the isomorphous point groups* $D_3 = \{E\ 2C_3\ 3C_2'\}$ *and* $C_{3v} = \{E\ 2C_3\ 3\sigma_v\}$.

The dashed line separates the vector representations, for which j is an integer, from the spinor representations, which correspond to half-integer values of j.

	D_3, C_{3v}	E	$2C_3$	$3C_2'$,$3\sigma_v$	j	D_3	C_{3v}
	$A_1\ \Gamma_1$	1	1	1	0	$\langle u_0^0\vert$	
	$A_2\ \Gamma_2$	1	1	−1	1	$\langle u_0^1\vert$	$\langle u_3\vert_+$
	$E\ \ \Gamma_3$	2	−1	0	1	$\langle u_1^1\ u_{-1}^1\vert$	
	$E_{1/2}\ \Gamma_4$	2	1	0	$1/2$	$\langle u_{1/2}^{1/2}\ \ u_{-1/2}^{1/2}\vert$	
$E_{3/2}$	${}^1E_{3/2}\ \Gamma_5$	1	−1	i	$3/2,\ -3/2$	$\langle u_{3/2}^{3/2}\vert_+$	$\langle u_{3/2}^{3/2}\vert_-$
	${}^2E_{3/2}\ \Gamma_6$	1	−1	−i	$3/2,\ -3/2$	$\langle u_{3/2}^{3/2}\vert_-$	$\langle u_{3/2}^{3/2}\vert_+$

and in this way the whole double group multiplication table could be derived by multiplication of quaternions. Matrix representatives of $\overline{g}_r$ are the negatives of those for g_r. Because D_3 and C_{3v} each contain three regular classes, their double groups will contain six classes, the characters for $\overline{\mathscr{C}}$ for spinor representations being the negatives of the characters for $\mathscr{C}$. Irregular classes contain both g_r and $\overline{g}_r$, and consequently the characters for these classes in spinor representations are necessarily zero. The use of different notation for the bases in Tables 12.7 and 12.8 is deliberate since both are in common use. The basis $\langle u_m^j|$ in function notation becomes, in Dirac notation, $\langle |j\ \ m\rangle|$, which we often abbreviate to $\langle m|$. For degenerate states, in the text m implies a list of the degenerate values of m and therefore a row of kets. In Table 12.7, the values of m are stated explicitly, as in $\langle 3/2\ \ -3/2|$ for example, which in an abbrevation for $\langle |3/2\ \ 3/2\rangle\ |3/2\ \ -3/2\rangle|$. (See also the caption to Table 12.7.)

Exercise 12.8-4 In D_3 reduce the following DPs into a direct sum of IRs: (i) $\Gamma_4 \otimes \Gamma_4$; (ii) $\Gamma_4 \otimes \Gamma_3$.

Answers to Exercises 12.8

Exercise 12.8-1 Consider $\lim_{a \to 0} a^n$. If n is finite, the limit is zero. But if n is zero, $a^0 = 1$ and the limit is 1.

Exercise 12.8-2 From Table 12.5, $C_{3z}^+\ C_{2b} = C_{2a}$ and $\sigma_e\ \sigma_f = C_{3z}^+$. FromTable 12.4,

$$\begin{aligned} &[1/2\ ;\ [0\ 0\ \sqrt{3}/2]]\,[0\ ;\ [-1/2\ \ \sqrt{3}/2\ \ 0]] \\ &= [0\ ;\ [-1/4\ \ \sqrt{3}/4\ \ 0] + [-3/4\ \ -\sqrt{3}/4\ \ 0]] \\ &= [0\ ;\ [-1\ 0\ 0]] = -1[0\ ;\ [1\ 0\ 0]]. \end{aligned}$$

Therefore $[C_{3z}^{+}\ ;\ C_{2\mathbf{b}}] = -1$. Using Table 12.4 once again,

$$\begin{aligned}&[0\ ;\ [-\sqrt{3}/2\quad -1/2\quad 0]]\,[0\ ;\ [\sqrt{3}/2\quad -1/2\quad 0]]\\ &= [-(-3/4 + 1/4)\ ;\ [0\quad 0\quad (\sqrt{3}/4 + \sqrt{3}/4)]]\\ &= [1/2\ ;\ [0\quad 0\quad \sqrt{3}/2]].\end{aligned}$$

Therefore $[\sigma_{\mathbf{e}}\ ;\ \sigma_{\mathbf{f}}] = 1$.

Exercise 12.8-3 Using Table 12.4,

$$\Gamma^{1/2}(\sigma_{\mathbf{d}}) = \begin{bmatrix} & -1\\ 1 & \end{bmatrix},\quad \Gamma^{1/2}(\sigma_{\mathbf{e}}) = \begin{bmatrix} & \varepsilon\\ -\varepsilon^* & \end{bmatrix},\quad \Gamma^{1/2}(\sigma_{\mathbf{f}}) = \begin{bmatrix} & \varepsilon^*\\ -\varepsilon & \end{bmatrix}.$$

From Table 12.7,

$$\Gamma^{1/2}(C_{3z}^{+})\ \Gamma^{1/2}(C_{2\mathbf{b}}) = \begin{bmatrix}\varepsilon^* & \\ & \varepsilon\end{bmatrix}\begin{bmatrix} & i\varepsilon\\ i\varepsilon^* & \end{bmatrix} = \begin{bmatrix} & i\\ i & \end{bmatrix} = -\Gamma^{1/2}(C_{2\mathbf{a}}),$$

in agreement with Tables 12.5 and 12.6. Similarly,

$$\Gamma^{1/2}(\sigma_{\mathbf{e}})^{1/2}(\sigma_{\mathbf{f}}) = \begin{bmatrix} & \varepsilon\\ -\varepsilon^* & \end{bmatrix}\begin{bmatrix} & \varepsilon^*\\ -\varepsilon & \end{bmatrix} = {}^{1/2}(C_{3z}^{+}),$$

as expected from Tables 12.5 and 12.6.

Exercise 12.8-4 In D_3, $\Gamma_4 \otimes \Gamma_4 = \{4\quad 1\quad 0\} = \Gamma_1 \oplus \Gamma_2 \oplus \Gamma_3$; $\Gamma_4 \otimes \Gamma_3 = \{4\quad -1\quad 0\} = \Gamma_4 \oplus \Gamma_5 \oplus \Gamma_6$.

Problems

12.1 Express the rotation matrix $\Gamma^1_{\mathbf{r}}(ab)$ in eq. (11.6.19) in terms of the quaternion parameters λ, $\mathbf{\Lambda}$.

12.2 This chapter has provided three ways of investigating the conjugation of g_i by g_k: (i) the direct calculation of $g_k\, g_i\, g_k^{-1}$; (ii) using eq. (12.6.30); and (iii) using eq. (12.5.24). Using the quaternion representation of a rotation, find the result of the conjugation of g_i by g_k by using all three of the above methods for $g_i = C_{2a}$ and (a) $g_k = C_{2\mathbf{b}}$ and (b) $g_k = C_{3z}^{+}$. (Note that the choice of h in Figure 12.10 satisfies the pole convention eq. (12.6.13), and the standardization condition, eq. (12.6.32), for the poles of binary rotations.)

12.3 Evaluate $C_{2z}\, C_{2x}\, C_{2z}^{-1}$ by method (iii) of Problem 12.2. Is this the expected result for BB rotations?

12.4 Show that the choice $0 \leq \theta < \pi$, $-\pi/2 \leq \varphi < \pi/2$, is not a suitable one for h for D_3.

12.5 Prove that the quaternion parameters for $S_m = \sigma_{\mathbf{n}} R(2\pi/m\ \mathbf{n})$ are those for $R(\pi-(2\pi/m),\ -\mathbf{n})$. Find the parameters for $S_{6\mathbf{z}}$ and $S_{3\mathbf{z}}$.

12.6 Prove the symmetry relation eq. (12.6.9). [*Hint*: Use the associativity relation (12.6.6) and the standardization condition (12.6.7).]

12.7 For the group D_3, with $g_i = C_{3\mathbf{z}}^{\ +}$, $g_j = C_{2\mathbf{a}}$, $g_k = C_{3\mathbf{z}}^{\ -}$ verify (i) the associativity of group elements $g_i(g_j\ g_k) = (g_i\ g_j)\ g_k$ and (ii) the associativity relation for PFs, eq. (12.6.6). [*Hint*: Use Tables 12.5 and 12.6.]

12.8 For the representation ${}^1E_{3/2}$ of D_3, verify eq. (12.6.37) with $g_i = C_{2\mathbf{a}}$, $g_k = C_{3\mathbf{z}}^{+}$.

12.9 Prove eq. (12.6.50).

12.10 The following operators were used in Chapter 7 as representative operators of the five classes of the cubic point group O: E, $R(2\pi/3\ [1\ 1\ 1])$, $R(\pi/2\ \mathbf{z})$, $R(\pi\ \mathbf{z})$, $R(\pi\ [1\ 1\ 0])$. Derive the standard representation for these operators and show that this representation is irreducible. [*Hint*: You may check your results by referring to the tables given by Altmann and Herzig (1994) or Onadera and Okasaki (1966).]

12.11 (a) Examine the splitting of the $j = 5/2$ atomic state in a crystal field of cubic symmetry O using both projective representations and the double group method. [*Hint*: Character tables need not be derived since they are known from Chapter 8.] What further splittings occur when the symmetry is lowered from O to C_{3v}? (b) Using only PRs verify that the transition $F_{5/2} \rightarrow E_{3/2}$ is allowed for E1 radiation in O symmetry. Find the allowed transitions and polarizations that originate from $F_{5/2} \rightarrow E_{3/2}$ when the symmetry is lowered from O to C_{3v}.

13 Time-reversal symmetry

Warning In the classification of IRs listed after eq. (13.4.12) and again after eq. (13.4.31) I have followed Altmann and Herzig (1994). In many other books and papers, the labels (b) and (c) are interchanged.

13.1 Time evolution

The invariance of transition probabilities under the action of a symmetry operator $\hat{T}$,

$$|\langle \hat{T}\varphi|\hat{T}\psi\rangle|^2 = |\langle\varphi|\psi\rangle|^2, \tag{1}$$

requires that either

$$\langle \hat{T}\varphi|\hat{T}\psi\rangle = \langle\varphi|\psi\rangle \tag{2}$$

or

$$\langle \hat{T}\varphi|\hat{T}\psi\rangle = \langle\varphi|\psi\rangle^*. \tag{3}$$

Operators that induce transformations in space satisfy eq. (2) and are therefore *unitary* operators with the property $\hat{T}^\dagger\hat{T} = 1$. An operator that satisfies eq. (3) is said to be *antiunitary*. In contrast to spatial symmetry operators, the time-reversal operator is antiunitary. Let $\hat{U}$ denote a unitary operator and let $\hat{T}$ denote an antiunitary operator.

(2) $$\langle \hat{U}\varphi|\hat{U}c\psi\rangle = \langle\varphi|c\psi\rangle = c\langle\varphi|\psi\rangle = c\langle\hat{U}\varphi|\hat{U}\psi\rangle; \tag{4}$$

(3) $$\langle \hat{T}\varphi|\hat{T}c\psi\rangle = \langle\varphi|c\psi\rangle^* = c^*\langle\varphi|\psi\rangle^* = c^*\langle\hat{T}\varphi|\hat{T}\psi\rangle. \tag{5}$$

Hence, unitary operators are *linear* operators, but an antiunitary operator is *antilinear*.

Time evolution in quantum mechanics is described, in the Schrödinger representation, by the Schrödinger time-dependent equation

$$\partial\psi/\partial t = -\mathrm{i}\hat{H}\psi. \tag{6}$$

For an infinitesimal increase δt in t from an initial time t_0 to $t_1 = t_0 + \delta t$,

(6) $$\psi(t_1) = \psi(t_0 + \delta t) = [1 - \mathrm{i}\hat{H}\delta t]\psi(t_0) = \hat{\mathscr{U}}(t_1 - t_0)\psi(t_0). \tag{7}$$

The operator in square brackets, which is designated by $\hat{\mathscr{U}}(t_1 - t_0)$, tells us how to calculate $\psi(t_0 + \delta t)$ from $\psi(t_0)$ and is therefore called the (infinitesimal) *time-evolution* operator. Since $\hat{\mathscr{U}}$ does not depend on t_0,

$$\psi(t_2) = \hat{\mathscr{U}}(t_2 - t_1)\psi(t_1) = \hat{\mathscr{U}}(t_2 - t_1)\hat{\mathscr{U}}(t_1 - t_0)\psi(t_0) = \hat{\mathscr{U}}(t_2 - t_0)\psi(t_0). \tag{8}$$

Equation (8) expresses the composition property of the time-evolution operator,

$$\hat{\mathscr{U}}(t_2 - t_0) = \hat{\mathscr{U}}(t_2 - t_1)\hat{\mathscr{U}}(t_1 - t_0). \tag{9}$$

When $\hat{H}$ is independent of t, $\hat{\mathscr{U}}(t - t_0)$, for a finite time interval, can be obtained by applying the composition property repeatedly to n successive time intervals each of length $\delta t = (t - t_0)/n$. Then

(9), (7)
$$\hat{\mathscr{U}}(t - t_0) = \lim_{\delta t \to 0}[1 - \mathrm{i}\hat{H}\delta t]^n = \lim_{n \to \infty}[1 - \mathrm{i}\hat{H}(t - t_0)/n]^n$$
$$= \exp[-\mathrm{i}\hat{H}(t - t_0)]. \tag{10}$$

13.2 Time reversal with neglect of electron spin

Provided $\hat{H}$ is real, which will be true at a level of approximation that neglects electron spin,

(13.1.10)
$$\hat{\mathscr{U}}(-(t - t_0)) = \exp(+\mathrm{i}\hat{H}(t - t_0)) = \hat{\mathscr{U}}(t - t_0)^*. \tag{1}$$

Therefore, at this level (with spin suppressed) the time-reversal operator is just the complex conjugation operator $\hat{\mathscr{K}}$ which replaces i by $-$i.

Example 13.2-1 The motion of a free particle (to the right, in the positive x direction) is described by the plane wave $\psi(x, t) = \exp[\mathrm{i}(kx - \omega t)]$. Then

$$\hat{\mathscr{K}}\psi(x, t) = \psi(x, t)^* = \exp[-\mathrm{i}(kx - \omega t)] = \exp[\mathrm{i}(k(-x) - \omega(-t))], \tag{2}$$

which represents a plane wave moving backward in time to the left, in the negative x direction. Note that the motion has been *reversed* by the operator $\hat{\mathscr{K}}$.

Let $\hat{M}$ denote a Hermitian operator; then the expectation value of the dynamical variable M in the time-reversed state $\hat{\mathscr{K}}\psi = \psi^*$ is

$$\langle\hat{\mathscr{K}}\psi|\hat{M}|\hat{\mathscr{K}}\psi\rangle = \langle\psi^*|\hat{M}|\psi^*\rangle = \langle\hat{M}\psi^*|\psi^*\rangle = \langle\psi^*|\hat{M}\psi^*\rangle^*$$
$$= \langle\psi|\hat{M}^*|\psi\rangle = \langle M^*\rangle. \tag{3}$$

Thus, real operators are unaffected by time reversal but linear and angular momenta, which have factors of i, change sign under time reversal. Therefore,

$$\hat{\mathscr{K}}\mathbf{r}\hat{\mathscr{K}}^{-1} = \mathbf{r}, \quad \hat{\mathscr{K}}\hat{\mathbf{p}}\hat{\mathscr{K}}^{-1} = -\mathbf{p}, \quad \hat{\mathscr{K}}\hat{\mathbf{J}}\hat{\mathscr{K}}^{-1} = -\hat{\mathbf{J}}. \tag{4}$$

$\hat{\mathscr{K}}$ is *antilinear* because

$$\hat{\mathscr{K}}(c\psi) = c^*\psi^* = c^*\hat{\mathscr{K}}(\psi), \tag{5}$$

whereas a linear operator $\hat{M}$ has the property $\hat{M}c\psi = c\hat{M}\psi$. Note that $\hat{\mathscr{K}}$ is also *antiunitary* because

$$\langle\hat{\mathscr{K}}\varphi|\hat{\mathscr{K}}\psi\rangle = \langle\varphi|\psi\rangle^*. \tag{6}$$

13.3 Time reversal with spin–orbit coupling

We now remove the restriction that $\hat{H}$ is real, introduce the symbol $\hat{\Theta}$ for the time-reversal symmetry operator, and choose $t_0 = 0$. Now $\hat{\Theta}\psi$ is the transformed function which has the same value at $-t$ as the original function ψ at time t,

(13.1.10) $$\hat{\Theta}\psi(-t) = \psi(t) = \exp[-\mathrm{i}\hat{H}t]\psi(0). \tag{1}$$

For the infinitesimal time interval δt,

(1) $$\hat{\Theta}\psi(-\delta t) = \psi(\delta t) = [1 - \mathrm{i}\hat{H}\delta t]\psi(0) = [1 - \mathrm{i}\hat{H}\delta t]\hat{\Theta}\psi(0). \tag{2}$$

The state described by $\psi(0)$ evolving backwards in time for the same time interval becomes one described by

(13.1.7) $$\psi(-\delta t) = [1 - \mathrm{i}\hat{H}(-\delta t)]\psi(0). \tag{3}$$

Operate on eq. (3) with Θ to obtain

(3) $$\hat{\Theta}\psi(-\delta t) = \hat{\Theta}[1 + \mathrm{i}\hat{H}\delta t]\psi(0); \tag{4}$$

(2), (4) $$\hat{\Theta}\mathrm{i}\hat{H} = -\mathrm{i}\hat{H}\hat{\Theta}. \tag{5}$$

If $\hat{\Theta}$ were a unitary operator,

(5) $$\hat{\Theta}\hat{H} = -\hat{H}\hat{\Theta}, \tag{6}$$

with the consequence that every stationary state ψ of the system with energy E would be accompanied by one $\hat{\Theta}\psi$ with energy $-$E . But time reversal reverses velocities, leaving E invariant, so $\hat{\Theta}$ cannot be a unitary operator but is antiunitary. Therefore

(13.1.5), (5) $$\hat{\Theta}\hat{H} = \hat{H}\hat{\Theta}, \tag{7}$$

which tells us that time reversal commutes with the Hamiltonian. Consequently, if ψ is an eigenstate of $\hat{H}$ then so is $\hat{\Theta}\psi$, with the same energy. This means that either ψ and $\hat{\Theta}\psi$ represent the same state, and so can differ only by a phase factor, or that they correspond to distinct (and therefore degenerate) states. Since two successive reversals of time leave all physical systems invariant,

$$\hat{\Theta}^2\psi = c\psi, \tag{8}$$

where c is the same phase factor for all systems. Because $\hat{\Theta}$ is antiunitary

$$\text{(13.1.5), (8)} \qquad \langle\hat{\Theta}\varphi|\psi\rangle = \langle\hat{\Theta}\psi|\hat{\Theta}^2\varphi\rangle = c\langle\hat{\Theta}\psi|\varphi\rangle = c\langle\hat{\Theta}\varphi|\hat{\Theta}^2\psi\rangle = c^2\langle\hat{\Theta}\varphi|\psi\rangle, \tag{9}$$

so that $c = \pm 1$. When $\phi = \psi$ and $c = -1$

$$\text{(9)} \qquad \langle\hat{\Theta}\psi|\psi\rangle = -\langle\hat{\Theta}\psi|\psi\rangle = 0. \tag{10}$$

In this case ($c = -1$) $\hat{\Theta}\psi$ and ψ are orthogonal and so correspond to different degenerate states.

13.3.1 *Determination of the time-reversal operator*

The product of two antiunitary operators is a unitary operator. Consequently,

$$\hat{\Theta}\hat{\mathscr{K}} = \hat{U}, \tag{11}$$

where $\hat{U}$ is unitary.

$$\text{(11)} \qquad \hat{\Theta} = \hat{U}\hat{\mathscr{K}}. \tag{12}$$

The linear Hermitian operators of quantum mechanics can be divided into two categories with respect to time reversal. In the first category are those operators $\hat{A}$ which correspond to dynamical variables that are either independent of t or depend on an even power of t. Let ψ_k be an eigenfunction of $\hat{A}$ with (real) eigenvalue a_k. Then $\hat{\Theta}\psi_k$ is also an eigenfunction of $\hat{A}$ with the same eigenvalue,

$$\hat{A}\hat{\Theta}\psi_k = a_k\hat{\Theta}\psi_k. \tag{13}$$

Any state φ is a linear superposition of the $\{\psi_k\}$, and since $\hat{\Theta}$ is antilinear

$$\hat{\Theta}\hat{A}\varphi = \hat{\Theta}\hat{A}\sum_k c_k\,\psi_k = \Theta\sum_k c_k\,a_k\,\psi_k = \sum_k c_k^*\,a_k\,\hat{\Theta}\psi_k; \tag{14}$$

$$\text{(13)} \qquad \hat{A}\hat{\Theta}\varphi = \hat{A}\hat{\Theta}\sum_k c_k\,\psi_k = \hat{A}\sum_k c_k^*\,\hat{\Theta}\psi_k = \sum_k c_k^*\,a_k\,\hat{\Theta}\psi_k; \tag{15}$$

$$\text{(14), (15)} \qquad \hat{A}\hat{\Theta} = \hat{\Theta}\hat{A}. \tag{16}$$

For operators $\hat{B}$ in the second class, which correspond to dynamical variables that depend on an odd power of t and for which

$$\hat{B}\psi_k = b_k\psi_k, \qquad \hat{B}\hat{\Theta}\psi_k = -b_k\hat{\Theta}\psi_k, \tag{17}$$

the same argument yields

$$\hat{B}\hat{\Theta} = -\hat{\Theta}\hat{B}. \tag{18}$$

$$\text{(16), (18)} \qquad \hat{\Theta}\,\mathbf{r}\,\hat{\Theta}^{-1} = \mathbf{r}, \quad \hat{\Theta}\,\hat{\mathbf{p}}\,\hat{\Theta}^{-1} = -\mathbf{p}, \quad \hat{\Theta}\,\hat{\mathbf{J}}\,\hat{\Theta}^{-1} = -\hat{\mathbf{J}}, \tag{19}$$

as already seen for systems in which electron spin is neglected and $\hat{\Theta}$ becomes $\hat{\mathscr{K}}$. Therefore the spin operators $\hat{s}_x$, $\hat{s}_y$, $\hat{s}_z$ which are represented by the Pauli spin matrices σ_1, σ_2, σ_3, anticommute with $\hat{\Theta}$. Because σ_1 and σ_3 are real (see eq. (11.6.3))

(12), (19) $$\hat{\Theta}\hat{s}_x = \hat{U}\hat{\mathscr{K}}\hat{s}_x = \hat{U}\hat{s}_x\hat{\mathscr{K}} = -\hat{s}_x\hat{U}\hat{\mathscr{K}}, \tag{20}$$

so that $\hat{s}_x$ anticommutes with $\hat{U}$. Similarly, $\hat{s}_z$ anticommutes with $\hat{U}$. But σ_2, which represents $\hat{s}_y$, is purely imaginary. Therefore,

(12), (19) $$\hat{\Theta}\hat{s}_y = \hat{U}\hat{\mathscr{K}}\hat{s}_y = -\hat{U}\hat{s}_y\hat{\mathscr{K}} = -\hat{s}_y\hat{U}\hat{\mathscr{K}}, \tag{21}$$

so that $\hat{s}_y$ commutes with $\hat{U}$.

(20), (21) $$\hat{U}\,\hat{s}_x = -\hat{s}_x\,\hat{U}; \quad \hat{U}\,\hat{s}_y = \hat{s}_y\,\hat{U}; \quad \hat{U}\,\hat{s}_z = -\hat{s}_z\,\hat{U}. \tag{22}$$

Using the commutation properties of the Pauli spin matrices, eqs. (22), determine $\hat{U}$ as $\hat{s}_y$, apart from a phase factor exp(iγ) which has no effect on eq. (22).

Exercise 13.3-1 Verify explicitly, by using the spin matrices from eq. (11.6.8), that the matrix representative (MR) of $\hat{U} = \hat{s}_y$ satisfies the matrix representation of eq. (22).

For an N-electron system, $\hat{U}$ is a product of the individual imaginary spin operators

$$\hat{\Theta} = \exp(\mathrm{i}\gamma)\prod_{k=1}^{N}\hat{s}_{yk}\hat{\mathscr{K}}. \tag{23}$$

Let φ denote any spinor function; then

(23) $$\hat{\Theta}^2\varphi = \left(\exp(\mathrm{i}\gamma)\prod_{k=1}^{N}\hat{s}_{yk}\hat{\mathscr{K}}\right)\left(\exp(\mathrm{i}\gamma)\prod_{l=1}^{N}\hat{s}_{yl}\hat{\mathscr{K}}\right)\varphi = (-1)^N\varphi = c\varphi, \tag{24}$$

where we have used

$$\hat{\mathscr{K}}\exp(\mathrm{i}\gamma) = \exp(-\mathrm{i}\gamma)\hat{\mathscr{K}};$$

$$\hat{\mathscr{K}}\prod_{l=1}^{N}\hat{s}_{yl} = (-1)^N\prod_{l=1}^{N}\hat{s}_{yl}\hat{\mathscr{K}}; \tag{25}$$

$$\hat{\mathscr{K}}^2\varphi = \varphi; \qquad \hat{s}_y^2 = \hat{E}.$$

For an even number of electrons N, $(-1)^N = 1$, $c = +1$, and there are no extra degeneracies. But for an odd number of electrons $(-1)^N = -1$, $c = -1$, and $\hat{\Theta}\psi$ and ψ correspond to different degenerate states (see eq. (10)). This conclusion assumes the absence of an external magnetic field. In the presence of a magnetic field, $\hat{H}$ contains terms linear in $\hat{\mathbf{L}}$ and $\hat{\mathbf{S}}$ and therefore no longer commutes with $\hat{\Theta}$ (eq. (19)). This means that $\hat{\Theta}$ is not a symmetry operator in the presence of an external magnetic field. However, the commutation of $\hat{\Theta}$ with $\hat{H}$ can be restored if the direction of the magnetic field is reversed along with the reversal of t. These results from eqs. (24) and (10) are embodied in Kramers' theorem (Kramers (1930)), which states that the energy levels of a system containing an odd number of electrons must be at least doubly degenerate provided there is no external magnetic field present to remove time-reversal symmetry.

Exercise 13.3-2 Show that the choice of phase factor $\exp(\mathrm{i}\gamma) = -\mathrm{i}$ makes $\hat{\Theta} = \hat{R}(\pi\mathbf{y})\hat{\mathscr{K}}$, where $R(\pi\ \mathbf{y})$ denotes a binary rotation about $\mathbf{y}$ in SU(2). [*Hint*: See eq. (11.6.11).]

Answers to Exercises 13.3

Exercise 13.3-1 With $\hat{U} = \hat{s}_y$ the matrix representation of eq. (22) is

$$\sigma_1\sigma_2 = \begin{bmatrix} & 1 \\ 1 & \end{bmatrix}\begin{bmatrix} & -\mathrm{i} \\ \mathrm{i} & \end{bmatrix} = \begin{bmatrix} \mathrm{i} & \\ & -\mathrm{i} \end{bmatrix},$$
$$\sigma_2\sigma_1 = \begin{bmatrix} & -\mathrm{i} \\ \mathrm{i} & \end{bmatrix}\begin{bmatrix} & 1 \\ 1 & \end{bmatrix} = \begin{bmatrix} -\mathrm{i} & \\ & \mathrm{i} \end{bmatrix} = -\sigma_1\sigma_2,$$

thus verifying $\hat{s}_y\hat{s}_x = -\hat{s}_x\hat{s}_y$. Similarly,

$$\sigma_3\sigma_2 = \begin{bmatrix} 1 & \\ & -1 \end{bmatrix}\begin{bmatrix} & -\mathrm{i} \\ \mathrm{i} & \end{bmatrix} = \begin{bmatrix} & -\mathrm{i} \\ -\mathrm{i} & \end{bmatrix},$$
$$\sigma_2\sigma_3 = \begin{bmatrix} & -\mathrm{i} \\ \mathrm{i} & \end{bmatrix}\begin{bmatrix} 1 & \\ & -1 \end{bmatrix} = \begin{bmatrix} & \mathrm{i} \\ \mathrm{i} & \end{bmatrix} = -\sigma_3\sigma_2,$$

verifying $\hat{s}_y\hat{s}_z = -\hat{s}_z\hat{s}_y$.

Exercise 13.3-2 If $\exp(-\mathrm{i}\gamma) = -\mathrm{i}$,

(11.6.3), (11.6.11) $$-\mathrm{i}\sigma_2 = \begin{bmatrix} & -1 \\ 1 & \end{bmatrix} = \Gamma^{1/2}(R(\pi\ \mathbf{y})), \quad -\mathrm{i}\ \hat{s}_y\hat{\mathscr{K}} = R(\pi\ \mathbf{y})\hat{\mathscr{K}}.$$

13.4 Co-representations

Consider the set of operators $\{R\} \oplus \{\Theta R\}$, where $\mathrm{H} = \{R\}$ is a group of unitary symmetry operators and $\{\Theta R\}$ is therefore a set of antiunitary operators. Since rotations and time reversal commute, the multiplication rules within this set are

$$\begin{aligned} RS &= T, \quad R, S, T \in \{R\}, \\ \Theta RS &= \Theta T, \\ S\Theta R &= \Theta SR = \Theta T',\ T' \in \{R\}, \\ \Theta R\Theta S &= cRS,\ c = \pm 1, \end{aligned} \tag{1}$$

(13.3.24)

where $c = +1$ for N even and $c = -1$ for N odd. These multiplication rules show that $\underline{\mathrm{G}} = \{R\} + \{\Theta R\}$ is a group, that the unitary operators $\{R\}$ form a normal (invariant) subgroup H of $\underline{\mathrm{G}}$, and that the antiunitary operators $\{\Theta R\}$ form a coset of H,

$$\underline{\mathrm{G}} = \{\mathrm{H}\} \oplus \Theta\{\mathrm{H}\}. \tag{2}$$

This conclusion is not unique to $\underline{G} = \{R, \Theta R\}$ but is true for any group $\underline{G} = \{H, AH\}$ that contains unitary $\{H\}$ and antiunitary $\{AH\}$ operators. Let $\{\psi_s\}$ be an orthonormal set of eigenfunctions which form a basis for H. Then $\{\hat{\Theta}\hat{R}\psi_s\}$ is also a set of eigenfunctions so that

(5.1.6) $$\hat{\Theta}\hat{R}\psi_s = \sum_p \psi_p \, \Gamma(\Theta R)_{ps}. \tag{3}$$

Since $\hat{\Theta}$ is antiunitary and $\hat{R}$ is unitary,

$$\langle \hat{\Theta}\hat{R}\psi_p | \hat{\Theta}\hat{R}\psi_s \rangle = \langle \hat{R}\psi_s | \hat{R}\psi_p \rangle = \langle \psi_s | \psi_p \rangle = \delta_{ps}; \tag{4}$$

(3), (4) $$\begin{aligned} \delta_{ps} &= \langle \hat{\Theta}\hat{R}\psi_p | \hat{\Theta}\hat{R}\psi_s \rangle = \langle \sum_q \psi_q \, \Gamma(\Theta R)_{qp} | \sum_r \psi_r \, \Gamma(\Theta R)_{rs} \rangle \\ &= \sum_{q,r} \Gamma(\Theta R)^*_{qp} \, \Gamma(\Theta R)_{rs} \langle \psi_q | \psi_r \rangle \\ &= \sum_q \Gamma(\Theta R)^*_{qp} \Gamma(\Theta R)_{qs} \\ &= [\Gamma(\Theta R)^\dagger \Gamma(\Theta R)]_{ps}. \end{aligned} \tag{5}$$

(5) $$\Gamma(\Theta R)^\dagger \Gamma(\Theta R) = \mathrm{E}, \tag{6}$$

which shows that $\Gamma\,(\Theta R)$ is a unitary matrix. However, because $\hat{\Theta}$ is antilinear,

(3) $$\begin{aligned} \hat{\Theta}\hat{S}\hat{\Theta}\hat{R}\psi_p &= \sum_q \hat{\Theta}\hat{S}\psi_q \, \Gamma(\Theta R)_{qp} \\ &= \sum_q \Gamma(\Theta R)^*_{qp} \, \hat{\Theta}\hat{S}\psi_q \\ &= \sum_q \Gamma(\Theta R)^*_{qp} \sum_r \psi_r \, \Gamma(\Theta S)_{rq}. \end{aligned} \tag{7}$$

But the LS of eq. (7) is $\hat{\Theta}\hat{S}\hat{\Theta}\hat{R}\psi_p = \sum_r \psi_r \, \Gamma(\Theta S \; \Theta R)_{rp}$. Therefore

(7), (1) $$\Gamma(\Theta S)\Gamma(\Theta R)^* = \Gamma(\Theta S \Theta R) = \Gamma(cSR) = \Gamma(cT'), c = \pm 1. \tag{8}$$

Similarly,

$$\Gamma(\Theta S)\Gamma(R)^* = \Gamma(\Theta SR) = \Gamma(\Theta T') \tag{9}$$

so that the MRs $\{\Gamma(R), \Gamma(\Theta R)\}$ do not form a representation of $\underline{G} = \{R, \Theta R\}$. Such sets of matrices where the complex conjugate (CC) of the second factor appears (as in eqs. (8) and (9)) when the first operator is antiunitary are called *co-representations* (Wigner (1959)).

Suppose that the set of eigenfunctions $\{\psi_p\}$ form a basis for one of the IRs of $G = \{R\}$ and define $\hat{\Theta}\psi_p = \overline{\psi}_p$. The inclusion of time reversal, which enlarges H to $\underline{G}$, introduces new degeneracies if the $\{\overline{\psi}_p\}$ are linearly independent (LI) of the $\{\psi_p\}$. Under the transformation induced by the symmetry operator R,

$$\begin{aligned} R\overline{\psi}_p = R\Theta\psi_p = \Theta R\psi_p = \hat{\Theta}\sum_q \psi_q \Gamma(R)_{qp} \\ = \sum_q \overline{\psi}_q \Gamma(R)^*_{qp}, \end{aligned} \tag{10}$$

which shows that if $\{\psi_p\}$ forms a basis for the representation Γ, then $\{\overline{\psi}_p\}$ forms a basis for Γ^*. If Γ is an IR, then the orthogonality theorem (OT) for the characters gives

$$g^{-1}\sum_R |\chi(R)|^2 = \langle\chi|\chi\rangle = 1, \tag{11}$$

where we have introduced the notation $|\chi\rangle$ to denote the character vector whose components are the normalized characters $\sqrt{c_k/g}\ \chi_k$ of the classes $\{\mathscr{C}_k\}$ and $\langle|\rangle$ is the Hermitian scalar product (HSP).

(11) $$\langle\chi|\chi\rangle = 1 = g^{-1}\sum_R \chi(R)^*\chi(R) = g^{-1}\sum_R \chi(R)\chi(R)^* = \langle\chi^*|\chi^*\rangle, \tag{12}$$

so that if Γ is an IR then so is Γ^*. There are three possibilities:

(a) Γ is equivalent to Γ^* ($\Gamma \approx \Gamma^*$), and they are equivalent to a real representation Γ';
(b) Γ, Γ^* are inequivalent (Γ is not $\approx \Gamma^*$);
(c) $\Gamma \approx \Gamma^*$ but they are not equivalent to a real representation.

If Γ is not $\approx \Gamma^*$ then the character system $\{\chi\}$ of Γ must be complex (that is, contain at least one complex character) since if $\{\chi\}$ is complex, Γ and Γ^* have different characters and so they cannot be equivalent.

Exercise 13.4-1 Prove the converse statement to that in the preceding sentence, namely that if $\{\chi\}$ is real, $\Gamma \approx \Gamma^*$.

If $\Gamma \approx \Gamma^*$ (cases (a) and (c)) then there exits a non-singular matrix Z such that

$$\Gamma(R)^* = \mathrm{Z}\Gamma(R)\mathrm{Z}^{-1} \qquad \forall R \in \mathrm{G}. \tag{13}$$

(13) $$\Gamma(R) = \mathrm{Z}^*\Gamma(R)^*(\mathrm{Z}^*)^{-1}; \tag{14}$$

(14), (13) $$\Gamma(R)\mathrm{Z}^*\mathrm{Z} = \mathrm{Z}^*\mathrm{Z}\Gamma(R). \tag{15}$$

Since $\mathrm{Z}^*\mathrm{Z}$ commutes with $\Gamma(R)$; $\forall R \in \mathrm{G}$, by Schur's lemma (Appendix A1.5) it is a multiple of the unit matrix,

$$\mathrm{Z}^*\mathrm{Z} = c(\mathrm{Z})\mathrm{E}, \tag{16}$$

where $c(\mathrm{Z})$ is real and non-zero. Consequently,

(16) $$\mathrm{Z}\mathrm{Z}^* = c(\mathrm{Z})\mathrm{E}. \tag{17}$$

Exercise 13.4-2 Verify the above statements about $c(\mathrm{Z})$, namely that it is real and non-zero.

Furthermore, any other matrix Z' that transforms Γ into Γ^* is a non-zero multiple of Z,

$$\mathrm{Z}' = a\mathrm{Z} \qquad a \neq 0, \tag{18}$$

(17), (18) $$\mathrm{Z}'\mathrm{Z}'^* = c(\mathrm{Z}')\mathrm{E} = |a|^2 c(\mathrm{Z})\mathrm{E}. \tag{19}$$

Consequently, if $c(\mathrm{Z})>0$, all possible $c(\mathrm{Z}')>0$. Similarly, if $c(\mathrm{Z})<0$, all the $c(\mathrm{Z}')<0$. We shall see that these two kinds of transformation ($c(\mathrm{Z})>0$ or $c(\mathrm{Z})<0$) will enable us to distinguish between cases (a) and (c) above. Suppose first that $c(\mathrm{Z})>0$. Then since eq. (13) is satisfied for any non-zero choice of a in eq. (18), it is in particular satisfied for $a=c(\mathrm{Z})^{-1/2}$, which makes

(19) $$\mathrm{Z}'\mathrm{Z}'^{*}=\mathrm{E}. \tag{20}$$

Construct a real representation Γ' equivalent to Γ with matrices

$$\Gamma'(R)=(\mathrm{Z}'+\mathrm{e}^{\mathrm{i}\gamma}\mathrm{E})^{-1}\Gamma(R)(\mathrm{Z}'+\mathrm{e}^{\mathrm{i}\gamma}\mathrm{E})\quad \forall R\in \mathrm{G}, \tag{21}$$

where the phase factor $\mathrm{e}^{\mathrm{i}\gamma}$ is chosen so that $\mathrm{Z}'+\mathrm{e}^{\mathrm{i}\gamma}\mathrm{E}$ is non-singular.

Exercise 13.4-3 What condition must $\mathrm{e}^{\mathrm{i}\gamma}$ satisfy in order that $\mathrm{Z}'+\mathrm{e}^{\mathrm{i}\gamma}\mathrm{E}$ shall be non-singular?

$$\begin{aligned}
&(21) && \Gamma'(R)^{*} =(\mathrm{Z}'^{*}+\mathrm{e}^{-\mathrm{i}\gamma}\mathrm{E})^{-1}\Gamma(R)^{*}(\mathrm{Z}'^{*}+\mathrm{e}^{-\mathrm{i}\gamma}\mathrm{E})\\
&(20),(13) && =[(\mathrm{Z}')^{-1}+\mathrm{e}^{-\mathrm{i}\gamma}\mathrm{E}]^{-1}(\mathrm{Z}')^{-1}\Gamma(R)\mathrm{Z}'[(\mathrm{Z}')^{-1}+\mathrm{e}^{-\mathrm{i}\gamma}\mathrm{E}]\\
& && =\mathrm{e}^{\mathrm{i}\gamma}(\mathrm{e}^{\mathrm{i}\gamma}\mathrm{E}+\mathrm{Z}')^{-1}\Gamma(R)(\mathrm{e}^{\mathrm{i}\gamma}\mathrm{E}+\mathrm{Z}')\mathrm{e}^{-\mathrm{i}\gamma}\\
& && =\Gamma'(R),
\end{aligned} \tag{22}$$

which verifies that $\Gamma'=\{\Gamma'(R)\}$ is a real representation. If $c(\mathrm{Z})>0$ then Γ (and therefore Γ^{*} which is $\approx\Gamma$) is equivalent to a real representation Γ'. Conversely, if Γ, Γ^{*} are equivalent to a real representation Γ', then there exists a transformation

$$\mathrm{Q}\Gamma(R)\mathrm{Q}^{-1}=\Gamma'(R)=\Gamma'(R)^{*}=\mathrm{Q}^{*}\Gamma(R)^{*}(\mathrm{Q}^{*})^{-1}\qquad \forall\, R\in \mathrm{G}. \tag{23}$$

(23), (13) $$\Gamma(R)^{*}=\mathrm{Q}^{*-1}\mathrm{Q}\Gamma(R)\mathrm{Q}^{-1}\mathrm{Q}^{*}=\mathrm{Z}\Gamma(R)\mathrm{Z}^{-1}; \tag{24}$$

(24), (17) $$\mathrm{Z}\,\mathrm{Z}^{\star}=(\mathrm{Q}^{\star-1}\mathrm{Q})(\mathrm{Q}^{-1}\mathrm{Q}^{*})=\mathrm{E}=c(\mathrm{Z})\mathrm{E}\qquad c(\mathrm{Z})=1>0. \tag{25}$$

We have thus established the theorem that if $c(\mathrm{Z})>0$, then Γ and Γ^{*} are equivalent to a real representation (case (a)) and also its converse, that if Γ, $\Gamma^{*}\approx\Gamma'=\Gamma'^{*}$, then $c(\mathrm{Z})>0$. Consequently, if $\Gamma\approx\Gamma^{*}$ but they are not equivalent to a real representation (case(c)) then $c(\mathrm{Z})$ must be <0. (Note that $c(\mathrm{Z})$ is non-zero.) We now have a criterion for deciding between the two cases (a) and (c), but it will be more useful in the form of a character test. For unitary, equivalent (IRs) Γ, Γ^{*} of dimension l, the OT takes the form

(A1.6.11) $$(l/g)\sum_{R}\Gamma(R)_{pq}\;\Gamma(R)_{rs}=\mathrm{Z}_{rp}(\mathrm{Z}^{-1})_{qs}. \tag{26}$$

Since Z in eq. (26) may be unitary (Section A1.6), and here is unitary,

(17) $$c(\mathrm{Z})=\pm 1. \tag{27}$$

Exercise 13.4-4 Show that if Z is unitary, then $c(\mathrm{Z}) = \pm 1$.

When Z is unitary

(26) $$(l/g)\sum_R \Gamma(R)_{pq}\ \Gamma(R)_{rs} = \mathrm{Z}_{rp}\mathrm{Z}^*_{sq}. \tag{28}$$

Set $p = s$, $q = r$, and sum over r and s:

(28), (27) $$g^{-1}\sum_R \chi(R^2) = l^{-1}\mathrm{Tr}(\mathrm{Z}\ \mathrm{Z}^*) = c(\mathrm{Z}) = \pm 1; \tag{29}$$

$c(\mathrm{Z}) = +1$ corresponds to case (a) and $c(\mathrm{Z}) = -1$ corresponds to case (c). When Γ, Γ^* are inequivalent, case (b),

(A1.5.32), (26) $$(l/g)\sum_R \Gamma(R)_{pq}\ \Gamma(R)_{rs} = 0; \tag{30}$$

(30) $$g^{-1}\sum_R \chi(R^2) = 0. \tag{31}$$

We have therefore established a diagnostic test (called the Frobenius–Schur test) which classifies the IRs Γ of a point group according to the three cases listed after eq. (12). (*Note*: See the warning at the beginning of this chapter.)

(a) If Γ and Γ^* are equivalent and they are equivalent to the same real representation, then $g^{-1}\sum\limits_R \chi(R^2) = +1$.

(b) If Γ is not $\approx \Gamma^*$, $g^{-1}\sum\limits_R \chi(R^2) = 0$.

(c) If $\Gamma \approx \Gamma^*$, but they are not equivalent to the same real representation, then $g^{-1}\sum\limits_R \chi(R^2) = -1$.

Exercise 13.4-5 Show that the dimension l of representations of the third kind (c) is an even number.

Recall that if $\{\psi_r\}$ forms a basis for Γ, then $\{\overline{\psi}_r\} = \{\Theta\psi_r\}$ forms a basis for Γ^*, and consider first the case when the number of electrons N is an even number. If Γ is not $\approx \Gamma^*$, case (b), $\{\psi_r\}$ and $\{\overline{\psi}_r\}$ are linearly independent (LI) and so time reversal causes a doubling of degeneracy. If $\Gamma \approx \Gamma^*$, cases (a) and (c), then there exists a non-singular matrix Z which transforms Γ into Γ^*,

$$\Gamma^* = \mathrm{Z}\Gamma(R)\mathrm{Z}^{-1} \quad \forall R \in \mathrm{G}, \tag{32}$$

where, from the remarks following eq. (25),

$$\mathrm{Z}\ \mathrm{Z}^\star = c(\mathrm{Z})\mathrm{E}, \quad \text{(a) } c(\mathrm{Z}) = +1, \quad \text{(c) } c(\mathrm{Z}) = -1. \tag{33}$$

Let $\{\psi_r\}$ denote an LI basis set of dimension l. Then if Θ does not introduce any new degeneracies,

$$\Theta\psi_r = \overline{\psi}_r = \sum_{s=1}^{l} \psi_s \mathrm{Z}_{sr}; \tag{34}$$

Table 13.1. *The effect of time-reversal symmetry on the degeneracy of quantum states.*

When the number of electrons N is even, the spin quantum number S is an integer, and, when N is odd, S is a half-integer. Γ' denotes a real representation.

Case	$g^{-1}\sum_R \chi(R^2)$	Equivalence of Γ, Γ^*	N even	N odd
(a)	+1	$\Gamma \approx \Gamma^* \approx \Gamma'$	no change	doubled
(b)	0	Γ not $\approx \Gamma^*$	doubled	doubled
(c)	−1	$\Gamma \approx \Gamma^*$ not $\approx \Gamma'$	doubled	no change

(34) $$\Theta^2\psi_r = \sum_s \Theta\psi_s \mathrm{Z}_{sr} = \sum_{s,q} \psi_q \mathrm{Z}_{qs} \mathrm{Z}_{sr}{}^* = \sum_q \psi_q(\mathrm{Z}\ \mathrm{Z}^*)_{qr}. \tag{35}$$

But since the ψ_r are LI,

$$(\mathrm{Z}\ \mathrm{Z}^*)_{qr} = \delta_{qr},\ \mathrm{Z}\ \mathrm{Z}^* = \mathrm{E}, \tag{36}$$

and Γ belongs to case (a). It follows also that if $\mathrm{Z}\ \mathrm{Z}^*$ is of type (b), $\overline{\psi}_r = \Theta\psi_r$ cannot be a linear combination of $\{\psi_r\}$, so $\{\psi_r,\ \overline{\psi}_r\}$ is of dimension $2l$ and Θ causes a doubling of degeneracy.

When N is an odd integer, for $\{\overline{\psi}_r\}$ to be linearly dependent on $\{\psi_r\}$,

$$\Theta\psi_r = \overline{\psi}_r = \sum_{s=1}^{l} \psi_s \mathrm{Z}_{sr} \tag{34}$$

for some non-singular matrix Z.

(13.3.24), (34) $$\Theta^2\psi_r = -\psi_r = \sum_s \Theta\psi_s \mathrm{Z}_{sr} = \sum_{s,q} \psi_q \mathrm{Z}_{qs} \mathrm{Z}_{sr}{}^* = \sum_q \psi_q(\mathrm{Z}\,\mathrm{Z}^*)_{qr}. \tag{37}$$

Since the ψ_r are LI,

(37) $$(\mathrm{Z}\,\mathrm{Z}^*)_{qr} = -\delta_{qr}, \quad \mathrm{Z}\,\mathrm{Z}^* = -\mathrm{E}\ (\text{case (c)}). \tag{38}$$

Therefore linear dependence leads only to case (c). But if the sets $\{\psi_r\}$, $\{\overline{\psi}_r\}$ are LI, this can arise only for cases (a) or (b). Thus, for N equal to an odd integer, time reversal leads to a doubling of degeneracy in cases (a) and (b). The consequences of time-reversal symmetry are summarized in Table 13.1.

Exercise 13.4-6 It might appear that the last line of Table 13.1 contravenes Kramers' theorem. Explain why this is not so.

When evaluating the character of the MR of the product $g_i g_j$ of two symmetry operators $g_i g_j \in \mathrm{G}$ for PRs, remember that $\Gamma(g_i)\,\Gamma(g_j) = [g_i\ ;\ g_j]\,\Gamma(g_k)$, where $g_k = g_i g_j$ and $[g_i\ ;\ g_j]$ is the appropriate projective factor (PF). For vector representations PF = 1 always.

Exercise 13.4-7 Determine if time-reversal symmetry introduces any additional degeneracies in systems with symmetry (1) C_3 and (2) C_4, for (i) N even and (ii) N odd. [*Hints*: Do not make use of tabulated PFs but calculate any PFs not already given in the examples in Section 12.4. Characters may be found in the character tables in Appendix A3.]

Answers to Exercises 13.4

Exercise 13.4-1 If $\{\chi\}$ is real, Γ and Γ^* have the same characters and they are therefore equivalent.

Exercise 13.4-2 Take the CC of eq. (16) and pre-multiply it by Z^{-1} and post-multiply by Z, giving $Z^{-1} Z Z^* Z = Z^{-1}, c(Z)^* E Z$, or $Z^* Z = c(Z)^* E$. Comparison with eq. (16) shows that $c(Z)^* = c(Z)$ so that $c(Z)$ is real.

$$\det Z Z^* = |\det Z|^2 = c(Z)^l \neq 0 \tag{17}$$

because Z is non-singular. Therefore $c(Z) \neq 0$.

Exercise 13.4-3 If $B = Z' + e^{i\gamma}E$ is a singular matrix, $\det B = 0$ and $e^{i\gamma}$ is then the negative of one of the eigenvalues of Z'. Therefore, a non-singular B can be ensured by having $-e^{i\gamma}$ not equal to any of the eigenvalues of Z'.

Exercise 13.4-4 Since Z is unitary, so are Z^* and $Z Z^*$. Equation (27) then gives $|c(Z)|^2 = 1$, and since, by Exercise 13.4-2, $c(Z)$ is real, $c(Z)$ can only be ± 1.

Exercise 13.4-5 From Exercise 13.4-2 and the remark after eq. (29), $|\det Z|^2 = c(Z)^l = (-1)^l$. But $|\det Z|^2 > 0$, so l can only be an even number.

Exercise 13.4-6 It was shown in Exercise 13.4-5 that the dimension l of representations of type (c) is an even integer. Therefore, even though time reversal introduces no new degeneracies, l is always at least 2 and Kramers' theorem is satisfied.

Exercise 13.4-7 (1) In C_3 the PF $[C_3^+ \,;\, C_3^+] = -1$, as shown in Example 12.6-2. For C_3^-, from Table 12.4, $[\lambda,\ \Lambda] = [{}^1\!/_2\ ,\ {}^{\sqrt{3}}\!/_2[0\ 0\ \bar{1}]]$. So, $[\lambda,\ \Lambda][\lambda,\ \Lambda] = [{}^1\!/_2,\ {}^{\sqrt{3}}\!/_2[0\ 0\ \bar{1}]]$ $[{}^1\!/_2,\ {}^{\sqrt{3}}\!/_2[0\ 0\ \bar{1}]] = -[{}^1\!/_2,\ {}^{\sqrt{3}}\!/_2[0\ 0\ 1]]$ and for PRs $[C_3^- \,;\, C_3^-] = -1$, $\Gamma(C_3^-)\Gamma(C_3^-) = -\Gamma(C_3^+)$. For A_1, $g^{-1}\sum_R \chi(R^2) = ({}^1\!/_3)[1 + 1 + 1] = 1$, so A_1 is of type (a). For ${}^1E$, $g^{-1}\sum_R \chi(R^2) = ({}^1\!/_3)[1 + \varepsilon + \varepsilon^*] = 1 + 2\cos(2\pi/3) = 0$, and similarly for ${}^2E$. Therefore these representations are of type (b). For ${}^1E_{1/2}, g^{-1}\sum_R \chi(R^2) = ({}^1\!/_3)[1 + \varepsilon + \varepsilon^*] = 0$, and similarly for ${}^2E_{1/2}$ so they are also of type (b). Note that although $(C_3^+)^2 = C_3^-$, for spinor representations $\chi((C_3^+)^2) = [C_3^+ \,;\, C_3^+]\chi(C_3^-) = -1(-\varepsilon^*) = \varepsilon^*$ for the class of C_3^+ in ${}^1E_{1/2}$, and similarly. For $B_{3/2}$, $g^{-1}\sum_R \chi(R^2) =$

Table 13.2.

Type	N even	N odd
(a)	none	doubled
(b)	doubled	none

$(1/3)[1+1+1]=1$, since $\chi((C_3^+)^2)=(-1)\chi(C_3^-)=(-1)(-1)=1$, and similarly for C_3^-. Therefore $B_{3/2}$ is of type (a).

(2) In C_4, $\{R^2\}=\{E\ \ C_2\ \ E\ \ C_2\}$. The PF $[E\ ;\ E]=1$, because of standardization. For C_2, $[\lambda\ ;\ \Lambda]=[0\ ;\ [0\ 0\ 1]]$ and $[\lambda\ ;\ \Lambda][\lambda\ ;\ \Lambda]=-1[1\ ;\ [0\ 0\ 0]]$ so that $[C_2\ ;\ C_2]=-1$. For $C_4^+[\lambda\ ;\ \Lambda]=[1/\sqrt{2}\ ;\ 1/\sqrt{2}[0\ 0\ 1]]$ and $[\lambda\ ;\ \Lambda][\lambda\ ;\ \Lambda]=[0\ ;\ [0\ 0\ 1]]$, so that for spinor as well as vector representations $\Gamma(C_4^+)\Gamma(C_4^+)=\Gamma(C_2)$. For C_4^-, $[\lambda\ ;\ \Lambda]=[1/\sqrt{2}\ ;\ 1/\sqrt{2}[0\ 0\ \bar{1}]]$ so that $[\lambda\ ;\ \Lambda][\lambda\ ;\ \Lambda]=-[0\ ;\ [0\ 0\ 1]]$ and $\Gamma(C_4^-)\,\Gamma(C_4^-)=-\Gamma(C_2)$ for the spinor representations. Thus, for spinor IRs, $\chi(E^2)=\chi(E)$, $\chi((C_4^+)^2)=\chi(C_2)$, $\chi((C_4^-)^2)=-\chi(C_2)$ and $\chi((C_2)^2)=-\chi(E)$. For the vector representations, all the PFs are $+1$. Therefore for A, B, $g^{-1}\sum_R \chi(R)^2=(1/4)[1+1+1+1]=1$, type (a). For ^{1}E, ^{2}E, $g^{-1}\sum_R \chi(R)^2=(1/4)[1-1+1-1]=0$, type (b). For ${}^1E_{1/2}$, ${}^2E_{1/2}$, $g^{-1}\sum_R \chi(R)^2=(1/4)[1-\mathrm{i}-1+\mathrm{i}]=0$, type (b). The change in degeneracy in states of C_3 and C_4 symmetry, that are induced by time-reversal symmetry are, therefore, as shown in Table 13.2.

Problems

13.1 Determine if time reversal introduces any further degeneracy into the quantum states of systems with N even and N odd and with point group symmetry D_2, D_3, and D_4.

13.2 Repeat Problem 13.1 using the double group $\bar{G}$ in place of G. [*Hint*: Remember that the multiplication rules in $\bar{G}$ are different from those of G.]

13.3 Prove that the number of inequivalent, real vector IRs of a symmetry group G is equal to the number of ambivalent classes of G. Test this theorem by referring to character tables for the point groups D_2, D_3, and T_h. [*Hints*: The inverse class $\mathscr{C}_{\bar{k}}$ of the class $\mathscr{C}_k=\{R\}$ is the class $\{R^{-1}\}$. An ambivalent class is one for which $\mathscr{C}_{\bar{k}}=\mathscr{C}_k$. You will need to use the orthogonality of the rows and of the columns of the character table.]

13.4 Consider the splitting of a state with $j=3/2$ in an electrostatic field of C_{3v} symmetry. [*Hint*: Assume that there is no external magnetic field.]

14 Magnetic point groups

14.1 Crystallographic magnetic point groups

Because the neutron has a magnetic moment, neutron diffraction can reveal not only the spatial distribution of the atoms in a crystal but also the orientation of the spin magnetic moments. Three main kinds of magnetic order can be distinguished. In ferromagnetic crystals (e.g. Fe, Ni, Co) the spin magnetic moments are aligned parallel to a particular direction. In antiferromagnetically ordered crystals, such as MnO, the spins on adjacent Mn atoms are antiparallel, so there is no net magnetic moment. In ferrimagnetic crystals (ferrites, garnets) the antiparallel spins on two sublattices are of unequal magnitude so that there is a net magnetic moment. In classical electromagnetism a magnetic moment is associated with a current, and consequently time reversal results in a reversal of magnetic moments. Therefore the point groups $\underline{\mathrm{G}}$ of magnetic crystals include *complementary* operators ΘR, where Θ is the time-reversal operator introduced in Chapter 13. The thirty-two crystallographic point groups, which were derived in Chapter 2, do not involve any complementary operators. In such crystals (designated as type I) the orientation of all spins is invariant under all $R \in \mathrm{G}$. In Shubnikov's (1964) description of the point groups, in which a positive spin is referred to as "black" and a negative spin as "white," so that the time-reversal operator Θ induces a "color change," these groups would be *singly colored*, either black or white. Diamagnetic or paramagnetic crystals, in which there is no net magnetic moment in the absence of an applied magnetic field, belong to one of the thirty-two type II "gray" groups which contain Θ explicitly, so that

$$\underline{\mathrm{G}} = \{R\} \oplus \Theta\{R\}. \tag{1}$$

Magnetic crystals with a net magnetic moment belong to one of the point groups $\underline{\mathrm{G}}$ which contain complementary operators, but for which $\Theta \notin \underline{\mathrm{G}}$. If $\mathrm{G} = \mathrm{H} + Q\mathrm{H}$, where H is a halving subgroup (invariant subgroup of index 2) of G, and $Q \in \mathrm{G}$ but $Q \notin \mathrm{H}$, then

$$\underline{\mathrm{G}} = \mathrm{H} + \Theta Q\mathrm{H} = \mathrm{H} + \Theta(\mathrm{G} - \mathrm{H}). \tag{2}$$

The distinguishing characteristic of the Shubnikov point groups are summarized in Table 14.1. A systematic determination of the fifty-eight type III magnetic point groups is summarized in Table 14.2, which shows $\underline{\mathrm{G}}$, G, H, Q, and the classes of G−H. The elements of $\underline{\mathrm{G}}$ are $\{\mathrm{H}\}$ and $\Theta\{\mathrm{G} - \mathrm{H}\}$. The elements of H can be identified from the character tables of the crystallographic point groups in Appendix A3, except that in the subgroup m or $\mathrm{C}_{1\mathrm{h}}$ of $\underline{mm}2$ the elements are $\{E\ \sigma_{\mathrm{y}}\}$ instead of $\{E\ \sigma_{\mathrm{z}}\}$ used for the point

Table 14.1. *The Shubnikov (or colored) point groups.*

$\mathbf{M}_i$ is the magnetic moment of the ith atom.

Type	Number	Color	Point group	Magnetic moment
I	32	singly colored	$\mathrm{G} = \{R\}$	$\mathbf{M}_i$ invariant under R
II	32	gray	$\mathrm{G} = \{R\} \otimes \{E\ \Theta\}$	$\mathbf{M} = \sum_i \mathbf{M}_i = 0$
III	58	black and white	$\mathrm{G} = \{\mathrm{H}\} \otimes \Theta\{\mathrm{G}-\mathrm{H}\}$	$\Theta\mathbf{M}_i = -\mathbf{M}_i$

Table 14.2. *The fifty-eight type III magnetic point groups.*

Underlines in the International notation for $\underline{\mathrm{G}}$ show which operators are complementary ones. Alternatively, these may be identified from the classes of G–H by multiplying each operator by Θ; G is the ordinary crystallographic point group from which $\underline{\mathrm{G}}$ was constructed by eq. (14.1.2); H is given first in International notation and then in Schönflies notation, in square brackets. Subscript **a** denotes the unit vector along [1 1 0].

No.	$\underline{\mathrm{G}}$	G	H	G–H	Q
1	$\underline{\bar{1}}$	C_i	$1[\mathrm{C}_1]$	I	I
2	$\underline{2}$	C_2	$1[\mathrm{C}_1]$	$C_{2\mathbf{z}}$	$C_{2\mathbf{z}}$
3	$\underline{m}$	C_s	$1[\mathrm{C}_1]$	$\sigma_{\mathbf{z}}$	$\sigma_{\mathbf{z}}$
4	$2/\underline{m}$	C_{2h}	$2[\mathrm{C}_2]$	$I,\ \sigma_{\mathbf{z}}$	I
5	$\underline{2}/m$	C_{2h}	$m[\mathrm{C}_{1h}]$	$I,\ C_{2\mathbf{z}}$	I
6	$\underline{2}/\underline{m}$	C_{2h}	$\bar{1}[\mathrm{C}_i]$	$C_{2\mathbf{z}},\ \sigma_{\mathbf{z}}$	$C_{2\mathbf{z}}$
7	$\underline{22}2$	D_2	$2[\mathrm{C}_2]$	$C_{2\mathbf{x}},\ C_{2\mathbf{y}}$	$C_{2\mathbf{x}}$
8	$\underline{mm}2$	C_{2v}	$2[\mathrm{C}_2]$	$\sigma_{\mathbf{x}},\ \sigma_{\mathbf{y}}$	$\sigma_{\mathbf{x}}$
9	$\underline{m}m\underline{2}$	C_{2v}	$m[\mathrm{C}_{1h}]$	$C_{2\mathbf{z}},\ \sigma_{\mathbf{x}}$	$\sigma_{\mathbf{y}}$
10	$\underline{mmm}$	D_{2h}	$222[\mathrm{D}_2]$	$I,\ \sigma_{\mathbf{x}},\ \sigma_{\mathbf{y}},\ \sigma_{\mathbf{z}}$	$C_{2\mathbf{z}}$
11	$mm\underline{m}$	D_{2h}	$mm2[\mathrm{C}_{2v}]$	$C_{2\mathbf{x}},\ C_{2\mathbf{y}},\ I,\ \sigma_{\mathbf{z}}$	I
12	$\underline{mm}m$	D_{2h}	$2/m[\mathrm{C}_{2h}]$	$C_{2\mathbf{x}},\ C_{2\mathbf{y}},\ \sigma_{\mathbf{x}},\ \sigma_{\mathbf{y}}$	$C_{2\mathbf{x}}$
13	$\underline{4}$	C_4	$2[\mathrm{C}_2]$	$C_{4\mathbf{z}}^+,\ C_{4\mathbf{z}}^-$	$C_{4\mathbf{z}}^+$
14	$\underline{\bar{4}}$	S_4	$2[\mathrm{C}_2]$	$S_{4\mathbf{z}}^-,\ S_{4\mathbf{z}}^+$	$S_{4\mathbf{z}}^-$
15	$4\underline{22}$	D_4	$4[\mathrm{C}_4]$	$2C_2',\ 2C_2''$	$C_{2\mathbf{x}}$
16	$\underline{4}2\underline{2}$	D_4	$222[\mathrm{D}_2]$	$2C_4,\ 2C_2''$	$C_{2\mathbf{a}}$
17	$4/\underline{m}$	C_{4h}	$4[\mathrm{C}_4]$	$I,\ S_{4\mathbf{z}}^-,\ \sigma_{\mathbf{z}},\ S_{4\mathbf{z}}^+$	I
18	$\underline{4}/\underline{m}$	C_{4h}	$\bar{4}[\mathrm{S}_4]$	$I,\ C_{4\mathbf{z}}^+,\ \sigma_{\mathbf{z}},\ C_{4\mathbf{z}}^-$	I
19	$\underline{4}/m$	C_{4h}	$2/m[\mathrm{C}_{2h}]$	$C_{4\mathbf{z}}^+,\ C_{4\mathbf{z}}^-,\ S_{4\mathbf{z}}^-,\ S_{4\mathbf{z}}^+$	$C_{4\mathbf{z}}^+$
20	$4\underline{mm}$	C_{4v}	$4[\mathrm{C}_4]$	$2\sigma_{\mathrm{v}},\ 2\sigma_{\mathrm{v}}'$	$\sigma_{\mathbf{x}}$
21	$\underline{4}m\underline{m}$	C_{4v}	$mm2[\mathrm{C}_{2v}]$	$2C_4,\ 2\sigma_{\mathrm{v}}'$	$\sigma_{\mathbf{a}}$
22	$\bar{4}\underline{2m}$	D_{2d}	$\bar{4}[\mathrm{S}_4]$	$2C_2',\ 2\sigma_{\mathrm{d}}$	$C_{2\mathbf{x}}$
23	$\underline{\bar{4}}2\underline{m}$	D_{2d}	$222[\mathrm{D}_2]$	$2S_{4\mathbf{z}},\ 2\sigma_{\mathrm{d}}$	$\sigma_{\mathbf{a}}$
24	$\underline{\bar{4}}m\underline{2}$	D_{2d}	$mm2[\mathrm{C}_{2v}]$	$2S_{4\mathbf{z}},\ 2C_2'$	$C_{2\mathbf{a}}$
25	$4/\underline{mmm}$	D_{4h}	$422[\mathrm{D}_4]$	$I,\ \sigma_{\mathbf{z}},\ 2S_{4\mathbf{z}},\ 2\sigma_{\mathrm{v}},\ 2\sigma_{\mathrm{d}}$	I
26	$4/\underline{m}mm$	D_{4h}	$4mm[\mathrm{C}_{4v}]$	$I,\ \sigma_{\mathbf{z}},\ 2S_{4\mathbf{z}},\ 2C_2',\ 2C_2''$	I
27	$\underline{4}/mm\underline{m}$	D_{4h}	$mmm\ [\mathrm{D}_{2h}]$	$2C_{4\mathbf{z}},\ 2C_2'',\ 2S_{4\mathbf{z}},\ 2\sigma_{\mathrm{d}}$	$C_{2\mathbf{a}}$
28	$\underline{4}/\underline{mm}m$	D_{4h}	$\bar{4}2m[\mathrm{D}_{2d}]$	$I,\ \sigma_{\mathbf{z}},\ 2C_{4\mathbf{z}},\ 2\sigma_{\mathrm{v}},\ 2C_2''$	I
29	$4/m\underline{mm}$	D_{4h}	$4/m[\mathrm{C}_{4h}]$	$2C_2',\ 2C_2'',\ 2\sigma_{\mathrm{v}},\ 2\sigma_{\mathrm{d}}$	$C_{2\mathbf{x}}$
30	$3\underline{2}$	D_3	$3[\mathrm{C}_3]$	$3C_2'$	$C_{2\mathbf{x}}$
31	$3\underline{m}$	C_{3v}	$3[\mathrm{C}_3]$	$3\sigma_{\mathrm{v}}$	$\sigma_{\mathbf{y}}$
32	$\underline{\bar{6}}$	C_{3h}	$3[\mathrm{C}_3]$	$\sigma_{\mathrm{h}},\ S_3^-,\ S_3^+$	σ_{h}

Table 14.2. (*cont.*)

No.	$\underline{\mathrm{G}}$	G	H	G−H	Q
33	$\bar{6}\underline{m2}$	$\mathrm{D_{3h}}$	$\bar{6}[\mathrm{C_{3h}}]$	$3C_2', 3\sigma_\mathrm{v}$	$C_{2\mathbf{x}}$
34	$\underline{\bar{6}}m\underline{2}$	$\mathrm{D_{3h}}$	$3m[\mathrm{C_{3v}}]$	$\sigma_\mathrm{h}, 2S_3, 3C_2'$	σ_h
35	$\underline{\bar{6}}\,\underline{m}2$	$\mathrm{D_{3h}}$	$32[\mathrm{D_3}]$	$\sigma_\mathrm{h}, 2S_3, 3\sigma_\mathrm{v}$	σ_h
36	$\underline{6}$	$\mathrm{C_6}$	$3[\mathrm{C_3}]$	C_6^+, C_6^-, C_2	$C_{2\mathbf{z}}$
37	$\underline{\bar{3}}$	$\mathrm{S_6}$	$3[\mathrm{C_3}]$	I, S_6^-, S_6^+	I
38	$\bar{3}\underline{m}$	$\mathrm{D_{3d}}$	$\bar{3}[\mathrm{S_6}]$	$3C_2', 3\sigma_\mathrm{d}$	$C_{2\mathbf{x}}$
39	$\underline{\bar{3}}m$	$\mathrm{D_{3d}}$	$3m[\mathrm{C_{3v}}]$	$I, 2S_6, 3C_2'$	I
40	$\underline{\bar{3}}\,\underline{m}$	$\mathrm{D_{3d}}$	$32[\mathrm{D_3}]$	$I, 2S_6, 3\sigma_\mathrm{d}$	I
41	$6\underline{22}$	$\mathrm{D_6}$	$6[\mathrm{C_6}]$	$3C_2', 3C_2''$	$C_{2\mathbf{x}}$
42	$\underline{62}2$	$\mathrm{D_6}$	$32[\mathrm{D_3}]$	$C_2, 2C_6, 3C_2''$	$C_{2\mathbf{z}}$
43	$6/\underline{m}$	$\mathrm{C_{6h}}$	$6[\mathrm{C_6}]$	$I, S_3^-, S_3^+, S_6^-, S_6^+, \sigma_\mathrm{h}$	I
44	$\underline{6}/\underline{m}$	$\mathrm{C_{6h}}$	$\bar{3}[\mathrm{S_6}]$	$C_6^+, C_6^-, \mathrm{C}_2, S_3^-, S_3^+, \sigma_\mathrm{h}$	$C_{2\mathbf{z}}$
45	$\underline{6}/m$	$\mathrm{C_{6h}}$	$\bar{6}[\mathrm{C_{3h}}]$	$I, S_6^-, S_6^+, C_2, C_6^+, C_6^-$	I
46	$6\underline{mm}$	$\mathrm{C_{6v}}$	$6[\mathrm{C_6}]$	$3\sigma_\mathrm{d}, 3\sigma_\mathrm{v}$	$\sigma_\mathbf{x}$
47	$\underline{6m}m$	$\mathrm{C_{6v}}$	$3m[\mathrm{C_{3v}}]$	$C_2, 2C_6, 3\sigma_\mathrm{v}$	$C_{2\mathbf{z}}$
48	$\underline{6}/mm\underline{m}$	$\mathrm{D_{6h}}$	$\bar{6}2m[\mathrm{D_{3h}}]$	$I, C_2, 2S_6, 2C_6, 3C_2'', 3\sigma_\mathrm{d}$	I
49	$\underline{6}/\underline{mm}m$	$\mathrm{D_{6h}}$	$\bar{3}m[\mathrm{D_{3d}}]$	$C_2, \sigma_\mathrm{h}, 2C_6, 2S_3, 3C_2'', 3\sigma_\mathrm{v}$	$C_{2\mathbf{z}}$
50	$6/\underline{mmm}$	$\mathrm{D_{6h}}$	$622[\mathrm{D_6}]$	$I, \sigma_\mathrm{h}, 2S_3, 2S_6, 3\sigma_\mathrm{d}, 3\sigma_\mathrm{v}$	I
51	$6/\underline{m}mm$	$\mathrm{D_{6h}}$	$6mm[\mathrm{C_{6v}}]$	$I, \sigma_\mathrm{h}, 2S_3, 2S_6, 3C_2', 3C_2''$	I
52	$6/m\underline{mm}$	$\mathrm{D_{6h}}$	$6/m[\mathrm{C_{6h}}]$	$3C_2', 3C_2'', 3\sigma_\mathrm{d}, 3\sigma_\mathrm{v}$	$C_{2\mathbf{x}}$
53	$\underline{m}3$	$\mathrm{T_h}$	$23[\mathrm{T}]$	$I, 4S_6^-, 4S_6^+, 3\sigma_\mathrm{h}$	I
54	$\underline{\bar{4}}3\underline{m}$	$\mathrm{T_d}$	$23[\mathrm{T}]$	$6S_4, 6\sigma_\mathrm{d}$	$\sigma_\mathbf{a}$
55	$\underline{4}3\underline{2}$	O	$23[\mathrm{T}]$	$6C_4, 6C_2'$	$C_{2\mathbf{a}}$
56	$\underline{m}3\underline{m}$	$\mathrm{O_h}$	$432[\mathrm{O}]$	$I, 8S_6, 3\sigma_\mathrm{h}, 6S_4, 6\sigma_\mathrm{d}$	I
57	$\underline{m}3m$	$\mathrm{O_h}$	$\bar{4}3m[\mathrm{T_d}]$	$I, 8S_6, 3\sigma_\mathrm{h}, 6C_2', 6C_4$	I
58	$m3\underline{m}$	$\mathrm{O_h}$	$m3[\mathrm{T_h}]$	$6C_4, 6C_2', 6S_4, 6\sigma_\mathrm{d}$	$C_{2\mathbf{a}}$

group m. In interpreting the International symbols for $\underline{\mathrm{G}}$, it is necessary to identify the appropriate symmetry elements from the positions of the symbols as given in the *International Tables for Crystallography* (Hahn (1983), (1992)). The total number of Shubnikov point groups, summarized in Table 14.1 is therefore $32 + 32 + 58 = 122$. International notation is used for $\underline{\mathrm{G}}$ (because it is more economical and more common) and Schönflies notation is used for G. Underlined elements in $\underline{\mathrm{G}}$ show which operators are complementary ones; removing the underlines would give G, which is why G is given separately only in Schönflies notation. The subgroup H is identified in both International and Schönflies notation. (Schönflies notation for G(H) is often used to identify both G and its subgroup H.)

14.2 Co-representations of magnetic point groups

Consider the group $\underline{\mathrm{G}} = \{\mathrm{H}, A\mathrm{H}\}$ that contains unitary $\{\mathrm{H}\}$ and antiunitary $\{A\mathrm{H}\}$ operators. H is necessarily an invariant subgroup of $\underline{\mathrm{G}}$ of index 2 and AH is a coset of H with coset representative A (which may be any one of the antiunitary operators of $\underline{\mathrm{G}}$) so that

$$\underline{\mathrm{G}} = \{\mathrm{H}\} \oplus A\{\mathrm{H}\}. \tag{1}$$

For example, A might be the time-reversal operator Θ (Section 13.4)

$$\underline{\mathrm{G}} = \{\mathrm{H}\} \oplus \Theta\{\mathrm{H}\}. \tag{2}$$

Another realization of eq. (1) is

$$\underline{\mathrm{G}} = \{\mathrm{H}\} \oplus \Theta Q\{\mathrm{H}\}, \tag{3}$$

with $A = \Theta Q$, and the unitary operator $Q \in \mathrm{G} - \mathrm{H}$. The corresponding unitary group is $\mathrm{G} = \{\mathrm{H}\} \oplus Q\{\mathrm{H}\}$. Equations (2) and (3) provided the basis for the derivation of types II and III magnetic point groups in Section 14.1. Let $R \in \mathrm{H}$ and suppose that $\langle\psi| = \langle\psi_p|$, $p = 1, 2, \ldots, l$ forms a basis for the unitary IR Γ of H, so that

$$\hat{R}\langle\psi| = \langle\psi|\Gamma(R). \tag{4}$$

Define $\hat{A}\psi_p$ as $\bar{\psi}_p$; then

$$\begin{aligned} \hat{R}\langle\overline{\psi}| &= \hat{R}\hat{A}\langle\psi| = \hat{A}(\hat{A}^{-1}\hat{R}\hat{A})\langle\psi| \\ &= \hat{A}\langle\psi|\,\Gamma(A^{-1}RA) \qquad A^{-1}RA \in \mathrm{H} \\ &= \langle\overline{\psi}|\,\Gamma(A^{-1}RA)^* = \langle\overline{\psi}|\,\overline{\Gamma}(R) \end{aligned} \tag{5}$$

($\hat{A}$ is anti-linear), where

$$\overline{\Gamma}(R) = \Gamma(A^{-1}RA)^*. \tag{6}$$

(4), (5)
$$\hat{R}\langle\psi\;\overline{\psi}| = \langle\psi\;\overline{\psi}|\begin{bmatrix}\Gamma(R) & 0\\ 0 & \overline{\Gamma}(R)\end{bmatrix} = \langle\psi\;\overline{\psi}|\,\underline{\Gamma}(R). \tag{7}$$

Let $B = AR$; then

(4)
$$\begin{aligned} \hat{B}\langle\psi| &= \hat{A}\hat{R}\langle\psi| = \hat{A}\langle\psi|\,\Gamma(R) \\ &= \langle\overline{\psi}|\,\Gamma(R)^* = \langle\overline{\psi}|\,\Gamma(A^{-1}B)^*; \end{aligned} \tag{8}$$

$$\hat{B}\langle\overline{\psi}| = \hat{B}\hat{A}\,\langle\psi| = \langle\psi|\,\Gamma(BA); \tag{9}$$

(8), (9)
$$\hat{B}\langle\psi\;\overline{\psi}| = \langle\psi\;\overline{\psi}|\begin{bmatrix}0 & \Gamma(BA)\\ \Gamma(A^{-1}B)^* & 0\end{bmatrix} = \langle\psi\;\overline{\psi}|\,\underline{\Gamma}(B). \tag{10}$$

Equations (7) and (10) confirm that $\langle\psi\;\overline{\psi}|$ forms a $2l$-dimensional basis for $\underline{\mathrm{G}}$. The representation $\underline{\Gamma}$ based on $\langle\psi\;\overline{\psi}|$ has matrix representatives (MRs)

(7), (10)
$$\underline{\Gamma}(R) = \begin{bmatrix}\Gamma(R) & 0\\ 0 & \overline{\Gamma}(R)\end{bmatrix}, \quad \underline{\Gamma}(B) = \begin{bmatrix}0 & \Gamma(BA)\\ \Gamma(A^{-1}B)^* & 0\end{bmatrix}. \tag{11}$$

However, the set of MRs $\{\underline{\Gamma}(R)\ \underline{\Gamma}(B)\}$ *do not form* an ordinary representation of $\underline{G}$. As confirmed in Problem 14.1, the matrices in eq. (11) obey the multiplication rules

$$\begin{aligned} \underline{\Gamma}(R)\ \underline{\Gamma}(S) &= \underline{\Gamma}(RS), \\ \underline{\Gamma}(R)\ \underline{\Gamma}(B) &= \underline{\Gamma}(RB), \\ \underline{\Gamma}(B)\ \underline{\Gamma}(R)^* &= \underline{\Gamma}(BR), \\ \underline{\Gamma}(B)\ \underline{\Gamma}(C)^* &= \underline{\Gamma}(BC), \end{aligned} \tag{12}$$

which hold for $\forall\, R, S \in \mathrm{H},\ \forall\, B, C \in A\mathrm{H}$. These equations demonstrate that when two of the $\underline{\Gamma}$ matrices are multiplied together, the second factor must be replaced by its complex conjugate (CC) when the first factor is the MR of an antiunitary operator. Such a set of MRs is called a *co-representation* (Wigner (1959)). Co-representations and their multiplication rules (eqs. (12)) have already been encountered in Section 13.4 for the particular case of $A = \Theta$. However, the derivation of the MRs is easier in that case because Θ commutes with R so that Γ for the basis $\langle\overline{\psi}|$ is just Γ^*. Equations (11) show that the matrices of the co-representation $\underline{\Gamma}$ can be expressed in terms of the MRs Γ of the unitary subgroup H.

Now consider the unitary transformation

$$\langle\psi'\ \overline{\psi}'| = \langle\psi\ \overline{\psi}|\mathrm{U}. \tag{13}$$

(13)
$$\begin{aligned} \hat{R}\langle\psi'\ \overline{\psi}'| &= \langle\psi'\ \overline{\psi}'|\ \underline{\Gamma}'(R) = \langle\psi\ \overline{\psi}|\ \mathrm{U}\ \underline{\Gamma}'(R) \\ &= \hat{R}\langle\psi\ \overline{\psi}|\ \mathrm{U} = \langle\psi\ \overline{\psi}|\ \underline{\Gamma}(R)\ \mathrm{U}; \end{aligned} \tag{14}$$

(14)
$$\underline{\Gamma}'(R) = \mathrm{U}^{-1}\ \underline{\Gamma}(R)\mathrm{U}; \tag{15}$$

(13)
$$\begin{aligned} \hat{B}\langle\psi'\ \overline{\psi}'| &= \langle\psi'\ \overline{\psi}'|\ \underline{\Gamma}'(B) = \langle\psi\ \overline{\psi}|\ \mathrm{U}\ \underline{\Gamma}'(B) \\ &= \hat{B}\langle\psi\ \overline{\psi}|\ \mathrm{U} = \langle\psi\ \overline{\psi}|\ \underline{\Gamma}(B)\ \mathrm{U}^* \end{aligned} \tag{16}$$

(B is antilinear),

(16)
$$\underline{\Gamma}'(B) = \mathrm{U}^{-1}\ \underline{\Gamma}(\mathrm{B})\ \mathrm{U}^*. \tag{17}$$

Therefore $\underline{\Gamma}$ is equivalent to $\underline{\Gamma}'$ ($\underline{\Gamma} \approx \underline{\Gamma}'$) if there is a unitary matrix U such that

$$\underline{\Gamma}'(R) = \mathrm{U}^{-1}\ \underline{\Gamma}(R)\ \mathrm{U},\ \underline{\Gamma}'(B) = \mathrm{U}^{-1}\ \underline{\Gamma}(B)\mathrm{U}^*,\ \forall\, R \in \mathrm{H}, \forall\, B \in A\mathrm{H}. \tag{18}$$

One might be concerned as to whether the equivalence of $\underline{\Gamma}'$ and $\underline{\Gamma}$ depends on the choice of A. But in fact, two co-representations $\underline{\Gamma}$, $\underline{\Gamma}'$ of $\underline{G}$ being equivalent depends only on the equivalence of the subduced representations Γ, Γ' of H and *not* on the choice of A in $\underline{G} = \mathrm{H} + A\mathrm{H}$ (Jansen and Boon (1967)). Note that $\overline{\Gamma}(R) = \Gamma(A^{-1}RA)^*$ may or may not be $\approx \Gamma$. Suppose first that $\overline{\Gamma}$ is not equivalent to Γ ($\overline{\Gamma}$ not $\approx$ Γ) and attempt the reduction of $\underline{\Gamma}$. Since $\underline{\Gamma}(R) = \Gamma(R) \oplus \overline{\Gamma}(R)$ (eq. (11)), any equivalent form must also be a direct sum, which means that U must be the direct sum $\mathrm{U}_1 \oplus \mathrm{U}_2$. But no such block-diagonal matrix can reduce $\underline{\Gamma}(B)$ in eq. (11), and so we conclude that if $\overline{\Gamma}$ is not $\approx \Gamma$ the co-representation $\underline{\Gamma}$, which consists of matrices of the form $\underline{\Gamma}(R)$, $\underline{\Gamma}(B)$ in eq. (11), must be irreducible.

Suppose next that $\overline{\Gamma} \approx \Gamma$; then there exists a unitary matrix Z such that

$$\Gamma(R) = \mathrm{Z}\ \overline{\Gamma}(R)\mathrm{Z}^{-1}. \tag{19}$$

But $A^2 \in \{R\}$, so

(19), (6) $$\Gamma(A^2) = \mathrm{Z}\ \overline{\Gamma}(A^2)\ \mathrm{Z}^{-1} = \mathrm{Z}\ \Gamma(A^2)^*\ \mathrm{Z}^{-1}; \tag{20}$$

(19), (6) $$\Gamma(R)^* = \mathrm{Z}^*\ \Gamma(A^{-1}RA)(\mathrm{Z}^{-1})^*; \tag{21}$$

(6), (19), (21) $$\begin{aligned}\overline{\Gamma}(R) &= \Gamma(A^{-1}RA)^* = \mathrm{Z}^*\ \Gamma(A^{-2}RA^2)(\mathrm{Z}^{-1})^* \\ &= \mathrm{Z}^*\ \Gamma^{-1}(A^2)\Gamma(R)\ \Gamma(A^2)(\mathrm{Z}^{-1})^*;\end{aligned} \tag{22}$$

(19), (22) $$\Gamma(R) = \mathrm{Z}\,\mathrm{Z}^*\ \Gamma^{-1}(A^2)\ \Gamma(R)\ \Gamma(A^2)(\mathrm{Z}^{-1})^*\mathrm{Z}^{-1},\ \forall R \in \mathrm{H}. \tag{23}$$

But Γ is an irreducible representation (IR) of H and so by Schur's lemma (Appendix A1.5) $\mathrm{ZZ}^*\ \Gamma^{-1}(A^2)$ is a multiple of the unit matrix, or

$$c(\mathrm{Z})\ \Gamma(A^2) = \mathrm{Z}\,\mathrm{Z}^*; \tag{24}$$

(24) $$\Gamma(A^2)^* = (c(\mathrm{Z})^*)^{-1}\ \mathrm{Z}^*\mathrm{Z}. \tag{25}$$

Equations (20), (24), and (25) show that $c(\mathrm{Z})$ is real. Moreover, since $\Gamma(A^2)$ and Z are unitary, $c(\mathrm{Z}) = \pm 1$, and

$$\mathrm{Z}\,\mathrm{Z}^* = \pm\Gamma(A^2). \tag{26}$$

Whether the co-representation $\underline{\Gamma}$ of $\underline{\mathrm{G}}$ (which is related to the IR Γ of H by eqs. (11) and (6)) is reducible or not, depends on which sign applies in eq. (26). We first of all generate the equivalent representation $\underline{\Gamma}'$ from $\underline{\Gamma}$ by

(11), (18), (19) $$\underline{\Gamma}'(R) = \begin{bmatrix}\mathrm{E} & 0 \\ 0 & \mathrm{Z}\end{bmatrix}\begin{bmatrix}\Gamma(R) & 0 \\ 0 & \overline{\Gamma}(R)\end{bmatrix}\begin{bmatrix}\mathrm{E} & 0 \\ 0 & \mathrm{Z}^{-1}\end{bmatrix} = \begin{bmatrix}\Gamma(R) & 0 \\ 0 & \Gamma(R)\end{bmatrix}, \tag{27}$$

(11), (18), (20) $$\underline{\Gamma}'(A) = \begin{bmatrix}\mathrm{E} & 0 \\ 0 & \mathrm{Z}\end{bmatrix}\begin{bmatrix}0 & \Gamma(\mathrm{A}^2) \\ \mathrm{E} & 0\end{bmatrix}\begin{bmatrix}\mathrm{E} & 0 \\ 0 & (\mathrm{Z}^*)^{-1}\end{bmatrix} = \begin{bmatrix}0 & \Gamma(\mathrm{A}^2)(\mathrm{Z}^*)^{-1} \\ \mathrm{Z} & 0\end{bmatrix}. \tag{28}$$

(Since eq. (18) holds $\forall B \in A\mathrm{H}$, it holds in particular when $B = A$.) Choosing the positive sign in eq. (26),

(28), (26) $$\underline{\Gamma}'(A) = \begin{bmatrix}0 & \mathrm{Z} \\ \mathrm{Z} & 0\end{bmatrix}, \tag{29}$$

which can be converted to diagonal form by the unitary transformation

$$\underline{\Gamma}''(A) = \mathrm{W}^{-1}\underline{\Gamma}'(A)\mathrm{W}, \tag{30}$$

with

$$\mathrm{W} = 2^{\frac{1}{2}}\begin{bmatrix} 1 & -1 \\ 1 & 1 \end{bmatrix}. \tag{31}$$

(29), (30), (31)
$$\underline{\Gamma}''(A) = \begin{bmatrix} \mathrm{Z} & 0 \\ 0 & -\mathrm{Z} \end{bmatrix}, \tag{32}$$

(27), (31)
$$\underline{\Gamma}''(R) = \mathrm{W}^{-1}\ \underline{\Gamma}'(R)\ \mathrm{W} = \underline{\Gamma}'(R). \tag{33}$$

Equations (32) and (33) show that $\underline{\Gamma}'$ has been reduced by the unitary transformation in eqs. (30) and (33). But if the negative sign is taken in eq. (26)

(28), (26)
$$\underline{\Gamma}'(A) = \begin{bmatrix} 0 & -\mathrm{Z} \\ \mathrm{Z} & 0 \end{bmatrix}, \tag{34}$$

which cannot be diagonalized by a unitary transformation, eq. (30), that preserves the diagonal form of $\underline{\Gamma}'(R)$ in eq. (27). B is *any one* of the antiunitary operators in $\underline{\mathrm{G}}$, i.e. $B \in \underline{\mathrm{G}} - \mathrm{H} = \mathrm{A}\{\mathrm{H}\} = \{\mathrm{H}\}A$. Therefore the above equations for B hold for $B = RA$. The co-representation $\underline{\Gamma}(B)$ of $B = RA$ is

(12)
$$\underline{\Gamma}(B) = \underline{\Gamma}(RA) = \underline{\Gamma}(R)\ \underline{\Gamma}(A) = \underline{\Gamma}(BA^{-1})\ \underline{\Gamma}(A). \tag{35}$$

Since the transformation from the unprimed $\underline{\Gamma}$ matrices to the $\underline{\Gamma}''$ set involves two successive unitary transformations,

(35)
$$\underline{\Gamma}'(B) = \underline{\Gamma}'(RA) = \underline{\Gamma}'(R)\ \underline{\Gamma}'(A) = \underline{\Gamma}'(BA^{-1})\ \underline{\Gamma}'(A), \tag{36}$$

(35)
$$\underline{\Gamma}''(B) = \underline{\Gamma}''(RA) = \underline{\Gamma}''(R)\ \underline{\Gamma}''(A) = \underline{\Gamma}''(BA^{-1})\ \underline{\Gamma}''(A), \tag{37}$$

(37), (33), (27), (35), (32)
$$\begin{aligned}\underline{\Gamma}''(B) &= \begin{bmatrix} \Gamma(BA^{-1}) & 0 \\ 0 & \Gamma(BA^{-1}) \end{bmatrix}\begin{bmatrix} \mathrm{Z} & 0 \\ 0 & -\mathrm{Z} \end{bmatrix} \\ &= \begin{bmatrix} \Gamma(\mathrm{BA}^{-1})\mathrm{Z} & 0 \\ 0 & -\Gamma(\mathrm{BA}^{-1})\mathrm{Z} \end{bmatrix},\end{aligned} \tag{38}$$

(36), (27), (34)
$$\begin{aligned}\underline{\Gamma}'(B) &= \begin{bmatrix} \Gamma(BA^{-1}) & 0 \\ 0 & \Gamma(BA^{-1}) \end{bmatrix}\begin{bmatrix} 0 & -\mathrm{Z} \\ \mathrm{Z} & 0 \end{bmatrix} \\ &= \begin{bmatrix} 0 & -\Gamma(\mathrm{BA}^{-1})\mathrm{Z} \\ \Gamma(\mathrm{BA}^{-1})\mathrm{Z} & 0 \end{bmatrix}.\end{aligned} \tag{39}$$

This completes the derivation of the irreducible co-representations $\underline{\Gamma}$ of $\underline{\mathrm{G}} = \mathrm{H} + A\mathrm{H}$ from the IRs Γ of H. The results may be summarized as follows. Z is the unitary matrix that transforms $\Gamma(R)$ into $\overline{\Gamma}(R) = \Gamma(A^{-1}RA)^*$ when $\Gamma \approx \overline{\Gamma}$, by

$$\mathrm{Z}^{-1}\Gamma(R)\ \mathrm{Z} = \overline{\Gamma}(R). \tag{19}$$

Case (a):
$$\Gamma(R) \approx \overline{\Gamma}(R), \quad c(\mathrm{Z}) = +1, \quad \mathrm{Z\,Z}^* = \Gamma(A^2),$$
$$\underline{\Gamma}''(R) = \Gamma(R), \tag{40}$$
$$\underline{\Gamma}''(B) = \pm\Gamma(BA^{-1})\ \mathrm{Z}.$$

Case (b):
$$\Gamma(R) \text{ not } \approx \overline{\Gamma}(R) = \Gamma(A^{-1}RA)^*,$$
$$\underline{\Gamma}(R) = \Gamma(R) \oplus \overline{\Gamma}(R), \tag{41}$$
$$\underline{\Gamma}(B) = \begin{bmatrix} 0 & \Gamma(\mathrm{BA}) \\ \Gamma(\mathrm{A}^{-1}\mathrm{B})^* & 0 \end{bmatrix}.$$

Case (c):
$$\Gamma(R) \approx \overline{\Gamma}(R),\ c(\mathrm{Z}) = -1,\ \mathrm{Z\,Z}^* = -\Gamma(A^2),$$
$$\underline{\Gamma}'(R) = \Gamma(R) \oplus \Gamma(R),\ \Gamma(R) = \mathrm{Z}\ \Gamma(A^{-1}RA)^*\mathrm{Z}^{-1}, \tag{42}$$
$$\underline{\Gamma}'(B) = \begin{bmatrix} 0 & -\Gamma(\mathrm{BA}^{-1})\mathrm{Z} \\ \Gamma(\mathrm{BA}^{-1})\mathrm{Z} & 0 \end{bmatrix}.$$

(See the warning at the beginning of Chapter 13 regarding the nomenclature used for (a), (b), and (c).)

Given the IRs Γ of H, all the irreducible co-representations $\underline{\Gamma}$ of $\underline{\mathrm{G}}$ can be determined from eqs. (40)–(42). Although the equivalence of Γ, $\overline{\Gamma}$ and the sign of $c(\mathrm{Z})$ provide a criterion for the classification of the co-representations of point groups with antiunitary operators, this will be more useful in the form of a character test.

$$\begin{aligned}\sum_{B\in A\mathrm{H}} \chi(B^2) &= \sum_{B\in A\mathrm{H}} \sum_{p} \Gamma(B^2)_{pp} = \sum_{R\in\mathrm{H}} \sum_{p} \Gamma(ARAR)_{pp} \\ &= \sum_{R\in\mathrm{H}} \sum_{p,q,r} \Gamma(A^2)_{pq}\ \Gamma(A^{-1}RA)_{qr}\ \Gamma(R)_{rp} \\ &= \sum_{p,q,r} \Gamma(A^2)_{pq} \sum_{R\in\mathrm{H}} \overline{\Gamma}(R)_{qr}{}^*\ \Gamma(R)_{rp}. \end{aligned} \tag{43}$$

If Γ is not $\approx \overline{\Gamma}$, then from the orthogonality theorem (OT)

(43), (A1.5.27)
$$\sum_{B\in A\mathrm{H}} \chi(B^2) = 0. \tag{44}$$

If $\Gamma \approx \overline{\Gamma}$, then, since Z is unitary,

(A1.6.11)
$$\begin{aligned}\sum_r (l/h) \sum_R \overline{\Gamma}(R)^*_{qr}\Gamma(R)_{rp} &= \sum_r \mathrm{Z}_{rq}\ \mathrm{Z}^{-1}_{rp} \\ &= \sum_r (\mathrm{Z}^*)^{-1}_{qr}\ \mathrm{Z}^{-1}_{rp} \\ &= ((\mathrm{Z}^*)^{-1}\ \mathrm{Z}^{-1})_{qp} = (\mathrm{Z\,Z}^*)^{-1}_{qp}, \end{aligned} \tag{45}$$

where h is the order of the unitary halving subgroup H. (Note that the definition of Z is the same in eqs. (19) and (A1.6.1), where here Γ^i is $\overline{\Gamma}$ and Γ^j is Γ.)

$$\text{(45), (43), (26)} \quad \sum_{B \in AH} \chi(B^2) = (h/l) \sum_{p,q} \Gamma(A^2)_{pq} \, (\mathrm{Z}\,\mathrm{Z}^*)^{-1}_{qp} = (h/l) \sum_{p=1}^{l} \pm \mathrm{E}_{pp} = \pm h, \qquad (46)$$

where the positive sign corresponds to case (a) and the negative sign corresponds to case (c). In summary, a diagnostic test (the Frobenius–Schur test) has been established which classifies the co-representations $\underline{\Gamma}$ of $\underline{\mathrm{G}} = \mathrm{H} + A\mathrm{H}$ according to the three cases listed in eqs. (40)–(42):

$$\text{(44)–(46)} \qquad h^{-1} \sum_{B \in AH} \chi(B^2) = \begin{cases} +1 & \text{case (a)} \\ 0 & \text{case (b)} \\ -1 & \text{case (c).} \end{cases} \qquad (47)$$

Equations (40)–(42) require Z: if Z cannot be determined by inspection, then

$$\text{(A1.5.30)} \qquad \mathrm{Z} = \sum_{R \in \mathrm{H}} \Gamma(R) \, \mathrm{X} \, \overline{\Gamma}(R^{-1}), \qquad (48)$$

where X is an arbitrary matrix, the purpose of which is to ensure that Z is unitary.

For type II magnetic point groups $A = \Theta$, so $B = AR$ becomes $B = \Theta R$ and $\sum_B \chi(B^2) = \Theta^2 \sum_R \chi(R^2)$, with $\Theta^2 = c = +1$ when N is even and -1 when N is odd. Thus eq. (47) in this case gives results identical with those in Table 13.1.

Example 14.2-1 Determine the type ((a), (b), or (c)) for the co-representations of the magnetic point group $\underline{\mathrm{G}} = \underline{m}3$ (which is #53 in Table 14.2).

For $\underline{\mathrm{G}} = \underline{m}3$, H = 23 or T and G = $m3$ or $\mathrm{T_h}$. This is the same example as that considered by Bradley and Cracknell (1972), although the method of solution used here is different. The character table of H = 23 is reproduced in Table 14.3, which also shows the determination of the type of representation for $\Gamma_1, \ldots, \Gamma_7$. Note that $A = \Theta I$ commutes with all $R \in \mathrm{H}$; therefore, for real vector representations, $\overline{\Gamma}(R) = \Gamma(R)^* = \Gamma(R)$, $\mathrm{Z} = \mathrm{E}$. For the complex representations ${}^{1,2}\mathrm{E}$, $\overline{\Gamma}(R) = \Gamma(R)^*$, which is not $\approx \Gamma(R)$. The case (b) co-representations may be written down from eq. (41). The same statement applies to the spinor representations ${}^{1,2}\mathrm{F}_{3/2}$ which are also case (b) (see Table 14.3). For ${}^{1,2}\mathrm{E}$,

$$\mathrm{Z} = \begin{bmatrix} 0 & 1 \\ \overline{1} & 0 \end{bmatrix} = \kappa$$

(Bradley and Cracknell (1972)). In fact, $\mathrm{Z} = \kappa$ also for the doubly degenerate case (a) spinor representations of type II magnetic point groups $\underline{\mathrm{G}} = \mathrm{H} + \Theta\mathrm{H}$ when H = $mm2(\mathrm{C_{2v}})$, $222(\mathrm{D_2})$, $32(\mathrm{D_3})$, $3m(\mathrm{C_{3v}})$, $422(\mathrm{D_4})$, $4mm(\mathrm{C_{4v}})$, $\overline{4}2m(\mathrm{D_{2d}})$, $622(\mathrm{D_6})$, $6mm(\mathrm{C_{6v}})$, $\overline{6}2m(\mathrm{D_{3h}})$, $432(\mathrm{O})$, and $\overline{4}3m(\mathrm{T_d})$, while for the $432(\mathrm{O})$ and $\overline{4}3m(\mathrm{T_d})$ type (a) spinor F representations $\mathrm{Z} = \kappa \oplus \kappa$. Other examples of Z may be found in Bradley and Cracknell (1972). For the solution of Example 14.2-1 by the double-group method, see Bradley and Cracknell (1972), pp. 626–9.

Table 14.3. *Character table of the invariant subgroup* H = *23 (or* T*) of* $\underline{G} = \underline{m}3$ (T_h).

$\varepsilon = \exp(-2\pi i/3)$; $\underline{G} = \{R\} \oplus \Theta I\{R\} = \{E\ 3C_2\ 4C_3^+\ 4C_3^-\} \oplus \Theta\ \{I\ 3\sigma_h\ 4S_6^-\ 4S_6^+\}$; $A = \Theta\ I$ commutes with $\forall R \in H$.

23, T	E	$3C_2$	$4C_3^+$	$4C_3^-$	$h^{-1}\sum_B \chi(B^2)$	(case)[a]
Γ_1 A	1	1	1	1	$(^1/_{12})[1+3+4+4]=1$	(a)
Γ_2 ^{1}E	1	1	ε^*	ε	$(^1/_{12})[1+3+4\varepsilon+4\varepsilon^*]=0$	(b)
Γ_3 ^{2}E	1	1	ε	ε^*	$(^1/_{12})[1+3+4\varepsilon^*+4\varepsilon]=0$	(b)
Γ_4 T	3	−1	0	0	$(^1/_{12})[3+9+0+0]=1$	(a)
Γ_5 $E_{1/2}$	2	0	1	1	$(-^1/_{12})[2-6-4-4]=1$	(a)[b,c]
Γ_6 $^1F_{3/2}$	2	0	ε^*	ε	$(-^1/_{12})[2-6-4\varepsilon-4\varepsilon^*]=0$	(b)[b,c]
Γ_7 $^2F_{3/2}$	2	0	ε	ε^*	$(-^1/_{12})[2-6-4\varepsilon^*-4\varepsilon]=0$	(b)[b,c]

[a] For the (a) representations the Z matrices are: Γ_1, E_1 ; Γ_4, E_3 ; Γ_5, κ (see text). The (a) and (b) representations may be found from eqs. (14.2.40) and (14.2.41).
[b] The minus sign outside the [ ] comes from $\Theta^2 = -1$ for spinor representations (with N an odd number).
[c] The minus signs within [ ] come from the Projective factors $[C_2\ ;\ C_2] = -1$, $[C_3^+\ ;\ C_3^+] = -1$, $[C_3^-\ ;\ C_3^-] = -1$, which may be verified by the methods of Chapter 12 or found in tables given by Altmann and Herzig (1994), p. 602. Recall that $[I\ ;\ I] = 1$ in the Pauli guage.

Exercise 14.2-1 Confirm that $Z^{-1}\Gamma(R)Z = \Gamma(R)^*$ for H = 23 (or T) for the MRs $\Gamma(R)$ with $R = C_{2x}$, $R = C_{2z}$ and $R = C_{31}^+$. [*Hint*: Use $Z = \kappa$ (see above).]

These MRs are (see, for example, Altmann and Herzig (1994)):

$$\Gamma(C_{2x}) = \begin{bmatrix} 0 & -i \\ -i & 0 \end{bmatrix},\ \Gamma(C_{2z}) = \begin{bmatrix} -i & 0 \\ 0 & i \end{bmatrix},\ \Gamma(C_{31}^+) = \begin{bmatrix} \varepsilon^* & -\varepsilon \\ \varepsilon^* & \varepsilon \end{bmatrix}.$$

Example 14.2-2 Find the co-representations of the magnetic point group $4\underline{mm}$ or $C_{4v}(C_{2v})$. Take $Q = \sigma_{\mathbf{a}}$, with **a** the unit vector along [110].

The character table of $2mm$ (C_{2v}) is given in Table 14.4, together with the determination of the type of co-representation of $\underline{4}m\underline{m}$ from $h^{-1}\sum_B \chi(B^2)$ and the projective factors (PFs) needed in the solution of this example. Since $B = AR = \Theta QR$, for projective representations (PRs)

$$\chi(B^2) = \Theta^2\ [QR\ ;\ QR]^{-1}\ [Q\ ;\ R]^{-2}\ \chi(QRQR). \tag{49}$$

The PFs are all ±1 so that $[Q\ ;\ R]^{-2} = +1$. For PRs, $\Theta^2 = -1$ (N odd) and so

(49) $$\chi(B^2) = (-1)\ [QR\ ;\ QR]^{-1}\chi(QRQR), \tag{50}$$

which enables us to write down $\sum_B \chi(B^2)$ in Table 14.4. The vector co-representations are $\underline{\Gamma}_1$ from Γ_1 (or A_1), $\underline{\Gamma}_2$ from Γ_2 (or A_2), which both belong to case (a), and $\underline{\Gamma}_3$ from Γ_3, Γ_4 (B_1, B_2), which is case (b). For the case (a) representations, from

Table 14.4. *Co-representations* $\underline{\Gamma}$ *of the magnetic point group* $4\underline{mm}$ [*or* $C_{4v}(C_{2v})$] *with* $Q = \sigma_{\mathbf{a}}$.

The derivation is explained in the text: see eqs. (40) and (51)–(55). Also given are the PFs required in the determination of the type of co-representation and in the derivation of $\underline{\Gamma}_5(B)$ (see eq. (55)). The matrices for the antiunitary operators in the spinor representation are to be multiplied by $\pm 2^{-1/2}$.

$4\underline{mm}$ C_{4v}	E	$R \in \{H\}$ $C_{2\mathbf{z}}$	$\sigma_{\mathbf{x}}$	$\sigma_{\mathbf{y}}$	$h^{-1} \sum_B \chi(B^2)$
$\underline{\Gamma}_1$ $\underline{A}_1$	1	1	1	1	1
$\underline{\Gamma}_2$ $\underline{A}_2$	1	1	−1	−1	1
$\underline{\Gamma}_3$ $\underline{B}_1$	$\begin{bmatrix} 1 & 0 \\ 0 & 1 \end{bmatrix}$	$\begin{bmatrix} -1 & 0 \\ 0 & -1 \end{bmatrix}$	$\begin{bmatrix} -1 & 0 \\ 0 & 1 \end{bmatrix}$	$\begin{bmatrix} 1 & 0 \\ 0 & -1 \end{bmatrix}$	0
$\underline{\Gamma}_5$ $\underline{E}_{1/2}$	$\begin{bmatrix} 1 & 0 \\ 0 & 1 \end{bmatrix}$	$\begin{bmatrix} -i & 0 \\ 0 & i \end{bmatrix}$	$\begin{bmatrix} 0 & -i \\ -i & 0 \end{bmatrix}$	$\begin{bmatrix} 0 & -1 \\ 1 & 0 \end{bmatrix}$	1
QR	$\sigma_{\mathbf{a}}$	$\sigma_{\mathbf{b}}$	$C_{4\mathbf{z}}^+$	$C_{4\mathbf{z}}^-$	
$[QR\ ;\ QR]$	−1	−1	1	−1	
$[Q\ ;\ R]$	1	1	−1	−1	
$[QR\ ;\ Q^{-1}]$	−1	1	1	1	
$[QR\ ;\ Q]$	−1	1	1	1	
$B = AR =$	$\Theta\sigma_{\mathbf{a}}$	$\Theta\sigma_{\mathbf{b}}$	$\Theta C_{4\mathbf{z}}^+$	$\Theta C_{4\mathbf{z}}^-$	
$\underline{\Gamma}_1$ $\underline{A}_1$	1	1	1	1	
$\underline{\Gamma}_2$ $\underline{A}_2$	1	1	−1	−1	
$\underline{\Gamma}_3$ $\underline{B}_1$	$\begin{bmatrix} 0 & 1 \\ 1 & 0 \end{bmatrix}$	$\begin{bmatrix} 0 & -1 \\ -1 & 0 \end{bmatrix}$	$\begin{bmatrix} 0 & 1 \\ -1 & 0 \end{bmatrix}$	$\begin{bmatrix} 0 & -1 \\ 1 & 0 \end{bmatrix}$	
$\underline{\Gamma}_5$ $\underline{E}_{1/2}$	$\begin{bmatrix} 1-i & 0 \\ 0 & 1+i \end{bmatrix}$	$\begin{bmatrix} -1-i & 0 \\ 0 & -1+i \end{bmatrix}$	$\begin{bmatrix} 0 & -1-i \\ 1-i & 0 \end{bmatrix}$	$\begin{bmatrix} 0 & 1-i \\ -1-i & 0 \end{bmatrix}$	

Table 14.5. *Rotation parameters ϕ **n**, quaternion parameters $[\lambda\ ;\ \Lambda]$ and Cayley–Klein parameters a, b for the point group 2mm (or C_{2v}).*

$a = \lambda - i\Lambda_z$, $b = -\Lambda_y - i\,\Lambda_x$. Also included are the values of $[\lambda\ ;\ \Lambda]$ for σ_a, σ_b since this information is needed in the evaluation of PFs in Example 14.2-2. In spinor representations, for improper rotations IR, the quaternion parameters $[\lambda\ ;\ \Lambda]$ for IR are the same as $[\lambda\ ;\ \Lambda]$ for R.

$2mm$, C_{2v}	ϕ	**n** or **m**	λ	$\boldsymbol{\Lambda}$	a	b
E	0	[0 0 0]	1	**0**	1	0
C_{2z}	π	[0 0 1]	0	[0 0 1]	$-i$	0
C_{2x}	π	[1 0 0]	0	[1 0 0]	0	$-i$
C_{2y}	π	[0 1 0]	0	[0 1 0]	0	-1
σ_x	π	[1 0 0]	0	[1 0 0]	0	$-i$
σ_y	π	[0 1 0]	0	[0 1 0]	0	-1
σ_a			0	$2^{-1/2}[1\ 1\ 0]$		
σ_b			0	$2^{-1/2}[1\ \bar{1}\ 0]$		

eq. (40), $\underline{\Gamma}''(R) = \Gamma(R)$, $\underline{\Gamma}''(B) = \pm\Gamma(BA^{-1})Z$. Since $Z = E_1$, on taking the positive sign (the negative sign merely gives an equivalent representation) $\underline{\Gamma}''(B) = \Gamma(BA^{-1}) = \Gamma(ARA^{-1}) = \Gamma(QRQ^{-1}) = \Gamma(R)$, for $\underline{\Gamma}_1$, $\underline{\Gamma}_2$ (since $\chi(\sigma_y) = \chi(\sigma_x)$ for these 1-D IRs – see Table 14.4). Thus we obtain the $\underline{\Gamma}_1$, $\underline{\Gamma}_2$ in Table 14.4. For Γ_3 (or B_1) $\overline{\Gamma}_3(R) = \Gamma_3(A^{-1}RA)^* = \Gamma_3(Q^{-1}RQ)^* = \Gamma_4(R)$. Therefore, from eq. (41), $\underline{\Gamma}_3(R) = \Gamma_3(R) \oplus \Gamma_4(R)$. Similarly, $\overline{\Gamma}_4(R)$ is $\Gamma_3(R)$, and so, had we started from $\Gamma_4(R)$, this would have given for the case (b) vector co-representation, $\Gamma_4(R) \oplus \Gamma_3(R)$, which is an equivalent representation to $\underline{\Gamma}_3(R)$. The significance of this remark is that the labels B_1 and B_2 are assigned arbitrarily (Mulliken notation giving no guide as to the assigning of priorities to equivalent vertical planes) and interchanging the labels B_1, B_2 results in apparently different but equivalent co-representations (see Bradley and Davies (1968)).

$$\underline{\Gamma}_3(B) = \begin{bmatrix} 0 & \Gamma_3(BA) \\ \Gamma_3(A^{-1}B)^* & 0 \end{bmatrix}, \tag{11}$$

where $\Gamma(BA) = \Gamma(ARA) = \Gamma(QRQ)$ for vector representations. Therefore $\Gamma_3(BA) = \Gamma_4(R)$ (Table 14.4). Further, $\Gamma(A^{-1}\,B)^* = \Gamma(A^{-1}\,AR)^* = \Gamma(R)$, since $\Gamma_3(R)$, $\Gamma_4\,(R)$ are real. So, $\underline{\Gamma}_3$ may now be written down, and is given in Table 14.4. For the projective (spinor) representation, which also belongs to case (a), $\underline{\Gamma}_5(R) = \Gamma_5(R)$ and $\underline{\Gamma}_5(B)$ is $\underline{\Gamma}_5''(B) = \pm\Gamma(BA^{-1})Z$ with

$$Z = \pm 2^{-1/2}\begin{bmatrix} 1-i & 0 \\ 0 & 1+i \end{bmatrix} \tag{51}$$

(see Bradley and Cracknell (1972), p. 631). The derivation of the $\underline{\Gamma}_5(R)$ matrices is summarized in Table 14.5. For spinor representations $\Gamma(IR) = \Gamma(R)$ (Pauli gauge), so $\Gamma^{1/2}(a, b)$ may be written down using eqs. (12.8.3) and (12.8.5). For example, for $R = \sigma_x$

$$\Gamma_5(\sigma_{\mathbf{x}}) = \begin{bmatrix} 0 & -\mathrm{i} \\ -\mathrm{i} & 0 \end{bmatrix} \begin{matrix} m' = {}^1\!/_2 \\ m' = -{}^1\!/_2. \end{matrix} \tag{52}$$

$$m = {}^1\!/_2 \quad -{}^1\!/_2$$

For the co-representations $\Gamma(B)$ (see eq. (40))

$$\begin{aligned}\Gamma(BA^{-1}) &= \Gamma(ARA^{-1}) = \Gamma(QRQ^{-1}) \\ &= \Gamma(Q)\Gamma(R)\Gamma(Q^{-1})[Q\ ;\ R]^{-1}[QR\ ;\ Q^{-1}]^{-1},\end{aligned} \tag{53}$$

$$\Gamma(Q^{-1})\ \Gamma(Q) = [Q^{-1}\ ;\ Q]\ \Gamma(E); \tag{54}$$

(53), (54) $$\Gamma(QRQ^{-1}) = \Gamma(Q)\Gamma(R)\Gamma(Q)^{-1}[Q\,;\,R]^{-1}\ [QR\,;\,Q^{-1}]^{-1}\ [Q^{-1}\,;\,Q]. \tag{55}$$

In this group, $Q = C_{2\mathbf{a}}$, so Q^{-1} within the PFs is $C_{2\mathbf{a}}$ and $[Q^{-1}\ ;\ Q] = -1$. The $\Gamma(B)$ matrices may now be obtained from eqs. (40), (51), (53) and (55), and are given in Table 14.2.

Answer to Exercise 14.2-1

This is straightforward matrix multiplication but a useful exercise nevertheless to confirm that the matrix $\mathbb{Z}$ is correctly given by κ.

14.3 Clebsch–Gordan coefficients

The inner direct product (DP) (or inner Kronecker product)

$$\Gamma^{ij}(\mathrm{H}) = \Gamma^i(\mathrm{H}) \boxtimes \Gamma^j(\mathrm{H}) \tag{1}$$

may be reducible,

$$\Gamma^{ij}(\mathrm{H}) = \sum_k c_{ij,\,k}\ \Gamma^k(\mathrm{H}), \tag{2}$$

where

$$c_{ij,\,k} = h^{-1} \sum_R\ \chi^i(R)\ \chi^j(R)\ \chi^k(R)^*, \quad R \in \mathrm{H}. \tag{3}$$

The $c_{ij,k}$ are called Clebsch–Gordan (CG) coefficients. They have the property

(3) $$c_{ij,\,k} = c_{ji,\,k}. \tag{4}$$

Example 14.3-1 From the character table of $3m$ (C_{3v}) in Appendix A3, $c_{11,1} = 1$, $c_{12,2} = 1$, $c_{13,3} = 1$, $c_{22,1} = 1$, $c_{23,3} = 1$, and $\Gamma_3 \boxtimes \Gamma_3 = \Gamma_1 \oplus \Gamma_2 \oplus \Gamma_3$, $c_{33,1} = 1$, $c_{33,2} = 1$, $c_{33,3} = 1$. All the other CG coefficients not given by eq. (4) are zero.

For magnetic point groups, the inner DP $\underline{\Gamma}^{ij}$ of the co-representations $\underline{\Gamma}^{i}$, $\underline{\Gamma}^{j}$ is

$$\underline{\Gamma}^{ij} = \underline{\Gamma}^{i} \boxtimes \underline{\Gamma}^{j} = \textstyle\sum_k d_{ij,\,k}\ \underline{\Gamma}^{k}. \tag{5}$$

The generalization of eq. (3) to magnetic groups (Bradley and Davis (1968); Karavaev (1965)) is

$$d_{ij,k} = \frac{h^{-1} \sum\limits_{R \in \mathrm{H}} \underline{\chi}^{i}(R)\underline{\chi}^{j}(R)\underline{\chi}^{k}(R)^{*}}{h^{-1} \sum\limits_{R \in \mathrm{H}} \underline{\chi}^{k}(R)\underline{\chi}^{k}(R)^{*}}, \tag{6}$$

where the $d_{ij,k}$ can be expressed in terms of the CG coefficients of the subgroup H. The normalization factor in the denominator of eq. (6) is only unity for case (a), since

(14.2.40)–(14.2.42)
$$\underline{\chi}^{k}(R) = \begin{cases} \chi^{k}(R), \text{ case (a)} \\ \chi^{k}(R) + \overline{\chi}^{k}(R), \text{ case (b)} \\ 2\chi^{k}(R), \text{ case (c)}, \end{cases} \tag{7}$$

where $\overline{\chi}^{k}(R)$ denotes the character of $\overline{\Gamma}^{k}(R)$. Consequently, in eq. (6)

(7)
$$h^{-1} \sum_{R \in \mathrm{H}} \underline{\chi}^{k}(R)\underline{\chi}^{k}(R)^{*} = \begin{cases} 1 \text{ in case (a)} \\ 2 \text{ in case (b)} \\ 4 \text{ in case (c)}. \end{cases} \tag{8}$$

Again using eqs. (14.2.40)–(14.2.42) for the numerator in eq. (6), the $d_{ij,k}$ can be expressed in terms of the $c_{ij,k}$. Because of the relation

(6)
$$d_{ij,\,k} = d_{ji,\,k}, \tag{9}$$

there are 3(3!) = 18 possible different combinations of $\underline{\Gamma}^{i}$, $\underline{\Gamma}^{j}$, $\underline{\Gamma}^{k}$. The results of using eqs. (7), (8) and (3) in eq. (6) are given in Table 14.6. For example, for the case aab in Table 14.6, if the representation Γ^{ij} contains Γ^{k} $c_{ij,k}$ times, it also contains $\overline{\Gamma}_{k}$ $c_{ij,k}$ times, and since the denominator is two for case (b), $d_{ij,k} = 2c_{ij,k}/2$. For the case aac, eqs. (7) and (8) require that $d_{ij,k} = 2c_{ij,k}/4 = \frac{1}{2}c_{ij,k}$. The rest of Table 14.6 can be completed in similar fashion.

Exercise 14.3-1 Write down the non-zero CG coefficients for the inner DPs of the point group *mm*2 (C_{2v}). [*Hints*: See Table 14.4. Recall that $\overline{\chi}_3(R)$ means $\chi(\overline{\Gamma}_3)$ and that for this group $\overline{\Gamma}_3 = \Gamma_4$.] Using Table 14.6 derive expressions for the non-zero CG d coefficients of the magnetic point group $4\underline{m}\underline{m}$ in terms of the $c_{ij,k}$ and evaluate these. Hence write down the CG decomposition for the Kronecker products of the IRs $\underline{\Gamma}$ of $4\underline{m}\underline{m}$.

Answer to Exercise 14.3-1

$$\begin{aligned} c_{11,1} &= c_{12,3} = c_{13,3} = c_{14,4} = c_{22,1} = c_{23,4} = c_{24,3} = c_{33,1} \\ &= c_{34,2} = c_{44,1} = c_{i5,5} = c_{55,i} = 1, \quad i = 1, 2, 3, 4. \end{aligned} \tag{10}$$

All other CG coefficients except those derived from eq. (10) using eq. (4) are zero. For the vector representations, from Tables 14.5 and 14.6 and eq. (10) we write down

Table 14.6. *Clebsch–Gordan coefficients for the inner direct products of irreducible co-representations.*

A barred suffix (e.g. $\bar{k}$) indicates a $\bar{\Gamma}$ representation, in this instance of $\bar{k}$, $\bar{\Gamma}^k$ with $\bar{\Gamma}^k(R) = \Gamma^k(A^{-1}RA^*)$.

$\underline{\Gamma}^i$	$\underline{\Gamma}^j$	$\underline{\Gamma}^k$	$d_{ij,k}$
a	a	a	$c_{ij,k}$
a	a	b	$c_{ij,k}$
a	a	c	$(1/2)c_{ij,k}$
a	b	a	$c_{ij,k} + c_{i\bar{j},k}$
a	b	b	$c_{ij,k} + c_{i\bar{j},k}$
a	b	c	$(1/2)(c_{ij,k} + c_{i\bar{j},k})$
a	c	a	$2c_{ij,k}$
a	c	b	$2c_{ij,k}$
a	c	c	$c_{ij,k}$
b	b	a	$c_{ij,k} + c_{i\bar{j},k} + c_{\bar{i}j,k} + c_{\bar{i}\bar{j},k}$
b	b	b	$c_{ij,k} + c_{i\bar{j},k} + c_{\bar{i}j,k} + c_{\bar{i}\bar{j},k}$
b	b	c	$(1/2)(c_{ij,k} + c_{i\bar{j},k} + c_{\bar{i}j,k} + c_{\bar{i}\bar{j},k})$
b	c	a	$2(c_{ij,k} + c_{i\bar{j},k})$
b	c	b	$2(c_{ij,k} + c_{i\bar{j},k})$
b	c	c	$c_{ij,k} + c_{\bar{i}j,k}$
c	c	a	$4c_{ij,k}$
c	c	b	$4c_{ij,k}$
c	c	c	$2c_{ij,k}$

After Bradley and Davies (1968), but recall the warning at the beginning of Chapter 13.

$$
\begin{aligned}
&d_{11,1} = c_{11,1} = 1, \quad d_{12,2} = c_{12,2} = 1, \quad d_{13,3} = c_{13,3} + c_{14,3} = 1 + 0 = 1,\\
&d_{22,1} = c_{22,1} = 1, \quad d_{23,3} = c_{23,3} + c_{24,3} = 0 + 1 = 1,\\
&d_{33,1} = c_{33,1} + c_{34,1} + c_{43,1} + c_{44,1} = 1 + 0 + 0 + 1 = 2,\\
&d_{33,2} = c_{33,2} + c_{34,2} + c_{43,2} + c_{44,2} = 0 + 1 + 1 + 0 = 2,
\end{aligned} \tag{11}
$$

(Note the use of eq. (4) in the last expression above.)

(10)
$$
\begin{aligned}
&\underline{\Gamma}_1 \boxtimes \underline{\Gamma}_1 = \underline{\Gamma}_1,\ \underline{\Gamma}_1 \boxtimes \underline{\Gamma}_2 = \underline{\Gamma}_2,\ \underline{\Gamma}_1 \boxtimes \underline{\Gamma}_3 = \underline{\Gamma}_3,\\
&\underline{\Gamma}_2 \boxtimes \underline{\Gamma}_2 = \underline{\Gamma}_1,\ \underline{\Gamma}_2 \boxtimes \underline{\Gamma}_3 = \underline{\Gamma}_3,\ \underline{\Gamma}_3 \boxtimes \underline{\Gamma}_3 = 2\underline{\Gamma}_1 + 2\underline{\Gamma}_2.
\end{aligned} \tag{12}
$$

For Kronecker products involving the spinor representation $\underline{\Gamma}_5$,

$$
\begin{aligned}
&d_{15,5} = c_{15,5} = 1, \quad d_{25,5} = c_{25,5} = 1, \quad d_{35,5} = c_{35,5} + c_{45,5} = 1 + 1 = 2,\\
&d_{55,k} = c_{55,k} = 1, \quad k = 1, 2, 3.
\end{aligned} \tag{13}
$$

All other $d_{ij,k}$ not derived from eqs. (11) and (13) by using eq. (9) are zero.

(12)
$$
\begin{aligned}
&\underline{\Gamma}_1 \boxtimes \underline{\Gamma}_5 = \underline{\Gamma}_5,\ \underline{\Gamma}_2 \boxtimes \underline{\Gamma}_5 = \underline{\Gamma}_5,\ \underline{\Gamma}_3 \boxtimes \underline{\Gamma}_5 = 2\underline{\Gamma}_5,\\
&\underline{\Gamma}_5 \boxtimes \underline{\Gamma}_5 = \underline{\Gamma}_1 \boxtimes \underline{\Gamma}_2 \boxtimes \underline{\Gamma}_3.
\end{aligned} \tag{14}
$$

14.4 Crystal-field theory for magnetic crystals

The splitting of atomic energy levels in a crystal field (CF) with the symmetry of one of the magnetic point groups has been considered in detail by Cracknell (1968). Consider an atomic ^{2}P level ($L = 1$) in an intermediate field of $2mm$ or C_{2v} symmetry and assume that $H_{\mathbf{S.L}} < H_{CF}$. The degenerate ^{2}P level is split into three components, $\Gamma_1 \oplus \Gamma_3 \oplus \Gamma_4$. But in a field of $4\underline{m}\underline{m}$ symmetry the two levels Γ_3 and Γ_4 "stick together," that is, are degenerate, since $\underline{\Gamma}_3 = \Gamma_3 \oplus \Gamma_4$ (case (b)), while Γ_1 is re-labeled as $\underline{\Gamma}_1$ (case (a)). (See Table 14.4 and the accompanying text.) Therefore,

$$\Gamma^L = \underline{\Gamma}_1 \oplus \underline{\Gamma}_3 \qquad (4\underline{m}\underline{m} \text{ symmetry}). \tag{1}$$

The possible splitting of the $\underline{\Gamma}_1$, $\underline{\Gamma}_3$ levels due to $H_{\mathbf{S.L}}$ is determined by

$$\underline{\Gamma}_i \boxtimes \underline{\Gamma}_j = \sum_k d_{ij,k}\, \underline{\Gamma}^k, \tag{2}$$

where $i = 1$ or 3 and $\underline{\Gamma}_j$ is $\underline{\Gamma}^S$, which is $\underline{E}_{1/2}$ or $\underline{\Gamma}_5$ in Table 14.4. The $c_{ij,k}$ must be determined first, remembering that $\underline{\Gamma}_3 = \underline{\Gamma}_3 \oplus \underline{\Gamma}_4$. Using Table 14.3,

$$\Gamma_1 \boxtimes \Gamma_5 = \Gamma_5,\ \Gamma_3 \boxtimes \Gamma_5 = \Gamma_5,\ \Gamma_4 \boxtimes \Gamma_5 = \Gamma_5. \tag{3}$$

(3) $$c_{15,5} = c_{35,5} = c_{45,5} = 1, \tag{4}$$

with $c_{ij,k}$ zero otherwise. From Table 14.6 (in the general case) or eq. (13) (for this example)

$$d_{15,5} = c_{15,5} = 1, \quad d_{35,5} = c_{35,5} + c_{45,5} = 2; \tag{5}$$

(4), (5) $$\underline{\Gamma}_1 \boxtimes \underline{\Gamma}_5 = \underline{\Gamma}_5,\ \underline{\Gamma}_3 \boxtimes \underline{\Gamma}_5 = \underline{\Gamma}_5 \oplus \underline{\Gamma}_5. \tag{6}$$

The splitting of the ^{2}P level in $\underline{G} = 4\underline{m}\underline{m}$ due to the CF and weak spin–orbit coupling is shown on the LS of Figure 14.1. When $H_{CF} < H_{\mathbf{S.L}}$ (the weak-field case) the crystal field acts on the components of the ^{2}P multiplet split by spin–orbit coupling. This is shown on the RS of Figure 14.1. In evaluating $\underline{\Gamma}^L \oplus \underline{\Gamma}^S$, make use of the fact that $\underline{\Gamma}^L$, $\underline{\Gamma}^S$, and $\underline{\Gamma}^J$ all

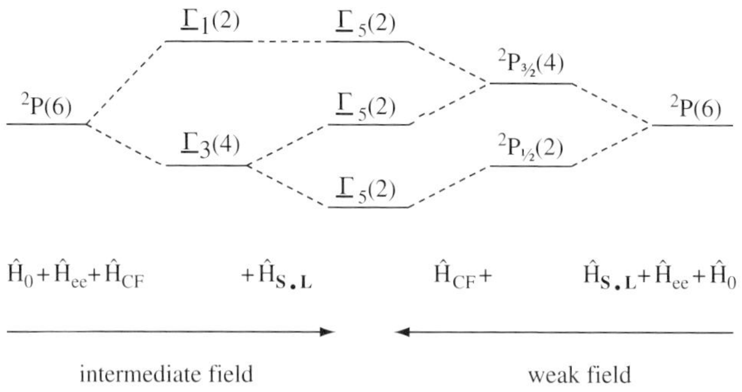

Figure 14.1. The splitting and re-labeling of a P^1: ^{2}P term in a crystal field of $4\underline{m}\underline{m}$ symmetry. On the LS, $H_{\mathbf{S.L}} < H_{CF}$ (see eqs. (14.3.15) and (14.3.20). The RS shows the weak-field case when $H_{CF} < H_{\mathbf{S.L}}$. The degeneracy of each state is shown in parentheses.

belong to case (a), so that $d_{ij,k} = c_{ij,k}$. The maximum degeneracy of a spinor representation in $4\underline{mm}$ symmetry is 2 (Table 14.4) so, in a field of this symmetry, the $\underline{\mathrm{F}}_{3/2}$ state, with character vector $|4\ 0\ 0\ 0\ 0\rangle$ splits into $\underline{\mathrm{E}}_{1/2} \oplus \underline{\mathrm{E}}_{1/2}$ or $\underline{\Gamma}_5 \oplus \underline{\Gamma}_5$.

For a slightly more complicated example, consider a $d^1 : {}^2\mathrm{D}$ term in a field of $\underline{m}3$ symmetry (see Table 14.3). The co-representations of $\underline{m}3$ are $\underline{\Gamma}_1$ (from Γ_1 of the subgroup 23), $\underline{\Gamma}_2$ (from Γ_2, Γ_3), $\underline{\Gamma}_4$ (from Γ_4), $\underline{\Gamma}_5$ (from Γ_5), and $\underline{\Gamma}_6$ (from Γ_6, Γ_7). The twenty-three subgroup levels Γ_2, Γ_3 "stick together" in the magnetic group $\underline{m}3$ to form the doubly degenerate $\underline{\Gamma}_2$ with $\underline{\Gamma}_2(R) = \Gamma_2(R) \oplus \Gamma_3(R)$ with character vector $|\underline{\chi}_2\rangle = |2\ 2\ -1\ -1\rangle$. Similarly, Γ_6, Γ_7 "stick together" to produce $\underline{\Gamma}_6$ with $\underline{\Gamma}_6(R) = \Gamma_6(R) \oplus \Gamma_7(R)$ and $|\underline{\chi}_6\rangle = |4\ 0\ -1\ -1\rangle$. In an intermediate field of $\underline{m}3$ symmetry, the ${}^2\mathrm{D}(10)$ term splits into ${}^2\underline{\Gamma}_2(4)$ and ${}^2\underline{\Gamma}_4(6)$. In $\underline{m}3$ the $\Gamma_{1/2}$ spinor forms a basis for $\underline{\Gamma}_5$ so that spin–orbit coupling is described by

$$\underline{\Gamma}_2 \otimes \underline{\Gamma}_5 = \underline{\Gamma}_6(4), \ \underline{\Gamma}_4 \otimes \underline{\Gamma}_5 = \underline{\Gamma}_5(2) \oplus \underline{\Gamma}_6(4). \tag{7}$$

In a weak $\underline{m}3$ field the ${}^2\mathrm{D}$ multiplet is split by spin–orbit coupling into its ${}^2\mathrm{D}_{3/2}(4)$ and ${}^2\mathrm{D}_{5/2}(6)$ components. The ${}^2\mathrm{D}_{3/2}$ state is $\underline{\Gamma}_6(4)$ while the ${}^2\mathrm{D}_{5/2}$ state splits into $\underline{\Gamma}_5(2)$ and $\underline{\Gamma}_6(4)$ components, which correlate with the intermediate field case in eq. (21).

Problems

14.1 Show that the matrices $\underline{\Gamma}(R)$, $\underline{\Gamma}(B)$ of eq. (14.2.11) obey the multiplication rules, eq. (14.2.12), for co-representations.

14.2 Find the type ((a) or (b), or (c)) of the co-representations of the magnetic point group $3\underline{m}$ (no. 31 in Table 14.2), which has the unitary subgroup 3. [*Hint:* $A = \Theta Q = \Theta\sigma_{\mathbf{d}}$. See Figure 12.10 for the definition of $\sigma_{\mathbf{d}}$.]

14.3 Find the type ((a) or (b), or (c)) of the co-representations of the magnetic point group $\underline{4}/m$ (no. 19 in Table 14.2) which has the unitary subgroup $2/m$. [*Hint:* $A = \Theta Q = \Theta C_{4\mathbf{z}}^{+}$.]

14.4 Derive the PFs given in Table 14.4.

14.5 Verify that the $\underline{\Gamma}_5$ matrices in Table 14.4 satisfy the multiplication rules for co-representations. It is not necessary to examine all possible products. Select, as a typical example, $R = \sigma_{\mathbf{x}}$ and $B = \Theta\sigma_{\mathbf{a}}$ in eq. (14.2.12). [*Hint*: Recall $\Theta^2 = -1$ for spinor representations.]

14.6 Write down the eight 2×2 matrices of the co-representation of $\underline{m}3$ that are derived from the 1×1 $\Gamma(R)$ matrices of the ${}^1\mathrm{E}$ or Γ_2 representation of $m3$. [*Hint:* Use the information in Table 14.3 and eqs. (14.2.40) or (14.2.41).]

14.7 Give a complete analysis of a ${}^2\mathrm{D}$ term in an $\underline{m}3$ field, justifying the various statements made in the solution sketched in the text. Include a correlation diagram (like Figure 14.1) that shows the splitting of the ${}^2\mathrm{D}$ level and indicate how the intermediate- and weak-field states shift with increasing strength of the crystal field. Perform a similar analysis for a ${}^3\mathrm{D}$ term in an $\underline{m}3$ field.

15 Physical properties of crystals

15.1 Tensors

I shall hardly do justice in this chapter to the tensor properties of crystals, a subject which has been discussed in several excellent treatises. Those I am familiar with include Bhagavantam (1966), Cracknell (1975), Nowick (1995), Nye (1957), and Wooster (1973). In addition, some books on group theory (for example, Lax (1974) and Lyubarskii (1960)) devote some attention to this topic. That group theory can be useful in this context is shown by *Neumann's principle* that the symmetry of every physical property of a crystal must include at least that of the point group of the crystal. The physical properties of crystals that are amenable to group-theoretical treatment are represented by tensors. I shall therefore begin with a brief introduction to tensors, sufficient for the main purpose of this chapter, which is the application of group theory to simplify the tensor description of the physical properties of crystals.

It will be more economical in the first two sections to label the coordinates of a point P by $\{x_1\, x_2\, x_3\}$. Symmetry operations transform points in space so that under a proper or improper rotation A, P$(x_1\ x_2\ x_3)$ is transformed into P$'(x_1{}'\ x_2{}'\ x_3{}')$. The matrix representation of this transformation is

$$|x_1{}'\ x_2{}'\ x_3{}'\rangle = \mathbb{A}|x_1\ x_2\ x_3\rangle. \tag{1}$$

A more concise representation is

$$x_i{}' = a_{ij}x_j, \tag{2}$$

in which the *Einstein summation convention* implies a sum over repeated indices. Since the a_{ij} are real, eq. (2) is an orthogonal transformation with

$$\mathbb{A}\mathbb{A}^{\mathrm{T}} = \mathbb{E},\ \det \mathbb{A} = \pm 1, \tag{3}$$

where $+1$ applies to proper rotations and -1 applies to improper rotations (Appendix A1.3). If two symmetry operators B, A are applied successively, then $\mathrm{P} \rightarrow \mathrm{P}' \rightarrow \mathrm{P}''$ and

(2) $$x_i{}'' = a_{ij}x_j{}' = a_{ij}b_{jk}x_k = c_{ik}x_k. \tag{4}$$

The symmetry operator $C = AB$ is represented by $\mathbb{C} = [c_{ik}] = \mathbb{A}\mathbb{B}$.

A quantity T that is invariant under all proper and improper rotations (that is, under all orthogonal transformations) so that $T' = T$, is a *scalar*, or *tensor of rank* 0, written $T(0)$. If T is invariant under proper rotations but changes sign on inversion, then it is a *pseudoscalar*.

Pseudoscalars with the property $T' = \pm T$ (where the positive sign applies to proper rotations and the negative sign applies to improper rotations) are also called *axial tensors of rank* 0, $T(0)^{ax}$. A quantity T with three components $\{T_1\ T_2\ T_3\}$ that transform like the coordinates $\{x_1\ x_2\ x_3\}$ of a point P, that is like the components of the position vector **r**, so that

$$T_i' = a_{ij}\ T_j,\ i,\ j = 1, 2, 3 \tag{5}$$

is a *polar vector* or *tensor of rank* 1, $T(1)$. If

$$T_i' = \pm\ a_{ij}\ T_j,\ i,\ j = 1, 2, 3, \tag{6}$$

where the positive sign applies to proper rotations $\{R\}$ and the negative sign applies to improper rotations $\{IR\}$, then T is a *pseudovector* or *axial vector*, or *axial (or pseudo-) tensor of rank* 1, $T(1)^{ax}$. Given two vectors **u** and **v**, the products of their coordinates $u_i\ v_j$ transform as

$$u_i'\ v_j' = a_{ik}\ a_{jl}\ u_k\ v_l. \tag{7}$$

Any set of nine quantities $\{T_{ij}\}$ that transform like the products of the components of two vectors in eq. (7), that is so that

$$T_{ij}' = a_{ik}\ a_{jl}\ T_{kl}, \tag{8}$$

is the set of components of a *tensor of rank* 2, $T(2)$. This definition is readily extended so that a *tensor of rank n*, $T(n)$, is a quantity with 3^n components that transform like

$$T'_{ijk\,\ldots} = a_{ip}\, a_{jq}\, a_{kr}\ \ldots\ T_{pqr\,\ldots}, \tag{9}$$

where $i, j, k, \ldots, p, q, r, \ldots = 1, 2$ or 3 and there are n subscripted indices on T and T'. If

$$T'_{ijk\,\ldots} = \pm\ a_{ip}\ a_{jq}\ a_{kr}\ \ldots T_{pqr\,\ldots}, \tag{10}$$

where the negative sign applies to improper rotations only, then T is an *axial tensor or pseudotensor of rank n*, $T(n)^{ax}$. If a $T(2)$ has

$$T_{ij} = T_{ji},\ \forall\ i,\ j, \tag{11}$$

then it is *symmetric* and consequently has only six independent components. But if

$$T_{ij} = -T_{ji},\ \forall\ i,\ j\ (\text{which entails } T_{ii} = 0, \forall\ i), \tag{12}$$

then it is *antisymmetric* with consequently only three independent components. By extension, any $T(n)$ that is invariant (changes sign) under the interchange $i \leftrightarrow j$ is symmetric (antisymmetric) with respect to these two indices i and j. For example, if $d_{ijk} = d_{ikj}, \forall\ i$, then the $T(3)$ d is symmetric with respect to the interchange of j and k. Symmetric and antisymmetric $T(2)$s will be indicated by $T(2)^{s}$ and $T(2)^{as}$, respectively. The symmetry of a tensor is an intrinsic feature of the physical property represented by the tensor, which is unaffected by any proper or improper rotation. Depending on the crystal class such symmetry operations may impose additional relations between the tensor components.

Symmetric tensors of rank 2 occur commonly in crystal physics. Consider the quadratic form

$$T_{ij}\ x_i\ x_j = 1 \tag{13}$$

of such a tensor, T. Then it is always possible to find a *principal axis transformation*

$$x_i{}' = a_{ij}\ x_j,\ \ T_{ij}{}' = a_{ij}\ a_{jl}\ T_{kl} \tag{14}$$

which results in T' being in diagonal form, with matrix representative (MR)

$$\mathrm{T}' = \begin{bmatrix} T_1{}' & 0 & 0 \\ 0 & T_2{}' & 0 \\ 0 & 0 & T_3{}' \end{bmatrix}. \tag{15}$$

The quadric eq. (13) referred to principal axes (on dropping primes in eq. (15)) is

$$\langle x_1\ x_2\ x_3|\mathrm{T}|x_1\ x_2\ x_3\rangle = T_1\ x_1{}^2 + T_2\ x_2{}^2 + T_3\ x_3{}^2 = 1. \tag{16}$$

If all three principal values are positive, the quadric surface is an ellipsoid with semiaxes $a_i = T_i^{-1/2}$, but if one or two of the principal values are negative the quadric surface is a hyperboloid. For example, the (relative) impermeability tensor β is defined by κ_0/κ, where κ is the permittivity and κ_0 is the permittivity of free space. As for any symmetric $T(2)$ the components of β define the representation quadric $\beta_{ij}x_ix_j = 1$, which here is called the *indicatrix* or optical index ellipsoid. Referred to principal axes the indicatrix is

$$\beta_1\ x_1{}^2 + \beta_2\ x_2{}^2 + \beta_3\ x_3{}^2 = 1, \tag{17}$$

where $\beta_i = \kappa_0/\kappa_i = 1/n_i^2 > 0$ and the n_i are the principal refractive indices.

A single-index notation for symmetric $T(2)$s introduced by Voigt is often very convenient. The pair of indices ij is contracted to the single index p according to the following scheme:

ij (or kl)	11	22	33	23 or 32	31 or 13	12 or 21
p (or q)	1	2	3	4	5	6

The Voigt notation may be extended to a symmetric $T(4)$ tensor when T_{ijkl} becomes T_{pq}.

Warnings (i) The $\{T_{pq}\}$ do not form a second-rank tensor and so unitary transformations must be carried out using the four-index notation T_{ijkl}. (ii) The contraction of T_{ijkl} may be accompanied by the introduction of numerical factors, for example when $T(4)$ is the elastic stiffness (Nye (1957)).

In order to apply group theory to the physical properties of crystals, we need to study the transformation of tensor components under the symmetry operations of the crystal point group. These tensor components form bases for the irreducible repsensentations (IRs) of the point group, for example $\{x_1\ x_2\ x_3\}$ for $T(1)$ and the set of infinitesimal rotations $\{R_x\ R_y\ R_z\}$ for $T(1)^{\mathrm{ax}}$. (It should be remarked that although there is no unique way of decomposing a finite rotation $R(\phi\ \mathbf{n})$ into the product of three rotations about the coordinate axes, *infinitesimal* rotations do commute and the vector $\phi\ \mathbf{n}$ can be resolved uniquely into its three projections ϕn_1, ϕn_2, ϕn_3 when ϕ is an infinitesimal angle.) The components of the basis vectors are *functions* of $x_1\ x_2\ x_3$, which transform in configuration space according to

$$\begin{aligned} R\langle \mathbf{e}_1\ \mathbf{e}_2\ \mathbf{e}_3 |\ x_1\ x_2\ x_3\rangle &= \langle \mathbf{e}_1\ \mathbf{e}_2\ \mathbf{e}_3\ |\Gamma(R)|\ x_1\ x_2\ x_3\rangle \\ &= \langle \mathbf{e}_1\ \mathbf{e}_2\ \mathbf{e}_3\ |x_1{}'\ x_2{}'\ x_3{}'\rangle. \end{aligned} \tag{18}$$

For tensors of higher rank we must ensure that the bases are properly normalized and remain so under the unitary transformations that correspond to proper or improper rotations. For a symmetric $T(2)$ the six independent components transform like binary products. There is only one way of writing $x_1\ x_1$, but since $x_1\ x_2 = x_2\ x_1$ the factors x_1 and x_2 may be combined in two equivalent ways. For the bases to remain normalized under unitary transformations the square of the normalization factor N for each tensor component is the number of combinations of the suffices in that particular product. For binary products of two unlike factors this number is two (namely ij and ji) and so $N^2 = 2$ and $x_i\ x_j$ appears as $\sqrt{2}x_i\ x_j$. The properly normalized orthogonal basis transforming like

$$\langle x_1{}^2\ x_2{}^2\ x_3{}^2\ \sqrt{2}x_2x_3\ \sqrt{2}x_3x_1\ \sqrt{2}x_1x_2\ | \tag{19}$$

is therefore

$$\langle \alpha_{11}\ \alpha_{22}\ \alpha_{33}\ \sqrt{2}\alpha_{23}\ \sqrt{2}\alpha_{31}\ \sqrt{2}\alpha_{12}| \tag{20}$$

or, in single suffix notation,

$$\langle \alpha_1\ \alpha_2\ \alpha_3\ \alpha_4\ \alpha_5\ \alpha_6|. \tag{21}$$

Equivalently, in a general $T(2)$ $x_i\ x_j$ is neither symmetric nor antisymmetric but may be symmetrized by $\mathscr{S}\ x_i\ x_j = 2^{-1/2}(x_i\ x_j + x_j\ x_i)$. But if the $T(2)$ is intrinsically symmetric (because of the property it represents) then $T_{ij} = T_{ji}$ and $2^{-1/2}(x_i\ x_j + x_j\ x_i)$ becomes $2^{1/2}x_i\ x_j$, as in eq. (19). In general, for a symmetric $T(n)$, the number of times the component $T_{ijk\ldots}$ occurs is the number of combinations of $ijk\ldots$, that is

$$P_n/P_{r_1}P_{r_2}\ldots = n!/r_1!\ r_2!\ \ldots, \tag{22}$$

where $P_n = n!$ is the number of permutations of n objects and there are n subscripts in all, r_1 alike of one kind, r_2 alike of another kind, and so on. For example, for a symmetric $T(3)$, T_{iii} occurs once ($N = 1$), T_{iij} occurs $3!/2! = 3$ times ($N = \sqrt{3}$), and T_{ijk} occurs $3! = 6$ times ($N = \sqrt{6}$). Therefore, the orthonormal symmetric $T(3)$ is

$$\langle \beta_1\ \beta_2\ \beta_3\ \beta_4\ \beta_5\ \beta_6\ \beta_7\ \beta_8\ \beta_9\ \beta_{10}|, \tag{23}$$

which transforms like

$$\langle x_1{}^3\ x_2{}^3\ x_3{}^3\ \sqrt{3}x_1{}^2x_2\ \sqrt{3}x_1{}^2x_3\ \sqrt{3}x_2{}^2x_3\ \sqrt{3}x_2{}^2x_1\ \sqrt{3}x_3{}^2x_1\ \sqrt{3}x_3{}^2x_2\ \sqrt{6}x_1x_2x_3|. \tag{24}$$

15.2 Crystal symmetry: the direct method

If a tensor T represents a physical property of a crystal, it must be invariant under the operations of the point group of the crystal. But if T is invariant under the generators of the point group, it is certainly invariant under any of the point group operators and so it will be sufficient to examine the effect of the group generators on T.

Example 15.2-1 Consider a crystal property that is described by a symmetric $T(2)$ for crystals of (a) D_{2h} and (b) C_2 symmetry. The transformation of $\{x_1\ x_2\ x_3\}$ and of $\{\alpha_1\ \alpha_2\ \alpha_2\ \alpha_4\ \alpha_5\ a_6\}$ under the group generators of D_{2h} is given in Table 15.1(a). Only $(\alpha_1\ \alpha_2\ \alpha_3)$ are invariant under $\{R\}$ and so they are the only non-zero elements of $T(2)$. The MR $\mathbb{T}$ of a symmetric $T(2)$ in (i) thus has the structure (ii) in D_{2h} symmetry:

$$\begin{matrix} \text{(i)} & \text{(ii)} & \text{(iii)} \\ \begin{bmatrix} \alpha_1 & \alpha_6 & \alpha_5 \\ \alpha_6 & \alpha_2 & \alpha_4 \\ \alpha_5 & \alpha_4 & \alpha_3 \end{bmatrix} & \begin{bmatrix} \alpha_1 & 0 & 0 \\ 0 & \alpha_2 & 0 \\ 0 & 0 & \alpha_3 \end{bmatrix} & \begin{bmatrix} \alpha_1 & \alpha_6 & 0 \\ \alpha_6 & \alpha_2 & 0 \\ 0 & 0 & \alpha_3 \end{bmatrix} \\ \text{triclinic crystal} & \mathrm{D_{2h}}\ \text{symmetry} & \mathrm{C_2}\ \text{symmetry} \end{matrix}. \tag{1}$$

The group $C_2 = \{E\ C_2\}$ has the generator C_2. Table 15.1(b) shows that $(\alpha_1\ \alpha_2\ \alpha_3\ \alpha_6)$ are invariant under $\{R\}$ and the structure of T is therefore as shown in (iii).

Example 15.2-2 Determine the non-zero elements of the elasticity tensor c_{ijkl} for a crystal of D_4 symmetry. The generalized form of Hooke's law is

$$\sigma_{ij} = c_{ijkl}\ \varepsilon_{kl},\ i, j, k, l = 1, 2, 3, \tag{2}$$

where both the stress σ and strain ε are symmetric $T(2)$s. They are *field tensors* that describe the applied forces and the resulting strains. The eighty-one *elastic stiffness constants* c_{ijkl} form a $T(4)$ which is symmetric with respect to the interchanges

$$i \leftrightarrow j,\ k \leftrightarrow l,\ ij \leftrightarrow kl. \tag{3}$$

These symmetries in eq. (3) reduce the number of independent tensor components for a triclinic crystal from eighty-one to twenty-one, which in Voigt notation form a symmetric

Table 15.1. *Transformation of the tensor components* $\{x_1\ x_2\ x_3\}$ *and* $\{\alpha_1\ \alpha_2\ \alpha_3\ \alpha_4\ \alpha_5\ \alpha_6\}$ *under (a) the generators of* D_{2h} *and (b) the generator of* C_2.

The definition of the components of α is given in eqs. (15.1.21) and (15.1.19).

	R	x_1	x_2	x_3	α_1	α_2	α_3	α_4	α_5	α_6
(a)	C_{2z}	$-x_1$	$-x_2$	x_3	α_1	α_2	α_3	$-\alpha_4$	$-\alpha_5$	α_6
	C_{2y}	$-x_1$	x_2	$-x_3$	α_1	α_2	α_3	$-\alpha_4$	α_5	$-\alpha_6$
	I	$-x_1$	$-x_2$	$-x_3$	α_1	α_2	α_3	α_4	α_5	α_6
(b)	C_2	$-x_1$	$-x_2$	x_3	α_1	α_2	α_3	$-\alpha_4$	$-\alpha_5$	α_6

6×6 matrix $[c_{pq}]$. The symmetry of $[c_{pq}]$ follows from thermodynamic arguments (see, for example, Nye (1957), pp. 136–7). The generators of D_4 are C_{4z}, C_{2x}, and the transformation of $[c_{pq}]$ in suffix notation is given in Tables 15.2(b), (c). Since $[c_{pq}]$ is invariant under the point group generators, if c_{pq} transforms into $-c_{pq}$, it must be zero, whereas if c_{pq} transforms into c_{rs}, then $c_{rs} = c_{pq}$. For example, Tables 15.2(b) and (c) show that $15 = -15 = 0 = 24$, that $14 = -25 = -14 = 0$, and so on. The resulting matrix of second-order elastic stiffness constants (in Voigt notation) is in Table 15.2(d).

A similar procedure may be followed for other point groups and for tensors representing other physical properties of crystals.

Table 15.2. *Table (a) shows the transforms $\{x_1'\ x_2'\ x_3'\}$ of $\{x_1\ x_2\ x_3\}$ under the group generators of* D_4. *Tables (b) and (c) give, in suffix notation, the transforms c'_{pq} of c_{pq} under* C_{2x} *and* C_{4z}. *Table (d) gives the upper half of the symmetric matrix $[c_{pq}]$ for crystals of* D_4 *symmetry. The tables on the right of (b) and (c) explain how the entries in the tables on the left are arrived at, using Table (a) to derive $i'j'$ (or $k'l'$).*

(a)

D_4	x_1'	x_2'	x_3'
C_{2x}	x_1	$-x_2$	$-x_3$
C_{4z}	$-x_2$	x_1	x_3

(b)

							p	ij	$i'j'$	p'
C_{2x}	11	12	13	14	−15	−16	1	11	11	1
		22	23	24	−25	−26	2	22	22	2
			33	34	−35	−36	3	33	33	3
				44	−45	−46	4	23	23	4
					55	56	5	13	−13	−5
						66	6	12	−12	−6
							q	kl	$k'l'$	q'

(c)

							p	ij	$i'j'$	p'
C_{4z}	22	12	23	25	−24	−26	1	11	22	2
		11	13	15	−14	−16	2	22	11	1
			33	35	−34	−36	3	33	33	3
				55	−45	−56	4	23	13	5
					44	46	5	13	−23	−4
						66	6	12	−12	−6
							q	kl	$k'l'$	q'

(d)

$[c_{pq}]$	11	12	13	0	0	0
		11	13	0	0	0
			33	0	0	0
				44	0	0
					44	0
						66

15.3 Group theory and physical properties of crystals

A new approach to the application of group theory in the study of the physical properties of crystals, which is more powerful than the direct method described in Section 15.2, has been developed by Nowick and is described fully in his book *Crystal Properties via Group Theory* (Nowick (1995)). A brief outline of Nowick's method will be given here. The equilibrium physical properties of crystals are described by constitutive relations which are Taylor expansions of some thermodynamic quantity Y_i in terms of a set of thermodynamic variables X_j. Usually, only the first term is retained giving the linear relations

$$Y_i = K_{ij}\, X_j. \tag{1}$$

Additional symmetries arise when the tensors X_j and/or Y_i are symmetric, and from crystal symmetry in accordance with Neumann's principle, as seen in Section 15.2. These symmetries are properties of the tensor and the crystal point group, and, if different physical properties may be represented by the same kind of tensor, it will exhibit the same structure, irrespective of the actual physical property under consideration.

In the linear-response region the fluxes J_i of matter and heat are related to the thermodynamic forces F_k by linear phenomenological relations

$$J_i = L_{ik}\, F_k. \tag{2}$$

The forces F_k involve gradients of intensive properties (temperature, electrochemical potential). The L_{ik} are called phenomenological coefficients and the fundamental theorem of the thermodynamics of irreversible processes, due originally to Onsager (1931a, b), is that when the fluxes and forces are chosen to satisfy the equation

$$T\ \mathrm{d}S/\mathrm{d}t = J_k F_k, \tag{3}$$

where $\mathrm{d}S/\mathrm{d}t$ is the rate of entropy production, then the matrix of phenomenological coefficients is symmetric,

$$L_{ik}(\mathbf{B}) = L_{ki}(-\mathbf{B}), \tag{4}$$

where $\mathbf{B}$ is magnetic induction. Onsager's theorem is based on the time-reversal symmetry of the equations of classical and quantum mechanics, and therefore if a magnetic field is present the sign of $\mathbf{B}$ must be reversed when applying the symmetry relation, eq. (4). Onsager's original demonstration of his reciprocity theorem, eq. (4), was based on the assumption that the regression of fluctuations in the thermodynamic parameters is governed by the same linear laws as are macroscopic processes (Callen (1960)). But there are difficulties in applying this hypothesis to continuous systems (Casimir (1945)), and the modern proof (De Groot and Mazur (1962)) employs time-correlation functions and the fluctuation–dissipation theorem. For the purpose of applying group theory to the physical properties of crystals, we may confidently assume the validity of the Onsager reciprocal relations (ORRs) for the linear-response coefficients (eq. (4)), which have also been verified in particular cases experimentally and by statistical-mechanical calculation of the L_{ik} (Allnatt and Lidiard (1993), chap. 6).

Table 15.3. *Transformation of the bases* $\{x\ y\ z\}$ *and* $\{R_x\ R_y\ R_z\}$ *under the generators* C_{4z} *and* σ_x *of the point group* C_{4v}.

See the text for an explanation of the determination of the orientation of the polar and axial vectors in the doubly degenerate E representation.

Basis	C_{4z}	σ_x	IRs	Orientation (for E basis)
xyz	$y\bar{x}z$	$\bar{x}yz$	$A_1 \oplus E$	$x,\ y$
$R_xR_yR_z$	$R_y\bar{R}_xR_z$	$R_x\bar{R}_y\bar{R}_z$	$A_2 \oplus E$	$R_y,\ -R_x$

The tensor components that form bases for the IRs of the point groups are given in character tables, usually for $T(1)$, $T(1)^{ax}$, and $T(2)$ only. In all other cases, one may use the projection operator P^{γ} in

$$P^{\gamma}\ X_i = \sum_R\ \chi^{\gamma}(R)^* \hat{R} X_i, \tag{5}$$

where X_i is a component of the tensor X and γ labels the IRs. For degenerate representations the projection must be continued with a second and, if necessary, third component until all the degenerate bases have been obtained. Many examples of finding these *symmetry coordinates* (or symmetry-adapted linear combinations, as they are often called) have been given in Chapters 6 and 9. An easier method of finding the IRs for bases that consist of products of components of tensors of lower rank is to form direct product (DP) representations. When more than one set of symmetry coordinates form bases for degenerate representations, it is advantageous to ensure that these sets are *similarly oriented*, that is that they transform in a corresponding fashion under the point group generators (which ensures that they will do so under all the operators of the point group). For example, the generators of C_{4v} are C_{4z} and σ_x. (Since the components of a polar vector are always labeled by $x\ y\ z$ rather than by $x_1\ x_2\ x_3$ in character tables, we revert now to this notation.) The Jones symbols for the bases $\{x\ y\ z\}$ and $\{R_x\ R_y\ R_z\}$ are given in Table 15.3 for the generators of C_{4v}. This table shows that the pairs $(x\ \ y)$ and $(R_y\ \ -R_x)$ transform in a corresponding manner. To see this, we note that under C_{4z}, R_y transforms into its partner namely $-R_x$ (just as $x \rightarrow y$) while $-R_x$ transforms into the negative of its partner, namely $-R_y$ (just as $y \rightarrow -x$). Again, under σ_x, R_y transforms into $-R_y$ (like x into $-x$) while $-R_x$ is invariant (and so is y). It is not usual in character tables to order the degenerate pairs (and triples) so as to preserve similarity of orientation, since this is not a consideration in other applications of group theory. Nevertheless, it may always be worked out in the above manner using the group generators. Alternatively, one may find tables which give similarly oriented bases in appendix E of Nowick (1995). In the groups C_{nv}, D_n, D_{nh}, and also D_{2d} and D_{3d}, it is advisable to determine similarity of orientation in E representations using $\alpha_4\ \alpha_5$, rather than by $x\ y$ (as in the C_{4v} example above) as this simplifies the analysis of stress and strain (Nowick and Heller (1965)). Henceforth in this chapter "symmetry coordinates" will imply similarly oriented bases. If it happens that X_1, X_2 form bases for

Table 15.4. *Matrix elements of* S *(see eq. (15.3.7)) for the group* C_{4v} *and a basis comprising the symmetrical T(2) tensor components* $\{\alpha_k\}$ *defined in eq. (15.1.19).*

S transforms $|\alpha_1\,\alpha_2\,\alpha_3\,\alpha_4\,\alpha_5\,\alpha_6\rangle$ into the symmetry coordinate basis $|X^{\gamma}_{dr}\rangle$. Normalization factors are included in S but omitted from $X^{\gamma}_{dr}\rangle$, as is usually the case in character tables. For the reason given in the text, in this group $\alpha_4\,\alpha_5$ are used to define similarity of orientation in the doubly degenerate E representation.

γ	r	d	X^{γ}_{dr}	S						α_k
A_1	1	1	$\alpha_1+\alpha_2$	$2^{-1/2}$	$2^{-1/2}$	0	0	0	0	α_1
A_1	2	1	α_3	0	0	1	0	0	0	α_2
B_1	1	1	$\alpha_1-\alpha_2$	$2^{-1/2}$	$2^{-1/2}$	0	0	0	0	α_3
B_2	1	1	α_6	0	0	0	0	0	1	α_4
E	1	1	α_4	0	0	0	1	0	0	α_5
		2	α_5	0	0	0	0	1	0	α_6

two 1-D representations, then it is necessary to check the transformation of $X_1 \pm X_2$ to see which linear combination forms a basis for each of the two IRs.

The first step in the group theoretical determination of the MR K of the physical property k defined by eq. (1), or its MR,

$$|Y\rangle = \mathrm{K}\,|X\rangle, \tag{6}$$

(1)

is to write down the unitary matrix S that transforms the tensor components $\{X_k\}$ and $\{Y_k\}$ into (oriented) symmetry coordinates. Note that $\{X_k\}$ and $\{Y_k\}$ denote one of the tensor bases $\{x\ y\ z\}$, $\{R_x\ R_y\ R_z\}$, $\{\alpha_k\}$ (eq. (15.1.21)), or $\{\beta_k\}$ (eq. (15.1.23)). This can be done from the information available in character tables, though a little extra work may be needed to determine the IRs spanned by $\{\beta_k\}$. Similarity of orientation must also be determined at this stage. Alternatively, the symmetry coordinate tables given by Nowick (1995) may be consulted. This transformation is for $\{X_k\}$ (in tensor notation)

$$X^{\gamma}_{dr} = S^{\gamma}_{dr,k}\ X_k, \tag{7}$$

where γ denotes one of the IRs spanned by $\{X_k\}$, d is the degeneracy index, and r is the repeat index: $r = 1, 2, \ldots$ identifies the basis when the same IR occurs more than once, and $d = 1,\ 2, \ldots$ enumerates the components of $\{X_k\}$ that form a basis for a degenerate representation. Since S is unitary, each row of S must be normalized and orthogonal to every other row. As an example, the matrix S for the group C_{4v} and basis $\{\alpha_k\}$ is given in Table 15.4.

The usefulness of group theory in establishing the non-zero elements of K is a consequence of the *fundamental theorem* (FT; Nowick (1995)), which may be stated as follows.

Provided the different sets of symmetry coordinates that form bases for a particular IR are similarly oriented, only symmetry coordinates in $\{Y_k\}$ and $\{X_k\}$ that belong to the same IR and have the same degeneracy index d are coupled (that is, connected by a non-zero coefficient in $\mathbb{K}$). For degenerate representations the different sets of symmetry coordinates are coupled by the same coefficient.

Suppose that the tensor bases in the relation

$$|Y\rangle = \mathbb{K}|X\rangle, \tag{6}$$

have already been expressed in (similarly oriented) symmetry coordinates. Then, under a symmetry operation R,

$$|X'\rangle = \Gamma^{(X)}(R)\ |X\rangle,\ \ |Y'\rangle = \Gamma^{(Y)}(R)\ |Y\rangle, \tag{8}$$

where the superscripts in parentheses distinguish the bases.

(6), (8)
$$\begin{aligned} |Y'\rangle &= \Gamma^{(Y)}(R)\ |Y\rangle = \Gamma^{(Y)}(R)\mathbb{K}\ |X\rangle \\ &= \mathbb{K}'\ |X'\rangle = \mathbb{K}'\Gamma^{(X)}(R)\ |X\rangle; \end{aligned} \tag{9}$$

(9)
$$\mathbb{K}' = \Gamma^{(Y)}(R)\ \mathbb{K}\ \Gamma^{(X)}(R)^{\dagger}. \tag{10}$$

But since $\mathbb{K}$ represents a physical property, $\mathbb{K}$ is invariant under R and

(10)
$$\Gamma^{(Y)}(R)\ \mathbb{K} = \mathbb{K}\ \Gamma^{(X)}(R), \tag{11}$$

in which $\Gamma^{(Y)}(R)$ and $\Gamma^{(X)}(R)$ are each a direct sum of MRs for particular IRs. By Schur's lemma (see Section A1.5) all the blocks (submatrices of $\mathbb{K}$) that connect the same IRs in $\Gamma^{(Y)}(R)$ and $\Gamma^{(X)}(R)$ are multiples of the unit matrix, while those that connect different IRs are zero. The importance of similarity of orientation lies in the fact that it guarantees that the blocks of $\Gamma^{(Y)}(R)$, $\Gamma^{(X)}(R)$ have the same form rather than just equivalent forms. It follows as a corollary to the FT that if $\{X\}$ and $\{Y\}$ have no IRs in common, then $\mathbb{K}$ is identically zero. For a triclinic crystal, all symmetry coordinates belong to the A representation and therefore are coupled, with consequently no zero entries in $\mathbb{K}$. In general, it follows from the FT that

(6)
$$Y^{\gamma}_{dr} = K^{\gamma}_{rs}\ X^{\gamma}_{ds}, \tag{12}$$

where the FT requires the same γ and d on X and Y. Note that $\mathbb{K}$ is invariant under $R \in G$ and that the superscript γ on K_{rs} serves as a reminder that it couples a Y and an X which form bases for the IRγ. If γ is 1-D, d is redundant, and if the repeat indices $r, s = 1$ they may also be dropped, giving

(12)
$$Y^{\gamma} = K^{\gamma}\ X^{\gamma}, \tag{13}$$

but if X_1, X_2 both form bases for the same 1-D representation ($s = 1, 2$) then

(12)
$$Y^{\gamma} = K^{\gamma}_1 X^{\gamma}_1 + K^{\gamma}_2 X^{\gamma}_2. \tag{14}$$

If γ is also repeated in Y, so that $r, s = 1, 2$,

(12)
$$\begin{aligned} Y_1^{\gamma} &= K_{11}^{\gamma} X_1^{\gamma} + K_{12}^{\gamma} X_2^{\gamma}, \\ Y_2^{\gamma} &= K_{21}^{\gamma} X_1^{\gamma} + K_{22}^{\gamma} X_2^{\gamma}. \end{aligned} \tag{15}$$

If there are no other IRs common to Y and X, then eqs. (15) show that there are only four non-zero K_{rs} coefficients in $\mathbb{K}$. Depending on the nature of the property it represents, $\mathbb{K}$ may be a symmetric matrix, and if this is true here $K_{12} = K_{21}$, leaving three independent coefficients. Equations (13)–(15) hold also for E or T representations since according to the FT the K coefficients that belong to different degeneracy indices are zero, while the coefficients for the same d are independent of d.

Exercise 15.3-1 Determine the number of independent K coefficients when Y is a vector and X is a symmetric tensor of rank 2, for crystals that belong to the point groups (a) C_{3v}, and (b) D_{2d}.

When Γ is not $\approx \Gamma^*$ (case (b) in Section 13.4), Γ, Γ^* form a degenerate pair which are generally labeled by ^1E, ^2E (sometimes with subscripts). Applying the projection operator $\sum\limits_R \chi(R)^* \hat{R}$ to a tensor component will therefore yield symmetry coordinates from ^1E, ^2E that are complex conjugates (CCs) of one another. Suppose that $X(^1E) = X_1 - iX_2$, $X(^2E) = X_1 + iX_2$, and similarly for the tensor Y. Then from the FT

$$Y_1 - iY_2 = K^E(X_1 - iX_2), \quad Y_1 + iY_2 = K^{E*}(X_1 + iX_2), \tag{16}$$

where

$$K^E = K^E_{Re} + iK^E_{Im}. \tag{17}$$

(16), (17)
$$Y_1 = K^E_{Re} X_1 + K^E_{Im} X_2, \; Y_2 = -K^E_{Im} X_1 + K^E_{Re} X_2, \tag{18}$$

(18)
$$\begin{bmatrix} Y_1 \\ Y_2 \end{bmatrix} = \begin{bmatrix} K^E_{Re} & K^E_{Im} \\ -K^E_{Im} & K^E_{Re} \end{bmatrix} \begin{bmatrix} X_1 \\ X_2 \end{bmatrix}, \tag{19}$$

so that there are two real independent coefficients K^E_{Re} and K^E_{Im}, unless $\mathbb{K}$ happens to be symmetric, in which case K^E_{Im} is zero. But if the conjugate bases for ^1E and ^2E are $X_1 \mp iX_2$, $Y_1 \pm iY_2$ then a similar analysis yields

$$K^E = \begin{bmatrix} K^E_{Re} & K^E_{Im} \\ K^E_{Im} & -K^E_{Re} \end{bmatrix} \tag{20}$$

with two independent coefficients.

The method described above is more powerful than earlier methods, especially for tensors of higher rank and for groups that have three-fold or six-fold principal axes.

Table 15.5.

γ	Basis $\{x\ y\ z\}$	$\{\alpha_k\}$
A_1	z	$\alpha_1+\alpha_2,\ \alpha_3$
E	$(y\ \ x)$	$(\alpha_1-\alpha_2,\ \alpha_6),\ (\alpha_4\ \ \alpha_5)$

Table 15.6.

E	S_4	$C_{2\mathbf{x}}$
xyz	$\bar{y}x\bar{z}$	$x\bar{y}\bar{z}$
$\alpha_4\alpha_5$	$\bar{\alpha}_5\alpha_4$	$\alpha_4\bar{\alpha}_5$

Answer to Exercise 15.3-1

(a) From the character table for C_{3v} the IR for which both $\{x\ y\ z\}$ and $\{\alpha_k\}$ form bases are given in Table 15.5. For both representations $r=1$, $s=1, 2$. Therefore eq. (14) holds for both IRs and there are four independent components. (Since $\alpha_4=2^{1/2}yz$, $\alpha_5=2^{1/2}zx$, and z is invariant under the generators of C_{3v}, similarity of orientation requires $(y\ x)$ as the E-basis in column 2 of Table 15.5).

(b) For D_{2d} the IRs B_2 and E each have one basis function from $\{x\ y\ z\}$ and $\{\alpha_k\}$, namely z and α_6 for B_2, and $(x\ \ y)$ and $(\alpha_4\ \ \alpha_5)$ for E. That $(x\ \ y)$ is similarly oriented to $(\alpha_4\ \ \alpha_5)$ is readily confirmed. The transformation of bases under the generators of D_{2d} are in Table 15.6. Equation (13) holds for $\gamma=B_2$, E, and there are therefore two independent coefficients.

15.4 Applications

The fundamental equation for the internal energy U of a thermodynamic system is

$$dU = X_i\ dY_i, \tag{1}$$

where the gradient of X_i is a generalized force, and the consequent change in Y_i is its conjugate response. For example, X_i might be the temperature T and Y_i might be the entropy S of the system. Experiment shows that, in general, a given response depends on all the forces and that when the forces are sufficiently small this dependence is linear, so

$$dY_i = K_{ij}\ dX_j. \tag{2}$$

For systems in equilibrium, eq. (2) yields

$$\Delta Y_i = K_{ij}\ \Delta X_j, \tag{3}$$

where $\Delta Y_i = Y_i - Y_{i0}$, $\Delta X_j = X_j - X_{j0}$, and the subscripts zero on Y and X indicate the initial state of the system. The integration that yields eq. (3) assumes that the K_{ij} are independent of the forces $\{X_j\}$ (the linear approximation). In the majority of cases it is convenient to take the reference values Y_{i0}, X_{j0} as zero, which amounts to saying, for example, that the crystal is not strained ($\varepsilon_{ij}=0$) when there is no applied stress ($\sigma_{ij}=0$). In such cases eq. (3) becomes

$$Y_i = K_{ij}\, X_j. \tag{4}$$

This is the form we have already used to describe the linear responses which define the properties of materials, but in some cases, notably for the temperature T, it is inconvenient to set the initial value T_0 to zero (this would require redefining the thermodynamic temperature scale), and so eq. (3) is used instead (see Table 15.7). In the particular example of a change in temperature, the conjugate response is

$$\Delta S = (C/T)\ \Delta T, \tag{5}$$

where C is the heat capacity of the system. Table 15.7 summarizes the names and symbols used for the equilibrium properties which determine the linear response ΔY_i to the forces ΔX_j, or Y_i to X_j when their values in their initial states are set to zero. The symmetry of the matrices $\mathbb{K}$ in any point group may be determined by the methods that were covered in Section 15.3.

Example 15.4-1 Obtain the $\mathbb{K}$ matrix for a $T(2)$ for a crystal that belongs to one of the uniaxial groups. [*Hint*: Take Y and X both as $T(1)$s.]

The uniaxial groups are of two kinds, those that contain case (b) ^{1}E, ^{2}E representations which are CCs of one another (called "lower symmetry" groups), and those with no case (b) representations (termed "upper symmetry groups"). For the latter, (x y) span an E irreducible representation while z is the basis of non-degenerate A or B representation. Consequently,

$$Y_1 = K^{\mathrm{E}}\, X_1,\quad Y_2 = K^{\mathrm{E}}\, X_2,\quad Y_3 = K^{\mathrm{A,B}}\, X_3, \tag{6}$$

and the dielectric permittivity matrix κ, for example, is

$$\kappa = \begin{bmatrix} \kappa_{11} & 0 & 0 \\ 0 & \kappa_{11} & 0 \\ 0 & 0 & \kappa_{33} \end{bmatrix}. \tag{7}$$

For the lower uniaxial groups,

$$(15.3.19)\qquad \mathbb{K} = \begin{bmatrix} K^{\mathrm{E}}_{\mathrm{Re}} & K^{\mathrm{E}}_{\mathrm{Im}} & 0 \\ -K^{\mathrm{E}}_{\mathrm{Im}} & K^{\mathrm{E}}_{\mathrm{Re}} & 0 \\ 0 & 0 & K_{\mathrm{A,B}} \end{bmatrix} \tag{8}$$

But for the dielectric permittivity, and any symmetric $T(2)$, $\mathbb{K}$ is symmetric ($K^{\mathrm{E}}_{\mathrm{Im}}=0$) and κ has the same form (eq. (7)) for all uniaxial groups.

Example 15.4-2 For a crystal to exhibit optical activity the gyration tensor $[g_{ij}]$ with $i, j = 1, 2, 3$, which is a symmetric axial second-rank tensor, must have at least one non-zero element. Determine the form of the gyration tensor for C_{4v} and D_{2d} symmetry.

Table 15.7. *Equilibrium crystal properties which determine the linear response ΔY_i to the forces ΔX_j.*

Each response is the sum of terms in that row (observing the summation convention). The coefficients are tensors, and the rank n of $T(n)$ is shown by the number of subscripts.

	To forces (ΔX_j)			
Response (ΔY_i)	Temperature change	Electric field	Magnetic field	Stress
ΔS, entropy	$(C/T)\,\Delta T$, heat capacity C, divided by T	$p_i\,E_i$, electrocaloric effect	$q_i\,H_i$, magnetocaloric effect	$\alpha_{ij}\,\sigma_{ij}$, piezocaloric effect
D_i, electric displacement	$p_i\,\Delta T$, pyroelectric effect	$\kappa_{ij}\,E_j$, permittivity	$\lambda_{ij}\,H_j$, magnetoelectric polarizability	$d_{ijk}\,\sigma_{jk}$, piezoelectric effect
B_i, magnetic induction	$q_i\,\Delta T$, pyromagnetic effect	$\lambda_{ji}\,E_j$, converse magnetoelectric effect	$\mu_{ij}\,H_j$, magnetic permeability	$Q_{ijk}\,\sigma_{jk}$, piezomagnetic effect
ε_{ij}, elastic strain	$\alpha_{ij}\,\Delta T$, thermal expansivity	$d_{kij}\,E_j$, converse piezoelectric effect	$Q_{kij}\,H_j$, converse piezomagnetic effect	$s_{ijkl}\,\sigma_{kl}$, elastic compliance
$\Delta\beta_{ij}$, impermeability change		$r_{ijk}\,E_k$, Pockels effect		$q_{ijkl}\,\sigma_{kl}$, photoelastic effect

For a symmetric $T(2)^{ax}$ we may work out the structure of $\mathbb{K}$ by taking Y as a $T(1)^{ax}$ (basis $\{R_x\ R_y\ R_z\}$) and X as a $T(1)$ (basis $\{x\ y\ z\}$) and then making $\mathbb{K}$ symmetric. From the character table for D_{2d} (on determining the similarity of orientation) we find that $(x\ y)$ and $(R_x\ -R_y)$ both form a basis for E while z and R_z belong to different IRs. Therefore, from the FT, $K_{33} = 0$ and

$$Y_1 = K_{11}\ X_1 + K_{12}\ X_2,$$
$$Y_2 = -K_{21}\ X_1 - K_{22}\ X_2. \tag{9}$$

Since $\mathbb{K}$ is symmetric, the off-diagonal elements are zero and

$$[g_{ij}] = \begin{bmatrix} g_{11} & 0 \\ 0 & -g_{11} \end{bmatrix}. \tag{10}$$

For C_{4v}, $(x\ y)$ and $(R_y\ -R_x)$ are bases for the E representation. Therefore,

$$Y_1 = -K\ X_2, \quad Y_2 = K\ X_1. \tag{11}$$

But since $\mathbb{K}$ is symmetric, $[g_{ij}]$ vanishes in this point group (and, in fact, in all the uniaxial C_{nv} point groups). (Note the importance of similarity of orientation in reaching the correct conclusions in this example.)

Generally the linear approximation suffices, but, because the refractive index can be measured with considerable precision, the change in the impermeability tensor due to stress and electric field should be written as

$$\Delta\beta_{ij} = r_{ijk}E_k + p_{ijkl}E_kE_l + q_{ijkl}\sigma_{kl}. \tag{12}$$

The $T(3)$ r_{ijk} gives the *linear electro-optic* (Pockels) effect, while the $T(4)$ p_{ijkl} is responsible for the *quadratic electro-optic* (Kerr) effect; q_{ijkl} is the photoelastic tensor.

To describe large deformations, the Lagrangian strains η_{ij} are defined by

$$\eta_{ij} = \frac{1}{2}\left(\frac{\partial u_i}{\partial x_j} + \frac{\partial u_j}{\partial x_i} + \frac{\partial u_k}{\partial x_i}\frac{\partial u_k}{\partial x_j}\right), \tag{13}$$

where u_i is the displacement in the ith direction in the deformed state. A Taylor expansion of the elastic strain energy in terms of the strains η_{ij} about a state of zero strain gives

$$U - U_0 = \tfrac{1}{2}c_{ijkl}\ \eta_{ij}\ \eta_{kl} + \tfrac{1}{6}C_{ijklmn}\ \eta_{ij}\ \eta_{kl}\ \eta_{mn} + \cdots \tag{14}$$

The thermodynamic tensions t_{ij} are defined by $(\partial U/\partial\eta_{ij})$ so that

(14) $$t_{ij} = (\partial U/\partial\,\eta_{ij}) = c_{ijkl}\ \eta_{kl} + \tfrac{1}{2}C_{ijklmn}\ \eta_{kl}\ \eta_{mn} + \cdots \tag{15}$$

In Voigt notation,

(15) $$t_p = (\partial U/\partial\,\eta_p) = c_{pq}\ \eta_q + \tfrac{1}{2}C_{pqr}\ \eta_q\ \eta_r + \cdots \tag{16}$$

The fifty-six components C_{pqr} (which do *not* constitute a tensor) are the *third-order elastic constants*: they are symmetric with respect to all interchanges of p, q, and r. The expansion of $U - U_0$ in eq. (14) can be carried to higher orders, but fourth- (and higher) order elastic constants are of limited application.

Example 15.4-3 Both the piezoelectric effect and the Pockels effect involve coupling between a vector and a symmetric *T*(2). The structure of $\mathbb{K}$ is therefore similar in the two cases, the only difference being that the 6×3 matrix $[r_{qi}]$ is the transpose of the 3×6 matrix $[d_{iq}]$ where $i = 1, 2, 3$ denote the vector components and $q = 1, \ldots, 6$ denote the components of the symmetric *T*(2) in the usual (Voigt) notation. Determine the structure of the piezoelectric tensor for a crystal of C_{3v} symmetry.

The allocation of vector and tensor components to IRs and similarity of orientation have already been determined for the point group C_{3v} in Exercise 15.3-1. Therefore the linear equations relating the vector $\{x\ y\ z\}$ and the symmetric *T*(2) $\{\alpha_k\}$ are

$$\begin{aligned} z &= 2^{-1/2} K_1^{A_1} (\alpha_1 + \alpha_2) + K_2^{A_1} \alpha_3 \\ y &= 2^{-1/2} K_1^{E} (\alpha_1 - \alpha_2) + K_2^{E} \alpha_4, \\ x &= K_1^{E} \alpha_6 + K_2^{E} \alpha_5. \end{aligned} \tag{17}$$

The factors $2^{-1/2}$ ensure normalization of the rows of the $\mathbb{S}$ matrix, and the subscripts 1, 2 on *K* are the values of the repeat index. Therefore, on writing $K_{15} = K_2^E$, $K_{22} = 2^{-1/2} K_1^E$, $K_{31} = 2^{-1/2} K_1^{A_1}$, $K_{33} = K_2^{A_1}$,

(17)
$$\mathbb{K} = \begin{bmatrix} 0 & 0 & 0 & 0 & K_{15} & \sqrt{2}K_{22} \\ K_{22} & -K_{22} & 0 & K_{15} & 0 & 0 \\ K_{31} & K_{31} & K_{33} & 0 & 0 & 0 \end{bmatrix}. \tag{18}$$

Group theory can tell us which elements of $\mathbb{K}$ are non-zero and about equalities between non-zero elements, but numerical factors (like $\sqrt{2}$ in the first row of $\mathbb{K}$) are simply a matter of how the K_{iq} are defined in terms of the constants $K_1^{A_1}$, $K_2^{A_1}$, K_1^E, and K_2^E, this being usually done in a way that reduces the number of numerical factors. In $LiNbO_3$ the electro-optic coefficient r_{33} is more than three times r_{13}, which gives rise to a relatively large difference in refractive index in directions along and normal to the optic (**z**) axis, thus making this material particularly useful in device applications.

15.4.1 *Thermoelectric effects*

In a crystal in which the only mobile species are electrons and there is no magnetic field present, the flux equations (15.3.2) for the transport of electrons and heat are

$$\mathbf{J}_e = -\alpha\ \nabla\bar{\mu} - (\beta/T)\ \nabla T, \tag{19}$$

$$\mathbf{J}_q = -\beta^\dagger\ \nabla\bar{\mu} - (\gamma/T)\ \nabla T, \tag{20}$$

where $\bar{\mu}$ is the electrochemical potential. To simplify the notation, α, β, and γ, which are *T*(2)s, have been used for the phenomenological coefficients L_{ee}, L_{eq}, and L_{qq}. (In this section the superscript † is used to denote the transpose even when the matrix is real, because in thermodynamics a superscript T is often used to denote constant temperature.) That the thermodynamic forces are $\nabla\bar{\mu}$ and $\nabla T/T$ follows from eq. (15.3.3) (for example Callen (1960), De Groot and Mazur (1962)). From the ORRs (eq. (15.3.4)), α and γ are symmetric $T(2)$s; β is real, but not symmetric, the ORRs being met by its transpose $\beta^{\dagger}$ appearing in eq. (20). Setting

$$\alpha^{-1}\beta/T = \Sigma, \tag{21}$$

$$-(\beta^{\dagger}\alpha^{-1}\beta - \gamma)/T = k, \tag{22}$$

(19)–(22)

$$\nabla\bar{\mu} = -\alpha^{-1}\mathbf{J}_e - \Sigma\nabla T, \tag{23}$$

$$\mathbf{J}_q = T\Sigma^{\dagger}\mathbf{J}_e - k\nabla T. \tag{24}$$

Σ is the *thermoelectric power* tensor. In a homogeneous isothermal system, $\nabla T = 0$, $\nabla\mu = \nabla\bar{\mu} + e\nabla\phi = 0$ (μ is the chemical potential), and the electrical current density $\mathbf{j}$ is given by

(23)

$$\mathbf{j} = -e\mathbf{J}_e = -\alpha e^2\nabla\phi = \alpha e^2\mathbf{E} = \sigma\mathbf{E}, \tag{25}$$

where $\mathbf{E} = -\nabla\phi$ is the electric field, $\sigma = \alpha e^2$ is the *electrical conductivity*, and $\sigma^{-1} = \rho$, the *resistivity*. Since α is a symmetric $T(2)$, so are σ and ρ. When there is no electric current, $\mathbf{J}_e = 0$ and

$$\mathbf{J}_q = -k\nabla T, \tag{26}$$

which shows that k is the *thermal conductivity*. The potential gradient produced by a temperature gradient under open circuit conditions is

(23)

$$\nabla\phi = (1/e)(\Sigma\nabla T + \nabla\mu). \tag{27}$$

If eq. (27) is integrated around a circuit from I in a metal b through metal a (with the b/a junction at T) to II also in metal b (at the same temperature as I) but with the a/b junction at $T + \Delta T$, then the potential difference $\Delta\phi = \phi^{\mathrm{II}} - \phi^{\mathrm{I}}$ (which is called the *Seebeck effect*) is

(27)

$$\Delta\phi + (1/e)(\Sigma^b - \Sigma^a)\Delta T. \tag{28}$$

Such an electrical circuit is a thermocouple, and $-\Delta\phi/\Delta T = \theta$ is the *thermoelectric power* of the thermocouple,

(28)

$$\theta = (-1/e)(\Sigma^b - \Sigma^a). \tag{29}$$

The sign convention adopted for metals (but not for ionic conductors) is that θ is positive if the hot electrode is negative, so that positive current flows from a to b at the hot junction.

Exercise 15.4-1 Why does the term $\nabla\mu$ vanish in going from eq. (27) to eq. (28)?

The thermoelectric power tensor Σ is not symmetric because $\Sigma = \alpha^{-1}\beta$, and although the ORRs require α to be symmetric, this is not true of the off-diagonal $T(2)$ β. Equation (24) shows that when there is no temperature gradient, a flow of electric current produces heat (the *Peltier effect*), the magnitude of which is determined by $\Sigma^{\dagger}$.

Table 15.8. *Symbols and names used for transport phenomena in crystals.*

The tensor components shown in columns 2–5 are obtained by expanding the tensor component in column 1 in powers of **B**, as in eqs. (15.4.34) and (15.4.35). The Hall tensor ρ_{ikl} is also commonly denoted by R_{ikl}. A dash means that there is no common name for that property.

Tensor component		Coefficient of		
		B_l	$B_l B_m$	$B_l B_m B_n$
Galvanomagnetic effects				
$\rho_{ik}(\mathbf{B})$	ρ_{ik}, electrical resistivity	ρ_{ikl}, Hall effect	ρ_{iklm}, magnetoresistance	ρ_{iklmn}, second-order Hall effect
$\Sigma^{\dagger}_{ik}(\mathbf{B})$	$\Sigma^{\dagger}_{ik}$, –	$\Sigma^{\dagger}_{ikl}$, Ettinghausen effect	$\Sigma^{\dagger}_{iklm}$, –	$\Sigma^{\dagger}_{iklmn}$, second-order Ettinghausen effect?
Thermomagnetic effects				
$k_{ik}(\mathbf{B})$	k_{ik}, thermal conductivity	k_{ikl}, Leduc–Righi effect	k_{iklm}, magnetothermal conductivity	k_{iklmn}, second-order Leduc–Righi effect
$\Sigma_{ik}(\mathbf{B})$	Σ_{ik}, thermoelectric power	Σ_{ikl}, Nernst effect	Σ_{iklm}, magneto-thermoelectric power	Σ_{iklmn}, second-order Nernst effect

15.4.2 *Galvanomagnetic and thermomagnetic effects*

Names and symbols used for galvanomagnetic and thermomagnetic effects in crystals are summarized in Table 15.8. In the presence of a magnetic field, crystal properties become functions of the magnetic induction **B**, and the ORRs, hitherto applied in the zero-field form $L_{ik} = L_{ki}$ are

$$L_{ik}(\mathbf{B}) = L_{ki}(-\mathbf{B}). \tag{15.3.4}$$

This means that the resistivity and thermal conductivity tensors are no longer symmetric. For example,

$$\rho_{ik}(\mathbf{B}) = \rho_{ki}(-\mathbf{B}). \tag{30}$$

However, any second-rank tensor can be written as the sum of symmetric and anti-symmetric parts, so

$$\rho_{ik}(\mathbf{B}) = \rho^{\mathrm{s}}_{ik}(\mathbf{B}) + \rho^{\mathrm{as}}_{ik}(\mathbf{B}), \tag{31}$$

where

$$\rho^{\mathrm{s}}_{ik}(\mathbf{B}) = \rho^{\mathrm{s}}_{ki}(\mathbf{B}),\ \ \rho^{\mathrm{as}}_{ik}(\mathbf{B}) = -\rho^{\mathrm{as}}_{ki}(\mathbf{B}), \tag{32}$$

(30)–(32) $$\rho^{\mathrm{s}}_{ik}(\mathbf{B}) + \rho^{\mathrm{as}}_{ik}(\mathbf{B}) = \rho^{\mathrm{s}}_{ki}(-\mathbf{B}) + \rho^{\mathrm{as}}_{ki}(-\mathbf{B}) = \rho^{\mathrm{s}}_{ik}(-\mathbf{B}) - \rho^{\mathrm{as}}_{ik}(-\mathbf{B}). \tag{33}$$

Equating symmetric and antisymmetric parts of the LS and RS of eq. (33) gives

$$\rho^{s}_{ik}(\mathbf{B}) = \rho^{s}_{ik}(-\mathbf{B}),\ \ \rho^{as}_{ik}(\mathbf{B}) = -\rho^{as}_{ik}(-\mathbf{B}), \tag{34}$$

(33)

which shows that ρ^{s}_{ik}, ρ^{as}_{ik} are, respectively, even and odd functions of **B**. We now expand ρ^{s}_{ik} in powers of **B**. Terms with even powers give the symmetric component, and those with odd powers of **B** provide the antisymmetric component, so that

$$\rho^{s}_{ik}(\mathbf{B}) = \rho_{ik} + \rho_{iklm}B_l B_m + \cdots \tag{35}$$

$$\rho^{as}_{ik}(\mathbf{B}) = \rho_{ikl}B_l + \rho_{iklm}B_l B_m B_n + \cdots \tag{36}$$

The thermal conductivity tensor k may likewise be split into symmetric and antisymmetric parts, with expansions in powers of **B** as in eqs. (35) and (36). But Σ is not necessarily a symmetric tensor at $\mathbf{B}=0$, and so the expansion of the antisymmetric part of Σ in an equation like eq. (36) is not applicable. Instead,

$$\Sigma_{ik}(\mathbf{B}) = \Sigma_{ik} + \Sigma_{ikl}\,B_l + \Sigma_{iklm}\,B_l B_m + \cdots \tag{37}$$

Example 15.4-4 Determine the structure of the Nernst tensor for cubic crystals.

Σ_{ik} is a measure of the ith component of the electric field produced by the kth component of the temperature gradient (eq. (27)) and Σ_{ikl} is a measure of the effect of the lth component of the magnetic induction B_l on Σ_{ik}. Therefore, Σ_{ikl} describes the coupling between a $T(2)$, Σ_{ik}, and an axial vector **B**, the components of which transform like $\{R_x\ R_y\ R_z\}$. The components $\{Y_k\}$ of a $T(2)$ transform like binary products of coordinates, that is like the nine quantities

(15.1.19) $$\{x_1\ x_2\ x_3\} \otimes \{x_1\ x_2\ x_3\}$$

$$= \{x_1{}^2\ x_2{}^2\ x_3{}^2\ 2^{-1/2}x_2x_3\ 2^{-1/2}x_3x_1\ 2^{-1/2}x_1x_2\ 2^{-1/2}x_3x_2\ 2^{-1/2}x_1x_3\ 2^{-1/2}x_2x_1\} \tag{38}$$

$$= \{x_ix_j\} = \{Y_{ij}\} = \{Y_k\},\ i,\ j = 1, 2, 3,\ k = 1, \ldots, 9, \tag{39}$$

where Y_{ij} transforms like x_ix_j and the factors of $2^{-1/2}$ ensure normalization. The set of nine components in eq. (39) may be separated into two subsets which are symmetric and antisymmetric with respect to $i \leftrightarrow j$.

(39)
$$\begin{aligned}\{Y_k\} &= \{Y_{11}\ Y_{22}\ Y_{33}\ 2^{-1/2}(Y_{23}+Y_{32})\ 2^{-1/2}(Y_{31}+Y_{13})\ 2^{-1/2}(Y_{12}+Y_{21})\}\\ &\quad \oplus \{2^{-1/2}(Y_{23}-Y_{32})\ 2^{-1/2}(Y_{31}-Y_{13})\ 2^{-1/2}(Y_{12}-Y_{21})\}\\ &= \{Y_1\ Y_2\ Y_3\ Y_4\ Y_5\ Y_6\} \oplus \{Y_7\ Y_8\ Y_9\} = \{Y^{s}\} \oplus \{Y^{as}\};\end{aligned} \tag{40}$$

$\{Y^{s}\}$ is just the symmetric $T(2)$ with basis $\{\alpha_k\}$, but the components of $\{Y^{as}\}$ transform like the axial vector $\{R_x\ R_y\ R_z\}$ (see eq. (40)). Therefore one needs to determine from character tables the IRs with bases that are components of the $T(1)^{ax}$ $\{R_x\ R_y\ R_z\}$ and the symmetric $T(2)$ $\{\alpha_k\}$. The first-order correction to Σ_{ik} in a magnetic field is

$$\Sigma^{(1)}_{ik} = \Sigma_{ikl}\,B_l, \tag{41}$$

or, in matrix notation, the supermatrix

$$\Sigma^{(1)} = \Sigma_{[ik,l]}\, B_l \tag{42}$$

where $\Sigma_{[ik,l]}$ denotes a 3×9 matrix consisting of three 3×3 blocks, each of the blocks describing the coupling to B_1, B_2, and B_3, respectively. The subscript l of the matrix element in eq. (42) tells us which component of **B** it will multiply. In the lower cubic groups T (23), T_h ($m3$), $(R_x\ \ R_y\ \ R_z)$ and $(\alpha_4\ \ \alpha_5\ \ \alpha_6)$ both form bases for a triply degenerate representation, T or T_g, respectively. Therefore, the $T(1)^{ax}$ $(X_1\ \ X_2\ \ X_3)$ – which here are $(B_1\ \ B_2\ \ B_3)$ – are coupled with both the symmetric components $(Y_4\ \ Y_5\ \ Y_6)$ and the antisymmetric components $(Y_7\ \ Y_8\ \ Y_9)$ of the $T(2)$ (which here is Σ_{ik}). Therefore, for $l = 1$,

$$Y_4 = 2^{-1/2}(Y_{23} + Y_{32}) = K_1^{\mathrm{T}} B_1, \tag{43}$$

$$Y_7 = 2^{-1/2}(Y_{23} - Y_{32}) = K_2^{\mathrm{T}} B_1, \tag{44}$$

and similarly for Y_5, Y_8, which couple with B_2, and Y_6, Y_9, which couple with B_3; $(Y_1\ Y_2\ Y_3)$ do not occur in this IR and so $Y_1 = Y_2 = Y_3 = 0$.

(43), (44) $$Y_{23} = 2^{-1/2}(K_1^{\mathrm{T}} + K_2^{\mathrm{T}})B_1, \quad Y_{32} = 2^{-1/2}(K_1^{\mathrm{T}} - K_2^{\mathrm{T}})\, B_1. \tag{45}$$

Re-writing eqs. (45) in notation appropriate to the current problem,

(45) $$\Sigma_{23}^{(1)} = \Sigma_{231}\, B_1, \quad \Sigma_{32}^{(1)} = \Sigma_{321}\, B_1. \tag{46}$$

Proceeding similarly for $l = 2$ and $l = 3$, and recalling that $K_1^{\mathrm{T}}, K_2^{\mathrm{T}}$ are independent of the degeneracy index so that the constants in eqs. (43)–(46) are independent of l, the $\Sigma_{[ik,l]}$ matrix is

$$\left[\begin{array}{ccc|ccc|ccc} 0 & 0 & 0 & 0 & 0 & 321 & 0 & 231 & 0 \\ 0 & 0 & 231 & 0 & 0 & 0 & 321 & 0 & 0 \\ 0 & 321 & 0 & 231 & 0 & 0 & 0 & 0 & 0 \end{array}\right], \tag{47}$$
$$\qquad l = 1 \qquad\qquad l = 2 \qquad\qquad l = 3$$

in which Σ is omitted, so that only the subscripts ikl are given; $l = 1, 2,$ and 3 mark the three 3×3 blocks. In the upper cubic groups T_d, O, O_h ($\bar{4}3m$, 432, $m3m$), no components of $\{Y^s\}$ share a representation with $T(1)^{ax}$ which forms a basis for T_1 or T_{1g}. Therefore $\{Y_1\ \ldots\ Y_6\}$ are zero and $T(1)^{ax}$ couples only with $\{Y^{as}\}$,

$$Y_4 = 2^{-1/2}(Y_{23} + Y_{32}) = 0, \tag{48}$$

$$Y_7 = 2^{-1/2}(Y_{23} - Y_{32}) = K^{\mathrm{T}} B_1, \tag{49}$$

(48), (49) $$Y_{23} = 2^{-1/2} K^{\mathrm{T}} B_1, \quad Y_{32} = -Y_{23}, \tag{50}$$

with similar results for $l = 2, 3$. In notation appropriate to the Nernst tensor,

(50) $$\Sigma_{23}^{1} = \Sigma_{231}\, B_1, \quad \Sigma_{32}^{(1)} = -\Sigma_{231}\, B_1, \tag{51}$$

and similarly, so that $\Sigma_{[ik,l]}$ takes the form

$$\begin{bmatrix} 0 & 0 & 0 & 0 & 0 & -231 & 0 & 231 & 0 \\ 0 & 0 & 231 & 0 & 0 & 0 & -231 & 0 & 0 \\ 0 & -231 & 0 & 231 & 0 & 0 & 0 & 0 & 0 \end{bmatrix}. \quad (52)$$
$$l = 1 \qquad\qquad l = 2 \qquad\qquad l = 3$$

For other point groups this analysis of the symmetry properties of a $T(3)^{ax}$ can be repeated, or alternatively tables given by Bhagavantam (1966) or Nowick (1995) may be consulted. The Hall tensor ρ_{ikl} (and likewise the Leduc–Righi tensor k_{ikl}) is also a $T(3)^{ax}$ tensor but differs from the Nernst tensor in that ρ_{ik} is symmetric and $\rho_{ikl}\, B_l = \rho_{kil}(-B_l)$ so that the blocks $\rho_{[ik,l]}$ are antisymmetric with respect to $i \leftrightarrow k$. This follows from the ORRs and is true in all point group symmetries. For cubic crystals and $l = 1$,

$$Y_7 = 2^{-1/2}(Y_{23} - Y_{32}) = K^T B_1. \quad (53)$$

But here $Y_{23} = -Y_{32}$ and so

$$Y_{23} = 2^{-1/2} K^T B_1 \text{ or } \rho_{23}^{(1)} = \rho_{231} B_1, \quad (54)$$

with $2^{1/2} K^T B_1 = \rho_{231}$. Similarly, for $l = 2$,

$$Y_8 = 2^{-1/2}(Y_{31} - Y_{13}) = K^T B_2. \quad (55)$$

On setting $Y_{31} = -Y_{13}$, because of the antisymmetry of $\rho_{ik}^{(1)}$

(55), (54) $$-Y_{13} = 2^{1/2} K^T B_2 \text{ or } \rho_{13}^{(1)} = -\rho_{132} B_2 = -2^{1/2} K^T B_2 = -\rho_{231} B_2, \quad (56)$$

and similarly for $Y_9 = K^T B_3$. Therefore the $T(3)^{ax}$ Hall tensor is

$$\begin{bmatrix} 0 & 0 & 0 & 0 & 0 & -231 & 0 & 231 & 0 \\ 0 & 0 & 231 & 0 & 0 & 0 & -231 & 0 & 0 \\ 0 & -231 & 0 & 231 & 0 & 0 & 0 & 0 & 0 \end{bmatrix} \quad (57)$$
$$B_1 \qquad\qquad B_2 \qquad\qquad B_3$$

The subscripts and signs of $\rho_{[ik,l]}$ may vary in some published tables, but such variations are purely conventional. Group theory gives us the *structure* of the MR of the tensor, that is it tells us which coefficients are zero and gives equalities between and relative signs of non-zero coefficients. In the lower cubic groups T and T_h, $T(1)^{ax}$ is not prevented by point group symmetry from coupling with $\{Y_4\ Y_5\ Y_6\}$, but the coupling coefficients have to be zero in order to satisfy the ORRs.

Answer to Exercise 15.4-1

I and II are points in two phases of identical composition at the same temperature. Therefore $\mu^a(\text{I}) = \mu^a(\text{II})$.

15.5 Properties of crystals with magnetic point groups

Two kinds of crystal properties have been considered in this chapter, namely properties of crystals in equilibrium and transport properties. The latter are associated with thermodynamically irreversible processes and are accompanied by an increase in entropy, $\Delta S > 0$. Such processes occur naturally, or spontaneously. Time reversal is not a permitted symmetry operation in systems undergoing irreversible processes because the operation $t \rightarrow -t$ would require the spontaneous process to be reversed and so contravene the second law of thermodynamics. Consequently, time reversal is limited to crystals in equilibrium. Most physical properties are unaffected by time reversal but because $\mathbf{B}(-t) = -\mathbf{B}(t)$ numerical values of some properties may be reversed in sign. Consequently Neumann's principle must be extended to include time-reversal symmetry as well as spatial symmetry. Tensors which change sign under Θ are called *c-tensors* and those which are invariant under Θ are called *i-tensors*. For crystals which belong to the type II "gray" groups defined by $\underline{\mathrm{G}} = \mathrm{H} + \Theta\mathrm{H}$, if both Y and X are either both symmetric or both antisymmetric under Θ, K is an i-tensor and its structure is the same as that obtained from the subgroup $\mathrm{H} = \{R\}$. But if only one of Y and X is an i-tensor then K is a c-tensor and

(15.3.11)
$$\mathrm{K} = \Gamma^{(Y)}(R)\ \mathrm{K}\ \Gamma^{(X)}(R)^{\dagger}, \quad R \in \mathbf{H}; \qquad (1)$$
$$\mathrm{K} = -\Gamma^{(Y)}(B)\ \mathrm{K}\ \Gamma^{(X)}(B)^{\dagger}, \quad B = \Theta R,$$

which means that when K is a c-tensor, it must be identically zero.

For magnetic crystals belonging to the type III (black and white) point groups

$$\underline{\mathrm{G}} = \mathrm{H} + \Theta(\mathrm{G} - \mathrm{H}) = \{R\} + A\{R\},\ A = \Theta Q,\ Q \in \mathrm{G} - \mathrm{H}. \qquad (2)$$

We may assume that the IRs of

$$\mathrm{G} = \{R\} + Q\{R\} \qquad (3)$$

are known (Appendix A3). Let Γ be one of these IRs and define the *complementary* IR by

$$\Gamma_{\mathrm{c}}(R) = \Gamma(R),\ \Gamma_{\mathrm{c}}(QR) = -\Gamma(QR),\ \forall R \in \mathrm{H},\ \forall QR \in \mathrm{G} - \mathrm{H}. \qquad (4)$$

Then the complementary matrices obey eq. (1) for the group $\underline{\mathrm{G}}$ and so can be used to determine the structure of K. The structure of K for magnetic crystals can thus be found from the representations of G by substituting the complementary representation Γ^j_{c} for Γ^j for c-tensors, and thus avoiding the necessity of actually determining the IRs of $\underline{\mathrm{G}}$. Most commonly, the c-tensor will be the magnetic induction $\mathbf{B}$ or the magnetic field $\mathbf{H}$, both of which are axial vectors transforming like $\{R_x\ R_y\ R_z\}$. Of course, proper orientation has to be determined, as explained in Section 15.3, or reference must be made to the tables given by Nowick (1995) for the thirty-two type I and fifty-eight type III magnetic point groups.

15.5.1 *Ferromagnetism*

A ferromagnetic material is one that possesses a magnetic moment $\mathbf{M}$ in the absence of a magnetic field. The magnetic induction is given by

Table 15.9. *The thirty-one magnetic point groups in which ferromagnetism is possible.*

Type I			Type III						
1	$\bar{1}$								
2	m	$2/m$	$\underline{2}$	$\underline{m}$	$\underline{2}/\underline{m}$	$2\underline{22}$	$\underline{mm}2$	$\underline{m}m\underline{2}$	$m\underline{mm}$
3	$\bar{3}$		$3\underline{2}$	$3\underline{m}$	$\bar{3}\underline{m}$				
4	$\bar{4}$	$4/m$	$4\underline{22}$	$4\underline{mm}$	$\bar{4}\underline{2m}$	$4/m\underline{mm}$			
6	$\bar{6}$	$6/m$	$6\underline{22}$	$6\underline{mm}$	$\bar{6}\underline{m2}$	$6/m\underline{mm}$			

$$\mathbf{B} = \mu_0\mathbf{H} + \mathbf{I}, \tag{5}$$

where the magnetic intensity **I** is the magnetic moment per unit volume. Since **B** and **H** are axial vectors, so is **I**. Therefore the magnetic point groups in which ferromagnetism is possible are those in which at least one of the components $\{R_x\ R_y\ R_z\}$ belongs to the totally symmetric representation. A systematic examination of the thirty-two type I and fifty-eight type III magnetic point groups reveals that this is so only for the thirty-one magnetic point groups listed in Table 15.9. Since ΔT and P are invariant under Θ for crystals in thermodynamic equilibrium, these magnetic point groups are also those which exhibit the magnetocaloric effect and the pyromagnetic effect and a particular case of the piezomagnetic effect in which the applied stress is a uniform pressure.

15.5.2 *Magnetoelectric polarizability*

The magnetoelectric polarizability $[\lambda_{ij}]$ of a crystalline material gives rise to a magnetic moment **I** (Y an axial vector, or $T(1)^{ax}$) when the crystal is placed in an electric field **E** (X a polar vector, or $T(1)$). Its transpose $[\lambda_{ji}]$ describes the converse magnetoelectric effect in which the roles of Y and X are interchanged. To find the structure of $\mathrm{K} = [\lambda_{ij}]$, we look for IRs which have components of both $\{x\ y\ z\}$ and $\{R_x\ R_y\ R_z\}$ as bases.

Example 15.5-1 Determine the structure of the matrix $[\lambda_{ij}]$ for the type I magnetic point group $4mm$ (C_{4v}) and the type III group $4\underline{mm}$ with $G = C_{4v}$ and $H = C_{2v}$.

From the character table of $4mm$ (C_{4v}), $\Gamma^{(r)} = A_1 \oplus E$, $\Gamma^{(\mathbf{R})} = A_2 \oplus E$. Orientation for the E basis, already determined in Section 15.3, is $(x\ y)$ and $(R_y\ {-R_x})$. Therefore

$$\lambda = \begin{bmatrix} 0 & \lambda_{12} & 0 \\ -\lambda_{12} & 0 & 0 \\ 0 & 0 & 0 \end{bmatrix}. \tag{6}$$

For $4\underline{mm}$, $H = \{E\ C_{2z}\ \sigma_{\mathbf{x}}\ \sigma_{\mathbf{y}}\}$ and $G - H = \{C_{4z}^{+}\ C_{4z}^{-}\ \sigma_{\mathbf{a}}\ \sigma_{\mathbf{b}}\}$, where $\sigma_{\mathbf{a}}$, $\sigma_{\mathbf{b}}$ bisect the angles between **x** and **y**. In $\underline{G} = \{H\} + \Theta\{G - H\}$ the complementary representations are obtained by replacing χ by $-\chi$ in the classes $\{C_{4z}^{+}\ C_{4z}^{-}\}$ and $\{\sigma_{\mathbf{a}}\ \sigma_{\mathbf{b}}\}$, which gives $\Gamma^{(\mathbf{R})} = B_2 \oplus E$ ($\Gamma^{(\mathbf{r})}$ is unaffected by Θ). Orientation must be re-determined. In $4mm$, under C_{4z}, $R_x \to R_y$, $R_y \to -R_x$ (Table 15.3); therefore in $4\underline{mm}$ under ΘC_{4z}^{+}, $R_x \to -R_y$, and $R_y \to R_x$. Consequently, in $4\underline{mm}$, the oriented E-bases are $(x\ y)$ and $(R_y\ R_x)$,

$$\lambda = \begin{bmatrix} 0 & \lambda_{12} & 0 \\ \lambda_{12} & 0 & 0 \\ 0 & 0 & 0 \end{bmatrix}. \tag{7}$$

The structure of λ (and therefore of any property represented by a $T(2)^{ax}$ tensor) for the fifty-eight magnetic groups in which λ is not identically zero, is given by Nowick (1995), p. 138.

15.5.3 Piezomagnetic effect

The coupling of the magnetic induction **B** (Y, a $T(1)^{ax}$) with an applied stress (X, a $T(2)^{s}$) gives rise to the piezomagnetic effect Q_{ijk}, which is a $T(3)^{ax}$ tensor, symmetric with respect to $j \leftrightarrow k$. Its converse Q_{kij} describes the coupling of the elastic strain with the magnetic field (Table 15.7). Using the single index notation for elastic stress or strain, Q is a 3 × 6 matrix, like that for the piezoelectric effect. The structure of Q is determined by finding representations common to $\Gamma_c^{(\mathbf{R})}$ and $\Gamma^{(\alpha)}$.

Example 15.5-2 Determine the structure of Q for the magnetic point groups $4mm$ and $4\underline{mm}$.

In $4mm$ (or C_{4v}) $\Gamma_c^{(\mathbf{R})} = A_2 \oplus E$ and $\Gamma^{(\alpha)} = A_1 \oplus B_1 \oplus B_2 \oplus E$, the orientation of components being such that R_x is coupled with α_4 and $-R_y$ with α_5. Therefore

$$Q = \begin{bmatrix} 0 & 0 & 0 & Q_{14} & 0 & 0 \\ 0 & 0 & 0 & 0 & -Q_{14} & 0 \\ 0 & 0 & 0 & 0 & 0 & 0 \end{bmatrix}. \tag{8}$$

In $4\underline{mm}$, $G - H = 2C_4 \oplus 2\sigma_d$, so that $\Gamma_c^{(\mathbf{R})} = B_2 \oplus E$. In $\Gamma_c^{(\mathbf{R})}$, R_z is a basis for $\mathbf{B}_2$ and is therefore coupled with α_6. In the degenerate representation, $(R_y \;\; R_x)$ couples with $(\alpha_5 \; \alpha_4)$. Consequently,

$$Q = \begin{bmatrix} 0 & 0 & 0 & Q_{14} & 0 & 0 \\ 0 & 0 & 0 & 0 & Q_{14} & 0 \\ 0 & 0 & 0 & 0 & 0 & Q_{36} \end{bmatrix}. \tag{9}$$

The piezomagnetic tensor Q for all magnetic point groups is given by Bhagavantam (1966), p. 173, and Nowick (1995) in his Table 8–3.

Problems

15.1 Determine the form of the matrix c of second-order elastic constants for crystals with the point group O.

15.2 Find the similarly oriented bases (symmetry coordinates) for the components of a $T(1)$, a $T(1)^{ax}$, and a symmetric $T(2)$, and a symmetric $T(3)$ for the point group D_{2d}.

15.3 In a piezoelectric crystal an applied stress σ_{jk} produces an electric polarization **P** so that $P_i = d_{ijk}\sigma_{jk}$. Prove that a crystal with a center of symmetry cannot exhibit the phenomenon of piezoelectricity.

15.4 Determine the point groups in which ferroelectricity is possible. [*Hint*: Check point-group character tables to see in which point groups at least one of x, y, or z form a basis for Γ_1.]

15.5 Find the structure of $\mathbb{K}$ when there is one complex Y coordinate $Y_1 \mp \mathrm{i}Y_2$ for $^1\mathrm{E}$, $^2\mathrm{E}$ but two X coordinates $X_1 \mp \mathrm{i}X_2$, $X_3 \mp \mathrm{i}X_4$. What changes arise when there are two Y coordinates but only one X coordinate for a degenerate pair of CC representations?

15.6 Obtain the permittivity matrix κ for orthorhombic and monoclinic crystals. What differences arise in the monoclinic case for a $T(2)$ that is not symmetric?

15.7 Obtain the MRs of the elastic constant tensor for an upper hexagonal crystal and a lower tetragonal crystal.

15.8 Find the 9×6 MR of the magnetothermoelectric power Σ_{ijkl} for a crystal of orthorhombic symmetry. Express your results in four-index notation, giving indices only.

16 Space groups

Crystals ... their essential virtues are but two; – the first is to be pure and the second is to be well shaped.

John Ruskin *The Ethics of the Dust* (1865)

16.1 Translational symmetry

A *crystal structure* is the spatial arrangement of all the atoms (ions, molecules) of which the crystal is composed. It is represented by the *crystal pattern*, which is a minimal set of points having the same symmetry as the crystal structure. These points are commonly shown in diagrams by small circles, although more elaborate figures (ornaments) may be used to emphasize particular symmetry elements. The essential characteristic of a crystal structure (and therefore of the crystal pattern) is its *translational symmetry*, which is described in terms of a space lattice (Figure 16.1). A space lattice (or lattice) is an array of points in space which, for clarity of representation, are joined by straight lines (Figure 16.1(a)). "Translational symmetry" means that the environment of a particular lattice point O is indistinguishable from that of any other lattice point reached from O by the lattice translation vector

$$\mathbf{a}_n = n_1\mathbf{a}_1 + n_2\mathbf{a}_2 + n_3\mathbf{a}_3 = \langle \mathbf{a}_1\ \mathbf{a}_2\ \mathbf{a}_3 | n_1\ n_2\ n_3 \rangle = \langle \mathbf{a} \,|\, n \rangle. \tag{1}$$

The integers n_1, n_2, and n_3 are the components of $\mathbf{a}_n$, and $\mathbf{a}_1$, $\mathbf{a}_2$, and $\mathbf{a}_3$ are the *fundamental lattice translation vectors*, which we shall often abbreviate to fundamental translation vectors or fundamental translations (see Bradley and Cracknell (1972)). There seems to be no generally agreed name for the $\{\mathbf{a}_i\}$ which have also been called, for example, the "primitive translation vectors of the lattice" (McWeeny (1963)), "primitive vectors" (Altmann 1977), "unit vectors" (Altmann (1991)), "basis translation vectors" (Evarestov and Smirnov (1997)) and "basis vectors" (Kim (1999)). In a particular lattice the location of a lattice point P is specified uniquely by the components of the position vector OP which are the coordinates of P(n_1 n_2 n_3) referred to the fundamental lattice translation vectors $\langle\, \mathbf{a}_1\, \mathbf{a}_2\, \mathbf{a}_3 \,|$. The parallelepiped defined by $\{\mathbf{a}_1\, \mathbf{a}_2\, \mathbf{a}_3\}$ is the *unit cell* of the lattice and it follows from eq. (1) that it is a space-filling polyhedron. Each fundamental translation vector may be resolved into its components along the Cartesian coordinate axes OX, OY, and OZ,

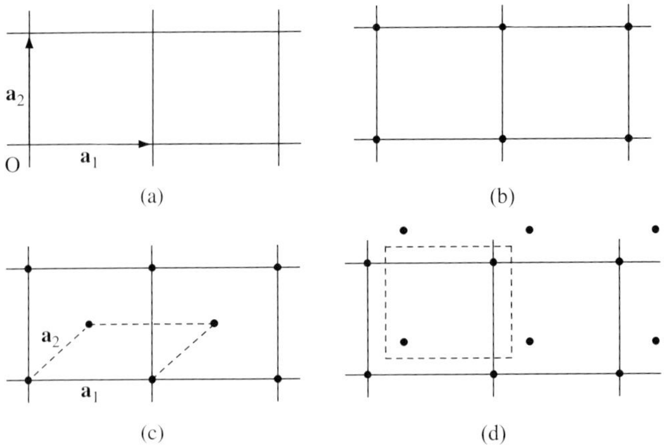

Figure 16.1. (a) Portion of a 2-D (space) lattice. (b) Lattice with a basis, primitive unit cell, $\kappa = [00]$. (c) Non-primitive unit cell, $\kappa_1 = [0\ \ 0]$, $\kappa_2 = [\frac{1}{2}\ \ \frac{1}{2}]$. All lattice points are translationally equivalent, as shown by the alternative choice of fundamental translations $\mathbf{a}_1$, $\mathbf{a}_2$ which give a primitive unit cell. (d) Non-primitive unit cell: no possible choice of fundamental translations exists which would make all pattern points translationally equivalent. Two possible unit cells are shown, which differ only in the choice of origin. Each unit cell of this lattice contains two pattern points, $s = 2$.

$$\mathbf{a}_i = \langle \mathbf{e}_1\ \mathbf{e}_2\ \mathbf{e}_3 \,|\, a_{ix}\ a_{iy}\ a_{iz} \rangle, \quad i = 1, 2, 3; \tag{2}$$

(1), (2)

$$\begin{aligned} \mathbf{a}_n &= \langle \mathbf{e}_1\ \mathbf{e}_2\ \mathbf{e}_3 | \begin{bmatrix} a_{1x} & a_{2x} & a_{3x} \\ a_{1y} & a_{2y} & a_{3y} \\ a_{1z} & a_{2z} & a_{3z} \end{bmatrix} \begin{bmatrix} n_1 \\ n_2 \\ n_3 \end{bmatrix} \\ &= \langle \mathbf{e}_1\ \mathbf{e}_2\ \mathbf{e}_3 | \mathbb{A} | n_1\ n_2\ n_3 \rangle. \end{aligned} \tag{3}$$

If every pattern point can be reached from O by a translation $\mathbf{a}_n$, the lattice is said to be *primitive*, in which case the unit cell contains just one pattern point and is also described as "primitive" (Figure 16.1). The location of a pattern point in a unit cell is specified by the vector

$$\mathbf{a}_\kappa = \kappa_1 \mathbf{a}_1 + \kappa_2 \mathbf{a}_2 + \kappa_3 \mathbf{a}_3 = \langle \mathbf{a}_1\ \mathbf{a}_2\ \mathbf{a}_3 \,|\, \kappa_1\ \kappa_2\ \kappa_3 \rangle = \langle \mathbf{a} \,|\, \kappa \rangle, \tag{4}$$

with κ_1, κ_2, $\kappa_3 < 1$. Every pattern point in a non-primitive cell is connected to O by the vector

$$\mathbf{a}_{n\kappa} = \mathbf{a}_n + \mathbf{a}_\kappa. \tag{5}$$

The pattern points associated with a particular lattice are referred to as the *basis* so that the description of a crystal pattern requires the specification of the space lattice by $\{\mathbf{a}_1\ \mathbf{a}_2\ \mathbf{a}_3\}$ and the specification of the basis by giving the location of the pattern points in one unit cell by κ_i, $i = 1, 2, \ldots, s$ (Figure 16.1(b), (c)). The choice of the fundamental translations is a matter of convenience. For example, in a face-centred cubic (*fcc*) lattice we could choose orthogonal fundamental translation vectors along OX, OY, OZ, in which case the unit cell contains $(^1/_8)8 + (^1/_2)6 = 4$ lattice points (Figure 16.2(a)). Alternatively, we might choose a primitive unit cell with the fundamental translations

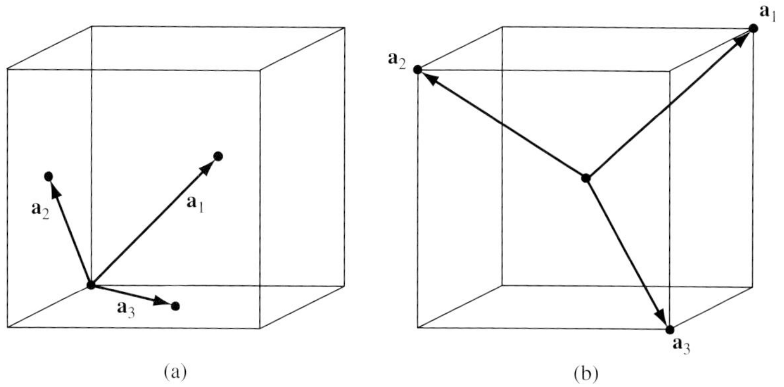

Figure 16.2. Conventional (non-primitive) unit cells of (a) the face-centered cubic and (b) the body-centered cubic lattices, showing the fundamental vectors $\mathbf{a}_1$, $\mathbf{a}_2$, and $\mathbf{a}_3$ of the primitive unit cells. (A *conventional* unit cell is one that displays the macroscopic symmetry of the crystal.)

$$\begin{aligned}\mathbf{a}_1 &= (a/2)\langle \mathbf{e}_1\ \mathbf{e}_2\ \mathbf{e}_3 \,|\, 0\ 1\ 1\rangle,\\ \mathbf{a}_2 &= (a/2)\langle \mathbf{e}_1\ \mathbf{e}_2\ \mathbf{e}_3 \,|\, 1\ 0\ 1\rangle,\\ \mathbf{a}_3 &= (a/2)\langle \mathbf{e}_1\ \mathbf{e}_2\ \mathbf{e}_3 \,|\, 1\ 1\ 0\rangle;\end{aligned} \tag{6}$$

(3), (6)

$$\mathbb{A} = (a/2)\begin{bmatrix} 0 & 1 & 1\\ 1 & 0 & 1\\ 1 & 1 & 0\end{bmatrix}. \tag{7}$$

Similarly, for the body-centered cubic (*bcc*) lattice one might choose an orthogonal set for the fundamental translations giving a non-primitive unit cell with two lattice points per cell (Figure 16.2(b)) or one could choose a primitive unit cell with the fundamental translations

$$\begin{aligned}\mathbf{a}_1 &= (a/2)\langle \mathbf{e}_1\ \mathbf{e}_2\ \mathbf{e}_3 \,|\, \bar{1}\ 1\ 1\rangle,\\ \mathbf{a}_2 &= (a/2)\langle \mathbf{e}_1\ \mathbf{e}_2\ \mathbf{e}_3 \,|\, 1\ \bar{1}\ 1\rangle,\\ \mathbf{a}_3 &= (a/2)\langle \mathbf{e}_1\ \mathbf{e}_2\ \mathbf{e}_3 \,|\, 1\ 1\ \bar{1}\rangle;\end{aligned} \tag{8}$$

(3), (8)

$$\mathbb{A} = (a/2)\begin{bmatrix} \bar{1} & 1 & 1\\ 1 & \bar{1} & 1\\ 1 & 1 & \bar{1}\end{bmatrix}. \tag{9}$$

A primitive centered unit cell, called the Wigner–Seitz cell, is particularly useful. To construct the Wigner–Seitz cell, draw straight lines from a chosen lattice point to all its near neighbors and bisect these lines perpendicularly by planes: then the smallest polyhedron enclosed by these planes is the Wigner–Seitz cell (Figure 16.3). A lattice direction is specified by its *indices* $[w_1\ w_2\ w_3]$, that is, the smallest set of integers in the same ratio as the components of a vector in that direction; $[[w_1\ w_2\ w_3]]$ denotes a set of equivalent directions. The orientation of a lattice plane is specified by its Miller indices $(h_1\ h_2\ h_3)$, which are the smallest set of integers in the same ratio as the reciprocals of the intercepts made by the plane on the vectors $\mathbf{a}_1$, $\mathbf{a}_2$, $\mathbf{a}_3$, in units of a_1, a_2, a_3; $((h_1\ h_2\ h_3))$ denotes a set of equivalent planes.

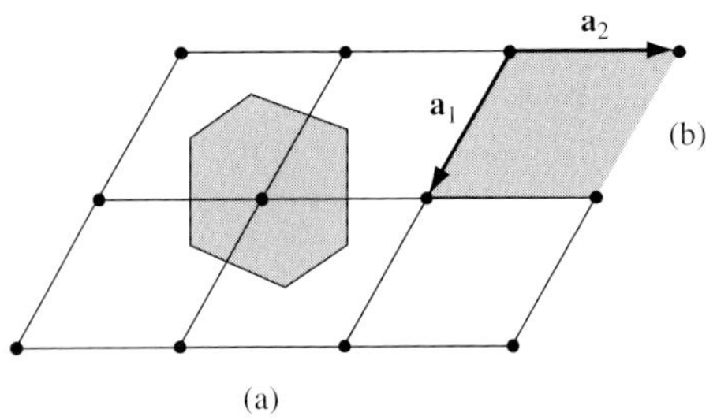

Figure 16.3. (a) Construction of the Wigner–Seitz cell in a 2-D hexagonal close-packed (*hcp*) lattice. (b) Primitive unit cell of the *hcp* lattice.

A crystal pattern may possess rotational symmetry as well as translational symmetry, although the existence of translational symmetry imposes restrictions on the order of the axes. The fundamental translations $\langle \mathbf{a} \mid$ in eq. (1) are the basis vectors of a linear vector space (LVS). Suppose that they are transformed into a new set $\langle \mathbf{a}' \vert$ by a unitary *transformation*

$$\langle \mathbf{a}' \vert = \langle \mathbf{a} \mid \mathrm{T}; \tag{10}$$

(10) $$R\langle \mathbf{a}' \mid = \langle \mathbf{a}' \mid \mathrm{R}' = R\langle \mathbf{a} \mid \mathrm{T} = \langle \mathbf{a} \vert\ \mathrm{RT} = \langle \mathbf{a}' \mid \mathrm{T}^{-1}\mathrm{RT}. \tag{11}$$

R is a unitary matrix, and any unitary matrix can be diagonalized by a unitary transformation,

(10), (11) $$\mathrm{R}' = \mathrm{T}^{-1}\mathrm{RT} = \mathrm{T}^{\dagger}\mathrm{RT} \quad (\mathrm{T}\ \text{unitary}). \tag{12}$$

But the trace of R is invariant under a similarity transformation and therefore

$$\mathrm{Tr}\ \mathrm{R} = \pm 1 + 2\cos\phi \tag{13}$$

whatever the choice of $\langle \mathbf{a} \vert$. A symmetry operator R transforms $\mathbf{a}_n$ into the lattice translation vector $\mathbf{a}_n{}'$, where both $\mid n\rangle$ and $\mid n'\rangle$ contain integers only.

(1) $$R\langle \mathbf{a} \vert n\rangle = \langle \mathbf{a}' \vert n\rangle = \langle \mathbf{a} \vert \mathrm{R} \vert n\rangle = \langle \mathbf{a} \vert n'\rangle. \tag{14}$$

Since $\mid n\rangle$ and $\mid n'\rangle$ consist of integers only, the diagonal form of R can consist only of integers and so

(13) $$\mathrm{Tr}\ \mathrm{R} = \pm 1 + 2\cos\phi = p, \tag{15}$$

where p is an integer.

(15) $$2\cos\phi = p \mp 1 = 0,\ \pm 1,\ \pm 2. \tag{16}$$

The values of $2\pi/\phi = n$ (where n is the order of the axis of rotation) that satisfy eq. (16) and therefore are compatible with translational symmetry, are shown in Table 16.1. It follows that the point groups compatible with translational symmetry are limited to the twenty-seven axial groups with $n = 1, 2, 3, 4$, or 6 and the five cubic groups, giving thirty-two *crystallographic point groups* (Table 2.9).

Table 16.1. *The orders of the axes of rotation, from eq. (16.1.16), that are compatible with translational symmetry.*

$\cos\phi$	0	$+\frac{1}{2}$	$-\frac{1}{2}$	$+1$	-1
ϕ	$\pi/2$	$\pi/3$	$2\pi/3$	2π	π
$n=2\pi/\phi$	4	6	3	1	2

It is not always possible to choose a unit cell which makes every pattern point translationally equivalent, that is, accessible from O by a translation $\mathbf{a}_n$. The maximum set of translationally equivalent points constitutes the *Bravais lattice* of the crystal. For example, the cubic unit cells shown in Figure 16.2 are the repeat units of Bravais lattices. Because n_1, n_2, and n_3 are integers, the inversion operator simply exchanges lattice points, and the Bravais lattice appears the same after inversion as it did before. Hence every Bravais lattice has inversion symmetry. The metric $\mathrm{M}=[\mathbf{a}_i \cdot \mathbf{a}_j]$ is invariant under the congruent transformation

$$\mathrm{M} = \mathrm{R}^{\dagger}\mathrm{M}\mathrm{R}, \tag{17}$$

where R is the matrix representative (MR) of the symmetry operator R. The invariance condition in eq. (17) for the metric imposes restrictions on both M and R, which determine the Bravais lattice (from M) and the crystallographic point groups ("crystal classes") from the group generators $\{R\}$. The results of a systematic enumeration of the Bravais lattices and the assignment of the crystal classes to the crystal systems (see, for example, Burns and Glazer (1963) and McWeeny (1978)) are summarized in Table 16.2. Unit cells are shown in Figure 16.4. Their derivation from eq. (17) is straightforward and so only one example will be provided here.

Example 16.1-1 Find the Bravais lattices, crystal systems, and crystallographic point groups that are consistent with a C_{3z} axis normal to a planar hexagonal net.

As Figure 16.5 shows, **z** is also a C_6 axis. From Figure 16.5, the *hexagonal crystal system* is defined by

$$\mathrm{a}_1 = \mathrm{a}_2 \neq \mathrm{a}_3, \quad \alpha_{12} = 2\pi/3, \quad \alpha_{23} = \alpha_{31} = \pi/2, \tag{18}$$

where α_{ij} is the angle between $\mathbf{a}_i$ and $\mathbf{a}_j$. Consequently,

$$\mathrm{M} = [\mathbf{a}_i \cdot \mathbf{a}_j] = a^2 \begin{bmatrix} 1 & -\frac{1}{2} & 0 \\ -\frac{1}{2} & 1 & 0 \\ 0 & 0 & c^2 \end{bmatrix}, \tag{19}$$

where $c = \mathrm{a}_3/a$. From Figure 16.5,

$$R\left(\frac{2\pi}{3}\ [0\ 0\ 1]\right)\langle \mathbf{a}_1\ \mathbf{a}_2\ \mathbf{a}_3| = \langle \mathbf{a}_2\ \mathbf{a}_4\ \mathbf{a}_3| = \langle \mathbf{a}_1\ \mathbf{a}_2\ \mathbf{a}_3| \begin{bmatrix} 0 & -1 & 0 \\ 1 & -1 & 0 \\ 0 & 0 & 1 \end{bmatrix}; \tag{20}$$

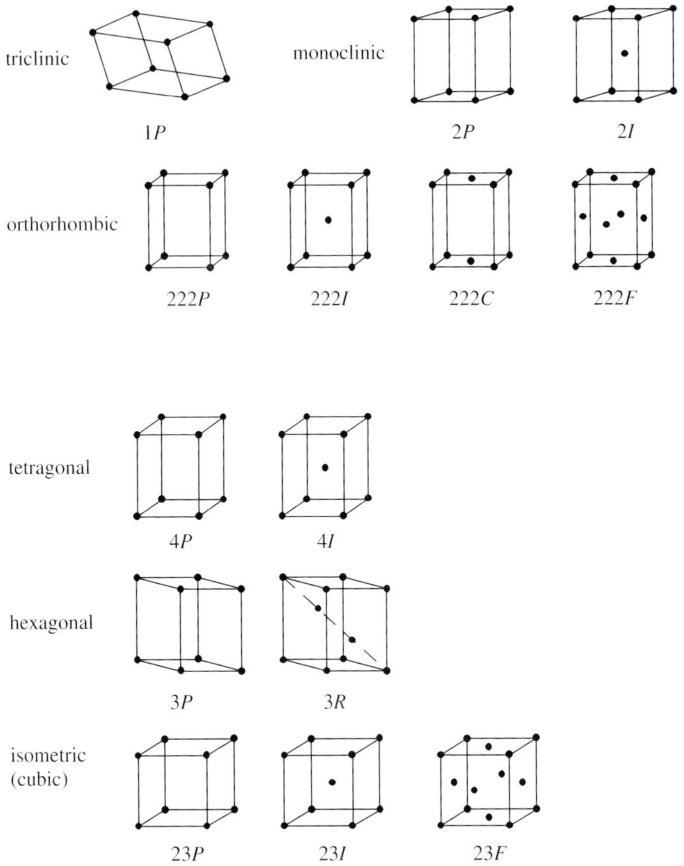

Figure 16.4. Unit cells of the fourteen space lattices classified into six crystal systems.

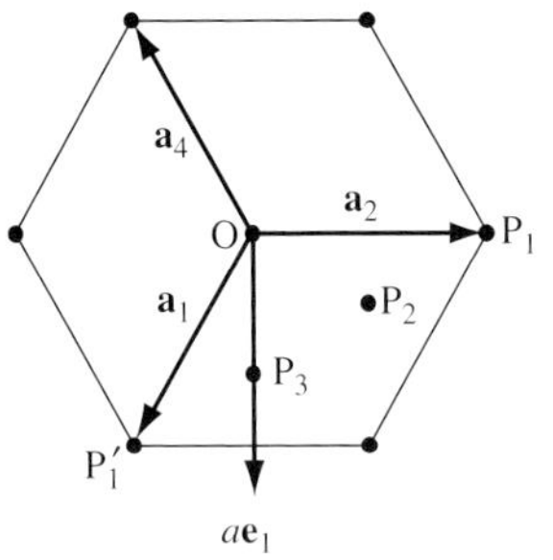

Figure 16.5. A hexagonal planar net is generated by the fundamental translations $\mathbf{a}_1$, $\mathbf{a}_2$ (each of length a) and $\alpha_{12} = 2\pi/3$. To generate a space lattice with three-fold rotational symmetry, the second and third layers must be translated so that P_1 lies over the points marked P_2 and P_3, respectively, that is at (1/3 2/3 1/3) and (2/3 1/3 2/3). If using hexagonal coordinates $\mathbf{a}_3$ is normal to the plane of $\mathbf{a}_1$, $\mathbf{a}_2$ and lies along $\mathbf{e}_3$, so that this unit cell (3R) contains three lattice points (Figure 16.4).

Table 16.2. *Crystal systems, space lattices, and crystallographic point groups.*

Unit cells are shown in Figure 16.4.

Crystal system	Unit cell	Lattice	Fundamental translations	Point groups
Triclinic	$1P$	$1P$	$a_1 \neq a_2 \neq a_3$ $\alpha_{12} \neq \alpha_{23} \neq \alpha_{31}$	$1, \bar{1}$
Monoclinic	$2P$	$2P$, $2I$	$a_1 \neq a_2 \neq a_3$ $\alpha_{31} \neq \alpha_{12} = \alpha_{23} = \pi/2^a$ $\alpha_{23} \neq \alpha_{12} = \alpha_{31} = \pi/2^b$	2, m, $2/m$
Orthorhombic	$222P$	$222P$, $222I$ $222C$, $222F$	$a_1 \neq a_2 \neq a_3$ $\alpha_{12} = \alpha_{23} = \alpha_{31} = \pi/2$	222, $2mm$ mmm
Tetragonal	$4P$	$4P$, $4I$	$a_1 = a_2 \neq a_3$ $\alpha_{12} = \alpha_{23} = \alpha_{31} = \pi/2$	4, $\bar{4}$, $4/m$, 422 $4mm$, $\bar{4}\,2m$, $4/mmm$
Hexagonal	$3P$	$3P$, $3R$	$a_1 = a_2 \neq a_3$ $\alpha_{12} = 2\pi/3$, $\alpha_{23} = \alpha_{31} = \pi/2$	3, $\bar{3}$, 32 $3m$, $\bar{3}m$
	$3P$	$3P$	$a_1 = a_2 \neq a_3$ $\alpha_{12} = 2\pi/3$, $\alpha_{23} = \alpha_{31} = \pi/2$	6, $\bar{6}$, $6/m$, 622, $6mm$, $\bar{6}m2$, $6/mmm$
Cubic	$23P$	$23P$, $23I$ $23F$	$a_1 = a_2 = a_3$ $\alpha_{12} = \alpha_{23} = \alpha_{31} = \pi/2$	23, $m3$, 432, $\bar{4}3m$, $m3m$

[a] First setting; [b] second setting.

$$(20)\quad \mathrm{R}^\dagger\,\mathrm{M}\,\mathrm{R} = a^2\begin{bmatrix} 0 & 1 & 0 \\ -1 & -1 & 0 \\ 0 & 0 & 1 \end{bmatrix}\begin{bmatrix} 1 & -\tfrac{1}{2} & 0 \\ -\tfrac{1}{2} & 1 & 0 \\ 0 & 0 & c^2 \end{bmatrix}\begin{bmatrix} 0 & -1 & 0 \\ 1 & -1 & 0 \\ 0 & 0 & 1 \end{bmatrix} = \mathrm{M}. \tag{21}$$

Here both M and R were deduced from the initial information that there is a C_{3z} axis, but eq. (21) is a useful consistency check. The next step is to check for possible C_2 axes normal to C_{3z},

$$R(\pi\ [1\ 0\ 0])\langle \mathbf{a}_1\ \mathbf{a}_2\ \mathbf{a}_3| = \langle \mathbf{a}_4\ \mathbf{a}_2\ -\mathbf{a}_3| = \langle \mathbf{a}_1\ \mathbf{a}_2\ \mathbf{a}_3| \begin{bmatrix} -1 & 0 & 0 \\ -1 & 1 & 0 \\ 0 & 0 & -1 \end{bmatrix}, \tag{22}$$

$$(22)\quad \mathrm{R}^\dagger\,\mathrm{M}\,\mathrm{R} = a^2\begin{bmatrix} -1 & -1 & 0 \\ 0 & 1 & 0 \\ 0 & 0 & -1 \end{bmatrix}\begin{bmatrix} 1 & -\tfrac{1}{2} & 0 \\ -\tfrac{1}{2} & 1 & 0 \\ 0 & 0 & c^2 \end{bmatrix}\begin{bmatrix} -1 & 0 & 0 \\ -1 & 1 & 0 \\ 0 & 0 & -1 \end{bmatrix} = \mathrm{M}. \tag{23}$$

Equations (21) and (23) show that M is invariant under C_{3z} and C_{2x}. Therefore the compatible point groups are those that contain a proper or improper three-fold axis, with or without proper or improper two-fold axes normal to the principal axis. These point groups are 3, $\bar{3}$, 32, $3m$, $\bar{3}m$ (or C_3, S_6, D_3, C_{3v}, D_{3d}). To generate a 3-D lattice with three-fold rotational symmetry, the second and third layers of the hexagonal planar net in Figure 16.5 must be translated so that P_1 lies over P_2 and P_3, respectively, i.e. at $(\tfrac{1}{3}\ \tfrac{2}{3}\ \tfrac{1}{3})$ and $(\tfrac{2}{3}\ \tfrac{1}{3}\ \tfrac{2}{3})$.

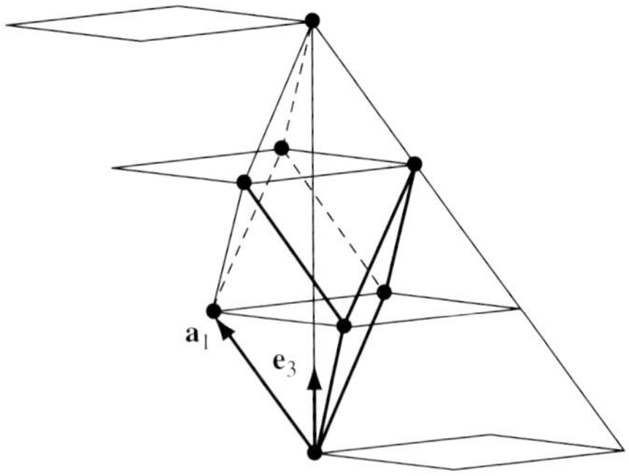

Figure 16.6. Primitive unit cell of the rhombohedral lattice $3R$. The three fundamental translations $\mathbf{a}_1$, $\mathbf{a}_2$, $\mathbf{a}_3$ are of equal length and make equal angles with $\mathbf{e}_3$. Hexagonal nets in four successive layers show how the rhombohedral cell may be constructed.

With $\mathbf{a}_3$ normal to the plane of $\mathbf{a}_1$, $\mathbf{a}_2$ at O, these fundamental translations generate a unit cell (named 3R) with two internal points (Figure 16.4). Figure 16.6 shows the primitive rhombohedral cell of this lattice. There is a third way of adding a second layer to the hexagonal net of Figure 16.5 which preserves the hexagonal symmetry of the initial net, and that is by the displacement $[0\ 0\ \mathbf{a}_3]$. Compatible symmetry operators that satisfy the invariance condition in eq. (17) are those associated with an inversion center, a horizontal or vertical mirror plane, or a two-fold axis giving the following seven point groups: 6, $\bar{6}$, $6/m$, 622, $6mm$, $\bar{6}m2$, and $6/mmm$ (C_6, C_{3h}, C_{6h}, D_6, C_{6v}, D_{3h}, and D_{6h}).

Exercise 16.1-1 Could a different lattice be generated by placing P_1' over P_3 and P_2, respectively, in the second and third layers of a hexagonal net?

Answer to Exercise 16.1-1

No. This lattice is equivalent to the first one because one may be converted into the other by a rotation through π about the normal to the plane containing $\mathbf{a}_1$ and $\mathbf{a}_2$ through the center of the rhombus with sides $\mathbf{a}_1$, $\mathbf{a}_2$.

16.2 The space group of a crystal

The space group G of a crystal is the set of all symmetry operators that leave the appearance of the crystal pattern unchanged from what it was before the operation. The most general kind of space-group operator (called a *Seitz operator*) consists of a point operator R (that is, a proper or improper rotation that leaves at least one point invariant) followed by a translation $\mathbf{v}$. For historical reasons the Seitz operator is usually written $\{R \mid \mathbf{v}\}$. However, we shall write it as $(R|\mathbf{v})$ to simplify the notation for sets of space-group operators. When a space-group operator acts on a position vector $\mathbf{r}$, the vector is transformed into

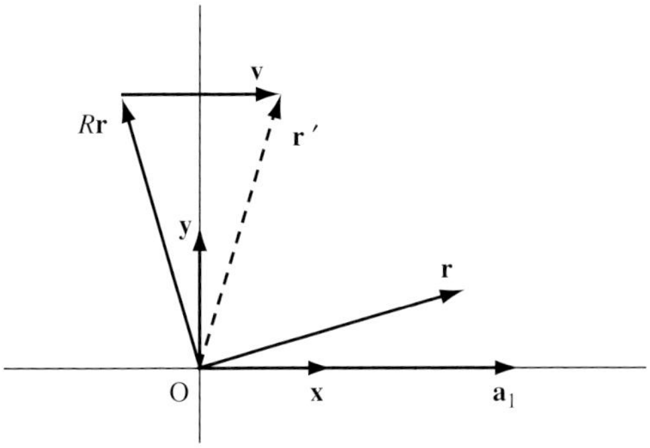

Figure 16.7. Action of the Seitz operator $(R|\mathbf{v})$ on the vector $\mathbf{r}$. In this example, R is $R(\pi/2\ \ [0\ 0\ 1])$ and $\mathbf{v} = \mathbf{a}_1/2$. The dashed line shows $\mathbf{r}' = R\mathbf{r}+\mathbf{v}$.

$$\mathbf{r}' = (R|\mathbf{v})\mathbf{r} = R\mathbf{r} + \mathbf{v} \tag{1}$$

(Figure 16.7). Note that $\mathbf{v}$ is not necessarily a lattice translation $\mathbf{t} \in \{\mathbf{a_n}\}$, though it must be if R is the identity E. Special cases of eq. (1) are as follows:

$$(E|\mathbf{t}),\ \text{a lattice translation;} \tag{2}$$

$$(E|\mathbf{0}),\ \text{the identity;} \tag{3}$$

$$(R|\mathbf{0}),\ \text{a point operator.} \tag{4}$$

In each of the equations (1)–(4) the crystal pattern appears the same after carrying out the operation signified. It follows from eq. (2) that the pattern, and therefore the subset of lattice translations

$$\mathrm{T} = \{(E|\mathbf{t})\}, \tag{5}$$

is infinite. This inconvenience may, however, be removed by a suitable choice of boundary conditions (see eq. (18) later).

The *product of two Seitz operators* is given by

(1)
$$(R_2|\mathbf{v}_2)(R_1|\mathbf{v}_1)\,\mathbf{r} = (R_2|\mathbf{v}_2)(R_1\mathbf{r} + \mathbf{v}_1) = R_2R_1\mathbf{r} + R_2\mathbf{v}_1 + \mathbf{v}_2$$
$$= (R_2R_1|R_2\mathbf{v}_1 + \mathbf{v}_2)\,\mathbf{r}; \tag{6}$$

(6)
$$(R_2|\mathbf{v}_2)(R_1|\mathbf{v}_1) = (R_2R_1|R_2\mathbf{v}_1 + \mathbf{v}_2). \tag{7}$$

The *inverse of a Seitz operator* is given by

(7)
$$(R|\mathbf{v})(R^{-1}|-R^{-1}\mathbf{v}) = (E|-\mathbf{v}+\mathbf{v}) = (\boldsymbol{E}|\mathbf{0}); \tag{8}$$

(8)
$$(R|\mathbf{v})^{-1} = (R^{-1}|-R^{-1}\mathbf{v}). \tag{9}$$

When a Seitz operator acts on configuration space all functions defined in that space are transformed, and the rule for carrying out this transformation is the same as that for rotation without translation. However, no special symbol is generally used in the case of the Seitz operator since it is clear when the corresponding function operator is intended. Thus $(R|\mathbf{v})f(\mathbf{r})$ implies

$$(R|\mathbf{v})f(\mathbf{r}) = f(R^{-1}\mathbf{r} - R^{-1}\mathbf{v}). \qquad (10)$$

The lattice translations form the *translation subgroup* of G.

(5), (7) $$(E|\mathbf{t}')(E|\mathbf{t}) = (E|\mathbf{t}+\mathbf{t}') = (E|\mathbf{t}'+\mathbf{t}) = (E|\mathbf{t})(E|\mathbf{t}'). \qquad (11)$$

Equation (11) shows that the set of lattice translations T form an Abelian subgroup of G. Moreover, T is an invariant subgroup of G, since

(7), (9) $$\begin{aligned}(R|\mathbf{v})(E|\mathbf{t})(R|\mathbf{v})^{-1} &= (R|R\mathbf{t}+\mathbf{v})(R^{-1}|-R^{-1}\mathbf{v}) \\ &= (E|-\mathbf{v}+R\mathbf{t}+\mathbf{v}) = (E|R\mathbf{t}) \in \mathbf{T}.\end{aligned} \qquad (12)$$

A lattice translation $\mathbf{t}$ is the sum of its projections along $\mathbf{a}_1$, $\mathbf{a}_2$, $\mathbf{a}_3$,

$$\mathbf{t} = \mathbf{t}_1 + \mathbf{t}_2 + \mathbf{t}_3 = n_1\mathbf{a}_1 + n_2\mathbf{a}_2 + n_3\mathbf{a}_3 \quad (n_1, n_2, n_3 \text{ integers}). \qquad (13)$$

(13), (11) $$(E|\mathbf{t}) = (E|\mathbf{t}_1)(E|\mathbf{t}_2)(E|\mathbf{t}_3), \qquad (14)$$

where $(E|\mathbf{t}_1)$ form a subgroup T_1 of T and similarly for T_2, T_3.

(14) $$T_1 \cap T_2 = (E|\mathbf{0}); \qquad (15)$$

(11) $$(E|\mathbf{t}_1)(E|\mathbf{t}_2) = (E|\mathbf{t}_2)(E|\mathbf{t}_1), \ \forall\ \mathbf{t}_1,\ \mathbf{t}_2; \qquad (16)$$

and similarly for T_2, T_3, and T_3, T_1.

(15), (16) $$T = T_1 \otimes T_2 \otimes T_3. \qquad (17)$$

We now remove the inconvenience of the translation subgroup, and consequently the Bravais lattice, being infinite by supposing that the crystal is a parallelepiped of sides $N_j\mathbf{a}_j$ where $\mathbf{a}_j$, $j = 1, 2, 3$, are the fundamental translations. The number of lattice points, $N_1N_2N_3$, is equal to the number of unit cells in the crystal, N. To eliminate surface effects we imagine the crystal to be one of an infinite number of replicas, which together constitute an infinite system. Then

$$(E\,|\,N_j\mathbf{a}_j) = (E\,|\,\mathbf{a}_j)^{Nj} = (E|\mathbf{0}),\ j = 1,\ 2,\ 3. \qquad (18)$$

Equation (18) is a statement of the Born and von Kármán *periodic boundary conditions*.

T is the direct product (DP) of three Abelian subgroups and so has $N_1N_2N_3$ 1-D representations. The MRs of T_1, $\Gamma(T_1)$, obey the same multiplication table as the corresponding operators, namely

(11) $$(E\,|\,n_1\,\mathbf{a}_1)(E\,|\,n_1{}'\,\mathbf{a}_1) = (E\,|\,(n_1 + n_1{}')\mathbf{a}_1); \qquad (19)$$

(19) $$\Gamma(E\,|\,n_1\,\mathbf{a}_1)\Gamma(E\,|\,n_1{}'\mathbf{a}_1) = \Gamma(E\,|\,(n_1 + n_1{}')\mathbf{a}_1). \qquad (20)$$

Equation (20) is satisfied by

$$\Gamma(E \mid n_1 \, \mathbf{a}_1) = \exp(-ik_1 \, n_1 \, \mathbf{a}_1). \tag{21}$$

Because of the DP in eq. (17), the representations of T obey the relation

(20)
$$\Gamma(E \mid n_1 \, \mathbf{a}_1 + n_2 \, \mathbf{a}_2 + n_3 \, \mathbf{a}_3) \; \Gamma(E \mid n_1{}' \, \mathbf{a}_1 + n_2{}' \, \mathbf{a}_2 + n_3{}' \, \mathbf{a}_3) \\ = \Gamma(E \mid (n_1 + n_1{}')\mathbf{a}_1 + (n_2 + n_2{}')\mathbf{a}_2 + (n_3 + n_3{}')\mathbf{a}_3); \tag{22}$$

(22), (13)
$$\Gamma(E|\mathbf{t}) \, \Gamma(E|\mathbf{t}') = \Gamma(E|\mathbf{t} + \mathbf{t}'); \tag{23}$$

(23), (21), (13)
$$\Gamma(E|\mathbf{t}) = \exp(-\mathrm{i} \, \mathbf{k} \cdot \mathbf{t}), \; \mathbf{t} \in \{\mathbf{a}_n\}, \tag{24}$$

which are the IRs of the translation group T. The MRs $\Gamma(E|\mathbf{t})$ of the translation operators $(E|\mathbf{t})$ are defined by

$$(E|\mathbf{t}) \, \psi_{\mathbf{k}}(\mathbf{r}) = \psi_{\mathbf{k}}(\mathbf{r}) \, \Gamma(E|\mathbf{t}); \tag{25}$$

(25), (24)
$$\psi_{\mathbf{k}} \, (\mathbf{r}) = \exp(\mathrm{i} \, \mathbf{k} \cdot \mathbf{r}) \, u_{\mathbf{k}}(\mathbf{r}). \tag{26}$$

The functions in eq. (26) are called *Bloch functions* and are plane waves modulated by the function $u_{\mathbf{k}}(\mathbf{r})$, which has the periodicity of the lattice,

$$u_{\mathbf{k}}(\mathbf{r}) = u_{\mathbf{k}}(\mathbf{r} + \mathbf{t}), \quad \forall \; \mathbf{t} \in \{\mathbf{a}_n\}. \tag{27}$$

We now confirm that eqs. (27) and (26) satisfy eq. (24):

$$(E|\mathbf{t}) \, \psi_{\mathbf{k}}(\mathbf{r}) = \psi_{\mathbf{k}}((E|\mathbf{t})^{-1}\mathbf{r}). \tag{28}$$

The configuration space operator on the RS of eq. (28) replaces $\mathbf{r}$ by $\mathbf{r} - \mathbf{t}$:

(26), (27), (28)
$$(E|\mathbf{t})\psi_{\mathbf{k}}(\mathbf{r}) = \exp[\mathrm{i} \, \mathbf{k} \cdot (\mathbf{r} - \mathbf{t})] \, u_{\mathbf{k}}(\mathbf{r} - \mathbf{t}) \\ = \exp(-\mathrm{i} \, \mathbf{k} \cdot \mathbf{t}) \, \psi_{\mathbf{k}}(\mathbf{r}), \tag{29}$$

in agreement with eq. (24).

Consider $\{(R|\mathbf{0})\} \subset$ G; then

$$(R|\mathbf{0})(R'|\mathbf{0}) = (RR'|\mathbf{0}) \in \mathrm{G}, \tag{30}$$

and so $\{(R|\mathbf{0})\} \subset$ G also form a subgroup of G, called the *point subgroup* P(G). A general space-group operator is represented in (modified) Seitz notation by $(R|\mathbf{v})$, where R is a point operator and $\mathbf{v}$ is a translation, though not necessarily a lattice translation $(E|\mathbf{t})$. If $(R|\mathbf{v}) = (S|\mathbf{w}) \in$ G, where $\mathbf{w} \notin \{\mathbf{t}\}$, then neither $(S|\mathbf{0})$ nor $(E|\mathbf{w})$ are $\in$ G. Here S is being used to indicate a special point operator that is associated with a particular, unique (non-lattice) translation $\mathbf{w} \notin \{\mathbf{t}\}$.

Exercise 16.2-1 Why are $(R|\mathbf{w})$ and $(S|\mathbf{t})$, $S \notin$ P(G), not $\in$ G?

(7) $$(R|\mathbf{0})(R'|\mathbf{0}) = (RR'|\mathbf{0}) \in \mathrm{G}, \ \forall R, \tag{31}$$

which demonstrates closure in the set $\{(R|\mathbf{0})\}$. Therefore, the set $\{(R|\mathbf{0})\}$, which is obtained from G by setting $\forall\ \mathbf{v} = \mathbf{0}$, and which may therefore contain some rotations which were associated in G with special, non-lattice translations **w**, is a group P called the *point group* or sometimes the *isogonal point group*, which avoids any possible confusion with the point subgroup P(G). The distinction between eq. (30) and eq. (31) is important: in the former equation all members of the set $\{(R|\mathbf{0})\} \in \mathrm{G}$, but in eq. (31) some of the set $\{(R|\mathbf{0})\}$ may not be $\in$ G. In general, the point group P is not a subgroup of G, unless G contains no operators of the form $(S|\mathbf{w})$ with $\mathbf{w} \notin \{\mathbf{t}\}$, in which case it will be the same as the point subgroup. Although $(S|\mathbf{0}) \notin \mathrm{G}$, it is a symmetry operator for the pattern that remains after removing all the pattern points that lie within the unit cell, but leaving all those which are the lattice points of the Bravais lattice. In other words, the Bravais lattice is invariant under the operations of the point group P, and P is therefore either the point group $\mathrm{P_{BL}}$ of the Bravais lattice or it is a subgroup of $\mathrm{P_{BL}}$. For example, the point subgroup of the *hcp* lattice is $\mathrm{D_{3h}}$, whereas the point group is $\mathrm{D_{6h}}$ (Figure 16.5).

There are two kinds of operators $(S|\mathbf{w})$.

(1) A *screw rotation* is one in which S is a rotation about a specified axis **n** and **w** is a translation along that axis. Screw rotations are described by the symbol n_p, in which n signifies a rotation through $2\pi/n$ about the screw axis **n**, followed by a translation $p\mathbf{t}/n$, where **t** is the translation between nearest-neighbor lattice points along **n** (Figure 16.8).
(2) A *glide reflection* is one in which S is a reflection in the glide plane followed by a translation **w**, not necessarily parallel to the reflection plane. The three possible types of translation **w** are described in Table 16.3.

It follows from the existence of operators $(S|\mathbf{w})$ that space groups G may be classified as either symmorphic or non-symmorphic. *Symmorphic* space groups consist only of operators of the type $(R|\mathbf{t})$, where $(R|\mathbf{0})$ and $(E|\mathbf{t})$ are members of the set of G. *Non-symmorphic* space groups contain besides operators of the type $(R|\mathbf{t})$ at least one operator$(S|\mathbf{w})$ in which neither $(S|\mathbf{0})$ nor $(E|\mathbf{w}) \in \mathrm{G}$, so that the list of symmetry elements contains one or more screw axes or glide planes.

The *coset expansion* of G on its invariant subgroup T is

$$\mathrm{G} = \sum_{\{R\}} (R|\mathbf{w})\mathrm{T}, \qquad \forall R \in \mathrm{P}, \tag{32}$$

where **w** is either the null vector **0** or the unique special vector associated with some screw axis or glide plane. (No coset representatives $(R|\mathbf{v})$ are necessary in eq. (32) because $(R|\mathbf{v})\mathrm{T} = (R|\mathbf{w})(E|\mathbf{t})\mathrm{T} = (R|\mathbf{w})\mathrm{T}$.) If there are no screw axes or glide planes, then there are no operators with $\mathbf{w} \neq \mathbf{0}$ in G and

(32) $$\mathrm{G} = \sum_{\{R\}} (R|\mathbf{0})\,\mathrm{T}. \tag{33}$$

In this case the point subgroup $\{(R|\mathbf{0})\}$ of G is identical with the point group P, and G may be written as the semidirect product

$$\mathrm{G} = \mathrm{T} \wedge \mathrm{P} \qquad (\mathrm{G\ symmorphic}). \tag{34}$$

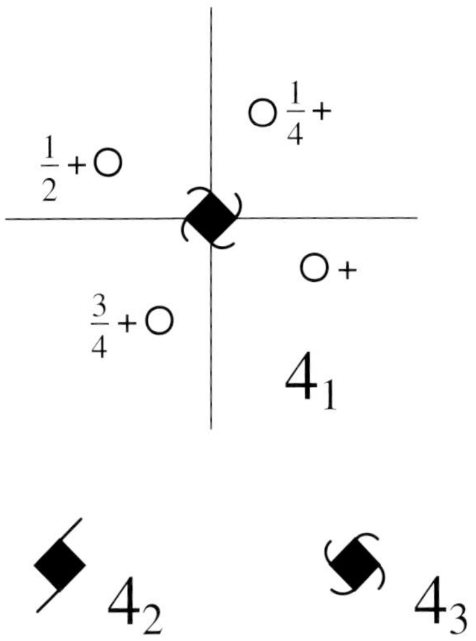

Figure 16.8. Equivalent positions (shown by open circles) generated by screw rotations about a 4_1 screw axis along **z**. The location of a screw axis n_p is specified by an n-sided filled polygon with extensions ("hooks") on q sides where qp/n is the smallest possible integer. The height of the point above the **xy** plane is shown by the symbol $^p/_n+$ which means $(^p/_n + z)$**t**. The curved "hooks" indicate the sign of the rotation through an angle less than π: anticlockwise, or positive, for $p < n/2$, but clockwise (negative) for $p > n/2$. For a 4_2 axis, $q = 2$, because $2(2/4) = 1$, but for both 4_1 and 4_3, $q = 4$ because $4(\frac{1}{4}) = 1$ and $4(\frac{3}{4}) = 3$.

Table 16.3.1 *Definition of glide planes.*

Notation: The fundamental translations are denoted in this book by $\mathbf{a}_1$, $\mathbf{a}_2$, $\mathbf{a}_3$. Superscript tc denotes tetragonal and cubic systems only.

Type	Symbol	Translation **w**
Axial	a, b, c	$\frac{1}{2}\mathbf{a}_1$, $\frac{1}{2}\mathbf{a}_2$, $\frac{1}{2}\mathbf{a}_3$
Diagonal	n	$\frac{1}{2}(\mathbf{a}_1 + \mathbf{a}_2)$, $\frac{1}{2}(\mathbf{a}_2 + \mathbf{a}_3)$, $\frac{1}{2}(\mathbf{a}_3 + \mathbf{a}_1)$
		$\frac{1}{2}(\mathbf{a}_1 + \mathbf{a}_2 + \mathbf{a}_3)^{tc}$
Diamond	d	$\frac{1}{4}(\mathbf{a}_1 \pm \mathbf{a}_2)$, $\frac{1}{4}(\mathbf{a}_2 \pm \mathbf{a}_3)$, $\frac{1}{4}(\mathbf{a}_3 \pm \mathbf{a}_1)$
		$\frac{1}{4}(\mathbf{a}_1 \pm \mathbf{a}_2 \pm \mathbf{a}_3)^{tc}$

In non-symmorphic space groups the point group P is not a subgroup of G and it is not possible to express G as a semidirect product.

Since T in eq. (32) is an invariant subgroup of G, the cosets $(R|\mathbf{w})$T form the factor group (Section 1.4)

$$\mathrm{F} = \mathrm{G/T} = \sum_{\{R\}} [(R|\mathbf{w})\,\mathrm{T}] \sim \mathrm{P}. \tag{35}$$

Each term in square brackets in eq. (35) is itself a set of elements, being T multiplied by the coset representative $(R|\mathbf{w})$. Therefore F is isomorphous with the point group P. The kernel

Table 16.4. *Coordinates of equivalent positions $x'\ y'\ z'$ and the space-group operators $(R|\mathbf{v})$ that transform $x\ y\ z$ into $x'\ y'\ z'$, for the Wyckoff position 8g of the space group 59 (Pmmn or D_{2h}^{13}).*

$x\ y\ z$	$\bar{x}\ \bar{y}\ z$	$\bar{x}\ y\ z$	$x\ \bar{y}\ z$
$(E\|\ 000)$	$(C_{2z}\|\ 000)$	$(\sigma_x\|\ 000)$	$(\sigma_y\|\ 000)$
$\frac{1}{2}-x,\ \frac{1}{2}-y,\ \bar{z}$	$\frac{1}{2}-x,\ \frac{1}{2}+y,\ \bar{z}$	$\frac{1}{2}+x,\ \frac{1}{2}+y,\ \bar{z}$	$\frac{1}{2}+x,\ \frac{1}{2}-y,\ \bar{z}$
$(I\|{}^1\!/_2\,{}^1\!/_2\,0)$	$(C_{2y}\|{}^1\!/_2\,{}^1\!/_2\,0)$	$(\sigma_z\|{}^1\!/_2\,{}^1\!/_2\,0)$	$(C_{2x}\|{}^1\!/_2\,{}^1\!/_2\,0)$

K of a homomorphism $G \rightarrow G'$ is the subset $K \subset G$ that is the fiber of E' in G' (Section 1.7). If the fibers of G' all have the same order, then G is an *extension* of G' by K with

(1.7.6) $$g = k\,g', \tag{36}$$

where g is the order of G and similarly. It follows from eq. (32) that G is the extension of the point group P by T. Firstly, the condition (36) is satisfied since

(31) $$g = N\,g(\mathrm{P}), \tag{37}$$

with N and g(P) the orders of T and P, respectively. Secondly, G is homomorphous to P, with T the kernel of the homomorphism. The mapping $(R|\mathbf{w}) \rightarrow R$ maintains the multiplication rules, since

$$(R'|\mathbf{w}')(R|\mathbf{w}) = (R'R|R'\mathbf{w} + \mathbf{w}') \tag{38}$$

maps on to $(R')(R) = R'R$, thus establishing the homomorphism (Section 1.7). Each one of the subgroup of translations $\mathrm{T} = \{(E|\mathbf{t})\}$ maps on to the identity so that T is indeed the kernel of the homomorphism.

The *matrix representation of the space-group operation*

$$\mathbf{r}' = (R|\mathbf{v})\mathbf{r} = R\,\mathbf{r} + \mathbf{v} \tag{39}$$

is

$$\begin{bmatrix} & \Gamma(R) & & \begin{matrix} v_1 \\ v_2 \\ v_3 \end{matrix} \\ 0 & 0 & 0 & 1 \end{bmatrix} \begin{bmatrix} x \\ y \\ z \\ 1 \end{bmatrix} = \begin{bmatrix} x' \\ y' \\ z' \\ 1 \end{bmatrix}, \tag{40}$$

where the 3×3 submatrix $\Gamma(R)$, or R, is the MR of the point symmetry operator R. The positions $x,\ y,\ z$ and $x',\ y',\ z'$ are called *equivalent positions*. They are given in the *International Tables of Crystallography* (Hahn (1983), and subsequently refered to as the ITC) for every space group, and from these coordinates the space-group operators can be determined. For example, for the space group 59, *Pmmn* or D_{2h}^{13}, we find in the ITC the coordinates of the eight equivalent positions in Table 16.4. Given below each set of coordinates is the space-group operator $(R|\mathbf{v})$ that transforms the general point $x,\ y,\ z$ into x', y', z'. In the second row of the table the space-group operators are of the form $(R|\mathbf{v})$, but in the fourth row each point operator is associated with the non-lattice translation $\mathbf{w} = [\frac{1}{2}\ \frac{1}{2}\ 0]$. This group is therefore a non-symmorphic space group. The corresponding symmetry elements are shown in the figures for space group 59 in the ITC. In addition to the equivalent positions of the general point $x\ y\ z$, the ITC also give equivalent positions for points in special positions on symmetry elements. The generating elements for the 230 space groups are listed by Bradley and Cracknell (1972) in their Table 3.7.

A set of points equivalent by symmetry form a *crystallographic orbit*. All the points in an orbit may be obtained from one, the generating point $\mathbf{q}$, by $(R|\mathbf{v})\mathbf{q}$, where, in general, $\mathbf{v} = \mathbf{w} + \mathbf{t}$. The *site symmetry* group $G_{\mathbf{q}}$, which is isomorphous to one of the crystallographic point groups, comprises the set of symmetry operators $(R|\mathbf{v})$ that leave $\mathbf{q}$ invariant. The site symmetry groups G_j of different points $\mathbf{q}_j$ of the same orbit are conjugate; that is, $G_j = g_j\ G_{\mathbf{q}}\ g_j^{-1}$. For a point $\mathbf{q}$ in a general position, $G_{\mathbf{q}} = C_1$, but for special points $G_{\mathbf{q}}$ has higher symmetry. All the symmetry points that have the same site symmetry group belong to a subset of the crystallographic orbits called a *Wyckoff position*. A particular Wyckoff position consists of only one orbit unless this contains one or more variable parameters. Wyckoff positions are labeled successively by lower case letters. For example, the Wyckoff position in Table 16.4 is called $8g$: 8 because there are eight equivalent positions that would be obtained from the general point $x\ y\ z$ by the symmetry operators $(R|\mathbf{v})$ listed in this table, and g because there are six Wyckoff positions labeled a–f which have lower site symmetry. The number of orbit points in the primitive unit cell is $g(\mathrm{P})/g(\mathbf{q})$, where $g(\mathrm{P})$ is the order of the point group P and $g(\mathbf{q})$ is the order of the site symmetry group $G_{\mathbf{q}}$. In this example, $g(\mathbf{q}) = 1$ (for C_1) and $g(\mathrm{P}) = 8$ (for D_{2h}) so there are eight equivalent positions for the Wyckoff position $8g$ of D_{2h}. But for a point $0\ y\ z$ in the mirror plane $x = 0$, the site symmetry group is m or $C_s = \{E\sigma_x\}$ with $g(\mathbf{q}) = 2$, and so there are $8/2 = 4$ equivalent positions in the unit cell. The ITC table for space group 59 confirms that there are four equivalent positions and gives their coordinates as

$$0\ y\ z;\quad 0\ \bar{y}\ z;\quad \tfrac{1}{2},\ \tfrac{1}{2}-y,\ \bar{z};\quad \tfrac{1}{2},\ \tfrac{1}{2}+y,\ \bar{z}.$$

Exercise 16.2-2 List the space-group operators $(R|\mathbf{v})$ which generate the above four positions from the point $0\ y\ z$.

In a space group the *choice of origin* is refered to as the *setting*. Suppose the coordinate axes O_1 X Y Z are shifted to O_2 X Y Z by a vector $\mathbf{t}_0$. Then any vector $\mathbf{r}_1$ referred to O_1 X Y Z becomes, in the new coordinate system,

$$\mathbf{r}_2 = \mathbf{r}_1 - \mathbf{t}_0. \tag{41}$$

The space-group operation

$$(R_1|\mathbf{v}_1)\,\mathbf{r}_1 = R_1\,\mathbf{r}_1 + \mathbf{v}_1 = \mathbf{r}_1{}' \tag{42}$$

becomes

$$(R_2|\mathbf{v}_2)\,\mathbf{r}_2 = R_2\,\mathbf{r}_2 + \mathbf{v}_2 = \mathbf{r}_2{}'; \tag{43}$$

(41), (42), (43) $$(R_2|\mathbf{v}_2) = (R_1|\mathbf{v}_1 + R_1\,\mathbf{t}_0 - \mathbf{t}_0). \quad (44)$$

For example, two different settings are used for space group 59 (*Pmmn* or D_{2h}^{13}). The first setting was used in Table 16.4; the second setting is related to the first by $\mathbf{t}_0 = (\mathbf{a}_1/4) - (\mathbf{a}_2/4)$. Therefore, in this second setting, the space group operator which replaces $(C_{2\mathbf{z}}|000)$ of the first setting is

(44) $$(C_{2\mathbf{z}}|-\tfrac{1}{2}\ \ \tfrac{1}{2}\ \ 0),\ \text{or}\ (C_{2\mathbf{z}}|\tfrac{1}{2}\ \ \tfrac{1}{2}\ \ 0), \quad (45)$$

on adding the fundamental translation $\mathbf{a}_1$ to simplify $(R_2|\mathbf{v}_2)$. From eq. (45), $(C_{2\mathbf{z}}|\tfrac{1}{2}\ \tfrac{1}{2}\ 0)$ $[x\ y\ z] = [\tfrac{1}{2}-x\ \ \tfrac{1}{2}-y\ \ z]$, which is one of the equivalent positions in Table 16.4. The other Seitz operators in the second setting may be written in a similar manner. The choice of a particular setting is arbitrary, and the settings chosen by various authors do not always agree with those in the ITC.

Example 16.2-1 List the symmetry operators of the space group 33, $Pna2_1$. What is the point group of this space group? Find the equivalent positions $[x'\ y'\ z']$ that result from applying these operators to the general point $[x\ y\ z]$.

The position of a symmetry element in the space-group symbol gives the unique direction associated with that element, namely the axis of a rotation or the normal to a reflection plane (see Table 16.5). Therefore, in $Pna2_1$ the symmetry elements are: a diagonal glide plane normal to [100]; an axial glide plane normal to [010] with glide direction [100]; and a 2_1 screw axis parallel to [001]. The symmetry operators of $Pna2_1$ are therefore $(E|000)$, $(\sigma_{\mathbf{x}}|0\ \tfrac{1}{2}\ \tfrac{1}{2})$, $(\sigma_{\mathbf{y}}|\tfrac{1}{2}\ 0\ 0)$, and $(C_{2\mathbf{z}}|0\ 0\ \tfrac{1}{2})$. The point group $P = \{E\ \sigma_{\mathbf{x}}\ \sigma_{\mathbf{y}}\ C_{2\mathbf{z}}\} = C_{2v}$. Since there are two improper C_2 axes normal to one another and to the proper C_2 axis, this space group belongs to the orthorhombic system. The space-group symbol tells us the symmetry elements but not their *location* in the unit cell. For this we must consult the ITC, which gives, for each space group, diagrams showing the location of equivalent points and space-group symmetry elements. Part of this information about space group 33 is in Figure 16.9; $(E|\,0\ 0\ 0)$ leaves the point $x\ y\ z$ at its original position. The MR of the Seitz operator $(\sigma_{\mathbf{x}}|\,0\ \tfrac{1}{2}\ \tfrac{1}{2})$ operating on $x\ y\ z$, with the glide plane at $x = \tfrac{1}{4}$, is

Table 16.5. *Conventions used to specify, in a space-group symbol, the unique direction associated with a space-group operator.*

	Unique direction for		
Crystal system	First position	Second position	Third position
Monoclinic and orthorhombic	[100]	[010]	[001]
Tetragonal and hexagonal	[001]	[100]	[1$\bar{1}$0]
Isometric (cubic)	[001]	[111]	[110]

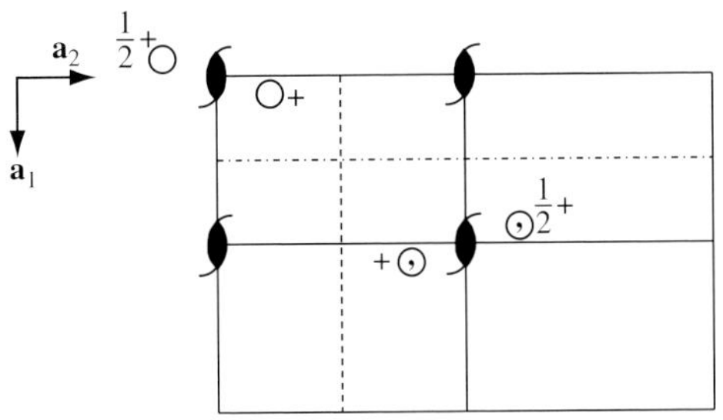

Figure 16.9. Location of some of the equivalent points and symmetry elements in the unit cell of space group $Pna2_1$. An open circle marked + denotes the position of a general point xyz, the + sign meaning that the point lies at a height z above the xy plane. Circles containing a comma denote equivalent points that result from mirror reflections. The origin is in the top left corner, and the filled digon with "tails" denotes the presence of a two-fold screw axis at the origin. Small arrows in this figure show the directions of $\mathbf{a}_1$, $\mathbf{a}_2$, which in an orthorhombic cell coincide with **x**, **y**. The dashed line shows the location of an axial glide plane at $y = ¼$ and the –·–· line shows the location of the diagonal glide plane at $x = ¼$.

$$\begin{bmatrix} \bar{1} & 0 & 0 & \tfrac{1}{2} \\ 0 & 1 & 0 & \tfrac{1}{2} \\ 0 & 0 & 1 & \tfrac{1}{2} \\ 0 & 0 & 0 & 1 \end{bmatrix} \begin{bmatrix} x \\ y \\ z \\ 1 \end{bmatrix} = \begin{bmatrix} \tfrac{1}{2} - x \\ \tfrac{1}{2} + y \\ \tfrac{1}{2} + z \\ 1 \end{bmatrix}. \tag{46}$$

The displacement by ½ in the first row and fourth column comes from the mirror reflection in the plane at $x = 1/4$. The ½ in the second and third rows of the fourth column are the components of the diagonal glide. The location of the transformed point is that marked by a comma (,) and ½+. The MR of the operation $(\sigma_{\mathbf{y}}|½\ 0\ 0)[xyz]$, when the axial glide plane lies at $y = ¼$, is

$$\begin{bmatrix} 1 & 0 & 0 & \tfrac{1}{2} \\ 0 & \bar{1} & 0 & \tfrac{1}{2} \\ 0 & 0 & 1 & 0 \\ 0 & 0 & 0 & 1 \end{bmatrix} \begin{bmatrix} x \\ y \\ z \\ 1 \end{bmatrix} = \begin{bmatrix} \tfrac{1}{2} + x \\ \tfrac{1}{2} - y \\ z \\ 1 \end{bmatrix}. \tag{47}$$

The ½ in the second row and fourth column is the displacement in y due to reflection in the axial a-glide plane at $y = ¼$. The location of the equivalent point resulting from this operation is that marked by a "," and ½ + in Figure 16.9. The MR of the operation of the 2_1 screw axis at the origin on the general point xyz, that is $(C_{2\mathbf{z}}|\ 0\ 0\ ½)(xyz)$, is

$$\begin{bmatrix} \bar{1} & 0 & 0 & 0 \\ 0 & \bar{1} & 0 & 0 \\ 0 & 0 & 1 & \tfrac{1}{2} \\ 0 & 0 & 0 & 1 \end{bmatrix} \begin{bmatrix} x \\ y \\ z \\ 1 \end{bmatrix} = \begin{bmatrix} -x \\ -y \\ \tfrac{1}{2} + z \\ 1 \end{bmatrix}, \tag{48}$$

and the resulting equivalent point is that marked by an open circle and ½+.

Answers to Exercises 16.2

Exercise 16.2-1 If $(R|\mathbf{0}) \in \mathrm{G}$, then R leaves the appearance of the crystal pattern indistinguishable from what it was before this operation. Therefore to maintain the crystal pattern any subsequent translation must be $\in \{\mathbf{t}\}$. Similarly, since $S \notin \mathrm{P(G)}$ it does not leave the pattern self-coincident and a subsequent translation $\mathbf{w}$ must therefore $\notin \{\mathbf{t}\}$, in order to restore self-coincidence.

Exercise 16.2-2 $(E|0\ 0\ 0)$, $(C_{2z}|0\ 0\ 0)$, $(C_{2x}|\tfrac{1}{2}\ \tfrac{1}{2}\ 0)$, $(C_{2y}|\tfrac{1}{2}\ \tfrac{1}{2}\ 0)$.

16.3 Reciprocal lattice and Brillouin zones

The *reciprocal lattice* is generated from the fundamental translations $\{\mathbf{b}_1\ \mathbf{b}_2\ \mathbf{b}_3\}$ defined by

$$\mathbf{b}_i \cdot \mathbf{a}_j = 2\pi\delta_{ij}, \quad i, j = 1, 2, 3, \tag{1}$$

where the fundamental translations $\{\mathbf{a}_j\}$ define a *primitive* unit cell of the Bravais lattice. The solutions to eq. (1) are given by

$$\mathbf{b}_1 = (2\pi/\mathrm{v}_a)(\mathbf{a}_2 \times \mathbf{a}_3), \quad \mathbf{b}_2 = (2\pi/\mathrm{v}_a)(\mathbf{a}_3 \times \mathbf{a}_1), \quad \mathbf{b}_3 = (2\pi/\mathrm{v}_a)(\mathbf{a}_1 \times \mathbf{a}_2). \tag{2}$$

$$\mathrm{v}_a = \mathbf{a}_1 \cdot \mathbf{a}_2 \times \mathbf{a}_3 \tag{3}$$

is the volume of the unit cell in the direct lattice. (In crystallography the reciprocal lattice is usually defined without the factor 2π in eq. (1), which, however, is invariably introduced in solid state physics.) Since the space lattice is primitive, then so is the reciprocal lattice, and each lattice point can be reached from O by a translation

$$\mathbf{b}_m = m_1\,\mathbf{b}_1 + m_2\,\mathbf{b}_2 + m_3\,\mathbf{b}_3 = \langle \mathbf{b}_1\ \mathbf{b}_2\ \mathbf{b}_3\,|\,m_1\ m_2\ m_3\rangle = \langle \mathbf{b}\,|\,m\rangle, \tag{4}$$

where m_1, m_2, and m_3 are integers.

(4), (16.1.3)
$$\mathbf{b}_m = \langle \mathbf{e}_1\ \mathbf{e}_2\ \mathbf{e}_3\,|\mathrm{B}|\,m_1\ m_2\ m_3\rangle; \tag{5}$$

$$\mathrm{B} = \begin{bmatrix} b_{1x} & b_{2x} & b_{3x} \\ b_{1y} & b_{2y} & b_{3y} \\ b_{1z} & b_{2z} & b_{3z} \end{bmatrix}. \tag{6}$$

The MR of the scalar product (SP) $\mathbf{b}_m \cdot \mathbf{a}_n$ (which conforms with the laws of matrix multiplication) is

$$\langle \mathbf{b}|m\rangle^{\mathrm{T}} \cdot \langle \mathbf{a}\,|\,n\rangle = \langle m_1\ m_2\ m_3|\ \mathbf{b}_1\mathbf{b}_2\mathbf{b}_3\rangle \cdot \langle \mathbf{a}_1\ \mathbf{a}_2\ \mathbf{a}_3\,|\,n_1\ n_2\ n_3\rangle$$

$$= \langle m_1\ m_2\ m_3\,|\,\mathrm{B}^{\mathrm{T}}\mathrm{A}|\ n_1\ n_2\ n_3\rangle \tag{7}$$

(1)
$$= \langle m_1\ m_2\ m_3|2\pi\mathrm{E}_3|n_1\ n_2\ n_3\rangle; \tag{8}$$

(8) $$\mathbf{b}_m \cdot \mathbf{a}_n = 2\pi p, \quad p = m_1 n_1 + m_2 n_2 + m_3 n_3 = \text{an integer}. \tag{9}$$

(7), (8) $$\mathrm{B} = 2\pi(\mathrm{A}^{-1})^{\mathrm{T}} = (2\pi/|\,\mathrm{A}\,|)(-1)^{r+s}\,[\alpha_{rs}], \tag{10}$$

where α_{rs} is the complementary minor of a_{rs} in $|\mathrm{A}| = |a_{rs}|$.

Example 16.3-1 Find the reciprocal lattice of the *fcc* direct lattice. From Figure 16.2(a),

$$\mathrm{A} = (a/2)\begin{bmatrix} 0 & 1 & 1 \\ 1 & 0 & 1 \\ 1 & 1 & 0 \end{bmatrix}; \tag{11}$$

$$\mathrm{B} = 2\pi(\mathrm{A}^{-1})^{\mathrm{T}} = (2\pi/a)\begin{bmatrix} -1 & 1 & 1 \\ 1 & -1 & 1 \\ 1 & 1 & -1 \end{bmatrix}, \tag{12}$$

so that the reciprocal lattice of the *fcc* lattice is *bcc* with cube edge $b = 4\pi/a$ (Figure 16.2(b)).

Example 16.3-2 Find the reciprocal lattice of the planar hexagonal net which has the primitive unit cell shown in Figure 16.10.

In the hexagonal net a lattice vector $\mathbf{a}_n = n_1\mathbf{a}_1 + n_2\mathbf{a}_2$, where $|\mathbf{a}_1| = |\mathbf{a}_2| = a$, and $\alpha_{12} = 2\pi/3$. From Figure 16.10,

$$\mathbf{a}_1 = \mathbf{e}_1\, a_{1x} + \mathbf{e}_2\, a_{1y} = \mathbf{e}_1\,(a\sqrt{3}/2) - \mathbf{e}_2\,(a/2), \tag{13}$$

$$\mathbf{a}_2 = \mathbf{e}_1\, a_{2x} + \mathbf{e}_2\, a_{2y} = \mathbf{e}_1\,(0) + \mathbf{e}_2\,(a). \tag{14}$$

(13), (14) $$[\mathbf{a}_1\ \mathbf{a}_2] = a\,[\mathbf{e}_1\ \mathbf{e}_2]\begin{bmatrix} \sqrt{3}/2 & 0 \\ -1/2 & 1 \end{bmatrix}. \tag{15}$$

(15) $$\mathrm{A} = a\begin{bmatrix} \sqrt{3}/2 & 0 \\ -1/2 & 1 \end{bmatrix}; \tag{16}$$

(16), (10) $$\mathrm{B} = (4\pi/a\sqrt{3})\begin{bmatrix} 1 & 1/2 \\ 0 & \sqrt{3}/2 \end{bmatrix}. \tag{17}$$

(17), (6) $$\mathbf{b}_1 = (4\pi/a\sqrt{3})\mathbf{e}_1; \quad \mathbf{b}_2 = (2\pi/a\sqrt{3})(\mathbf{e}_1 + \sqrt{3}\mathbf{e}_2), \tag{18}$$

so that the reciprocal lattice is also a planar hexagonal net.

Exercise 16.3-1 Confirm that $\mathrm{B}^{\mathrm{T}}\mathrm{A} = 2\pi\mathrm{E}_2$.

The direct lattice and reciprocal lattice unit cells are marked on the crystal pattern of a planar hexagonal net in Figure 16.10, using eqs. (13), (14), and (18). The scales chosen for

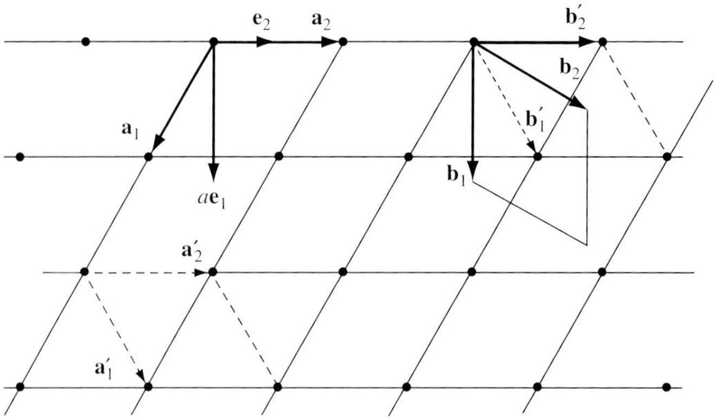

Figure 16.10. Hexagonal net with primitive lattice vectors $\mathbf{a}_1$, $\mathbf{a}_2$. The reciprocal lattice vectors $\mathbf{b}_1$, $\mathbf{b}_2$ also generate a hexagonal net which, at first sight, does not appear to be the same as the original net. However, the two nets may be brought into coincidence by using the alternative choice of fundamental lattice translation vectors $\mathbf{a}_1'$, $\mathbf{a}_2'$ (shown by the unit cell marked by dashed lines) and rotating $\mathbf{b}_1$, $\mathbf{b}_2$ through an angle of $\pi/6$. The two nets would then be proportional but have been made to coincide through the choice of the scales used in the drawing of the two nets.

the two lattices in Figure 16.10 are $a = b = \sqrt{2\pi}$. From eq. (17) the angle between $\mathbf{b}_1$ and $\mathbf{b}_2$ is $\pi/3$, but the two nets may be brought into coincidence by a rotation through an angle of $\pi/6$, as emphasized by the rotated unit cell defined $\mathbf{b}_1{}'$, $\mathbf{b}_2{}'$. For any crystal pattern there may be alternative choices of fundamental translations, for example $\{\mathbf{a}_1{}', \mathbf{a}_2{}'\}$ and $\{\mathbf{a}_1, \mathbf{a}_2\}$ in Figure 16.10. The fact that the original net and the reciprocal lattice net coincide, rather than scale, is a consequence of our deliberate choice of scales making $b = a = \sqrt{2\pi}$. Any other choice of a would have resulted in a reciprocal net that (after rotation) scaled, rather than coincided, with the original net.

The crystal lattice and the reciprocal lattice representations have different purposes. The crystal lattice describes, and enables us to visualize, the crystal structure. The reciprocal lattice will provide a means of describing electron states and phonon states in crystals.

Applying *periodic boundary conditions*, eq. (16.2.18), to the Bloch functions of eq. (16.2.26) yields

$$\exp(\mathrm{i}\,\mathbf{k}\cdot r)\,u_{\mathbf{k}}(\mathbf{r}) = \exp\left[\mathrm{i}\,\mathbf{k}\cdot\left(\mathbf{r} + \sum_{j=1}^{3} N_j\mathbf{a}_j\right)\right]\mathbf{u}_{\mathbf{k}}(\mathbf{r}). \tag{19}$$

(19), (1)
$$\mathbf{k} = \sum_{j=1}^{3}(m_j/N_j)\,\mathbf{b}_j = \sum_{j=1}^{3} k_j\mathbf{b}_j, \qquad m_j = 0, \pm 1, \pm 2\ldots, \tag{20}$$

showing that $\mathbf{k}$ is a vector in the reciprocal lattice with components $k_j = m_j/N_j$.

(19), (20)
$$\mathbf{k}\cdot\left(\sum_{j=1}^{3} N_j\mathbf{a}_j\right) = 2\pi\sum_{j=1}^{3} m_j = 2\pi p, \qquad p \text{ an integer}, \tag{21}$$

which confirms that eq. (19) is satisfied by eq. (20). Equation (20) tells us that the number of $\mathbf{k}$ vectors allowed by the periodic boundary conditions (PBC) is $N_1N_2N_3 = N$, the number of unit cells in the crystal lattice, and that these **k** vectors just fill a volume of the reciprocal lattice equal to v_b, the volume of the primitive unit cell of the reciprocal lattice. This volume is the first *Brillouin zone* (BZ). It could be chosen in various ways, but it is usual to take the Wigner–Seitz cell of the reciprocal lattice as the first BZ, except for monoclinic and triclinic space groups where the primitive unit cell is used instead. The number of **k** vectors per unit volume of the reciprocal lattice is

$$N(\mathbf{k}) = N/v_b = V/8\pi^3, \tag{22}$$

where $V = Nv_a$ is the volume of the crystal.

The MR $\Gamma(E|\,\mathbf{t})$ of the translation operator $(E|\,\mathbf{t})$ is given by eq. (16.2.24) as $\exp(-i\,\mathbf{k}\cdot\mathbf{t})$, where **t** is a translation $\mathbf{a}_n$. As **k** runs over its N allowed values in eq. (20) it generates the N irreducible representations (IRs) of the translation group T, which we therefore label by **k**, as in $\Gamma_{\mathbf{k}}\,(E|\mathbf{t})$.

(16.2.24), (7)–(9) $$\Gamma_{\mathbf{k}+\mathbf{b}_m}(E|\mathbf{a}_n) = \exp[-i(\mathbf{k}+\mathbf{b}_m)\cdot\mathbf{a}_n)] = \Gamma_{\mathbf{k}}(E|\mathbf{a}_n). \tag{23}$$

Therefore, $\mathbf{k}+\mathbf{b}_m$ and **k** label the same representation and are said to be *equivalent* ($\cong$). By definition, no two interior points can be equivalent but every point on the surface of the BZ has at least one equivalent point. The $\mathbf{k} = \mathbf{0}$ point at the center of the zone is denoted by Γ. All other internal high-symmetry points are denoted by capital Greek letters. Surface symmetry points are denoted by capital Roman letters. The elements of the point group which transform a particular **k** point into itself or into an equivalent point constitute the *point group of the wave vector* (or *little co-group* of **k**) $P(\mathbf{k}) \subseteq P$, for that **k** point.

We now describe a general method for the construction of the BZ. It is a consequence of the SP relation eqs. (7)–(9) that every reciprocal lattice vector $\mathbf{b}_m$ is normal to a set of planes in the direct lattice. In Figure 16.11(a), $\mathbf{b}_m$ is a reciprocal lattice vector that connects lattice point O to some other lattice point P_1. Let 1 be the plane through P_1 that is normal to $\mathbf{b}_m$ and let 0 be the plane parallel to 1 through O. Let $\mathbf{a}_n$ be the lattice vector from O to some

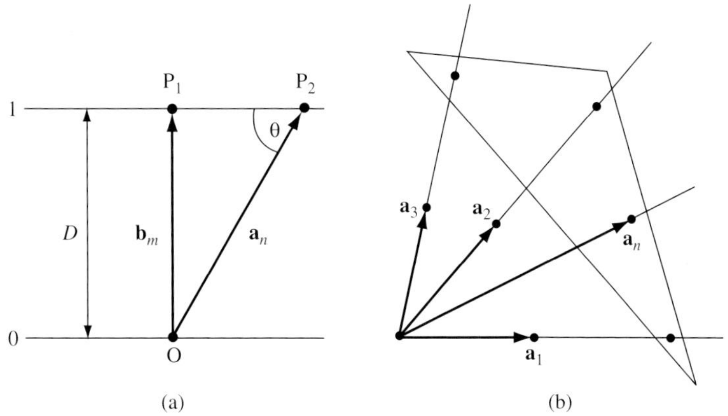

Figure 16.11. (a) $\mathbf{b}_m$ is a vector from the origin O to a lattice point P_1 in the reciprocal lattice representation, and plane 1 is normal to $\mathbf{b}_m$. The lattice translation $\mathbf{a}_n$ is a vector from O to another lattice point P_2 on plane 1. Plane 0 is parallel to plane 1 through O. (b) $\mathbf{a}_n$ intersects plane 1 at one of the other lattice points in plane 1. If $\mathbf{a}_n$ lies along $\mathbf{a}_1$, n_2 and n_3 are zero and $\mathbf{a}_n = n_1\mathbf{a}_1$. Similarly for $\mathbf{a}_2$, $\mathbf{a}_3$.

other lattice point P_2 on plane 1; $\mathbf{a}_n$ makes an angle θ with plane 1. The distance D between these planes is given by

(9) $$D = |\mathbf{a}_n| \cos\left(\frac{\pi}{2} - \theta\right) = \langle \mathbf{b}_m | \mathbf{a}_n \rangle / |\mathbf{b}_m| = 2\pi p / |\mathbf{b}_m|, \tag{24}$$

where $p = m_1 n_1 + m_2 n_2 + m_3 n_3$ is an integer. The intercept made by this plane 1 on $\mathbf{a}_1$ is $n_1 \mathbf{a}_1$. When $\mathbf{a}_n$ lies along $\mathbf{a}_1$, n_2 and n_3 are zero and $n_1 = p / m_1$. Similarly the intercepts $n_2 \mathbf{a}_2$, $n_3 \mathbf{a}_3$ along $\mathbf{a}_2$, $\mathbf{a}_3$ are given by $n_2 = p / m_2$, $n_3 = p / m_3$ respectively. Therefore the Miller indices of this plane, and so of a whole stack of parallel planes, are proportional to $((m_1\ m_2\ m_3))$. Removal of any common factor gives the Miller indices $((h_1\ h_2\ h_3))$. The spacing between each pair of adjacent planes is given by

(24) $$D/p = d_m = 2\pi / |\mathbf{b}_m|. \tag{25}$$

The plane normal to $\mathbf{b}_m$ with the same Miller indices, but located at a distance $|\mathbf{b}_m|/2$ from O, is one of the faces of the BZ. The equation $k\ (x\ y\ z)$ to this face is

$$\mathbf{k} \cdot \mathbf{u}_m = |\mathbf{b}_m|/2, \tag{26}$$

where $\mathbf{k}$ is a vector in the reciprocal lattice from O to this face, $\mathbf{u}_m$ is a unit vector normal to the face and $|\mathbf{b}_m|/2$ is the length of the normal from O to this face. A similar analysis for another vector $|\mathbf{b}_m|$ from O to a nearby lattice point yields another face of the BZ, and so on, until the equations to all the faces not connected by symmetry have been obtained and the whole BZ has thus been determined. Equation (23) emphasizes the primary importance of the first BZ. Consequently, BZ, or Brillouin zone, when unqualified, means the first BZ. The use of successively larger reciprocal lattice vectors $\mathbf{b}_m$ in eq. (25) gives the second, third, ... BZs (see, for example, Landsberg (1969)). For example, the smallest volume (lying outside the first zone), and enclosed by the next set of planes that satisfy eq. (25), forms the second BZ, and so on.

Exercise 16.3-2 Show that if $\mathbf{k}$ is the wave vector of incident radiation (X-ray or neutron) or the wave vector of a particle or quasiparticle, then eq. (25) leads to the Bragg diffraction condition.

Example 16.3-3 Construct the BZs of the primitive cubic and *fcc* lattices.

For the primitive cubic lattice $\mathbb{A} = a\mathrm{E}_3$, so that the reciprocal lattice is also primitive cubic with cube edge $b = 2\pi / a$. The shortest vectors from O to its near neighbors are

$$\begin{aligned} &\pm (2\pi/a)\,[1\ 0\ 0], \quad \pm (2\pi/a)\,[0\ 1\ 0], \quad \pm (2\pi/a)\,[0\ 0\ 1], \quad \text{or} \\ &\pm (2\pi/a)\,[[1\ 0\ 0]] \end{aligned} \tag{27}$$

where the double brackets signify the set of vectors equivalent by symmetry. The planes which bisect these vectors perpendicularly at $\pm(\pi/a)$ [[1 0 0]] determine the zone boundaries so that the BZ is a cube with cube edge $2\pi/a$. Symmetry points are marked in Figure 16.12(a), and their coordinates and point groups are given in Table 16.6.

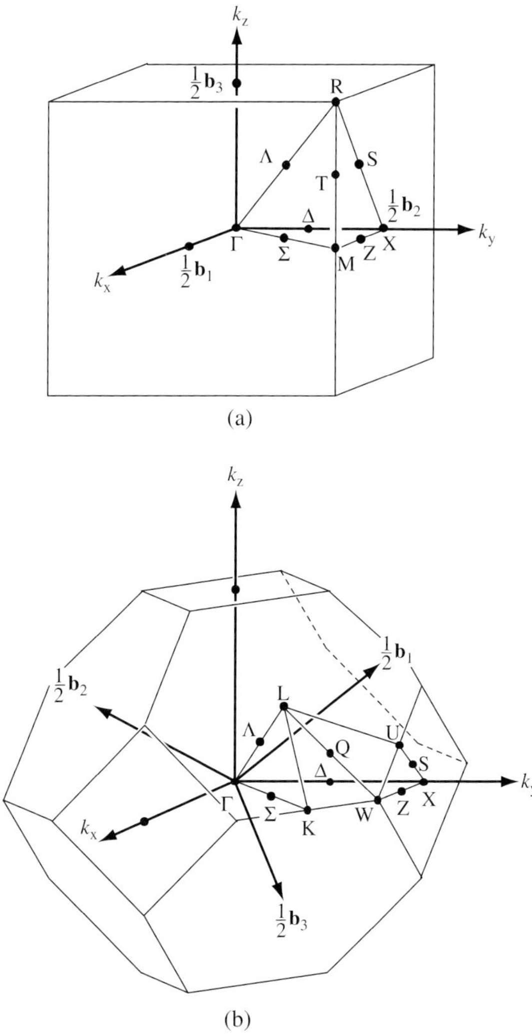

Figure 16.12. Brillouin zones, with symmetry points marked, of (a) the primitive cubic Bravais lattice and (b) the cubic close-packed or *fcc* Bravais lattice.

The reciprocal lattice of the *fcc* lattice (see eq. (12)) is *bcc*, and the fourteen planes which bisect the shortest vectors to near-neighbor reciprocal lattice points have Miller indices

$$\pm((111)), \quad \pm((200)), \tag{28}$$

which are, respectively, the eight hexagonal faces and the six square faces shown in Figure 16.12(b), which also shows the principal symmetry points. In these straightforward examples it was hardly necessary to find the equations for the faces of the BZ by the general method outlined earlier (see eq. (25)). The BZ for each of the fourteen Bravais lattices is shown in Bradley and Cracknell (1972), who also list the symmetry points and their point groups P(**k**). In their Table 3.7 they give the generating elements for all the 230 space groups. The principal

Table 16.6. *Symmetry points in the Brillouin zones (see Figure 16.12) of the reciprocal lattices of (a) the primitive cubic space lattice (simple cubic, sc), for which the reciprocal lattice is also sc, and (b) the fcc space lattice, which has a bcc reciprocal lattice.*

$\mathbf{k} = \alpha\mathbf{b}_1 + \beta\mathbf{b}_2 + \gamma\mathbf{b}_3 = \xi\mathbf{e}_1 + \eta\mathbf{e}_2 + \zeta\mathbf{e}_3$. In column 2, the coordinates of the point **k** are the components of the vector **k** with respect to the basis $\langle\mathbf{b}_1\ \mathbf{b}_2\ \mathbf{b}_3|$, in units of $2\pi/a$ for both (a) and (b). For the *sc* lattice, $\mathrm{B} = \mathrm{E}_3$ and $[\xi\ \eta\ \zeta] = [\alpha\ \beta\ \gamma]$. In (b), $[\xi\ \eta\ \zeta]$ are the Cartesian coordinates of the point **k** in units of $2\pi/a$, so to obtain $[\xi\ \eta\ \zeta]$ for **k** in units of the cube edge $4\pi/a$, divide the values in column 5 by two.

(a)	Point	Coordinates	P(**k**)		
	Γ	[0 0 0]	$m3m$	O_h	
	X	[0 ½ 0]	$4/mmm$	D_{4h}	
	M	[½ ½ 0]	$4/mmm$	D_{4h}	
	R	[½ ½ ½]	$m3m$	O_h	
	Δ(ΓX)	[0 β 0]	$4mm$	C_{4v}	
	Σ(ΓM)	[α α 0]	$mm2$	C_{2v}	
	Λ(ΓR)	[α α α]	$3m$	C_{3v}	
	S(XR)	[α ½ α]	$mm2$	C_{2v}	
	Z(XM)	[α ½ 0]	$mm2$	C_{2v}	
	T(MR)	[½ ½ γ]	$4mm$	C_{4v}	
(b)	**Point**	**Coordinates**	**P(k)**		**[ξ η ζ]**
	Γ	[0 0 0]	$m3m$	O_h	[0 0 0]
	X	[½ 0 ½]	$4/mmm$	D_{4h}	[0 1 0]
	L	[½ ½ ½]	$\bar{3}m$	D_{3d}	[½ ½ ½]
	W	[½ ¼ ¾]	$\bar{4}2m$	D_{2d}	[½ 1 0]
	K	[⅜ ⅜ ¾]	$mm2$	C_{2v}	[¾ ¾ 0]
	Δ(ΓX)	[α 0 α]	$4mm$	C_{4v}	[0 η 0]
	Σ(ΓK)	[α α 2α]	$mm2$	C_{2v}	[ξ ξ 0]
	Λ(ΓL)	[α α α]	$3m$	C_{3v}	[ξ ξ ξ]
	S(XU)	[½ + α, 2α, ½ + α]	$mm2$	C_{2v}	[ξ 1 ξ]
	Z(XW)	[½, α, ½ + α]	$mm2$	C_{2v}	[ξ 1 0]
	Q(LW)	[½, ½ − α, ½ + α]	2	C_2	[½, ½ + η, ½ − η]

symmetry points in the BZs shown in Figure 16.12 are listed in Table 16.6, together with their coordinates and point groups P(**k**). The notation Δ(ΓX), for example, means any point on the axis of symmetry ΓX (excluding the end points when they have different symmetry).

Answers to Exercises 16.3

Exercise 16.3-1 From eqs. (17) and (16),

$$\mathrm{B}^{\mathrm{T}}\,\mathrm{A} = (2b/\sqrt{3})a\begin{bmatrix} 1 & 0 \\ 1/2 & \sqrt{3}/2 \end{bmatrix}\begin{bmatrix} \sqrt{3}/2 & 0 \\ -1/2 & 1 \end{bmatrix}$$

$$= (ba)(2/\sqrt{3})\begin{bmatrix} \sqrt{3}/2 & 0 \\ 0 & \sqrt{3}/2 \end{bmatrix} = 2\pi\,\mathrm{E}_2. \tag{16}$$

Exercise 16.3-2 From eqs. (25) and (26),

$$\mathbf{k}\cdot\mathbf{u}_m = (2\pi/\lambda)\sin\theta = \pi/d_m, \qquad \lambda = 2\,d_m\ \sin\theta. \tag{29}$$

16.4 Space-group representations

The action of the space-group operator $(R|\mathbf{v}) \in \mathrm{G}$ (with $R \in \mathrm{P}$), the point group of the space group G on the Bloch function $\psi_{\mathbf{k}}(\mathbf{r})$ gives the transformed function $\psi'_{\mathbf{k}}(\mathbf{r})$. To find the transformed wave vector $\mathbf{k}'$ we need the eigenvalue $\exp\,(-\mathrm{i}\mathbf{k}'\cdot\mathbf{t})$ of the translation operator $(E|\mathbf{t})$.

$$(E|\mathbf{t})[(R|\mathbf{v})\,\psi_k(\mathbf{r})] = (R|\mathbf{v})\,(E|R^{-1}\mathbf{t})\,\psi_{\mathbf{k}}(\mathbf{r}) \tag{1}$$

$$= (R|\mathbf{v})\exp(-\mathrm{i}\,\mathbf{k}\cdot R^{-1}\mathbf{t})\,\psi_{\mathbf{k}}(\mathbf{r}) \tag{2}$$

$$= \exp(-\mathrm{i}\,R\mathbf{k}\cdot\mathbf{t})(R|\mathbf{v})\,\psi_{\mathbf{k}}(\mathbf{r}). \tag{3}$$

The space-group operator $(R|\,\mathbf{v})$ acts on functions of $\mathbf{r}$, and therefore the exponential factor in eq. (2), which is not a function of $\mathbf{r}$, is unaffected by $(R|\,\mathbf{v})$.

Exercise 16.4-1 (a) Verify the equality of the operator products on each side of eq. (1). [*Hint*: Use the multiplication rule for Seitz operators.] (b) Verify the equality of the RS of eqs. (2) and (3). (c) Find the transformed Bloch function $\psi'_{\mathbf{k}}(\mathbf{r})$ when $(R\,|\,\mathbf{v})$ is $(I\,|\,\mathbf{0})$.

Equation (3) shows that the space-group operator $(R\,|\,\mathbf{v})$ transforms a Bloch function with wave vector $\mathbf{k} \in$ BZ into one with wave vector $R\,\mathbf{k}$, which either also lies in the BZ or is equivalent to ($\cong$) a wave vector $\mathbf{k}'$ in the first BZ. (The case $\mathbf{k}' = \mathbf{k}$ is not excluded.) Therefore, as R runs over the whole $\{R\} = \mathrm{P}$, the isogonal point group of G, it generates a basis $\langle\psi_{\mathbf{k}}|$ for a representation of the space group G,

(3) $$(R|\mathbf{v})\,\psi_{\mathbf{k}} = \psi_{R\,\mathbf{k}}, \qquad \forall\ \mathbf{k} \in \mathrm{BZ}, \forall R \in \mathrm{P}, R\mathbf{k} \in \mathrm{BZ}\ \text{or} \cong \mathbf{k}' \in \mathrm{BZ}. \tag{4}$$

On introducing the notation $\langle\mathbf{k}|$, meaning the whole set of Bloch functions that form a basis for a representation of G,

(4) $$(R|\mathbf{v})\langle\mathbf{k}| = \langle R\ \mathbf{k}|. \tag{5}$$

Because $R\,\mathbf{k} \cong \mathbf{k}' \in \langle\mathbf{k}|, (R|\mathbf{v})\langle\mathbf{k}|$ simply reorders the basis as eq. (5) implies.

Exercise 16.4-2 Prove that two bases $\langle\mathbf{k}|$, $\langle\mathbf{k}'|$ either have no $\mathbf{k}$ vector in common or they are identical.

The point group $\mathrm{P} \subseteq \mathrm{P_{BL}}$ and when $\mathrm{P} = \mathrm{P_{BL}}$ (which is so for a holosymmetric space group) the points and lines of symmetry mark out the *basic domain* Ω of the Brillouin zone. When

$P \subset P_{BL}$ the points and lines of symmetry define the *representation domain* $\Phi \supseteq \Omega$, such that $\Sigma R\Phi, \forall R \in P$, is equal to the whole BZ. If $R\mathbf{k}_1 = \mathbf{k}_1 + \mathbf{b}_m$, where the reciprocal lattice vector $\mathbf{b}_m$ may be the null vector $\mathbf{0}$, so that $R\mathbf{k}_1 = \mathbf{k}_1'$ is either identical ($\equiv$) or $\cong \mathbf{k}_1$, then $R \in P(\mathbf{k}_1)$, the *point group of the wave vector* $\mathbf{k}_1$. But if $R\mathbf{k}_1 = \mathbf{k}_2$, where $\mathbf{k}_2$ is not $\equiv$ or $\cong \mathbf{k}_1$, then $\mathbf{k}_1, \mathbf{k}_2, \ldots \in {}^*\mathbf{k}_1$, called the *star* of $\mathbf{k}_1$. That is, the star of $\mathbf{k}_1$ is the set of *distinct* (*inequivalent*) $\mathbf{k}$ vectors $\in \{R\mathbf{k}_1\}$. The $\{R\}$ for which $R\mathbf{k}_1$ is not $\equiv$ or $\cong \mathbf{k}_1$ are called the generators of ${}^*\mathbf{k}_1$.

The *little group* (or group of the wave vector) $G(\mathbf{k})$ is the space group

$$G(\mathbf{k}) = \sum (R_j|\mathbf{w}_j)\,T, \quad \forall R_j \in P(\mathbf{k}), \tag{6}$$

where $\mathbf{w}_j$ is either the null vector $\mathbf{0}$ or the special non-lattice vector associated with some screw axis or glide plane. The product of two coset representatives in eq. (6)

(6) $$(R_i|\,\mathbf{w}_i)(R_j|\,\mathbf{w}_j) = (R_i\,R_j|R_i\,\mathbf{w}_j + \mathbf{w}_i) = (E\,|\mathbf{t}_{ij})(R_k|\,\mathbf{w}_k) \tag{7}$$

$$= (R_k|\,\mathbf{w}_k)\,(E|\,R_k^{-1}\,\mathbf{t}_{ij}), \tag{8}$$

where

$$\mathbf{t}_{ij} = \mathbf{w}_i + R_i\,\mathbf{w}_j - \mathbf{w}_k \in T. \tag{9}$$

Equations (8) and (9) show that $\{(R_j|\mathbf{w}_j)\}$ is not, in general, closed since only for symmorphic groups is $\mathbf{t}_{ij} = \mathbf{0}$, $\forall\, i, j$. They also establish the multiplication rule for the cosets as

$$[(R_i\,|\,\mathbf{w}_i)\,T]\,[R_j\,|\,\mathbf{w}_j)\,T] = [(R_k\,|\,\mathbf{w}_k)\,T], \tag{10}$$

and that $\{(R_j|\mathbf{w}_j)T\}$ is closed, thus confirming that $G(\mathbf{k})$ is a group. T is an invariant subgroup of $G(\mathbf{k})$, and so the *little factor group*

(6) $$F(\mathbf{k}) = G(\mathbf{k})/T = \{(R_j\,|\,\mathbf{w}_j)T\}, \forall\, R_j \in P(\mathbf{k}). \tag{11}$$

Equation (10) confirms that $F(\mathbf{k})$ is a group and that $F(\mathbf{k}) \sim P(\mathbf{k})$. In view of the mappings of the factor group

$$F = G/T = \{(R_j\,|\,\mathbf{w}_j)T\}, \quad \forall\, R_j \in P, \tag{12}$$

on to P, $F \leftrightarrow P$, and of the little factor group $F(\mathbf{k}) \leftrightarrow P(\mathbf{k})$,

(12), (11) $$F = G/T = \sum_j\,(R_j\,|\,\mathbf{w}_j)\,F(\mathbf{k}), \quad R_j\mathbf{k} \in {}^*\mathbf{k}. \tag{13}$$

(13) $$P \sim \sum_j R_jP(\mathbf{k}),\ j = 1, \ldots, s(\mathbf{k}); \tag{14}$$

(14) $$s(\mathbf{k}) = g(P)/p(\mathbf{k}), \tag{15}$$

where $g(P)$ is the order of P and $s(\mathbf{k})$, $p(\mathbf{k})$ are the orders of ${}^*\mathbf{k}$ and of $P(\mathbf{k})$.

Equation (11) represents a major simplification in the problem of determining the IRs of $G(\mathbf{k}) \subset G$. Because it contains the translation subgroup $T = \{(E\,|\,\mathbf{t})\}$, $\mathbf{t} \in \{\mathbf{a_k}\}$, $G(\mathbf{k})$ is a very large group. $F(\mathbf{k})$ is a much smaller group than $G(\mathbf{k})$, but the elements of $F(\mathbf{k})$ are the

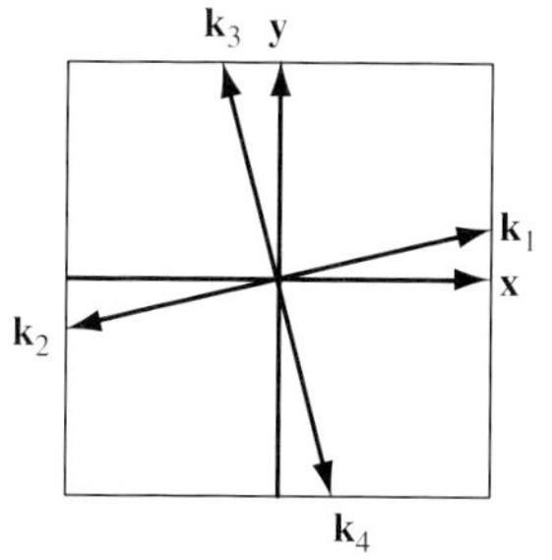

Figure 16.13. Brillouin zone of the reciprocal lattice of the strictly 2-D square lattice, demonstrating the construction of $^*\mathbf{k}_1 : E\mathbf{k}_1 = \mathbf{k}_1,\ C_{2z}\mathbf{k}_1 = \mathbf{k}_2,\ C^+_{4z}\mathbf{k}_1 = \mathbf{k}_3, C^-_{4z}\mathbf{k}_1 = \mathbf{k}_4$.

cosets $\{(R_j|\mathbf{w}_j)\mathrm{T}\}$ and it is easier to work with its isomorph, the point group P(**k**), which is therefore also known as *the little co-group*. For symmorphic space groups, or for internal points of non-symmorphic space groups, P(**k**) (or F(**k**), see, for example, Cornwell (1984)) will provide a satisfactory route to the space-group representations. The disadvantage is that not only the translations for which $\exp(-\mathrm{i}\,\mathbf{k}\cdot\mathbf{t}) = 1$, but the whole set $\{(E|\mathbf{t})\}$ is mapped on to the identity. For points that lie on surface lines of symmetry, or for surface symmetry points, this results in a loss of information which can, however, be restored by finding projective representations (PRs) of P(**k**), although for symmetry points that lie on the surface of the BZ an alternative method proposed by Herring (1942) is generally somewhat easier to apply (Section 16.7).

The star of **k** was defined as the set of distinct (inequivalent) **k** vectors that are $\subset \{R_j\,\mathbf{k}\}$, the $R_j\,\mathbf{k}$ that are $\equiv$ or $\cong$ **k** being $\in$ P(**k**). Equation (14) establishes the equivalent definition

$$^*\mathbf{k} = \{R_j\,\mathbf{k}\}, \forall\, R_j \text{ in } \mathrm{P} \sim \Sigma R_j\,\mathrm{P}(\mathbf{k}), j = 1, \ldots, s(\mathbf{k}). \tag{16}$$

The isomorphism in eq. (14) is a sufficient requirement, but in fact the equality often holds, as it does in Example 16.4-2 below. An example of the isomorphism is provided by Altmann (1977), p. 208.

Example 16.4-1 Construct the star of $\mathbf{k}_1$ in the BZ of the reciprocal lattice of the strictly 2-D square lattice shown in Figure 16.13.

The point group P is $\mathrm{C_{4v}} = \{E\ C_2\ 2C_4\ 2\sigma_\mathrm{v}\ 2\sigma_\mathrm{d}\}$, of order $g(\mathrm{P}) = 8$. Shown in Figure 16.13 are the vectors of $^*\mathbf{k}_1$, comprising the inequivalent vectors $\mathbf{k}_1$, $\mathbf{k}_2$, $\mathbf{k}_3$, $\mathbf{k}_4$ generated by $R\,\mathbf{k}_1$ with $R \in \{E\ C_{2z}\ C^+_{4z}\ C^-_{4z}\} = \mathrm{C_4}$.

Exercise 16.4-3 Show graphically that the vectors produced by the remaining operators of $\mathrm{C_{4v}}$ are equivalent to a member of $^*\mathbf{k}_1 = \{\mathbf{k}_1\ \mathbf{k}_2\ \mathbf{k}_3\ \mathbf{k}_4\}$.

Example 16.4-2 Write a coset expansion of the point group P, for the strictly 2-D square lattice, on P(**k**), where **k** is the vector from the origin to point Z in Figure 16.14.

Here P is $\mathrm{C_{4v}}$ and P(**k**) at Z is $\mathrm{C}_s = \{E\ \sigma_\mathrm{x}\}$ because $\mathrm{Z}' \cong \mathrm{Z}$, but no other $R \in \mathrm{P}$ give $R\,\mathbf{k} \equiv \mathrm{Z}$ or $\cong$ Z. Therefore, $p(\mathbf{k}) = 2$, and from eq. (15) the number of elements in $^*\mathrm{Z}$ is

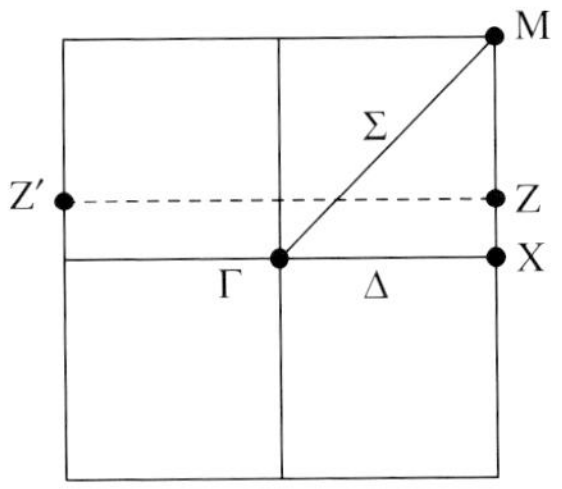

Figure 16.14. Brillouin zone of the reciprocal lattice of the strictly 2-D square lattice. The symmetry points mark out the basic domain; Z′ and Z are equivalent points.

$s(\mathrm{Z}) = 8/2 = 4$. From Example 16.4-1, $^*\mathrm{Z}$ is $R\,\mathbf{k}_1$ with $\{R\} = \{E\ C_{2z}\ C_{4z}^{+}\ C_{4z}^{-}\}$, and therefore the coset expansion on P(**k**) is

(14) $$E\{E\ \sigma_x\} \oplus C_{2z}\{E\ \sigma_x\} \oplus C_{4z}^{+}\{E\ \sigma_x\} \oplus C_{4z}^{-}\{E\ \sigma_x\}$$

$$= \{E\ \sigma_x\ C_{2z}\ \sigma_y\ C_{4z}^{+}\ \sigma_a\ C_{4z}^{-}\ \sigma_b\} = C_{4v} = P,$$

where $\mathbf{a} = 2^{-1/2}[1\ 1\ 0]$, $\mathbf{b} = 2^{-1/2}[\bar{1}\ 1\ 0]$.

16.4.1 Representations of the little group

Let $\{\widetilde{\Gamma}_{\mathbf{k}}(R\,|\,\mathbf{w})\}$ be the set of MRs of the coset representatives in eq. (6).

(7), (9) $$\widetilde{\Gamma}_{\mathbf{k}}(R_i\,|\,\mathbf{w}_i)\,\widetilde{\Gamma}_{\mathbf{k}}(R_j\,|\,\mathbf{w}_j) = \exp(-\mathrm{i}\,\mathbf{k}\cdot\mathbf{t}_{ij})\,\widetilde{\Gamma}_{\mathbf{k}}(R_k\,|\,\mathbf{w}_k). \tag{17}$$

Define

$$\Gamma_{\mathbf{k}}(R) = \exp(\mathrm{i}\,\mathbf{k}\cdot\mathbf{w})\,\widetilde{\Gamma}_{\mathbf{k}}(R\,|\,\mathbf{w}). \tag{18}$$

(17), (18) $$\widetilde{\Gamma}_{\mathbf{k}}(R_i)\,\widetilde{\Gamma}_{\mathbf{k}}(R_j) = \exp[-\mathrm{i}\,\mathbf{k}\cdot(\mathbf{t}_{ij} - \mathbf{w}_i - \mathbf{w}_j + \mathbf{w}_k)]\,\widetilde{\Gamma}_{\mathbf{k}}(R_k)$$

(9) $$= \exp[-\mathrm{i}\,\mathbf{k}\cdot(R_i\mathbf{w}_j - \mathbf{w}_j)]\,\widetilde{\Gamma}_{\mathbf{k}}(R_k)$$

$$= \exp(\mathrm{i}\,\mathbf{k}\cdot\mathbf{w}_j)\exp(-\mathrm{i}R_i^{-1}\mathbf{k}\cdot\mathbf{w}_j)\,\Gamma_{\mathbf{k}}(R_k). \tag{19}$$

$R_i^{-1}\mathbf{k} \cong \mathbf{k}$ and so can differ from **k** only by a reciprocal lattice vector,

$$R_i^{-1}\mathbf{k} = \mathbf{k} + \mathbf{b}_i. \tag{20}$$

In eq. (20) $\mathbf{b}_i$ is different from **0** only if **k** lies on the surface of the BZ.

(19), (20) $$\Gamma_{\mathbf{k}}(R_i)\,\Gamma_{\mathbf{k}}(R_j) = \exp(-\mathrm{i}\,\mathbf{b}_i\cdot\mathbf{w}_j)\,\Gamma_{\mathbf{k}}(R_k). \tag{21}$$

In eq. (21) $\mathbf{w}_j$ is either **0** or the non-lattice translation associated with R_j. Equations (18) and (21) show that the MRs $\Gamma_{\mathbf{k}}(R|\mathbf{w})$ of the coset representatives $(R\,|\,\mathbf{w})$ in the little group of the **k** vector G(**k**) can be written as the product of an exponential factor $\exp(-\mathrm{i}\,\mathbf{k}\cdot\mathbf{w})$ and the

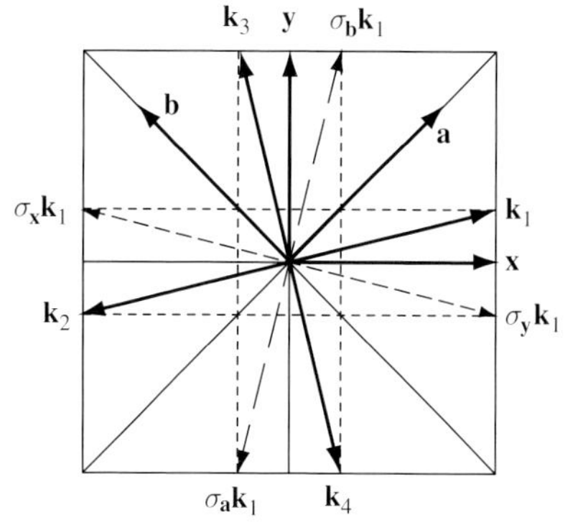

Figure 16.15. This figure, in conjunction with Figure 16.13, proves that $\{R\}$, $R\mathbf{k}_1 = {}^*\mathbf{k}_1$, is C_4, since $\sigma_y\mathbf{k}_1 \cong \mathbf{k}_2$, $\sigma_x\mathbf{k}_1 \cong \mathbf{k}_1$, $\sigma_a\mathbf{k}_1 \cong \mathbf{k}_3$, $\sigma_b\mathbf{k}_1 \cong \mathbf{k}_4$.

MR $\Gamma_{\mathbf{k}}(R)$, where the set of matrices $\{\Gamma_{\mathbf{k}}(R)\}$ form a PR of the point-group operators R that form the point group P($\mathbf{k}$) of the little group G($\mathbf{k}$).

Exercise 16.4-4 Explain why the $\{\Gamma_{\mathbf{k}}(R)\}$ are just vector representations for symmorphic space groups.

An alternative method due to Herring (1942) (see also Altmann (1977) and Bradley and Cracknell (1972)) avoids the use of PRs and instead involves finding the ordinary vector representations of a point group of order greater than that of P($\mathbf{k}$). Herring's method will be illustrated in Section 16.7.

Answers to Exercises 16.4

Exercise 16.4-1 (a) $(R\,|\,\mathbf{v})(E\,|\,R^{-1}\mathbf{t}) = (R\,|\,\mathbf{t}+\mathbf{v}) = (R\,|\,\mathbf{v}+\mathbf{t}) = (E\,|\,\mathbf{t})(R\,|\,\mathbf{v})$. (b) $\mathbf{k}$, $\mathbf{t}$ are vectors in the same space, the space of the crystal pattern and the Bravais lattice. The scalar product of two vectors is invariant under a rigid rotation of the LVS, and so

$$\mathbf{k} \cdot R^{-1}\mathbf{t} = R\mathbf{k} \cdot RR^{-1}\mathbf{t} = R\,\mathbf{k} \cdot \mathbf{t}.$$

(c) $(I\,|\mathbf{0})\psi_{\mathbf{k}}(\mathbf{r}) = \psi_{I\mathbf{k}}(\mathbf{r}) = \psi_{-\mathbf{k}}(\mathbf{r})$

Exercise 16.4-2 $R\mathbf{k} \in \langle\mathbf{k}|$, $R\,\mathbf{k}' \in \langle\mathbf{k}'\,|$, $\forall\, R, R' \in$ P. Suppose that $R\mathbf{k} = R'\mathbf{k}'$; then $\mathbf{k}' = R'^{-1}$ $R\mathbf{k} = R''\mathbf{k}$ (closure in P). Therefore $\mathbf{k}' \in \langle\mathbf{k}|$, and consequently $R\mathbf{k}' \in \langle\mathbf{k}|$, $\forall\, R' \in$ P, so that $\langle\mathbf{k}'| \equiv \langle\mathbf{k}|$. Therefore two bases either have no $\mathbf{k}$ vector in common, or they are identical.

Exercise 16.4-3 Figure 16.13 showed the result $R\mathbf{k}_1$ when $R \in C_4$. The remaining elements of C_{4v} (namely, σ_x, σ_y, σ_a, σ_b) yield vectors (shown by dashed lines in Figure 16.15) which are $\cong$ to members of the ${}^*\mathbf{k}_1$.

Exercise 16.4-4 For symmorphic space groups $\mathbf{w}_j = \mathbf{0}$, $\forall\,(R_j\,|\,\mathbf{w}_j)$ and the projective factors (PFs) $\exp(-\,\mathbf{b}_i \cdot \mathbf{w}_j)$ are all unity.

16.5 The covering group

The problem of finding space-group representations of the little group G(**k**) has been reduced in Section 16.4 to that of finding PRs of the point group P(**k**), as defined by eq. (16.4.18). The way that we shall do this (Altmann (1977); but *cf.* Hurley (1966) and Kim (1999)) is by finding the *vector representations* of the *covering group* P(**k**)′. The construction of P(**k**)′ involves finding the central extension of P(**k**) by a cyclic Abelian subgroup Z_n. Since we always use a factor system that is normalized and standardized, the PFs (which are known from eqs. (16.4.21) and (16.4.20)) are all *n*th roots ofunity,

$$[i\,;\,j] = \exp(2\pi\,\mathrm{i}\,n^{-1}z_{ij}), \tag{1}$$

where n and z_{ij} are integers. The cyclic group Z_n, of order $z(n)$, is defined by

$$Z_n = \{z\},\ z = 1, 2, \ldots, n = 0, \tag{2}$$

with binary composition defined as addition modulo n. The covering group P(**k**)′, of order $z(n)p(\mathbf{k})$, is then

$$\mathrm{P}(\mathbf{k})' = \{(R_i, z_l)\},\ \forall\ R_i \in \mathrm{P}(\mathbf{k}),\ \forall\ z_l \in Z_n, \tag{3}$$

with binary composition defined by

$$\{(R_i, z_l)\}\{(R_j, z_m)\} - \{(R_iR_j, z_l + z_m + z_{ij})\}, \text{addition modulo } n. \tag{4}$$

The set $\{(R_i, z_l)\}$ has the following properties:

(i) it is *closed*, which follows from eq. (4);
(ii) it contains the *identity* $(E, 0)$;
(iii) each element (R_i, z_l) has an inverse

$$(R_i, z_l)^{-1} = (R_i^{-1}, -z_l - z_{ii}); \tag{5}$$

(iv) it exhibits *associativity*,

$$[(R_i, z_l)(R_i, z_m)](R_k, z_n) = (R_i, z_l)[(R_i, z_m)(R_k, z_n)]. \tag{6}$$

Exercise 16.5-1 Verify the above assertions concerning the inverse of (R_i, z_l) and the associative property of $\{(R_i, z_l)\}$.

Exercise 16.5-2 Prove that $z_{EE} = z_{ER_i} = z_{R_iE} = n, \mathrm{mod}\ n$. Hence verify that (E, z_m) commutes with (R_i, z_l).

That P(**k**)′ defined by eq. (3) is a central extension of P(**k**) by Z_n is readily established. Firstly, $(R_i, z_l) \to R_i, \forall\, z_l \in Z_n$, so that (R_i, z_l) is the pre-image of R_i, while eq. (4) shows that binary composition is preserved. Secondly, $(R_i, z_l) \to R_i$ is a homomorphous mapping, and since $(E, z_l) \to E, \forall\ z_l \in Z_n$, is the kernel of the homomorphism. Lastly (E, z_l) commutes with $\forall\,(R_i, z_l) \in \mathrm{P}(\mathbf{k})'$ so that (E, z_l) is the *center* of $\mathrm{P}(\mathbf{k})'$. Therefore, $\mathrm{P}(\mathbf{k})'$ is a group which is the *central extension* of $\mathrm{P}(\mathbf{k})$ by Z_n, and the further condition

$$(R_i, 0) \to R_i \tag{7}$$

ensures that $\mathrm{P}(\mathbf{k})'$ is a covering group of $\mathrm{P}(\mathbf{k})$. This is so because of a general theorem that, given a group G and its central extension G′ by $\mathrm{H}' \subset \mathrm{Z}(\mathrm{G}')$, G′ is a covering group of G, with the property that the irreducible vector representations of G′ are irreducible projective representations of G, provided that there exists a rule (such as eq. (7)) that establishes the pre-image of g_i, $\forall\, g_i \in \mathrm{G}$. Moreover this procedure gives all the IRs of G for a given factor system. I shall not reproduce the rather lengthy proof of this theorem here: it is given, for example, in Altmann (1977), pp. 86–8, 95–6, and Bradley and Cracknell (1972), pp. 181–3.

Answers to Exercises 16.5

Exercise 16.5-1

$$(R_i, z_l)(R_i^{-1}, -z_l - z_{ii}) = (E, -z_l - z_{ii} + z_l + z_{ii}) = (E, 0).$$

On substituting eq. (1) in eq. (12.6.6), $z_{ij} + z_{ij,k} = z_{i,jk} + z_{j,k}$ and eq. (6) follows.

Exercise 16.5-2 $[E, E] = 1 = \exp(2\pi i n^{-1} z_{EE})$; therefore $z_{EE} = n$ and similarly for the other two relations. $(E, z_m)(R_i, z_l) = (ER_i, z_m + z_l + z_{ER_i}) = (R_i, z_m + z_l + n)$, $(R_i, z_l)(E, z_m) = (R_iE, z_l + z_m + z_{R_iE}) = (R_i, z_m + z_l + n)$.

16.6 The irreducible representations of G

Now that we have the PRs of $\mathrm{P}(\mathbf{k})$, the small IRs $\widetilde{\Gamma}_{\mathbf{k}}\,(R\,|\,\mathbf{w})$ of the little group $\mathrm{G}(\mathbf{k})$ follow from

(16.4.18) $$\widetilde{\Gamma}_{\mathbf{k}}\,(R|\mathbf{w}) = \exp(-i\,\mathbf{k}\cdot\mathbf{w})\,\Gamma_{\mathbf{k}}(R). \tag{1}$$

The final step going from the small IRs of the little group $\mathrm{G}(\mathbf{k})$ to the IRs of G requires the theory of induced representations (Section 4.8). At a particular $\mathbf{k}$ in the representation domain, the left coset expansion of G on the little group $\mathrm{G}(\mathbf{k})$ is

$$\mathrm{G} = \Sigma(R|\mathbf{w})\ \mathrm{G}(\mathbf{k}),\quad \forall R \in {}^*\mathbf{k}. \tag{2}$$

(2), (4.8.18) $$\Gamma_{\mathbf{k}}(R|\mathbf{v})_{[r\ s]} = \widetilde{\Gamma}_{\mathbf{k}}[(R_r|\mathbf{w}_r)^{-1}(R|\mathbf{v})(R_s|\mathbf{w}_s)]$$

$$= \widetilde{\Gamma}_{\mathbf{k}}(R_u|\mathbf{v}_{\mathbf{q}}),\quad (R_u|\mathbf{v}_q) \in \mathrm{G}(\mathbf{k}) \tag{3}$$

$$= 0,\quad (R_u|\mathbf{v}_q) \notin \mathrm{G}(\mathbf{k}), \tag{4}$$

$$(3) \qquad R_u = R_r^{-1} R\, R_s, \qquad (5)$$

$$(3) \qquad \mathbf{v}_q = R_r^{-1}(R\mathbf{w}_s + \mathbf{v} - \mathbf{w}_r). \qquad (6)$$

Exercise 16.6-1 Verify eqs. (5) and (6) by evaluating the product of space-group operators in eq. (3).

Let $\mathbf{w}_u$ be the non-lattice translation vector that belongs to R_u in the coset expansion, eq. (2). Then

$$(R_u|\mathbf{v}_q) = (R_u|\mathbf{w}_u)(E| - R_u^{-1}\mathbf{w}_u + R_u^{-1}\mathbf{v_q}) \qquad (7)$$

$$= (R_u|\mathbf{w}_u)(E| - R_u^{-1}\mathbf{w}_u)(E|R_u^{-1}\mathbf{v_q}). \qquad (8)$$

But the only coset representative in $\{(R|\mathbf{w})\}$ with $R = E$ is $(E|\mathbf{0})$; therefore, the second and third space-group operators on the RS of eq. (8) must be $\in$ G($\boldsymbol{k}$). Define $\{R\}'$ as $\{R\}$ in eq. (2) *with the exception of E*. Then from eqs. (3) and (8), $(R_u|\mathbf{v}_q) \in$ G($\mathbf{k}$) implies that $(R_u|\mathbf{w}_u) \in$ G($\mathbf{k}$) and therefore that $R_u \notin \{R\}'$. Conversely, $(R_u|\mathbf{v}_q) \notin$ G($\mathbf{k}$) implies that $R_u \in \{R\}'$. Therefore, this criterion $R_u \in$ or $\notin \{R\}'$ enables us to decide whether the $[r\,s]$ element of the supermatrix $\Gamma_{\mathbf{k}}(R|\mathbf{v})$ in the induced representation $\Gamma_{\mathbf{k}}(R|\mathbf{w}) = \tilde{\Gamma}_{\mathbf{k}} \uparrow$ G is to be replaced by the null matrix or by $\tilde{\Gamma}_{\mathbf{k}}(R_u|\mathbf{v}_q)$ with R_u, $\mathbf{v}_q$ given by eqs. (5) and (6). Therefore,

$$(3) \qquad \Gamma_{\mathbf{k}}(R|\mathbf{v}_{[r\,s]}) = \tilde{\Gamma}_{\mathbf{k}}(R_u|\mathbf{v}_q), \quad R_u \notin \{R\}' \qquad (9)$$

$$(4) \qquad = 0, \quad R_u \in \{R\}'. \qquad (10)$$

This method of finding all the IRs of the space group G at any particular value of $\mathbf{k}$ in the representation domain of the BZ will now be summarized.

(i) Form the little co-group

$$\mathrm{P}(\mathbf{k}) = \{R\},\ R\mathbf{k} = \mathbf{k} + \mathbf{b}_m,\ R \in \mathrm{P}. \qquad (11)$$

(ii) Find the ${}^*\mathbf{k}$, that is the set of distinct, inequivalent $\mathbf{k}$ vectors $\in \{R\mathbf{k}\}$, $R \in$ P.

(iii) Write down the coset expansion

$$\textstyle\sum R\ \mathrm{P}(\mathbf{k}),\ R\mathbf{k} \in {}^*\mathbf{k}. \qquad (12)$$

The set $\{(R|\mathbf{w})\}$, with R from eq. (12), label the rows and columns of the supermatrix $\Gamma_{\mathbf{k}}(R|\mathbf{w})$.

(iv) Find the factor system for the little co-group P($\mathbf{k}$),

$$[(R_i|\mathbf{w}_i)\ ;\ (R_j|\mathbf{w}_j)] = \exp(-\mathrm{i}\mathbf{b}_i \cdot \mathbf{w}_j),\ \mathbf{b}_i = R_i^{-1}\mathbf{k} - \mathbf{k},\ \forall\ R \in \mathrm{P}(\mathbf{k}). \qquad (13)$$

(v) Find all the $\Gamma(R)$ matrices of the projective IRs of P($\mathbf{k}$) with the factor system, eq. (13); hence write down the small representations, that is, the IRs of the group of the $\mathbf{k}$ vector

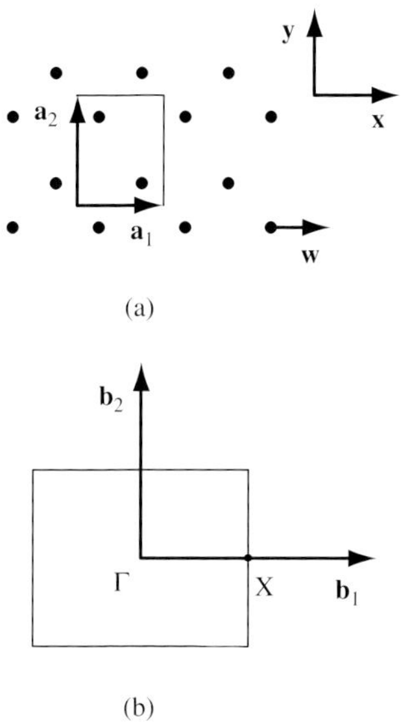

Figure 16.16. (a) Crystal pattern of a 2-D non-symmorphic space group.(b) Brillouin zone of the reciprocal lattice; $\Gamma = [0\ 0]$, $\mathrm{X} = [½\ 0]$.

$$\tilde{\Gamma}_{\mathbf{k}}(R|\mathbf{w}) = \Gamma_{\mathbf{k}}(R)\exp(-\mathrm{i}\,\mathbf{k}\cdot\mathbf{w}). \tag{14}$$

When $\mathbf{w} = \mathbf{0}$ the small representations are vector representations, $\Gamma(R|\mathbf{0}) = \Gamma(R)$.

(vi) Use the $\tilde{\Gamma}_{\mathbf{k}}(R|\mathbf{w})$ (including those with $\mathbf{w} = \mathbf{0}$) to determine the induced representations $\Gamma_{\mathbf{k}}(R|\mathbf{w})$ from eqs. (9) and (10). These matrices, when multiplied by $\Gamma_{\mathbf{k}}(E|\,\mathbf{t})$, give the space-group representations.

Repeat steps (i)–(vi) for each $\mathbf{k}$ vector for which the space-group representations $\Gamma_{\mathbf{k}}(R|\mathbf{w})\ \Gamma_{\mathbf{k}}(E|\mathbf{t})$ are required.

Special cases

The above procedure is simplified in either of two special cases. For symmorphic groups $\mathbf{w}_j = \mathbf{0}$, $\forall\, g_j \in \mathrm{G}$; hence the PFs are all unity and the representations of the little co-group are all vector representations. For internal points $\mathbf{k}$ there are no other points equivalent to $\mathbf{k}$, and so from eq. (16.4.20) $\mathbf{b}_i = 0$ and, again, there are no PRs, so that all one needs to do is to find the IRs of the little co-group.

Example 16.6-1 The pattern of a strictly 2-D space group is shown in Figure 16.16. There is a C_{2z} axis normal to the plane at the intersection of $\mathbf{a}_1$ and $\mathbf{a}_2$. The vector $\mathbf{w} = \mathbf{a}_1/2 = a_1[½\ 0]$. Find the IRs of G at X[½ 0].

$$\mathrm{G} = \{(E|\mathbf{0})\mathrm{T} \quad (C_{2z}|\mathbf{0})\mathrm{T} \quad (\sigma_{\mathrm{y}}|\mathbf{w})\mathrm{T} \quad (\sigma_{\mathrm{x}}|\mathbf{w})\mathrm{T}\}, \tag{15}$$

Table 16.7. *(a) Projective factors* $[(R_i|\mathbf{w}_i)\ ;\ (R_j|\ \mathbf{w}_j)]$ *calculated from eq.(16.6.13). (b) Multiplication table* $R_i\ R_j$. *(c) Table of values of* z_{ij}.

In (a) and (b) R_i labels the rows and R_j labels the columns. In (c), i and j label rows and columns, respectively.

(a)	$R_i\backslash R_j$	E	C_{2z}	σ_y	σ_x
	E	1	1	1	1
	C_{2z}	1	1	−1	−1
	σ_y	1	1	1	1
	σ_x	1	1	−1	−1
(b)	$R_i\backslash R_j$	E	C_{2z}	σ_y	σ_x
	E	E	C_{2z}	σ_y	σ_x
	C_{2z}	C_{2z}	E	σ_x	σ_y
	σ_y	σ_y	σ_x	E	C_{2z}
	σ_x	σ_x	σ_y	C_{2z}	E
(c)	$i\backslash j$	E	C_{2z}	σ_y	σ_x
	E	0	0	0	0
	C_{2z}	0	0	1	1
	σ_y	0	0	0	0
	σ_x	0	0	1	1

$$\mathrm{P} = \{E\ C_{2z}\ \sigma_y\ \sigma_x\}. \tag{16}$$

Exercise 16.6-2 Does the set of coset representatives in (15) form a group?

(i) At X, $\mathbf{k} = \mathbf{b}_1/2$ (Figure 16.16(b)) and

$$\mathrm{P}(\mathbf{k}) = \{E\ C_{2z}\ \sigma_y\ \sigma_x\} = \mathrm{C}_{2v}. \tag{17}$$

(ii) At X,

$$E\mathbf{k} = \mathbf{k}, C_{2z}\mathbf{k} = -\mathbf{k} = \mathbf{k} - \mathbf{a}_1 \cong \mathbf{k},\ \sigma_y\ \mathbf{k} = \mathbf{k},\ \sigma_x\ \mathbf{k} = -\mathbf{k} \cong \mathbf{k}; \tag{18}$$

(17)

$$^*\mathbf{k} = \{\mathbf{k}\},\ s(\mathbf{k}) = 1,\ p(\mathbf{k}) = g(\mathrm{P})/s(\mathbf{k}) = 4\,. \tag{19}$$

(iii)

$$E\mathrm{P}(\mathbf{k}) = \mathrm{P}. \tag{20}$$

(iv) The factor table follows from eq. (13) and is given in Table 16.7, which also shows the multiplication table of C_{2v}.

(v) From Table 16.7(a), the PFs are all roots of unity of order 2 so that $n = 2$ in eq. (16.5.1), and

Table 16.8. *Multiplication table for the covering group* G′ *of* G = C_{2v} *showing the isomorphism of* G′ *with* D_4.

D_4		E	C_{2x}	C_{4z}^{+}	C_{4z}^{-}	C_{2x}	C_{2y}	C_{2a}	C_{2b}
	G′	$E, 0$	$E, 1$	$\sigma_x, 0$	$\sigma_x, 1$	$C_{2z}, 0$	$C_{2z}, 1$	$\sigma_y, 0$	$\sigma_y, 1$
E	$E, 0$	$E, 0$	$E, 1$	$\sigma_x, 0$	$\sigma_x, 1$	$C_{2z}, 0$	$C_{2z}, 1$	$\sigma_y, 0$	$\sigma_y, 1$
C_{2z}	$E, 1$	$E, 1$	$E, 0$	$\sigma_x, 1$	$\sigma_x, 0$	$C_{2z}, 1$	$C_{2z}, 0$	$\sigma_y, 1$	$\sigma_y, 0$
C_{4z}^{+}	$\sigma_x, 0$	$\sigma_x, 0$	$\sigma_x, 1$	$E, 1$	$E, 0$	$\sigma_y, 0$	$\sigma_y, 1$	$C_{2z}, 1$	$C_{2z}, 0$
C_{4z}^{-}	$\sigma_x, 1$	$\sigma_x, 1$	$\sigma_x, 0$	$E, 0$	$E, 1$	$\sigma_y, 1$	$\sigma_y, 0$	$C_{2z}, 0$	$C_{2z}, 1$
C_{2x}	$C_{2z}, 0$	$C_{2z}, 0$	$C_{2z}, 1$	$\sigma_y, 1$	$\sigma_y, 0$	$E, 0$	$E, 1$	$\sigma_x, 1$	$\sigma_x, 0$
C_{2y}	$C_{2z}, 1$	$C_{2z}, 1$	$C_{2z}, 0$	$\sigma_y, 0$	$\sigma_y, 1$	$E, 1$	$E, 0$	$\sigma_x, 0$	$\sigma_x, 1$
C_{2a}	$\sigma_y, 0$	$\sigma_y, 0$	$\sigma_y, 1$	$C_{2z}, 0$	$C_{2z}, 1$	$\sigma_x, 0$	$\sigma_x, 1$	$E, 0$	$E, 1$
C_{2b}	$\sigma_y, 1$	$\sigma_y, 1$	$\sigma_y, 0$	$C_{2z}, 1$	$C_{2z}, 0$	$\sigma_x, 1$	$\sigma_x, 0$	$E, 1$	$E, 0$

Table 16.9. *Character table of the abstract group* G_8^4, *showing the corresponding classes of the covering group* C'_{2v} *and its isomorph* D_4.

G_8^4	E	P^2	P, P^3	Q, P^2Q	PQ, P^3Q
C'_{2v}	$(E, 0)$	$(E, 1)$	$(\sigma_x, 0)$ $(\sigma_x, 1)$	$(C_{2z}, 0)$ $(C_{2z}, 1)$	$(\sigma_y, 0)$ $(\sigma_y, 1)$
D_4	E	C_{2z}	$C_{4z}^{\pm}$	C_{2x}, C_{2y}	C_{2a}, C_{2b}
Γ_1	1	1	1	1	1
Γ_2	1	1	1	−1	−1
Γ_3	1	1	−1	1	−1
Γ_4	1	1	−1	−1	1
Γ_5	2	−2	0	0	0

$$[i\ ;\ j] = \exp(2\pi\mathrm{i}\ n^{-1} z_{ij}), \tag{21}$$

from which we calculate the values of z_{ij} shown in Table 16.7(c). C_{2v} is Abelian and therefore has four 1-D vector IRs. Since $n = 2$, the order of G′ = C'_{2v} is $8 = 4(1)^2 + 2^2$, showing that there is one 2-D PR, Γ_5. When $n = 2$, $Z_2 = \{1, 0\}$, and the elements of G′ are $\{(g_j, 0)\ (g_j, 1)\}$ with $g_j \in C_{2v}$. The multiplication table of G′ in Table 16.8 is now readily calculated from eq. (16.5.4) using the values of z_{ij} in Table 16.8.

Exercise 16.6-3 Evaluate the products $(C_{2z}, 1)(\sigma_y, 0)$ and $(\sigma_x, 1)\ (\sigma_y, 1)$.

The multiplication table for G′ = C'_{2v} shows that it is isomorphous with the abstract group G_8^4 defined by the generating relations $P^4 = E$, $Q^2 = E$, $QP = P^3Q$. From Table 16.8, $(\sigma_x, 0)^4 = (E, 1)^2 = (E, 0) \rightarrow E$ and $(C_{2z}, 0)^2 = (E, 0) \rightarrow E$, which show that $(\sigma_x, 0) \rightarrow P$ and $(C_{2z}, 0) \rightarrow Q$. As a check: $QP = (C_{2z}, 0)(\sigma_x, 0) = (\sigma_y, 1)$, and $P^3Q = (\sigma_x, 0)^3\ (C_{2z}, 0) = (\sigma_x, 1)\ (C_{2\ z}, 0) = (\sigma_y, 1)$. The point group D_4 is also isomorphous with G_8^4 with $C_{4z}^{+} \rightarrow P$ and

Table 16.10. *(a) Matrices* $\Gamma(R)$ *of the PR of* C_{2v}, *which is the vector representation* Γ_5 *of* $D_4 \sim C'_{2v}$. *(b) Matrices* $\Gamma_{\mathbf{k}}(R|\mathbf{w})$ *of the 2-D space-group representation at* $\mathbf{k} = [1/2\ 0]$. *(c) Values of* R_u *and* $\mathbf{v}_q$ *for* $R_u \in \{R\}'$ *and the PF* $\exp(-i\,\mathbf{k}\cdot\mathbf{w})$ *used in calculating the matrices in (b).*

The column headings in (a) identify the operators $(g_j, 0) \in C'_{2v}$, which map on to $g_j \in C_{2v}$, and also the corresponding operators of D_4 used in the calculation of these matrices. As the supermatrix, (b), has only one row and column, it follows from eqs. (3), (23), and (24) that the space-group representation $\Gamma_{\mathbf{k}}(R|\mathbf{w})$ is identical with the small representation $\tilde{\Gamma}_{\mathbf{k}}(R|\mathbf{w})$ of the little group given by eq. (16.6.14).

	C'_{2v} D_4	$(E, 0)$ E	$(C_{2z}, 0)$ C_{2x}	$(\sigma_y, 0)$ C_{2a}	$(\sigma_x, 0)$ C^+_{4z}
(a)	$\Gamma(R)$	$\begin{bmatrix} 1 & 0 \\ 0 & 1 \end{bmatrix}$	$\begin{bmatrix} 1 & 0 \\ 0 & -1 \end{bmatrix}$	$\begin{bmatrix} 0 & 1 \\ 1 & 0 \end{bmatrix}$	$\begin{bmatrix} 0 & -1 \\ 1 & 0 \end{bmatrix}$
	G	$(E\|\mathbf{0})$	$(C_{2z}\|\mathbf{0})$	$(\sigma_y\|\mathbf{w})$	$(\sigma_x\|\mathbf{w})$
(b)	$\Gamma_{\mathbf{k}}(R\|\mathbf{w})$	$\begin{bmatrix} 1 & 0 \\ 0 & 1 \end{bmatrix}$	$\begin{bmatrix} 1 & 0 \\ 0 & -1 \end{bmatrix}$	$\begin{bmatrix} 0 & -i \\ -i & 0 \end{bmatrix}$	$\begin{bmatrix} 0 & i \\ -i & 0 \end{bmatrix}$
(c)	R_u	C_{2z}	σ_y	σ_x	
	$\mathbf{v}_q$	$\mathbf{0}$	$\mathbf{w}$	$\mathbf{w}$	
	$\exp(-i\,\mathbf{k}\cdot\mathbf{w})$	1	$-i$	$-i$	

$C_{2x} \to Q$. Therefore, C'_{2v} is isomorphous with D_4, with the mappings $(C_{2z}, 0) \to C_{2x}$, $(\sigma_x, 0) \to C^+_{4z}$. Corresponding elements of G′ and D_4 are shown in Table 16.8. For G^4_8, $g = 8$, $n_c = 5$, $n_v = 5$. The character table of G^4_8 is shown in Table 16.9, in which column headings include the corresponding classes of $G' = C'_{2v}$ and D_4. Note that C'_{2v} has five vector representations: Γ_1 to Γ_4 are identical with those of C_{2v}. Since it is the operators $(g_j, 0)$ of G' which map on to the operators g_j of $G = C_{2v}$, the column headed $(E, 1)$ is omitted, because it is in C'_{2v} but not in C_{2v}. Therefore, Γ_5 is the PR of C_{2v}. (Check: $\sum |\chi_j|^2 = 4 = g$.) Since $(g_j, 0) \to g_j$, we choose a suitable basis for Γ_5 in C'_{2v} and calculate the MRs of $(g_j, 0)$. The simplest basis to use is $\langle \mathbf{e}_1\ \mathbf{e}_2|$. For example, using the correspondence of $(\sigma_y, 0)$ with C_{2a},

$$C_{2\mathbf{a}}\langle \mathbf{e}_1\ \mathbf{e}_2\,| = \langle \mathbf{e}_2\ \mathbf{e}_1\,| = \langle \mathbf{e}_1\ \mathbf{e}_2\,|\begin{bmatrix} 0 & 1 \\ 1 & 0 \end{bmatrix}. \tag{22}$$

The MRs of the $(g_j, 0)$ operators calculated in this way are given in Table 16.10(b). There is only one coset representative $(R_r|\mathbf{w}_r) = (R_s|\mathbf{w}_s) = (E|\mathbf{0})$ and so

(5) $$R_u = R, \tag{23}$$

(6) $$\mathbf{v}_q = R_r^{-1}(R\,\mathbf{w}_s + \mathbf{w} - \mathbf{w}_r) = E^{-1}(R\,\mathbf{0} + \mathbf{w} - \mathbf{0}) = \mathbf{w}. \tag{24}$$

In this way we derive the values of $\mathbf{v}_q$ shown in Table 16.10(c). Since $\mathbf{k} = \mathbf{b}_1/2 = \pi/\mathbf{a}_1$ and $\mathbf{w} = \mathbf{a}_1/2$ (Figure 16.16), $\mathbf{k} \cdot \mathbf{w} = (\pi/2)$ and $\exp(-\mathrm{i}\,\mathbf{k} \cdot \mathbf{w}) = -\mathrm{i}$, as entered in Table 16.10(c). In this example the supermatrix has only one row and column, so that

$$\Gamma_{\mathbf{k}}(R|\mathbf{w}) = \tilde{\Gamma}_{\mathbf{k}}(R)\exp(-\mathrm{i}\,\mathbf{k} \cdot \mathbf{w}).$$

Exercise 16.6-4 Using the MRs $\Gamma_{\mathbf{k}}(R|\mathbf{w})$ in Table 16.10 evaluate the product $\Gamma(C_{2\mathbf{z}}|\mathbf{0})$ $\Gamma(\sigma_{\mathbf{y}}|\mathbf{w})$. Evaluate also the corresponding operator products in C_{2v}, in C'_{2v}, and in D_4.

Finally, the matrices $\Gamma_{\mathbf{k}}(R|\mathbf{w})$ in Table 16.10 have to be multiplied by the appropriate representation $\Gamma_{\mathbf{k}}(E|\mathbf{t})$ of the translation subgroup to give the space-group representations $\Gamma_{\mathbf{k}}(R|\mathbf{v})$. At X,

$$\mathbf{k} \cdot \mathbf{t} = \tfrac{1}{2}\mathbf{b}_1.(n_1\mathbf{a}_1 + n_2\mathbf{a}_2) = \pi\, n_1, \tag{25}$$

(25)
$$\Gamma_{\mathrm{k}}(E|\mathbf{t}) = \exp(-\mathrm{i}\ \mathbf{k}\ \cdot\ \mathbf{t})\mathrm{E}_2 \begin{cases} = -\mathrm{E}_2\ (n_1\ \text{odd}) \\ = \ \ \mathrm{E}_2\ (n_1\ \text{even}) \end{cases} \tag{26}$$

and similarly at any other **k** point.

Answers to Exercises 16.6

Exercise 16.6-1 $(\mathbf{R_r}|\mathbf{w_r})^{-1}(\mathbf{R}|\mathbf{v})(\mathbf{R_s}|\mathbf{w_s}) = (R_r^{-1}|-R_r^{-1}\mathbf{w}_r)(R\,R_s|R\,\mathbf{w}_s + \mathbf{v})$

$$= (R_r^{-1}R\,R_s|R_r^{-1}(R\,\mathbf{w}_s + \mathbf{v}) - R_r^{-1}\,\mathbf{w}_r)$$

$$= (R_r^{-1}R\,R_s|R_r^{-1}(R\mathbf{w}_s + \mathbf{v} - \mathbf{w}_r))$$

$$= (R_u|\mathbf{v}_q).$$

Exercise 16.6-2 No, because it is not closed. For example, $(\sigma_{\mathbf{y}}|\mathbf{w})(\sigma_{\mathbf{y}}|\mathbf{w} = (\sigma_{\mathbf{y}}{}^2|\mathbf{w} + \mathbf{w}) = (E|2\mathbf{w}) = (E|\mathbf{a}_1)$.

Exercise 16.6-3 $(C_{2\mathbf{z}}, 1)(\sigma_{\mathbf{y}}, 0) = (\sigma_{\mathbf{x}}, 1 + 0 + 1) = (\sigma_{\mathbf{x}}, 0)$
$(\sigma_{\mathbf{x}}, 1)(\sigma_{\mathbf{y}}, 1) = (C_{2\mathbf{z}}, 1 + 1 + 1) = (C_{2\mathbf{z}}, 1).$

Exercise 16.6-4 $(C_{2\mathbf{z}}|\mathbf{0})(\sigma_{\mathbf{y}}|\mathbf{w}) = (\sigma_{\mathbf{x}}| - \mathbf{w} + \mathbf{0}) = (E| - \mathbf{a}_1)(\sigma_{\mathbf{x}}|\mathbf{w})$;

$$\Gamma(C_{2\mathbf{z}}|\mathbf{0})\Gamma(\sigma_{\mathbf{y}}|\mathbf{w}) = -\mathrm{i}\begin{bmatrix} 1 & 0 \\ 0 & -1 \end{bmatrix}\begin{bmatrix} 0 & 1 \\ 1 & 0 \end{bmatrix} = -\mathrm{i}\begin{bmatrix} 0 & 1 \\ -1 & 0 \end{bmatrix} = -1\begin{bmatrix} 0 & \mathrm{i} \\ -\mathrm{i} & 0 \end{bmatrix}$$

$$= -\Gamma(\sigma_{\mathbf{x}}|\ \mathbf{w}),$$

and so does $\Gamma(E|-a_1)\Gamma(\sigma_x|\mathbf{w}) = -\Gamma(\sigma_x|\mathbf{w})$. In C_{2v}, $C_{2z}\ \sigma_y = \sigma_x$; in C'_{2v}, $(C_{2z}, 0)$ $(\sigma_y, 0) = (\sigma_x,\ 1) \rightarrow\ C^-_{4z}$ in D_4. In D_4, $C_{2x}\ C_{2a} = C^-_{4z}$.

16.7 Herring method for non-symmorphic space groups

In Sections 16.4–16.6 the problem of finding the representations of a space group G at any particular **k** point was solved by reducing the size of the group of the wave vector

$$G(\mathbf{k}) = \sum (R|\mathbf{w})T, \ \ \forall\ R \in P(\mathbf{k}), \tag{1}$$

by forming its factor group with respect to T,

$$F(\mathbf{k}) = G(\mathbf{k})/T = \{(R|\mathbf{w})T\}, \ \ \forall\ R \in P(\mathbf{k}), \tag{2}$$

which is isomorphous with the point group of the wave vector $P(\mathbf{k})$. The whole of T is thereby mapped on to the identity in $F(\mathbf{k})$, so that all the translations are represented by the unit matrix E. This device results in a loss of information, which may be restored by finding PRs of $P(\mathbf{k})$ as vector representations of its central extension $P(\mathbf{k})'$. In Herring's method one uses instead an isomorph of the factor group of $G(\mathbf{k})$ with respect to a *subgroup* of T, $T(\mathbf{k})$, defined by

$$T(\mathbf{k}) = \{(E|\mathbf{t})\}, \ \ \exp(-i\,\mathbf{k}\cdot\mathbf{t}) = 1, \tag{3}$$

so that the elements of $T(\mathbf{k})$ *are* represented by the unit matrix. This avoids the loss of information referred to above, and consequently the representations of $G(\mathbf{k})$ may be found via vector representations of a group of order larger than that of $P(\mathbf{k})$, called the Herring group $\mathscr{H}(\mathbf{k})$. For points of high symmetry the use of Herring's method may involve isomorphisms with abstract groups of rather large order, but character tables for all the abstract groups required in deriving the representations of the 230 space groups have been given by Bradley and Cracknell (1972).

(1), (3)
$$G(\mathbf{k}) = \sum_{(R|\mathbf{w})} \sum_{(E|\mathbf{t})} (R|\mathbf{w})\ (E|\mathbf{t})\ T(\mathbf{k}), \tag{4}$$

where

$$\{(E|\mathbf{t})\} = (E|\mathbf{0}) \oplus \forall\ (E|\mathbf{t}) \notin T(\mathbf{k})$$

$$= (E|\mathbf{0}) \oplus \forall\ (E|\mathbf{t}) \text{ for which } \exp(-i\,\mathbf{k}\cdot\mathbf{t}) \neq 1 \tag{5}$$

$$= \mathscr{T}(\mathbf{k}). \tag{6}$$

The Herring factor group is

(4)–(6)
$$G(\mathbf{k})/T(\mathbf{k}) = \{(R|\mathbf{w})\ \mathscr{T}(\mathbf{k})\ T(\mathbf{k})\}, \tag{7}$$

which is isomorphous with the Herring group

(4)–(6)
$$\mathscr{H}(\mathbf{k}) = \{(R|\mathbf{w})\}\ \{(E|\mathbf{t})\}, \ \ \forall\ R \in P(\mathbf{k}), \ \ \forall (E|\mathbf{t}) \in \mathscr{T}(\mathbf{k}). \tag{8}$$

The law of binary composition in $\mathscr{H}(\mathbf{k})$ is the Herring multiplication rule, which is the Seitz rule for the multiplication of space-group elements supplemented by the additional condition that $(E|\mathbf{t})$ is to be replaced by $(E|\mathbf{0})$ whenever $\exp(-\mathrm{i}\,\mathbf{k}\cdot\mathbf{t}) = +1$.

The steps involved in Herring's method for a non-symmorphic space group will now be summarized as was done for the PR method in Section 16.6.

(i) Form the little co-group for any **k** in the representation domain of the BZ,

$$\mathrm{P}(\mathbf{k}) = \{R\}, \qquad R\mathbf{k} = \mathbf{k} + \mathbf{b}_m. \tag{9}$$

(ii) Find $^*\mathbf{k}$.

(iii) Write down the coset expansion

$$\sum R\ \mathrm{P}(\mathbf{k}),\ R\mathbf{k} \in {}^*\mathbf{k}. \tag{10}$$

The coset representatives $\{R\}$ are the star generators found in (ii).

(iv) Write down the set (which is not necessarily a group)

$$\{(R|\mathbf{w})\},\ \ \forall\ R \in \mathrm{P}(\mathbf{k}). \tag{11}$$

(v) Construct the translation group $\mathrm{T}(\mathbf{k})$ (eq. (3)).

(vi) Construct the group $\mathscr{T}(\mathbf{k})$ (eq. (5)).

(vii) Write down the Herring group $\mathscr{H}(\mathbf{k})$ (eq. (8)).

(viii) The IRs of $\mathscr{H}(\mathbf{k})$ give the small representations, but many of the representations of $\mathscr{H}(\mathbf{k})$ will duplicate information in the vector representations of the little co-group, so the only ones *required* are those for which

$$\Gamma_{\mathbf{k}}(E|\mathbf{t}) = \exp(-\mathrm{i}\,\mathbf{k}\cdot\mathbf{t})\ \mathrm{E}_{l(\mathbf{k})},\quad l(\mathbf{k}) = \text{dimension of } \Gamma_{\mathbf{k}}. \tag{12}$$

On using eqs. (16.6.9) and (16.6.10) the matrices of the required representations found in (viii) give the elements $\Gamma_{\mathbf{k}}$ of the supermatrix as in step (vi) of Section 16.6, and these matrices, when multiplied by $\Gamma_{\mathbf{k}}(E|\mathbf{t})$, are the space-group representations.

Comment The "required representations" are also referred to in the literature as "permitted" or "allowed" representations.

Example 16.7-1 Herring's method will be illustrated by re-working Example 16.6-1 on a 2-D non-symmorphic space group. As before,

$$\mathrm{P} = \{E\ C_{2z}\ \sigma_y\ \sigma_x\} = C_{2v}. \tag{13}$$

(i) At X, $\mathbf{k} = [\tfrac{1}{2}\ 0]$, and the little co-group

$$\mathrm{P}(\mathbf{k}) = \{E\ C_{2z}\ \sigma_y\ \sigma_x\} = C_{2v}. \tag{14}$$

(ii) Therefore, the only star generator is E and $^*\mathbf{k}$ is $E\mathbf{k}$.

(iii) The coset expansion on P(**k**) is

$$E\ \mathrm{C}_{2v} = \mathrm{P}. \tag{15}$$

The supermatrix therefore consists of just one submatrix $\Gamma_{[(E|\mathbf{0})(E|\mathbf{0})]}$.

(iv) The set

$$\{(R|\mathbf{w})\} = \{(E|\mathbf{0})(C_{2z}|\mathbf{0})(\sigma_y|\mathbf{w})(\sigma_x|\mathbf{w})\}. \tag{16}$$

(v)
$$\mathrm{T}(\mathbf{k}) = \{(E|\,n_1\mathbf{a}_1 + n_2\,\mathbf{a}_2), \quad n_1 \text{ even}, \ \forall n_2. \tag{17}$$

(n_1 must be an even integer in order to satisfy eq. (3).)

(vi) $(E|\mathbf{a}_1)$ is excluded from T(**k**) because $\exp(-\mathrm{i}\,\mathbf{k}\cdot\mathbf{a}_1) = \exp(-\mathrm{i}\pi) = -1$. Therefore

$$\mathscr{T}(\mathbf{k}) = \{(E|\mathbf{0})(E|\mathbf{a}_1)\}. \tag{18}$$

(vii)

$$\begin{aligned}\mathscr{H}(\mathbf{k}) &= \{(E|\mathbf{0})(C_2|\mathbf{0})(\sigma_y|\mathbf{w})(\sigma_x|\mathbf{w})\}\{(E|\mathbf{0})(E|\mathbf{a}_1)\}\\ &= \{(E|\mathbf{0})(C_{2z}|\mathbf{0})(\sigma_y|\mathbf{w})(\sigma_x|\mathbf{w})\} \oplus \{(E|\mathbf{a}_1)(C_{2z}|\mathbf{a}_1)(\sigma_y|\mathbf{w}+\mathbf{a}_1)(\sigma_x|\mathbf{w}+\mathbf{a}_1)\}.\end{aligned} \tag{19}$$

The direct sum in eq. (19) would not close without the Herring multiplication rule. For example, $(\sigma_x|\mathbf{w})(E|\mathbf{a}_1) = (\sigma_x|\mathbf{w} + \sigma_x\mathbf{a}_1) = (\sigma_x|\mathbf{w} - \mathbf{a}_1) = (\sigma_x|\mathbf{w} + \mathbf{a}_1)$, because $\exp(-\mathrm{i}\,\mathbf{k}\cdot 2\mathbf{a}_1) = +1$.

(viii) $\mathscr{H}(\mathbf{k})$ is isomorphous with G_8^4, which has the generating relations $P^4 = E = Q^2$, $QP = P^3Q$. In $\mathscr{H}(\mathbf{k})$, $(C_{2z}|\mathbf{0})^2 = (E|\mathbf{0}) \to E$, so that $(C_{2z}|\mathbf{0}) \to Q$; $(\sigma_y|\mathbf{w})^2 = (E|2\mathbf{w}) = (E|\mathbf{a}_1) \to (E, 1)$ of C'_{2v} and $(E|\mathbf{a}_1)^2 = (E|2\mathbf{a}_1) = (E|\mathbf{0}) \to E$, so that $(\sigma_y|\mathbf{w}) \to P$. (Note the difference between C'_{2v} and $\mathscr{H}(\mathbf{k})$: in the former it is $(\sigma_x, 0)$ that maps on to P.) Therefore, the character table of $\mathscr{H}(\mathbf{k})$ is the same as that of D_4. It is the *required* representations of $\mathscr{H}(\mathbf{k})$ that are small representations of G(**k**). These required representations are those which satisfy eq. (12) with $\mathbf{k} = ½\mathbf{b}_1$. Only the 2-D representation Γ_5 of D_4 does this for the class $(E|\mathbf{a}_1)$.

Exercise 16.7-1 Confirm that $QP = P^3\,Q$ in the isomorphism $\mathscr{H}(\mathbf{k}) \sim \mathrm{G}_8^4$.

Table 16.11. *Corresponding classes in the isomorphisms of* $\mathscr{H}(\mathbf{k})$ *for the 2-D non-symmorphic space group of Figure 16.7(a).*

	$\mathscr{C}_1$	$\mathscr{C}_2$	$\mathscr{C}_3$	$\mathscr{C}_4$	$\mathscr{C}_5$
G_8^4	E	P^2	P, P^3	Q, P^2Q	PQ, P^3Q
$\mathscr{H}(\mathbf{k})$	$(E\|\mathbf{0})$	$(E\|\mathbf{a}_1)$	$(\sigma_y\|\mathbf{w})$ $(\sigma_y\|\mathbf{w}+\mathbf{a}_1)$	$(C_{2z}\|\mathbf{0})$ $(C_{2z}\|\mathbf{a}_1)$	$(\sigma_x\|\mathbf{w})$ $(\sigma_x\|\mathbf{w}+\mathbf{a}_1)$
D_4	E	C_{2z}	$C_{4z}^{\pm}$	C_{2x}, C_{2y}	C_{2a}, C_{2b}
C'_{2v}	$(E, 0)$	$(E, 1)$	$(\sigma_x, 0)$ $(\sigma_x, 1)$	$(C_{2z}, 0)$ $(C_{2z}, 1)$	$(\sigma_y, 0)$ $(\sigma_y, 1)$

(ix) The four 2×2 matrices for $\{(R|\mathbf{w})\} = \{(E|\mathbf{0})(C_{2z}|\mathbf{0})\ (\sigma_y|\mathbf{w})(\sigma_x|\mathbf{w})\}$ in eq. (19) have already been given in Table 16.10(a) (with the difference noted above). The remaining four matrices for $\{(R|\mathbf{w})(E|\mathbf{a}_1)\}$ are obtained by multiplying those for $\{(R|\mathbf{w})\}$ by that for $(E|\mathbf{a}_1) = \exp(-\mathrm{i}\,\mathbf{k}\cdot\mathbf{a}_1)\,\mathrm{E}_2 = -\mathrm{E}_2$. Table 16.11 clarifies the isomorphisms of $\mathscr{H}(\mathbf{k})$ with G_8^4, D_4, and C'_{2v}.

Example 16.7-2 A more substantial example is provided by the problem of finding the representations of the space group 227 ($Fd3m$ or O_h^7), which is the space group of the diamond structure. As specific examples, the space-group representations will be constructed at the surface points W and X (Figure 16.12(b), Table 16.6(b)). For $Fd3m$ the little group

$$\mathrm{G}(\mathbf{k}) = \sum_{\{A\}}(A|\mathbf{0})\mathrm{T} + \sum_{\{B\}}(B|\mathbf{w})\mathrm{T},\ \ \mathbf{w} = [{}^1\!/\!{}_4\ {}^1\!/\!{}_4\ {}^1\!/\!{}_4], \tag{20}$$

where $A \in \mathrm{P}(\mathbf{k}) \cap \mathrm{T_d}$, $\mathrm{B} \in \{\mathrm{P}(\mathbf{k})\}$ that are not in $\mathrm{T_d}$. For example, at W, for which $\mathbf{k} = {}^1\!/\!{}_2\mathbf{b}_1 + {}^1\!/\!{}_4\,\mathbf{b}_2 + {}^3\!/\!{}_4\mathbf{b}_3 = {}^1\!/\!{}_2\mathbf{e}_1 + \mathbf{e}_2 + 0\mathbf{e}_3, \mathrm{P}(\mathbf{k}) = \mathrm{D_{2d}}$ and

$$\{A\} = \{E\ C_{2x}\ S_{4x}^+\ S_{4x}^-\},\quad \{B\} = \{C_{2c}\ C_{2\mathbf{d}}\ \sigma_y\ \sigma_z\}. \tag{21}$$

(21), (20) $\{(R|\mathbf{w})\} = \{(E|\mathbf{0})(C_{2x}|\mathbf{0})(S_{4x}^+|\mathbf{0})(S_{4x}^-|\mathbf{0})(C_{2c}|\mathbf{w})(C_{2\mathbf{d}}|\mathbf{w})(\sigma_y|\mathbf{w})(\sigma_z|\mathbf{w})\}$; (22)

(5), (6) $\mathscr{T}(\mathbf{k}) = \{(E|\mathbf{0})(E|\mathbf{a}_3)(E|\mathbf{a}_1)(E|\mathbf{a}_2)\}$ (23)

with $\exp(-\mathrm{i}\,\mathbf{k}\cdot\mathbf{t}) = 1, \mathrm{i}, -1, -\mathrm{i}$, respectively. (The translations in eq. (23) are the fundamental translations for the *fcc* lattice.)

Table 16.12. *Character table for the required representations of the Herring group at the symmetry point* W *in the BZ (Figure 16.6(b)) for space group* 227 ($Fd3m$ *or* O_h^7).

$\mathrm{W}_1 = \mathrm{R}_{11}$ and $\mathrm{W}_2 = \mathrm{R}_{12}$, where R_{11}, R_{12} are representations of the abstract group G_{32}^4 which is isomorphous with $\mathscr{H}(\mathbf{k})$. Column headings for the classes are the coset representatives $\{(R|\mathbf{w})\}$. Time-reversal symmetry is of type a for both representations.

	$\mathscr{C}_1$	$\mathscr{C}_2$	$\mathscr{C}_3$	$\mathscr{C}_4$	$\mathscr{C}_5$	$\mathscr{C}_6$	$\mathscr{C}_7$
W	$(E\|\mathbf{0})$	$(E\|\mathbf{a}_3)$	$(E\|\mathbf{a}_1)$	$(E\|\mathbf{a}_2)$	$(C_{2x}\|\mathbf{0})$ $(C_{2x}\|\mathbf{a}_3)$	$(C_{2x}\|\mathbf{a}_1)$ $(C_{2x}\|\mathbf{a}_2)$	$(S_{4x}^+\|\mathbf{0})$ $(S_{4x}^-\|\mathbf{a}_2)$
W_1	2	2i	−2	−2i	0	0	1 − i
W_2	2	2i	−2	−2i	0	0	−1 + i

	$\mathscr{C}_8$	$\mathscr{C}_9$	$\mathscr{C}_{10}$	$\mathscr{C}_{11}$	$\mathscr{C}_{12}$	$\mathscr{C}_{13}$	$\mathscr{C}_{14}$
W	$(S_{4x}^+\|\mathbf{a}_3)$ $(S_{4x}^-\|\mathbf{0})$	$(S_{4x}^+\|\mathbf{a}_1)$ $(S_{4x}^-\|\mathbf{a}_3)$	$(S_{4x}^+\|\mathbf{a}_2)$ $(S_{4x}^-\|\mathbf{a}_1)$	$(\sigma_y\|\mathbf{w}+\mathbf{a}_3)$ $(\sigma_y\|\mathbf{w}+\mathbf{a}_2)$ $(\sigma_z\|\mathbf{w})$ $(\sigma_z\|\mathbf{w}+\mathbf{a}_1)$	$(\sigma_y\|\mathbf{w})$ $(\sigma_y\|\mathbf{w}+\mathbf{a}_1)$ $(\sigma_z\|\mathbf{w}+\mathbf{a}_3)$ $(\sigma_z\|\mathbf{w}+\mathbf{a}_2)$	$(C_{2\mathbf{d}}\|\mathbf{w}+\mathbf{a}_3)$ $(C_{2\mathbf{d}}\|\mathbf{w}+\mathbf{a}_2)$ $(C_{2c}\|\mathbf{w})$ $(C_{2c}\|\mathbf{w}+\mathbf{a}_1)$	$(C_{2\mathbf{d}}\|\mathbf{w})$ $(C_{2\mathbf{d}}\|\mathbf{w}+\mathbf{a}_1)$ $(C_{2c}\|\mathbf{w}+\mathbf{a}_3)$ $(C_{2c}\|\mathbf{w}+\mathbf{a}_2)$
W_1	1 + i	−1 + i	−1 − i	0	0	0	0
W_2	−1 − i	1 − i	1 + i	0	0	0	0

Table 16.13. *Classes of the abstract group* G^4_{32}.

The classes of $\mathscr{H}(\mathbf{k})$ in Table 16.12 may be checked from these classes of G^4_{32} and the generators in eq. (24).

G^4_{32}	G^4_{32}
$\mathscr{C}_1$	E
$\mathscr{C}_2$	Q
$\mathscr{C}_3$	Q^2
$\mathscr{C}_4$	Q^3
$\mathscr{C}_5$	P^2, P^2Q^2
$\mathscr{C}_6$	P^2Q, P^2Q^3
$\mathscr{C}_7$	P, P^3Q^3
$\mathscr{C}_8$	PQ, P^3
$\mathscr{C}_9$	PQ^2, P^3Q
$\mathscr{C}_{10}$	PQ^3, P^3Q^2
$\mathscr{C}_{11}$	$PQR, P^3R, PQ^3R, P^3Q^2R$
$\mathscr{C}_{12}$	$PR, PQ^2R, P^3QR, P^3Q^3R$
$\mathscr{C}_{13}$	QR, Q^3R, P^2R, P^2Q^2R
$\mathscr{C}_{14}$	R, Q^2R, P^2QR, P^2Q^3R

Table 16.14. *Matrix representatives of the Herring translations* $\mathscr{T}(\mathbf{k})$.

E_l is the $l \times l$ unit matrix.

$(E\|\mathbf{0})$	$(E\|\mathbf{a}_3)$	$(E\|\mathbf{a}_1)$	$(E\|\mathbf{a}_2)$
E_2	iE_2	$-E_2$	$-iE_2$

Table 16.15. *Generating matrices for the required representations of* $\mathscr{H}(\mathbf{k}) \sim G^4_{32}$ *at* W.

	P	Q	R
R_{11}	p^*	iq	r
R_{12}	ip	iq	r

$$p = \begin{bmatrix} 1 & 0 \\ 0 & 1 \end{bmatrix}; q = \begin{bmatrix} 1 & 0 \\ 0 & 1 \end{bmatrix}; r = \begin{bmatrix} 0 & 1 \\ 1 & 0 \end{bmatrix}.$$

Exercise 16.7-2 Verify the values of $\exp(-i\,\mathbf{k}\cdot\mathbf{t})$ at $\mathbf{k} = [\frac{1}{2}\ \frac{1}{4}\ \frac{3}{4}](2\pi/a)$.

The Herring group $\mathscr{H}(\mathbf{k}) = \{(R|\mathbf{w})\}\ \{\mathscr{T}(\mathbf{k})\}$ at W is the set product of eqs. (22) and (23) and is isomorphous with the abstract group G^4_{32}, with generators

$$P = (S^+_{4\mathbf{x}}|\mathbf{0}),\ Q = (E|\mathbf{a}_3),\ R = (C_{2\mathbf{d}}|\mathbf{w}). \tag{24}$$

Table 16.16. *Character table for the required IRs of the Herring group for the space group 227 (Fd3m or O_h^7) at the symmetry point X in the BZ (see Figure 16.12 and Table 16.6).*

R_{10}, R_{11}, R_{13}, R_{14} are representations of the abstract group $G_{32}^2 \sim \mathscr{H}(\mathbf{k})$.

	$X_1(R_{10})$	$X_2(R_{11})$	$X_3(R_{13})$	$X_4(R_{14})$
$\mathscr{C}_1(E\|\mathbf{0})$	2	2	2	2
$\mathscr{C}_2(C_{2y}\|\mathbf{0})$	2	−2	2	−2
$\mathscr{C}_3(E\|\mathbf{a}_1)$	−2	−2	−2	−2
$\mathscr{C}_4(C_{2y}\|\mathbf{a}_1)$	−2	2	−2	2
$\mathscr{C}_5(I\|\mathbf{w}), (I\|\mathbf{w}+\mathbf{a}_1)$	0	0	0	0
$\mathscr{C}_6(\sigma_y\|\mathbf{w}), (\sigma_y\|\mathbf{w}+\mathbf{a}_1)$	0	0	0	0
$\mathscr{C}_7(\sigma_e\|\mathbf{0}), (\sigma_f\|\mathbf{0})$	2	0	−2	0
$\mathscr{C}_8(\sigma_e\|\mathbf{a}_1), (\sigma_f\|\mathbf{a}_1)$	−2	0	2	0
$\mathscr{C}_9(C_{2e}\|\mathbf{w}+\mathbf{a}_1), (C_{2f}\|\mathbf{w})$	0	2	0	−2
$\mathscr{C}_{10}(C_{2e}\|\mathbf{w}), (C_{2f}\|\mathbf{w}+\mathbf{a}_1)$	0	−2	0	2
$\mathscr{C}_{11}(C_{2z}\|\mathbf{0}), (C_{2x}\|\mathbf{0})\ (C_{2z}\|\mathbf{a}_1), (C_{2x}\|\mathbf{a}_1)$	0	0	0	0
$\mathscr{C}_{12}(\sigma_z\|\mathbf{w}), (\sigma_z\|\mathbf{w}+\mathbf{a}_1), (\sigma_x\|\mathbf{w}), (\sigma_x\|\mathbf{w}+\mathbf{a}_1)$	0	0	0	0
$\mathscr{C}_{13}(S_{4y}^{\mp}\|\mathbf{0}), (S_{4y}^{\mp}\|\mathbf{a}_1)$	0	0	0	0
$\mathscr{C}_{14}(C_{4y}^{\pm}\|\mathbf{w}), (C_{4y}^{\pm}\|\mathbf{w}+\mathbf{a}_1)$	0	0	0	0
Θ	a	a	a	a

The character table of G_{32}^4 is given by Bradley and Cracknell 1972, p. 241. The required representations in which $(E|\mathbf{a}_3)$ is represented by $\mathrm{i}E_l$ are R_{11} and R_{12}, with $l=2$. The character table of the Herring group $\mathscr{H}(\mathbf{k})$ at W is given in Table 16.12. It is customary to label the IRs Γ_1, Γ_2, ... by substituting the label for the symmetry point for Γ, so here W_1, W_2, ... The classes of G_{32}^4 are given in Table 16.13 and the MRs of $T(\mathbf{k})$ are given in Table 16.14. The generating matrices for the required representations of $\mathscr{H}(\mathbf{k}) \sim G_{32}^4$ at W are given in Table 16.15, and the generators P, Q, R are defined in eq. (24).

At X, $\mathbf{k} = {}^1\!/\!_2\mathbf{b}_1 + {}^1\!/\!_2\mathbf{b}_3$ and $\mathbf{k}\cdot\mathbf{a}_1 = (2\pi/a)[{}^1\!/\!_2\ \ 0\ \ {}^1\!/\!_2]\,.\,a[1\ \ 0\ \ 0] = \pi$, so that $\exp(-\mathrm{i}\,\mathbf{k}\cdot\mathbf{a}_1) = \exp(-\mathrm{i}\pi) = -1$. Therefore $\mathscr{T}(\mathbf{k}) = \{(E|\mathbf{0})(E|\mathbf{a}_1)\}$. The Herring group $\mathscr{H}(\mathbf{k})$ is the set product of $\{(R|\mathbf{w})\}$ and $\mathscr{T}(\mathbf{k})$. At X, $P(\mathbf{k}) = D_{4h} = D_4 \otimes C_i$, and therefore

$$\mathscr{H}(\mathbf{k}) = \{\sum_{\{A\}}(A|\mathbf{0}) + \sum_{\{A\}}(IA|\mathbf{w}) + \sum_{\{B\}}(B|\mathbf{w}) + \sum_{\{B\}}(IB|\mathbf{0})\}\{(E|\mathbf{0})(E|\mathbf{a}_1)\}, \tag{25}$$

where $A \in D_4 \cap T_d$, $B \in D_4$ but $\notin T_d$.

$$(25)\quad \mathscr{H}(\mathbf{k}) = \{(E|\mathbf{0})(C_{2y}|\mathbf{0})(C_{4y}^{\pm}|\mathbf{w})(C_{2z}|\mathbf{0})(C_{2x}|\mathbf{0})(C_{2e}|\mathbf{w})(C_{2f}|\mathbf{w})(I|\mathbf{w})$$

$$(\sigma_y|\mathbf{w})(S_{4y}^{\mp}|\mathbf{0})(\sigma_z|\mathbf{w})(\sigma_x|\mathbf{w})(\sigma_e|\mathbf{0})(\sigma_f|\mathbf{0})\}\{(E|\mathbf{0})(E|\mathbf{a}_1)\}. \tag{26}$$

$\mathscr{H}(\mathbf{k})$ is isomorphous with the abstract group G_{32}^2 with generators

$$P = (\sigma_x|\mathbf{w}), \quad Q = (S_{4y}^{+}|\mathbf{0}), \quad R = (C_{2x}|\mathbf{0}). \tag{27}$$

Table 16.17. *Generating matrices for the representations* X_1 *to* X_4.

	P	Q	R
X_1	p	r	r
X_2	p	−p	r
X_3	p	−r	r
X_4	p	p	r

$$\mathrm{p}=\begin{bmatrix}0 & -1\\ 1 & 0\end{bmatrix};\quad \mathrm{r}=\begin{bmatrix}0 & 1\\ 1 & 0\end{bmatrix}.$$

The classes of this realization of G_{32}^2 are given in Table 16.16. The extra translation apart from $(E|\mathbf{0})$ in $\mathscr{T}(\mathbf{k})$ is $(E|\mathbf{a}_1)$, which forms the class $\mathscr{C}_3$. The required representations of $\mathscr{H}(\mathbf{k})$ are those with character $\exp(-\mathrm{i}\,\mathbf{k}\,.\,\mathbf{a}_1)l=-l$, for $\chi(\mathscr{C}_3)$. (l is the dimension of the representation.) These are the representations R_{10}, R_{11}, R_{13}, R_{14} (with $l=2$) in the character table of G_{32}^2 (Bradley and Cracknell (1972), p. 240; see also Jones (1975), Table 44, in which $(E|\mathbf{a}_1)$ is $\mathscr{C}_{14}$). Table 16.16 is a partial character table of $\mathscr{H}(\mathbf{k})$, giving only the four required representations. Matrix representatives can be obtained from those of the generators in Table 16.17, using eq. (27) and the classes of G_{32}^2 which are given in Problem 16.8.

Tables of space-group representations are given by Bradley and Cracknell (1972), Kovalev (1993), Miller and Love (1967), Zak (1969). Stokes and Hatch (1988) describe various errors in these compilations and discuss the different settings and labels used.

Answers to Exercises 16.7

Exercise 16.7-1

$$P^3Q=(E|\mathbf{a}_1)(\sigma_y|\mathbf{w})(C_2|\mathbf{0})=(E|\mathbf{a}_1)(\sigma_x|\mathbf{w});$$

$$QP=(C_2|\mathbf{0})(\sigma_y|\mathbf{w})=(\sigma_x\,|-\mathbf{w})=(E|\mathbf{a}_1)(\sigma_x|\mathbf{w})=P^3Q.$$

Exercise 16.7-2

$$\exp(-\mathrm{i}\,\mathbf{k}\cdot\mathbf{t})=\exp(-\mathrm{i}(2\pi/a)[{}^1\!/_2\ {}^1\!/_4\ {}^3\!/_4].a[0\quad 0\quad 0]=1;$$

$$\exp(-\mathrm{i}(2\pi/a)[{}^1\!/_2\ {}^1\!/_4\ {}^3\!/_4].a[0\quad 0\quad 1])=\exp(-\mathrm{i}(3\pi/2)=\mathrm{i},$$

$$\exp(-\mathrm{i}(2\pi/a)[{}^1\!/_2\ {}^1\!/_4\ {}^3\!/_4].a[1\quad 0\quad 0])=\exp(-\mathrm{i}\pi)=-1,$$

$$\exp(-\mathrm{i}(2\pi/a)[{}^1\!/_2\ {}^1\!/_4\ {}^3\!/_4].a[0\quad 1\quad 0])=\exp(-\mathrm{i}\pi/2)=-\mathrm{i}.$$

Alternatively, the same results could be obtained using Cartesian axes.

16.8 Spinor representations of space groups

We have seen that the determination of space-group representations involves the study of point-group representations, albeit sometimes of rather large order, either by finding PRs or by the Herring method. The results for ordinary Bloch functions $\psi(\mathbf{r})$, which may be used when electron spin is neglected, must be generalized when the basis functions are two-component spinors. This may be done either by replacing the groups G, P and P(**k**) by the corresponding double groups $\overline{\mathrm{G}}$, $\overline{\mathrm{P}}$, and $\overline{\mathrm{P}}(\mathbf{k})$ (Chapter 8) or by using PRs (Chapter 12). Double space groups corresponding to $\mathrm{O}_\mathrm{h}^1, \mathrm{O}_\mathrm{h}^5, \mathrm{O}_\mathrm{h}^7$ and O_h^9 were first studied by Elliott (1954) and an account of double space-group representations has been given by Bradley and Cracknell (1972). Here, I shall show, by means of a few examples, how to derive the projective spinor representations of a space group at particular symmetry points. I shall use as an example the space group 219 ($F\overline{4}3c$ or T_d^5) because its double group representations have been discussed by Bradley and Cracknell (1972), thus affording the reader an opportunity of comparing the two methods. The method of deriving spinor representations has been described in Chapter 12, for the point groups D_3 and $\mathrm{C}_{3\mathrm{v}}$, and in Chapter 14 for $\mathrm{C}_{2\mathrm{v}}$. It involves the following steps.

(i) For any required point symmetry operator R write down the rotation parameters ϕ **n**.
(ii) Write down the quaternion parameters $[\lambda, \Lambda]$ for R.
(iii) Calculate the Cayley–Klein parameters a, b.
(iv) Write down the MRs $\Gamma^j(R)$ using eqs. (12.8.3) and (12.8.5). For improper rotations use the Pauli gauge, $\Gamma(IR) = \Gamma(R)$.
(v) Sum the diagonal elements to obtain the characters of $\{\Gamma(R)\}$.

If only characters but not MRs are required, steps (i)–(iv) need only be carried out for one member of each class. In this case the usual checks for normalization and orthogonality of the character systems for each spinor representation should be applied. For point groups of large order the tables of Altmann and Herzig (1994) may be consulted. Otherwise, one may use the representations of double groups given in Chapter 6 of Bradley and Cracknell (1972), omitting the information that relates to double group operators $(\overline{R}|\mathbf{w})$. However, multiplication rules (from PFs) will be required if products of group elements are to be evaluated.

Example 16.8-1 Determine the spinor representations for space group 219 ($F\overline{4}3c$ or T_d^5) at the symmetry points X and W.

The lattice is again *fcc* with the BZ shown in Figure 16.12. At X,

$$\mathbf{k} = [{}^1\!/\!_2\ 0\ {}^1\!/\!_2], \quad \mathrm{P}(\mathbf{k}) = \mathrm{D}_{2\mathrm{d}}, \quad \mathscr{T}(\mathbf{k}) = \{(E|\mathbf{0}), (E|\mathbf{a}_1)\}. \tag{1}$$

Exercise 16.8-1 Explain why $(E|\mathbf{a}_1)$ could be replaced by $(E|\mathbf{a}_3)$ in eq. (1).

The set $\{(R|\mathbf{w})\}$ is a subgroup of $\mathscr{H}(\mathbf{k})$ and is $\sim \mathrm{G}_8^4 \sim \mathrm{D}_{2\mathrm{d}}$ with generators

$$P = (S_{4\mathbf{y}}^+|\mathbf{w}), \; Q = (C_{2\mathrm{x}}|\mathbf{0}), \tag{2}$$

Table 16.18. *Rotation parameters ϕ **n** (or ϕ **m**), quaternion parameters $[\lambda, \Lambda]$, and Cayley–Klein parameters a, b for the elements of subgroup G_8^4 of $\mathscr{H}$(X).*

$a = \lambda - i\Lambda_z$, $b = -\Lambda_y - i\Lambda_x$. Parameters for *IR* are the same as for *R* (Pauli gauge).

	ϕ	**n** or **m**	λ	Λ	a	b
E	0	[0 0 0]	1	[0 0 0]	1	0
$S_{4\mathbf{y}}^-$	$\pi/2$	[0 1 0]	$1/\sqrt{2}$	$1/\sqrt{2}$ [0 1 0]	$1/\sqrt{2}$	$-1/\sqrt{2}$
$S_{4\mathbf{y}}^+$	$\pi/2$	$[0\ \bar{1}\ 0]$	$1/\sqrt{2}$	$1/\sqrt{2}$ $[0\ \bar{1}\ 0]$	$1/\sqrt{2}$	$1/\sqrt{2}$
$C_{2\mathbf{y}}$	π	[0 1 0]	0	[0 1 0]	0	-1
$C_{2\mathbf{x}}$	π	[1 0 0]	0	[1 0 0]	0	$-i$
$C_{2\mathbf{z}}$	π	[0 0 1]	0	[0 0 1]	$-i$	0
$\sigma_\mathbf{e}$	π	$1/\sqrt{2}$ [1 0 1]	0	$1/\sqrt{2}$ [1 0 1]	$-i/\sqrt{2}$	$-i/\sqrt{2}$
$\sigma_\mathbf{f}$	π	$1/\sqrt{2}$ $[1\ 0\ \bar{1}]$	0	$1/\sqrt{2}$ $[1\ 0\ \bar{1}]$	$i/\sqrt{2}$	$-i/\sqrt{2}$

Table 16.19. *Matrix representatives for elements of the subgroup G_8^4 of $\mathscr{H}$(X) calculated from eqs. (12.8.3) and (12.8.5) using the Cayley–Klein parameters in Table 16.18 for the symmetrized bases.*

$$\langle\, u\ v| = \langle |{}^1\!/_2\ \ {}^1\!/_2\rangle\ |{}^1\!/_2\ \ -{}^1\!/_2\rangle|,\ \langle v^*\ \ -u^*|\ = \langle |{}^1\!/_2\ -{}^1\!/_2\rangle^*\ \ -|{}^1\!/_2\ \ {}^1\!/_2\rangle^*|.$$

$\langle v^*\ \ -u^*|$ is the ungerade spinor from Chapter 12.

	E	$S_{4\mathbf{y}}^-$	$S_{4\mathbf{y}}^+$	$C_{2\mathbf{y}}$
$E^{1/2}$	$\begin{bmatrix}1 & 0\\ 0 & 1\end{bmatrix}$	$1/\sqrt{2}\begin{bmatrix}1 & -1\\ 1 & 1\end{bmatrix}$	$1/\sqrt{2}\begin{bmatrix}1 & 1\\ -1 & 1\end{bmatrix}$	$\begin{bmatrix}0 & -1\\ 1 & 0\end{bmatrix}$
$E^{3/2}$	$\begin{bmatrix}1 & 0\\ 0 & 1\end{bmatrix}$	$1/\sqrt{2}\begin{bmatrix}-1 & 1\\ -1 & -1\end{bmatrix}$	$1/\sqrt{2}\begin{bmatrix}-1 & -1\\ 1 & -1\end{bmatrix}$	$\begin{bmatrix}0 & -1\\ 1 & 0\end{bmatrix}$
	$C_{2\mathbf{x}}$	$C_{2\mathbf{z}}$	$\sigma_\mathbf{e}$	$\sigma_\mathbf{f}$
$E^{1/2}$	$\begin{bmatrix}0 & -i\\ -i & 0\end{bmatrix}$	$\begin{bmatrix}-i & 0\\ 0 & i\end{bmatrix}$	$1/\sqrt{2}\begin{bmatrix}-i & -i\\ -i & i\end{bmatrix}$	$1/\sqrt{2}\begin{bmatrix}i & -i\\ -i & -i\end{bmatrix}$
$E^{3/2}$	$\begin{bmatrix}0 & -i\\ -i & 0\end{bmatrix}$	$\begin{bmatrix}-i & 0\\ 0 & i\end{bmatrix}$	$1/\sqrt{2}\begin{bmatrix}i & i\\ i & -i\end{bmatrix}$	$1/\sqrt{2}\begin{bmatrix}-i & i\\ i & i\end{bmatrix}$

where $\mathbf{w} = {}^1\!/_2\mathbf{a}_1 + {}^1\!/_2\mathbf{a}_2 + {}^1\!/_2\mathbf{a}_3$ and $\{\mathbf{a}_1\ \mathbf{a}_2\ \mathbf{a}_3\}$ are the fundamental vectors of the *fcc* lattice. The group elements of $\{(R|\mathbf{w})\} \sim G_8^4$ are

$$E = (E|\mathbf{0}),\ P = (S_{4\mathbf{y}}^+|\mathbf{w}),\ P^2 = (C_{2\mathbf{y}}|\mathbf{0}),\ P^3 = (S_{4\mathbf{y}}^-|\mathbf{w}),$$
$$Q = (C_{2\mathbf{x}}|\mathbf{0}),\ PQ = (\sigma_\mathbf{e}|\mathbf{w}),\ P^2Q = (C_{2\mathbf{z}}|\mathbf{0}),\ P^3Q = (\sigma_\mathbf{f}|\mathbf{w}). \tag{3}$$

The character table of D_{2d} shows the five vector representations A_1, A_2, B_1, B_2, E which we re-label as X_1, X_2, X_3, X_4, X_5. Matrix representatives of *P*, *Q* for X_5 in G_8^4 are

$$P = \begin{bmatrix}0 & 1\\ -1 & 0\end{bmatrix},\ Q = \begin{bmatrix}1 & 0\\ 0 & -1\end{bmatrix}. \tag{4}$$

Table 16.20. *Character table for the spinor representations of the Herring subgroup of the space group* 219 ($F\bar{4}3c$ *or* T_d^5) *at the symmetry point* X.

These characters are real and therefore no additional degeneracies are to be expected in crystals exhibiting time-reversal symmetry.

	E	$(C_{2y}\|\mathbf{0})$	$(S_{4y}^-\|\mathbf{w})$ $(S_{4y}^+\|\mathbf{w})$	$(C_{2x}\|\mathbf{0})$ $(C_{2z}\|\mathbf{0})$	$(\sigma_e\|\mathbf{w})$ $(\sigma_f\|\mathbf{w})$	Θ
$E^{1/2}$	2	0	$\sqrt{2}$	0	0	a
$E^{3/2}$	2	0	$-\sqrt{2}$	0	0	a

Exercise 16.8-2 The generating relations for G_8^4 are $P^4 = E = Q^2$, $QP = P^3Q$. Verify these relations using (a) P, Q defined in eq. (2), and (b) the MRs, $\mathbb{P}$, $\mathbb{Q}$ defined in eqs. (4).

Rotation parameters, quaternion parameters, and Cayley–Klein parameters for the point operators R in $\{(R|\mathbf{w})\}\sim D_{2d}$ are given in Table 16.18. Table 16.19 shows the MRs for the PRs of G_8^4. Since $G_8^4 \sim D_{2d}$, this information could have been obtained from the tables of Altmann and Herzig (1994) but MRs have been worked out here to illustrate the method used in deriving PRs. The number of PRs (namely two) is equal to the number of regular classes (that is, those with no bilateral binary (BB) rotations), which is two in this group. Remember that the MR of the ungerade spinor changes sign on inversion so that for improper rotations only, the MR in $E^{3/2}$ is the negative of that in $E^{1/2}$. Summing the diagonal elements gives the characters in Table 16.20.

At the symmetry point W, $\mathbf{k} = [\frac{1}{2}\ \frac{1}{4}\ \frac{3}{4}]$, $P(\mathbf{k}) = S_4$, and $\mathscr{H}(\mathbf{k})$ is the DP $G_4^1 \otimes \mathscr{T}(k)$, where

$$\mathscr{T}(\mathbf{k}) = \{(E|\mathbf{0})\ (E|\mathbf{a}_2)\ (E|2\mathbf{a}_2)\ (E|3\mathbf{a}_2)\}. \tag{5}$$

The generator of G_4^1 is $P = (S_{4x}|\mathbf{w})$ with $\mathbf{w} = [\frac{1}{2}\ \frac{1}{2}\ \frac{1}{2}]$; G_4^1 is isomorphous with the point group S_4 and therefore has four 1-D representations $W_1 = A$, $W_2 = {}^2E$, $W_3 = B$, and $W_4 = {}^1E$. These IRs are of type b, and the pairs W_1, W_3 and W_2, W_4 become degenerate under time-reversal symmetry. Matrix representatives for the spinor bases $\langle u\ v|$, $\langle v^*\ -u^*|$ are in Table 16.21. The character systems for these representations are $\{2\ \pm2\ 0\ \pm2\}$ so that for both representations, $g^{-1}\sum|\chi|^2 = 3 \neq 1$. These MRs may be reduced by the transformation $\mathbb{S}^{-1}\mathbb{M}\mathbb{S} = \mathbb{M}'$, where

$$\mathbb{S} = 2^{-1/2}\begin{bmatrix} 1 & 1 \\ 1 & \bar{1} \end{bmatrix} = \mathbb{S}^{-1}. \tag{6}$$

The IRs and their characters are given in Table 16.21. The space-group representations are obtained by multiplying these 1-D IRs in Table 16.21 by the MRs of $\mathscr{T}(\mathbf{k})$.

Exercise 16.8-3 Write down the quaternion parameters for $R \in S_4$ and hence verify the matrices in Table 16.21.

Table 16.21. *Spinor representations and irreducible spinor representations for the Herring subgroup of the space group* 219 ($F\bar{4}3c$ *or* T_d^5) *at the symmetry point* W.

$\varepsilon = \exp(-i\pi/4)$. To obtain the space-group representations, multiply the 1-D irreducible representations by the matrix representatives of $\mathscr{T}(\mathbf{k})$. The spinor basis $\langle u\ v| = \left\langle |{}^1\!/_2\ \ {}^1\!/_2\rangle\ |{}^1\!/_2\ \ \overline{{}^1\!/_2}\rangle \right|$.

	E	$(S_{4x}^-\|\mathbf{w})$	$(C_{2x}\|\mathbf{0})$	$(S_{4x}^+\|\mathbf{w})$	Basis	Θ
$E_{1/2}$	$\begin{bmatrix}1 & 0\\ 0 & 1\end{bmatrix}$	${}^1\!/\!\sqrt{2}\begin{bmatrix}1 & -i\\ -i & 1\end{bmatrix}$	$\begin{bmatrix}0 & -i\\ -i & 0\end{bmatrix}$	${}^1\!/\!\sqrt{2}\begin{bmatrix}1 & i\\ i & 1\end{bmatrix}$	$\langle u\ v\|$	
$E_{3/2}$	$\begin{bmatrix}1 & 0\\ 0 & 1\end{bmatrix}$	${}^1\!/\!\sqrt{2}\begin{bmatrix}-1 & i\\ i & -1\end{bmatrix}$	$\begin{bmatrix}0 & -i\\ -i & 0\end{bmatrix}$	${}^1\!/\!\sqrt{2}\begin{bmatrix}-1 & -i\\ -i & -1\end{bmatrix}$	$\langle v^* - u^*\|$	
${}^1E_{1/2}$	1	ε	$-i$	ε^*	u	b
${}^2E_{1/2}$	1	ε^*	i	ε	v	b
${}^1E_{3/2}$	1	$-\varepsilon$	$-i$	$-\varepsilon^*$	$-u^*$	b
${}^2E_{3/2}$	1	$-\varepsilon^*$	i	$-\varepsilon$	v^*	b

Answers to Exercises 16.8

Exercise 16.8-1 $\exp(-i\,\mathbf{k}\cdot\mathbf{0}) = 1$; $\mathbf{k}\cdot\mathbf{a}_1 = (2\pi/a)[{}^1\!/_2\ 0\ {}^1\!/_2]\,.\,a[1\ 0\ 0] = \pi$, $\exp(-i\pi) = -1$; $\mathbf{k}\cdot\mathbf{a}_3 = (2\pi/a)[{}^1\!/_2\ 0\ {}^1\!/_2]\,.\,a[0\ 0\ 1] = \pi$, as for $\mathbf{a}_1$. The MRs are $\exp(-i\,\mathbf{k}\cdot\mathbf{t})$ (here 1 or -1) multiplied by the unit matrix E_l.

Exercise 16.8-2 $P = (S_{4y}^+|\mathbf{w})$ with

$$\mathbf{w} = \tfrac{1}{2}\mathbf{a}_1 + \tfrac{1}{2}\mathbf{a}_2 + \tfrac{1}{2}\mathbf{a}_3 = \tfrac{1}{2}[\mathbf{e}_1 + \mathbf{e}_2 + \mathbf{e}_3].$$

$$P^2 = (S_{4y}^+|\mathbf{w})(S_{4y}^+|\mathbf{w}) = (C_{2y}|\mathbf{e}_3) = (C_{2y}|\mathbf{0}),$$

$$P^3 = (S_{4y}^+|\mathbf{w})(C_{2y}|\mathbf{0}) = (S_{4y}^-|\mathbf{w}),$$

$$P^4 = P^2P^2 = (C_{2y}|\mathbf{0})^2 = (E|\mathbf{0}),$$

$$Q^2 = (C_{2x}|\mathbf{0})^2 = (E|\mathbf{0}),$$

$$QP = (C_{2x}|\mathbf{0})(S_{4y}^+|\mathbf{w}) = (\sigma_{\mathbf{f}}|{}^1\!/_2\mathbf{e}_1 - {}^1\!/_2\mathbf{e}_2 - {}^1\!/_2\mathbf{e}_3) = (\sigma_{\mathbf{f}}|\mathbf{w} - \mathbf{e}_2 - \mathbf{e}_3) = (\sigma_{\mathbf{f}}|\mathbf{w}),$$

$$P^3Q = (S_{4y}^-|\mathbf{w})(C_{2x}|\mathbf{0}) = (\sigma_{\mathbf{f}}|\mathbf{w}),$$

$$\mathrm{P}^2 = \begin{bmatrix}0 & 1\\ \bar{1} & 0\end{bmatrix}\begin{bmatrix}0 & 1\\ \bar{1} & 0\end{bmatrix} = \begin{bmatrix}\bar{1} & 0\\ 0 & \bar{1}\end{bmatrix},$$

$$\mathrm{P}^3 = \begin{bmatrix}0 & 1\\ \bar{1} & 0\end{bmatrix}\begin{bmatrix}\bar{1} & 0\\ 0 & \bar{1}\end{bmatrix} = \begin{bmatrix}0 & \bar{1}\\ 1 & 0\end{bmatrix}.$$

Table 16.22. *Quaternion and Cayley–Klein parameters for the symmetry operators of the point group* S_4.

R	λ	Λ	a	b
E	1	[0 0 0]	1	0
$S_{4\mathbf{x}}^-$	$2^{-1/2}$	$2^{-1/2}$[1 0 0]	$2^{-1/2}$	$-2^{-1/2}\,\mathrm{i}$
$C_{2\mathbf{x}}$	0	[1 0 0]	0	$-\mathrm{i}$
$S_{4\mathbf{x}}^+$	$2^{-1/2}$	$2^{-1/2}$[$\bar{1}$ 0 0]	$2^{-1/2}$	$2^{-1/2}\,\mathrm{i}$

$$\mathrm{P}^3\mathrm{Q} = \begin{bmatrix} 0 & \bar{1} \\ 1 & 0 \end{bmatrix}\begin{bmatrix} 1 & 0 \\ 0 & \bar{1} \end{bmatrix} = \begin{bmatrix} 0 & 1 \\ 1 & 0 \end{bmatrix},$$

$$\mathrm{QP} = \begin{bmatrix} 1 & 0 \\ 0 & \bar{1} \end{bmatrix}\begin{bmatrix} 0 & 1 \\ \bar{1} & 0 \end{bmatrix} = \begin{bmatrix} 0 & 1 \\ 1 & 0 \end{bmatrix} = \Gamma(\sigma_{\mathbf{f}}).$$

Exercise 16.8-3 Quaternion and Cayley–Klein parameters are given in Table 16.22. Using eq. (12.8.3), $\Gamma(R) = \begin{bmatrix} a & b \\ -b^* & a^* \end{bmatrix}$, and the MRs in Table 16.21 follow.

Problems

16.1 Find the Bravais lattice and crystallographic point groups that are compatible with a C_2 axis. [*Hint*: Use eq. (16.1.17).]

16.2 Demonstrate, by drawing unit cells, that a $4F$ space lattice is equivalent to type $4I$.

16.3 Write down the matrix representation of eq. (16.2.1). Hence find the coordinates (x' y' z') of a general point (x y z) after the following symmetry operations:

(a) a screw rotation 4_2 about the [001] axis;

(b) a diagonal glide operation $(\sigma_{\mathbf{z}}|\tfrac{1}{2}\mathbf{a}_1 + \tfrac{1}{2}\mathbf{a}_2)$ in a cubic lattice.

16.4 Find the space-group operators for space group 59 in the second setting, in which the origin is displayed by $-[\tfrac{1}{4}\ \tfrac{1}{4}\ 0]$. Find, for Wyckoff position $8g$, the points equivalent to (x y z) in the second setting. Check your working by referring to the *International Tables for Crystallography* (Hahn (1983), (1992)).

16.5 (i) For space group 33 draw separate diagrams showing the location of (a) symmetry elements and (b) equivalent points. [*Hint*: Do not forget translational symmetry!]

(ii) For space group 33 prove that the product of the two glide reflections gives the screw rotation, using (a) Seitz operators, and (b) MRs.

16.6 Prove that the reciprocal lattice of the *bcc* lattice is a *fcc* lattice. Find the equations to the faces of the BZ and sketch the BZ.

16.7 The primitive rhombohedral cell in Figure 16.6 can be specified by giving the length $a = |\mathbf{a}_1|$ and the angle α between any pair of the fundamental translation vectors $\mathbf{a}_1$, $\mathbf{a}_2$, $\mathbf{a}_3$. Choose $\mathbf{e}_1$ along the projection of $\mathbf{a}_1$ in the $\mathbf{xy}$ plane; θ is the angle made by $\mathbf{a}_1$ with $\mathbf{e}_3$.

Find the matrix $\mathbb{A}$ in terms of a and θ, and hence find an expression for θ in terms of α. Prove that the reciprocal lattice of a rhombohedral lattice is also rhombohedral. Take $|\mathbf{b}_1| = b$, the angle between any pair of $\mathbf{b}_1$, $\mathbf{b}_2$, $\mathbf{b}_3$ as β, and the angle between $\mathbf{b}_1$ and $\mathbf{e}_3$ as ϕ, and find expressions for b and β in terms of a and α. Find also the equations that determine the faces of the BZ.

16.8 To determine the classes of a point group not given in the usual compilations of point-group character tables (such as Appendix A.3) one would calculate its multiplication table from the group generators (or group elements) and then find the classes by the methods in Chapter 1. However, character tables of all the abstract groups required in the calculation of space-group representations have been given in the book by Bradley and Cracknell (1972). To find the classes of any particular realization of an abstract group one needs only to evaluate their expressions for the classes using the law of binary composition for that group. Verify the classes of space group 227 at X, which have been given in Table 16.16.

The group generators for this realization of G_{32}^2 are $P = (\sigma_{\mathbf{x}}|\mathbf{w})$, $Q = (S_{4\mathbf{y}}^{+}|\mathbf{0})$, $R = (C_{2\mathbf{x}}|\mathbf{0})$.

The classes of G_{32}^2 are

$$\mathscr{C}_1 = P^4 = Q^4 = R^2;\quad \mathscr{C}_2 = Q^2;\quad \mathscr{C}_3 = P^2;\quad \mathscr{C}_4 = P^2Q^2;\quad \mathscr{C}_5 = P^3R, PR;$$
$$\mathscr{C}_6 = P^3Q^2R, PQ^2R;\quad \mathscr{C}_7 = QR, Q^3R;\quad \mathscr{C}_8 = P^2Q^3R, P^2QR;\quad \mathscr{C}_9 = PQ, P^3Q^3;$$
$$\mathscr{C}_{10} = PQ^3, P^3Q;\quad \mathscr{C}_{11} = R, P^2R, Q^2R, P^2Q^2R;\quad \mathscr{C}_{12} = P, P^3, PQ^2, P^3Q^2;$$
$$\mathscr{C}_{13} = Q, P^2Q, Q^3, P^2Q^3;\quad \mathscr{C}_{14} = PQR, P^3QR, PQ^3R, P^3Q^3R.$$

16.9 For the space group 225 ($Fm3m$ or O_h^5) write down the coset expansion of the little group G($\mathbf{k}$) on T. Hence write down an expression for the small representations. State the point group of the $\mathbf{k}$ vector P($\mathbf{k}$) at the symmetry points L(½ ½ ½) and $\Sigma(\alpha\ \alpha\ 2\alpha)$. Work out also the Cartesian coordinates of L and Σ. Finally, list the space-group representations at L and Σ.

16.10 Find the representations of the space group 227 ($Fd3m$ or O_h^7) at the surface point B(½ + β, $\alpha+\beta$, ½ + α), point group $\mathrm{C_s} = \{E\ \sigma_{\mathbf{y}}\}$. [*Hints*: Use the method of induced representations. Look for an isomorphism of C_s' with a cyclic point group of low order. The multiplication table of C_s' will be helpful.]

16.11 Determine the space-group representations of 227 ($Fd3m$ or O_h^7) at L using the Herring method. The Herring group contains twenty-four elements and is the DP $\mathrm{C_{3v}} \otimes \{(E|\mathbf{0})\ (I|\mathbf{w})\} \otimes \mathrm{T}_1$, where T_1 consists of two translations. [*Comment*: Even though this is a surface point of the BZ and a non-symmorphic space group, its representations turn out to have the same characters as those of a symmorphic space group ($Fm3m$), as may sometimes happen at particular symmetry points.]

16.12 Find the spinor representations of the space group 219 ($F\bar{4}3c$ or T_d^5) at Δ and at Σ. Comment on whether time-reversal symmetry introduces any extra degeneracy.

17 Electronic energy states in crystals

17.1 Translational symmetry

Within the adiabatic and one-electron approximations, when electron spin is neglected, electron states in crystals are described by the eigenfunctions and their corresponding eigenvalues, which are the solutions of the Schrödinger equation,

$$\hat{H}\psi = -(\hbar^2/2m_e)\nabla^2\psi + V\psi = -i\hbar\,\partial\psi/\partial t, \tag{1}$$

in which the potential energy V has the periodicity of the crystal lattice. Surface effects may be eliminated by the choice of periodic boundary conditions (PBCs). If we make the simplest possible assumption that the potential energy may be approximated by a constant value V_0 inside the crystal and set $V_0 = 0$ by our choice of the arbitrary energy zero, then

(1) $$-(\hbar^2/2m_e)\nabla^2\psi = -i\hbar\,\partial\psi/\partial t, \tag{2}$$

which has plane-wave solutions

$$\psi_{\mathbf{k}}(\mathbf{r}, t) = \psi_{\mathbf{k}}(\mathbf{r})\phi(t) = V^{-1/2}\exp[i(\mathbf{k}\cdot\mathbf{r} - \omega t)], \tag{3}$$

$$E_{\mathbf{k}} = \hbar^2 k^2/2m_e, \tag{4}$$

where V is the volume of the crystal. The assumption that V may be replaced by V_0 is called the *free-electron approximation*. In reality, $V \neq V_0$ but has the periodicity of the crystal lattice described by its translational symmetry. The translational symmetry operators $\hat{T}$ commute with the Hamiltonian and, after separating out the time-dependence of $\psi(\mathbf{r}, t)$, the common eigenfunctions of $\hat{T}$ and $\hat{H}$ are the Bloch functions $\{\psi_{\mathbf{k}}(\mathbf{r})\}$, where

(16.2.26) $$\psi_{\mathbf{k}}(\mathbf{r}) = \exp(i\,\mathbf{k}\cdot\mathbf{r})\,u_{\mathbf{k}}(\mathbf{r}), \tag{5}$$

in which $u_{\mathbf{k}}(\mathbf{r})$ has the periodicity of the lattice. The eigenfunctions $\psi_{\mathbf{k}}(\mathbf{r})$ and eigenvalues $E_{\mathbf{k}}$ depend on the wave vector $\mathbf{k}$ and are therefore labeled by the subscript $\mathbf{k}$. The effect of a periodic potential is to introduce discontinuities in E at zone boundaries so that the continuous series of states implied by eq. (4) is broken up into *bands* separated by energy *gaps*.

17.2 Time-reversal symmetry

So far, electron spin has been neglected. Up to Chapter 11, electron spin was accounted for at an elementary level by recognizing that the existence of spin angular momentum resulted in a doubling of single-electron states and the presence of an extra term $\hat{H}_{\mathbf{S}.\mathbf{L}}$ in the Hamiltonian to account for spin–orbit coupling. In Section 11.8 the two-component spinor $|\mathrm{u\ v}\rangle$ was introduced to describe the $j = \pm\frac{1}{2}$ states. In the non-relativistic limit ($\mathbf{v}/c \to 0$) of the Dirac equation (where $\mathbf{v}$ is the electron velocity and c is the speed of light) and removing the self-energy $m_e c^2$ of the electron by our choice of energy zero, the Hamiltonian is just $\hat{T} + \hat{V} + \hat{H}_{\mathbf{S}.\mathbf{L}}$, which operates on a two-component eigenvector $\psi(\mathbf{r})|\mathrm{u\ v}\rangle$, where, in a crystal, the $\psi(\mathbf{r})$ are the Bloch functions $\psi_{\mathbf{k}}(\mathbf{r})$. Applying the time-reversal operator (Chapter 13)

$$\hat{\Theta}\psi_{\mathbf{k}}(\mathbf{r})|\mathrm{u\ v}\rangle = \hat{\sigma}_2 \hat{\mathscr{K}} \psi_{\mathbf{k}}(\mathbf{r})|\mathrm{u\ v}\rangle = \psi_{-\mathbf{k}}(\mathbf{r})|-\mathrm{i}\mathrm{v}^*\ \mathrm{i}\mathrm{u}^*\rangle. \tag{1}$$

Apart from the phase factor $-$ i, the transformed spinor will be recognized as the ungerade spinor of Chapter 11. The original and time-reversed states are orthogonal and therefore degenerate, and consequently

$$\mathrm{E}_{\mathbf{k}} = \mathrm{E}_{-\mathbf{k}}. \tag{2}$$

Equation (2) is clearly true for the free-electron model and is true in general if G($\mathbf{k}$) contains the inversion operator $(I|\mathbf{0})$ (Exercise 16.4-1), but eq. (2) shows that the energy curves $\mathrm{E}_{\mathbf{k}}$ are always symmetrical about $\mathbf{k} = 0$ and so need only be displayed for $\mathbf{k} > 0$.

17.3 Translational symmetry in the reciprocal lattice representation

Because of the translational symmetry of the reciprocal lattice (Section 16.3) and the definition of the Brillouin zone (BZ), the BZ faces occur in pairs separated by a reciprocal lattice vector. For example, the cubic faces of the first BZ of the simple cubic (*sc*) lattice occur in pairs separated by the reciprocal lattice vectors $\mathbf{b} = (2\pi/a)[[1\ 0\ 0]]$ (see eq. (16.3.27)). In general, for every $\mathbf{k}$ vector that terminates on a BZ face there exists an equivalent vector $\mathbf{k}'$ (Figure 17.1) such that

$$\mathbf{k}' = \mathbf{k} - \mathbf{b}_m, \quad |\mathbf{k}'| = |\mathbf{k}|. \tag{1}$$

Equations (1) are the von Laue conditions, which apply to the reflection of a plane wave in a crystal. Because of eqs. (1), the momentum normal to the surface changes abruptly from $\hbar\mathbf{k}_\perp$ to the negative of this value when $\mathbf{k}$ terminates on a face of the BZ (Bragg reflection). At a general point in the BZ the wave vector $\mathbf{k} + \mathbf{b}_m$ cannot be distinguished from the equivalent wave vector $\mathbf{k}$, and consequently

$$\mathrm{E}(\mathbf{k} + \mathbf{b}_m) = \mathrm{E}(\mathbf{k}). \tag{2}$$

Imagine $\mathbf{k}$ increasing along a line from the zone center Γ to the face center at $\frac{1}{2}\mathbf{b}_m$. When $\mathbf{k} = \frac{1}{2}\mathbf{b}_m + \boldsymbol{\delta}$, where $\boldsymbol{\delta}$ is a small increment in $\mathbf{k}$ normal to the face,

(17.2.2), (2) $$\mathrm{E}(\tfrac{1}{2}\mathbf{b}_m + \boldsymbol{\delta}) = \mathrm{E}(-\tfrac{1}{2}\mathbf{b}_m - \boldsymbol{\delta}) = \mathrm{E}(\tfrac{1}{2}\mathbf{b}_m - \boldsymbol{\delta}). \tag{3}$$

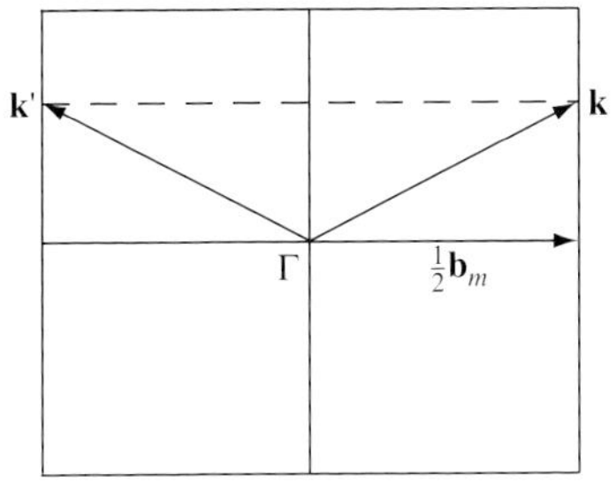

Figure 17.1. Illustration of the relation $\mathbf{k}' = \mathbf{k} - \mathbf{b}_m$.

It follows from eq. (3) that the gradient of $E(\mathbf{k})$ normal to the zone boundary vanishes at a face center,

$$\nabla_{\mathbf{k}\perp} E(\mathbf{k}) = \mathbf{0}, \quad \mathbf{k} = {}^1\!/_2\mathbf{b}_m \quad \text{(face center)}. \tag{4}$$

Equation (4) holds generally at the face center but is valid over the whole face if the crystal point group contains a reflection plane through the zone center that is parallel to the face. It also holds for all **k** vectors that terminate on a line in the BZ face that is parallel to a binary axis. The $E(\mathbf{k})$ may be described either by a single-valued function of **k** (with $\mathbf{k} \geq \mathbf{0}$), which is called the *extended zone scheme*, or by a multivalued function of **k** within the first BZ, the *reduced zone scheme* (see Figure 17.2).

Exercise 17.3-1 Show that the gradient of $E(\mathbf{k})$ normal to a BZ face vanishes over a face when there is a symmetry plane through the origin that is parallel to this face.

Answer to Exercise 17.3-1

Let $\mathbf{k} \in$ BZ and let $\boldsymbol{\delta}$ be a small increment in **k** normal to the face which is parallel to the symmetry plane through Γ. The perpendicular distance from Γ to the center of the face is $½\mathbf{b}_m$. Then

$$E(\mathbf{k} - \boldsymbol{\delta}) = E(\sigma(\mathbf{k} - \boldsymbol{\delta})) = E(\sigma(\mathbf{k} - \boldsymbol{\delta}) + \mathbf{b}_m) = E(\mathbf{k} + \boldsymbol{\delta}),$$

$$\nabla_{\mathbf{k}\perp} E(\mathbf{k}) = \lim_{\boldsymbol{\delta}\to\mathbf{0}} \{[E(\mathbf{k} + \boldsymbol{\delta}) - E(\mathbf{k} - \boldsymbol{\delta})]/2\boldsymbol{\delta}\} = \mathbf{0}.$$

17.4 Point group symmetry

At any symmetry point in the BZ,

$$(16.4.3) \qquad (R|\mathbf{v})\psi_{\mathbf{k}}(\mathbf{r}) = (E|\mathbf{t})(R|\mathbf{w})\psi_{\mathbf{k}}(\mathbf{r}) = \exp(-\mathrm{i}R\mathbf{k}\cdot\mathbf{t})\psi_{R\mathbf{k}}(\mathbf{r}), \;\; \forall R \in P(\mathbf{k}). \tag{1}$$

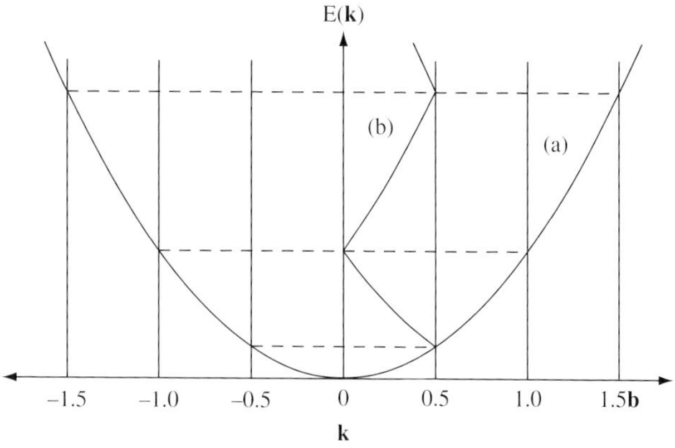

Figure 17.2. E(**k**) for the 1-D free electron model in (a) the extended zone scheme and (b) the reduced zone scheme.

Since the physical system (crystal) is indistinguishable from what it was before the application of a space-group operator, and a translational symmetry operator only changes the phase of the Bloch function without affecting the corresponding energy E(**k**),

$$E(R\,\mathbf{k}) = E(\mathbf{k}), \quad \forall\, R \in P(\mathbf{k}). \tag{2}$$

Equation (2) shows that the band energy has the symmetry of the point group of the wave vector. As R runs over the whole $\{R\} = P(\mathbf{k})$, it generates a set of degenerate eigenfunctions $\{\psi_{R\mathbf{k}}(\mathbf{r})\}$ which belong to the eigenvalue E(**k**) and which form a basis for a representation of G(**k**). The point group of the wave vector P(**k**) may be determined by inspection of the representation domain of the first BZ, but most of the P(**k**) likely to be encountered have already been given in the literature (see, for example, Table 3.5 in Bradley and Cracknell (1972)). The BZ of the *sc* lattice is in Figure 16.12(a) with symmetry points marked using the notation of Bouckaert *et al.* (1936), henceforth abbreviated to BSW. The point group symmetry at X and at M is $D_{4h} = D_4 \otimes C_i$. At X the four-fold axis is along k_y and at M it is parallel to k_z. Table 17.1 lists the symmetry of the states at X and the corresponding bases for each irreducible representation (IR). At M replace X by M and perform the cyclic permutation $y \to z$, $z \to x$, and $x \to y$ on the bases. The IRs are labeled by three sets of commonly used notation due to Mulliken (M), Bethe (B), and Bouckaert *et al.* (BSW). The basis functions of the primed representations are antisymmetric with respect to I, which transforms **k** into −**k**, as does the time-reversal operator Θ, and this is denoted by a prime in BSW notation. Besides the familiar *s*- , *p*- , and *d*- like functions, the bases include higher-order Cartesian tensors which may be verified either by forming direct product (DP) representations (Chapter 5) or by projecting a suitable function into the appropriate subspace (Chapters 6 and 9).

Exercise 17.4-1 Confirm the bases given in Table 17.1 for X_4, X_2', X_3', and X_1'. [*Hint*: Use the character table of D_{4h} and form the necessary DP representations.]

At the point Δ along ΓX, P(**k**) is C_{4v} which is isomorphous with D_4. The character table (Appendix A3) is therefore the same as the upper left quadrant of D_{4h} but the basis

Table 17.1. *Irreducible representations and basis functions for the symmetry point* X *in the BZ of the sc Bravais lattice.*

IRs are labeled by three sets of commonly used notation due to Mulliken (M), Bethe (B), and Bouckaert *et al.* (1936) (BSW). The character table for D_{4h} is in Appendix A3.

State				State			
M	B	BSW	Basis	M	B	BSW	Basis
A_{1g}	Γ_1^+	X_1	1	A_{1u}	Γ_1^-	X_1'	$xyz(x^2-z^2)$
A_{2g}	Γ_2^+	X_4	$zx(z^2-x^2)$	A_{2u}	Γ_2^-	X_4'	y
B_{1g}	Γ_3^+	X_2	(z^2-x^2)	B_{1u}	Γ_3^-	X_2'	xyz
B_{2g}	Γ_4^+	X_3	zx	B_{2u}	Γ_4^-	X_3'	$y(z^2-x^2)$
E_g	Γ_5^+	X_5	$(xy,\ yz)$	E_u	Γ_5^-	X_5'	$(z,\ x)$

Table 17.2. *The common classes of* C_{4v} *and* D_{4h}.

These basis functions are the appropriate ones when the principal axis is along k_y as it is for Δ in Figure 16.12(a). Binary axes **e** and **f** are defined in Figure 2.12. For T the principal axis is parallel to k_z and the bases would therefore be as given in Appendix A3. The representations at X that are compatible with (have the same characters as) those at Δ show how the BSW notation for the representations at Δ is derived.

D_{4h} (X)	C_{4v} (Δ, T)								
BSW	M	B	BSW	E	$C_{4y}^{\pm}$	C_{4y}^2	σ_z, σ_x	σ_e, σ_f	Basis
X_1	A_1	Γ_1	Δ_1	1	1	1	1	1	$1, y$
X_1'	A_2	Γ_2	Δ_1'	1	1	1	−1	−1	$zx(z^2-x^2)$
X_2	B_1	Γ_3	Δ_2	1	−1	1	1	−1	z^2-x^2
X_2'	B_2	Γ_4	Δ_2'	1	−1	1	−1	1	zx
X_5	E	Γ_5	Δ_5	2	0	−2	0	0	$(z,\ x)$

functions are different (Table 17.2) and depend on the choice of principal axis. (Jones (1962) takes Δ on k_z and BSW have Δ on k_x.) In Figure 16.12(a) Δ is on k_y because it is easier to visualize the symmetry points in the representation domain while maintaining the usual directions for the (right-handed) x, y, and z axes.

Exercise 17.4-2 Derive the basis function for Δ_1' in Table 17.2 by expressing Δ_1' as a DP representation.

When a lowering in symmetry occurs, an IR of the higher symmetry group is generally either re-labeled or, if it is not irreducible in the subgroup of lower symmetry, it forms a direct sum of the IRs of the subgroup,

$$\Gamma_i = \textstyle\sum_j c_{i,j}\ \Gamma_j, \tag{3}$$

Table 17.3. *Irreducible representations of the point group* $O_h = O \otimes C_i$ *and their Cartesian tensor bases (for principal axis along* k_y*).*

The character table of O_h is in Appendix A3. The BSW notation for the IRs depends on compatibility relations which are derived in Table 17.5.

IR			
M	B	BSW	Basis
A_{1g}	Γ_1^+	Γ_1	$1,\ x^2+y^2+z^2$
A_{2g}	Γ_2^+	Γ_2	$z^4(x^2-y^2)+x^4(y^2-z^2)+y^4(z^2-x^2)$
E_g	Γ_3^+	Γ_{12}	$\{z^2-x^2,\ 2y^2-(z^2+x^2)\}$
T_{1g}	Γ_4^+	Γ'_{15}	$\{zx(z^2-x^2),\ xy(x^2-y^2),\ yz(y^2-z^2)\}$
T_{2g}	Γ_5^+	Γ'_{25}	$\{zx,\ xy,\ yz\}$
A_{1u}	Γ_1^-	Γ_1'	$xyz[z^4(x^2-y^2)+x^4(y^2-z^2)+y^4(z^2-x^2)]$
A_{2u}	Γ_2^-	Γ_2'	xyz
E_u	Γ_3^-	Γ'_{12}	$\{xyz(z^2-x^2),\ xyz[2y^2-(z^2+x^2)]\}$
T_{1u}	Γ_4^-	Γ_{15}	$\{z,\ x,\ y\}$
T_{2u}	Γ_5^-	Γ_{25}	$\{y(z^2-x^2),\ z(x^2-y^2),\ x(y^2-z^2)\}$

(3)
$$\{\chi_i\} = \textstyle\sum_j c_{i,j}\ \{\chi_j\}, \tag{4}$$

where $\{\chi_i\}$ denotes the character set of the ith IR of P(**k**) at some symmetry point (say, K) and the $\{\chi_j\}$ are the character sets of the IRs in the point group (which is a subgroup of P(**k**) at K) at a point on a line of symmetry terminating at K. Representations that either change their labels or split into two or more IRs are said to be *compatible*. Only classes that are common to the subgroup and its parent group occur in the character sets in eq. (4). For example, if K is X, with P(**k**) = D_{4h}, then its subgroup at Δ on ΓX is C_{4v}. The characters of the classes common to C_{4v} and D_{4h} are given in Table 17.2. As these character sets show, the compatibility relations for D_{4h} and C_{4v} are

$$X_1 = \Delta_1,\ X_1' = \Delta_1',\ X_2 = \Delta_2,\ X_2' = \Delta_2',\ X_5 = \Delta_5, \tag{5}$$

which explain the reason for the BSW notation for the IRs of C_{4v}. Equals signs in eq. (5) and other compatibility relations means equality of the character sets for these IRs. The point group symmetery at Γ and at R is O_h. Table 17.3 lists the IRs of O_h in the three sets of principal notation used, together with the Cartesian tensors that form bases for these IRs. Since we wish to examine the lowering in symmetry that occurs along the line $\Gamma\Delta$X, the principal axis has been chosen along k_y instead of the more usual k_z.

Exercise 17.4-3 Find a basis for the IR Γ'_{12} of O_h by forming the DP of two other representations in Table 17.3.

Exercise 17.4-4 Verify the basis given for Γ'_{15} in Table 17.4 by projecting a fourth-order polynomial into the Γ'_{15} subspace. [*Hint*: Use the substitutions provided in Table17.4.]

Table 17.4. *Jones symbols for the transformation of functions (that is, R^{-1} {x y z}) for the twenty-four operations $R \in$ O.*

E	xyz
$3C_4^2$	$\bar{x}\bar{y}z$, $x\bar{y}\bar{z}$, $\bar{x}y\bar{z}$
$4C_3^{\pm}$	zxy, yzx, $z\bar{x}\bar{y}$, $\bar{y}\bar{z}x$, $\bar{z}\bar{x}y$, $\bar{y}z\bar{x}$, $\bar{z}x\bar{y}$, $y\bar{z}\bar{x}$
$3C_4^{\pm}$	$\bar{y}xz$, $y\bar{x}z$, $x\bar{z}y$, $xz\bar{y}$, $zy\bar{x}$, $\bar{z}yx$
$6C_2$	$yx\bar{z}$, $z\bar{y}x$, $\bar{x}zy$, $\bar{y}\bar{x}\bar{z}$, $\bar{z}\bar{y}\bar{x}$, $\bar{x}\bar{z}\bar{y}$

Table 17.5. *Characters of the IRs Γ_{12}, Γ'_{15}, and Γ'_{25} of O_h for classes common to O_h and C_{4v}.*

In the left-hand column are the direct sums of representations of C_{4v} which yield the same character sets as the IR of O_h in column 7. The $\sigma_{\mathbf{z}}$, $\sigma_{\mathbf{x}}$ are *vertical* planes in C_{4v} because they contain the four-fold axis but are *horizontal* planes in O_h because they are normal to the C_4 axes along **z** and **x**, respectively. The planes $\sigma_{\mathbf{e}}$ and $\sigma_{\mathbf{f}}$ in C_{4v} are two of the six dihedral planes in O_h (Figure 2.12).

C_{4v}	E	C_{4y}^2	$C_{4y}^{\pm}$	$\sigma_{\mathbf{z}}$, $\sigma_{\mathbf{x}}$	$\sigma_{\mathbf{e}}$, $\sigma_{\mathbf{f}}$	
O_h	E	$3C_4^2$	$3C_4^{\pm}$	$3\sigma_h$	$6\sigma_d$	
IRs of C_{4v}						IRs of O_h
$\Delta_1 \oplus \Delta_2$	2	2	0	2	0	Γ_{12}
$\Delta'_1 \oplus \Delta_5$	3	−1	1	−1	−1	Γ'_{15}
$\Delta'_2 \oplus \Delta_5$	3	−1	1	−1	1	Γ'_{25}

Compatibility relations for C_{4v} and O_h are derived in Table 17.5. This table shows again, as does eq. (5), that the primes and subscripts in BSW notation come from compatibility relations. Here:

$$\Delta_1 \oplus \Delta_2 = \Gamma_{12}; \quad \Delta'_1 \oplus \Delta_5 = \Gamma'_{15}; \quad \Delta'_2 \oplus \Delta_5 = \Gamma'_{25}. \tag{6}$$

Tables of compatibility relations for the simple cubic structure have been given by Jones (1962, 1975), and similar tables can be compiled for other structures, as shown by the examples in Tables 17.2 and 17.5. Compatibility relations are extremely useful in assigning the symmetry of electronic states in band structures. Their use in correlation diagrams in crystal-field theory was emphasized in Chapters 7 and 8, although there it is not so common to use BSW notation, which was invented to help describe the symmetry of electronic states in energy bands in crystals (Bouckaert *et al.* (1936)).

Compatibility relations between states at points on symmetry axes and states at end points of these axes are independent of the particular choice made from a set of equivalent axes. For example, it would make no difference to the compatibility relations in eqs. (5) and (6) if X were to be chosen on k_z or k_x instead of on k_y as in Figure 16.12(b). But there is another kind of compatibility relation which governs states on symmetry axes that lie in a plane and which can only be described in relation to a particular choice of coordinate axes.

Table 17.6. *Compatibility relations for the symmetry plane $k_z = 0$ in the simple cubic structure.*

Symmetric	Antisymmetric
Σ_1, Σ_4	Σ_2, Σ_3
$\Delta_1, \Delta_2, \Delta_5$	$\Delta'_1, \Delta'_2, \Delta_5$
Z_1, Z_3	Z_2, Z_4

For example, the symmetry points Λ, Σ, and T all lie in the $k_x = k_y$ plane. Therefore, basis functions for Λ states that are antisymmetric with respect to reflection in this plane are only compatible (because of continuity within the BZ) with basis functions for Σ and T states that are also antisymmetric with respect to reflection in this plane. Similarly, Z, T, and S all lie in the $k_z = b/2 = \pi/a$ plane. Compatibility relations for the plane $k_z = 0$ in the simple cubic structure are in Table 17.6. For example, for Δ'_1 and Δ_2 (see Table 17.2),

$$\sigma_z \; zx(z^2 - x^2) = -zx(z^2 - x^2), \quad \sigma_z \; (z^2 - x^2) = z^2 - x^2 \tag{7}$$

so that these bases are, respectively, antisymmetric and symmetric with respect to reflection in the plane $z = 0$. To resolve questions of compatibility due to symmetry planes one needs to know the necessary basis functions expressed as Cartesian tensors (Altmann and Herzig (1994); Jones (1962, 1975)).

Exercise 17.4-5 The Σ point lies on [1 1 0] between Γ and M in the $k_z = 0$ plane (Figure 16.12(b)). What is P(**k**) at Σ? List the basis functions for the IRs, naming them in both Mulliken and BSW notation. Note that Σ, Λ, and T all lie in the (1 1 0) plane through Γ defined by $x - y = 0$. Can the states Λ_2, T'_1, and T_2 exist in the same energy band as a Σ_2 state? What other Σ state is compatible with these Λ and T states? [*Hint*: These basis functions will differ from those usually seen in character tables with vertical planes $x = 0$, $y = 0$; here the vertical planes are $z = 0$, $x - y = 0$.]

Answers to Exercises 17.4

Exercise 17.4-1 The basis of a DP representation $A \otimes B$ is the DP of the bases of A and B. Therefore, in D_{4h} (Table 17.1),

DP	basis
$X_2 \otimes X_3 = X_4$,	$zx(z^2 - x^2)$,
$X_3 \otimes X_4' = X_2'$,	xyz,
$X_2 \otimes X_4' = X_3'$,	$y(z^2 - x^2)$,
$X_2 \otimes X_2' = X_1'$,	$xyz(z^2 - x^2)$.

Exercise 17.4-2 $\chi(\Delta'_2 \otimes \Delta_2) = \{1 \; 1 \; 1 \; -1 \; -1\} = \chi(\Delta'_1)$. Therefore, the basis of Δ'_1 is $zx(z^2 - x^2)$.

Table 17.7. *Character table for* C_{2v} *with principal axis along* **a**.

C_{2v} (Σ) Jones symbols	E xyz	$C_{2\mathbf{a}}$ $yx\bar{z}$	σ_z $xy\bar{z}$	$\sigma_{\mathbf{b}}$ yxz	
A_1 Σ_1	1	1	1	1	
A_2 Σ_2	1	1	-1	-1	$z(x-y)$
B_1 Σ_3	1	-1	-1	1	z
B_2 Σ_4	1	-1	1	-1	$x-y$

Exercise 17.4-3 From the character table for O_h, $\Gamma_{12} \otimes \Gamma'_2 = \Gamma'_{12}$. Therefore, a Cartesian tensor basis for Γ'_{12} is

$$\{xyz\} \otimes \{z^2-x^2,\ 2y^2-(z^2+x^2)\} = \{xyz(z^2-x^2),\ xyz[2y^2-(z^2+x^2)]\}.$$

The principal axis has been taken along k_y because of our interest in the line $\Gamma\Delta X$.

Exercise 17.4-4 x^4 does not provide a basis for Γ'_{15} but (after removing any unnecessary common factor c)

$$c^{-1}\sum_R \chi(\Gamma_{15}')^* \hat{R}(x^3y) = c^{-1}\{x^3y[3-1(1-1-1\}]$$
$$+xy^3[1(-1-1)-1(1+1)]\} = xy(x^2-y^2).$$

The other two independent functions $yz(y^2-z^2)$ and $zx(z^2-x^2)$ follow by cyclic permutation of this result.

Exercise 17.4-5 At Σ, P(**k**) is C_{2v} the character table for which is shown in Table 17.7. The basis functions shown are those for the IRs of C_{2v} when the principal axis is along **a**. Table 17.8 contains the character table for C_{3v} with basis functions for a choice of principal axis along [1 1 1]. The easiest way to transform functions is to perform the substitutions shown by the Jones symbols in these two tables. The states Σ_2 and Σ_4 are antisymmetric with respect to $\sigma_{\mathbf{b}}$, which interchanges x and y (see the Jones symbol for $\sigma_{\mathbf{b}}$ in Table 17.7). Note that $\sigma_{\mathbf{b}}$ is also one of the three vertical planes at Λ and, as Table 17.8 shows, Λ_2 is antisymmetric with respect to $\sigma_{\mathbf{b}}$. The plane $x=y$ is one of the dihedral planes in C_{4v} and from Table 17.2 we see that the bases for T'_1 and T_2 are antisymmetric with respect to $\sigma_{\mathbf{b}}$. (The four-fold axis at T is parallel to k_z, and carrying out the permutation $y \to z$, $z \to x$, $x \to y$ on the bases for Δ'_1 and Δ_2 gives $xy(x^2-y^2)$ and x^2-y^2 for the bases of T'_1 and T_2, which are antisymmetric with respect to $\sigma_{\mathbf{b}}$.)

17.5 Energy bands in the free-electron approximation: symmorphic space groups

Substituting eq. (17.1.5)

$$\psi_{\mathbf{k}}(\mathbf{r}) = \exp(\mathrm{i}\ \mathbf{k}\cdot\mathbf{r})\ u_{\mathbf{k}}(\mathbf{r}) \tag{1}$$

Table 17.8. *Character table for* C_{3v} *with principal axis along* [1 1 1].

C_{3v} (Λ)	E	$2C_3$	$3\sigma_v$	
Jones symbols	xyz	$zxy,\ yzx$	$yxz,\ zyx,\ xzy$	
$A_1\ \Lambda_1$	1	1	1	
$A_2\ \Lambda_2$	1	1	−1	$xy(x-y)+yz(y-z)+zx(z-x)$
$E\ \Lambda_3$	2	−1	0	$(x-z,\ y-z)$

(which satisfies the PBCs) into the Schrödinger one-electron time-independent equation

$$-(\hbar^2/2m_e)\nabla^2\psi(\mathbf{r}) = [E - V(\mathbf{r})]\psi(\mathbf{r}) \tag{2}$$

gives

$$\nabla^2 u_{\mathbf{k}}(\mathbf{r}) + 2i\mathbf{k}\cdot\nabla u_{\mathbf{k}}(\mathbf{r}) + (2m_e/\hbar^2)[E(\mathbf{k}) - (\hbar^2k^2/2m_e) - V(\mathbf{r})]\,u_{\mathbf{k}}(\mathbf{r}) = 0, \tag{3}$$

which must be solved self-consistently because of the difficulty involved in finding a satisfactory approximation to $V(\mathbf{r})$. However, useful insight into the form of the energy bands in a crystal may be gained by setting $V(\mathbf{r})=0$, which is called the free-electron (FE) approximation. With $V(\mathbf{r})=0$,

(3) $$u_{\mathbf{k}}(\mathbf{r}) = \exp(-i\,\mathbf{b}_m\cdot\mathbf{r}), \tag{4}$$

$$E_m(\mathbf{k}) = (\hbar^2/2m_e)|\mathbf{k}-\mathbf{b}_m|^2, \tag{5}$$

(1), (4) $$\psi_{m\mathbf{k}}(\mathbf{r}) = \exp[i(\mathbf{k}-\mathbf{b}_m)\cdot\mathbf{r}]. \tag{6}$$

In eqs. (5) and (6) the energy $E_m(\mathbf{k})$ and eigenfunctions $\psi_{m\mathbf{k}}(\mathbf{r})$ carry the subscript m because, in general, at any particular symmetry point, there may be several different values of $[m_1\ m_2\ m_3]$ which give the same energy. When degeneracy due to symmetry occurs, the appropriate eigenfunctions at that point are linear combinations of the $\psi_{m\mathbf{k}}(\mathbf{r})$. Such linear combinations of the correct symmetry may be determined by the use of projection operators (Chapter 5). Although historical usage dictates the continued use of the "free-electron" approximation, this is not perhaps the best description since the electron eigenfunctions are still required to obey the PBCs due to the translational symmetry of the lattice. Shockley (1937) used the phrase "empty lattice" in order to describe the test made by setting $V=0$, which was used in connection with the Wigner–Seitz cellular method for calculating wave functions in crystals. While "empty lattice approximation" would be a more accurate description in the present context, in view of historical precedence and familiarity I have continued to use "free electron" to describe the approximation of setting $V=0$. Accidental degeneracies not due to symmetry can occur in the FE approximation, but these are often removed when a more realistic potential is imposed.

When calculating FE energy states along particular directions in the BZ it is often convenient to work in Cartesian coordinates, that is to use the $\langle \mathbf{e}|$ basis rather than the $\langle \mathbf{b}|$ basis. The matrix representation of a reciprocal lattice vector $\mathbf{b}_m$ is

$$\langle \mathbf{b}|m\rangle = \langle \mathbf{e}|\mathrm{B}|m\rangle = \langle \mathbf{e}_1\ \mathbf{e}_2\ \mathbf{e}_3|m_x\ m_y\ m_z\rangle. \tag{7}$$

For the simple cubic lattice B is E_3 and so $|m_x\ m_y\ m_z\rangle$ is identical to $|m_1\ m_2\ m_3\rangle$ for this lattice only. But, in general, we must use eq. (7) to find $|m_x\ m_y\ m_z\rangle$ at a point whose coordinates $|m_1\ m_2\ m_3\rangle$ in the reciprocal lattice are given in the $\langle \mathbf{b}|$ basis.

(7), (16.3.6)
$$\mathrm{B}|m\rangle = \begin{bmatrix} b_{1x} & b_{2x} & b_{3x} \\ b_{1y} & b_{2y} & b_{3y} \\ b_{1z} & b_{2z} & b_{3z} \end{bmatrix} \begin{bmatrix} m_1 \\ m_2 \\ m_3 \end{bmatrix} = \begin{bmatrix} m_x \\ m_y \\ m_z \end{bmatrix}; \tag{8}$$

(8)
$$\begin{aligned} m_x &= b_{1x}m_1 + b_{2x}m_2 + b_{3x}m_3, \\ m_y &= b_{1y}m_1 + b_{2y}m_2 + b_{3y}m_3, \\ m_z &= b_{1z}m_1 + b_{2z}m_2 + b_{3z}m_3. \end{aligned} \tag{9}$$

The symmetrized eigenfunctions at $\mathbf{k}$ form bases for the group of the wave vector (the "little group") $\mathrm{G}(\mathbf{k}) \subset \mathrm{G}$. These IRs may be constructed in a number of ways, two of which were described in Chapter 16, namely via the central extension $\mathrm{P}(\mathbf{k})'$ of the point group of the wave vector, $\mathrm{P}(\mathbf{k})$, or by constructing the Herring group $\mathscr{H}(\mathbf{k})$. For symmorphic space groups, or for non-symmorphic space groups at internal points of symmetry (including points on lines of symmetry), there are no projective representations (PRs), and the IRs of $\mathrm{G}(\mathbf{k})$ are just vector representations of $\mathrm{P}(\mathbf{k})$ multiplied by the representations $\Gamma_{\mathbf{k}}(E|\mathbf{t})$ of the translation subgroup. To find the IRs of $\mathrm{G}(\mathbf{k})$ for non-symmorphic space groups at points that are on surface lines of symmetry, we use instead of $\mathrm{P}(\mathbf{k})$ its central extension $\mathrm{P}(\mathbf{k})'$. For surface points of symmetry we use either $\mathrm{P}(\mathbf{k})'$ or its isomorph the Herring group $\mathscr{H}(\mathbf{k})$, constructed from $\mathrm{P}(\mathbf{k})$. In these cases $\mathrm{P}(\mathbf{k})'$ and $\mathscr{H}(\mathbf{k})$ are isomorphous with an abstract group G_g^n, or with $\mathrm{G}_g^n \otimes \mathrm{T}_p$, or with $\mathrm{G}_g^n \otimes \mathrm{T}_p \otimes \mathrm{T}_r$. Here T_p and T_r are low-order Abelian subgroups which consist of translations $(E|\mathbf{t})$ and which may therefore be ignored because they affect only the phase of the Bloch functions $\psi_{m\mathbf{k}}$. The abstract group G_g^n may be, but is not necessarily, isomorphous with a crystallographic point group. The superscript denotes the ordinal number n in the list of abstract groups of order g. For example, for the 2-D space group of Sections 16.6 and 16.7, at the surface symmetry point X, $\mathrm{P}(\mathbf{k})$ is C_{2v} and $\mathrm{P}(\mathbf{k})'$ and $\mathscr{H}(\mathbf{k})$ are isomorphous with the abstract group $\mathrm{G}_8^4 \sim \mathrm{D}_4$.

In applying the projection operator method for the calculation of symmetrized linear combinations of eigenfunctions, we shall need the effect of a space-group function operator $(R|\mathbf{v})$ on the FE eigenfunction $\psi_{m\mathbf{k}}(\mathbf{r})$, which is

$$\begin{aligned} (R|\mathbf{v})\psi_{m\mathbf{k}}(\mathbf{r}) &= \psi_{m\mathbf{k}}((R|\mathbf{v})^{-1}\mathbf{r}) = \psi_{m\mathbf{k}}(R^{-1}\mathbf{r} - R^{-1}\mathbf{v}) \\ &= \exp[\mathrm{i}(\mathbf{k} - \mathbf{b}_m)\cdot(R^{-1}\mathbf{r} - R^{-1}\mathbf{v})] \\ &= \exp(\mathrm{i}\ \mathbf{b}_m \cdot R^{-1}\mathbf{v})\exp(-\mathrm{i}\ \mathbf{k}\cdot R^{-1}\mathbf{v})\exp[\mathrm{i}(\mathbf{k} - \mathbf{b}_m)\cdot R^{-1}\mathbf{r}]. \end{aligned} \tag{10}$$

In eq. (10), $\mathbf{v} = \mathbf{0}, \forall R$, for symmorphic space groups; $\mathbf{r}$ and $\mathbf{v}$ are vectors in the space of the crystal lattice and are measured in units of a lattice constant a of that lattice, so that $\mathbf{r}/a = x\mathbf{e}_1 + y\mathbf{e}_2 + z\mathbf{e}_3 = [x\ y\ z]$, where $[x\ y\ z]$ means a vector whose Cartesian components x, y, and z are dimensionless numbers. Similarly, when $\mathbf{v} = \mathbf{w} \neq \mathbf{0}$, $[w_1\ w_2\ w_3] = w_1\mathbf{e}_1 + w_2\mathbf{e}_2 + w_3\mathbf{e}_3$ means the vector whose Cartesian components are w_1, w_2, and w_3, in units of a.

Example 17.5-1 A crystal belonging to the symmorphic space group 221 (O_h^1 or $Pm3m$) has the simple cubic lattice. The unit cell is a cube with cube edge of length a. The reciprocal lattice defined by $\mathrm{B} = b\mathrm{E}_3$ is also simple cubic with cube edge of the unit cell $b = 2\pi/\mathrm{a}$. Expressing $\mathbf{k}$ in units of b gives $\mathbf{k}/b = [\xi\ \eta\ \zeta]$, where the components of $\mathbf{k}/b$, $[\xi\ \eta\ \zeta]$, are the coordinates of symmetry point K, $\mathbf{k}$ being the vector from the origin to K.

(5) $$(\hbar^2/2m_e a^2)^{-1}\mathrm{E}_m(\mathrm{K}) = \varepsilon_m(\mathrm{K}) = (\xi - m_x)^2 + (\eta - m_y)^2 + (\zeta - m_z)^2; \qquad (11)$$

(6) $$\psi_m(\mathrm{K}) = \exp\{2\pi\mathrm{i}[(\xi - m_x)x + (\eta - m_y)y + (\zeta - m_z)z]\}. \qquad (12)$$

ψ_m is a function of $\mathbf{r}$, as eq. (6) shows, but the notation $\psi_m(\mathrm{K})$ is used to convey that it is $\psi(\mathbf{r})$ at the symmetry point $\mathrm{K}[\xi\ \eta\ \zeta]$, for this $\mathbf{b}_m$, the components of which enter parametrically into eq. (12). Likewise, $\varepsilon_m(\mathrm{K})$ is the dimensionless energy at K.

At $\Gamma[0\ 0\ 0]$, $\mathrm{P}(\mathbf{k}) = \mathrm{O_h}$ and

(11) $$\varepsilon_m(\Gamma) = m_x{}^2 + m_y{}^2 + m_z{}^2, \qquad (13)$$

(12) $$\psi_m(\Gamma) = \exp[-2\pi\mathrm{i}(m_x x + m_y y + m_z z)]. \qquad (14)$$

The lowest energy at Γ occurs for $\mathbf{b}_m = \mathbf{0}$ and is $\varepsilon_0(\Gamma) = 0$. The corresponding (unnormalized) eigenfunction is $\psi_0(\Gamma) = 1$, and the symmetry of this state is therefore Γ_1 (or A_{1g}). Note that Δ is a point on ΓX, which is along k_y, so that $\Delta[0\ \eta\ 0]$. At Δ, $\mathrm{P}(\mathbf{k})$ is C_{4v}. In the lowest band ($m_x = m_y = m_z = 0$),

(11), (12) $$\varepsilon_0(\Delta) = \eta^2,\ \psi_0(\Delta) = \exp[2\pi\mathrm{i}\eta y],\ 0 < \eta < {}^1\!/_2, \qquad (15)$$

which is of symmetry Δ_1 in C_{4v} (Table 17.2). This first band ($m_y = 0$) ends at the zone boundary ($\eta = {}^1\!/_2$) where the second band ($m_y = 1$) starts. Therefore, at X,

$$\varepsilon_0(\mathrm{X}) = {}^1\!/_4,\quad \psi_0(\mathrm{X}) = \exp(\mathrm{i}\pi y),\ m_y = 0; \qquad (16)$$

$$\varepsilon_{[0\,1\,0]}(\mathrm{X}) = {}^1\!/_4,\quad \psi_{[010]}(\mathrm{X}) = \exp(-\mathrm{i}\pi y),\ m_y = 1. \qquad (17)$$

We now use the projection operator method for finding the linear combinations of the degenerate eigenfunctions $\psi_0(\mathrm{X})$, $\psi_{[0\ 1\ 0]}(\mathrm{X})$ that form bases for the IRs of $\mathrm{P}(\mathbf{k}) = D_{4h}$. For $j = X_1$ (or A_{1g}), which is compatible with Δ_1,

(5.2.10)
$$\begin{aligned} P^j\psi_0(X) &= c^{-1}\sum_R \chi^j(R)^*\hat{R}\psi_0(X), \quad R \in D_{4h} \\ &= \tfrac{1}{2}[\exp(i\pi y) + \exp(-i\pi y)] \\ &= \cos(\pi y), \end{aligned} \tag{18}$$

where c is simply a common factor that is cancelled in order to achieve a result without unnecessary numerical factors. Similarly for $j = X_4'$ (or A_{2u})

$$P^j\psi_{[010]}(X) = \sin(\pi y). \tag{19}$$

Along ΓX, the second band ($m_y = 1$) is given by

(11)
$$\varepsilon_{[010]}(\Delta) = (1-\eta)^2, \quad \psi_{[010]}(\Delta) = \exp[2\pi i(\eta - 1)y], \tag{20}$$

which forms a basis for Δ_1. The second band ends at Γ, where $m_y = 1$ gives

(13)
$$\varepsilon_{[010]}(\Gamma) = 1. \tag{21}$$

This energy is six-fold degenerate since the states with [[± 1 0 0]] all have the same energy $\varepsilon_m(\Gamma) = 1$. The eigenfunctions are linear combinations of the $\psi_m(\Gamma)$ in eq. (14) which are of the correct symmetry. To find these IRs we need to know the subspaces spanned by the m basis, which consists of the six permutations of $m = [1\ 0\ 0]$, and then use projection operators. But actually we have already solved this problem in Section 6.4 in finding the molecular orbitals of an ML_6 complex ion. There we found the IRs of $\{\sigma_1, \ldots, \sigma_6\}$, which map on to the six permutations of [1 0 0], to be A_{1g}, E_g and T_{1u} (or Γ_1, Γ_{12}, and Γ_{15}) and also the symmetrized bases listed in eqs. (6.4.19)–(6.4.25).

(14), (5.2.10)
$$P^j\psi_{[010]}(\Gamma) = c^{-1}\sum_R \chi^j(R)^*\exp(-2\pi i R^{-1}y), \ j = \Gamma_1; \tag{22}$$

(6.4.19)
$$\psi(\Gamma_1) = \cos(2\pi x) + \cos(2\pi y) + \cos(2\pi z). \tag{23}$$

Proceeding similarly for Γ_{12} and Γ_{15} yields

(6.4.20)
$$\psi_I(\Gamma_{12}) = \cos(2\pi z) - \cos(2\pi x), \tag{24}$$

(6.4.21)
$$\psi_{II}(\Gamma_{12}) = 2\cos(2\pi y) - [\cos(2\pi z) + \cos(2\pi x)], \tag{25}$$

(6.4.22)
$$\psi_I(\Gamma_{15}) = \sin(2\pi x), \ \psi_{II}(\Gamma_{15}) = \sin(2\pi y), \ \psi_{III}(\Gamma_{15}) = \sin(2\pi z). \tag{26}$$

At Δ, P(**k**) is C_{4v}, with the principal four-fold axis along k_y. The coordinates at Δ are $[0\ \eta\ 0]$, $0 < \eta < \tfrac{1}{2}$. The possible values of m are the six permutations of [1 0 0]. The energies in this band are

$$\begin{aligned} &\varepsilon_{[0\ 1\ 0]} = (1-\eta)^2, \quad \varepsilon_{[0\ \bar{1}\ 0]} = (1+\eta)^2, \\ &\varepsilon_{[1\ 0\ 0]} = \varepsilon_{[\bar{1}\ 0\ 0]} = \varepsilon_{[0\ 0\ 1]} = \varepsilon_{[0\ 0\ \bar{1}]} = 1 + \eta^2. \end{aligned} \tag{27}$$

One of these, namely $\varepsilon_{[0\ 1\ 0]} = (1-\eta)^2$, is the band described by eq. (20). The band

(11) $$\varepsilon_{[0\,\bar{1}\,0]}(\Delta) = (1+\eta)^2, \quad \psi_{[0\,\bar{1}\,0]}(\Delta) = \exp[2\pi \mathrm{i}(1+\eta)y], \tag{28}$$

is also of symmetry Δ_1 since $\psi_{[0\,\bar{1}\,0]}$ is invariant under the operations of C_{4v}. The four bands that correspond to the remaining four permutations [±1 0 0], [0 0 ±1] are degenerate with $\varepsilon_m = 1 + \eta^2$. For these bands, the character set $\chi(\Gamma_m)$ of the permutation representation is

$$\begin{array}{lllllll} C_{4v} = \{E & 2C_{4y}^{\pm} & C_{2y} & \sigma_z, \sigma_x & \sigma_e, \sigma_f\} \\ \chi(\Gamma_m) = \{4 & 0 & 0 & 2 & 0\}, \end{array} \tag{29}$$

the classes of C_{4v} being shown on the previous line. Therefore

$$\Gamma_m = \Delta_1 + \Delta_2 + \Delta_5 = A_1 + B_1 + E; \tag{30}$$

(12) $$\psi_{[0\,0\,1]}(\Delta) = \exp[2\pi \mathrm{i}(\eta y - z)], \; \psi_{[1\,0\,0]}(\Delta) = \exp[2\pi \mathrm{i}(\eta y - x)]. \tag{31}$$

Projecting the first of these into the Δ_1, Δ_2, and Δ_5 subspaces, and projecting the second one into the Δ_5 subspace, yields

$$\psi(\Delta_1) = \exp(2\pi \mathrm{i}\eta y)\,[\cos 2\pi z + \cos 2\pi x], \tag{32}$$

$$\psi(\Delta_2) = \exp(2\pi \mathrm{i}\eta y)\,[\cos 2\pi z - \cos 2\pi x], \tag{33}$$

$$\psi_{\mathrm{I}}(\Delta_5) = \exp(2\pi \mathrm{i}\eta y)\,\sin 2\pi z, \quad \psi_{\mathrm{II}}(\Delta_5) = \exp(2\pi \mathrm{i}\eta y)\sin 2\pi x. \tag{34}$$

At X, $\eta = 1/2$, and the energy of the four degenerate bands Δ_1, Δ_2, Δ_5 (eqs. (32)–(34)) is $\varepsilon_m(\mathrm{X}) = 5/4$. Since ξ and ζ are zero,

$$\varepsilon_m(\mathrm{X}) = m_x{}^2 + (m_y - 1/2)^2 + m_z{}^2 = 5/4, \tag{35}$$

which is satisfied by

$$[m_x\; m_y\; m_z] = [\pm 1\;0\;0], [0\;0\;\pm 1], [\pm 1\;1\;0], [0\;1\;\pm 1]. \tag{36}$$

P(**k**) at X is D_{4h}, and the character set of the permutation in eq. (36) is therefore given by

$$\begin{array}{lllllllllll} D_{4h} = \{E & C_{4y}^{\pm} & C_{4y}^{2} & C_{2z}, C_{2x} & C_{2e}, C_{2f} & I & \sigma_y & S_{4y}^{\mp} & \sigma_z, \sigma_x & \sigma_e, \sigma_f\} \\ \chi(\Gamma_m) = \{8 & 0 & 0 & 0 & 0 & 0 & 0 & 0 & 4 & 0\}, \end{array}$$

whence

$$\begin{aligned} \Gamma_m &= A_{1g} \oplus B_{1g} \oplus A_{2u} \oplus B_{2u} \oplus E_g \oplus E_u \\ &= X_1 \oplus X_2 \oplus X_4{}' \oplus X_3{}' \oplus X_5 \oplus X_5{}'. \end{aligned} \tag{37}$$

Classes of D_{4h} for X on k_y are specified above the character set for Γ_m. The translation into BSW notation in eq. (37) may be checked from Table 17.1.

Exercise 17.5-1 Verify the direct sum in eq. (37).

The symmetrized linear combinations that form bases for these IRs are now determined using the projection operator

Table 17.9. *Jones symbols $R^{-1}\{x\,y\,z\}$ for the set $\{R\}$ of the point group* $D_{4h} = \{R\} \oplus \{IR\}$.

Jones symbols for the set $\{IR\}$ are obtained by changing the sign of the symbols for $\{R\}$. The principal axis has been chosen along y; for the choices z or x, use cyclic permutations of $\{x\,y\,z\}$, or derive afresh, using the appropriate projection diagram.

	E	$C_{4y}^{\pm}$	C_{4y}^{2}	C_{2z}, C_{2x}	C_{2e}, C_{2f}
Jones symbol	xyz	$zy\bar{x}$ $\bar{z}yx$	$\bar{x}y\bar{z}$	$\bar{x}\bar{y}z$ $x\bar{y}\bar{z}$	$z\bar{y}x$ $\bar{z}\bar{y}\bar{x}$

(12) $$\psi^j = P^j\psi_{[100]}(X) = c^{-1}\sum_R \chi^j(R)^* \exp[-2\pi i R^{-1}(x + \tfrac{1}{2}y)], \tag{38}$$

where j denotes one of the IRs in eq. (37). The simplest way to effect the substitutions $R^{-1}(x + \frac{1}{2}y)$ is to first make a list of the Jones symbols. For degenerate states, project a different function in eq. (38). The symmetrized linear combinations obtained from eq. (38) using Table 17.9 are as follows:

$$\psi(X_1) = \cos \pi y\ [\cos 2\pi z + \cos 2\pi x], \tag{39}$$

$$\psi(X_2) = \cos \pi y\ [\cos 2\pi z - \cos 2\pi x], \tag{40}$$

$$\psi(X_3{}') = \sin \pi y\ [\cos 2\pi z - \cos 2\pi x], \tag{41}$$

$$\psi(X_4{}') = \sin \pi y\ [\cos 2\pi z + \cos 2\pi x], \tag{42}$$

$$\psi(X_5) = \{\sin \pi y \sin 2\pi z,\quad \sin \pi y \sin 2\pi x\}, \tag{43}$$

$$\psi(X_5{}') = \{\cos \pi y \sin 2\pi z,\quad \cos \pi y \sin 2\pi x\}. \tag{44}$$

Figure 17.3 shows the energy bands along $\Gamma\Delta X$ in the BZ of the *sc* lattice. Degeneracy and compatibility are satisfied at Γ and at the zone boundary. Since from eq. (11), $\varepsilon(\Gamma)$ in the next band is 2, $\varepsilon(\Delta) = 2 - \eta^2$ and m is one of the four permutations of [1 1 0] which have $m_y = 1$. From eq. (30) the IRs in this band are Δ_1, Δ_2, and Δ_5, as shown in Figure 17.3.

Example 17.5-2 Free-electron energy bands for the face-centered cubic (*fcc*) lattice. For the *fcc* lattice,

(9) $$\begin{aligned} m_x &= \tfrac{1}{2}b\ (-m_1 + m_2 + m_3), \\ m_y &= \tfrac{1}{2}b\ (m_1 - m_2 + m_3), \\ m_z &= \tfrac{1}{2}b\ (m_1 + m_2 - m_3), \end{aligned} \tag{45}$$

where $b = 4\pi/a$ and a is the cube edge of the unit cell in the *fcc* lattice. The dimensionless energy at K $[\xi\ \eta\ \zeta]\ (2\pi/a)$ is

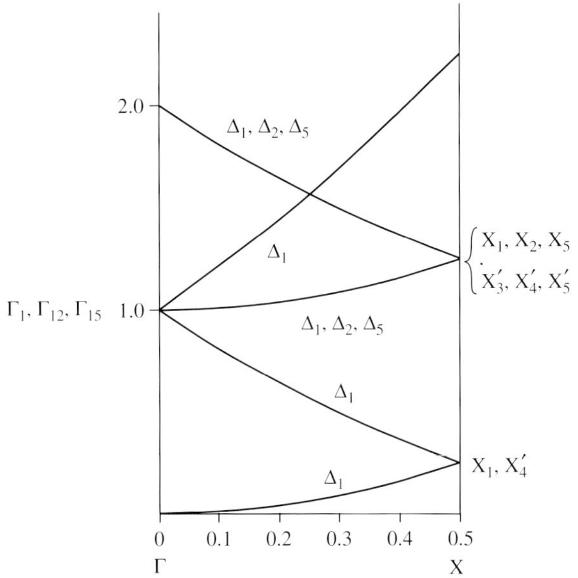

Figure 17.3. Energy bands for the simple cubic Bravais lattice in the free-electron approximation at Δ on ΓX. The symmetry of the eigenfunctions at Γ and at X given in the diagram satisfy compatibility requirements (Koster *et al.* (1963)). Degeneracies are not marked, but may be easily calculated from the dimensions of the representations.

$$\varepsilon_m(\mathrm{K}) = (\xi - m_x)^2 + (\eta - m_y)^2 + (\zeta - m_z)^2, \tag{46}$$

and the corresponding FE eigenfunction is given by

$$\psi_m(\mathrm{K}) = \exp\{2\pi \mathrm{i}[(\xi - m_x)x + (\eta - m_y)y + (\zeta - m_z)z]\}. \tag{47}$$

At Γ,

(46) $$\varepsilon_m(\Gamma) = m_x^{\ 2} + m_y^{\ 2} + m_z^{\ 2}; \tag{48}$$

(47) $$\psi_m(\Gamma) = \exp[-2\pi \mathrm{i}(m_x x + m_y y + m_z z)]. \tag{49}$$

The lowest energy state, $\varepsilon_0(\Gamma) = 0$, has $\psi_m(\Gamma) = 1$ and is of symmetry Γ_1 or A_{1g}. The next highest level is

$$\varepsilon_{\{111\}}(\Gamma) = 3, \quad \psi_{\{111\}}(\Gamma) = \exp[-2\pi \mathrm{i}(\pm x \pm y \pm z)], \tag{50}$$

where the subscript $\{111\}$ indicates that m is one of the eight permutations of [1 1 1]. These points lie at the corners of a cube. The permutation representation is

$$\Gamma_{\{111\}} = A_{1g} \oplus A_{2u} \oplus T_{2g} \oplus T_{1u} = \Gamma_1 \oplus \Gamma_2{}' \oplus \Gamma_{25}{}' \oplus \Gamma_{15}; \tag{51}$$

Γ_m may also be derived by reducing the representation spanned by the ligand orbitals in a cubic ML_8 molecule or configuration, as occurs, for example, when an ion such as In^+ is a substitutional impurity in CsCl.

Table 17.10. *Basis functions* $\phi_1 \dots \phi_8$, *and their variables, for the IRs of the permutation representation of* [1 1 1].

These same Jones symbols are used in the derivation of the ligand orbitals of an ML_8 molecule.

ϕ_1	xyz	ϕ_5	$\bar{x}\bar{y}\bar{z}$
ϕ_2	$\bar{x}\bar{y}z$	ϕ_6	$xy\bar{z}$
ϕ_3	$x\bar{y}\bar{z}$	ϕ_7	$\bar{x}yz$
ϕ_4	$\bar{x}y\bar{z}$	ϕ_8	$x\bar{y}z$

Exercise 17.5-2 Verify the permutation representation Γ_m in eq. (51).

(50) $$\psi_{[111]}(\Gamma) = \exp[-2\pi i(x + y + z)] = \phi_1. \tag{52}$$

The sets of variables (coordinates) for the complete list of ϕ functions are in Table 17.10. Basis functions of the correct symmetry may be found by using projection operators. For the jth IR in eq. (51),

$$\psi^j = c^{-1}\sum_R \chi^j(R)^* \hat{R}\ \psi_{[111]}(\Gamma). \tag{53}$$

To reduce writing we give only the transformed coordinates (Jones symbols) which are to be substituted for $\{x\ y\ z\}$ in eq. (52), rather than the actual functions, using parentheses to separate the classes of O_h. For $j = \Gamma_1$,

(53) $$\begin{aligned}\psi(\Gamma_1) = c^{-1}[&(xyz) + (\bar{x}\bar{y}z + x\bar{y}\bar{z} + \bar{x}y\bar{z})\\ &+ (zxy + yzx + \bar{z}x\bar{y} + y\bar{z}\bar{x} + \bar{y}\bar{z}x + z\bar{x}\bar{y} + \bar{y}z\bar{x} + \bar{z}\bar{x}y)\\ &+ (\bar{y}xz + y\bar{x}z + x\bar{z}y + xz\bar{y} + zy\bar{x} + \bar{z}yx)\\ &+ (yx\bar{z} + \bar{y}\bar{x}\bar{z} + z\bar{y}x + \bar{z}\bar{y}\bar{x} + \bar{x}zy + \bar{x}\bar{y}\bar{z})\\ &+ IR^{-1}(xyz)], \quad R \in O;\end{aligned} \tag{54}$$

(54) $$\psi(\Gamma_1) = xyz + \bar{x}yz + x\bar{y}z + \bar{x}\bar{y}z + xy\bar{z} + \bar{x}y\bar{z} + x\bar{y}\bar{z} + \bar{x}\bar{y}\bar{z}. \tag{55}$$

Recall that the notation in eqs. (54) and (55) is evocative rather than literal, so that "$+\bar{x}yz$", for example, means "$+\exp[-2\pi i(-x + y + z)]$". Alternatively, from eq. (55) and Table 17.10,

$$\psi(A_{1g}) = \phi_1 + \phi_2 + \phi_3 + \phi_4 + \phi_5 + \phi_6 + \phi_7 + \phi_8. \tag{56}$$

The ϕ notation for the basis functions is often convenient since it offers considerable economy in writing down the bases for each IR.

(55), or (56) $$\psi(\Gamma_1) = \cos(2\pi x)\ \cos(2\pi y)\ \cos(2\pi z). \tag{57}$$

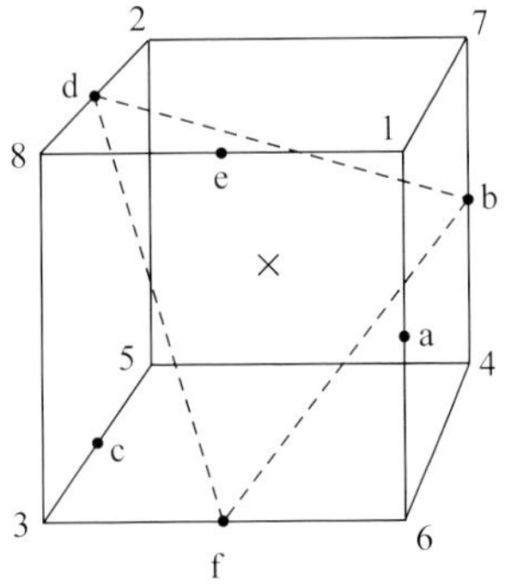

Figure 17.4. Labeling of the symmetry elements of a cube as used in the derivation of Γ_m and of the symmetrized bases. The points 1, 2, 3, and 4 label the three-fold axes. The poles of the two-fold axes are marked a, b, c, d, e, and f. The points 1, 2, 3, 4, 5, 6, 7, and 8 provide a graphical representation of the permutations of [1 1 1] (cf. Table 17.10).

Exercise 17.5-3 Derive eq. (56) directly from $\psi^j = c^{-1}\sum_R \chi^j(R)^* \hat{R}\phi_1$, where $\{\phi_i\}$ labels the points $i = 1, 2, \ldots, 8$ in Figure 17.4.

In the same way, for $j = \Gamma_2'$ (or A_{2u}), for which $\{\chi^j(\mathscr{C}_k)\} = \{1 \;\; 1 \;\; 1 \;\; -1 \;\; -1\}$ for the five classes of O (in the same order as in the O_h table in Appendix A3)

(53) $$\psi(\Gamma_2') = \sin(2\pi x)\,\sin(2\pi y)\,\sin(2\pi z). \tag{58}$$

For $j = \Gamma_{25}'$, with $\{\chi^j(\mathscr{C}_k)\} = \{3 \;\; -1 \;\; 0 \;\; -1 \;\; 1\}$,

$$\psi(\Gamma_{25}') = 3\phi_1 - (\phi_2 + \phi_3 + \phi_4) - (\phi_8 + \phi_7 + \phi_6) + 3\phi_5 \equiv \psi_1. \tag{59}$$

Projecting, in turn, ϕ_2, ϕ_3, ϕ_4 gives

$$\psi_2 = 3\phi_2 - (\phi_3 + \phi_4 + \phi_5) - (\phi_1 + \phi_8 + \phi_7) + 3\phi_6, \tag{60}$$

$$\psi_3 = 3\phi_3 - (\phi_4 + \phi_5 + \phi_6) - (\phi_2 + \phi_1 + \phi_8) + 3\phi_7, \tag{61}$$

$$\psi_4 = 3\phi_4 - (\phi_5 + \phi_6 + \phi_7) - (\phi_3 + \phi_2 + \phi_1) + 3\phi_8. \tag{62}$$

Of course, it is not necessary to carry out the actual projections because ψ_2, ψ_3, and ψ_4 are simply written down as cyclic permutations of ψ_1. There are only three linearly independent bases for a T_{2g} representation and we choose these to be $\psi_{\mathrm{I}} = ½[\psi_1 + \psi_3]$, $\psi_{\mathrm{II}} = ½[\psi_1 + \psi_4]$, $\psi_{\mathrm{III}} = ½[\psi_1 + \psi_2]$. On using Table 17.10 and eq. (52),

(59), (61) $$\psi_{\mathrm{I}}(\Gamma_{25}') = \cos(2\pi x)\sin(2\pi y)\sin(2\pi z), \tag{63}$$

(59), (62) $$\psi_{\mathrm{II}}(\Gamma_{25}') = \sin(2\pi x)\cos(2\pi y)\sin(2\pi z), \tag{64}$$

(59), (60) $$\psi_{\mathrm{III}}(\Gamma_{25}') = \sin(2\pi x)\sin(2\pi y)\cos(2\pi z). \tag{65}$$

Alternatively, ½$[\psi_1+\psi_2]=\psi_{\rm III}=\sin(2\pi x)\sin(2\pi y)\cos(2\pi z)$, and cyclic permutation of xyz gives $\psi_{\rm I}$ and $\psi_{\rm II}$. The three symmetrized bases for Γ'_{25} are symmetric under inversion, as they should be. In a similar way, the bases for $\Gamma_{15}(T_{1u})$ are obtained by projecting ϕ_1, ϕ_2, ϕ_3, ϕ_4 into the Γ_{15} subspace, with the following result:

(54) $$\psi_{\rm I}(\Gamma_{15})=\sin(2\pi x)\cos(2\pi y)\cos(2\pi z). \tag{66}$$

Cyclic permutation of $\{xyz\}$ in eq. (66) gives

(66) $$\psi_{\rm II}(\Gamma_{15})=\cos(2\pi x)\sin(2\pi y)\cos(2\pi z), \tag{67}$$

(66) $$\psi_{\rm III}(\Gamma_{15})=\cos(2\pi x)\cos(2\pi y)\sin(2\pi z). \tag{68}$$

At symmetry point L, with coordinates [½ ½ ½]$(2\pi/a)$,

(46) $$\varepsilon_m(\mathrm{L})=(½-m_x)^2+(½-m_y)^2+(½-m_z)^2, \tag{69}$$

(47) $$\psi_m(\mathrm{L})=\exp\{2\pi\mathrm{i}[(½-m_x)x+(½-m_y)y+(½-m_z)z]\}. \tag{70}$$

The lowest energy, which occurs for $m=0$ and $m=[1\ 1\ 1]$, is

$$\varepsilon_0(\mathrm{L})=3/4. \tag{71}$$

This $\varepsilon(\mathrm{L})=$ ¾ state is degenerate, with

(70) $$\psi_0(\mathrm{L})=\exp[\mathrm{i}\pi(x+y+z)], \tag{72}$$

(70) $$\psi_{[1\ 1\ 1]}(\mathrm{L})=\exp[-\mathrm{i}\pi(x+y+z)]. \tag{73}$$

P(**k**) at L is $D_{3d}=\{R\}\oplus\{IR\}$, where $\{R\}=\{E\ R(\pm(2\pi/3)[1\ 1\ 1])\ C_{2\mathbf{b}}\ C_{2\mathbf{d}}\ C_{2\mathbf{f}}\}$ (Figure 17.4). The Jones symbols describing the effect of these operators on a function $f(x, y, z)$ are in Table 17.11.

Exercise 17.5-4 Derive the permutation representation $\Gamma_m(\mathrm{L})$. [*Hint*: See Figure 17.5.]

Projecting $\psi_0(\mathrm{L})$ into the L_1, $L'_2(A_{1g}, A_{2u})$ subspaces yields

(72) $$\psi(\mathrm{L}_1)=\cos[\pi(x+y+z)], \tag{74}$$

$$\psi(\mathrm{L}'_2)=\sin[\pi(x+y+z)]. \tag{75}$$

At $\varepsilon(\mathrm{L})=2\,3/4$

(47) $$\psi_{[1\ 1\ \bar{1}]}(\mathrm{L})=\exp[-\mathrm{i}\pi(x+y-3z)]. \tag{76}$$

Projecting into the L_1, L'_2, L_3, L'_3 subspaces gives

$$\psi_{[1\ 1\ \bar{1}]}(\mathrm{L}_1)=\cos[\pi(x+y-3z)]+\cos[\pi(y+z-3x)]+\cos[\pi(z+x-3y)]; \tag{77}$$

$$\psi_{[1\ 1\ \bar{1}]}(\mathrm{L}'_2)=\sin[\pi(x+y-3z)]+\sin[\pi(y+z-3x)]+\sin[\pi(z+x-3y)]; \tag{78}$$

$$\psi_{[1\ 1\ \bar{1}]}(\mathrm{L}_3)=\cos[\pi(x+y-3z)]-\cos[\pi(y+z-3x)]; \tag{79}$$

Table 17.11. *Jones symbols $R^{-1}(xyz)$ for $R \in D_{3d}$ and character table for* D_{3d}.

Each of the three C_2 operators transforms ϕ_1 into ϕ_5 (see Figure 17.4).

	E	$C_{31}^{\pm}$	$C_{2\mathbf{b}}, C_{2\mathbf{d}}, C_{2\mathbf{f}}$	I	$S_{61}^{\pm}$	$\sigma_{\mathbf{b}}, \sigma_{\mathbf{d}}, \sigma_{\mathbf{f}}$
	xyz	zxy	$\overline{x}\overline{y}\overline{z}$	$\overline{x}\overline{y}\overline{z}$	$\overline{z}\overline{x}\overline{y}$	xyz
		yzx	$\overline{x}\overline{y}\overline{z}$		$\overline{y}\overline{z}\overline{x}$	xyz
			$\overline{x}\overline{y}\overline{z}$			xyz
L_1 A_{1g}	1	1	1	1	1	1
L_1' A_{1u}	1	1	1	−1	−1	−1
L_2 A_{2g}	1	1	−1	1	1	−1
L_2' A_{2u}	1	1	−1	−1	−1	1
L_3 E_g	2	−1	0	2	−1	0
L_3' E_u	2	−1	0	−2	1	0

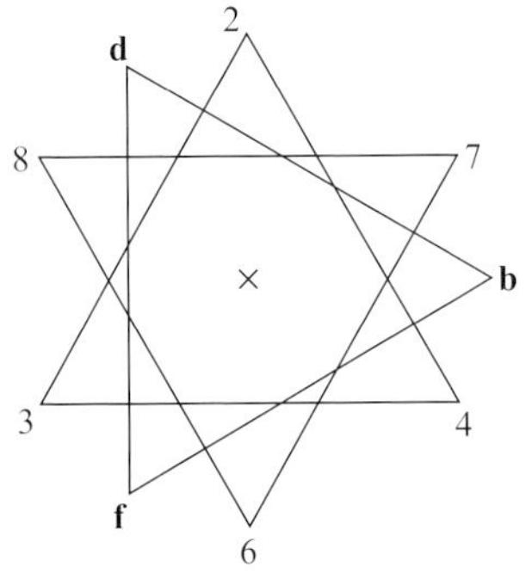

Figure 17.5. View looking down the [111] axis in Figure 17.4. The triangle containing 6, 7, 8 lies above the plane of the triangle containing the binary axes labeled **b**, **d**, and **f**, while that containing 2, 3, and 4 lies below this plane. Points marked 1 and 5 in Figure 17.4 lie on the axis through the center of this figure.

$$\psi_{[1\,1\,\overline{1}]}(L_3)' = \cos[\pi(z + x - 3y)] - \cos[\pi(y + z - 3x)]; \tag{80}$$

$$\psi_{[1\,1\,\overline{1}]}(L_3') = \sin[\pi(x + y - 3z)] - \sin[\pi(y + z - 3x)]; \tag{81}$$

$$\psi_{[1\,1\,\overline{1}]}(L_3')' = \sin[\pi(z + x - 3y)] - \sin[\pi(y + z - 3x)]. \tag{82}$$

These results can all be derived very easily with the aid of Table 17.11.

Exercise 17.5-5 Confirm the above results for the L_3 representation. For a point Λ on ΓL,

(47) $$\varepsilon_m(\Lambda) = (\xi - m_x)^2 + (\xi - m_y)^2 + (\xi - m_z)^2, \quad 0 < \xi < 1/2, \tag{83}$$

whence the free-electron energy bands along Λ are readily calculated, the representations being determined from compatibility requirements.

Answers to Exercises 17.5

Exercise 17.5-1 Use the character table for D_{4h} and eq. (4.4.20).

Exercise 17.5-2 Using Figure 17.4, the character set Γ_m is given by

$$\begin{array}{lcccccccccc} & E & 3C_4^2 & 4C_3^{\pm} & 3C_4^{\pm} & 6C_2 & I & 3\sigma_h & 4S_6^{\mp} & 3S_4^{\mp} & 6\sigma_d \\ \chi(\Gamma_m) = \{ & 8 & 0 & 2 & 0 & 0 & 0 & 0 & 0 & 0 & 4\}. \end{array}$$

Γ_m then follows from the character table for O_h and eq. (4.4.20). (The easiest way of determining $\chi(\Gamma_m)$ is to look at the figure representing the set of points to be permuted (in this case, a cube) and determine the number of points that are unshifted under one operation from each class. This number is the character for that class in the permutation representation Γ_m. For example, each of the dihedral planes contains four points which are invariant under reflections in that plane.)

Exercise 17.5-3

$$\begin{aligned} c^{-1}\sum_{R\in O}\chi^{\Gamma_1}(R)^*\hat{R}\phi_1 = c^{-1}[&(\phi_1) + (\phi_2+\phi_3+\phi_4) \\ &+ (\phi_1+\phi_1+\phi_3+\phi_4+\phi_4+\phi_2+\phi_2+\phi_3) \\ &+ (\phi_8+\phi_7+\phi_6+\phi_8+\phi_7+\phi_6) \\ &+ (\phi_6+\phi_5+\phi_8+\phi_5+\phi_7+\phi_5)]. \end{aligned}$$

The order of the operators here is the same as that used in the derivation of eq. (54). Each ϕ function occurs three times in this list, and consequently occurs another three times in the sum from IR, $R \in O$. Therefore, with $c = 6$, $\psi(\Gamma_1) = \phi_1+\phi_2+\phi_3+\phi_4+\phi_5+\phi_6+\phi_7+\phi_8$.

Exercise 17.5-4 From Figure 17.5,

$$\begin{array}{lccccccc} & & E & C_{31}^{\mp} & 3C_2 & I & S_{61}^{\pm} & 3\sigma_d \\ \chi(\Gamma_m) & = & \{6 & 0 & 0 & 0 & 0 & 2\}; \end{array}$$

$$\Gamma_m = A_{1g} \oplus A_{2u} \oplus E_g + E_u = L_1 \oplus L_2' \oplus L_3 \oplus L_3'$$

then follows from the character table for D_{3d}, either by inspection or by use of eq. (4.4.20).

Exercise 17.5-5 Projecting eq. (76) into the L_3 subspace yields

$$\begin{aligned} \psi_1(L_3) &= c^{-1}\sum_{R\in D_{3d}}\chi^{L_3}(R)^*\hat{R}\exp[-i\pi(x+y-3z)] \\ &= 2\cos[\pi(x+y-3z)] - \cos[\pi(z+x-3y)] - \cos[\pi(y+z-3x)], \end{aligned}$$

$$\psi_2(L_3) = 2\cos[\pi(y+z-3x)] - \cos[\pi(x+y-3z)] - \cos[\pi(z+x-3y)],$$

$$\psi_3(L_3) = 2\cos[\pi(z+x-3y)] - \cos[\pi(y+z-3x)] - \cos[\pi(x+y-3z)].$$

These three functions are not linearly independent, but on choosing linear combinations to give the simplest form,

$$\psi_{\text{I}}(\text{L}_3) = [\psi_1(\text{L}_3) - \psi_3(\text{L}_3)]/3 = \cos[\pi(x + y - 3z)] - \cos[\pi(y + z - 3x)],$$

$$\psi_{\text{II}}(\text{L}_3) = [\psi_3(\text{L}_3) - \psi_2(\text{L}_3)]/3 = \cos[\pi(z + x - 3y)] - \cos[\pi(y + z - 3x)].$$

17.6 Free-electron states for crystals with non-symmorphic space groups

For internal points of the BZ there are no PRs and the procedure for constructing space-group representations from P(**k**) is the same as that for symmorphic space groups. But for points on surface lines of symmetry we must use instead of P(**k**) its central extension P(**k**)′. For surface symmetry points it is generally easier to construct the Herring group and to use $\mathscr{H}(\mathbf{k})$ instead of P(**k**)′. Note that $\mathscr{H}(\mathbf{k})$ or P(**k**)′ is isomorphous either with an abstract group G_g^n or with the DP of an abstract group with one or more low-order translation groups; G_g^n may or may not be isomorphous with a crystallographic point group (Table 17.12).

The space group of diamond, and also silicon and germanium, is 227 (*Fd*3*m* or O_h^7). The point group of *Fd*3*m* is T_d, and the little group

$$\text{(16.7.20)} \qquad G(\mathbf{k}) = \sum_A (A|\mathbf{0})T + \sum_B (B|\mathbf{w})T, \qquad (1)$$

in which $\mathbf{w} = [\frac{1}{4}\ \frac{1}{4}\ \frac{1}{4}]$ in units of the cube edge a, $\{A\} = \text{P}(\mathbf{k}) \cap \text{T}_d$, and $B \in \text{P}(\mathbf{k})$ but $\notin \text{T}_d$. In the III/V semiconductor InSb, the Ge atom at [000] is replaced by an In atom and that at **w** by an Sb atom. The point group **P** is again T_d but P(**k**) at Γ is T_d instead of O_h. The point group of the wave vector P(**k**) and the Herring group $\mathscr{H}(\mathbf{k})$ for the InSb and Si structures at the symmetry points K = Γ, Δ, and X are given in Table 17.12. The derivation of the Herring groups from their corresponding point groups has been described in Section 16.7, where it was illustrated by the derivation of $\mathscr{H}(\mathbf{k}) \sim G_{32}^2$ for space group 227 at X. The other $\mathscr{H}(\mathbf{k})$ in Table 17.12 may be obtained similarly, using eq. (1). For space group 216 at X, P(**k**) is D_{2d} and the Herring translation group $\mathscr{T}(\mathbf{k})$ is $T_2 = \{(E|\mathbf{0})\ (E|\mathbf{a}_1 \text{ or } \mathbf{a}_3)\}$, where $\{\mathbf{a}_1, \mathbf{a}_2, \mathbf{a}_3\}$ are the fundamental translations of the *fcc* lattice. In this case, $\mathscr{H}(\mathbf{k})$ is isomorphous with the DP $D_{2d} \otimes T_2$. The advantage of identifying $\mathscr{H}(\mathbf{k})$ or P(**k**′) with an abstract group G_g^n (which, if inspection fails, requires a determination of the symmetry elements and of the group generators of $\mathscr{H}(\mathbf{k})$ or P(**k**′), and a comparison of generating relations with those of the abstract groups G_g to find n) is that when G_g^n is not isomorphous with a crystallographic point group, its classes and character table are then available from the literature (e.g. Bradley and Cracknell (1972)) and need not be worked out each time (cf. Jones (1962), (1975), Morgan (1969)).

Example 17.6-1 Free-electron energy bands along ΓΔX for space groups 216 and 227.

For InSb (or some other AB compound with the same structure) the point group at Γ is T_d and

$$\text{(17.5.45), (17.5.46)} \qquad \varepsilon_m(\Gamma) = m_x^2 + m_y^2 + m_z^2 = 0, 3, 4, \ldots \qquad (2)$$

Table 17.12. *The point group of the wave vector* P(**k**) *at the symmetry points* K = Γ, Δ, *and* X *for the InSb and Si structures.*

The third column gives the abstract group G_g^n which is isomorphous with P(**k**) or, for the surface point X, with the Herring group $\mathscr{H}(\mathbf{k})$. The group symbol for the abstract group is followed by the group generators of this realization of G_g^n (in the order $P, Q, R, \ldots$). This list includes the generator of the translation subgroup when $\mathscr{H}(\mathbf{k})$ is isomorphous with the direct product of an abstract group and a low-order Abelian translation subgroup (as happens for InSb at X). $\{\mathbf{a}_1\mathbf{a}_2\mathbf{a}_3\}$ are the fundamental translations of the *fcc* lattice, $\mathrm{T}_2 = \{(E|\mathbf{0})(E|\mathbf{a}_1)\}$, and $\mathbf{w} = [{}^1\!/\!_4\ {}^1\!/\!_4\ {}^1\!/\!_4]a$. Axes 1, 2, ..., 8 and **a**, ..., **f** are defined in Figure 17.4.

K	P(**k**)	~P(**k**) or $\mathscr{H}(\mathbf{k})$	Generators of G_g^n or $\mathrm{G}_g^n \otimes \mathrm{T}_2$
InSb, 216 ($F\bar{4}3m$ or T_d^2)			
Γ	$\mathrm{T_d}$	$\mathrm{G}_{24}^7 \sim \mathrm{T_d}$	$\{(C_{31}^-|\mathbf{0})\ (C_{2\mathbf{z}}|\mathbf{0})\ (C_{2\mathbf{x}}|\mathbf{0})\ (\sigma_\mathbf{a}|\mathbf{0})\}$
Δ	$\mathrm{C_{2v}}$	$\mathrm{G}_4^2 \sim \mathrm{C_{2v}}$	$\{(C_{2\mathbf{y}}|\mathbf{0})\ (\sigma_\mathbf{e}|\mathbf{0})\}$
X	$\mathrm{D_{2d}}$	$\mathrm{G}_8^4(\sim\mathrm{D_{2d}}) \otimes \mathrm{T}_2$	$\{(S_{4\mathbf{y}}^+|\mathbf{0})\ (C_{2\mathbf{x}}|\mathbf{0})\ (E|\mathbf{a}_1 \text{ or } \mathbf{a}_3)\}$
Si, 227 ($Fd3m$ or O_h^7)			
Γ	$\mathrm{O_h}$	$\mathrm{G}_{48}^7 \sim \mathrm{O_h}$	$\{(S_{61}^-|\mathbf{w})\ (\sigma_\mathbf{x}|\mathbf{w})\ (\sigma_\mathbf{z}|\mathbf{w})\ (C_{2\mathbf{e}}|\mathbf{w})\}$
Δ	$\mathrm{C_{4v}}$	$\mathrm{G}_8^4 \sim \mathrm{C_{4v}}$	$\{(C_{4\mathbf{y}}^+|\mathbf{w})\ (\sigma_\mathbf{e}|\mathbf{0})\}$
X	$\mathrm{D_{2d}}$	G_{32}^2	$\{(\sigma_\mathbf{x}|\mathbf{w})(S_{4\mathbf{y}}^+|\mathbf{0})\ (C_{2\mathbf{x}}|\mathbf{0})\}$

Generating relations for G_g^n

G_4^2 $\quad P^2 = Q^2 = E, \quad PQ = QP$

G_8^4 $\quad P^4 = Q^2 = E, \quad P^3Q = QP$

G_{24}^7 $\quad P^3 = Q^2 = R^2 = S^2 = E, \quad QP = PR, \quad RP = PQR, \quad RQ = QR, \quad SP = P^2RS, \quad SQ = QS$

G_{32}^2 $\quad P^4 = Q^4 = R^2 = E, \quad QP = P^3Q^3, \quad QP^2 = P^2Q, \quad RP = P^3R, \quad RQ = Q^3R$

G_{48}^7 $\quad P^6 = Q^2 = R^2 = S^2 = E, \quad QP = PR, \quad QP^3 = P^3Q, \quad RP = P^4QR, \quad RQ = QR, \quad SP = P^2QS, \quad SQ = RS, \quad SR = QS$

Table 17.13. *Free-electron band energies ε_m for InSb (space group 216) at Γ, Δ, and X, the BZ of the fcc Bravais lattice (Figure 16.12(b)).*

The BSW symbol for K is followed by its Cartesian coordinates $[\xi\,\eta\,\zeta]$ in units of $2\pi/a$, so that the zone boundary at X lies at $\eta = 1$ (see Table 16.6). K is either a symmetry point (Γ, X) or a point on a line of symmetry (Δ); $[m_x\,m_y\,m_z]$ are the Cartesian components of $\mathbf{b}_m$ in eqs. (17.5.11) and (17.5.12); l is the dimension of the permutation representation Γ_m; $\{m_x\,m_y\,m_z\}$ means all possible permutations of $[m_x\,m_y\,m_z]$, but components of $\mathbf{m}$ in bold-face type are not permuted. For example, $\{1\ \mathbf{1}\ 1\}$ means the $l = 4$ permutations $[\pm 1\ \ 1\ \ \pm 1]$. The IRs that comprise the direct sum Γ_m are given in the notation of Mulliken and Koster *et al.* (1963). Character tables may be found in Table 17.14, Appendix A3, Bouckaert *et al.* (1936), Altmann and Herzig (1994), or Koster *et al.* (1963).

m	ε_m	l	Mulliken	Koster *et al.* (1963)
$\Gamma[0\ 0\ 0]$, $P(\mathbf{k}) = T_d = \{E\ 3C_2\ 4C_3^{\pm}\ 3S_4^{\pm}\ 6\sigma_d\} \sim G_{24}^7$				
[0 0 0]	0	1	A_1	Γ_1
{1 1 1}	3	8	$2A_1 \oplus 2T_2$	$2\Gamma_1 \oplus 2\Gamma_5$
{0 2 0}	4	6	$A_1 \oplus E \oplus T_2$	$\Gamma_1 \oplus \Gamma_3 \oplus \Gamma_5$
$\Delta[0\ \eta\ 0]$, $P(\mathbf{k}) = C_{2v} = \{E\ C_{2y}\ \sigma_e\ \sigma_f\} \sim G_4^2$				
[0 0 0]	η^2	1	A_1	Γ_1
{1 **1** 1}	$(\eta-1)^2+2$	4	$2A_1 \oplus B_1 \oplus B_2$	$2\Gamma_1 \oplus \Gamma_2 \oplus \Gamma_4$
{1 $\bar{\mathbf{1}}$ 1}	$(\eta+1)^2+2$	4	$2A_1 \oplus B_1 \oplus B_2$	$2\Gamma_1 \oplus \Gamma_2 \oplus \Gamma_4$
[0 2 0]	$(\eta-2)^2$	1	A_1	Γ_1
[0 $\bar{2}$ 0]	$(\eta+2)^2$	1	A_1	Γ_1
{2 **0** 0}	η^2+4	4	$A_1 \oplus A_2 \oplus B_1 \oplus B_2$	$\Gamma_1 \oplus \Gamma_2 \oplus \Gamma_3 \oplus \Gamma_4$
$X[0\ 1\ 0]$, $\mathscr{H}(\mathbf{k}) = D_{2d} \otimes T_2$, $D_{2d} = \{E\ S_{4y}^{\pm}\ C_{2y}\ C_{2z}\ C_{2x}\ \sigma_e\ \sigma_f\} \sim G_8^4$				
[0 0 0], [0 2 0]	1	2	$A_1 \oplus B_2$	$\Gamma_1 \oplus \Gamma_4$
{1 **1** 1}	2	4	$A_1 \oplus B_1 \oplus E$	$\Gamma_1 \oplus \Gamma_3 \oplus \Gamma_5$
{2 0 0}, {2 **2** 0}	5	8	$A_1 \oplus A_2 \oplus B_1 \oplus B_2 \oplus 2E$	$\Gamma_1 \oplus \Gamma_2 \oplus \Gamma_3 \oplus \Gamma_4 \oplus 2\Gamma_5$

for $m = [0\,0\,0]$, $[\pm 1\ \pm 1\ \pm 1]$, $[\pm 2\ \ 0\ \ 0]$, ..., with degeneracies of 1, 8, 6, ... The eight permutations of [1 1 1] can be represented by the points lying at the corners of a cube. At Γ, $T_d = \{E\ 3C_2\ 4C_3^{\pm}\ 3S_4^{\pm}\ 6\sigma_d\}$ and $\chi(\Gamma_m) = \{8\ 0\ 2\ 0\ 4\}$, whence $\Gamma_m = 2A_1 \oplus 2T_2 = 2\Gamma_1 \oplus 2\Gamma_5$. The six permutations of [2 0 0] can be represented by the six points lying at the vertices of an octahedron; $\chi(\Gamma_m) = \{6\,2\,0\,0\,2\}$ and $\Gamma_m = A_1 \oplus E \oplus T_2 = \Gamma_1 \oplus \Gamma_3 \oplus \Gamma_5$. Band energies calculated from eq. (17.5.46) are in Table 17.13. The symmetrized eigenfunctions are obtained by using projection operators. For example, for $\varepsilon_m(\Delta) = (\eta - 1)^2 + 2$,

$$\psi_m^j(\Delta) = c^{-1}\sum_R \chi^j(R)^* \hat{R}\psi_{[1\ 1\ 1]}(\Delta),\ \ m = [\pm 1\ \ 1\ \ \pm 1],\quad R \in C_{2v} \tag{3}$$

since, as Table 17.14 shows, there are no non-lattice translations. Projecting $\psi_{[1\ 1\ 1]}$ into the Δ_1 subspace gives, with the aid of Table 17.14,

(17.5.47)
$$\psi_{[1\ 1\ 1]}(\Delta) = \exp[2\pi i(-x + (\eta - 1)y - z)]; \tag{4}$$

Table 17.14. *Jones symbols and character tables for InSb (space group* 216) *at* Δ *and* X.

Symmetry transformations may be deduced from the transformation of the points 1, 6, 4, and 7 in Figure 17.4. The reflecting plane of σ_e contains the points 6 and 7, and that of σ_f contains the points 1 and 4. In C_{2v}, the Mulliken designations B_1 and B_2 are arbitrary since they depend on which plane is chosen to determine the subscript. Only $B_1 \oplus B_2$ appears in the direct sums in Table 17.13. BSW(M) signifies the designations of IRs given by Morgan (1969).

$\Delta[0\ \eta\ 0]$, $P(\mathbf{k}) \sim C_{2v}$ $G_4^2 \sim C_{2v}$			$(E\|\mathbf{0})$	$(C_{2y}\|\mathbf{0})$	$(\sigma_f\|\mathbf{0})$	$(\sigma_e\|\mathbf{0})$
			xyz	$\bar{x}y\bar{z}$	zyx	$\bar{z}y\bar{x}$
M	BSW(M)	K	$xy\bar{z}$	$\bar{x}yz$	$zy\bar{x}$	$\bar{z}yx$
A_1	Δ_1	Γ_1	1	1	1	1
A_2	Δ_2	Γ_3	1	1	−1	−1
B_1	Δ_3	Γ_2	1	−1	1	1
B_2	Δ_4	Γ_4	1	−1	−1	1

$X[0\ 1\ 0]$, $\mathscr{H}(\mathbf{k}) = D_{2d} \otimes T_2$ $G_8^4 \sim D_{2d}$			$(E\|\mathbf{0})$	$(S_{4y}^{\pm}\|\mathbf{0})$	$(C_{2y}\|\mathbf{0})$	$(C_{2z}\|\mathbf{0}), (C_{2x}\|\mathbf{0})$	$(\sigma_e\|\mathbf{0}), (\sigma_f\|\mathbf{0})$
M	BSW(M)	K	xyz	$z\bar{y}\bar{x}, \bar{z}\bar{y}x$	$\bar{x}y\bar{z}$	$\bar{x}\bar{y}z, x\bar{y}\bar{z}$	$\bar{z}y\bar{x}, zyx$
A_1	X_1	Γ_2	1	1	1	1	1
A_2	X_4	Γ_2	1	1	1	−1	−1
B_1	X_2	Γ_3	1	−1	1	1	−1
B_2	X_3	Γ_4	1	−1	1	−1	1
E	X_5	Γ_5	2	0	−2	0	0

(3), (4)
$$\psi_{\text{I}}(\Delta_1) = \exp[2\pi\text{i}(-x + (\eta - 1)y - z)] + \exp[2\pi\text{i}(x + (\eta - 1)y + z)]$$
$$= \exp[2\pi\text{i}(\eta - 1)y]\cos[2\pi(x + z)]. \tag{5}$$

A second function of Δ_1 symmetry may be obtained by projecting $\psi_{[1\ 1\ \bar{1}]}(\Delta) = \exp[2\pi\text{i}(-x + (\eta - 1)y + z)]$ (Table 17.14), which gives

$$\psi_{\text{II}}(\Delta_1) = \exp[2\pi\text{i}(\eta - 1)y]\cos[2\pi(x - z)]. \tag{6}$$

On projecting $\psi_{[1\ 1\ 1]}(\Delta)$ with $j = \Delta_3$ and $\chi(\Delta_3) = \{1\ \ -1\ \ \ 1\ \ -1\}$,

$$\psi(\Delta_3) = \exp[2\pi\text{i}(\eta - 1)y]\sin[2\pi(x + z)]. \tag{7}$$

Since $\chi(\Delta_4) = \{1\ \ -1\ \ -1\ \ \ 1\}$, it can be seen from Table 17.14 that the projection of $\psi_{[1\ 1\ 1]}$ into the Δ_4 subspace will be zero. The results in eqs. (5)–(7) already obtained at Δ suggest that we should project $\psi_{[1\ 1\ \bar{1}]}(\Delta)$, and this yields

$$\psi(\Delta_4) = \exp[2\pi\text{i}(\eta - 1)y]\sin[2\pi(x - z)]. \tag{8}$$

Table 17.15. *Character table and Jones symbols for the rotational part R of the space group operator* $(R|\mathbf{v}) \in Fd3m$ (227 *or* O_h^7) *at* Δ.

Because $(R|\mathbf{v})\mathbf{r} = R\mathbf{r} + \mathbf{v}$, the non-lattice translation $\mathbf{w} = [\frac{1}{4}\ \frac{1}{4}\ \frac{1}{4}]a$ is to be added to the rotated vector $\mathbf{r}'$ when $\mathbf{v} \neq \mathbf{0}$. Characters for $(R|\mathbf{w})$ are obtained by multiplying $\chi(R)$ by $\exp(-\mathrm{i}\mathbf{k}\cdot\mathbf{v})$, when $\mathbf{v} = \mathbf{w} \neq \mathbf{0}$ (see eq. (16.6.14)).

C_{4v} $\mathscr{H}(\mathbf{k})$	$(E\|\mathbf{0})$ xyz	$(C_{4y}^{\pm}\|\mathbf{w})$ $\bar{z}yx, zy\bar{x}$	$(C_{2y}\|\mathbf{0})$ $\bar{x}y\bar{z}$	$(\sigma_z\|\mathbf{w})$, $(\sigma_x\|\mathbf{w})$ $xy\bar{z}, \bar{x}yz,$	$(\sigma_e\|\mathbf{0})$, $(\sigma_f\|\mathbf{0})$ $\bar{z}y\bar{x}, zyx$
A_1 Δ_1	1	1	1	1	1
A_2 Δ_1'	1	1	1	−1	−1
B_1 Δ_2	1	−1	1	1	−1
B_2 Δ_2'	1	−1	1	−1	1
E Δ_5	2	0	−2	0	0
$\exp(-\mathrm{i}\mathbf{k}.\mathbf{v})$	1	$\exp(-\mathrm{i}\mathbf{k}\cdot\mathbf{w})$	1	$\exp(-\mathrm{i}\mathbf{k}\cdot\mathbf{w})$	1

Exercise 17.6-1 Confirm the derivation of eqs. (6), (7), and (8).

The additional symmetry elements in the silicon structure are given in Table 17.12. At Δ, $P(\mathbf{k}) \sim C_{4v}$, which is isomorphous with G_8^4. Jones symbols and the character table are in Table 17.15.

Exercise 17.6-2 The group generators in this realization of the abstract group G_8^4 are $P = (C_{4y}^{+}|\mathbf{w})$, $Q = (\sigma_e|\mathbf{0})$. Confirm the generating relations $P^4 = Q^2 = E$, $QP = P^3Q$.

The permutation representation at Δ is

$$\chi(\Gamma_m) = \{4\ 0\ 0\ 0\ 2\},\ \Gamma_m = \Delta_1 \oplus \Delta_2' \oplus \Delta_5. \tag{9}$$

On projecting $\psi_{[1\ 1\ 1]}(\Delta)$, symmetrized eigenfunctions of Δ_1, Δ'_2, and Δ_5 symmetry are given by

$$\psi(\Delta_1) = c^{-1}\sum_R \chi^{\Delta_1}(R|\mathbf{v})^* \exp[2\pi\mathrm{i}(\eta - 1)y] \exp[-2\pi\mathrm{i}\{R^{-1}(z + x) - (R^{-1}\mathbf{v})y\}],$$

$$(R|\mathbf{v}) \in G_8^4,\ \mathbf{v} = \mathbf{0} \text{ or } \mathbf{w}$$

$$= \exp[2\pi\mathrm{i}(\eta - 1)y]\cos 2\pi[(z + x) + \mathrm{i}\cos(z - x)]; \tag{10}$$

$$\psi(\Delta_2') = \exp[2\pi\mathrm{i}(\eta - 1)y]\,[\cos\{2\pi(z + x)\} - \mathrm{i}\cos\{2\pi(z - x)\}]; \tag{11}$$

$$\psi_{\mathrm{I}}(\Delta_5) = \exp[2\pi\mathrm{i}(\eta - 1)y]\,[\sin\{2\pi(z + x)\} + \mathrm{i}\sin\{2\pi(z - x)\}]. \tag{12}$$

The second eigenfunction for Δ_5,

$$\psi_{\mathrm{II}}(\Delta_5) = \exp[2\pi\mathrm{i}(\eta - 1)y]\,[\sin\{2\pi(z + x)\} - \mathrm{i}\sin\{2\pi(z - x)\}], \tag{13}$$

is obtained by projecting $\psi_{[\bar{1}\ 1\ 1]}(\Delta)$. Equations (12) and (13) are sometimes replaced by their linear combinations. All these symmetrized eigenfunctions for Si at Δ are complex, and so their degeneracy is doubled under time reversal.

Answers to Exercises 17.6

Exercise 17.6-1 For the Δ_3 basis, in the short-hand notation introduced earlier in this chapter, projecting $\psi_{[1\ 1\ 1]}$ gives (with the aid of the second and sixth lines of Table 17.14)

$$(xyz) - (\bar{x}y\bar{z}) + (zyx) - (\bar{z}y\bar{x}),$$

$$\psi(\Delta_3) = \exp[2\pi i(\eta - 1)y]\ \sin[2\pi(x + z)]. \tag{7}$$

The Jones symbols which provide the substitutions for $\psi_{[1\ 1\ \bar{1}]}(\Delta)$ are in the third line of Table 17.14. On making use of these substitutions, a second basis for the Δ_1 representation is provided by

$$(xy\bar{z}) + (\bar{x}yz) + (zy\bar{x}) + (\bar{z}yx),$$

$$\psi_{\text{II}}(\Delta_1) = \exp[2\pi i(\eta - 1)y]\ \cos[2\pi(x - z)]. \tag{6}$$

Similarly, for the Δ_4 representation,

$$(xy\bar{z}) - (\bar{x}yz) - (zy\bar{x}) + (\bar{z}yx),$$

$$\psi(\Delta_4) = \exp[2\pi i(\eta - 1)y]\ \sin[2\pi(x - z)]. \tag{8}$$

Exercise 17.6-2 $P^2 = [(C_{4y}^+|\mathbf{w})]^2 = (C_{2y}|[\tfrac{1}{4}\ \tfrac{1}{4}\ -\tfrac{1}{4}] + [\tfrac{1}{4}\ \tfrac{1}{4}\ \tfrac{1}{4}])$

$$= (C_{2y}|[\tfrac{1}{2}\ \tfrac{1}{2}\ 0]) = (C_{2y}|\mathbf{a}_3);$$

$$P^4 = [(C_{2y}|\mathbf{a}_3)]^2 = (E|[-\tfrac{1}{2}\ \tfrac{1}{2}\ 0] + [\tfrac{1}{2}\ \tfrac{1}{2}\ 0]) = (E|[0\ 1\ 0]) = (E|\mathbf{0});$$

$$Q^2 = [(\sigma_\mathbf{e}|\mathbf{0})]^2 = (E|\mathbf{0}), \qquad P^4 = Q^2 = E;$$

$$P^3 = (C_{4y}^+|\mathbf{w})(C_{2y}|[\tfrac{1}{2}\ \tfrac{1}{2}\ 0]) = (C_{4y}^-|[0\ \tfrac{1}{2}\ -\tfrac{1}{2}] + \mathbf{w}) = (C_{4y}^-|\mathbf{w});$$

$$P^3Q = (C_{4y}^-|\mathbf{w})(\sigma_\mathbf{e}|\mathbf{0}) = (\sigma_\mathbf{z}|\mathbf{w});$$

$$QP = (\sigma_\mathbf{e}|\mathbf{0})(C_{4y}^+|\mathbf{w}) = (\sigma_\mathbf{z}|-\tfrac{1}{4}\ \tfrac{1}{4}\ -\tfrac{1}{4}) = (\sigma_\mathbf{z}|\mathbf{w} - \mathbf{a}_2) = (\sigma_\mathbf{z}|\mathbf{w}).$$

17.7 Spinor representations

We have seen that the calculation of FE energy bands at a point **k** in the BZ depends on the determination of the symmetrized eigenfunctions that form bases for the representations of

P($\mathbf{k}$), P($\mathbf{k}$)′, or the Herring group $\mathscr{H}(\mathbf{k})$, using plane-wave solutions (17.5.6) of the Schrödinger wave equation. This procedure is applicable only when electron spin is neglected. When spin is allowed for, the group representations include the j = half-integer spinor representations. These may be calculated using Bethe's double groups $\overline{\mathrm{G}}(\mathbf{k})$ and Opechewski's rules. However, this procedure would leave us without a group multiplication table. The alternative is to find the spinor representations as PRs of $\mathbf{P}(\mathbf{k})$, $\mathbf{P}(\mathbf{k})'$, or $\mathscr{H}(\mathbf{k})$ by calculating projective factors (PFs) using the quaternion representation. Examples of the calculation of spinor representations for space groups were given in Section 16.8. Here we shall consider again the line ΓΔX in the BZs of crystals with the silicon and indium antimonide structures, which belong to space groups 227 ($Fd3m$ or O_h^7) and 216 ($F\bar{4}3m$ or T_d^2), respectively. These are non-symmorphic space groups and so, for surface points, the Herring group should be used in place of $\mathbf{P}(\mathbf{k})$. Table 17.12 shows that for space groups 216 and 227, $\mathbf{P}(\mathbf{k})$ at Γ and Δ is isomorphous with a point group so that its projective spinor representations can be obtained from compilations of character tables of point groups, as in Appendix A3. At X for InSb, $\mathscr{H}(\mathbf{k}) \sim \mathrm{G}_8^4 \otimes \mathrm{T}_2$, where $\mathrm{G}_8^4 \sim \mathrm{C}_{4v}$, so again tables of spinor representations can be used. But for Si at X, $\mathscr{H}(\mathbf{k}) \sim \mathrm{G}_{32}^2$, which is not isomorphous with a crystallographic point group. The characters of the four required vector representations were given in Table 16.16. In such cases, the character systems of spinor representations can be determined either by the methods given in Chapter 12 or by extracting the characters for R from the spinor representations of the double group $\{R \oplus \bar{R}\}$. At X the double group is isomorphous with the abstract group G_{64}^2. The character table of G_{64}^2 is given by Bradley and Cracknell (1972, p. 262), and this includes the spinor representations of G_{32}^2. There are five regular classes in G_{32}^2 ($\mathscr{C}_1$, $\mathscr{C}_3$, $\mathscr{C}_5$, $\mathscr{C}_{13}$, $\mathscr{C}_{14}$ in Table 17.16) and therefore five spinor representations of dimensions 2, 2, 2, 2, and 4. (Check: $\sum_s l_s^2 = 4(2^2) + 1(4^2) = 32 = g$.) Their characters are included in the representations R_{15} to R_{19} in the character table of G_{64}^2. Only the spinor representation with $l = 4$ satisfies the Herring requirement that $\chi(\mathscr{C}_3) = \exp(-i\mathbf{k}\,.\,\mathbf{a}_1)l = -l$. When $\chi(\mathscr{C}_1) = \chi(E) = -4$ and $\chi(\mathscr{C}_3) = -4$, normalization of the character system of this row requires that the characters for all the other classes of G_{64}^2 be zero. Orthogonality of the rows requires that $\chi(\mathscr{C}_3) = +2$ for the four 2-D IRs, which shows that the four 2-D representations are not required representations. Therefore, at the point X in the first BZ of Si, all the states are four-fold degenerate.

17.8 Transitions between electronic states

Electron states in a crystal with space group G are described by state functions that belong to one of the IRs of G

$$\Gamma_i^p \equiv \Gamma^p[\mathrm{G}(\mathbf{k}_i)] \uparrow \mathrm{G}, \tag{1}$$

where $\Gamma^p[\mathrm{G}(\mathbf{k}_i)]$ is the pth small representation of the little group $\mathrm{G}(\mathbf{k}_i)$ and eq. (1) defines the abbreviated notation that will be used in this discussion of transitions between electronic states in crystals. In Section 5.4 it was shown that the matrix element $\langle \psi^q | \hat{\mathrm{Q}}^s | \psi^p \rangle$, which governs the transition probability from an initial state ψ^p to the final state ψ^q induced

Table 17.16. *Classes of* $\mathscr{H}(\boldsymbol{k}) \sim G_{32}^2$ *at X for the space group of Si* (227, O_h^7, *or Fd3m*).

The group generators for this realization of G_{32}^2 are $P = (\sigma_x|\mathbf{w}), Q = (S_{4y}^+|\mathbf{0}), R = (C_{2x}|\mathbf{0})$.

$\mathscr{C}_1 = P^4 = Q^4 = R^2 = (E|\mathbf{0})$
$\mathscr{C}_2 = Q^2 = (C_{2y}|\mathbf{0})$
$\mathscr{C}_3 = P^2 = (E|\mathbf{a}_1)$
$\mathscr{C}_4 = P^2Q^2 = (C_{2y}|\mathbf{a}_1)$
$\mathscr{C}_5 = \{P^3R \ \ PR\} = \{(I|\mathbf{w}+\mathbf{a}_1) \ \ (I|\mathbf{w})\}$
$\mathscr{C}_6 = \{P^3Q^2R \ \ PQ^2R\} = \{(\sigma_y|\mathbf{w}+\mathbf{a}_1) \ \ (\sigma_y|\mathbf{w})\}$
$\mathscr{C}_7 = \{QR \ \ Q^3R\} = \{(\sigma_f|\mathbf{0}) \ \ (\sigma_e|\mathbf{0})\}$
$\mathscr{C}_8 = \{P^2Q^3R \ \ P^2QR\} = \{(\sigma_e|\mathbf{a}_1) \ \ (\sigma_f|\mathbf{a}_1)\}$
$\mathscr{C}_9 = \{PQ \ \ P^3Q^3\} = \{(C_{2f}|\mathbf{w}) \ \ (C_{2e}|\mathbf{w}+\mathbf{a}_1)\}$
$\mathscr{C}_{10} = \{PQ^3 \ \ P^3Q\} = \{(C_{2e}|\mathbf{w}) \ \ (C_{2f}|\mathbf{w}+\mathbf{a}_1)\}$
$\mathscr{C}_{11} = \{R \ \ P^2R \ \ Q^2R \ \ P^2Q^2R\} = \{(C_{2x}|\mathbf{0}) \ \ (C_{2x}|\mathbf{a}_1) \ \ (C_{2z}|\mathbf{0}) \ \ (C_{2z}|\mathbf{a}_1)\}$
$\mathscr{C}_{12} = \{P \ \ P^3 \ \ PQ \ \ P^3Q^2\} = \{(\sigma_x|\mathbf{w}) \ \ (\sigma_x|\mathbf{w}+\mathbf{a}_1) \ \ (\sigma_z|\mathbf{w}) \ \ (\sigma_z|\mathbf{w}+\mathbf{a}_1)\}$
$\mathscr{C}_{13} = \{Q \ \ P^2Q \ \ Q^3 \ \ P^3Q^3\} = \{(S_{4y}^+|\mathbf{0}) \ \ (S_{4y}^+|\mathbf{a}_1) \ \ (S_{4y}^-|\mathbf{0}) \ \ (S_{4y}^-|\mathbf{a}_1)\}$
$\mathscr{C}_{14} = \{PQR \ \ P^3QR \ \ PQ^3R \ \ P^3Q^3R\} = \{(C_{4y}^+|\mathbf{w}) \ \ (C_{4y}^+|\mathbf{w}+\mathbf{a}_1) \ \ (C_{4y}^-|\mathbf{w}) \ \ (C_{4y}^-|\mathbf{w}+\mathbf{a}_1)\}$

by the operator $\hat{Q}^s$, is zero unless the DP $\Gamma^p \otimes \Gamma^q \otimes \Gamma^s$ contains the totally symmetric representation Γ_1. An equivalent statement (see section 5.4) is that this matrix element is zero unless the DP of any two of the representations Γ^p, Γ^q, Γ^s, contains the third one. In a crystal, the transition from a state ψ_i^p (which forms a basis for Γ_i^p) to a state ψ_j^q under the operator Q_l^s is forbidden unless the Clebsch–Gordan (CG) coefficient $c_{ij,l}^{pq,s}$ in the DP

$$\Gamma_i^p \otimes \Gamma_j^q = \sum_{l,s} c_{ij,l}^{pq,s} \Gamma_l^s \tag{2}$$

is non-zero. The coefficient $c_{ij,l}^{pq,s}$, which is the number of times the representation Γ_l^s appears in the direct sum on the RS of eq. (2), is called the *frequency* of Γ_l^s in $\Gamma_i^p \otimes \Gamma_j^q$. It is vital to appreciate that ψ_j^q is not determined by just the single wave vector $\mathbf{k}_j$, for ψ_j^q is a linear combination of the Bloch functions of the *whole star* of $\mathbf{k}_j$ (and similarly, of course, for ψ_j^p). Thus the situation is a great deal more complicated than for molecules. One possible approach would be to use the little groups $G(\mathbf{k}_i)$ and $G(\mathbf{k}_j)$ but with the possibility that transitions between other members ("prongs") of the stars of $\mathbf{k}_i$ and $\mathbf{k}_j$ could be missed unless eq. (2) was evaluated for all possible initial and final states. Fortunately this turns out not to be necessary, since Bradley (1966) has shown how the problem of determining the $c_{ij,l}^{pg,s}$ frequency coefficients may be solved by using the induction of the little groups $G(\mathbf{k}_i)$, $G(\mathbf{k}_j)$, $G(\mathbf{k}_l)$ and on to the space group G. This requires the use of double cosets.

17.8.1 *Double cosets*

If H_1 and H_2 are subgroups of a group G and $d_a \in G$ then

$$G = \sum_a H_1 d_a H_2 \quad \text{(with repetitions ignored)} \tag{3}$$

is the *double coset expansion* of G on H_1, H_2. In eq. (3) each term $H_1 d_a H_2$ is counted *once* only. The $\{d_a\}$ are the *double coset representatives* of G with respect to its subgroups H_1, H_2. The d_a are not unique, but the expansion is unique since the double cosets are all distinct.

Example 17.8-1 Write a double coset expansion of S(3) on $H_1 = \{P_0\ P_3\}$, $H_2 = \{P_0\ P_4\}$.

$$H_1 P_0 H_2 = \{P_0 P_3\}\{P_0 P_4\} = \{P_0\ P_4\ P_3\ P_1\};$$

$$H_1 P_5 H_2 = \{P_0 P_3\}\{P_5 P_2\} = \{P_5\ P_2\ P_2\ P_5\}$$
$$= \{P_2 P_5\} \text{ (repetitions ignored)}.$$

$H_1 P_0 H_2 \oplus H_1 P_5 H_2 = G$ and the double coset representatives are $\{P_0\ P_5\}$.

Bradley (1966) has shown that

$$c_{ij,l}^{pq,s} = \sum_a \sum_b g_{ab}^{-1} \sum_{(R_c|\mathbf{v}_c)\in G_{ab}} \chi_i^p((R_b|\mathbf{v}_b)^{-1}(R_c|\mathbf{v}_c)(R_b|\mathbf{v}_b))$$
$$\times \chi_j^q((R_a|\mathbf{v}_a)^{-1}(R_c|\mathbf{v}_c)(R_a|\mathbf{v}_a))\chi_l^s(R_c|\mathbf{v}_c)^*, \tag{4}$$

where g_{ab} is the order of

$$G_{ab} = G(\mathbf{k}_{bi}) \cap G(\mathbf{k}_{aj}); \tag{5}$$

Γ_i^p, Γ_j^q, and Γ_l^s are IRs of the little groups $G(\mathbf{k}_i)$, $G(\mathbf{k}_j)$, and $G(\mathbf{k}_l)$, and $\Gamma_i^p \uparrow G$, $\Gamma_j^q \uparrow G$, $\Gamma_l^s \uparrow G$ are representations of G induced by these small representations; $G(\mathbf{k}_{bi})$ is the little group of $R_b\mathbf{k}_i$ and Γ_{bi}^p is a small representation of $G(\mathbf{k}_{bi})$; $(R_c|\mathbf{v}_c) \in G(\mathbf{k}_{bi})$ so that

$$\Gamma_{bi}^p((R_c|\mathbf{v}_c)) = \Gamma_i^p((R_b|\mathbf{v}_b)^{-1}(R_c|\mathbf{v}_c)(R_b|\mathbf{v}_b)) \tag{6}$$

because representations at different prongs of a star are conjugate. $(R_a|\mathbf{v}_a)$ and $(R_b|\mathbf{v}_b)$ are double coset representatives in

$$G = \sum_a G(\mathbf{k}_l)(R_a|\mathbf{v}_a)\, G(\mathbf{k}_j), \tag{7}$$

$$G = \sum_b [G(\mathbf{k}_{aj}) \cap G(\mathbf{k}_l)](R_b|\mathbf{v}_b)G(\mathbf{k}_j). \tag{8}$$

Not all such pairs R_a, R_b will do but only those pairs for which

$$R_b\mathbf{k}_i + R_a\mathbf{k}_j = \mathbf{k}_l. \tag{9}$$

On summing over translations, it follows from the orthogonolity theorem for the characters of the IRs of the translation group that

$$c_{ij,l}^{pq,s} = \sum_a \sum_b g_{ab}^{-1} \sum_{(R_c|\mathbf{v}_c)\in G_{ab}/T} \chi_i^p[(R_b|\mathbf{v}_b)^{-1}(R_c|\mathbf{v}_c)(R_b|\mathbf{v}_b)]$$
$$\times \chi_j^q((R_a|\mathbf{v}_a)^{-1}(R_c|\mathbf{v}_c)(R_a|\mathbf{v}_a))\chi_{l'}^s(R_c|\mathbf{v}_c)^*, \tag{10}$$

where the two sums over a and b are restricted to those terms for which eq. (9) is satisfied. We may expect that relatively few terms in the triple sum in eq. (10) will survive this restriction, which emphasizes the considerable simplification introduced. These results were first given by Bradley (1966) and are quoted here without proof, which may be found in his 1966 paper and also in books by Altmann (1977) and Bradley and Cracknell (1972).

To summarize, the procedure for evaluating the $c_{ij,l}^{pq,s}$ CG coefficients is as follows.

(i) Determine the star of $\mathbf{k}_j$, that is the vectors $R_a\ \mathbf{k}_j$ with $R_a \in \mathrm{P}(\mathbf{k}_j)$.
(ii) Determine the star of $\mathbf{k}_i$, that is the vectors $R_b\ \mathbf{k}_i$ with $R_b \in \mathrm{G}(\mathbf{k}_i) \cap \mathrm{G}(\mathbf{k}_j)$.
(iii) With the required R_a, R_b established from eq. (9), evaluate from eq. (10) the $c_{ij,l}^{pq,s}$ frequency coefficients.

Example 17.8-2 Find the allowed transitions from a B_1 M-state to a B_2 M-state in a crystal with space group 195. This is an example used by Bradley (1966) of a case in which more than one term survives the sums over a and b in eq. (10). The Bravais lattice of the symmorphic space group 195 ($P23$ or T_1) is simple cubic and the point group P of G = T_1 is T = $\{E\ 3C_{2\mathbf{m}}\ 4C_{3\mathbf{j}}\}$, $\mathbf{m}$ = **x**, **y**, **z**; $\mathbf{j}$ = 1, 2, 3, 4 (Figure 2.12). The BZ is in Figure 16.12(a), from which we see that the stars of Γ, M are

$$\Gamma = [0\ 0\ 0], \tag{11}$$

$$\mathrm{M} = [\tfrac{1}{2}\ \tfrac{1}{2}\ 0], \qquad \mathrm{M}_2 = C_{31}{}^{+}\mathrm{M} = [0\ \tfrac{1}{2}\ \tfrac{1}{2}], \qquad \mathrm{M}_3 = C_{31}{}^{-}\mathrm{M} = [\tfrac{1}{2}\ 0\ \tfrac{1}{2}]. \tag{12}$$

Here both $\mathbf{k}_i$ and $\mathbf{k}_j$ refer to the symmetry point M. We thus need to see which possible prongs of the star of M in eqs. (12) will satisfy eq. (9). This is summarized in Table 17.17. The column headed $\cong \mathbf{k}$ shows that the only possible **k** vectors $\mathbf{k}_l$ are Γ and M. (A star is fully determined by any one of its members, say M, so that the pairs R_a, R_b that give M_2, M_3

Table 17.17 *Determination of the pairs of wave vectors $R_b\mathbf{k}_i$ and $R_a\mathbf{k}_j$ that satisfy eq. (17.8.9) when $\mathbf{k}_i = \mathbf{k}_j = [\tfrac{1}{2}\ \tfrac{1}{2}\ 0]$.*

The column headed $\cong \mathbf{k}$ lists all possible wave vectors that could result from summing a member of the star of $\mathbf{k}_i$ with one from the star of $\mathbf{k}_j$. Not all of these yield $\mathbf{k}_l$ because of the restriction imposed by eq. (17.8.9), that is by translational symmetry.

Prong of star of						
$\mathbf{k}_i$	$\mathbf{k}_j$	R_b	R_a	$R_b\mathbf{k}_i + R_a\mathbf{k}_j$	$\cong \mathbf{k}$	$\mathbf{k}_l$
M	M	E	E	[1 1 0]	Γ	Γ
	M_2	E	C_{31}^{+}	[½ 1 ½]	M_3	–
	M_3	E	C_{31}^{-}	[1 ½ ½]	M_2	–
M_2	M	$C_{31}{}^{+}$	E	[½ 1 ½	M_3	–
	M_2	$C_{31}{}^{+}$	C_{31}^{+}	[0 1 1]	Γ	–
	M_3	$C_{31}{}^{+}$	C_{31}^{-}	[½ ½ 1]	M	M
M_3	M	C_{31}^{-}	E	[1 ½ ½]	M_2	–
	M_2	C_{31}^{-}	C_{31}^{+}	[½ ½ 1]	M	M
	M_3	C_{31}^{-}	C_{31}^{-}	[1 0 1]	Γ	–

need not be considered.) The little groups $G(M_2)$, $G(M_3)$ are $\sim$ G(M) and the representations Γ^{M_2}, Γ^{M_3} are conjugate representations of Γ^{M}. For $\mathbf{k}_l = \Gamma$, the double coset expansion of G on its subgroups $G(\mathbf{k}_l) \equiv G(\Gamma)$ and $G(\mathbf{k}_j) \equiv G(M)$ is

$$\text{(7)} \qquad G = \sum_a G(\Gamma)(R_a|\mathbf{w}_a)\, G(M), \quad (R_a|\mathbf{w}_a) = (E|\mathbf{0}). \qquad (13)$$

Only a single term survives in the sum over a because of the "no repetitions" rule and the fact that G(M) is a subgroup of G(Γ).

$$\text{(8)} \qquad G = \sum_b [G(M) \cap G(\Gamma)](R_b|w_b)\, G(M) \qquad (14)$$
$$= \sum_b G(M)(R_b|\mathbf{w}_b)\, G(M), \quad (R_b|\mathbf{w}_b) = (E|\mathbf{0}), (C_{31}^{+}|\mathbf{0}), (C_{31}^{-}|\mathbf{0}).$$

Of these possibilities (as Table 17.17 shows) only the pair $d_a = E$, $d_b = E$ are double coset representatives that satisfy (eq. 9),

$$\text{(9), (13), (14)} \qquad R_b\mathbf{k}_i + R_a\mathbf{k}_j = EM + EM = \Gamma, \quad E \in \{R_a\}, E \in \{R_b\}. \qquad (15)$$

Although $C_{31}^{+}M + C_{31}^{+}M = M_2 + M_2 = \Gamma$, $C_{31}^{+} \notin \{R_a\}$. For $\mathbf{k}_l = M$,

$$\text{(7)} \qquad G = \sum_a G(M)(R_a|\mathbf{w}_a)\, G(M), \quad (R_a|\mathbf{w}_a) = (E|\mathbf{0}), (C_{31}^{+}|\mathbf{0}), (C_{31}^{-}|\mathbf{0}), \qquad (16)$$

$$\text{(8)} \quad G = \sum_b [G(M) \cap G(M)](R_b|\mathbf{w}_b)G(M), \quad (R_b|\mathbf{w}_b) = (E|\mathbf{0}), (C_{31}^{+}|\mathbf{0}), (C_{31}^{-}|\mathbf{0}). \qquad (17)$$

Table 17.17 shows that only the pairs R_a, $R_b = C_{31}^{+}$; C_{31}^{-} or C_{31}^{-}, C_{31}^{+} yield M.

17.8.2 *Transition probabilities*

$$\text{(2)} \qquad \Gamma_M^{B_1} \otimes \Gamma_M^{B_2} = \sum_s c_{MM,\,\Gamma}^{B_1B_2,\,s}\, \Gamma_\Gamma^{s} \oplus \sum_u c_{MM,\,M}^{B_1B_2,\,u}\, \Gamma_M^{u}. \qquad (18)$$

Since $\mathbf{k}_l$ is Γ or M, the transition probability $M(B_1) \to M(B_2)$ is allowed if $c_{MM,\,\Gamma}^{B_1B_2,\,s}$ and/or $c_{MM,\,M}^{B_1B_2,\,u}$ are non-zero, that is if the perturbation (electromagnetic radiation, for example) contains a term that transforms according to the sth IR at Γ and/or the uth IR at M. We have established in Table 17.17 that at Γ, $R_a = E$, $R_b = E$. Therefore,

$$G(R_b\mathbf{k}_i) \cap G(R_a\mathbf{k}_j) \cap G(\mathbf{k}_l) = G(M) \cap G(M) \cap G(\Gamma) = G(M), \qquad (19)$$

which is of order 4.

$$\text{(18), (13), (19)} \qquad c_{MM,\,\Gamma}^{B_1B_2,\,s} = \tfrac{1}{4} \sum_{R_c} \chi^{B_1}(R_c)\chi^{B_2}(R_c)\chi_\Gamma^{s}(R_c)^* = \boldsymbol{\delta}_{s,\,T}; \qquad (20)$$

$$G(\mathbf{k}_{b_i}) = G(M_2) \sim G(M), \; G(\mathbf{k}_{a_j}) = G(M_3) \sim G(M), \qquad (21)$$

$$\text{(21)} \qquad G(\mathbf{k}_{b_i}) \cap G(\mathbf{k}_{a_j}) \cap G(\mathbf{k}_l) = G(M), \qquad (22)$$

which is of order 4.

(18), (22)
$$c_{\mathrm{MM,M}}^{\mathrm{B_1B_2},u} = \tfrac{1}{4}\Big[\sum_{R_c}\chi_{\mathrm{M_2}}^{\mathrm{B_1}}(R_c)\chi_{\mathrm{M_3}}^{\mathrm{B_2}}(R_c)\chi_{\mathrm{M}}^{u}(R_c)^* + \sum_{R_c}\chi_{\mathrm{M_3}}^{\mathrm{B_1}}(R_c)\chi_{\mathrm{M_2}}^{\mathrm{B_2}}(R_c)\chi_{\mathrm{M}}^{u}(R_c)^*\Big]. \tag{23}$$

Representations at different prongs of the same star are conjugate; therefore since $\mathrm{M_2} = R_b\mathrm{M} = C_{31}{}^+\mathrm{M}$,

$$\Gamma_{\mathrm{M_2}}^{\mathrm{B_1}}(R_c) = \Gamma_{\mathrm{M}}^{\mathrm{B_1}}(R_b^{-1}R_cR_b) = \Gamma_{\mathrm{M}}^{\mathrm{B_1}}(C_{31}{}^-R_cC_{31}{}^+) = \Gamma_{\mathrm{M}}^{\mathrm{B_3}}(R_c). \tag{24}$$

Likewise,

$$\Gamma_{\mathrm{M_2}}^{\mathrm{B_2}} = \Gamma_{\mathrm{M}}^{\mathrm{B_1}},\quad \Gamma_{\mathrm{M_3}}^{\mathrm{B_1}} = \Gamma_{\mathrm{M}}^{\mathrm{B_2}},\quad \text{and } \Gamma_{\mathrm{M_3}}^{\mathrm{B_2}} = \Gamma_{\mathrm{M}}^{\mathrm{B_3}}. \tag{25}$$

Exercise 17.8-1 Verify the conjugate representations given in eq. (25). [*Hint*: $\mathrm{M_3} = C_{31}^-\mathrm{M}$.]

(24)
$$c_{\mathrm{MM,\,M}}^{\mathrm{B_1B_2},\,s} = \tfrac{1}{4}\sum_{R_c}\chi_{\mathrm{M}}^{\mathrm{B_3}}(R_c)\chi_{\mathrm{M}}^{\mathrm{B_3}}(R_c)\chi_{\mathrm{M}}^{u}(R_c)^* = \boldsymbol{\delta}_{s,\,\mathrm{A}}; \tag{26}$$

(25)
$$c_{\mathrm{MM,\,M}}^{\mathrm{B_1B_2},\,u} = \tfrac{1}{4}\sum_{R_c}\chi_{\mathrm{M}}^{\mathrm{B_2}}(R_c)\chi_{\mathrm{M}}^{\mathrm{B_1}}(R_c)\chi_{\mathrm{M}}^{u}(R_c)^* = \boldsymbol{\delta}_{u,\,\mathrm{B_3}}. \tag{27}$$

Therefore a transition between an electronic state described by the state function $\psi_{\mathrm{M}}^{\mathrm{B_1}}$ to one described by $\psi_{\mathrm{M}}^{\mathrm{B_2}}$ is symmetry-forbidden unless the perturbing operator belongs to the IR T at Γ or to A or $\mathrm{B_3}$ at M (see eqs. (20), (26), and (27)).

Bradley's (1966) work has removed the uncertainties about the subgroup method. The only comprehensive alternative seems to be a method described by Birman (1962, 1963) that uses the full group G. The effect of time-reversal symmetry on selection rules in crystals has been described by Lax (1962, 1965).

Answer to Exercise 17.8-1

$$C_{31}^-C_{2x}C_{31}^+\langle \mathbf{e}_1\,\mathbf{e}_2\,\mathbf{e}_3| = C_{31}^-C_{2x}\langle \mathbf{e}_2\,\mathbf{e}_3\,\mathbf{e}_1| = C_{31}^-\langle \bar{\mathbf{e}}_2\,\bar{\mathbf{e}}_3\,\mathbf{e}_1| = \langle \bar{\mathbf{e}}_1\,\bar{\mathbf{e}}_2\,\mathbf{e}_3|$$
$$= C_{2z}\langle \mathbf{e}_1\,\mathbf{e}_2\,\mathbf{e}_3|;$$

$$C_{31}^-C_{2y}C_{31}^+\langle \mathbf{e}_1\,\mathbf{e}_2\,\mathbf{e}_3| = C_{31}^-C_{2y}\langle \mathbf{e}_2\,\mathbf{e}_3\,\mathbf{e}_1| = C_{31}^-\langle \mathbf{e}_2\,\bar{\mathbf{e}}_3\,\bar{\mathbf{e}}_1| = \langle \mathbf{e}_1\,\bar{\mathbf{e}}_2\,\bar{\mathbf{e}}_3|$$
$$= C_{2x}\langle \mathbf{e}_1\,\mathbf{e}_2\,\mathbf{e}_3|;$$

$$C_{31}^-C_{2z}C_{31}^+\langle \mathbf{e}_1\,\mathbf{e}_2\,\mathbf{e}_3| = C_{31}^-C_{2z}\langle \mathbf{e}_2\,\mathbf{e}_3\,\mathbf{e}_1| = C_{31}^-\langle \bar{\mathbf{e}}_2\,\mathbf{e}_3\,\bar{\mathbf{e}}_1| = \langle \bar{\mathbf{e}}_1\,\mathbf{e}_2\,\bar{\mathbf{e}}_3|$$
$$= C_{2y}\langle \mathbf{e}_1\,\mathbf{e}_2\,\mathbf{e}_3|.$$

Deduction of the conjugate representations is summarized in Table 17.18.

Table 17.18. *Character table for the small representations of* M *and for the required conjugate representations.*

M	E	C_{2x}	C_{2y}	C_{2z}	
A	1	1	1	1	
B_1	1	−1	−1	1	
B_2	1	−1	1	−1	
B_3	1	1	−1	−1	
M_2	E	C_{2z}	C_{2x}	C_{2y}	
$B_1(M_2)$	1	1	−1	−1	$B_3(M)$
$B_2(M_2)$	1	−1	−1	1	$B_1(M)$
M_3	E	C_{2y}	C_{2z}	C_{2x}	
$B_1(M_3)$	1	−1	1	−1	$B_2(M)$
$B_2(M_3)$	1	1	−1	−1	$B_3(M)$

Problems

17.1 Prove that $\nabla_{\mathbf{k}\perp}E(\mathbf{k}) = \mathbf{0}$ for all $\mathbf{k}$ vectors which end on a line in a BZ face which is parallel to a binary axis. [*Hint*: See Exercise 17.3-1.]

17.2 Prove that $\nabla_{\mathbf{k}}E(\mathbf{k})$ vanishes at the center of the BZ. [*Hint*: A cusp is impossible. Why?]

17.3 For the BZ of the *sc* lattice, find compatibility relations between IRs at Z (which lies on the line XM) and at the end-points X and M.

17.4 Verify the eigenfunctions at X that are given in eqs. (17.5.39)–(17.5.44).

17.5 Give a detailed derivation of the FE eigenfunctions which form bases for the Γ_{15} representation at Γ in the BZ of the *sc* lattice. Check your results against eqs. (17.5.66)–(17.5.68).

17.6 Confirm the generating relations for the abstract groups G_4^2, G_{24}^7, and G_{48}^7 (which are isomorphous with Herring groups for InSb and Si at Γ and Δ) using the realizations of these groups that are given in Table 17.12.

17.7 Write down the Jones symbols and character table for the Herring group at the point Σ in the BZ of the Si structure. Calculate the FE energy at Σ for the first three bands and find the symmetrized FE eigenfunctions in these bands. Plot the FE energy as a function of $\mathbf{k}$, marking in your figure the IRs of these symmetrized bases.

17.8 Find expressions for, and sketch in the reduced zone scheme, FE energy bands along ΓH in the reciprocal lattice of the *bcc* lattice.

17.9 Find expressions for, and sketch in the reduced zone scheme, FE energy bands along ΓM in the reciprocal lattice of the *hcp* lattice.

18 Vibration of atoms in crystals

18.1 Equations of motion

In the harmonic approximation the potential energy Φ of a crystal in which the atoms are vibrating about their equilibrium positions differs from Φ_0, the potential energy with each atom on its equilibrium site, by

$$\Phi - \Phi_0 = \tfrac{1}{2} \sum_{n\kappa\alpha} \sum_{n'\kappa'\beta} \phi_{\alpha\beta}(n\kappa, n'\kappa') \; u_\alpha(n\kappa) u_\beta(n'\kappa'), \tag{1}$$

where $u_\alpha(n\kappa)$ is the αth Cartesian component of the displacement $\mathbf{u}(n\kappa)$ of the κth atom ($\kappa = 1, 2, \ldots, s$) in the nth unit cell from its equilibrium position, $\mathbf{a}_{n\kappa} = \mathbf{a}_n + \mathbf{a}_\kappa$ (Section 16.1). The

$$\phi_{\alpha\beta}(n\kappa, n'\kappa') = \left[\frac{\partial^2 \Phi}{\partial u_\alpha(n\kappa) \partial u_\beta(n'\kappa')} \right]_0 \tag{2}$$

are atomic force constants, and the zero subscript means that the second derivatives are to be evaluated with the atoms at their equilibrium sites. Equation (1) is the result of a Taylor expansion of Φ about Φ_0. There is no term linear in the displacements because there is no net force on each atom at equilibrium. Truncation of the Taylor expansion at terms quadratic in the displacements constitutes the *harmonic approximation*, the usefulness of which is due to the fact that the displacements are generally small in comparison with the interatomic distances. Since Φ is a continuous function of the atom displacements, with continuous partial derivatives, the first-order partial derivatives on the RS of eq. (2) commute, with the result that the force constants are symmetric with respect to the interchange of the indices $(n\kappa\alpha)$ with $(n'\kappa'\beta)$,

$$\phi_{\alpha\beta}(n\kappa, n'\kappa') = \phi_{\beta\alpha}(n'\kappa', n\kappa). \tag{3}$$

Because the potential energy is invariant under an arbitrary displacement of the whole crystal, the force constants obey the sum rule

$$\sum_{n'\kappa'} \phi_{\alpha\beta}(n\kappa, n'\kappa') = 0. \tag{4}$$

From translational symmetry, with $l \in \{n\}$,

$$\phi_{\alpha\beta}(n\kappa, n'\kappa') = \phi_{\alpha\beta}(n+l, \kappa \; ; \; n'+l, \kappa'). \tag{5}$$

In particular, if $l = -n'$, or $l = -n$,

$$(5) \qquad \phi_{\alpha\beta}(n\kappa, n'\kappa') = \phi_{\alpha\beta}(n-n', \kappa\ ;\ 0, \kappa'), \tag{6}$$

$$(5) \qquad \phi_{\alpha\beta}(n\kappa, n'\kappa') = \phi_{\alpha\beta}(0, \kappa\ ;\ n'-n, \kappa'). \tag{7}$$

Summing each side of eq. (7) over n' yields

$$(7) \qquad \sum_{n'} \phi_{\alpha\beta}(n\kappa, n'\kappa') = \sum_{n'} \phi_{\alpha\beta}(0, \kappa\ ;\ n'-n, \kappa'). \tag{8}$$

Replacing the summation index n' on the RS of eq. (8) by $2n-n'$ yields

$$(8) \qquad \sum_{n'} \phi_{\alpha\beta}(n\kappa, n'\kappa') = \sum_{n'} \phi_{\alpha\beta}(0, \kappa;\ n-n', \kappa'); \tag{9}$$

$$(9), (7) \qquad \sum_{n'} \phi_{\alpha\beta}(n\kappa, n'\kappa') = \sum_{n'} \phi_{\alpha\beta}(n'\kappa, n\kappa'). \tag{10}$$

The equations of motion are

$$(1) \qquad M_\kappa \ddot{u}_\alpha(n\kappa) = -\frac{\partial \Phi}{\partial u_\alpha(n\kappa)} = -\sum_{n'\kappa'\beta} \phi_{\alpha\beta}(n\kappa, n'\kappa')\ u_\beta(n'\kappa'). \tag{11}$$

Equation (11) describes the time dependence of the displacements. In the harmonic approximation the displacements are plane waves

$$(11) \qquad u_\alpha(n\kappa) = M_\kappa^{-1/2} u_\alpha(\kappa) \exp[\mathrm{i}(\mathbf{q}\cdot\mathbf{a}_n - \omega t)], \tag{12}$$

where $\mathbf{q}$ is the wave vector; $u_\alpha(\kappa)$ is the amplitude of the vibration of the κth atom, and it is independent of both n and the time t.

$$(12), (11)\ \omega^2 u_\alpha(\kappa) = \sum_{\kappa'\beta} u_\beta(\kappa')(M_\kappa M_{\kappa'})^{-1/2} \sum_{n'} \phi_{\alpha\beta}(n\kappa,\ n'\kappa') \exp[-\mathrm{i}\mathbf{q}\cdot(\mathbf{a}_n - \mathbf{a}_{n'})], \tag{13}$$

$$(13) \qquad \omega^2 u_\alpha(\kappa) = \sum_{\kappa'\beta} \mathrm{D}_{\alpha\beta}(\kappa\kappa'|\mathbf{q})\ u_\beta(\kappa'). \tag{14}$$

The $\mathrm{D}_{\alpha\beta}(\kappa\kappa'|\mathbf{q})$, which are defined by eqs. (13) and (14) as

$$(13), (14) \qquad \mathrm{D}_{\alpha\beta}(\kappa\kappa'|\mathbf{q}) = (M_\kappa M_{\kappa'})^{-1/2} \sum_{n'} \phi_{\alpha\beta}(n\kappa, n'\kappa') \exp[-\mathrm{i}\mathbf{q}\cdot(\mathbf{a}_n - \mathbf{a}_{n'})] \tag{15}$$

are the elements of the Fourier-transformed dynamical matrix $\mathrm{D}(\mathbf{q})$ and they are independent of n because of translational symmetry.

Exercise 18.1-1 Show that $\mathrm{D}(\mathbf{q})$ is a Hermitian matrix.

The $3s$ linear homogeneous equations (14) have non-trivial solutions if

$$|\mathrm{D}_{\alpha\beta}(\kappa\kappa'|\mathbf{q}) - \omega^2 \delta_{\alpha\beta}\delta_{\kappa\kappa'}| = 0. \tag{16}$$

Equation (16) is an algebraic equation of degree $3s$ in ω^2, and the solutions $\{\omega_j^2(\mathbf{q})\}$, $j = 1, 2, \ldots, 3s$, are the *eigenvalues* of $\mathrm{D}(\mathbf{q})$. Since $\mathrm{D}(\mathbf{q})$ is a Hermitian matrix, its eigenvalues are real. But the frequencies $\{\omega_j(\mathbf{q})\}$ must also be real if eq. (12) is to describe a vibrational motion, so the eigenvalues $\{\omega_j^2(\mathbf{q})\}$ are positive as well as real. At each $\mathbf{q}$ the $3s$ values of $\omega_j(\mathbf{q})$ form the *branches* of the multivalued function

$$\omega(\mathbf{q}) = \omega_j(\mathbf{q}), \quad j = 1, 2, \ldots, 3s, \tag{17}$$

which is the *dispersion relation*. Three of the $3s$ branches (the *acoustic* modes) tend to zero as $\mathbf{q} \to \mathbf{0}$. The other $3s - 3$ modes are *optic* modes. Substituting the $\omega_j^2(\mathbf{q})$, one at a time, in eq. (14) gives the eigenvector components $u_\alpha(\kappa)$, which will now be written as $\mathrm{e}_\alpha(\kappa|\mathbf{q}j)$ to emphasize their correspondence with the $3s$ eigenvalues $\omega_j^2(\mathbf{q})$ at each wave vector $\mathbf{q}$. With this notational change,

(14)
$$\omega_j(\mathbf{q})^2 \mathrm{e}_\alpha(\kappa|\mathbf{q}j) = \sum_{\kappa'\beta} \mathrm{D}_{\alpha\beta}(\kappa\kappa'|\mathbf{q}) \mathrm{e}_\beta(\kappa'|\mathbf{q}j). \tag{18}$$

Because these equations are homogeneous, the eigenvector components $\mathrm{e}_\alpha(\kappa|\mathbf{q}j)$ are determined only to within a constant factor, which may be chosen to satisfy the orthonormal

$$\sum_{\kappa\alpha} \mathrm{e}_\alpha(\kappa|\mathbf{q}j)^* \mathrm{e}_\alpha(\kappa|\mathbf{q}j') = \delta_{jj'} \tag{19}$$

and closure

$$\sum_{j} \mathrm{e}_\beta(\kappa'|\mathbf{q}j)^* \mathrm{e}_\alpha(\kappa|\mathbf{q}j) = \delta_{\alpha\beta}\delta_{\kappa\kappa'} \tag{20}$$

conditions. (Closure follows from the completeness property of the eigenvectors of a Hermitian matrix.) In eq. (18), we replace $\mathbf{q}$ by $-\mathbf{q}$ and take the complex conjugate:

(18), (15)
$$\omega_j^2(-\mathbf{q}) \mathrm{e}_\alpha(\kappa|-\mathbf{q}j)^* = \sum_{\kappa'\beta} \mathrm{D}_{\alpha\beta}(\kappa\kappa'|\mathbf{q}) \mathrm{e}_\beta(\kappa'|-\mathbf{q}j)^*. \tag{21}$$

Therefore, $\omega_j^2(\mathbf{q})$ and $\omega_j^2(-\mathbf{q})$ are eigenvalues of the same matrix $\mathrm{D}(\mathbf{q})$,

(18), (21)
$$\omega_j^2(\mathbf{q}) = \omega_j^2(-\mathbf{q}). \tag{22}$$

Thus the dispersion relation is symmetric about $\mathbf{q} = \mathbf{0}$. It follows from eqs. (18) and (21) that the components of $\mathbf{e}^*(-\mathbf{q}\,j)$ satisfy the same set of $3s$ linear homogeneous equations as the components of the eigenvector $\mathbf{e}(\mathbf{q}\,j)$. Therefore, if degeneracy is absent, $\mathbf{e}(\mathbf{q}\,j)$ and $\mathbf{e}^*(-\mathbf{q}\,j)$ can only differ by a phase factor (which preserves normalization). The physical properties of the system are independent of the choice of this phase factor, which we take to be

$$\exp(\mathrm{i}\delta) = +1 \tag{23}$$

(Born and Huang (1954), but see also Liebfried (1955)). With this convention,

$$\mathrm{e}_\alpha(\kappa|-\mathbf{q}j)^* = \mathrm{e}_\alpha(\kappa|\mathbf{q}j), \tag{24}$$

which can always be ensured, even when there is degeneracy.

Since

$$\mathbf{a}_n \cdot \mathbf{b}_m = 2\pi p, \tag{25}$$

where $\mathbf{b}_m$ is a reciprocal lattice vector and p is an integer,

(15), (25) $$\mathrm{D}_{\alpha\beta}(\kappa\kappa'|\mathbf{q}+\mathbf{b}_m) = \mathrm{D}_{\alpha\beta}(\kappa\kappa'|\mathbf{q}), \tag{26}$$

and $\mathrm{D}_{\alpha\beta}(\kappa\kappa'|\mathbf{q})$ has the periodicity of the reciprocal lattice. Therefore the vibrational frequencies and the eigenvectors also have the periodicity of the reciprocal lattice:

$$\omega_j(\mathbf{q}+\mathbf{b}_m) = \omega_j(\mathbf{q}), \tag{27}$$

$$\mathrm{e}_\alpha(\kappa|\mathbf{q}+\mathbf{b}_m,\ j) = \mathrm{e}_\alpha(\kappa|\mathbf{q}j). \tag{28}$$

Consequently, the dispersion relation is usually displayed by plotting $\omega_j(\mathbf{q})$ along high-symmetry directions in the Brilliouin zone. If $\omega_j(\mathbf{q})$ is degenerate, then the RS of eq. (27) should read $\omega_{j'}(\mathbf{q})$, where j' labels one of the modes degenerate with $\omega_j(\mathbf{q})$. However, the choice $j' = j$ is a convenient one that ensures that points of degeneracy can be treated in the same way as points where degeneracy is absent. However, eq. (28) would then be true only to within a phase factor, so that in this form eq. (28) implies that this phase factor has been chosen to be unity (Maradudin *et al.* (1971)).

Answer to Exercise 18.1-1

(15) $$\mathrm{D}_{\beta\alpha}(\kappa'\kappa|\mathbf{q})^* = (M_{\kappa'}M_\kappa)^{-1/2}\sum_{n'}\phi_{\beta\alpha}(n\kappa', n'\kappa)\exp[-\mathrm{i}\mathbf{q}\cdot(\mathbf{a}_{n'}-\mathbf{a}_n)]$$

(3) $$= (M_\kappa M_{\kappa'})^{-1/2}\sum_{n'}\phi_{\alpha\beta}(n'\kappa, n\kappa')\exp[-\mathrm{i}\mathbf{q}\cdot(\mathbf{a}_{n'}-\mathbf{a}_n)]. \tag{29}$$

On replacing the summation variable n' by $2n-n'$,

(29), (10) $$\mathrm{D}_{\beta\alpha}(\kappa'\kappa|\mathbf{q})^* = (M_\kappa M_{\kappa'})^{-1/2}\sum_{n'}\phi_{\alpha\beta}(n\kappa, n'\kappa')\exp[-\mathrm{i}\mathbf{q}\cdot(\mathbf{a}_n-\mathbf{a}_{n'})]$$

$$= \mathrm{D}_{\alpha\beta}(\kappa\kappa'|\mathbf{q}). \tag{30}$$

Therefore $\mathrm{D}(\mathbf{q})$ is a Hermitian matrix.

18.2 Space-group symmetry

The symmetry operators of the space-group G of a crystal are of the form

$$(R|\mathbf{v}) = (R|\mathbf{w}+\mathbf{t}),\ \ \mathbf{t} \in \{\mathbf{a}_n\}. \tag{1}$$

In eq. (1), $\mathbf{v}$ is not necessarily a lattice translation $\mathbf{t}$, since $\mathbf{w}$ may be either the null vector $\mathbf{0}$ or the particular non-lattice translation associated with some screw axis or glide plane. If $\mathbf{v} \in \{\mathbf{a}_n\}\ \forall\ R$, then there are no screw axes or glide planes among the symmetry elements of the crystal and G is a symmorphic space-group (Section 16.2). When $(R|\mathbf{v})$ acts on the position vector $\mathbf{a}_{n\kappa}$ of the κth atom in the nth cell,

$$(R|\mathbf{w}+\mathbf{t})\mathbf{a}_{n\kappa} = R\mathbf{a}_{n\kappa} + \mathbf{w} + \mathbf{t} = \mathbf{a}_{NK}, \tag{2}$$

in which (following Maradudin and Vosko (1968)) the capital letters (NK) are used to label the transformed vector (or site). Since $(R|\mathbf{v})$ is a symmetry operator the site (NK) is one occupied by an atom of the same chemical species as that at the site $(n\kappa)$. When $(R|\mathbf{v})$ acts on the crystal pattern, in the active representation as in eq. (2), any function $f(\mathbf{a}_{n\kappa})$ that depends on the atom positions is transformed into the function

$$(R|\mathbf{w}+\mathbf{t})f(\mathbf{a}_{n\kappa}) = f'(\mathbf{a}_{n\kappa}) = f(R^{-1}\mathbf{a}_{n\kappa} - R^{-1}\mathbf{w} - R^{-1}\mathbf{t}). \tag{3}$$

Note that in eq. (3), as in Chapter 16, no special symbol is used to signify when $(R|\mathbf{w}+\mathbf{t})$ is a space-group *function operator* since this will always be clear from the context. It will often be convenient (following Venkataraman *et al.* (1975)) to shorten the notation for a space-group operator to

$$\Re_l \equiv (R|\mathbf{w}+\mathbf{a}_l). \tag{4}$$

For example,

(4), (2)
$$\Re_l\mathbf{a}_{n\kappa} = R\mathbf{a}_{n\kappa} + \mathbf{w} + \mathbf{a}_l = \mathbf{a}_{NK}. \tag{5}$$

In addition, to minimize the need for multiple subscripts, $\mathbf{a}_{n\kappa}$ will now be denoted by the alternative (and completely equivalent) notation $\mathbf{a}(n\kappa)$ and similarly $\mathbf{r}_{n\kappa}$ will be denoted by $\mathbf{r}(n\kappa)$.

(4), (5)
$$\Re_l\mathbf{r}(n\kappa) = \Re_l[\mathbf{a}(n\kappa) + \mathbf{u}(n\kappa)] = R\mathbf{a}(n\kappa) + R\mathbf{u}(n\kappa) + \mathbf{w} + \mathbf{a}_l$$
$$= \Re_l\mathbf{a}(n\kappa) + R\mathbf{u}(n\kappa) = \mathbf{a}(NK) + \mathbf{u}(NK) = \mathbf{r}(NK); \tag{6}$$

(6)
$$\mathbf{u}(NK) = R\mathbf{u}(n\kappa). \tag{7}$$

Thus, because of the space-group symmetry, the displacement at (NK) is equal to $R\mathbf{u}(n\kappa)$, the rotated displacement from the equivalent site (nk).

(7)
$$u_A(NK) = \sum_\alpha \mathrm{R}_{A\alpha}\, u_\alpha(n\kappa), \tag{8}$$

where R is the 3×3 matrix representative (MR) of the point symmetry operator R. The potential energy Φ is invariant during any space-group operation $\Re_l$,

(6)
$$\Phi(\mathbf{r}(n\kappa)) = \Phi(\Re_l\ \mathbf{r}(n\kappa)) = \Phi(\mathbf{r}(NK)); \tag{9}$$

(9), (6), (7)
$$\Phi(\mathbf{r}(n\kappa)) = \Phi(\mathbf{a}(n\kappa) + \mathbf{u}(n\kappa)) = \Phi(\mathbf{a}(NK) + R\mathbf{u}(n\kappa)). \tag{10}$$

Expanding the LS of eq. (10) in powers of $\mathbf{u}(n\kappa)$ and the RS in powers of $\mathbf{u}(NK) = R\mathbf{u}(n\kappa)$ gives (to terms of second order in the displacements)

(8)
$$\sum_{n\kappa,n'\kappa'}\sum_{\alpha,\beta} u_\alpha(n\kappa)\ \phi_{\alpha\beta}(n\kappa,n'\kappa')\ u_\beta(n'\kappa')$$
$$=\sum_{n\kappa,n'\kappa'}\sum_{A,B} u_A(NK)\ \phi_{AB}(NK,N'K')\ u_B(N'K')$$
$$=\sum_{n\kappa,\ n'\kappa'}\sum_{A,B}\sum_{\alpha,\beta}\mathrm{R}_{A\alpha}u_\alpha(n\kappa)\ \phi_{AB}(NK,N'K')\ \mathrm{R}_{B\beta}u_\beta(n'\kappa')$$
$$=\sum_{n\kappa,n'\kappa'}\sum_{\alpha,\beta} u_\alpha(n\kappa)\left[\sum_{A,B}\mathrm{R}^{\mathrm{T}}_{\alpha A}\ \phi_{AB}(NK,N'K')\,\mathrm{R}_{B\beta}\right] u_\beta(n'\kappa')$$
$$=\sum_{n\kappa,n'\kappa'}\sum_{\alpha\beta} u_\alpha(n\kappa)[\mathrm{R}^{\mathrm{T}}\phi(NK,N'K')\mathrm{R}]_{\alpha\beta}\ u_\beta(n'\kappa'). \tag{11}$$

On equating coefficients of the arbitrary displacements $u_\alpha(n\kappa)\,u_\beta(n'\kappa')$,

(11)
$$\phi_{\alpha\beta}(n\kappa,n'\kappa')=[\mathrm{R}^{\mathrm{T}}\phi(NK,N'K')\mathrm{R}]_{\alpha\beta},\quad \forall\ \alpha,\beta, \tag{12}$$

(12)
$$\phi(n\kappa,n'\kappa')=\mathrm{R}^{\mathrm{T}}\phi(NK,N'K')\mathrm{R}, \tag{13}$$

where $\phi(n\kappa,n'\kappa')$ is the 3×3 matrix with elements $\phi_{\alpha\beta}(n\kappa,n'\kappa')$. Since the MR of the point symmetry operator R is a 3×3 orthogonal matrix,

(13)
$$\phi(NK,N'K')=\mathrm{R}\phi(n\kappa,n'\kappa')\mathrm{R}^{\mathrm{T}}. \tag{14}$$

The result in eq. (14) is not limited to the harmonic approximation because coefficients of like powers of the displacements on each side of eq. (10) are equal, irrespective of the order to which the Taylor expansions are made.

If $\mathfrak{R}_l$ is a point group operator ($\mathbf{w}=\mathbf{0}$, $\mathbf{a}_l=\mathbf{0}$) then $(NK)=(n\kappa)$ and

(14)
$$\phi(n\kappa,n'\kappa')=\mathrm{R}\phi(n\kappa,n'\kappa')\mathrm{R}^{\mathrm{T}}. \tag{15}$$

Equation (15), together with the permutation symmetry condition

(18.1.3)
$$\phi_{\alpha\beta}(n\kappa,n'\kappa')=\phi_{\beta\alpha}(n'\kappa',n\kappa) \tag{16}$$

determine the non-zero elements of the force constant matrix $\phi(n\kappa,n'\kappa')$ (Chapter 15). Similarly, if $\mathfrak{R}_l$ is a space-group operator that interchanges the sites $(n\kappa)$ and $(n'\kappa')$

(14)
$$\phi(n'\kappa',n\kappa)=\mathrm{R}\phi(n\kappa,n'\kappa')\mathrm{R}^{\mathrm{T}}, \tag{17}$$

and eq. (17) with eq. (16) determine the non-zero elements of $\phi(n\kappa,n'\kappa')$.

18.2.1 Periodic boundary conditions

Suppose the crystal is a parallelepiped of sides $N_j\mathbf{a}_j$, where $\mathbf{a}_j$, $j=1, 2, 3$, are the fundamental lattice translations (Section 16.1). To eliminate surface effects, we imagine the crystal to be one of an infinite number of replicas, so that

(16.2.18)
$$(E|N_j\mathbf{a}_j)=(E|\mathbf{a}_j)^{N_j}=(E|\mathbf{0}),\ j=1,2,3. \tag{18}$$

Equation (18) is a statement of the Born–von Kármán periodic boundary conditions; $N = N_1 N_2 N_3$ is the number of unit cells in the crystal lattice.

$$\text{(18)} \qquad \mathbf{u}(n+N, \kappa) = \mathbf{u}(n, \kappa), \tag{19}$$

where the LS of eq. (19) is the displacement of the κth atom in the $(n+N)$th unit cell which is connected to the origin by the vector

$$\mathbf{a}(n+N) = \langle \mathbf{a}_1 \ \mathbf{a}_2 \ \mathbf{a}_3 | n_1+N_1 \quad n_2+N_2 \quad n_3+N_3 \rangle; \tag{20}$$

$$\text{(19), (18.1.12)} \qquad \exp\left[\mathrm{i}\mathbf{q} \cdot \sum_{j=1}^{3} N_j \mathbf{a}_j\right] = 1. \tag{21}$$

Equation (21) shows that $\mathbf{q}$ is a vector in the reciprocal lattice,

$$\begin{aligned} \mathbf{q} = \sum_{j=1}^{3} (m_j/N_j)\mathbf{b}_j, \quad & m_j = 0, \pm 1, \pm 2, \ldots, \pm(N_j - 1)/2, \text{if } N_j \text{ is odd}, \\ & m_j = 0, \pm 1, \pm 2, \ldots, \pm(N_j - 2)/2, \ N_j/2, \text{if } N_j \text{ is even}. \end{aligned} \tag{22}$$

Exercise 18.2-1 Show that eq. (21) is satisfied by eq. (22).

The N $\mathbf{q}$ vectors allowed by the boundary conditions just fill the first Brillouin zone (BZ) of volume equal to v_b, the volume of the primitive unit cell of the reciprocal lattice. Because of this dense, uniform distribution of $\mathbf{q}$ vectors it is possible to treat $\mathbf{q}$ as a continuous variable and thus replace

$$\sum_{\mathbf{q}} f(\mathbf{q}) \quad \text{by} \quad \frac{V}{8\pi^3} \int_{\mathrm{v}_b} f(\mathbf{q}) \, \mathrm{d}\mathbf{q}, \tag{23}$$

where $V = N\mathrm{v}_a$ is the volume of the crystal and the integration is over the volume of the first BZ. But the BZ is invariant under the $g(\mathrm{P})$ operations of the point group P so that an *irreducible volume* of the first BZ of volume $\mathrm{v}_b/g(\mathrm{P})$ can be defined, such that the $g(\mathrm{P})$ operators of P will generate from this irreducible volume the whole BZ. Thus the range of integration can be limited to this irreducible volume.

Answer to Exercise 18.2-1

$$\exp\left[\mathrm{i}\mathbf{q} \cdot \sum_{j=1}^{3} N_j \mathbf{a}_j\right] = \exp\left[\mathrm{i} \sum_{j=1}^{3} m_j \, \mathbf{b}_j \cdot \mathbf{a}_j\right] = \exp(2\pi \, \mathrm{i} p) = 1,$$

since p is an integer or zero.

18.3 Symmetry of the dynamical matrix

The αth component of the displacement $\mathbf{u}(n\kappa)$ of the κth atom in the nth unit cell due to the normal mode of vibration $(\mathbf{q}j)$ is a function of $\mathbf{a}_n$ and time t,

(18.1.12), (18.1.14), (18.1.17) *et seq.* $$u_\alpha(n\kappa) = \mathrm{e}_\alpha(\kappa|\mathbf{q}j)M_\kappa^{-1/2}\exp[\mathrm{i}(\mathbf{q}\cdot\mathbf{a}_n - \omega_j(\mathbf{q})t)]. \tag{1}$$

Under the pure translation $\Re_l = (E|\mathbf{a}_l)$,

(1) $$(E|\mathbf{a}_l)\,u_\alpha(n\kappa) = u_\alpha[(E|\mathbf{a}_l)^{-1}\mathbf{a}_n,\ t] = \exp(-\mathrm{i}\mathbf{q}\cdot\mathbf{a}_l)\,u_\alpha(n\kappa). \tag{2}$$

The wave vector $\mathbf{q}$ in the transformed displacement $(E|\mathbf{a}_l)u_\alpha(n\kappa)$ appears in the scalar product with $\mathbf{a}_l$ in the eigenvalue of the function operator $(E|\mathbf{a}_l)$, and is unaffected by a pure translation, as we should expect. Under the pure rotation $(R|\mathbf{0})$, the rotated displacement

$$Ru_\alpha(n\kappa) = u_\alpha(R^{-1}\mathbf{a}_n, t)$$

$$= \mathrm{e}_\alpha(\kappa|\mathbf{q}j)M_\kappa^{-1/2}\exp[\mathrm{i}(\mathbf{q}\cdot R^{-1}\mathbf{a}_n - \omega_j(\mathbf{q})t)] \tag{3}$$

$$= \mathrm{e}_\alpha(\kappa|\mathbf{q}j)M_\kappa^{-1/2}\exp[\mathrm{i}(R\mathbf{q}\cdot\mathbf{a}_n - \omega_j(\mathbf{q})t)]. \tag{4}$$

Equation (4) follows from eq. (3) because the scalar product (SP) in eq. (3) is invariant under the rigid rotation R.

(2), (4) $$(E|\mathbf{a}_l)\,Ru_\alpha(n\kappa) = (E|\mathbf{a}_l)\,(R|\mathbf{0})\,u_\alpha(n\kappa) = \Re_l u_\alpha(n\kappa) \tag{5}$$

(5), (4), (2) $$= \exp(-\mathrm{i}R\mathbf{q}\cdot\mathbf{a}_l)R\,u_\alpha(n\kappa). \tag{6}$$

Equation (6) verifies that under the rotation $(R|\mathbf{0})$ the wave vector $\mathbf{q}$ is rotated into $R\mathbf{q}$, just as we might have anticipated in the active representation. The transformed function

(1), (5), (6) $$\begin{aligned} u_\alpha(NK) &= \Re_l u_\alpha(n\kappa) = u_\alpha(\Re_l^{-1}\mathbf{a}(n\kappa), t) \\ &= \exp(-\mathrm{i}R\mathbf{q}\cdot\mathbf{a}_l)Ru_\alpha(n\kappa). \end{aligned} \tag{7}$$

Since $\Re_l$ is a symmetry operator, $(n\kappa)$ and (NK) are equivalent sites occupied by the same kind of atom, so that $M_K = M_\kappa$. It follows from eq. (7) that the Fourier-transformed dynamical matrix $\mathrm{D}(\mathbf{q})$ is transformed by $\Re_l$ into $\mathrm{D}(R\mathbf{q})$. The AK, BK' element of $\mathrm{D}(R\mathbf{q})$ is

(18.1.7) $$\mathrm{D}_{AB}(KK'|R\mathbf{q})$$

$$= (M_K M_{K'})^{-1/2}\sum_{N'}\phi_{AB}(NK, N'K')\exp[-\mathrm{i}R\mathbf{q}\cdot(\mathbf{a}_N - \mathbf{a}_{N'})] \tag{8}$$

$$= (M_K M_{K'})^{-1/2}\sum_{N'}\phi_{AB}(NK, N'K')\exp[-\mathrm{i}\mathbf{q}\cdot R^{-1}(\mathbf{a}_N - \mathbf{a}_{N'})], \tag{9}$$

where eq. (9) follows from eq. (8) because of the invariance of the SP under rotations.

$$\Re_l \mathbf{a}(n\kappa) = (R|\mathbf{w} + \mathbf{a}_l)(\mathbf{a}_n + \mathbf{a}_\kappa) = R\mathbf{a}_n + R\mathbf{a}_\kappa + \mathbf{w} + \mathbf{a}_l = \mathbf{a}(NK) = \mathbf{a}_N + \mathbf{a}_K; \quad (10)$$

(10)
$$R^{-1}\mathbf{a}_N = \mathbf{a}_n + \mathbf{a}_\kappa - \Re_l^{-1}\mathbf{a}_K. \quad (11)$$

(9), (11), (18.2.14)
$$\mathrm{D}_{AB}(KK'|R\mathbf{q}) = (M_K M_{K'})^{-1/2} \sum_{\alpha\beta n'} \mathrm{R}_{A\alpha} \exp[\mathrm{i}\mathbf{q}\cdot(\Re_l^{-1}\mathbf{a}_K - \mathbf{a}_\kappa)] \times \phi_{\alpha\beta}(n\kappa, n'\kappa') \exp[-\mathrm{i}\mathbf{q}\cdot(\mathbf{a}_n - \mathbf{a}_{n'})]\mathrm{R}^{\mathrm{T}}_{\beta B} \exp[-\mathrm{i}\mathbf{q}\cdot(\Re_l^{-1}\mathbf{a}_{K'} - \mathbf{a}_{\kappa'})] \quad (12)$$

$$= \sum_{\alpha\beta} \tilde{\Gamma}_{A\alpha}(K\kappa|\mathbf{q}, \Re_l)\mathrm{D}_{\alpha\beta}(\kappa\kappa'|\mathbf{q})\tilde{\Gamma}_{\beta B}(\kappa' K'|\mathbf{q}, \Re_l)^\dagger, \quad (13)$$

where

$$\tilde{\Gamma}_{A\alpha}(K\kappa|\mathbf{q}, \Re_l) = \mathrm{R}_{A\alpha} \exp[\mathrm{i}\mathbf{q}\cdot(\Re_l^{-1}\mathbf{a}_K - \mathbf{a}_\kappa)]\delta(K, (R|\mathbf{w})\kappa), \quad (14)$$

(13)
$$\mathrm{D}(R\mathbf{q}) = \tilde{\Gamma}(\mathbf{q}, \Re_l)\mathrm{D}(\mathbf{q})\tilde{\Gamma}(\mathbf{q}, \Re_l)^\dagger. \quad (15)$$

In eq. (13), the first factor in the sum is the $(AK, \alpha\kappa)$ element of $\tilde{\Gamma}(\mathbf{q}, \Re_l)$ and the last factor is the $(\beta\kappa', BK')$ element of $\tilde{\Gamma}(\mathbf{q}, \Re_l)^\dagger$, that is the $(BK', \beta\kappa')$ element of $\tilde{\Gamma}(\mathbf{q}, \Re_l)^*$. (Note that A, B, α, and β label the Cartesian components x, y, or z, and that κ, K, κ', and K' label the atoms in the unit cell.) The factor $\delta(K, (R|\mathbf{w})\kappa)$ ensures that it is an atom on the sublattice occupied by atom K that results from applying the symmetry operator $\Re_l$ to the atom at $\mathbf{a}_\kappa$. (The sublattice labeling by $\kappa, K, \ldots$ is invariant under lattice translations.) R is the 3×3 orthogonal MR of R, and $\tilde{\Gamma}$ is the $3s \times 3s$ unitary matrix that transforms $\mathrm{D}(\mathbf{q})$ into $\mathrm{D}(R\mathbf{q})$ by the unitary transformation, eq. (15).

Exercise 18.3-1 Prove that the matrix $\tilde{\Gamma}$ defined by eq. (14) is unitary.

Exercise 18.3-2 Show that $\omega_j(R\mathbf{q})^2 = \omega_j(\mathbf{q})^2$.

Suppose that $\Re_l = \Re_i\, \Re_j$, where $\Re_l, \Re_i, \Re_j \in$ the space group G and

$$\mathbf{a}_K = \Re_l \mathbf{a}_\kappa = \Re_i \Re_j \mathbf{a}_\kappa = \Re_i \mathbf{a}_{\kappa_1}. \quad (16)$$

(14)
$$\tilde{\Gamma}_{A\alpha}(K\kappa|\mathbf{q}, \Re_l) = \mathrm{R}_{A\alpha} \exp[\mathrm{i}\mathbf{q}.(\Re_j^{-1}\Re_i^{-1}\mathbf{a}_K - \mathbf{a}_\kappa)]\ \delta(K, (R|\mathbf{w})\kappa), \quad (17)$$

where

$$\delta(K, (R|\mathbf{w})\kappa) = \sum_{\kappa_1} \delta(K, (R_i|\mathbf{w}_i)\kappa_1)\ \delta(\kappa_1, (R_j|\mathbf{w}_j)\kappa). \quad (18)$$

The choice of $\Re_i\, \Re_j$, and therefore of κ_1, is immaterial as long as eq. (16) is satisfied. The exponent in eq. (17) is

$$\mathrm{i}\mathbf{q}.(\Re_j^{-1}\Re_i^{-1}\mathbf{a}_K - \mathbf{a}_\kappa) = \mathrm{i}\Re_j\mathbf{q}\cdot(\Re_i^{-1}\mathbf{a}_\kappa - \mathbf{a}_{\kappa_1}) + \mathrm{i}\mathbf{q}\cdot(\Re_j^{-1}\mathbf{a}_{\kappa_1} - \mathbf{a}_\kappa), \quad (19)$$

in which $\Re_j^{-1}\mathbf{a}_{\kappa_1}$ has been subtracted and added to the RS of eq. (19).

$$(17),(18),(19) \qquad \tilde{\Gamma}_{A\alpha}(K\kappa|\mathbf{q},\Re_l) = \sum_{\gamma\kappa_1} \mathrm{R}_{A\gamma}\exp[\mathrm{i}R_j\mathbf{q}\cdot(\Re_i^{-1}\mathbf{a}_k - \mathbf{a}_{\kappa_1})$$
$$\times\, \delta(K,(R_i|\mathbf{w}_i)\kappa_1)\mathrm{R}_{\gamma\alpha}\exp[\mathrm{i}\mathbf{q}\cdot(\Re_\varphi^{-1}\mathbf{a}_{\kappa_1} - \mathbf{a}_\kappa)]\delta(\kappa_1,(\mathbf{P}_\varphi|\mathbf{w}_\varphi)\kappa); \tag{20}$$

$$(20) \qquad \tilde{\Gamma}(\mathbf{q},\Re_l) = \tilde{\Gamma}(\Re_j\mathbf{q},\Re_i)\,\tilde{\Gamma}(\mathbf{q},\Re_j). \tag{21}$$

Because $\mathbf{q}$ is replaced by $\Re_j\mathbf{q}$ in the first factor on the RS of eq. (21), $\{\tilde{\Gamma}(\mathbf{q},\Re_j)\}$ does not form an MR of the space group G. But if $\Re_l$, $\Re_i$, $\Re_j \in \mathrm{G}(\mathbf{q})$ (the little group, or group of the wave vector $\mathbf{q}$) then

$$R\mathbf{q} = \mathbf{q} - \mathbf{b}, \tag{22}$$

where $\mathbf{b}$ is a reciprocal lattice vector that is non-zero only if $\mathbf{q}$ terminates on the surface of the first BZ. On using eq. (22) the first exponential in eq. (20) may be written as

$$\mathrm{R}_{A\alpha}\exp[\mathrm{i}\mathbf{q}\cdot\Re_\iota^{-1}(\mathbf{a_K} - \mathbf{a}_{\kappa_1})]\exp[\mathrm{i}\mathbf{b}_\varphi\cdot(\Re_\iota^{-1}\mathbf{a_K} - \mathbf{a}_{\kappa_1})], \tag{23}$$

where (as described by the Kronecker δ in eq. (20)) $\Re_j^{-1}\,\mathbf{a}_K - \mathbf{a}_{\kappa 1}$ is a lattice translation (see eq. (14) *et seq*.). Consequently, the second factor in eq. (23) is unity and

$$(21),(23) \qquad \tilde{\Gamma}(\mathbf{q},\Re_i\,\Re_j) = \tilde{\Gamma}(\mathbf{q},\Re_i)\,\tilde{\Gamma}(\mathbf{q},\Re_j),\ \ \Re_i,\Re_j \in \mathrm{G}(\mathbf{q}). \tag{24}$$

Thus $\{\tilde{\Gamma}(\mathbf{q},\Re_l)\}$ do form a unitary MR of dimension $3s$ of the little space group G($\mathbf{q}$). Define

$$\Gamma(\mathbf{q},R) = \exp[\mathrm{i}\mathbf{q}\cdot(\mathbf{w}+\mathbf{a}_l)]\;\tilde{\Gamma}(\mathbf{q},\Re_l); \tag{25}$$

$$\begin{aligned}
(25)\quad \Gamma(\mathbf{q},R_i)\Gamma(\mathbf{q},R_j) &= \exp[\mathrm{i}\mathbf{q}\cdot(\mathbf{w}_i+\mathbf{a}_i)]\exp[\mathrm{i}\mathbf{q}\cdot(\mathbf{w}_j+\mathbf{a}_j)]\tilde{\Gamma}(\mathbf{q},\Re_i)\tilde{\Gamma}(\mathbf{q},\Re_j)\\
(24)\quad &= \exp[\mathrm{i}\mathbf{q}\cdot(\mathbf{w}_i+\mathbf{a}_i)]\exp[\mathrm{i}\mathbf{q}\cdot(\mathbf{w}_j+\mathbf{a}_j)]\exp[-\mathrm{i}\mathbf{q}\cdot R_i(\mathbf{w}_j+\mathbf{a}_j)+\mathbf{w}_i+\mathbf{a}_i]\Gamma(\mathbf{q},R_iR_j)\\
&= \exp[\mathrm{i}(\mathbf{q}-R_i^{-1}\mathbf{q})\cdot(\mathbf{w}_j+\mathbf{a}_j)]\Gamma(\mathbf{q},R_iR_j)\\
(22)\quad &= \exp[\mathrm{i}\mathbf{b}_i\cdot\mathbf{w}_j]\Gamma(\mathbf{q},R_iR_j).
\end{aligned} \tag{26}$$

Equation (26) shows that $\{\Gamma(\mathbf{q},R)\}$ forms a unitary projective (or multiplier) representation of $\{R\} = \mathrm{P}(\mathbf{q})$. Only for non-symmorphic groups with $\mathbf{b}$ different from zero (that is, when $\mathbf{q}$ lies on the surface of the BZ) are the projective factors exp $[\mathrm{i}\mathbf{b}_i.\mathbf{w}_j]$ in eq. (26) different from unity.

$$(18.1.26) \qquad \mathrm{D}(\mathbf{q}-\mathbf{b}) = \mathrm{D}(\mathbf{q}),\ \ \mathbf{b}\in\{\mathbf{b}_m\}; \tag{27}$$

$$(15),(22),(27) \qquad \mathrm{D}(\mathbf{q}) = \tilde{\Gamma}(\mathbf{q},\Re_l)\mathrm{D}(\mathbf{q})\tilde{\Gamma}(\mathbf{q},\Re_l)^\dagger. \tag{28}$$

Equation (28) shows that the unitary matrices $\tilde{\Gamma}(\mathbf{q},\Re_l)$ commute with the dynamical matrix,

$$(28) \qquad \mathrm{D}(\mathbf{q})\tilde{\Gamma}(\mathbf{q},\Re_l) = \tilde{\Gamma}(\mathbf{q},\Re_l)\mathrm{D}(\mathbf{q}), \tag{29}$$

(29), (25)
$$\mathrm{D}(\mathbf{q})\Gamma(\mathbf{q}, R) = \Gamma(\mathbf{q}, R)\mathrm{D}(\mathbf{q}), \tag{30}$$

so that the $\Gamma(\mathbf{q}, R)$ matrices also commute with $\mathrm{D}(\mathbf{q})$.

Answers to Exercises 18.3

Exercise 18.3-1 From eqs. (12), (13), and (14),

$$[\tilde{\Gamma}(K\kappa|\mathbf{q}, \Re_l)\tilde{\Gamma}(K'\kappa|\mathbf{q}, \Re_l)^{\dagger}]_{AB} = \sum_{\alpha} \mathrm{R}_{A\alpha} \exp[\mathrm{i}\mathbf{q}.(\Re_l^{-1}\mathbf{a}_K - \mathbf{a}_\kappa)]$$
$$\times\, \delta(K, (R|\mathbf{w})\kappa)\mathrm{R}^{\mathrm{T}}_{\alpha B} \exp[-\mathrm{i}\mathbf{q}\cdot(\Re_l^{-1}\mathbf{a}_{K'} - \mathbf{a}_\kappa)]\delta(K', (R|\mathbf{w})\kappa). \tag{31}$$

The second delta function on the RS of eq. (31) is zero unless $K' = K$, when the exponential factors cancel and

(31)
$$\sum_{\alpha} \mathrm{R}_{A\alpha}\mathrm{R}^{\mathrm{T}}_{\alpha B} = (\mathrm{R}\mathrm{R}^{\mathrm{T}})_{AB} = \delta_{AB} \tag{32}$$

since R is an orthogonal matrix. Therefore, $\mathrm{R}\mathrm{R}^{\mathrm{T}} = \mathrm{E}$ and $\tilde{\Gamma}(\mathbf{q}, \Re_l)$ is a unitary matrix.

Exercise 18.3-2 $\mathrm{D}(R\mathbf{q})$ and $\mathrm{D}(\mathbf{q})$ are related by the unitary transformation, eq. (15), and since the eigenvalues of a matrix are invariant under a unitary transformation, $\omega_j(R\mathbf{q})^2 = \omega_j(\mathbf{q})^2$.

18.4 Symmetry coordinates

The determination of the eigenvalues $\omega_j(\mathbf{q})^2$ may be simplified by an orthogonal transformation to "symmetry coordinates," which are linear combinations of the Cartesian displacements of the atoms which represent the actual displacements of the atoms in the unit cell. Simultaneously, the eigenvectors undergo the same orthogonal transformation (see Section 9.4, especially eqs. (9.4.4) and (9.4.6)). In matrix notation,

(18.1.18)
$$\mathrm{D}(\mathbf{q})|\mathbf{e}(\mathbf{q}j)\rangle = \omega_j(\mathbf{q})^2|\mathbf{e}(\mathbf{q}j)\rangle. \tag{1}$$

Multiplying each side of eq. (1) by $\Gamma(\mathbf{q}, R)$, $R \in \mathrm{P}(\mathbf{q})$, and using the commutation of $\mathrm{D}(\mathbf{q})$ with $\Gamma(\mathbf{q}, R)$,

(1), (18.3.30)
$$\mathrm{D}(\mathbf{q})\Gamma(\mathbf{q}, R)|\mathbf{e}(\mathbf{q}j)\rangle = \omega_j(\mathbf{q})^2\Gamma(\mathbf{q}, R)|\mathbf{e}(\mathbf{q}j)\rangle. \tag{2}$$

The eigenvalues $\omega_j(\mathbf{q})^2$ in eq. (2) are not necessarily all distinct, so the index j will now be replaced by the double index $\sigma\lambda$ where σ labels the distinct eigenvalues of $\mathrm{D}(\mathbf{q})$ and $\lambda = 1, 2, \ldots, l(\sigma)$ labels the linearly independent (LI) eigenvectors associated with the degenerate eigenvalue σ. In this notation,

(1)
$$\mathrm{D}(\mathbf{q})|\mathbf{e}(\mathbf{q}\sigma\lambda)\rangle = \omega_\sigma(\mathbf{q})^2|\mathbf{e}(\mathbf{q}\sigma\lambda)\rangle. \tag{3}$$

Equation (2) shows that $\Gamma(\mathbf{q}, R)|\mathbf{e}(\mathbf{q}\sigma\lambda)\rangle$ must be a linear combination of the LI eigenvectors of $\mathrm{D}(\mathbf{q})$ with eigenvalue $\omega_\sigma(\mathbf{q})^2$,

$$(2),(3) \qquad \Gamma(\mathbf{q},R)|\mathbf{e}(\mathbf{q}\sigma\lambda)\rangle = \sum_{\lambda'=1}^{l(\sigma)} \overline{\Gamma}^{\sigma}_{\lambda\lambda'}(\mathbf{q},R)|\mathbf{e}(\mathbf{q}\sigma\lambda')\rangle, \quad \forall R \in \mathrm{P}(\mathbf{q}), \tag{4}$$

where the coefficients $\overline{\Gamma}^{\sigma}_{\lambda\lambda'}(\mathbf{q},R)$ are the elements of the λth row of the MR $\overline{\Gamma}^{\sigma}(\mathbf{q},R)$ of R in the irreducible representation $\overline{\Gamma}^{\sigma}$. Multiply both sides of eq. (4) by $\Gamma(\mathbf{q}, R')$, $R' \in \mathrm{P}(\mathbf{q})$,

$$(4) \qquad \Gamma(\mathbf{q},R')\Gamma(\mathbf{q},R)|\mathbf{e}(\mathbf{q}\sigma\lambda)\rangle = \sum_{\lambda'=1}^{l(\sigma)}\sum_{\lambda''=1}^{l(\sigma)} \overline{\Gamma}^{\sigma}_{\lambda\lambda'}(\mathbf{q},R)\overline{\Gamma}^{\sigma}_{\lambda'\lambda''}(\mathbf{q},R')|\mathbf{e}(\mathbf{q}\sigma\lambda'')\rangle. \tag{5}$$

Since $\{\Gamma(\mathbf{q}, R)\}$, with $R, R', R'R \in \mathrm{P}(\mathbf{q})$, forms a projective representation (PR) of $\mathrm{P}(\mathbf{q})$, the LS of eq. (5) is given by

$$(5) \qquad [R';R]\Gamma(\mathbf{q},R'R)|\mathbf{e}(\mathbf{q}\sigma\lambda)\rangle = [R';R]\sum_{\lambda''=1}^{l(\sigma)} \overline{\Gamma}^{\sigma}_{\lambda\lambda''}(\mathbf{q},R'R)|\mathbf{e}(\mathbf{q}\sigma\lambda'')\rangle. \tag{6}$$

It follows from eqs. (5) and (6) that $\{\overline{\Gamma}^{\sigma}(\mathbf{q},R)\}$ forms a PR of $\mathrm{P}(\mathbf{q})$ with the same PFs as the PR $\{\Gamma(\mathbf{q}, R)\}$. The $\{\Gamma(\mathbf{q}, R)\}$ form a unitary PR of $\mathrm{P}(\mathbf{q})$, therefore,

$$\Gamma^{\sigma}(\mathbf{q},R)^{\dagger}\Gamma^{\sigma'}(\mathbf{q},R) = \mathrm{E}\ \delta(\sigma\sigma'). \tag{7}$$

Multiply this $l(\sigma) \times l(\sigma)$ square matrix from the left by $\langle\mathbf{e}(\mathbf{q}\sigma\lambda)|$ and from the right by $|\mathbf{e}(\mathbf{q}\sigma'\lambda')\rangle$, where $\langle\mathbf{e}(\mathbf{q}\sigma\lambda)|$ is the row matrix containing the $l(\sigma)$ eigenvectors $\mathbf{e}(\mathbf{q}\sigma\lambda)$, $\lambda = 1, 2 \ldots, l(\sigma)$. Then

$$(7),(18.1.18) \qquad \begin{aligned}&\langle\mathbf{e}(\mathbf{q}\sigma\lambda)|\Gamma^{\sigma}(\mathbf{q},R)^{\dagger}\Gamma^{\sigma'}(\mathbf{q},R)|\mathbf{e}(\mathbf{q}\sigma'\lambda')\rangle \\ &\quad = \langle\mathbf{e}(\mathbf{q}\sigma\lambda)|\mathbf{e}(\mathbf{q}\sigma'\lambda')\rangle = \delta(\sigma,\sigma')\delta(\lambda,\lambda').\end{aligned} \tag{8}$$

Multiply each side of eq. (8) by $l(\sigma)/p(\mathbf{q})$, where $p(\mathbf{q})$ is the order of $\mathrm{P}(\mathbf{q})$, and $\sum\limits_{R\in\mathrm{P}(\mathbf{q})}$. The LS of eq. (8) then becomes

$$(8),(4) \qquad \begin{aligned}&\langle\mathbf{e}(\mathbf{q}\sigma\lambda)|\sum_{\lambda'=1}^{l(\sigma)}\sum_{\lambda''=1}^{l(\sigma)}\sum_{R\in\mathrm{P}(\mathbf{q})}[l(\sigma)/p(\mathbf{q})]\overline{\Gamma}^{\sigma}_{\lambda'\lambda}(\mathbf{q},R)\overline{\Gamma}^{\sigma'}_{\lambda\lambda''}((\mathbf{q},R))|\mathbf{e}(\mathbf{q}\sigma'\lambda')\rangle \\ &= \sum_{\lambda'=1}^{l(\sigma)}\sum_{\lambda''=1}^{l(\sigma)}\langle\mathbf{e}(\mathbf{q}\sigma\lambda)|\delta(\sigma,\sigma')\delta(\lambda',\lambda)\delta(\lambda,\lambda'')|\mathbf{e}(\mathbf{q}\sigma'\lambda')\rangle \\ &= \langle\mathbf{e}(\mathbf{q}\sigma\lambda)|\delta(\sigma,\sigma')\delta(\lambda,\lambda')|\mathbf{e}(\mathbf{q}\sigma'\lambda')\rangle = \delta(\sigma,\sigma')\delta(\lambda,\lambda') = \text{RS of eq. (8)}\end{aligned} \tag{9}$$

iff the $\{\overline{\Gamma}^{\sigma}(\mathbf{q},R)\}$ forms an irreducible unitary PR of $\mathrm{P}(\mathbf{q})$. Consequently, the eigenvalues and eigenvectors of $\mathrm{D}(\mathbf{q})$ may be classified (and labeled) by the irreducible PRs $\{\overline{\Gamma}^{\sigma}(\mathbf{q},R)\}$ of $\mathrm{P}(\mathbf{q})$. Reduction of the reducible representation $\Gamma(\mathbf{q}, R)$ is effected in the usual way from

$$c^{\sigma} = [p(\mathbf{q})]^{-1}\sum_{R}\chi(\mathbf{q},R)\overline{\chi}^{\sigma}(\mathbf{q},R)^{*}, \tag{10}$$

where c^{σ} is the number of times that the IR $\overline{\Gamma}^{\sigma}(\mathbf{q},R)$ occurs in the reduction of $\Gamma(\mathbf{q}, R)$; $\chi(\mathbf{q}, R)$ is the character of $\Gamma(\mathbf{q}, R)$; $\overline{\chi}^{\sigma}(\mathbf{q},R)$ is the character of $\overline{\Gamma}^{\sigma}(\mathbf{q},R)$; and $p(\mathbf{q})$ is the order of $\mathrm{P}(\mathbf{q})$.

Since there may be more than one IR $\overline{\Gamma}^{\sigma}(\mathbf{q}, R)$ of the same symmetry, an additional index $\mu = 1, 2, \ldots, c^{\sigma}$ may be needed to label the different eigenvalues of the same symmetry σ:

$$\text{(1)} \qquad \mathrm{D}(\mathbf{q})|\mathbf{e}(\mathbf{q}\sigma\mu\lambda)\rangle = \omega_{\sigma\mu}(\mathbf{q})^2|\mathbf{e}(\mathbf{q}\sigma\mu\lambda)\rangle. \qquad (11)$$

Equation (4) holds for each value of μ so that

$$\text{(4)} \qquad \Gamma(\mathbf{q}, R)|\mathbf{e}(\mathbf{q}\sigma\mu\lambda)\rangle = \sum_{\lambda'=1}^{l(\sigma)} \overline{\Gamma}^{\sigma}_{\lambda\lambda'}(\mathbf{q}, R)|\mathbf{e}(\mathbf{q}\sigma\mu\lambda')\rangle. \qquad (12)$$

Symmetry coordinates are linear combinations of the eigenvectors $\mathbf{e}(\mathbf{q}\sigma\mu\lambda)$ that describe the displacement patterns of the atoms in the unit cell (see Section 9.4). These patterns observe the symmetry of the IRs $\{\overline{\Gamma}^{\sigma}(\mathbf{q}, R)\}$ and they may be calculated by applying the projection operator

$$\mathrm{P}^{\sigma}_{\lambda\lambda'}(\mathbf{q}) = [l(\sigma)/p(\mathbf{q})]\sum_{R}\overline{\Gamma}^{\sigma}_{\lambda\lambda'}(\mathbf{q},\ R)^{*}\Gamma(\mathbf{q},\ R) \qquad (13)$$

to an arbitrary vector $|\psi\rangle$ with $3s$ rows to give

$$|\psi(\mathbf{q}\sigma\lambda)\rangle = \mathrm{P}^{\sigma}_{\lambda\lambda'}(\mathbf{q})|\psi\rangle, \qquad (14)$$

which transforms according to the λth row of the MR of R in the σth representation $\overline{\Gamma}^{\sigma}(\mathbf{q}, R)$,

$$\Gamma(\mathbf{q}, R)|\psi(\mathbf{q}\sigma\lambda)\rangle = \sum_{\lambda'=1}^{l(\sigma)} \overline{\Gamma}^{\sigma}_{\lambda\lambda'}(\mathbf{q},\ R)|\psi(\mathbf{q}\sigma\mu\lambda')\rangle, \qquad (15)$$

as do the $|\mathbf{e}(\mathbf{q}\sigma\mu\lambda)\rangle$. Therefore

$$\text{(9), (7)} \qquad |\psi(\mathbf{q}\sigma\lambda)\rangle = \sum_{\mu=1}^{c^{\sigma}} c_{\mu}|\mathbf{e}(\mathbf{q}\sigma\mu\lambda)\rangle. \qquad (16)$$

The coefficients c_{μ} are not determined by symmetry but depend on the particular form of $\mathrm{D}(q)$ (see the molecular case in Section 9.4). To find the c^{σ} LI vectors $|\psi(\mathbf{q}\sigma\mu\lambda)\rangle$ in the $\sigma\lambda$ subspace, set $\lambda' = \lambda$ and obtain the $3s$ vectors:

$$\text{(13)} \qquad |\psi(\mathbf{q}\sigma a\lambda)\rangle = [l(\sigma)/p(\mathbf{q})]\sum_{R}\overline{\Gamma}^{\sigma}_{\lambda\lambda}(\mathbf{q}, R)^{*}\Gamma_{a}(\mathbf{q}, R), \quad a = 1, 2, \ldots, 3s, \qquad (17)$$

where $\Gamma_a(\mathbf{q},\ R)$ is the ath column of $\Gamma(\mathbf{q},\ R)$ (Worlton and Warren (1972)). From these $3s$ vectors c^{σ} LI orthonormal vectors $\{|\psi(\mathbf{q}\sigma\mu\lambda)\rangle\}$, which are the symmetry coordinates, can be obtained by the Schmidt procedure (see, for example, Margenau and Murphy (1943), Section 10.8). These vectors $|\psi(\mathbf{q}\sigma\mu\lambda)\rangle$ transform according to eq. (15) for each value of μ so that

$$\Gamma(\mathbf{q}, R)|\psi(\mathbf{q}\sigma\mu\lambda)\rangle = \sum_{\lambda'=1}^{l(\sigma)} \overline{\Gamma}^{\sigma}_{\lambda\lambda'}(\mathbf{q}, R)|\psi(\mathbf{q}\sigma\mu\lambda')\rangle. \qquad (18)$$

The matrix elements of $\mathrm{D}(\mathbf{q})$, $\overline{\mathrm{D}}(\mathbf{q}) = \langle(\Gamma^{\sigma})^{-1}|\mathrm{D}(\mathbf{q})|\Gamma^{\sigma}\rangle$ (where $\langle|\rangle$ denotes the Hermitian SP) are

$$\overline{\mathrm{D}}(\mathbf{q}|\sigma\mu\lambda,\sigma'\mu'\lambda') = \langle\psi(\mathbf{q}\sigma\mu\lambda)|\Gamma^{\sigma\dagger}\mathrm{D}(\mathbf{q})\Gamma^{\sigma}|\psi(q\sigma'\mu'\lambda')\rangle$$
$$= \Big\langle \sum_{\lambda_1'=1}^{l(\sigma)} \overline{\Gamma}^{\sigma}_{\lambda\lambda_1'}(\mathbf{q},\, R)\psi(\mathbf{q}\sigma\mu\lambda_1')|\mathrm{D}(\mathbf{q})| \sum_{\lambda_2'=1}^{l(\sigma')} \overline{\Gamma}^{\sigma'}_{\lambda'\lambda_2'}(\mathbf{q},\, R)\psi(\mathbf{q}\sigma'\mu'\lambda_2')\Big\rangle. \tag{19}$$

Divide each side of eq. (19) by $p(\mathbf{q})$ and sum over R; then from the orthogonality theorem for the MRs $\overline{\Gamma}^{\sigma}$

$$\overline{\mathrm{D}}(\mathbf{q}|\mu,\mu') = [l(\sigma)]^{-1}\langle\psi(\mathbf{q}\sigma\mu\lambda)|\mathrm{D}(\mathbf{q})|\psi(\mathbf{q}\sigma\mu'\lambda)\rangle. \tag{20}$$

Thus the matrix $\overline{\mathrm{D}}$ has been transformed into one with a block-diagonal structure.The diagonal blocks labeled $\overline{\mathrm{D}}^{\sigma}(\mathbf{q})$ are matrices of dimensions $c^{\sigma}\times c^{\sigma}$, with elements

$$\begin{aligned}\overline{\mathrm{D}}^{\sigma}(\mathbf{q}|\mu,\mu') &= \langle\psi(\mathbf{q}\sigma\mu\lambda)|\mathrm{D}|\psi(\mathbf{q}\sigma\mu'\lambda)\rangle\\ &= \sum_{\alpha\kappa}\sum_{\alpha'\kappa'}\langle\psi_{\alpha}(\kappa|\mathbf{q}\sigma\mu\lambda)|\mathrm{D}_{\alpha\alpha'}(\kappa\kappa'|\mathbf{q})|\psi_{\alpha'}(\kappa'|\mathbf{q}\sigma\mu'\lambda)\rangle.\end{aligned} \tag{21}$$

Each $\overline{\mathrm{D}}^{\sigma}(\mathbf{q})$ block appears c^{σ} times along the diagonal. The eigenvalues of $\overline{\mathrm{D}}^{\sigma}(\mathbf{q})$ are $\omega^2_{\sigma\mu}$ and their degeneracy is $l(\sigma)$, the dimension of the IR σ. This completes the solution to the problem of finding the frequencies and the eigenvectors of the dynamical matrix $\overline{\mathrm{D}}(\mathbf{q})$, except for a consideration of extra degeneracies that may arise from time-reversal symmetry.

18.5 Time-reversal symmetry

It was shown in Secion 13.2 that for motion in which spin is neglected, the time-reversal operator Θ is just the complex conjugation operator $\mathscr{K}$. Therefore

(18.1.12) $$\mathscr{K}\, u_{\alpha}(n\kappa) = M_{\kappa}^{-1/2} u_{\alpha}(\kappa)\exp[\mathrm{i}(-\mathbf{q}\cdot\mathbf{a}_n + \omega t)] \tag{1}$$

so that t is replaced by $-t$ and $\mathbf{q}$ by $-\mathbf{q}$. The equations of motion in the time-reversed state are therefore

(18.4.1) $$\mathrm{D}(-\mathbf{q})|\mathbf{e}(-\mathbf{q}j)\rangle = \omega_j(-\mathbf{q})^2|\mathbf{e}(-\mathbf{q}j)\rangle, \tag{2}$$

(18.1.15) $$\mathrm{D}(-\mathbf{q}) = \mathrm{D}(\mathbf{q})^*, \tag{3}$$

(18.4.1) $$\mathrm{D}(\mathbf{q})^*\mathbf{e}(\mathbf{q}j)^* = \omega_j(\mathbf{q})^2\mathbf{e}(\mathbf{q}j)^*, \tag{4}$$

(2), (3) $$\mathrm{D}(\mathbf{q})^*\mathbf{e}(-\mathbf{q}j) = \omega_j(-\mathbf{q})^2\mathbf{e}(-\mathbf{q}j), \tag{5}$$

(4), (5) $$\omega_j(-\mathbf{q})^2 = \omega_j(\mathbf{q})^2, \tag{6}$$

(4), (5) $$\mathbf{e}(-\mathbf{q}j) = \mathbf{e}(\mathbf{q}j)^*, \tag{7}$$

with the appropriate choice of phase in eq. (7). Thus the symmetry of the normal mode frequencies in the BZ about $\mathbf{q} = \mathbf{0}$ is a consequence of time-reversal symmetry. (A similar situation was encountered with the symmetry of $E(\mathbf{k})$ about $\mathbf{k} = \mathbf{0}$ (see Section 17.2).) If ψ is an arbitrary vector that is a linear combination of the $\mathbf{e}(\mathbf{q})$, then

$$\mathscr{K}\psi = \psi^*. \tag{8}$$

Exercise 18.5-1 (a) Find the operator $\mathscr{K}_0{}^{-1}$. (b) Prove that $\mathscr{K}_0$ is an antiunitary operator. (c) Show that

$$\mathscr{K}_0 \mathrm{D} \mathscr{K}_0^{-1} = \mathrm{D}^*. \tag{9}$$

Time-reversal symmetry may be responsible for additional degeneracies beyond that stated in eq. (6). These arise when $\mathrm{P}(\mathbf{q})$ contains an operator Q such that

$$Q\mathbf{q} = -\mathbf{q} \in {}^*\mathbf{q}\,, \quad Q \in \mathrm{P}(\mathbf{q}). \tag{10}$$

In this case $\mathrm{P}(\mathbf{q})$ is an invariant subgroup of index 2 of

(10) $$\mathrm{P}(\mathbf{q}, -\mathbf{q}) = \mathrm{P}(\mathbf{q}) + \mathscr{K}Q\mathrm{P}(\mathbf{q}) = \mathrm{P}(\mathbf{q}) + A\ \mathrm{P}(\mathbf{q}), \tag{11}$$

where A is antiunitary. The MRs of $\mathrm{P}(\mathbf{q})$ are $\Gamma(\mathbf{q}, R)$, and those of $A\ \mathrm{P}(\mathbf{q})$ are

$$\Gamma_{\alpha\beta}(\kappa\kappa'|\mathbf{q}, AR) = \exp[i\mathbf{q}\cdot(\mathbf{w}(QR) + \mathbf{a}_l)]\mathscr{K}\Gamma_{\alpha\beta}(\kappa\kappa'|\mathbf{q}, (\mathscr{Q}\mathfrak{R})_l), \tag{12}$$

where

$$(\mathscr{Q}\mathfrak{R})_l = (QR|\mathbf{w}(QR) + \mathbf{a}_l). \tag{13}$$

We note the following (Maradudin and Vosko (1968)).

(i) $\Gamma(\mathbf{q}, AR)$ is antiunitary,

(13.1.3) $$\langle\Gamma(\mathbf{q}, AR)\varphi|\Gamma(\mathbf{q}, AR)\psi\rangle = \langle\psi|\varphi\rangle. \tag{14}$$

(ii) Since $\Gamma(\mathbf{q}, R)$ and $\mathscr{K}$ both commute with $\mathrm{D}(\mathbf{q})$, the $\{\Gamma(\mathbf{q}, AR)\}$ commute with $\mathrm{D}(\mathbf{q})$. This is referred to as the time-reversal invariance of the dynamical matrix.

(iii) Multiplying each side of eq. (18.4.11) by $\Gamma(\mathbf{q}, AR)$ and using (ii) gives

(18.4.11) $$\mathrm{D}(\mathbf{q})|\Gamma(\mathbf{q}, AR)|\mathbf{e}(\mathbf{q}\sigma\mu\lambda)\rangle = \omega_{\sigma\mu}(\mathbf{q})^2|\Gamma(\mathbf{q}, AR)|\mathbf{e}(\mathbf{q}\sigma\mu\lambda)\rangle. \tag{15}$$

Equations (15) and (18.4.11) state that $|\mathbf{e}(\mathbf{q}\sigma\mu\lambda)\rangle$ and $|\Gamma(\mathbf{q}, AR)\ |\mathbf{e}(\mathbf{q}\sigma\mu\lambda)\rangle$ are eigenvectors of $\mathrm{D}(\mathbf{q})$ with the same eigenvalue $\omega_{\sigma\mu}(\mathbf{q})^2$. Whether or not this involves any extra degeneracy depends on whether or not these eigenvectors are LI. This question can be answered by applying the Frobenius–Schur test described in Section 13.4 (see also the warning given at the beginning of Chapter 13). This test depends on the value of

$$\mathrm{W}(\mathbf{q}) = p(\mathbf{q})^{-1} \sum_{R \in \mathrm{P}(\mathbf{q})} [A\,;A]\chi^{\sigma}(\mathbf{q}, A^2), \tag{16}$$

(16), (18.3.26) $$[A\,;A] = \exp[-i(\mathbf{q} + A^{-1}\mathbf{q})\cdot\mathbf{w}(A)] \tag{17}$$

(a) If $W(\mathbf{q}) = +1$, $|\mathbf{e}(\mathbf{q}\sigma\mu\lambda)\rangle$ and $\Gamma(\mathbf{q}, AR)|\mathbf{e}(\mathbf{q}\sigma\mu\lambda)\rangle$ are not LI and there is no extra degeneracy.
(b) If $W(\mathbf{q}) = 0$, then two frequencies $\omega_{\sigma\mu}$ and $\omega_{\sigma'\mu'}$ $(\sigma' \neq \sigma)$, which correspond to two *different* irreducible PRs, $\Gamma^{\sigma}(\mathbf{q})$ and $\Gamma^{\sigma'}(\mathbf{q})$, are degenerate through time-reversal symmetry. $\Gamma^{\sigma}(\mathbf{q})$ and $\Gamma^{\sigma'}(\mathbf{q})$ occur in pairs.
(c) $W(\mathbf{q}) = -1$, then $\omega^2_{\sigma\mu}(\mathbf{q}) = \omega^2_{\sigma\mu'}(\mathbf{q})$ $(\mu' \neq \mu)$; that is, two frequencies corresponding to two different occurrences of the same irreducible PR $\Gamma^{\sigma}(\mathbf{q})$ are degenerate. In this case $\Gamma^{\sigma}(\mathbf{q})$ will occur an even number of times.

There are two instances in which the antiunitary operator is just $\mathscr{K}$ rather than $\mathscr{K}Q$, namely at $\mathbf{q} = \mathbf{0}$ and $\mathbf{q} = \frac{1}{2}\mathbf{b}_m$.

$$D(\tfrac{1}{2}\mathbf{b}_m) = D(\tfrac{1}{2}\mathbf{b}_m - \mathbf{b}_m) = D(-\tfrac{1}{2}\mathbf{b}_m) = D^*(\tfrac{1}{2}\mathbf{b}_m); \tag{18}$$

(9), (18)
$$D^*(\tfrac{1}{2}\mathbf{b}_m) = \mathscr{K} D(\tfrac{1}{2}\mathbf{b}_m)\mathscr{K} = D(\tfrac{1}{2}\mathbf{b}_m); \tag{19}$$

(16), (17)
$$W(\tfrac{1}{2}\mathbf{b}_m) = \sum_{R\in P\{\mathbf{q}\}} \exp\left[-\tfrac{1}{2}\mathrm{i}(\mathbf{b}_m + R^{-1}\mathbf{b}_m)\cdot\mathbf{w}(R)\right]\chi^{\sigma}\left(\tfrac{1}{2}\mathbf{b}_m, R^2\right). \tag{20}$$

Similarly, at $\mathbf{q} = \mathbf{0}$,

$$W(\mathbf{0}) = \sum_{R\in P(\mathbf{q})} \chi^{\sigma}(\mathbf{0}, R^2). \tag{21}$$

This rather brief sketch of the consequences of time reversal in lattice dynamics may be amplified by fuller accounts to be found in an article by Maradudin and Vosko (1968) and the books by Lax (1974), Maradudin *et al.* (1971), and Venkataraman *et al.* (1975).

Answer to Exercise 18.5-1

(a) $\mathscr{K}^2\psi = \mathscr{K}\mathscr{K}\psi = E\psi$, $\mathscr{K}^{-1} = \mathscr{K}$.
(b) $\langle\mathscr{K}\psi_1|\mathscr{K}\psi_2\rangle = \langle\psi_1|\psi_2\rangle^*$ (see eq. (13.1.3)).
(c) Using (a), $\mathscr{K}D\mathscr{K}^{-1}\psi = \mathscr{K}D\psi^* = D^*\psi$, therefore $\mathscr{K}D\mathscr{K}^{-1} = D^*$.

18.6 An example: silicon

Silicon, and also germanium, have the diamond structure, which is face-centered cubic. With cube edge a, the fundamental translations are $(a/2)[0\ 1\ 1]$, $(a/2)[1\ 0\ 1]$, $(a/2)[1\ 1\ 0]$. There are two atoms per unit cell with $\kappa_1 = [0\ 0\ 0]$, $\kappa_2 = (a/4)[1\ 1\ 1]$. The space group of diamond (and of silicon and germanium) is 227 ($Fd3m$ or O_h^7), which is non-symmorphic. The reciprocal lattice is body-centered cubic with cube edge $b = 4\pi/a$ and reciprocal lattice vectors $(b/2)[\bar{1}\ 1\ 1]$, $(b/2)[1\ \bar{1}\ 1]$, and $(b/2)[1\ 1\ \bar{1}]$. The free-electron band structure along $[0\ 1\ 0]$ was described in Section 17.6. The nearest neighbors are the pair of atoms κ_1 and κ_2 separated by a distance of $a/4$ along $[1\ 1\ 1]$. (Sometimes an origin displaced to the midpoint between κ_1 and κ_2 may be convenient.) The symmetry group of the nearest-neighbor pair interaction (also called the "group of the bond" (Lax (1974)) is $C_{3v} = \{E\ C_3^{\pm}\ \sigma_{\mathbf{b}}\ \sigma_{\mathbf{d}}\}$. Note that $\sigma_{\mathbf{b}}$ interchanges x and y, so that (in tensor notation) $xy = yx$. The two nearest-neighbors are invariant under the C_3 rotations $R(\pm 2\pi/3\ [1\ 1\ 1])$ which produce the transformations $(xyz) \rightarrow (yzx)$, (zxy). Therefore the three diagonal terms in the matrix representation of the nearest-neighbor force-constant tensor are equal. Again, because of the C_3 axis $xy = yz = zx$ and $yx = zx$. Therefore, the nearest-neighbor force constant matrix is

$$\begin{bmatrix} \alpha & \beta & \beta \\ \beta & \alpha & \beta \\ \beta & \beta & \alpha \end{bmatrix}.$$

For brevity, the eigenvectors $\mathbf{e}(\kappa_1)$ and $\mathbf{e}(\kappa_2)$ will now be denoted by $\mathbf{e}_1$ and $\mathbf{e}_2$. The point group of $Fd3m$ is T_d and the little group

$$G(\mathbf{q}) = \sum_A (A|\mathbf{0})T + \sum_B (B|\mathbf{w})T, \tag{1}$$

in which $\{A\} = P(\mathbf{q}) \cap T_d$ and $B \in P(\mathbf{q})$ but $\notin T_d$. The non-lattice translation $\mathbf{w}$ interchanges κ_1 and κ_2. Therefore the characters of the 6×6 MRs of $(R|\mathbf{v})$ are

$$\chi(A|0) = 2\chi^{(\mathbf{e})}(A), \quad \chi(B|\mathbf{w}) = 0, \tag{2}$$

where $\chi^{(\mathbf{e})}(A)$ is the character of the MR of A for the polar vector basis $|\mathbf{e}\rangle = |\mathbf{e}_x\ \mathbf{e}_y\ \mathbf{e}_z\rangle$. (Compare with Section 9.1; here $N_R = 2$. Because of using a column vector basis $|\mathbf{e}\rangle$ to symmetrize D, this MR $\Gamma^{(\mathbf{e})}$ is the transpose of our usual MR for a polar vector basis $\Gamma^{(\mathbf{r})}$ or R. However, $\chi^{(\mathbf{e})}(A) = \chi^{(\mathbf{r})}(A)$.) Note that $\mathbf{e}_1$ and $\mathbf{e}_2$ each have three Cartesian components $\mathbf{e}_x$, $\mathbf{e}_y$, $\mathbf{e}_z$ so that $|\mathbf{e}_1\ \mathbf{e}_2\rangle$ is a 6×1 column matrix.

The point group of the wave vector $P(\mathbf{q})$ at Γ is O_h. The classes of the factor group $\{(R|\mathbf{v})\}$ at Γ are given in Table 18.1. The characters are not reprinted here since at $\Gamma[0\ 0\ 0]$ they are the characters of the point group O_h. When $(R|\mathbf{v})$ is $(A|\mathbf{0})$, the characters for the basis $|\mathbf{e}_1\ \mathbf{e}_2\rangle$ are those for A with the polar basis $|\mathbf{e}\rangle$, multiplied by two because $N_R = 2$ in the notation of Chapter 9. When $(R|\mathbf{v})$ is $(B|\mathbf{w})$, the characters are zero since $N_R = 0$ and the submatrices $\Gamma^{(\mathbf{e})}(B)$ occupy off-diagonal positions because the translation $\mathbf{w}$ exchanges the two atoms in the unit cell at κ_1 and κ_2. Hence the characters $\chi[(R|\mathbf{v})]$ at Γ are either $2\chi(R, T_{1u})$ or zero and so can be obtained from the character table for the point group O_h. This representation is reducible and, as shown in Table 18.1,

$$\Gamma[(R|\mathbf{v})] = T_{1u} \oplus T_{2g} = \Gamma_{15} \oplus \Gamma'_{25}. \tag{3}$$

The displacements at Γ are linear combinations of $\mathbf{e}_1$ and $\mathbf{e}_2$ obtained by projecting an arbitrary, but suitably chosen, vector $\mathbf{e}(\mathbf{q})$ into the T_{1u} and T_{2g} subspaces using

$$\mathbf{e}(\mathbf{q},\ \sigma) = \sum_R \chi^{\sigma}(\mathbf{q}, R)(R|\mathbf{q})\ \mathbf{e}(\mathbf{q}), \tag{4}$$

$$\chi^{\sigma}(\mathbf{q}, R) = \chi(R)\exp(-i\mathbf{q}\cdot\mathbf{w}) = \chi(R), \quad R \in O_h, \tag{5}$$

Table 18.1. *Classes, and a typical element of each class, for the factor group* F(**q**) = G(**q**)/T *at* Γ [0 0 0].

Since $\varepsilon = \exp(-i\mathbf{q}\cdot\mathbf{w}) = 1$, the characters of $(A \mid \mathbf{0})$ for the basis $|e_1\ e_2\rangle$ are just those of A in P(**q**) = O_h multiplied by two (see the O_h character table in Appendix A3). The characters of $(B|\mathbf{w})$ are zero. Axes are defined in Figure 2.12.

Class	$(E\|\mathbf{0})$	$3(C_4^2\|\mathbf{0})$	$4(C_3^{\pm}\|\mathbf{0})$	$3(C_4^{\pm}\|\mathbf{w})$	$6(C_2\|\mathbf{w})$
Element	$(E\|\mathbf{0})$	$(C_{4y}^2\|\mathbf{0})$	$(C_{31}^{+}\|\mathbf{0})$	$(C_{4y}^{+}\|\mathbf{w})$	$(C_{2\mathbf{d}}\|\mathbf{w})$
$\chi\,[(R\|\,\mathbf{v}),\,\|\,\mathbf{e}_1\ \mathbf{e}_2\rangle]$	6	−2	0	0	0
Class	$(I\|\,\mathbf{w})$	$3(\sigma_h\|\mathbf{w})$	$4(S_6^{\mp}\|\mathbf{w})$	$3(S_4^{\mp}\|\mathbf{0})$	$6(\sigma_d\|\mathbf{0})$
Element	$(I\|\,\mathbf{w})$	$(\sigma_y\|\mathbf{w})$	$(S_{61}^{-}\|\mathbf{w})$	$(S_{4y}^{-}\|\mathbf{0})$	$(\sigma_{\mathbf{d}}\|\mathbf{0})$
$\chi\,[(R\|\,\mathbf{v}),\,\|\,\mathbf{e}_1\ \mathbf{e}_2\rangle]$	0	0	0	−2	2

$\chi[(R|\mathbf{v}), |\mathbf{e}_1\mathbf{e}_2\rangle = \chi(\mathrm{T}_{1u}) = \chi(\mathrm{T}_{2g}) = \chi(\Gamma_{15}) + \chi(\Gamma'_{25})$

since $\mathbf{q} = 0$ at Γ. Choosing $\mathbf{e}(\mathbf{q}) = |\mathbf{e}_1\rangle = |\mathbf{e}_{1x}\ \mathbf{e}_{1y}\ \mathbf{e}_{1z}\rangle$, where $\mathbf{e}_{1x}$ is the component of $\mathbf{e}_1$ along **x**, and making use of the character set for T_{1u} from the character table for O_h,

$$|\mathbf{e}(\Gamma, \mathrm{T}_{1u}))\rangle = \tfrac{1}{2}|\mathbf{e}_1 + \mathbf{e}_2; \mathbf{e}_1 + \mathbf{e}_2\rangle. \tag{6}$$

Similarly, we find

$$|\mathbf{e}(\Gamma, \mathrm{T}_{2g}))\rangle = \tfrac{1}{2}|\mathbf{e}_1 - \mathbf{e}_2; \mathbf{e}_2 - \mathbf{e}_1\rangle. \tag{7}$$

Note that the factor $l(\sigma)p(\mathbf{q})$ is omitted from the projection operator in eq. (4) and that common factors may also be omitted in the penultimate step of the derivations of eqs. (7) and (8) because the final eigenvector is always renormalized.

In eq. (6) the two atoms in the unit cell vibrate in phase as in an *acoustic* mode, while in eq. (7) the two atoms in the unit cell vibrate in antiphase so that it is called an *optic* mode.

18.6.1 Vibrational modes at Δ [0 η 0]

At Δ, the point group of the wave vector P(**q**) is C_{4v}. The character table of the factor group G(**q**)/T = {(*R*|**v**)}, $R \in C_{4v}$, is shown in Table 18.2.

Exercise 18.6-1 Verify the following compatibility relations at Γ [0 0 0].

$$\text{Acoustic modes:}\quad \Gamma_{15} = \Delta_1 \oplus \Delta_5; \tag{8}$$

$$\text{optic modes:}\quad \Gamma'_{25} = \Delta'_2 \oplus \Delta_5. \tag{9}$$

Re-write the relations (8) and (9) in Mulliken notation.

Projecting the general vector

$$|\mathbf{e}_1\ \mathbf{e}_2\rangle = |\mathbf{e}_{1x}\ \mathbf{e}_{1y}\ \mathbf{e}_{1z}; \mathbf{e}_{2x}\ \mathbf{e}_{2y}\ \mathbf{e}_{2z}\rangle \tag{10}$$

Table 18.2. *Character table of the factor group* G($\mathbf{q}$)/T *at* Δ [0 η 0], *together with the corresponding classes of* G($\mathbf{q}$)/T *at* Γ *and the Jones symbols* $R(xyz)$, *where* (xyz) *is an abbreviation for* $(\mathbf{e}_x\ \mathbf{e}_y\ \mathbf{e}_z)$.

The phase factor $\varepsilon = \exp[-i\mathbf{q}\cdot\mathbf{w}] = \exp[-i\pi\eta]$. Without this phase factor, the characters would be those of the point group C_{4v}.

G(Γ)/T G(Δ)/T $R(xyz)$	$(E\|\mathbf{0})$ $(E\|\mathbf{0})$ xyz	$3(C_4^2\|\mathbf{0})$ $(C_{2y}\|\mathbf{0})$ $\bar{x}y\bar{z}$	$3(C_4^\pm\|\mathbf{w})$ $(C_{4y}^\pm\|\mathbf{w})$ $\bar{z}yx$, $zy\bar{x}$	$3(\sigma_h\|\mathbf{w})$ $(\sigma_z\|\mathbf{w})$, $(\sigma_x\|\mathbf{w})$ $xy\bar{z}$, $\bar{x}yz$	$6(\sigma_d\|\mathbf{0})$ $(\sigma_e\|\mathbf{0})$, $(\sigma_f\|\mathbf{0})$ $\bar{z}y\bar{x}$, zyx
A_1 Δ_1	1	1	ε	ε	1
A_2 Δ_1'	1	1	ε	$-\varepsilon$	-1
B_1 Δ_2	1	1	$-\varepsilon$	ε	-1
B_2 Δ_2'	1	1	$-\varepsilon$	$-\varepsilon$	1
E Δ_5	2	-2	0	0	0

into the $\Delta_1 = A_1$ subspace and utilizing the Jones symbols in Table 18.2 yields

$$|0, \mathbf{e}_{1y} + \mathbf{e}_{2y},\ 0;\ 0,\ \mathbf{e}_{1y} + \mathbf{e}_{2y}, 0\rangle. \tag{11}$$

But $\mathbf{e}_{1y}$, $\mathbf{e}_{2y}$ are both unit vectors along $\mathbf{y}$, and so the normalized eigenvector $\mathbf{e}(\mathbf{q}, \sigma)$ (in the usual notation for an eigenvector when the crystal has two atoms per unit cell (see, for example, Lax (1974)) is

(11) LA: $$\mathbf{e}(\Delta, A_1) = |0\ 1\ 0;\ 0\ 1\ 0\rangle. \tag{12}$$

(It is customary not to give the normalization factor explicitly in such expressions.) Equation (12) describes a longitudinal acoustic (LA) mode; it is an acoustic mode because the two atoms in the unit cell vibrate in phase, and it is a longitudinal mode because the direction of the displacements is along the wave vector, that is along $\mathbf{y}$. Similarly, the characters for Δ_2' (or B_2) in C_{4v} require that the displacements of the two atoms in the unit cell be out of phase by π, so that on projecting eq. (10) into the B_2 subspace, we find the longitudinal optic (LO) mode

LO: $$\mathbf{e}(\Delta, B_2) = |0\ 1\ 0;\ 0\ \bar{1}\ 0\rangle. \tag{13}$$

Projecting the vector, eq. (10), into the Δ_5 (or E) subspace gives for $\mathbf{e}(\mathbf{q}\ \sigma\ \mu\ \lambda)$

TA[1 0 1]: $$\mathbf{e}(\Delta, E, 1, 1) = |1\ 0\ 1;\ 1\ 0\ 1\rangle. \tag{14}$$

where TA [1 0 1] signifies a transverse acoustic mode polarized along the [1 0 1] direction. The degenerate Δ_5 mode with the same frequency is an optic mode, since eqs. (8) and (9) tell us that there is one acoustic and one optic mode of Δ_5 symmetry. The eigenvector of this transverse optic (TO) mode is orthogonal to eq. (14) and therefore polarized along [1 0 $\bar{1}$], so that it is

TO[1 0 $\bar{1}$]: $$\mathbf{e}(\Delta, E, 2, 1) = |1\ 0\ \bar{1};\ \bar{1}\ 0\ 1\rangle. \tag{15}$$

Table 18.3 *Projection of the vector* $|1\ 1\ 1;\ 0\ 0\ 0\rangle$ *into the* E_g (X_5) *and* E_u (X_5') *subspaces.*

The projected eigenvectors $\mathbf{e}(\mathbf{q}\sigma\lambda)$ in eqs. (21) and (22) are obtained by addition of these transformed vectors, multiplied by the character for $(R|\mathbf{v})$ (which is the character for R in D_{4h}, multiplied by $\exp(-i\mathbf{q}\cdot\mathbf{v})$). All the classes with non-zero characters have only one member.

Class	Character of $(R\|\mathbf{v})$ in E_g	Transformed vector	Character of $(R\|\mathbf{v})$ in E_u
$(E\|\mathbf{0})$	2	$\|1\,1\,1\ ;\ 0\,0\,0\rangle$	2
$(C_{2y}\|\mathbf{0})$	−2	$\|\bar{1}\,1\,\bar{1}\ ;\ 0\,0\,0\rangle$	−2
$(I\|\mathbf{w})$	$-2i$	$\|0\,0\,0\ ;\ \bar{1}\,\bar{1}\,\bar{1}\rangle$	$2i$
$(\sigma_y\|\mathbf{w})$	$2i$	$\|0\,0\,0\ ;\ 1\,\bar{1}\,1\rangle$	$-2i$

The IR $\sigma = E$ is two-fold degenerate, and since the $\lambda = 1$, $\lambda = 2$ eigenvectors are mutually orthogonal, the remaining two eigenvectors at Δ are

$$\text{TA}[1\ 0\ \bar{1}]: \qquad \mathbf{e}(\Delta, \text{E}, 1, 2) = |1\ 0\ \bar{1};\ 1\ 0\ \bar{1}\rangle, \tag{16}$$

$$\text{TO}[1\ 0\ 1]: \qquad \mathbf{e}(\Delta, \text{E}, 2, 2) = |1\ 0\ 1;\ \bar{1}\ 0\ \bar{1}\rangle. \tag{17}$$

18.6.2 Vibrational modes at X [0 ½ 0]

The point group P(**q**) at X is $D_{4h} = D_4 \otimes C_i$; D_4 is isomorphous with C_{4v}, so the compatibility relations at X (at which $\omega = \exp(-i\pi/2) = -i$) are, in both Mulliken and BSW notation,

$$\Delta_1 = X_1\ (\text{or } A_1),\ \ \Delta_2' = X_1\ (\text{or } A_1), \tag{18}$$

$$\Delta_5 = X_5\ (\text{or } E_g),\ \ \Delta_5 = X_5'\ (\text{or } E_u). \tag{19}$$

Therefore the Δ_1 (LA) and Δ_2' (LO) modes become degenerate at X where they are both labeled by X_1 (Weber (1977)). There is no group-theoretical reason for the TA and TO Δ_5 modes to become degenerate at the BZ surface, and in the older literature (for example, Bilz and Kress (1979)) these modes are often referred to as X_3 and X_4. The space-group of silicon is non-symmorphic and the non-lattice translation **w** in the space-group operators $(B|\mathbf{w}$ has two effects: (i) it interchanges the atoms at the sites κ_1 and κ_2; and (ii) it introduces a phase factor $\exp(-i\mathbf{q}.\mathbf{w})$ into the characters of $\{(B|\mathbf{w})\}$. At X, this factor is $-i$. The results of projecting the vector $|1\ 1\ 1;\ 0\ 0\ 0\rangle$ into the E_g and E_u subspaces are given in Table 18.3. Multiplying each of these projections (for the four classes with non-zero characters) by the character for $(R|\mathbf{v})$ given in Table 18.3 and adding the results gives the $\mathbf{e}(\mathbf{q}\sigma\lambda)$ in eqs. (21) and (22). These show that the mode of symmetry $E_g(X_5)$ is the acoustic mode and that $E_u(X_5')$ is the optic mode,

$$\text{TA} \qquad \mathbf{e}(\text{X}, \text{E}_g, 1) = |1\ 0\ 1;\ \text{i}\ 0\ \text{i}\rangle, \tag{20}$$

TO $$\mathbf{e}(\mathrm{X}, \mathrm{E}_u, 1) = |1\ 0\ 1;\ \bar{\mathrm{i}}\ 0\ \bar{\mathrm{i}}\rangle. \tag{21}$$

(Each of these representations E_g, E_u occurs once only so that the index μ is redundant here.) The degenerate $\lambda = 2$ eigenvectors are orthogonal to these, and therefore are given by

(20) TA $$\mathbf{e}(\mathrm{X}, \mathrm{E}_g, 2) = |1\ 0\ \bar{1};\ \mathrm{i}\ 0\ \bar{\mathrm{i}}\rangle, \tag{22}$$

(21) TO $$\mathbf{e}(\mathrm{X}, \mathrm{E}_u, 2) = |1\ 0\ \bar{1};\ \bar{\mathrm{i}}\ 0\ \mathrm{i}\rangle. \tag{23}$$

Note that phase factors are included in the four eigenvectors in eqs. (20)–(23) so that the relative magnitudes of the displacements in $\mathbf{e}(\mathrm{X}, \mathrm{E}_g, 2)$, for example, are given by

(23) TA $$[1\ 0\ \bar{1}] \quad |1\ 0\ \bar{1};\ 1\ 0\ \bar{1}\rangle. \tag{24}$$

However, eq. (24) does not convey the information that the displacement of atom 2 in the unit cell is $\pi/2$ out of phase with that of atom 1.

I have not described the calculation of the eigenvalues, which requires the solution of the equations of motion and therefore a knowledge of the force constants. The shell model for ionic crystals, introduced by Dick and Overhauser (1958), has proved to be extremely useful in the development of empirical crystal potentials for the calculation of phonon dispersion and other physical properties of perfect and imperfect ionic crystals. There is now a considerable literature in this field, and the following references will provide an introduction: Catlow *et al.* (1977), Gale (1997), Grimes *et al.* (1996), Jackson *et al.* (1995), Sangster and Attwood (1978). The shell model can also be used for polar and covalent crystals and has been applied to silicon and germanium (Cochran (1965)).

Answer to Exercise 18.6-1

At Γ, the phase factors ω in χ are unity, so the characters of the relevant direct sums for the classes of the factor group at Γ that occur also at Δ, are as given in Table 18.4. (See the C_{4v} and O_h character tables in Appendix A3.) Therefore $\Delta_1 + \Delta_5 = \Gamma_{15}$ (or $\mathrm{A}_1 + \mathrm{E} = \mathrm{T}_{1u}$) and $\Delta_2' + \Delta_5 = \Gamma_{25}'$ (or $\mathrm{B}_2 + \mathrm{E} = \mathrm{T}_{2g}$).

Table 18.4. *Corresponding classes of the factor group* G/T *at* Γ *and* Δ *in the BZ of silicon and characters for the direct sums at* Δ *that are compatible with IRs at* Γ.

G(Γ)/T G(Δ)/T	$(E\|\mathbf{0})$ $(E\|\mathbf{0})$	$3(C_4^2\|\mathbf{0})$ $(C_{2y}\|\mathbf{0})$	$3(C_4^{\mp}\|\mathbf{w})$ $(C_{4y}^{\pm}\|\mathbf{w})$	$3(\sigma_h\|\mathbf{w})$ $(\sigma_z\|\mathbf{w}), (\sigma_x\|\mathbf{w})$	$6(\sigma_d\|\mathbf{0})$ $(\sigma_e\|\mathbf{0}), (\sigma_f\|\mathbf{0})$
$\Delta_1 + \Delta_5$	3	−1	1	1	1
$\Delta_2' + \Delta_5$	3	−1	−1	−1	1

Problems

18.1 In some crystals (NaCl is an example) every atom is at a center of symmetry. Show that when this is true

$$\phi(0, \kappa;\, n' - n, \kappa') = \phi(0, \kappa;\, N - N', \kappa').$$

18.2 In the diamond and fluorite structures, for example, the inversion operator interchanges like atoms on different sublattices, so that $K = \kappa'$, $K' = \kappa$, where κ, κ' signify atoms on different sublattices.

(a) Show that for such crystals, $\phi(n\kappa, n'\,\kappa')$ is a symmetric matrix.

(b) Show, when $\mathfrak{R}_l$ is the space-group operator that exchanges the pair of atoms on the same sublattice at $(n\kappa)$, $(n'\kappa)$ for the pair at $(N\kappa)$, $(N'\kappa)$, respectively, where $(n' - n) = -(N' - N) = \Lambda$, that

$$\phi(0, \kappa;\, \Lambda, \kappa) = \phi(\Lambda, \kappa;\, 0, \kappa).$$

18.3 Show that in silicon at Σ, the acoustic modes are of $\Sigma_1 \oplus \Sigma_3 \oplus \Sigma_4$ symmetry and the optic modes are of $\Sigma_1 \oplus \Sigma_3 \oplus \Sigma_4$ symmetry. Write down the character table for the factor group (including phase factors) and deduce the eigenvectors of the modes at Σ. Make a rough sketch of the $\omega(\mathbf{q})$ dispersion curves along $\Gamma\Sigma$X. Discuss the classification into acoustic and optic modes along Σ. [*Hint*: The group-theoretical classification of the modes according to their IRs is a fundamental property imposed by the symmetry of the system, but a classification into LA, TA, LO, and TO modes is not always possible at points other than Γ.]

1 Appendices

A1 Determinants and matrices

A1.1 Determinants

A *determinant* det A or $|A|$ is an $n \times n$ array of elements

$$|A| = |a_{rs}| = \begin{vmatrix} a_{11} & a_{12} & \cdots & \cdots & a_{1n} \\ a_{21} & a_{22} & \cdots & \cdots & a_{2n} \\ \cdots & \cdots & \cdots & \cdots & \cdots \\ \cdots & \cdots & \cdots & \cdots & \cdots \\ a_{n1} & a_{n2} & \cdots & \cdots & a_{nn} \end{vmatrix}, \tag{1}$$

where a_{rs} is the element common to the rth row and sth column. The *complementary minor* of a_{rs}, $M^{rs}(A)$, is the $(n-1) \times (n-1)$ determinant obtained by deleting the rth row and sth column of $|A|$. The *co-factor* of a_{rs}, A^{rs}, is obtained from $M^{rs}(A)$ by attaching the sign $(-1)^{r+s}$, so that

$$A^{rs} = (-1)^{r+s}\, M^{rs}(A). \tag{2}$$

The determinant of A is evaluated by the following rules:

(i) expansion down a column (s constant)

$$|A| = \sum_{r=1}^{n} a_{rs} A^{rs} = \sum_{r=1}^{n} (-1)^{r+s} a_{rs} M^{rs}(A); \tag{3}$$

(ii) expansion across a row (r constant)

$$|A| = \sum_{s=1}^{n} a_{rs} A^{rs} = \sum_{s=1}^{n} (-1)^{r+s} a_{rs} M^{rs}(A), \tag{4}$$

(3) or (4)

$$\partial|A|/\partial a_{rs} = (-1)^{r+s} M^{rs}(A). \tag{5}$$

A1.1.1 Product of determinants

If $|C| = |A||B|$,

$$c_{ik} = \sum_{j=1}^{n} a_{ij}\, b_{jk}. \tag{6}$$

This is the same rule as the "row by column" rule for the multiplication of two matrices.

A1.1.2 Properties of determinants

(a) The value of $|\mathrm{A}|$ is unchanged if rows and columns are transposed. This follows from points (i) and (ii) above.

(b) If two rows or columns of $|\mathrm{A}|$ are interchanged, the sign of $|\mathrm{A}|$ is reversed. This also follows from the expansion rules for evaluating $|\mathrm{A}|$.

Exercise A1.1-1 Given

$$|\mathrm{A}| = \begin{vmatrix} a_1 & b_1 & c_1 \\ a_2 & b_2 & c_2 \\ a_3 & b_3 & c_3 \end{vmatrix},$$

express $|\mathrm{A}|$ as the sum of three 2×2 determinants by expanding across the first row. Form $|\mathrm{A}|'$ by interchanging the first and second rows of $|\mathrm{A}|$ and prove that $|\mathrm{A}|' = -|\mathrm{A}|$. [*Hint*: Expand across the second row of $|\mathrm{A}|'$.]

(c) If two rows or columns of $|\mathrm{A}|$ are identical, then $|\mathrm{A}| = 0$.

(d) If each element of any row (or column) is the sum of two (or more) elements, the determinant may be written as the sum of two (or more) determinants of the same order. This follows from the rules for evaluating $|\mathrm{A}|$.

(e) If a linear combination of any number of rows or (columns) is added to a particular row (or column), that is if a_{ij} is replaced by

$$a'_{ij} = a_{ij} + \sum_{k=1}^{n} \sum_{j=1}^{n} c_k \, a_{kj}, \; j = 1, 2, \ldots, n, \tag{7}$$

then the value of $|\mathrm{A}|$ is unchanged.

(7)

$$|\mathrm{A}'| = \sum_{k=1}^{n} \sum_{j=1}^{n} (a_{ij} + c_k a_{kj}) \mathrm{A}^{ij}$$

$$= \sum_{j=1}^{n} a_{ij} \mathrm{A}^{ij} + \sum_{k=1}^{n} c_k \sum_{j=1}^{k} a_{kj} \mathrm{A}^{ij} = |\mathrm{A}|. \tag{8}$$

The second sum in the second term in eq. (8) is zero because A^{ij} includes the kth row, and so this term is the expansion of a determinant in which two rows, are identical.

Answer to Exercise A1.1-1

$$|\mathrm{A}| = \begin{vmatrix} a_1 & b_1 & c_1 \\ a_2 & b_2 & c_2 \\ a_3 & b_3 & c_3 \end{vmatrix} = a_1 \begin{vmatrix} b_2 & c_2 \\ b_3 & c_3 \end{vmatrix} - b_1 \begin{vmatrix} a_2 & c_2 \\ a_3 & c_3 \end{vmatrix} + c_1 \begin{vmatrix} a_2 & b_2 \\ a_3 & b_3 \end{vmatrix};$$

$$|A|' = \begin{vmatrix} a_2 & b_2 & c_2 \\ a_1 & b_1 & c_1 \\ a_3 & b_3 & c_3 \end{vmatrix} = -a_1 \begin{vmatrix} b_2 & c_2 \\ b_3 & c_3 \end{vmatrix} + b_1 \begin{vmatrix} a_2 & c_2 \\ a_3 & c_3 \end{vmatrix} - c_1 \begin{vmatrix} a_2 & b_2 \\ a_3 & b_3 \end{vmatrix} = -|A|.$$

A1.2 Definitions and properties of matrices

A matrix $\mathbb{A}$ is a rectangular array of elements

$$\mathbb{A} = [a_{rs}] = \begin{bmatrix} a_{11} & a_{12} & \cdots & \cdots & a_{1n} \\ a_{21} & a_{22} & \cdots & \cdots & a_{2n} \\ \cdots & \cdots & \cdots & \cdots & \cdots \\ \cdots & \cdots & \cdots & \cdots & \cdots \\ a_{m1} & a_{m2} & \cdots & \cdots & a_{mn} \end{bmatrix}. \quad (1)$$

$(\mathbb{A})_{rs}$, usually written a_{rs} or $\mathbb{A}_{rs}$, is a typical element of $\mathbb{A}$, and the subscripts indicate that it is the element common to the rth row and sth column. In this matrix there are m rows ($r = 1, 2, \ldots, m$) and n columns ($s = 1, 2, \ldots, n$). If $m = n$, $\mathbb{A}$ is a *square matrix*. The set of elements with $r = s$ in a square matrix are the *diagonal elements*. A square matrix with $a_{rs} = \delta_{rs}$ is a unit matrix, $\mathbb{E}$. The dimensions of $\mathbb{E}$ may usually be understood from the context, but when necessary the unit matrix of dimensions $n \times n$ will be denoted by $\mathbb{E}_n$.

If $m = 1$, $\mathbb{A}$ is a *row matrix*

$$\langle a| = \langle a_1, \ a_2, \ \ldots, \ a_n|. \quad (2)$$

If $n = 1$, $\mathbb{A}$ is the *column matrix*

$$|a\rangle = |a_1, \ a_2, \ \ldots, \ a_m\rangle. \quad (3)$$

For example, the components of a vector **r** in configuration space may be represented by the column matrix $|x \ y \ z\rangle$.

A1.2.1 Rules of matrix algebra

(1) $\mathbb{A} = \mathbb{B}$ if $a_{rs} = b_{rs}$, $\forall r, s$.
(2) If $\mathbb{C} = \mathbb{A} + \mathbb{B}$, then $c_{rs} = a_{rs} + b_{rs}$, $\forall \ r, s$. It follows that $k \ \mathbb{A} = [k \ a_{rs}]$. Clearly, in rules (1) and (2) the matrices $\mathbb{A}$ and $\mathbb{B}$ must contain the same numbers of rows and columns.
(3) $\mathbb{A}$ and $\mathbb{B}$ are conformable for the product $\mathbb{AB}$ if the number of columns in $\mathbb{A}$ is equal to the number of rows in $\mathbb{B}$. This product is then $\mathbb{C} = [c_{ik}]$, where

$$c_{ik} = \sum_j a_{ij} b_{jk}. \quad (4)$$

This is the "row × column" law of *matrix multiplication*.

Example A1.2-1 If the row matrix $\langle \mathbf{e}|$ contains the 3-D configuration space basis vectors $\{\mathbf{e}_1 \ \mathbf{e}_2 \ \mathbf{e}_3\}$ and the column matrix $|r\rangle$ contains the components $|x \ y \ z\rangle$ of a vector **r**, then the *matrix representative* (MR) of the vector **r** is

$$\langle \mathbf{e}|r\rangle = \langle \mathbf{e}_1\ \mathbf{e}_2\ \mathbf{e}_3|x\ y\ z\rangle \qquad (4)$$
$$= \mathbf{e}_1 x + \mathbf{e}_2 y + \mathbf{e}_3 z. \qquad (5)$$

(4) Matrix multiplication is *associative*,

$$\mathrm{A(BC) = (AB)C}. \qquad (6)$$

(5) Matrix multiplication is not necessarily *commutative*, but if AB = BA the matrices A and B are said to commute.

(6) Division by a matrix A is defined as multiplication by the *inverse* of A, written A^{-1}, and defined by the relation

$$\mathrm{A}^{-1}\mathrm{A} = \mathrm{AA}^{-1} = \mathrm{E}. \qquad (7)$$

Because the multiplication of determinants and of matrices obey the same "row × column" rule

$$|\mathrm{AB}| = |\mathrm{A}||\mathrm{B}|, \qquad (8)$$

where $|\mathrm{A}|$ signifies the determinant of the matrix A,

$$(8), (7) \qquad |\mathrm{A}^{-1}\mathrm{A}| = |\mathrm{A}^{-1}||\mathrm{A}| = |\mathrm{E}| = 1. \qquad (9)$$

It follows from eq. (9) that A has an inverse only if

$$|\mathrm{A}| = |a_{rs}| \neq 0. \qquad (10)$$

If a matrix A has a determinant $|\mathrm{A}| = 0$, it is said to be *singular*. Consequently, A has an inverse A^{-1} only if eq. (10) is satisfied and A is *non-singular*. The *trace* of a square matrix A is the sum of the diagonal elements of A,

$$\mathrm{Tr\ A} = \sum_j a_{jj}, \qquad (11)$$

$$(11), (4) \qquad \mathrm{Tr\ AB} = \sum_i \sum_j a_{ij} b_{ji} = \sum_j \sum_i b_{ji} a_{ij} = \mathrm{Tr\ BA}. \qquad (12)$$

If Q is a non-singular matrix and $\mathrm{B} = \mathrm{QAQ}^{-1}$, B and A are said to be related by a *similarity transformation*. The trace of a matrix is invariant under a similarity transformation, for

$$(12) \qquad \mathrm{Tr\ QAQ}^{-1} = \mathrm{Tr\ Q(AQ}^{-1}) = \mathrm{Tr\ (AQ}^{-1})\mathrm{Q} = \mathrm{Tr\ A}. \qquad (13)$$

The determinant of A is also invariant under a similarity transformation, for

$$(8), (9) \qquad |\mathrm{QAQ}^{-1}| = |\mathrm{Q}||\mathrm{A}||\mathrm{Q}^{-1}| = |\mathrm{Q}||\mathrm{A}||\mathrm{Q}|^{-1} = |\mathrm{A}|. \qquad (14)$$

The transpose A^{T} of a matrix $\mathrm{A} = [a_{ij}]$ is obtained by interchanging rows and columns of A, so that

$$\mathrm{A}^{\mathrm{T}} = [a_{ji}]. \qquad (15)$$

If $A^T = A$, the matrix A is *symmetric*; if $A^T = -A$, A is *skew-symmetric*.

$$(AB)^T_{ik} = (AB)_{ki} = \sum_j a_{kj} b_{ji} = \sum_j b^T_{ij} a^T_{jk} = (B^T A^T)_{ik}, \ \forall i, k,$$

so that

$$(AB)^T = B^T A^T. \tag{16}$$

The *complex conjugate* (CC) matrix of $A = [a_{ij}]$ is $A^* = [a_{ij}^*]$. The *adjoint* matrix of A is $A^\dagger = (A^T)^*$. If $A^\dagger = A$, A is *Hermitian*, but if $A^\dagger = -A$, then A is *skew-Hermitian*.

Exercise A1.2-1 Show that $(AB)^\dagger = B^\dagger A^\dagger$.

Define the matrix $\bar{A}$ as the transpose of the matrix of co-factors of $|A|$ so that $\bar{A}_{ij} = (A^{ij})^T$, where A^{ij} is the co-factor of a_{ij} in $|A|$. From the expansion property of $|A|$,

(A1.1.4) $$\sum_{s=1}^{n} a_{is} A^{js} = |A| \delta_{ij}, \tag{17}$$

(4), (17) $$A\bar{A} = B = [b_{ij}], \ b_{ij} = \sum_s a_{is} (A^{sj})^T = \sum_s a_{is} A^{js} = |A| \delta_{ij}, \tag{18}$$

(18) $$A\bar{A} = B = |A| E, \tag{19}$$

(19) $$A^{-1} = \bar{A}/|A|. \tag{20}$$

Clearly, the inverse A^{-1} of A may only be evaluated if A is non-singular.

Exercise A1.2–2 Show that $(AB)^{-1} = B^{-1} A^{-1}$. [*Hint*: Use the definition of the inverse of AB, namely that $(AB)^{-1}$ is the matrix which on multiplying AB gives the unit matrix.]

If $A^{-1} = A^T$, A is an *orthogonal* matrix.

Exercise A1.2–3 Show that the product of two orthogonal matrices A and B is an orthogonal matrix.

Exercise A1.2–4 Show that if A is an orthogonal matrix, then

$$\sum_k a_{ij} \ a_{jk} = \delta_{ij}, \tag{21}$$

$$\sum_k a_{ki} \ a_{kj} = \delta_{ij}. \tag{22}$$

[*Hint*: Make use of the definition of the inverse matrix, eq. (7), and the property of the transposed matrix.]

Equations (21) and (22) state that the rows or columns of an orthogonal matrix are orthonormal. A is a *unitary* matrix if

Table A1.1. *Names, symbols, and defining relations for various special matrices.*

E is the $n \times n$ unit matrix and a_{ij} is the ijth element of A.

Name of matrix	Notation used	Definition
Diagonal	D	$a_{ij} = a_i\ \delta_{ij}$
Inverse	A^{-1}	$A^{-1}\ A = E$
Transpose	A^T	$(A^T)_{ij} = a_{ji}$
Symmetric		$A^T = A$
Skew-symmetric		$A^T = -A$
Complex conjugate	A^*	$(A^*)_{ij} = a^*_{ij}$
Adjoint	$A^\dagger$	$A^\dagger = (A^T)^*$
Orthogonal		$A^T A = E$
Unitary		$A^\dagger A = E$
Hermitian		$A^\dagger = A$
Skew-Hermitian		$A^\dagger = -A$
Normal		$AA^\dagger = A^\dagger A$
Permutation, Pseudo-permutation		*see text*

$$A^{-1} = A^\dagger = (A^T)^*. \tag{23}$$

If A is unitary,

(23) $$|A^\dagger A| = |A^T|^*|A| = |A|^*|A| = 1, \tag{24}$$

so that the determinant of a unitary matrix is a complex number of modulus unity.

Exercise A1.2–5 Prove that, if A is a unitary matrix,

$$\sum_k a_{ik}\ a^*_{jk} = \delta_{ij}, \tag{25}$$

$$\sum_k a^*_{ki}\ a_{kj} = \delta_{ij}. \tag{26}$$

Equations (25) and (26) show that the rows or the columns of a unitary matrix are orthonormal when the scalar product is defined to be the Hermitian scalar product.

A *permutation* (*pseudo-permutation*) matrix is one in which every element in each row and column is equal to zero, except for one element which is +1 (−1 or +1). For convenience of reference the defining relations for special matrices are summarized in Table A1.1.

Exercise A1.2–6 Show that

$$\begin{bmatrix} 0 & 0 & 1 \\ 1 & 0 & 0 \\ 0 & 1 & 0 \end{bmatrix} \begin{bmatrix} x_1 \\ x_2 \\ x_3 \end{bmatrix}$$

produces a cyclic permutation of the elements of the column matrix $|x_1\ x_2\ x_3\rangle$. Construct the permutation matrix that on multiplying $|x_1\ x_2\ x_3\rangle$ produces the permutation $|x_2\ x_3\ x_1\rangle$.

A *normal* matrix is one that commutes with its adjoint, $\mathrm{AA}^{\dagger} = \mathrm{A}^{\dagger}\mathrm{A}$. Normal matrices include diagonal, real symmetric, orthogonal, unitary, Hermitian (self-adjoint), permutation, and pseudo-permutation matrices.

Answers to Exercises A1.2

Exercise A1.2-1 $(\mathrm{AB})^{\dagger} = ((\mathrm{AB})^{\mathrm{T}})^{*} = (\mathrm{B}^{\mathrm{T}}\mathrm{A}^{\mathrm{T}})^{*} = \mathrm{B}^{\dagger}\mathrm{A}^{\dagger}$.

Exercise A1.2-2 $\mathrm{B}^{-1}\mathrm{A}^{-1}\,\mathrm{AB} = \mathrm{E}$; therefore $(\mathrm{AB})^{-1} = \mathrm{B}^{-1}\mathrm{A}^{-1}$.

Exercise A1.2-3 Since A, B are orthogonal, $(\mathrm{AB})^{-1} = \mathrm{B}^{-1}\mathrm{A}^{-1} = \mathrm{B}^{\mathrm{T}}\mathrm{A}^{\mathrm{T}} = (\mathrm{AB})^{\mathrm{T}}$.

Exercise A1.2-4 If A is an orthogonal matrix, $\mathrm{AA}^{-1} = \mathrm{AA}^{\mathrm{T}} = \mathrm{E}$. Therefore $\sum\limits_k a_{ik}\ a^{\mathrm{T}}_{kj} = \sum\limits_k a_{ik}\ a_{jk} = \delta_{ij}$, which proves that the rows of an orthogonal matrix A are orthonormal. Again, $\mathrm{A}^{\mathrm{T}}\mathrm{A} = \mathrm{E}$, and $\sum\limits_k a^{\mathrm{T}}_{ik}\ a_{kj} = \sum\limits_k a_{ki}\ a_{kj} = \delta_{ij}$, showing that the columns of an orthogonal matrix are orthonormal.

Exercise A1.2-5 If A is unitary, $\mathrm{AA}^{\dagger} = \mathrm{A}(\mathrm{A}^{\mathrm{T}})^{*} = \mathrm{E}$, $\sum\limits_k a_{ik}\ a^{*}_{jk} = \delta_{ij}$, and the two vectors (in a unitary linear vector space) whose components are the elements in two rows of a unitary matrix, are orthonormal. Similarly, $\mathrm{A}^{\dagger}\mathrm{A} = (\mathrm{A}^{\mathrm{T}})^{*}\mathrm{A} = \mathrm{E}$, $\sum\limits_k a^{*}_{ki}\ a_{kj} = \delta_{ij}$, and vectors whose components are the elements in two columns of a unitary matrix are orthonormal. These results are important in applications of group theory where symmetry operators are represented by unitary matrices. The reader has no doubt noted that a real unitary matrix is an orthogonal matrix.

Exercise A1.2-6

$$\begin{bmatrix} 0 & 0 & 1 \\ 1 & 0 & 0 \\ 0 & 1 & 0 \end{bmatrix} \begin{bmatrix} x_1 \\ x_2 \\ x_3 \end{bmatrix} = \begin{bmatrix} x_3 \\ x_1 \\ x_2 \end{bmatrix};$$

$|x_3\ x_1\ x_2\rangle$ is a cyclic permutation of $|x_1\ x_2\ x_3\rangle$. Similarly,

$$\begin{bmatrix} 0 & 1 & 0 \\ 0 & 0 & 1 \\ 1 & 0 & 0 \end{bmatrix} \begin{bmatrix} x_1 \\ x_2 \\ x_3 \end{bmatrix} = \begin{bmatrix} x_2 \\ x_3 \\ x_1 \end{bmatrix}.$$

A1.3 Eigenvalues and eigenvectors; diagonalization

Generally, when a square matrix A multiplies a column matrix $|x\rangle$ from the left, it changes $|x\rangle$ into a new column matrix $|x'\rangle$. In the particular case that $|x'\rangle$ is just $|x\rangle$ multiplied by a constant a,

$$\mathrm{A}|x\rangle = a|x\rangle, \tag{1}$$

$|x\rangle$ is an *eigenvector* of A and a is the corresponding *eigenvalue*. The set of homogeneous linear equations (1) has non-trivial solutions only if

$$|\mathrm{A} - a\mathrm{E}| = 0. \tag{2}$$

Equation (2) is called the *characteristic* (or *secular*) equation of A, and its roots are the eigenvalues of A, $\{a_k\}$. The problem of finding the eigenvalues of a matrix is intimately connected with its conversion to diagonal form. For if A were a diagonal matrix, then its characteristic equation would be

(2) $$\prod_{i=1}^{n} (a_{ij} - a_j)\delta_{ij} = 0, \tag{3}$$

and its eigenvalues a_i would be given by the diagonal elements, a_{ii}, $i = 1, \ldots, n$. Two matrices A_1 and A_2 of the same dimensions are *equivalent*, $\mathrm{A}_1 \approx \mathrm{A}_2$, if they are related by a *similarity transformation*, that is there exists a non-singular matrix S such that

$$\mathrm{A}_2 = \mathrm{S}\ \mathrm{A}_1\ \mathrm{S}^{-1}. \tag{4}$$

A matrix is *diagonalizable* if it is equivalent to a diagonal matrix D. The characteristic equation of A is invariant under a similarity transformation, for

$$|\mathrm{S\ A\ S}^{-1} - a\mathrm{E}| = |\mathrm{S}(\mathrm{A} - a\mathrm{E})\mathrm{S}^{-1}| = |\mathrm{S}|\ |(\mathrm{A} - a\mathrm{E})|\ |\mathrm{S}^{-1}| = |(\mathrm{A} - a\mathrm{E})|, \tag{5}$$

and this means that the eigenvalues of a matrix are invariant under a similarity transformation. A matrix A can be diagonalized by the unitary matrix S,

$$\mathrm{S\ A\ S}^{-1} = \mathrm{D}, \tag{6}$$

iff (meaning if and only if) A is a *normal* matrix.

Proof

(6) $$\mathrm{A\ S}^{-1} = \mathrm{S}^{-1}\mathrm{D}; \tag{7}$$

(7) $$(\mathrm{S}^{-1})^{\dagger}\mathrm{A}^{\dagger} = \mathrm{D}^{*}(\mathrm{S}^{-1})^{\dagger}; \tag{8}$$

(8) $$\mathrm{S\ A}^{\dagger} = \mathrm{D}^{*}\mathrm{S} \quad (\mathrm{S}\ \text{unitary}); \tag{9}$$

(9), (7) $$\mathrm{S\ A}^{\dagger}\ \mathrm{A\ S}^{-1} = \mathrm{D}^{*}\mathrm{S\ S}^{-1}\mathrm{D} = \mathrm{D}^{*}\mathrm{D}; \tag{10}$$

(9), (7) $$\mathrm{A\ S}^{-1}\mathrm{S\ A}^{\dagger} = \mathrm{S}^{-1}\mathrm{D\ D}^{*}\mathrm{S} = \mathrm{AA}^{\dagger}; \tag{11}$$

(11) $$\mathrm{S\ A\ A}^{\dagger}\ \mathrm{S}^{-1} = \mathrm{D\ D}^{*} = \mathrm{D}^{*}\ \mathrm{D}; \tag{12}$$

(12), (10) $$\mathrm{A\ A}^{\dagger} = \mathrm{A}^{\dagger}\ \mathrm{A}, \tag{13}$$

which proves that $\mathbb{A}$ can be diagonalized by a similarity transformation with a unitary matrix $\mathbb{S}$, iff $\mathbb{A}$ is a normal matrix.

In quantum mechanics special importance attaches to Hermitian matrices, which have real eigenvalues. Multiply each side of eq. (1) from the left by the adjoint of $|x\rangle$, $\langle x^*|$

$$\langle x^*|\mathbb{A}|x\rangle = a\langle x^*|x\rangle. \tag{14}$$

Take the adjoint of each side of eq. (3), using the result in Exercise A1.2-1 that the adjoint of a product is the product of the adjoints, in reverse order.

(14) $$\langle x^*|\mathbb{A}^\dagger|x\rangle = a^*\langle x^*|x\rangle, \tag{15}$$

since the adjoint of the complex number a is a^*. But if $\mathbb{A}$ is Hermitian or skew-Hermitian, $\mathbb{A}^\dagger = \pm\mathbb{A}$, and so

(14), (15) $$a = \pm a^*. \tag{16}$$

Therefore, the eigenvalues of a Hermitian matrix are real, and the eigenvalues of a skew-Hermitian matrix are pure imaginary. Now consider the eigenvectors $|x\rangle$ and $|x'\rangle$ belonging to two different eigenvalues a, a' of a self-adjoint matrix $\mathbb{A}$.

$$\mathbb{A}|x\rangle = a|x\rangle,\ \mathbb{A}|x'\rangle = a'|x'\rangle; \tag{17}$$

(17) $$\langle x'^*|\mathbb{A}|x\rangle = a\langle x'^*|x\rangle; \tag{18}$$

(17) $$\langle x^*|\mathbb{A}|x'\rangle = a'\langle x^*|x'\rangle; \tag{19}$$

(10) $$\langle x^*|\mathbb{A}^\dagger|x'\rangle = a^*\langle x^*|x'\rangle; \tag{20}$$

(12), (8) $$\langle x^*|\mathbb{A}|x'\rangle = a\langle x^*|x'\rangle \quad (\mathbb{A}\text{ Hermitian}); \tag{21}$$

(19), (21) $$(a - a')\langle x^*|x'\rangle = 0. \tag{22}$$

Thus, if a and a' are distinct eigenvalues, their eigenvectors $|x\rangle$, $|x'\rangle$ are orthogonal, and since they may always be normalized

$$\langle x^*|x'\rangle = \delta_{x,x'} \quad \text{or} \quad \sum_j x_j^* x_j' = \delta_{x,x'}. \tag{23}$$

If the eigenvalues are degenerate, $a = a'$, then the eigenvectors $|x\rangle$, $|x'\rangle$ are not necessarily orthogonal, but a mutually orthogonal set may always be found (Schmidt orthogonalization). Consequently, a Hermitian matrix of order n has n mutually orthogonal and normalizable eigenvectors $|x\rangle$. The same result holds for skew-Hermitian matrices.

For orthogonal matrices,

$$\mathbb{A}|x\rangle = a|x\rangle; \tag{24}$$

(24) $$\langle x|\mathbb{A}^{\mathrm{T}} = a\langle x|; \tag{25}$$

(24), (25) $$\langle x|\mathbb{A}^{\mathrm{T}}\mathbb{A}|x\rangle = a^2\langle x|x\rangle = \langle x|x\rangle \quad (\mathbb{A} \text{ orthogonal}). \tag{26}$$

Equation (26) implies that the *real eigenvectors* of an orthogonal matrix must correspond to eigenvalues with $a^2 = 1$. For real orthogonal matrices, $\mathbb{A} = \mathbb{A}^*$,

(24) $$\mathbb{A}^*|x^*\rangle = a^*|x^*\rangle, \;\; \mathbb{A}|x^*\rangle = a^*|x^*\rangle. \tag{27}$$

Equations (24) and (27) show that the eigenvalues and eigenvectors of real orthogonal matrices occur in CC pairs. Furthermore, if $|x\rangle$ is real, eqs. (24) and (27) imply that $a = a^*$, that is that the eigenvalues that correspond to real eigenvectors of real, orthogonal matrices are also real. In fact, because $a^2 = 1$,

$$a = \pm 1 \quad \text{(for real eigenvectors of real orthogonal matrices)}. \tag{28}$$

The MRs of proper and improper rotations in 3-D configuration space $\Re^3$ are 3×3 real orthogonal matrices (see eqs. (3.2.11) and (3.2.12)). There are, therefore, three eigenvalues and therefore only two possibilities: either there is one real eigenvector and one CC pair, or there are three real eigenvectors. If there is one real eigenvector then the eigenvalues are ± 1, ω, and ω^*.

(3.2.11), (3.2.12) $$|\mathbb{A}| = \pm 1, \tag{29}$$

where $|\mathbb{A}| = +1$ corresponds to a *proper rotation* $R(\phi\ \mathbf{n})$ and $|\mathbb{A}| = -1$ corresponds to an *improper rotation* $S(\phi\ \mathbf{n})$. *Binary rotations* $\phi = \pi$ are excluded since, in this case, the eigenvalues of $\mathbb{A}$ are all real. The four possible cases that can arise when a 3×3 real orthogonal matrix has three real eigenvalues are summarized in Table A1.2.

For a unitary matrix $\mathbb{A}$,

$$\mathbb{A}|x\rangle = a|x\rangle; \tag{30}$$

(30) $$\langle x^*|\mathbb{A}^\dagger = a^*\langle x^*|; \tag{31}$$

(31), (30) $$\langle x^*|\mathbb{A}^\dagger\mathbb{A}|x\rangle = aa^*\langle x^*|x\rangle = aa^*\sum_k |x_k|^2; \tag{32}$$

(32) $$aa^* = 1, \;\; |a| = 1 \;\; (\mathbb{A} \text{ unitary}), \tag{33}$$

Table A1.2. *Real eigenvalues of a real* 3×3 *orthogonal matrix* $\mathbb{A}$ *with real eigenvectors.*

Symmetry operation	Eigenvalues			Reference to MR	$\lvert\mathbb{A}\rvert$
	a_1	a_2	a_3		
Identity	1	1	1	Example 3.2–1	+1
Inversion	−1	−1	−1	Equation (2.1.4)	−1
Binary rotation	1	−1	−1	Equation (3.2.11)	+1
Reflection	1	1	−1	Equation (3.2.15)	−1

since eq. (32) reminds us that $\langle x^* | x \rangle$ is positive definite. The eigenvalues of a unitary matrix are therefore complex numbers of modulus unity. If $\mathbb{A}$ is real, a real unitary matrix becomes a real orthogonal matrix, and $\mathbb{A}^T\mathbb{A} = \mathrm{E}$ (see eq. (3.2.14) *et seq.*).

In the general case, if $\mathbb{A}$ is not a normal matrix, then it is not necessarily diagonalizable. However, it is diagonalizable if the characteristic equation has n distinct roots.

(2) $$|\mathbb{A} - a\mathrm{E}| = \sum_{\nu=0}^{n}(-1)^{\nu}\sigma_{\nu}a^{\nu} = 0, \tag{34}$$

where σ_ν is the sum of all the $(n-\nu)$-rowed principal minors of A. In particular,

$$\sigma_0 = |\mathbb{A}|,\ \sigma_{n-1} = \mathrm{Tr}\ \mathbb{A},\ \sigma_n = 1. \tag{35}$$

(A $(n-\nu)$-rowed principal minor of a matrix $\mathbb{A}$ of order n is the determinant of the matrix formed by removing from $\mathbb{A}$ any ν rows and the corresponding ν columns with the same indices.) Since $|\mathbb{A} - a\mathrm{E}|$ and the eigenvalues of A are invariant under a similarity transformation, the σ_ν are similarly invariant, as we already know for Tr $\mathbb{A}$ and $|\mathbb{A}|$, eqs. (A1.2.13) and (A1.2.14). The n roots of eq. (34) are the n eigenvalues of A. Substituting one of these eigenvalues (say a_k) in the homogeneous linear equations (1) gives the eigenvector $|x_k\rangle$. This contains an arbitrary constant which can be removed by normalization. Repeat for all the other eigenvalues in turn, and then construct the square matrix X, the columns of which are the eigenvectors $|x_k\rangle$:

$$\mathrm{X} = \langle|x_1\rangle\ |x_2\rangle \ldots |x_k\rangle \ldots |x_n\rangle|. \tag{36}$$

(36), (1) $$\mathbb{A}\ \mathrm{X} = \langle a_1|x_1\rangle\ a_2|x_2\rangle \ldots a_k|x_k\rangle \ldots a_n|x_n\rangle| = \mathrm{X}[a_i\delta_{ik}]; \tag{37}$$

(37) $$\mathrm{X}^{-1}\ \mathbb{A}\ \mathrm{X} = [a_i\delta_{ik}]. \tag{38}$$

This reduction of $\mathbb{A}$ to diagonal form is unique, except for the order in which the eigenvalues occur on the diagonal. When the roots are not all distinct, it may not be possible to convert $\mathbb{A}$ to diagonal form. However, $\mathbb{A}$ may then be reduced to Jordan canonical form in which the eigenvalues occur on the diagonal, with the position immediately above each eigenvalue occupied by unity or zero and with a zero everywhere else.

If a number of normal matrices $\mathbb{A}_1, \mathbb{A}_2, \ldots$ commute with one another then they can all be diagonalized by a similarity transformation with the same unitary matrix S. If $\mathbb{A}_1, \mathbb{A}_2$ are both diagonalized by S, then

$$\mathrm{S}\ \mathbb{A}_1\ \mathrm{S}^{-1} = \mathrm{D}_1,\ \mathrm{S}\ \mathbb{A}_2\mathrm{S}^{-1} = \mathrm{D}_2; \tag{39}$$

(39) $$\mathrm{S}\ \mathbb{A}_1\ \mathbb{A}_2\ \mathrm{S}^{-1} = \mathrm{S}\ \mathbb{A}_1\ \mathrm{S}^{-1}\ \mathrm{S}\ \mathbb{A}_2\ \mathrm{S}^{-1} = \mathrm{D}_1\ \mathrm{D}_2; \tag{40}$$

(39) $$\mathrm{S}\ \mathbb{A}_2\ \mathbb{A}_1\ \mathrm{S}^{-1} = \mathrm{D}_2\ \mathrm{D}_1 = \mathrm{D}_1\ \mathrm{D}_2; \tag{41}$$

(40), (41) $$\mathbb{A}_1\ \mathbb{A}_2 = \mathbb{A}_2\ \mathbb{A}_1, \tag{42}$$

showing that the condition that A_1, A_2 commute is necessary. It is also sufficient, for if A_1 is diagonalized by S and A_1, A_2 commute,

$$\begin{aligned} \mathrm{S}\ \mathrm{A}_1\ \mathrm{A}_2\ \mathrm{S}^{-1} &= \mathrm{D}_1\ \mathrm{S}\ \mathrm{A}_2\ \mathrm{S}^{-1} = \mathrm{D}_1\ \mathrm{B}, \text{ say}, \\ \mathrm{S}\ \mathrm{A}_2\ \mathrm{A}_1\ \mathrm{S}^{-1} &= \mathrm{S}\ \mathrm{A}_2\ \mathrm{S}^{-1}\ \mathrm{D}_1 = \mathrm{B}\ \mathrm{D}_1. \end{aligned} \tag{43}$$

$\mathrm{S}\ \mathrm{A}_2\ \mathrm{S}^{-1} = \mathrm{B}$ is a matrix that commutes with a diagonal matrix D_2 and is therefore also diagonal. For if $\mathrm{BD} = \mathrm{DB}$, with D diagonal,

$$(\mathrm{DB})_{ik} = \sum_j d_i\ \delta_{ij}\ b_{jk} = d_i\ b_{ik} = (\mathrm{BD})_{ik} = \sum_j b_{ij}\ d_j\ \delta_{jk} = b_{ik}\ d_k. \tag{44}$$

So $\mathrm{BD} = \mathrm{DB}$ requires that b_{ik} be zero unless $i = k$ and therefore $\mathrm{B} = \mathrm{S}\ \mathrm{A}_2\ \mathrm{S}^{-1}$ *is* a diagonal matrix, say D_2.

A1.4 Matrix representations

The *direct sum* of two matrices A_1, A_2 is the block-diagonal matrix

$$\mathrm{A} = \mathrm{A}_1 \oplus \mathrm{A}_2 = \begin{bmatrix} \mathrm{A}_1 & \\ & \mathrm{A}_2 \end{bmatrix}. \tag{1}$$

A matrix A is *reducible* if it is equivalent to a direct sum of two or more matrices. A matrix system of order g is a set of g matrices

$$\{\mathrm{A}\} = \{\mathrm{A}_1\ \mathrm{A}_2\ \ldots\ \mathrm{A}_g\}. \tag{2}$$

Two matrix systems of the same order g are equivalent if there exists a non-singular matrix S such that

$$\mathrm{A}_i' = \mathrm{S}\mathrm{A}_i\mathrm{S}^{-1}, \quad i = 1, 2 \ldots, g. \tag{3}$$

The matrix system $\{\mathrm{A}\} = \{\mathrm{A}_1\ \mathrm{A}_2\ \ldots\ \mathrm{A}_g\}$ is *reducible* if it is equivalent to a direct sum of matrix systems,

$$\{\mathrm{A}\} = \{\mathrm{A}^1\} \oplus \{\mathrm{A}^2\} \oplus \ldots \tag{4}$$

If each of the blocks in the matrices comprising the matrix system $\{\mathrm{A}\}$ cannot be reduced further, the matrix system has been reduced completely and each of the matrix systems $\{\mathrm{A}^1\}$, $\{\mathrm{A}^2\}$, ... in the direct sum is said to be *irreducible*. Matrix systems that are isomorphous to a group G are called *matrix representations* (Chapter 4). *Irreducible representations* (IRs) are of great importance in applications of group theory in physics and chemistry. A matrix representation in which the matrices are unitary matrices is called a *unitary representation*. Matrix representations are not necessarily unitary, but any representation of a finite group that consists of non-singular matrices is equivalent to a unitary representation, as will be demonstrated in Section A1.5.

A1.5 Schur's lemma and the orthogonality theorem

This section contains the proofs of the orthogonality theorem (OT) and three other results that are required in the proof of the OT (Wigner (1959)). They need not be studied in detail by readers willing to accept the orthogonality relations embodied in the OT, eq. (29). The implications and applications of these orthogonality relations are discussed in the text, beginning at Section 4.3. The proofs are for the most part standard ones and follow arguments presented in the classic book on group theory by Wigner (1959).

Lemma 1 Any matrix representation consisting of non-singular matrices is equivalent to a unitary representation.

Proof

Denote by $\Gamma(R)$, $\Gamma(S)$, ... the matrices representing the group elements R, S, ... in the group G, and construct a Hermitian matrix H by

$$\mathrm{H} = \sum_R \Gamma(R)\Gamma(R)^\dagger = \mathrm{H}^\dagger. \tag{1}$$

A Hermitian matrix H is an example of a normal matrix and can therefore be diagonalized by the similarity transformation $\mathrm{U\ H\ U}^{-1}$ with U a unitary matrix (see eq. (A1.3.6)).

(1)
$$\begin{aligned}\mathrm{D} = \mathrm{U\ H\ U}^{-1} &= \sum_R \mathrm{U}\ \Gamma(R)\ \Gamma(R)^\dagger\ \mathrm{U}^{-1} \\ &= \sum_R \mathrm{U}\ \Gamma(R)\ \mathrm{U}^{-1}\ \mathrm{U}\ \Gamma(R)^\dagger\ \mathrm{U}^{-1} = \sum_R \Gamma'(R)\ \Gamma'(R)^\dagger,\end{aligned} \tag{2}$$

where the primed matrices are members of a representation $\{\Gamma'(R)\} \approx \{\Gamma(R)\}$. The unitary property of U, namely that $\mathrm{U}^{-1} = \mathrm{U}^\dagger$, has been used in writing the last equality in eq. (2). The diagonal elements of $\mathrm{D} = [d_i\ \delta_{ij}]$ are the (real) eigenvalues of H, and the definition H ensures that these d_i are positive real numbers. We may therefore construct the real positive diagonal matrices

$$\mathrm{D}^{1/2} = [d_i^{1/2}\ \delta_{ij}], \quad \mathrm{D}^{-1/2} = [d_i^{-1/2}\ \delta_{ij}], \tag{3}$$

where $d_i^{1/2}$ is the positive square root of d_i. Since diagonal matrices commute,

(2)
$$\mathrm{D}^{-1/2}\left[\sum_R \Gamma'(R)\ \Gamma'(R)^\dagger\right]\mathrm{D}^{-1/2} = \mathrm{E}. \tag{4}$$

Now define the set of matrices $\{\Gamma''(R)\}$ by

$$\Gamma''(R) = \mathrm{D}^{-1/2}\ \Gamma'(R)\ \mathrm{D}^{1/2}. \tag{5}$$

It will be shown that the set $\{\Gamma''(R)\}$ is a unitary representation.

$$
\begin{aligned}
&(5)\quad \Gamma''(R)\ \Gamma''(R)^\dagger = \mathrm{D}^{-1/2}\ \Gamma'(R)\ \mathrm{D}^{1/2}\ \mathrm{D}^{1/2}\ \Gamma'(R)^\dagger\ \mathrm{D}^{-1/2} \\
&(4)\qquad\qquad\qquad\ = \mathrm{D}^{-1/2}\ \Gamma'(R)\ \mathrm{D}^{1/2}\ \mathrm{D}^{-1/2}\left[\sum_S \Gamma'(S)\ \Gamma'(S)^\dagger\right]\mathrm{D}^{-1/2}\ \mathrm{D}^{1/2}\ \Gamma'(R)^\dagger\ \mathrm{D}^{-1/2} \\
&\qquad\qquad\qquad\qquad = \mathrm{D}^{-1/2}\left[\sum_S \Gamma'(R)\ \Gamma'(S)\ \Gamma'(S)^\dagger\ \Gamma'(R)^\dagger\right]\mathrm{D}^{-1/2} \\
&(4), (RS = T)\qquad\ = \mathrm{D}^{-1/2}\left[\sum_T \Gamma'(T)\ \Gamma'(T)^\dagger\right]\mathrm{D}^{-1/2} = \mathrm{E}.
\end{aligned}
\tag{6}
$$

Therefore, given a representation $\{\Gamma(R)\}$, an equivalent unitary representation $\{\Gamma''(R)\}$ can always be constructed by forming

$$\Gamma''(R) = \mathrm{D}^{-1/2}\ \mathrm{U}\ \Gamma(R)\ \mathrm{U}^{-1}\ \mathrm{D}^{1/2}, \tag{7}$$

where U and D are defined in eq. (2).

Lemma 2 (Schur's lemma) A matrix which commutes with every matrix of an IR must be a multiple of the unit matrix. Schur's lemma provides a criterion for the reducibility of a matrix representation. For if a matrix can be found which commutes with $\Gamma(R)$, $\forall R \in \mathrm{G}$, and which is not a multiple of E, then $\Gamma = \{\Gamma(R)\}$ is a reducible representation. Because of Lemma 1, we may restrict our considerations to unitary representations. Let M be a matrix that commutes with $\Gamma(R)$, $\forall R \in \mathrm{G}$. Then

$$\mathrm{M}\ \Gamma(R) = \Gamma(R)\ \mathrm{M}. \tag{8}$$

(8)
$$\Gamma(R)^\dagger\ \mathrm{M}^\dagger = \mathrm{M}^\dagger\ \Gamma(R)^\dagger; \tag{9}$$

(9), ($\Gamma(R)$ unitary)
$$\mathrm{M}^\dagger\ \Gamma(R) = \Gamma(R)\ \mathrm{M}^\dagger. \tag{10}$$

Equation (10) shows that $\mathrm{M}^\dagger$ also commutes with the unitary matrices of Γ, and so therefore do the two Hermitian matrices

$$\mathrm{H}_1 = (\mathrm{M} + \mathrm{M}^\dagger) \tag{11}$$

and

$$\mathrm{H}_2 = \mathrm{i}(\mathrm{M} - \mathrm{M}^\dagger). \tag{12}$$

But a Hermitian matrix H (which could be either H_1 or H_2) can always be diagonalized by a unitary transformation

$$\mathrm{U}\ \mathrm{H}\ \mathrm{U}^{-1} = \mathrm{D}. \tag{13}$$

Define $\Gamma'(R)$ by

$$\Gamma'(R) = \mathrm{U}\ \Gamma(R)\ \mathrm{U}^{-1}. \tag{14}$$

(14), (13)
$$\begin{aligned}\Gamma'(R)\ \mathrm{D} &= \mathrm{U}\ \Gamma(R)\ \mathrm{U}^{-1}\ \mathrm{U}\ \mathrm{H}\ \mathrm{U}^{-1}\\ &= \mathrm{U}\ \mathrm{H}\ \Gamma(R)\ \mathrm{U}^{-1}\quad (\Gamma(R)\ \text{commutes with H})\\ &= \mathrm{U}\ \mathrm{H}\ \mathrm{U}^{-1}\ \mathrm{U}\ \Gamma(R)\ \mathrm{U}^{-1}\\ &= \mathrm{D}\ \Gamma'(R);\end{aligned}\tag{15}$$

(15)
$$\Gamma'(R)_{ij}\ d_j\ \delta_{jk} = d_i\ \delta_{ij}\ \Gamma'(R)_{jk}.\tag{16}$$

Summing each side of eq. (16) over j gives

(16)
$$\Gamma'(R)_{ik}(d_k - d_i) = 0.\tag{17}$$

The subscripts $i, k = 1, 2, \ldots, l$ fall into two sets (one of which may be empty): (i) $d_k = d_i$, $\forall\ k \in \{a\}$; (ii) $d_j \neq d_i$, $\forall\ j \in \{b\}$, $\forall\ i \in \{a\}$. When $d_j \neq d_i$ (so that D is not a multiple of the unit matrix), eq. (17) requires that

$$\Gamma'(R)_{ij} = \Gamma'(R)_{ji} = 0,\quad \forall\ i \in \{a\},\ \forall\ j \in \{b\},\ \forall\ R \in \mathrm{G}.\tag{18}$$

In this case, the transformation of Γ to Γ' has brought $\{\Gamma'(R)\}$ into block-diagonal form and the matrix representation Γ was therefore reducible. But if Γ is irreducible, then $d_k = d_i$, $\forall\ k$, and D is a constant matrix, that is the constant d_i times the unit matrix. But if UHU^{-1} is a multiple of the unit matrix, then so is H. And if H_1 and H_2 are multiples of the unit matrix, then so also is $\mathrm{M} = \frac{1}{2}(\mathrm{H}_1 - i\mathrm{H}_2)$, which proves Schur's lemma.

Lemma 3 If Γ^i, Γ^j are two IRs of the same group G of dimensions l_i, l_j, respectively, and if there exists a rectangular matrix M such that

$$\mathrm{M}\ \Gamma^i(R) = \Gamma^j(R)\ \mathrm{M},\quad \forall\ R \in \mathrm{G},\tag{19}$$

then there are two possibilities: (i) if $l_i \neq l_j$, then $\mathrm{M} = 0$; (ii) if $l_i = l_j$, either $\mathrm{M} = 0$ or $|\mathrm{M}| \neq 0$. In this latter case, M has an inverse and

(19)
$$\Gamma^j(R) = \mathrm{M}\ \Gamma^i(R)\ \mathrm{M}^{-1},\quad \forall\ R \in \mathrm{G},\tag{20}$$

so that $\Gamma^j \approx \Gamma^i$. Matrix M is called the *intertwining* matrix, and Lemma 3 asserts that two IRs Γ^i and Γ^j can be intertwined only trivially, that is either $\mathrm{M} = 0$ or $\Gamma^j \approx \Gamma^i$.

Proof

From Lemma 1, we may take the representations Γ^i, Γ^j as unitary. Assume, without loss in generality, that $l_i \leq l_j$.

(19)
$$\Gamma^i(R)^\dagger\ \mathrm{M}^\dagger = \mathrm{M}^\dagger\ \Gamma^j(R)^\dagger;\tag{21}$$

(21)
$$\Gamma^i(R^{-1})\ \mathrm{M}^\dagger = \mathrm{M}^\dagger\ \Gamma^j(R^{-1});\tag{22}$$

(22)
$$\mathrm{M}\ \Gamma^i(R^{-1})\ \mathrm{M}^\dagger = \mathrm{M}\ \mathrm{M}^\dagger\ \Gamma^j(R^{-1});\tag{23}$$

(23), (19) $$\Gamma^j(R^{-1})\ \mathrm{M}\ \mathrm{M}^\dagger = \mathrm{M}\ \mathrm{M}^\dagger\ \Gamma^j(R^{-1}),\ \forall R \in \mathrm{G}. \qquad (24)$$

Therefore $\mathrm{M}\ \mathrm{M}^\dagger$ commutes with all the matrices of Γ^j, and so by Schur's lemma it must be a multiple of the unit matrix,

$$\mathrm{M}\ \mathrm{M}^\dagger = c\mathrm{E}. \qquad (25)$$

If $l_i = l_j = l$, M is a square matrix and

$$|\mathrm{M}\ \mathrm{M}^\dagger| = |\mathrm{M}||\mathrm{M}^\dagger| = |\det\ \mathrm{M}|^2 = c^l. \qquad (26)$$

If $c \neq 0$, $|\mathrm{M}| \neq 0$ and M has an inverse, so that $\Gamma^i \approx \Gamma^j$. If $c = 0$, $\mathrm{M}\mathrm{M}^\dagger = 0$ and the prth element of $\mathrm{M}\ \mathrm{M}^\dagger$

$$\sum_{q=1}^{l} \mathrm{M}_{pq}\ \mathrm{M}^*_{rq} = 0, \quad \forall\ p, r. \qquad (27)$$

For the diagonal elements $r = p$

$$\sum_{q=1}^{l} |\mathrm{M}_{pq}|^2 = 0, \quad \forall\ p = 1, 2, \ldots, l, \qquad (28)$$

which necessitates that $\mathrm{M}_{pq} = 0$, $\forall\ q = 1, 2, \ldots, l$, that is that $\mathrm{M} = 0$. The other possibility is that $l_i < l_j$ so that M has l_i columns and l_j rows. A square matrix M' of dimensions $l_j \times l_j$ can be constructed from M by adding $l_j - l_i$ columns of zeros. Since $|\mathrm{M}'| = 0$ and $\mathrm{M}'\ \mathrm{M}'^\dagger = \mathrm{M}\ \mathrm{M}^\dagger$, $|\mathrm{M}\ \mathrm{M}^\dagger| = 0$. Therefore eq. (25) requires that $c = 0$ and consequently that $\mathrm{M} = 0$ (from the above argument that includes eqs. (27) and (28)). This completes the proof of Lemma 3.

The orthogonality theorem The inequivalent irreducible unitary matrix representations of a group G satisfy the orthogonality relations

$$\sum_R \sqrt{l_i/g}\ \Gamma^i(R)^*_{pq} \sqrt{l_j/g}\ \Gamma^j(R)_{rs} = \delta_{ij}\ \delta_{pr}\ \delta_{qs}, \qquad (29)$$

where l_i is the dimension of Γ^i and g is the order of the group.

Proof: Define

$$\mathrm{M} = \sum_R \Gamma^j(R)\ \mathrm{X}\ \Gamma^i(R^{-1}), \qquad (30)$$

where X is an arbitrary matrix with l_j rows and l_i columns. Then

$$\begin{aligned}\Gamma^j(S)\ \mathrm{M} &= \sum_R \Gamma^j(S)\ \Gamma^j(R)\ \mathrm{X}\ \Gamma^i(R^{-1})\\ &= \sum_R \Gamma^j(S)\ \Gamma^j(R)\ \mathrm{X}\ \Gamma^i(R^{-1})\ \Gamma^i(S^{-1})\ \Gamma^i(S)\\ &= \left[\sum_R \Gamma^j(SR)\ \mathrm{X}\ \Gamma^i((SR)^{-1})\right]\Gamma^i(S)\\ &= \left[\sum_R \Gamma^j(R)\ \mathrm{X}\ \Gamma^i(R^{-1})\right]\Gamma^i(S)\end{aligned} \qquad (31)$$

since the sum is over all group elements. Therefore

(31), (30) $$\Gamma^j(S)\ \mathrm{M} = \mathrm{M}\ \Gamma^i(S), \quad \forall\ S \in \mathrm{G}. \tag{32}$$

Since Γ^i, Γ^j are inequivalent representations (Γ^i not $\approx \Gamma^j$), Lemma 3 requires that M is the null matrix. Therefore,

(30) $$\mathrm{M}_{rp} = \sum_R \sum_s \sum_q \Gamma^j(R)_{rs}\ \mathrm{X}_{sq}\ \Gamma^i(R^{-1})_{qp} = 0, \tag{33}$$

the second equality holding when Γ^i is not $\approx \Gamma^j$. But X is an arbitrary matrix, and we may choose the elements of X in whatever way we please. If we choose $\mathrm{X}_{sq} = 1$ and all the other elements of X to be zero, then

(33) $$\sum_R \Gamma^j(R)_{rs}\ \Gamma^i(R^{-1})_{qp} = 0. \tag{34}$$

But since Γ^i is a unitary representation,

$$\sum_R \Gamma^i(R)^*_{pq}\ \Gamma^j(R)_{rs} = 0\ (\Gamma^i \approx \Gamma^j). \tag{35}$$

Now suppose that $i = j$, then

$$\mathrm{M} = \sum_R \Gamma^i(R)\ \mathrm{X}\ \Gamma^i(R^{-1}) \tag{36}$$

commutes with all the other $\Gamma^i(S)$, $S \in \mathrm{G}$, because none of the steps from eq. (30) to eq. (32) preclude $j = i$. By Schur's lemma, $\mathrm{M} = c\mathrm{E}$ and so

(36) $$\mathrm{M}_{rp} = \sum_{t,u} \sum_R \Gamma^i(R)_{rt}\ \mathrm{X}_{tu}\ \Gamma^i(R^{-1})_{up} = c\ \delta_{pr}, \quad r, p = 1, 2, \ldots, l_i. \tag{37}$$

Choosing all the $\mathrm{X}_{tu} = 0$ except for $\mathrm{X}_{sq} = 1$ reduces eq. (37) to

$$\sum_R \Gamma^i(R)_{rs}\ \Gamma^i(R^{-1})_{qp} = c\ \delta_{pr}. \tag{38}$$

Set $r = p$ and sum over p,

$$\sum_R \sum_p \Gamma^i(R^{-1})_{qp}\ \Gamma^i(R)_{ps} = c\ \sum_p \delta_{pr}; \tag{39}$$

(39) $$\sum_R \Gamma^i(R^{-1}R)_{qs} = c\ l_i, \tag{40}$$

(40) $$\sum_R \Gamma^i(E)_{qs} = g\ \delta_{qs} = c\ l_i, \tag{41}$$

(41) $$c = g\ \delta_{qs}/l_i, \tag{42}$$

(38), (42) $$\sum_R \Gamma^i(R)_{rs}\ \Gamma^i(R^{-1})_{qp} = (g/l_i)\delta_{pr}\ \delta_{qs}, \tag{43}$$

(43) $$\sum_R \Gamma^i(R)^*_{pq}\ \Gamma^i(R)_{rs} = (g/l_i)\delta_{pr}\ \delta_{qs}. \tag{44}$$

Combining eqs. (35) and (44) yields the general form, eq. (29), of the OT for unitary IRs that are inequivalent when $i \neq j$ and identical when $i = j$. This is the situation met in most practical applications of group theory using character tables that contain the characters of the inequivalent IRs of a group of spatial symmetry operators. Moreover, eqs. (38) and (42) provide the generalization for non-unitary representations. However, in developing a test to determine whether extra degeneracies are introduced by time-reversal symmetry, we need an orthogonality theorem for *equivalent* IRs. This is derived in section A1.6.

A1.6 Orthogonality theorem for equivalent irreducible representations

When Γ^i is not $\approx \Gamma^j$ we have shown in Section A1.5 that

$$\Gamma^j(R)\ \mathrm{M} = \mathrm{M}\ \Gamma^i(R), \quad \forall\ R \in \mathrm{G}, \tag{A1.5.32}$$

leads to

$$\sum_R \Gamma^j(R)_{rs}\ \Gamma^i(R^{-1})_{qp} = 0, \tag{A1.5.34}$$

which, for unitary representations, becomes

$$\sum_R \Gamma^i(R)^*_{pq}\ \Gamma^j(R)_{rs} = 0. \tag{A1.5.35}$$

Now suppose that $\Gamma^i \approx \Gamma^j$ and let Z be the non-singular matrix that transforms Γ^j into Γ^i,

$$\Gamma^i(R) = \mathrm{Z}\ \Gamma^j(R)\ \mathrm{Z}^{-1}, \quad \forall\ R \in \mathrm{G}, \tag{1}$$

(A1.5.32), (1) $$\Gamma^j(R)\ \mathrm{M} = \mathrm{M}\ \mathrm{Z}\ \Gamma^j(R)\ \mathrm{Z}^{-1}, \tag{2}$$

(2) $$\Gamma^j(R)\ \mathrm{M}\ \mathrm{Z} = \mathrm{M}\ \mathrm{Z}\ \Gamma^j(R). \tag{3}$$

From Schur's lemma (Lemma 2 of Section A1.5)

$$\mathrm{M}\ \mathrm{Z} = c\ \mathrm{E}. \tag{4}$$

Since M depends linearly on X, eq. (A1.5.30), c must also depend linearly on X, its most general form being

$$c = \sum_{s=1}\sum_{q=1} C_{sq}\ \mathrm{X}_{sq}, \tag{5}$$

where the coefficients C_{sq} are independent of X.

(A1.5.34), (4), (5) $$\begin{aligned} \mathrm{M}_{rp} &= c(\mathrm{Z}^{-1})_{rp} = \sum_{s,q} C_{sq}\ \mathrm{X}_{sq}\ (\mathrm{Z}^{-1})_{rp} \\ &= \sum_R \sum_{s'} \sum_{q'} \Gamma^j(R)_{rs'}\ \mathrm{X}_{s'q'}\ \Gamma^i(R^{-1})_{q'p}. \end{aligned} \tag{6}$$

But X is an arbitrary matrix, so on equating coefficients of X_{sq},

$$C_{sq}\ (Z^{-1})_{rp} = \sum_R \Gamma^j(R)_{rs}\ \Gamma^i(R^{-1})_{qp}. \tag{7}$$

Multiply each side of eq. (7) from the right by Z_{pr} and sum over p:

(7), (1)
$$C_{sq}\ E_{rr} = \sum_R \Gamma^j(R)_{rs}\ [\Gamma^i(R^{-1})\ Z]_{qr}. \tag{8}$$

On summing over r,

(8)
$$C_{sq}\ l_j = g\ Z_{qs}, \tag{9}$$

(9), (7)
$$Z_{qs}\ (Z^{-1})_{rp} = (l_j/g)\sum_R \Gamma^j(R)_{rs}\ \Gamma^i(R^{-1})_{qp}. \tag{10}$$

For unitary representations $\Gamma(R^{-1}) = \Gamma(R)^{\dagger}$, and

(10)
$$(l_i/g)\sum_R\ \Gamma^i(R)^*_{pq}\ \Gamma^j(R)_{rs} = Z_{qs}\ (Z^{-1})_{rp}. \tag{11}$$

Equations (10) and (11) hold when $\Gamma^i \approx \Gamma^j$; if $i = j$, then eq. (1) shows that Z is a multiple of the unit matrix and

(11)
$$\sum_R \sqrt{l_j/g}\ \Gamma^j(R)^*_{pq}\ \sqrt{l_j/g}\ \Gamma^j(R)_{rs} = \delta_{pr}\ \delta_{qs} \quad (i = j), \tag{12}$$

in agreement with (A1.5.29) when $i = j$.

We shall also need the following corollary to Schur's lemma: the complete set of matrices $\{Z\}$ that transforms Γ^j into Γ^i is cZ, where Z is one such matrix and $c \neq 0$; if Γ^i and Γ^j are unitary then $\{Z\}$ contains unitary matrices and the complete set of unitary matrices that transform Γ^j into Γ^i is $\{e^{i\gamma}Z\}$.

Proof

If $Z' \in \{Z\}$, then

(1)
$$\Gamma^i(R) = Z'\ \Gamma^j(R)\ Z'^{-1} = Z\ \Gamma^j(R)\ Z^{-1}, \quad \forall\ R \in G \tag{13}$$

(13)
$$\Gamma^j(R) = Z'^{-1}\ Z\Gamma^j(R)\ Z^{-1}\ Z' = Z'^{-1}\ Z\ \Gamma^j(R)(Z'^{-1}Z)^{-1} \tag{14}$$

so that $Z'^{-1}\ Z$ commutes with $\Gamma^j(R)$, $\forall\ R \in G$. By Schur's lemma (Section A1.5) $Z'^{-1}\ Z$ is a multiple (say, c^{-1}) of the unit matrix and so $Z' = c\ Z$ and $\{c\ Z\}$ is the complete set that transforms Γ^j into Γ^i. Now suppose Γ^i and Γ^j to be unitary.

(1)
$$\Gamma^i(R)^{\dagger} = (Z^{-1})^{\dagger}\ \Gamma^j(R)^{\dagger}\ Z^{\dagger}, \tag{15}$$

(15)
$$\Gamma^i(R^{-1}) = (Z^{-1})^{\dagger}\ \Gamma^j(R^{-1})\ Z^{\dagger}, \tag{16}$$

$$(16),\ (R^{-1} \in \mathrm{G}) \qquad \Gamma^i(R) = (\mathrm{Z}^{-1})^\dagger\ \Gamma^j(R)\ \mathrm{Z}^\dagger. \tag{17}$$

Equations (1) and (17) show that Z is unitary and therefore that $\{e^{i\gamma}\mathrm{Z}\}$ is the complete set of unitary matrices that transform Γ^j into Γ^i.

A1.7 Direct product matrices

If A, B are two square matrices of dimensions $m \times m$ and $n \times n$, respectively, then the direct product (DP) of A and B is

$$\mathrm{A} \otimes \mathrm{B} = \begin{bmatrix} a_{11}\mathrm{B} & a_{12}\mathrm{B} & \dots & \dots & a_{1m}\mathrm{B} \\ a_{21}\mathrm{B} & a_{22}\mathrm{B} & \dots & \dots & a_{2m}\mathrm{B} \\ \dots & \dots & \dots & \dots & \dots \\ \dots & \dots & \dots & \dots & \dots \\ a_{m1}\mathrm{B} & a_{m2}\mathrm{B} & \dots & \dots & a_{mm}\mathrm{B} \end{bmatrix}. \tag{1}$$

$\mathrm{A} \otimes \mathrm{B}$ is of dimensions $mn \times mn$. The general term is denoted by

$$[\mathrm{A} \otimes \mathrm{B}]_{pr,\,qs} = a_{pq}\ b_{rs}. \tag{2}$$

Note that in eq. (2) the first two subscripts denote the two row indices, p and r, while the second pair of indices denote the two column indices q and s.

Example A1.7–1 Find $\mathrm{A} \otimes \mathrm{B}$ when A and B are both 2×2 matrices:

$$\mathrm{A} = \begin{bmatrix} a_{11} & a_{12} \\ a_{21} & a_{22} \end{bmatrix}; \quad \mathrm{B} = \begin{bmatrix} b_{11} & b_{12} \\ b_{21} & b_{22} \end{bmatrix}. \tag{3}$$

$$\mathrm{A} \otimes \mathrm{B} = \begin{bmatrix} a_{11}b_{11} & a_{11}b_{12} & a_{12}b_{11} & a_{12}b_{12} \\ a_{11}b_{21} & a_{11}b_{22} & a_{12}b_{21} & a_{12}b_{22} \\ a_{21}b_{11} & a_{21}b_{12} & a_{22}b_{11} & a_{22}b_{12} \\ a_{21}b_{21} & a_{21}b_{22} & a_{22}b_{21} & a_{22}b_{22} \end{bmatrix} \tag{4}$$

column indices ($q\ s$) 11 12 21 22

For example, the element $[\mathrm{A} \otimes \mathrm{B}]_{21,21}$ of the DP matrix is $a_{22}b_{11}$. An alternative notation, based on the definition of the DP matrix in eq. (1), is to represent one block of the supermatrix $\mathrm{A} \otimes \mathrm{B}$ by

$$[\mathrm{A} \otimes \mathrm{B}]_{[pq]} = a_{pq}\mathrm{B}. \tag{5}$$

A *supermatrix* is a matrix, each element of which is itself a matrix. The subscripts $[pq]$ on the LS of eq. (5) denote that the general term $[pq]$ of the supermatrix $\mathrm{A} \otimes \mathrm{B}$ is the pqth term of the first matrix a_{pq} mutiplying the second matrix B. For example, the upper right-hand block of $\mathrm{A} \otimes \mathrm{B}$ in eq. (4) is

$$[\mathtt{A} \otimes \mathtt{B}]_{[12]} = a_{12}\mathtt{B}. \tag{6}$$

The product of two DPs is the direct product DP of the product of the first members of each DP, and the product of the second members of each DP,

$$(\mathtt{A} \otimes \mathtt{B})(\mathtt{C} \otimes \mathtt{D}) = \mathtt{AC} \otimes \mathtt{BD}. \tag{7}$$

The proof of this formula is an application of the second notation for a DP, introduced in eq. (5):

$$\begin{aligned} [(\mathtt{A} \otimes \mathtt{B})(\mathtt{C} \otimes \mathtt{D})]_{[rs]} &= \sum_t a_{rt}\ \mathtt{B}\ c_{ts}\ \mathtt{D} \\ &= (\mathtt{A}\ \mathtt{C})_{rs}\ \mathtt{B}\ \mathtt{D} \\ &= [\mathtt{A}\ \mathtt{C} \otimes \mathtt{B}\ \mathtt{D}]_{[rs]}. \end{aligned} \tag{8}$$

A2 Class algebra

A2.1 The Dirac character

The jth class $\mathscr{C}_j$ of a group G is the set of all the elements of G that are conjugate to g_j, so that

$$\mathscr{C}_j = \{g_k\ g_j\ g_k^{-1}\},\ k = 1, 2, \ldots, g, \text{ with repetitions deleted.} \tag{1}$$

Because binary composition is unique, the classes of G are all disjoint, with no elements in common. For example, the classes of the permutation group S_3 are $\mathscr{C}_1 = \{P_0\}$, $\mathscr{C}_2 = \{P_1\ P_2\}$, $\mathscr{C}_3 = \{P_3\ P_4\ P_5\}$ (see Section 1.2). The *Dirac character* Ω of a class (sometimes called the "class sum") is the sum of the elements in a class

$$\Omega_j = \sum_{j=1}^{c_j} g_j \tag{2}$$

(1.8.14), (1.8.15)
$$= \sum_{r=1}^{c_j} g_r\ g_j\ g_r^{-1}, \tag{3}$$

where g_r, $r = 1, 2\ \ldots, c_j$ (with $g_1 = E$) are the c_j coset representatives of the centralizer of g_j, $Z_j = Z\,(g_j|\mathrm{G})$, which is the set (of order z) of all the elements of G which commute with g_j. The $\{g_r\}$ are found from the coset expansion of G on Z_j,

(1.8.13), (1.8.15)
$$\mathrm{G} = \sum_{r=1}^{c_j} g_r\ Z_j,\ g_1 = E. \tag{4}$$

Equation (3) determines $\mathscr{C}_j = \{g_j\}, j = 1, 2, \ldots, c_j$, without repetitions.

The *inverse class* of $\mathscr{C}_j$, written $\mathscr{C}_{\bar{j}}$, is the set of elements of G that are the inverses of the elements of the class $\mathscr{C}_j$, so that

$$\mathscr{C}_{\bar{j}} = \{g_j^{-1}\},\ j = 1, 2, \ldots, c_j,\ g_j \in \mathscr{C}_j. \tag{5}$$

It follows from the definition of the inverse class that the mapping $\mathscr{C}_j \rightarrow \mathscr{C}_{\bar{j}}$ is isomorphous, so that $c_{\bar{j}} = c_j$.

Exercise A2.1-1 Show that if $g_j \in \mathscr{C}_j$, then $\{g_j^{-1}\}$ form a class with $c_{\bar{j}} = c_j$.

From the multiplication table for S(3), Table 1.3,

$$\mathscr{C}_{\bar{1}} = \{P_0\} = \mathscr{C}_1,\ \mathscr{C}_{\bar{2}} = \{P_2\ P_1\} = \mathscr{C}_2,\ \mathscr{C}_{\bar{3}} = \{P_3\ P_4\ P_5\} = \mathscr{C}_3. \tag{6}$$

$\mathscr{C}_{\bar{1}} = E = \mathscr{C}_1$, always. It often happens, as in S(3), that a class $\mathscr{C}_j$ is the same as its inverse $\mathscr{C}_{\bar{j}}$, in which case the class $\mathscr{C}_j = \mathscr{C}_{\bar{j}}$ is said to be *ambivalent*. Any group G contains at least one ambivalent class, namely $\mathscr{C}_1 = E$. Equation (6) shows that all three classes of S(3) are ambivalent. A class $\mathscr{C}_j$ may be ambivalent because each element is equal to its inverse, $g_j^{-1} = g_j,\ \forall\, j = 1, 2, \ldots, c_j$ as is true for $\mathscr{C}_3$ in S_3. Or a class might be ambivalent with $g_j^{-1} \neq g_j,\ \forall\, g_j \in \mathscr{C}_j$, as is true for $\mathscr{C}_2$ in S_3, in which $P_1^{-1} = P_2$. The Dirac character of the inverse class $\mathscr{C}_{\bar{j}}$ is

$$\Omega_{\bar{j}} = \sum_{j=1}^{c_j} g_j^{-1}. \tag{7}$$

Answer to Exercise A2.1-1

Suppose that $g_l \in \mathscr{C}_j = \{g_j\ g_l \ldots\}$. Then for some $g_k \in$ G, $g_l = g_k\ g_j\ g_k^{-1}$ and $g_l^{-1} = (g_k\, g_j\, g_k^{-1})^{-1} = g_k\, g_j^{-1}\, g_k^{-1}$, so that g_j^{-1}, g_l^{-1} are in the same class $\mathscr{C}_{\bar{j}} = \{g_j^{-1}\ g_l^{-1} \ldots\}$. From the definition of the inverse class, $\mathscr{C}_{\bar{j}}$ contains only those elements $\{g_j^{-1}\}$ that are the inverses of the elements $g_j \in \mathscr{C}_j$. Since each of the $\mathscr{C}_j$ elements $g_j \in \mathscr{C}_j$ has a unique inverse g_j^{-1}, and these are all in the same class $\mathscr{C}_{\bar{j}}$, the order $c_{\bar{j}}$ of $\mathscr{C}_{\bar{j}}$ is the same as the order c_j of $\mathscr{C}_j$.

A2.2 Properties of the Dirac characters (class sums)

(a) The Dirac characters commute with every element g_k of G. Suppose that the transform of g_j by g_k is g_l,

$$g_k\ g_j\ g_k^{-1} = g_l, \quad g_j,\ g_l \in \mathscr{C}_j; \quad g_k \in \text{G}. \tag{1}$$

Since binary composition is unique, as $j = 1, 2, \ldots, c_j$ with the same g_k, a different element g_l is generated for each $g_j \in \mathscr{C}_j$, and so there are c_j different elements in all. Each of these c_j elements $g_l \in \mathscr{C}_j$ and so applying the procedure, eq. (1), to $\forall\, g_j \in \mathscr{C}_j$ in turn simply regenerates the elements of $\mathscr{C}_j$, albeit usually in a different order. Therefore,

$$g_k\ \Omega_j\ g_k^{-1} = \sum_{j=1}^{c_j} g_k\ g_j\ g_k^{-1} = \sum_{l=1}^{c_j} g_l = \Omega_j, \tag{2}$$

(2)

$$\Omega_j\ g_k = g_k\ \Omega_j. \tag{3}$$

(b) The Dirac characters commute with each other. The elements of $\mathscr{C}_j$ each commute with Ω_i and so

$$\Omega_i\ \Omega_j = \sum_{j=1}^{c_j} \Omega_i\ g_j = \sum_{j=1}^{c_j} g_j\ \Omega_i = \Omega_j\ \Omega_i. \tag{4}$$

In the last equality we have made use of the fact that classes are disjoint.

(c) The product of two Dirac characters is a linear combination of Dirac characters. Suppose that $\mathscr{C}_i$, $\mathscr{C}_j$ are two of the N_c classes of G. Then,

$$\Omega_i \ \Omega_j = \sum_{i=1}^{c_i} \sum_{j=1}^{c_j} g_i \ g_j. \tag{5}$$

By closure, a particular one $g_i \ g_j$ of the $c_i \ c_j$ terms on the RS of eq. (5) is an element of G which must be the inverse of some element of G, $g_{\bar{k}}^{-1}$,

$$g_i \ g_j = g_{\bar{k}}^{-1} \in \mathscr{C}_{\bar{k}}. \tag{6}$$

Every element of $\mathscr{C}_{\bar{k}}$ occurs equally often on the RS of eq. (5). Let $g_{\bar{l}}^{-1} \in \mathscr{C}_{\bar{k}}$; then $g_{\bar{l}}^{-1}$ is related to $g_{\bar{k}}^{-1}$ by a similarity transformation and consequently, for some $g_p \in$ G,

$$g_{\bar{l}}^{-1} = g_p \ g_{\bar{k}}^{-1} \ g_p^{-1} = g_p \ g_i \ g_j \ g_p^{-1} = g_p \ g_i \ g_p^{-1} \ g_p \ g_j \ g_p^{-1} = g_q \ g_s, \tag{7}$$

where $g_q \in \mathscr{C}_i$, $g_s \in \mathscr{C}_j$. Thus for every product $g_i \ g_j = g_{\bar{k}}^{-1}$ on the RS of eq. (5), there occur also the products $g_q \ g_s = g_{\bar{l}}^{-1} = g_p \ g_{\bar{k}}^{-1} \ g_p^{-1}$, which are in the same class as $g_{\bar{k}}^{-1}$. Therefore, if $g_{\bar{k}}^{-1} \in \mathscr{C}_{\bar{k}}$ occurs C_{ij}^k times in the sum $\sum_{i=1}^{c_i} \sum_{j=1}^{c_j} g_i \ g_j$, *every* element of $\mathscr{C}_{\bar{k}}$ must appear the same number of times, and

$$\Omega_i \ \Omega_j = \sum_{k=1}^{N_c} C_{ij}^k \ \Omega_{\bar{k}}. \tag{8}$$

The C_{ij}^k are integers, or zero, called *class constants*.

Remark The reader will no doubt have noticed that the above argument could be carried out equally well for the classes $\mathscr{C}_m$ instead of for the inverse classes $\mathscr{C}_{\bar{k}}$. Why then have we used the seemingly more complicated route of expressing the RS of eq. (8) as a linear combination of *inverse classes* rather than as a linear combination of *classes*, as is done, for example, in the books by Hall (1959) and Jansen and Boon (1967)? It is because the symmetry properties of the class constants defined by eq. (8) are more extensive than they would have been had the product $\Omega_i \ \Omega_j$ been expressed as a linear combination of classes. Of course, each class has an inverse class and so the same terms will occur on the RS of eq. (8), but their ordering by the index k will differ unless all classes of G are ambivalent.

(d) The product $g_i \ g_j$ on the RS of eq. (5) can equal $g_1 = E$ only if $\mathscr{C}_j = \mathscr{C}_{\bar{i}}$, and then E occurs c_i times so that with $k = 1$

(8)
$$C_{ij}^1 = c_i \ \delta_{\bar{i}j}. \tag{9}$$

Exercise A2.2-1 Show that

$$C_{\bar{i}j}^1 = c_i \ \delta_{ij}. \tag{9'}$$

[*Hint*: Consider the product $\Omega_{\bar{i}} \ \Omega_j$.]

Table A2.1. *Multiplication table for the Dirac characters (class sums) of* S(3).

Since
$\Omega_{\bar{1}} = P_0^{-1} = P_0 = E = \Omega_1$, $\Omega_{\bar{2}} = P_1^{-1} + P_2^{-1} = P_2 + P_1 = \Omega_2$, $\Omega_{\bar{3}} = P_3^{-1} + P_4^{-1} + P_5^{-1} = P_3 + P_4 + P_5 = \Omega_3$, all three classes of S(3) are ambivalent.

	Ω_1	Ω_2	Ω_3
Ω_1	$\Omega_{\bar{1}}$	$\Omega_{\bar{2}}$	$\Omega_{\bar{3}}$
Ω_2	$\Omega_{\bar{2}}$	$2\Omega_{\bar{1}} + \Omega_{\bar{2}}$	$2\Omega_{\bar{3}}$
Ω_3	$\Omega_{\bar{3}}$	$2\Omega_{\bar{3}}$	$3\Omega_{\bar{1}} + 3\Omega_{\bar{2}}$

Example A2.2-1 Develop the multiplication table for the Dirac characters of S(3).

Using the multiplication table for S(3) in Table 1.3, we find the multiplication table for the Dirac characters in Table A2.1. The entries in this table illustrate eq. (8). The class constants C_{ij}^k form a 3-D cubical array. For fixed k the $C_{ij}^k, i = 1, 2, \ldots, N_c$, $j = 1, 2, \ldots, N_c$ may be arranged in an $N_c \times N_c$ square matrix, these square matrices being successive "slices" of the 3-D array for $k = 1, 2, \ldots, N_c$.

Remark If we were to take in eq. (8) a linear combination of classes, instead of inverse classes, the slices of the 3-D array might, in general, occur in a different order and this would result in some loss of symmetry in the class constants for groups with non-ambivalent classes.

For S(3),

$$\begin{array}{ccc} k=1 & k=2 & k=3 \\ C_{ij}^1 = \begin{bmatrix} 1 & 0 & 0 \\ 0 & 2 & 0 \\ 0 & 0 & 3 \end{bmatrix} & C_{ij}^2 = \begin{bmatrix} 0 & 1 & 0 \\ 1 & 1 & 0 \\ 0 & 0 & 3 \end{bmatrix} & C_{ij}^3 = \begin{bmatrix} 0 & 0 & 1 \\ 0 & 0 & 2 \\ 1 & 2 & 0 \end{bmatrix}. \end{array} \tag{10}$$

Exercise A2.2-2 Evaluate the class constants C_{ij}^k for $k = 1$ for the group S(3) from eq. (9) and check your results against the matrix C_{ij}^1 in eq. (10).

Exercise A2.2-3 Show that $c_i \; c_j = \sum_{k=1}^{N_c} c_k \; C_{ij}^k$. [*Hint*: Consider the number of terms on each side of eq. (8).] Demonstrate the validity of this result for (i) $i = 2, j = 3$, and (ii) $i = 3, j = 3$ for the group S(3).

(e) The symmetry properties of the class constants C_{ij}^k may be summarized by the statement that $c_k \; C_{ij}^k$ is invariant under (i) a permutation of indices and (ii) when all the classes are inverted.

Proof (i) The triple product

(8), (9) $$\Omega_i\ \Omega_j\ \Omega_k = \left(\sum_l C^l_{ij}\ \Omega_{\bar{l}}\right)\Omega_k = c_k\ C^k_{ij}\ \Omega_1 + \cdots. \tag{11}$$

The second equality in eq. (11) follows because $\Omega_1 = E$ and the identity, which can only arise when $l = k$, is repeated c_k times. Because the Dirac characters commute (eq. (4)), the triple product in eq. (11) is invariant under any permutation of i, j, k, so that

$$c_k\ C^k_{ij} = c_i\ C^i_{jk} = c_j\ C^j_{ki} = c_j\ C^j_{ik} = c_i\ C^i_{kj} = c_k\ C^k_{ji}. \tag{12}$$

The numerical factor c_k multiplying C^k_{ij} in eq. (12) may be inconvenient in some applications. It can be avoided by defining the *average class sums* by

$$\overline{\Omega}_i = \Omega_i / c_i, \quad i = 1, 2, \ldots, N_c, \tag{13}$$

and a new set of class constants c^k_{ij} by

$$c^k_{ij} = C^k_{ij} / c_i c_j, \quad i, j, k = 1, 2, \ldots, N_c. \tag{14}$$

(ii) $$\Omega_i\ \Omega_j = \sum_{i=1}^{c_i} \sum_{j=1}^{c_j} g_i\ g_j. \tag{5}$$

Each term g_k^{-1} of the double sum is a member of some class (the $\bar{k}$th) of G. From eq. (8) the number of terms ν_k that are in the $\bar{k}$th class is $c_k\ C^{\bar{k}}_{ij}$. But if $g_i\ g_j = g_k^{-1}$, then $g_j^{-1}\ g_i^{-1} = g_k$ and so ν_k is also the number of terms that are in the kth class in the product $\Omega_{\bar{j}}\ \Omega_{\bar{i}} = \Omega_{\bar{i}}\ \Omega_{\bar{j}}$, namely $c_k\ C^{k}_{\bar{i}\bar{j}}$. Therefore,

$$C^k_{ij} = C^{\bar{k}}_{\bar{i}\bar{j}}. \tag{15}$$

Exercise A2.2-4 Show that $\sum_l C^l_{ij} C^m_{\bar{l}k} = \sum_n C^n_{jk} C^m_{i\bar{n}}$. [*Hint*: Evaluate the triple product in eq. (10) in two different ways, making use of the fact that the multiplication of group elements, and therefore of the Dirac characters, is associative.]

Answers to Exercises A2.2

Exercise A2.2-1

(8) $$\Omega_{\bar{i}}\ \Omega_j = C^1_{\bar{i}j}\ \Omega_{\bar{1}} + \cdots.$$

The terms on the RS can be $\Omega_{\bar{1}} = E$ only when $j = i$, and then $\Omega_{\bar{i}}\ \Omega_i = c_i\ E$, so that

$$C^1_{\bar{i}j} = c_i\ \delta_{ij}. \tag{9'}$$

Exercise A2.2-2 From eq. (9), $C_{ij}^1 = c_i\ \delta_{\bar{i}j} = c_i\ \delta_{ij}$ for S(3), since all classes are ambivalent. Therefore, all non-diagonal elements of C_{ij}^1 are zero, so that $C_{ij}^1 = 0$, if $j \neq i$. If $i = 1$, $C_{11}^1 = c_1 = 1$; for $i = 2$, $C_{22}^1 = c_2 = 2$; and with $i = 3$, $C_{33}^1 = 3$, in agreement with eq. (10).

Exercise A2.2-3 The number of terms $g_i\ g_j = g_k^{-1}$ on the LS of eq. (3) is $c_i\ c_j$. Each of these belongs to some class $\bar{k}$ of G, and the number of terms in the $\bar{k}$th class is the number of times the $\bar{k}$th class occurs, C_{ij}^k, multiplied by the order of the $\bar{k}$th class c_k. Summing over all classes of G gives the total number of terms. For the group S(3), for which all three classes are ambivalent, the number of terms in $\Omega_2\ \Omega_3$ is, from the LS of eq. (8), $c_2\ c_3 = (2)(3) = 6$. From the RS it is $\sum\limits_{k=1}^{3} c_k\ C_{23}^k = 1(C_{23}^1) + 2(C_{23}^2) + 3(C_{23}^3) = 6$. Similarly, for $i = 3$, $j = 3$, $\text{LS} = (3)(3) = 9, \text{RS} = 1C_{33}^1 + 2C_{33}^2 + 3C_{33}^3 = 1(3) + 2(3) + 3(0) = 9$.

Exercise A2.2-4

$$(\Omega_i\ \Omega_j)\Omega_k = \sum_l C_{ij}^l\ \Omega_{\bar{l}}\ \Omega_k = \sum_{l,m} C_{ij}^l\ C_{\bar{l}k}^m\ \Omega_{\overline{m}},$$

$$\Omega_i(\Omega_j\ \Omega_k) = \sum_n C_{jk}^n\ \Omega_i\ \Omega_{\bar{n}} = \sum_{n,m} C_{jk}^n\ C_{i\bar{n}}^m\ \Omega_{\overline{m}}.$$

Equating coefficients of $\Omega_{\overline{m}}$,

$$\sum_l C_{ij}^l\ C_{\bar{l}k}^m = \sum_n C_{jk}^n\ C_{i\bar{n}}^m.$$

A2.3 The class algebra as a vector space

Multiplication of the Dirac characters produces a linear combination of Dirac characters (see eq. (4.2.8)), as do the operations of addition and scalar multiplication. The Dirac characters therefore satisfy the requirements of a linear associative algebra in which the elements are linear combinations of Dirac characters. Since the classes are disjoint sets, the N_c Dirac characters in a group G are linearly independent, but any set of $N_c + 1$ vectors made up of sums of group elements is necessarily linearly dependent. We need, therefore, only a satisfactory definition of the inner product for the class algebra to form a vector space. The inner product of two Dirac characters Ω_i, Ω_j is defined as the coefficient of the identity C_{ij}^1 in the expansion of the product $\Omega_i\ \Omega_j$ in eq. (A2.2.8),

(2.2.9′) $$\langle \Omega_i | \Omega_j \rangle = C_{ij}^1 = c_i\ \delta_{ij}. \qquad (1)$$

This definition satisfies all the requirements of an inner product (see, for example, Cornwell (1984), p. 274). Therefore, the Dirac characters form a set of orthogonal but not normalized vectors. An orthonormal basis can be defined by $\{(c_i)^{-1/2}\ \Omega_i\}$, for then

$$\langle (c_i)^{-1/2}\ \Omega_i | (c_j)^{-1/2}\ \Omega_j \rangle = (c_i\ c_j)^{-1/2}\ C_{\bar{i}j}^1 = \delta_{ij}. \qquad (2)$$

A general vector in this space is

$$\mathbf{X} = \sum_i (c_i)^{-1/2}\ \Omega_i\ x_{\bar{i}}, \tag{3}$$

where $x_{\bar{i}}$ is the $\bar{i}$th component of $\mathbf{X}$. The inner product of two vectors $\mathbf{X}$ and $\mathbf{Y}$ is then

$$\langle \mathbf{X}|\mathbf{Y}\rangle = \sum_{i,j} x_{\bar{i}}^*\ y_j\ \delta_{ij} = \sum_j x_{\bar{j}}^*\ x_j. \tag{4}$$

Equation (A2.2.8),

$$\Omega_j\ \Omega_k = \sum_{l=1}^{N_c} C_{jk}^l\ \Omega_{\bar{l}}, \tag{5}$$

may now be re-interpreted in terms of a vector space in which the basis vectors are the normalized Dirac characters $\{(c_i)^{-1/2}\ \Omega_i\}$. If Ω_k in eq. (5) is a basis vector, then Ω_j is an operator that acts on Ω_k (the operation being that of multiplication of Dirac characters) to produce a linear combination of basis vectors $\Omega_{\bar{l}}$. In vector notation,

$$\Omega_j|\Omega_k\rangle = \sum_{l=1}^{N_c} C_{jk}^l|\Omega_{\bar{l}}\rangle, \tag{6}$$

(6), (2), (A2.2.11)

$$\langle\Omega_i|\Omega_j|\Omega_k\rangle = \sum_{l=1}^{N_c} C_{jk}^l\langle\Omega_i|\Omega_{\bar{l}}\rangle = \sum_{l=1}^{N_c} C_{jk}^l\delta_{il}$$
$$= c_i C_{jk}^i = c_j C_{ki}^j = c_k C_{ij}^k. \tag{7}$$

This is the matrix element of Ω_j, and according to eq. (A2.2.11) it is invariant under any permutation of the indices i, j, k.

A2.4 Diagonalization of the Dirac characters

In Section A1.4 we proved that each one of a set of normal matrices can be diagonalized by a similarity transformation with the same unitary matrix S provided that they commute. In Chapter 4 a matrix representation of a group G was defined as a set of matrices that form a group isomorphous with G. Such a matrix system is reducible when it is equivalent (Section A1.4) to a direct sum of matrix systems of smaller dimensions and irreducible when it cannot be reduced any further into a direct sum of matrix systems of smaller dimensions. Each of the matrix systems in this direct sum forms an irreducible representation (IR). The character system of an IR Γ^m is the set of N_c numbers $\{\chi_j^m\}$, $j = 1, 2, \ldots, N_c$, called the characters, where χ_j^m is the sum of the diagonal elements (the trace) of the matrix representative (MR) of any member of the jth class in the mth IR. The character χ_j^m is the same for all members of the same class, and is therefore a *class property*. For symmetry groups the MRs are unitary. In general, matrix representations are not necessarily unitary, but any representation of a finite group consisting of non-singular matrices is equivalent to a unitary representation (Section A1.5). Consequently we may confine our attention to unitary representations. Since the Dirac character of the ith class is the sum of the group elements in that class, the MR of the Dirac character of any class is the sum of the MRs of the elements comprising that class,

$$\Omega_i = \sum_{i=1}^{c_i} g_i, \quad \Gamma(\Omega_i) = \sum_{i=1}^{c_i} \Gamma(g_i). \tag{1}$$

Whereas the MRs of group elements do not necessarily commute, the MRs of the Dirac characters do commute with one another and therefore also with their adjoints. They therefore form a set of normal matrices which can all be diagonalized by similarity transformations with the same unitary matrix S. The Dirac character matrices therefore have N_c (not necessarily distinct) eigenvalues and N_c corresponding eigenvectors.

In general, when a Dirac character multiplies a Dirac character, as in eq. (A2.2.8), it produces a linear combination of Dirac characters (vectors). But for a particular linear combination of Dirac characters

$$\Lambda = \sum_j y_j \; \Omega_{\bar{j}}, \tag{2}$$

that is an eigenvector of Ω_i,

$$\Omega_i \; \Lambda = \lambda_i \; \Lambda. \tag{3}$$

In this case, when Ω_i operates on Λ the result is just Λ multiplied by a constant λ_i, the eigenvalue of Ω_i.

(2), (3)
$$\Omega_i \sum_j y_j \; \Omega_{\bar{j}} = \lambda_i \sum_j y_j \; \Omega_{\bar{j}}; \tag{4}$$

(4), (A2.2.8)
$$\sum_{j,k} y_j \; C_{i\bar{j}}^{k} \; \Omega_{\bar{k}} = \lambda_i \sum_{j,k} y_j \; \delta_{jk} \; \Omega_{\bar{k}}; \tag{5}$$

(5)
$$\sum_j (C_{i\bar{j}}^{k} - \lambda_i \delta_{jk}) y_j = 0, \quad k = 1, 2, \ldots, N_c. \tag{6}$$

The set of equations (6) is a set of linear homogeneous equations for the y_j (which are the components of Λ_i in the $\{\Omega_i\}$ basis). The $N_r = N_c$ eigenvalues for the ith class, $\{\lambda_i^p\}$, $p = 1, 2, \ldots, N_c$, are the roots of the characteristic equation

$$|C_{i\bar{j}}^{k} - \lambda_i \delta_{jk}| = 0, \quad j, k = 1, 2, \ldots N_c, \quad i = 1, 2, \ldots N_c. \tag{7}$$

Remark One would normally expect to complete the solution of an eigenvalue problem by substituting the eigenvalues λ_i^p one at a time into eq. (6) and solving for the ratios of the coefficients of $\Omega_{\bar{j}}$ and $\Omega_1, y_j^p / y_1^p$. This would give the components of the pth eigenvector Λ^p, apart from an arbitrary constant which could be fixed by normalization. We shall not need to do this here, however, since our aim of determining the characters of all the IRs is satisfied by finding the eigenvalues.

Since the MRs of the pth IR of G form a group isomorphous with G, it follows from the definition of a class that

(A2.2.2)
$$\Gamma^p(g_k) \; \Gamma^p(\Omega_i) \; \Gamma^p(g_k^{-1}) = \Gamma^p(\Omega_i), \tag{8}$$

where $\Gamma^p\ (\Omega_i)$ is the sum of the MRs of the members of the ith class of G in the pth representation.

(8) $$\Gamma^p(g_k)\ \Gamma^p(\Omega_i) = \Gamma^p(\Omega_i)\ \Gamma^p(g_k)\ , \quad \forall\ g_k \in \mathrm{G}. \tag{9}$$

By Schur's lemma (see Section A1.5)

(9) $$\Gamma^p(\Omega_i) = \lambda_i^p\ \Gamma(E), \tag{10}$$

where $\Gamma(E)$ is the unit matrix of dimension l_p and λ_i^p is a scalar. If the MR $\Gamma^p\ \Omega_i$ of an operator Ω_i is a scalar times the unit matrix, then that scalar is one of the eigenvalues of Ω_i, and that is why the scalar in eq. (10) is written as λ_i^p.

(10) $$\mathrm{Tr}\ \Gamma^p(\Omega_i) = \sum_{g_j \in C_i} \Gamma^p(g_j) = c_i\ \chi_i^p = \lambda_i^p\ l_p; \tag{11}$$

(10), (11) $$\Gamma^p(\Omega_i) = (c_i\ \chi_i^p/l_p)\mathrm{E}; \tag{12}$$

(12) $$\Omega_i^p = (c_i\ \chi_i^p/l_p) = \lambda_i^p. \tag{13}$$

Therefore the characters χ_i^p in the pth representation are just the eigenvalues of Ω_i^p multiplied by l_p/c_i. Thus the calculation of the characters involves two steps: (i) the calculation of the eigenvalues λ_i^p by finding the roots of the characteristic equation, eq. (7), and (ii) the calculation of the characters χ_i^p from eq. (13). If the dimensions l_p of the IRs are not known, as for example when there is not a unique solution to

$$\sum_{p=1}^{N_r} l_p^2 = g, \tag{14}$$

then the $\{l_p\}$ must be determined first. This may be done from the normalization condition

$$g^{-1} \sum_{i=1}^{N_c} c_i |\chi_i^p|^2 = 1, \tag{15}$$

(15), (13) $$l_p^2 = g \left[\sum_{i=1}^{N_c} c_i^{-1} |\lambda_i^p|^2 \right]^{-1}. \tag{16}$$

Since $l_p \geq 1$, only the positive square root of the RS of eq. (16) is physically significant. This method yields the eigenvalues $\{\lambda_i^p\}$, $p = 1, 2, \ldots, N_c$, of the ith class in an arbitrary order. Since the characters are given by

(13) $$\chi_i^p = (l_p/c_i)\lambda_i^p, \tag{17}$$

we need to determine this order, that is to settle on the values for $p = 1, 2\ \ldots, N_c$ in any one of the N_c sets $\{\lambda_i^p\}$. This can usually be decided by satisfying normalization and orthogonality requirements.

Example A2.4-1 Determine the character table for the permutation group S(3).

From the multiplication table for S(3) (Table 1.3) we have, as already determined in eq. (A2.1.6), $g = 6$, $N_c = 3$, $\mathscr{C}_1 = \{P_0\}$, $\mathscr{C}_2 = \{P_1 P_2\}$, $\mathscr{C}_3 = \{P_3\, P_4\, P_5\}$, so that $\{c_i\} = \{1\ 2\ 3\}$.

Table A2.2. *Character table of* S(3) *deduced from the diagonalization of the MRs of the Dirac characters.*

S(3)	$\mathscr{C}_1$	$\mathscr{C}_2$	$\mathscr{C}_3$
Γ_1	1	1	1
Γ_2	1	1	−1
Γ_3	2	−1	0

All three classes are ambivalent $\mathscr{C}_{\bar{j}} = \mathscr{C}_j$, $j = 1, 2, 3$. The multiplication table for the Dirac characters is in Table A2.1. The entries in this table are $\sum_{k=1}^{N_c} C_{i\bar{j}}^{k}\, \Omega_{\bar{k}}$ so that the characteristic polynomials $|C_{i\bar{j}}^{k} - \lambda_i \delta_{jk}|$ for $i = 1, 2, 3$ are:

$$\begin{array}{cccccccccc} & & i=1 & & & i=2 & & & i=3 & \\ k = & 1 & 2 & 3 & 1 & 2 & 3 & 1 & 2 & 3 \end{array}$$

$$\begin{vmatrix} 1-\lambda_1 & 0 & 0 \\ 0 & 1-\lambda_1 & 0 \\ 0 & 0 & 1-\lambda_1 \end{vmatrix} \begin{vmatrix} -\lambda_2 & 1 & 0 \\ 2 & 1-\lambda_2 & 0 \\ 0 & 0 & 2-\lambda_2 \end{vmatrix} \begin{vmatrix} -\lambda_3 & 0 & 1 \\ 0 & -\lambda_3 & 2 \\ 3 & 3 & -\lambda_3 \end{vmatrix}. \tag{18}$$

$i = 1, 2, 3$ labels the classes, $k = 1, 2, 3$ labels the columns of the determinants, and the rows are labeled by $\bar{j} = 1, 2, 3$. Equating the three determinants to zero and solving for the roots yields the eigenvalues for the ith class in the pth representation

$$\begin{array}{ccccc} p\backslash i & 1 & 2 & 3 & l_p \\ 1 & 1 & 2 & 3 & 1 \\ 2 & 1 & 2 & -3 & 1 \\ 3 & 1 & -1 & 0 & 2 \end{array} \tag{19}$$

l_p is the dimension of the pth representation. Here we know the values of l_p since $\sum_{p=1}^{N_c} l_p^2 = g = 6$ has only one solution $l_1 = 1$, $l_2 = 1$, $l_3 = 2$. Nevertheless, we will illustrate the procedure to be used when $\{l_p\}$ is not known. From eq. (16), $l_1^2 = 6[1^2 + (\tfrac{1}{2})2^2 + (\tfrac{1}{3})3^2]^{-1}$, $l_1 = 1$; similarly, $l_2 = 1$, and $l_3^2 = 6[1^2 + (\tfrac{1}{2})(-1)^2]^{-1} = 4, l_3 = 2$. Finally, using eq. (17), we deduce from the eigenvalue table (19), the character table shown in Table A2.2, which is in complete agreement with that of its isomorph C_{3v}.

Example A2.4-2 Deduce the character table for the quaternion group Q defined in Chapter 12.

For Q, $g = 8$, $N_c = 5$, then, $\mathscr{C}_1 = 1$, $\mathscr{C}_2 = -1$, $\mathscr{C}_3 = \{q_1\ -q_1\}$, $\mathscr{C}_4 = \{q_2\ -q_2\}$, $\mathscr{C}_5 = \{q_3\ -q_3\}$. Again, each of these classes is ambivalent. From the multiplication table for Q, we deduce the multiplication table for the Dirac characters (Table A2.3). All the entries in

Table A2.3. *Multiplication table for the Dirac characters of the quaternion group* Q.

All five classes of Q are ambivalent.

Q	Ω_1	Ω_2	Ω_3	Ω_4	Ω_5
Ω_1	Ω_1	Ω_2	Ω_3	Ω_4	Ω_5
Ω_2	Ω_2	Ω_1	Ω_3	Ω_4	Ω_5
Ω_3	Ω_3	Ω_3	$2\Omega_1 + 2\Omega_2$	$2\Omega_5$	$2\Omega_4$
Ω_4	Ω_4	Ω_4	$2\Omega_5$	$2\Omega_1 + 2\Omega_2$	$2\Omega_3$
Ω_5	Ω_5	Ω_5	$2\Omega_4$	$2\Omega_3$	$2\Omega_1 + 2\Omega_2$

Table A2.4 *Characteristic determinants obtained in the diagonalization of the Dirac characters for the quaternion group* Q.

The rows of these determinants are labeled by $\bar{j} = 1, \ldots, 5$. Null entries are all zero.

$i = 1$

$k=$	1	2	3	4	5
	$1-\lambda_1$				
		$1-\lambda_1$			
			$1-\lambda_1$		
				$1-\lambda_1$	
					$1-\lambda_1$

$i = 2$

$k=$	1	2	3	4	5
	$-\lambda_2$	1			
	1	$-\lambda_2$			
			$1-\lambda_2$		
				$1-\lambda_2$	
					$1-\lambda_2$

$i = 3$

$k=$	1	2	3	4	5
	$-\lambda_3$		1		
		$-\lambda_3$	1		
	2	2	$-\lambda_3$		
				$-\lambda_3$	2
				2	$-\lambda_3$

$i = 4$

$k=$	1	2	3	4	5
	$-\lambda_4$			1	
		$-\lambda_4$		1	
			$-\lambda_4$		2
	2	2		$-\lambda_4$	
			2		$-\lambda_4$

$i = 5$

$k=$	1	2	3	4	5
	$-\lambda_5$				1
		$-\lambda_5$			1
			$-\lambda_5$	2	
			2	$-\lambda_5$	
	2	2			$-\lambda_5$

this table are $\sum_{k=1}^{N_c} C_{i\bar{j}}^{k} \Omega_{\bar{k}}$ but they are entered as sums of Ω_k because all classes of Q are ambivalent.

From the entries in Table A2.3 we find the characteristic determinants $|C_{i\bar{j}}^{k} - \lambda_i \delta_{jk}|$ given in Table A2.4. The roots of the characteristic equations yield the eigenvalues λ_i^p in Table A2.5.

Table A2.5. *Eigenvalues λ_i^p for the ith class in the pth IR of* Q *calculated from the diagonalization of the MRs of the Dirac characters.*

The degeneracy l_p of the pth representation is given in the right-hand column. The order c_i of the ith class is given in the bottom row.

$p\backslash i$	1	2	3	4	5	l_p
1	1	1	2	2	2	1
2	1	1	2	−2	−2	1
3	1	1	−2	2	−2	1
4	1	1	−2	−2	2	1
5	1	−1	0	0	0	2
c_i	1	1	2	2	2	

Table A2.6. *Character table for the quaternion group* Q *found by the diagonalization of the MRs of the Dirac characters.*

Q	$\mathscr{C}_1$	$\mathscr{C}_2$	$\mathscr{C}_3$	$\mathscr{C}_4$	$\mathscr{C}_5$
Γ_1	1	1	1	1	1
Γ_2	1	1	1	−1	−1
Γ_3	1	1	−1	1	−1
Γ_4	1	1	−1	−1	1
Γ_5	2	−2	0	0	0

The degeneracies l_p in Table A2.5 were calculated from eq. (16). Thus

(16) $$l_1^2 = 8[1 + 1 + 2 + 2 + 2]^{-1}, \tag{20}$$

so that $l_1 = 1$. Similarly, $l_2 = l_3 = l_4 = 1$, but $l_5^2 = 8[1 + 1]^{-1}$, $l_5 = 2$. Finally, the character table for Q calculated from eq. (17) is given in Table A2.6. The order of the eigenvalues λ_i^p is not determined by solving the characteristic equations. But considerations of normalization and orthogonality require the character table for Q to be as shown in Table A2.6 apart from the labels Γ_2, Γ_3, Γ_4, the order of which is arbitrary, but conventional.

There is available a variant of this method of determining character tables by diagonalization of the Dirac characters which is completely unambiguous (apart from the ordering of the rows of the character table, which is arbitrary) but which involves rather more work.

(6) $$|\mathrm{L} - \lambda \mathrm{E}| = 0; \tag{21}$$

(6), (21) $$\mathrm{L}_{jk} = \sum_{i=1}^{N_c} y_j \, C_{\bar{i}j}^{k} \, . \tag{22}$$

Here the determination of $|\lambda_i\rangle$, the column of eigenvalues for the ith class, from eqs. (21) and (22) provides a consistent ordering of the rows without the need to appeal to normalization and orthogonality conditions.

Example A2.4-3 Determine the character table for the quaternion group Q from eqs. (21) and (22).

Using Table A2.3 to determine the $C_{i\bar{j}}^{k}$ (see eq. (A2.3.5))

$$\text{(22)} \qquad \mathrm{L} = \begin{bmatrix} y_1 & y_2 & y_3 & y_4 & y_5 \\ y_2 & y_1 & y_3 & y_4 & y_5 \\ 2y_3 & 2y_3 & y_1 + y_2 & 2y_5 & 2y_4 \\ 2y_4 & 2y_4 & 2y_5 & y_1 + y_2 & 2y_3 \\ 2y_5 & 2y_5 & 2y_4 & 2y_3 & y_1 + y_2 \end{bmatrix}. \tag{23}$$

Multiplying out the determinant $|\mathrm{L}|$ and solving the fifth-order characteristic equation $|\mathrm{L}| = 0$ may be accomplished using one of the mathematical packages available. Using *Mathematica*, I obtained

$$|\lambda^p\rangle = \begin{bmatrix} 1 & 1 & 2 & 2 & 2 \\ 1 & 1 & 2 & \bar{2} & \bar{2} \\ 1 & 1 & \bar{2} & 2 & \bar{2} \\ 1 & 1 & \bar{2} & \bar{2} & 2 \\ 1 & \bar{1} & 0 & 0 & 0 \end{bmatrix} \begin{bmatrix} y_1 \\ y_2 \\ y_3 \\ y_4 \\ y_5 \end{bmatrix}. \tag{24}$$

The square matrix in eq. (24) is $[\lambda_i^p]$, where p labels the representations (rows) and i labels the classes (columns). For Q,

$$\langle c_i| = \langle 1\ 1\ 2\ 2\ 2|, \tag{25}$$

$$|l_p\rangle = |1\ 1\ 1\ 1\ 2\rangle, \tag{26}$$

$$\text{(25), (26)} \qquad |l_p\rangle\langle (c_i)^{-1}| = \begin{bmatrix} 1 & 1 & ½ & ½ & ½ \\ 1 & 1 & ½ & ½ & ½ \\ 1 & 1 & ½ & ½ & ½ \\ 1 & 1 & ½ & ½ & ½ \\ 2 & 2 & 1 & 1 & 1 \end{bmatrix}. \tag{27}$$

Multiplying each element in the square matrix $[\lambda_i^p]$ in eq. (24) by the corresponding element in eq. (27) (see eq. (17)) yields Table A2.6 without the need to use normalization and orthogonality conditions.

A3 Character tables for point groups

This appendix contains tables of characters for vector and spinor representations of the point groups G that are encountered most commonly in practical applications of group theory in chemical physics. Correlation tables are given separately in Appendix A4.

The character tables are grouped together in the following sections. *page*

A3.1 The proper cyclic groups C_n ($n = 1, 2, 3, 4, 5, 6$); 450
A3.2 The improper cyclic groups C_i, C_s, S_n ($n = 4, 6, 8$); 452
A3.3 The dihedral groups D_n ($n = 2, 3, 4, 5, 6$); 453
A3.4 The C_{nh} groups ($n = 2, 3, 4, 5, 6$); 455
A3.5 The C_{nv} groups ($n = 2, 3, 4, 5, 6, \infty$); 457
A3.6 The D_{nh} groups ($n = 2, 3, 4, 5, 6, \infty$); 459
A3.7 The D_{nd} groups ($n = 2, 3, 4, 5, 6$); 461
A3.8 The cubic groups T, T_h, T_d, O, O_h; 463
A3.9 The icosahedral groups Y, Y_h. 465

There are no tables for double groups $\overline{G}$ since these are made unnecessary by the inclusion of spinor representations. However, there is enough information in Chapters 8 and 11, and in these character tables, for readers who insist on using double groups to construct their own tables for the characters of double group representations very easily. The construction of the additional classes in $\overline{G} = \{R, \overline{R}\}$ is explained in Chapter 8. For vector representations, which are the irreducible representations of symmetry groups for systems with integral values of the total angular momentum quantum number j, the characters of classes containing *only* $\{\overline{R}_i\}$ are the same as the characters of the corresponding classes that contain $\{R_i\}$. For spinor representations (systems with half-integral total angular momentum quantum number j) the characters of classes that contain *only* $\{\overline{R}_i\}$ are the *negatives* of the characters of the corresponding classes that contain $\{R_i\}$. As an example, the classes and characters of the double group $\overline{C}_{3v}$ are given in Table A3.1. To derive multiplication rules in$\overline{G}$, or to multiply two matrix representatives R_i, R_j, requires projective factors $[R_i\,;\,R_j]$. Projective factors are not given explicitly because they may be calculated from the quaternion parameters $[\lambda\ \Lambda]$ (Chapter 12), which are obtained from the rotation parameter $(\phi\ \mathbf{n})$. (See the definition of a rotation $R(\phi\ \mathbf{n})$ in the "Notation and conventions" section, pp. xiii–xx).

Except for doubly degenerate complex conjugate representations, vector representations are named according to the Mulliken rules, which are explained in Section 4.5. For spinor representations, and the complex conjugate pairs of vector representations (which are degenerate through time-reversal symmetry), the notation follows that of Altmann and Herzig (1994), except that a non-degenerate representation with $j = 3/2$ and a character

Table A3.1. *Characters of the classes of the double group* $\overline{C}_{3v}$.

$\overline{C}_{3v}$	E	$2C_3$	$3\sigma_v$	$\overline{E}$	$2\overline{C}_3$	$3\overline{\sigma}_v$
A_1	1	1	1	1	1	1
A_2	1	1	−1	1	1	−1
E	2	−1	0	2	−1	0
$E_{1/2}$	2	1	0	−2	−1	0
$^1E_{3/2}$	1	−1	i	−1	1	−i
$^2E_{3/2}$	1	−1	−i	−1	1	i

Table A3.2. *Time-reversal classification of representations* Γ, *as listed in the column headed TR in the character tables.*

Vector representations correspond to integral values of the angular momentum quantum number *j* and therefore to systems with an even number of electrons. Spinor representations correspond to systems with half-integral *j* and therefore to systems with an odd number of electrons. Note that Γ^* is the complex conjugate of Γ.

	If Γ and Γ^* are	Extra degeneracy for vector IRs	Extra degeneracy for spinor IRs
a	real and equal	none	doubled
b	complex and inequivalent	doubled	doubled
c	complex and equivalent	doubled	none

of − 1 for the rotation C_n^+ about the principal axis is called $B_{3/2}$. In addition to the characters of vector and spinor representations, the tables include (in the column headed "TR") the time-reversal classification of the representations (*a*, *b*, or *c*) using the Altmann and Herzig (1994) criteria. Since the symbols *b* and *c* are interchanged in many other books and publications, this classification is repeated here in Table A3.2. The character tables include the bases of the vector representations, namely the infinitesimal rotations R_x, R_y, R_z, the *p* and *d* functions, and Cartesian tensors of rank 2. (The atomic *s* function is invariant under any proper or improper rotation and so always forms a basis for the totally symmetric representation.) Complete tables of spinor bases are given by Altmann and Herzig (1994), and it is recommended that the reader refers to these tables if spinor bases are required. Vector representations may include complex conjugate pairs named ^1E, ^2E. These pairs are bracketed together and the real bases given are those of $^1E \oplus {}^2E$. It may be dangerous to use more than one set of tables at a time. Though internally consistent, different sets of tables may differ in the naming of IRs. These tables have been checked against those in Altmann and Herzig (1994) and agree with them except in the definition of ε and the naming of $B_{3/2}$. The derivation and naming of IRs for cyclic groups do not necessarily conform with eq. (4.7.7), but results are equivalent.

My aim has been to give in these tables only the most commonly required information. For character tables for $n > 6$, Cartesian tensor bases of rank 3, spinor bases, rotation parameters, tables of projective factors, Clebsch–Gordan coefficients, direct product

Table A3.3. *Rotation parameter* (ϕ **n**), *quaternion parameters* [λ Λ], *and Cayley–Klein parameters a, b for the point group* C_3 ($b = 0$).

C_3	$R(\phi\ \mathbf{n})$	λ	Λ_z	a
E	$R(0\ [0\ 0\ 0])$	1	0	1
C_{3z}^+	$R(2\pi/3\ [0\ 0\ 1])$	$\cos(\pi/3)$	$\sin(\pi/3)$	$\exp(-\mathrm{i}\pi/3)$
C_{3z}^-	$R(-2\pi/3\ [0\ 0\ 1])$	$\cos(\pi/3)$	$-\sin(\pi/3)$	$\exp(\mathrm{i}\ \pi/3)$

representations, multiplication tables, and matrix representatives one should refer to the extensive compilation by Altmann and Herzig (1994). Although correlation tables (compatibility relations) are easily calculated from character tables, it is nevertheless useful in some applications (such as descent of symmetry) to have correlation tables available, and so correlation tables are included in Appendix A4. Character tables for groups of larger order (up to $n = 10$), direct product representations, and Cartesian bases of rank 3 are also given by Harris and Bertolucci (1978). However, those tables contain only vector representations. Other useful sources are: Atkins *et al.* (1970), Flurry (1980), Kim (1999), Koster *et al.* (1963) and Lax (1974).

The character χ^j (R) is the sum of the diagonal elements of the matrix representative $\Gamma^j(R)$ of the rotation $R(\phi\ \mathbf{n})$,

$$\chi^j(R) = \sum_m \Gamma^j_{mm}(R), \tag{1}$$

where

(11.8.43)
$$\Gamma^j_{mm}(a,b) = \sum_{k=0}^{j+m} \frac{(j+m)!(j-m)!\ a^{j+m-k}(a^*)^{j-m-k} b^k(-b^*)^k}{(j+m-k)!(j-m-k)!(k!)^2}, \tag{2}$$

in which

$$(-n)! = \infty, \quad \text{if } n > 0; \ (-n)! = 1, \quad \text{if } n = 0. \tag{3}$$

(12.5.21)
$$a = \lambda - \mathrm{i}\Lambda_z,\ b = -\Lambda_y - \mathrm{i}\Lambda_x, \tag{4}$$

(12.5.18)
$$\lambda = \cos\frac{1}{2}\phi, \quad \mathbf{\Lambda} = \left(\sin\frac{1}{2}\phi\right)\mathbf{n}. \tag{5}$$

If j is an integer, for improper rotations IR, multiply Γ^j (a, b) by $(-1)^j$. If j is a half-integer, $\Gamma(IR) = \Gamma(R)$ (Pauli gauge). For cyclic groups,

(12.8.3)
$$\Gamma^j_{mm}(a,0) = a^{j+m}\ (a^*)^{j-m}. \tag{6}$$

Example A3.1 Find the character table for the cyclic group C_3.

The calculation of a is in Table A3.3. The irreducible representations are 1-D, and $\chi(R)$ follows from eq. (6). From (6), $\chi(R) = a^{j+m}\ (a^*)^{j-m}$. For $j = 0$, $m = 0$, $\chi(R) = 1$,

$\forall R$, so that the basis $|0\ 0\rangle$ gives the totally symmetric representation A. For the basis $|1\ 1\rangle$, $\Gamma^j_{mm}(a) = a^2$, therefore $\chi(C_3^+) = \varepsilon$, $\chi(C_3^-) = \varepsilon^*$, where $\varepsilon = \exp(-\text{i}\ 2\pi/3)$. Similarly, for a basis $|1\ \bar{1}\rangle$, $\Gamma^j_{mm}(a) = (a^*)^2$, $\chi(C_3^+) = \varepsilon^*$, $\chi(C_3^-) = \varepsilon$. For the spinor basis $|{}^1\!/_2\ \overline{{}^1\!/_2}\rangle$, $\Gamma^j_{mm}(a) = a^*$, $\chi(C_3^+) = \exp(\text{i}\ \pi/3) = -\varepsilon$, $\chi(C_3^-) = -\varepsilon^*$; for $|{}^1\!/_2\ {}^1\!/_2\rangle$, $\Gamma^j_{mm}(a) = a$, $\chi(C_3^+) = \exp(-\text{i}\ \pi/3) = -\varepsilon^*$, $\chi(C_3^-) = -\varepsilon$. For $|{}^3\!/_2\ {}^3\!/_2\rangle$, $\Gamma^j_{mm}(a) = a^3$, $\chi(C_3^+) = \exp(-3\text{i}\ \pi/3) = -1$, and $\chi(C_3^-) = \exp(\text{i}\ \pi) = -1$.

A3.1 The proper cyclic groups C_n

1 C_1

C_1	E	TR	Bases
A	1	a	any $f(x, y, z)$
$A_{1/2}$	1	a	

2 C_2

C_2	E	C_2	TR	Bases
A	1	1	a	$z, R_z, x^2, y^2, z^2, xy$
B	1	−1	a	x, y, R_x, R_y, yz, zx
${}^1E_{1/2}$	1	i	b	
${}^2E_{1/2}$	1	−i	b	

3 C_3

C_3	E	C_3^+	C_3^-	TR	Bases
A	1	1	1	a	z, R_z, x^2+y^2, z^2
${}^1E$	1	ε	ε^*	b	} $(x, y), (R_x, R_y), (yz, zx), (xy, x^2-y^2)$
${}^2E$	1	ε^*	ε	b	
${}^1E_{1/2}$	1	$-\varepsilon$	$-\varepsilon^*$	b	
${}^2E_{1/2}$	1	$-\varepsilon^*$	$-\varepsilon$	b	
$B_{3/2}$	1	−1	−1	a	

$\varepsilon = \exp(-\text{i}2\pi/3)$.

4 C_4

C_4	E	C_4^+	C_2	C_4^-	TR	Bases
A	1	1	1	1	a	z, R_z, x^2+y^2, z^2
B	1	−1	1	−1	a	xy, x^2-y^2
${}^1E$	1	−i	−1	i	b	} $(x, y), (R_x, R_y), (yz, zx)$
${}^2E$	1	i	−1	−i	b	

Table 4 (*cont.*)

C_4	E	C_4^+	C_2	C_4^-	TR	Bases
$^1E_{1/2}$	1	ε	$-i$	ε^*	b	
$^2E_{1/2}$	1	ε^*	i	ε	b	
$^1E_{3/2}$	1	$-\varepsilon$	$-i$	$-\varepsilon^*$	b	
$^2E_{3/2}$	1	$-\varepsilon^*$	i	$-\varepsilon$	b	

$\varepsilon = \exp(-i2\pi/4)$.

5 C_5

C_5	E	C_5^+	C_5^{2+}	C_5^{2-}	C_5^-	TR	Bases
A	1	1	1	1	1	a	z, R_z, x^2+y^2, z^2
$^1E_1$	1	δ	ε	ε^*	δ^*	b	$(x, y), (R_x, R_y), (yz, zx)$
$^2E_1$	1	δ^*	ε^*	ε	δ	b	
$^1E_2$	1	ε	δ^*	δ	ε^*	b	(xy, x^2-y^2)
$^2E_2$	1	ε^*	δ	δ^*	ε	b	
$^1E_{1/2}$	1	$-\varepsilon$	δ^*	δ	$-\varepsilon^*$	b	
$^2E_{1/2}$	1	$-\varepsilon^*$	δ	δ^*	$-\varepsilon$	b	
$^1E_{3/2}$	1	$-\delta^*$	ε^*	ε	$-\delta$	b	
$^2E_{3/2}$	1	$-\delta$	ε	ε^*	$-\delta^*$	b	
$B_{5/2}$	1	-1	1	1	-1	a	

$\delta = \exp(-i2\pi/5)$; $\varepsilon = \exp(-i4\pi/5)$.

6 C_6

C_6	E	C_6^+	C_3^+	C_2	C_3^-	C_6^-	TR	Bases
A	1	1	1	1	1	1	a	z, R_z, x^2+y^2, z^2
B	1	-1	1	-1	1	-1	a	
$^1E_1$	1	$-\varepsilon^*$	ε	-1	ε^*	$-\varepsilon$	b	$(x, y), (R_x, R_y), (yz, zx)$
$^2E_1$	1	$-\varepsilon$	ε^*	-1	ε	$-\varepsilon^*$	b	
$^1E_2$	1	ε^*	ε	1	ε^*	ε	b	(xy, x^2-y^2)
$^2E_2$	1	ε	ε^*	1	ε	ε^*	b	
$^1E_{1/2}$	1	$-i\varepsilon^*$	$-\varepsilon$	i	$-\varepsilon^*$	$i\varepsilon$	b	
$^2E_{1/2}$	1	$i\varepsilon$	$-\varepsilon^*$	$-i$	$-\varepsilon$	$-i\varepsilon^*$	b	
$^1E_{3/2}$	1	$-i$	-1	i	-1	i	b	
$^2E_{3/2}$	1	i	-1	$-i$	-1	$-i$	b	
$^1E_{5/2}$	1	$-i\varepsilon$	$-\varepsilon^*$	i	$-\varepsilon$	$i\varepsilon^*$	b	
$^2E_{5/2}$	1	$i\varepsilon^*$	$-\varepsilon$	$-i$	$-\varepsilon^*$	$-i\varepsilon$	b	

$\varepsilon = \exp(-i2\pi/3)$.

A3.2 The improper cyclic groups C_i, C_s, S_n

$\bar{1}$ C_i

C_i	E	I	TR	Bases
A_g	1	1	a	$R_x, R_y, R_z, x^2, y^2, z^2, xy, yz, zx$
A_u	1	−1	a	x, y, z
$A_{1/2, g}$	1	1	a	
$A_{1/2, u}$	1	−1	a	

m C_s

C_s	E	σ_h	TR	Bases
A'	1	1	a	$x, y, R_z, x^2, y^2, z^2, xy$
A''	1	−1	a	z, R_x, R_y, yz, zx
$^1E_{1/2}$	1	i	b	
$^2E_{1/2}$	1	−i	b	

$\bar{4}$ S_4

S_4	E	S_4^-	C_2	S_4^+	TR	Bases
A	1	1	1	1	a	R_z, x^2+y^2, z^2
B	1	−1	1	−1	a	z, xy, x^2-y^2
^1E	1	−i	−1	i	b	$(x, y), (R_x, R_y), (yz, zx)$
^2E	1	i	−1	−i	b	
$^1E_{1/2}$	1	ε	−i	ε^*	b	
$^2E_{1/2}$	1	ε^*	i	ε	b	
$^1E_{3/2}$	1	$-\varepsilon$	−i	$-\varepsilon^*$	b	
$^2E_{3/2}$	1	$-\varepsilon^*$	i	$-\varepsilon$	b	

$\varepsilon = \exp(-i\pi/4)$.

$\bar{6}$ S_6

S_6	E	C_3^+	C_3^-	I	S_6^-	S_6^+	TR	Bases
A_g	1	1	1	1	1	1	a	R_z, x^2+y^2, z^2
^1E_g	1	ε	ε^*	1	ε	ε^*	b	$(R_x, R_y), (yz, zx), (xy, x^2-y^2)$
^2E_g	1	ε^*	ε	1	ε^*	ε	b	
A_u	1	1	1	−1	−1	−1	a	z
^1E_u	1	ε	ε^*	−1	$-\varepsilon$	$-\varepsilon^*$	b	(x, y)
^2E_u	1	ε^*	ε	−1	$-\varepsilon^*$	$-\varepsilon$	b	
$^1E_{1/2, g}$	1	$-\varepsilon$	$-\varepsilon^*$	1	$-\varepsilon$	$-\varepsilon^*$	b	

Table $\overline{6}$ (*cont.*)

S_6	E	C_3^+	C_3^-	I	S_6^-	S_6^+	TR	Bases
$^2E_{1/2,\,g}$	1	$-\varepsilon^*$	$-\varepsilon$	1	$-\varepsilon^*$	$-\varepsilon$	b	
$B_{3/2,\,g}$	1	-1	-1	1	-1	-1	a	
$^1E_{1/2,\,u}$	1	$-\varepsilon$	$-\varepsilon^*$	-1	ε	$-\varepsilon^*$	b	
$^2E_{1/2,\,u}$	1	$-\varepsilon^*$	$-\varepsilon$	-1	$-\varepsilon^*$	ε	b	
$B_{3/2,\,u}$	1	-1	-1	-1	1	1	a	

$\varepsilon = \exp(-i2\pi/3)$.

$\overline{8}$ S_8

S_8	E	S_8^{3-}	C_4^+	S_8^-	C_2	S_8^+	C_4^-	S_8^{3+}	TR	Bases
A	1	1	1	1	1	1	1	1	a	R_z, x^2+y^2, z^2
B	1	-1	1	-1	1	-1	1	-1	a	z
$^1E_1$	1	$-\varepsilon$	$-i$	ε^*	-1	ε	i	$-\varepsilon^*$	b	$(x, y), (R_x, R_y)$
$^2E_1$	1	$-\varepsilon^*$	i	ε	-1	ε^*	$-i$	$-\varepsilon$	b	
$^1E_2$	1	$-i$	-1	i	1	$-i$	-1	i	b	(xy, x^2-y^2)
$^2E_2$	1	i	-1	$-i$	1	i	-1	$-i$	b	
$^1E_3$	1	ε	$-i$	$-\varepsilon^*$	-1	$-\varepsilon$	i	ε^*	b	(yz, zx)
$^2E_3$	1	ε^*	i	$-\varepsilon$	-1	$-\varepsilon^*$	$-i$	ε	b	
$^1E_{1/2}$	1	δ^*	ε^*	$i\delta$	i	$-i\delta^*$	ε	δ	b	
$^2E_{1/2}$	1	δ	ε	$-i\delta^*$	$-i$	$i\delta$	ε^*	δ^*	b	
$^1E_{3/2}$	1	$-i\delta^*$	$-\varepsilon^*$	$-\delta$	i	$-\delta^*$	$-\varepsilon$	$i\delta$	b	
$^2E_{3/2}$	1	$i\delta$	$-\varepsilon$	$-\delta^*$	$-i$	$-\delta$	$-\varepsilon^*$	$-i\delta^*$	b	
$^1E_{5/2}$	1	$i\delta^*$	$-\varepsilon^*$	δ	i	δ^*	$-\varepsilon$	$-i\delta$	b	
$^2E_{5/2}$	1	$-i\delta$	$-\varepsilon$	δ^*	$-i$	δ	$-\varepsilon^*$	$i\delta^*$	b	
$^1E_{7/2}$	1	$-\delta^*$	ε^*	$-i\delta$	i	$i\delta^*$	ε	$-\delta$	b	
$^2E_{7/2}$	1	$-\delta$	ε	$i\delta^*$	$-i$	$-i\delta$	ε^*	$-\delta^*$	b	

$\delta = \exp(-i\pi/8)$; $\varepsilon = \exp(-i\pi/4)$.

A3.3 The dihedral groups D_n

222 D_2

D_2	E	C_{2z}	C_{2x}	C_{2y}	TR	Bases
A	1	1	1	1	a	x^2, y^2, z^2
B_1	1	1	-1	-1	a	z, R_z, xy
B_2	1	-1	-1	1	a	y, R_y, zx
B_3	1	-1	1	-1	a	x, R_x, yz
$E_{1/2}$	2	0	0	0	c	

32 D_3

D_3	E	$2C_3$	$3C_2'$	TR	Bases
A_1	1	1	1	a	x^2+y^2, z^2
A_2	1	1	−1	a	z, R_z
E	2	−1	0	a	$(x, y), (R_x, R_y), (xy, x^2-y^2), (yz, zx)$
$E_{1/2}$	2	1	0	c	
$^1E_{3/2}$	1	−1	i	b	
$^2E_{3/2}$	1	−1	−i	b	

422 D_4

D_4	E	$2C_4$	C_2	$2C_2'$	$2C_2''$	TR	Bases
A_1	1	1	1	1	1	a	x^2+y^2, z^2
A_2	1	1	1	−1	−1	a	z, R_z
B_1	1	−1	1	1	−1	a	$x-y^2$
B_2	1	−1	1	−1	1	a	xy
E	2	0	−2	0	0	a	$(x, y), (R_x, R_y), (yz, zx)$
$E_{1/2}$	2	$\sqrt{2}$	0	0	0	c	
$E_{3/2}$	2	$-\sqrt{2}$	0	0	0	c	

52 D_5

D_5	E	$2C_5$	$2C_5^2$	$5C_2'$	TR	Bases
A_1	1	1	1	1	a	x^2+y^2, z^2
A_2	1	1	1	−1	a	z, R_z
E_1	2	$2c_5^2$	$2c_5^4$	0	a	$(x,y), (R_x, R_y), (yz, zx)$
E_2	2	$2c_5^4$	$2c_5^2$	0	a	(xy, x^2-y^2)
$E_{1/2}$	2	$-2c_5^4$	$2c_5^2$	0	c	
$E_{3/2}$	2	$-2c_5^2$	$2c_5^4$	0	c	
$^1E_{5/2}$	1	−1	1	i	b	
$^2E_{5/2}$	1	−1	1	−i	b	

$c_n^m = \cos(m\pi/n)$. This economical notation was used by Altmann and Herzig (1994) and I have adopted it in order to reduce column within the tables.

622 D_6

D_6	E	$2C_6$	$2C_3$	C_2	$3C_2'$	$3C_2''$	TR	Bases
A_1	1	1	1	1	1	1	a	x^2+y^2, z^2
A_2	1	1	1	1	−1	−1	a	z, R_z
B_1	1	−1	1	−1	1	−1	a	

Table 622 (*cont.*)

D_6	E	$2C_6$	$2C_3$	C_2	$3C_2'$	$3C_2''$	TR	Bases
B_2	1	−1	1	−1	−1	1	*a*	
E_1	2	1	−1	−2	0	0	*a*	$(x, y), (R_x, R_y), (yz, zx)$
E_2	2	−1	−1	2	0	0	*a*	(xy, x^2-y^2)
$E_{1/2}$	2	$\sqrt{3}$	1	0	0	0	*c*	
$E_{3/2}$	2	0	−2	0	0	0	*c*	
$E_{5/2}$	2	$-\sqrt{3}$	1	0	0	0	*c*	

A3.4 The C_{nh} groups

2/*m* C_{2h}

C_{2h}	E	C_2	I	σ_h	TR	Bases
A_g	1	1	1	1	*a*	x^2, y^2, z^2, xy, R_z
B_g	1	−1	1	−1	*a*	R_x, R_y, yz, zx
A_u	1	1	−1	−1	*a*	z
B_u	1	−1	−1	1	*a*	x, y
$^1E_{1/2,g}$	1	i	1	i	*b*	
$^2E_{1/2,g}$	1	−i	1	−i	*b*	
$^1E_{1/2,u}$	1	i	−1	−i	*b*	
$^2E_{1/2,u}$	1	−i	−1	i	*b*	

3/*m* C_{3h}

C_{3h}	E	C_3^+	C_3^-	σ_h	S_3^+	S_3^-	TR	Bases
A'	1	1	1	1	1	1	*a*	R_z, x^2+y^2, z^2
$^1E'$	1	ε	ε^*	1	ε	ε^*	*b*	$(x, y), (xy, x^2-y^2)$
$^2E'$	1	ε^*	ε	1	ε^*	ε	*b*	
A''	1	1	1	−1	−1	−1	*a*	z
$^1E''$	1	ε	ε^*	−1	$-\varepsilon$	$-\varepsilon^*$	*b*	$(R_x, R_y), (yz, zx)$
$^2E''$	1	ε^*	ε	−1	$-\varepsilon^*$	$-\varepsilon$	*b*	
$^1E_{1/2}$	1	$-\varepsilon$	$-\varepsilon^*$	i	$i\varepsilon$	$-i\varepsilon^*$	*b*	
$^2E_{1/2}$	1	$-\varepsilon^*$	$-\varepsilon$	−i	$-i\varepsilon^*$	$i\varepsilon$	*b*	
$^1E_{3/2}$	1	−1	−1	i	1	−i	*b*	
$^2E_{3/2}$	1	−1	−1	−i	−1	i	*b*	
$^1E_{5/2}$	1	$-\varepsilon^*$	$-\varepsilon$	i	$i\varepsilon^*$	$-i\varepsilon$	*b*	
$^2E_{5/2}$	1	$-\varepsilon$	$-\varepsilon^*$	−i	$-i\varepsilon$	$i\varepsilon^*$	*b*	

$\varepsilon = \exp(-i2\pi/3)$.

4*m* C_{4h}

C_{4h}	E	C_4^+	C_2	C_4^-	I	S_4^-	σ_h	S_4^+	TR	Bases
A_g	1	1	1	1	1	1	1	1	a	R_z, x^2+y^2, z^2
B_g	1	−1	1	−1	1	−1	1	−1	a	xy, x^2-y^2
${}^1E_g$	1	−i	−1	i	1	−i	−1	i	b	$(R_x, R_y), (yz, zx)$
${}^2E_g$	1	i	−1	−i	1	i	−1	−i	b	
A_u	1	1	1	1	−1	−1	−1	−1	a	z
B_u	1	−1	1	−1	−1	1	−1	1	a	
${}^1E_u$	1	−i	−1	i	−1	i	1	−i	b	(x, y)
${}^2E_u$	1	i	−1	−i	−1	−i	1	i	b	
${}^1E_{1/2,g}$	1	ε	−i	ε^*	1	ε	−i	ε^*	b	
${}^2E_{1/2,g}$	1	ε^*	i	ε	1	ε^*	i	ε	b	
${}^1E_{3/2,g}$	1	$-\varepsilon$	−i	ε^*	1	$-\varepsilon$	−i	ε^*	b	
${}^2E_{3/2,g}$	1	$-\varepsilon^*$	i	ε	1	$-\varepsilon^*$	i	ε	b	
${}^1E_{1/2,u}$	1	ε	−i	ε^*	−1	$-\varepsilon$	i	$-\varepsilon^*$	b	
${}^2E_{1/2,u}$	1	ε^*	i	ε	−1	$-\varepsilon^*$	−i	$-\varepsilon$	b	
${}^1E_{3/2,u}$	1	$-\varepsilon$	−i	$-\varepsilon^*$	−1	ε	i	ε^*	b	
${}^2E_{3/2,u}$	1	$-\varepsilon^*$	i	$-\varepsilon$	−1	ε^*	−i	ε	b	

$\varepsilon = \exp(-i\pi/4)$.

5/*m* C_{5h}

C_{5h}	E	C_5^+	C_5^{2+}	C_5^{2-}	C_5^-	σ_h	S_5^+	S_5^{2+}	S_5^{2-}	S_5^-	TR	Bases
A'	1	1	1	1	1	1	1	1	1	1	a	R_z, x^2+y^2, z^2
${}^1E_1'$	1	δ	ε	ε^*	δ^*	1	δ	ε	ε^*	δ^*	b	(x, y)
${}^2E_1'$	1	δ^*	ε^*	ε	δ	1	δ^*	ε^*	ε	δ	b	
${}^1E_2'$	1	ε	δ^*	δ	ε^*	1	ε	δ^*	δ	ε^*	b	(xy, x^2-y^2)
${}^2E_2'$	1	ε^*	δ	δ^*	ε	1	ε^*	δ	δ^*	ε	b	
A''	1	1	1	1	1	−1	−1	−1	−1	−1	a	z
${}^1E_1''$	1	δ	ε	ε^*	δ^*	−1	$-\delta$	$-\varepsilon$	$-\varepsilon^*$	$-\delta^*$	b	$(R_x, R_y), (yz, zx)$
${}^2E_1''$	1	δ^*	ε^*	ε	δ	−1	$-\delta^*$	$-\varepsilon^*$	$-\varepsilon$	$-\delta$	b	
${}^1E_2''$	1	ε	δ^*	δ	ε^*	−1	$-\varepsilon$	$-\delta^*$	$-\delta$	$-\varepsilon^*$	b	
${}^2E_2''$	1	ε^*	δ	δ^*	ε	−1	$-\varepsilon^*$	$-\delta$	$-\delta^*$	$-\varepsilon$	b	
${}^1E_{1/2}$	1	$-\varepsilon$	δ^*	δ	$-\varepsilon^*$	i	$i\varepsilon$	$-i\delta^*$	$i\delta$	$-i\varepsilon^*$	b	
${}^2E_{1/2}$	1	$-\varepsilon^*$	δ	δ^*	$-\varepsilon$	−i	$-i\varepsilon^*$	$i\delta$	$-i\delta^*$	$i\varepsilon$	b	
${}^1E_{3/2}$	1	$-\delta^*$	ε^*	ε	$-\delta$	i	$i\delta^*$	$-i\varepsilon^*$	$i\varepsilon$	$-i\delta$	b	
${}^2E_{3/2}$	1	$-\delta$	ε	ε^*	$-\delta^*$	−i	$-i\delta$	$i\varepsilon$	$-i\varepsilon^*$	$i\delta^*$	b	
${}^1E_{5/2}$	1	−1	1	1	−1	i	i	−i	i	−i	b	
${}^2E_{5/2}$	1	−1	1	1	−1	−i	−i	i	−i	i	b	
${}^1E_{7/2}$	1	$-\delta$	ε	ε^*	$-\delta^*$	i	$i\delta$	$-i\varepsilon$	$i\varepsilon^*$	$-i\delta^*$	b	
${}^2E_{7/2}$	1	$-\delta^*$	ε^*	ε	$-\delta$	−i	$-i\delta^*$	$i\varepsilon^*$	$-i\varepsilon$	$i\delta$	b	
${}^1E_{9/2}$	1	$-\varepsilon^*$	δ	δ^*	$-\varepsilon$	i	$i\varepsilon^*$	$-i\delta$	$i\delta^*$	$-i\varepsilon$	b	
${}^2E_{9/2}$	1	$-\varepsilon$	δ^*	δ	$-\varepsilon^*$	−i	$-i\varepsilon$	$i\delta^*$	$-i\delta$	$i\varepsilon^*$	b	

$\delta = \exp(-i2\pi/5)$; $\varepsilon = \exp(-i4\pi/5)$.

$6/m$ C_{6h}

C_{6h}	E	C_6^+	C_3^+	C_2	C_3^-	C_6^-	I	S_3^-	S_6^-	σ_h	S_6^+	S_3^+	TR	Bases
A_g	1	1	1	1	1	1	1	1	1	1	1	1	a	R_z, x^2+y^2, z^2
B_g	1	−1	1	−1	1	−1	1	−1	1	−1	1	−1	a	
$^1E_{1g}$	1	ε^*	$-\varepsilon$	−1	$-\varepsilon^*$	ε	1	ε^*	$-\varepsilon$	−1	$-\varepsilon^*$	ε	a	$(R_x, R_y), (yz, zx)$
$^2E_{1g}$	1	ε	$-\varepsilon^*$	−1	$-\varepsilon$	ε^*	1	ε	$-\varepsilon^*$	−1	$-\varepsilon$	ε^*	a	
$^1E_{2g}$	1	$-\varepsilon$	$-\varepsilon^*$	1	$-\varepsilon$	$-\varepsilon^*$	1	$-\varepsilon$	$-\varepsilon^*$	1	$-\varepsilon$	$-\varepsilon^*$	a	(xy, x^2-y^2)
$^2E_{2g}$	1	$-\varepsilon^*$	$-\varepsilon$	1	$-\varepsilon^*$	$-\varepsilon$	1	$-\varepsilon^*$	$-\varepsilon$	1	$-\varepsilon^*$	$-\varepsilon$	a	
A_u	1	1	1	1	1	1	−1	−1	−1	−1	−1	−1	a	z
B_u	1	−1	1	−1	1	−1	−1	1	−1	1	−1	1	a	
$^1E_{1u}$	1	ε^*	$-\varepsilon$	−1	$-\varepsilon^*$	ε	−1	$-\varepsilon^*$	ε	1	ε^*	$-\varepsilon$	a	(x,y)
$^2E_{1u}$	1	ε	$-\varepsilon^*$	−1	$-\varepsilon$	ε^*	−1	$-\varepsilon$	ε^*	1	ε	$-\varepsilon^*$	a	
$^1E_{2u}$	1	$-\varepsilon$	$-\varepsilon^*$	1	$-\varepsilon$	$-\varepsilon^*$	−1	ε	ε^*	−1	ε	ε^*	a	
$^2E_{2u}$	1	$-\varepsilon^*$	$-\varepsilon$	1	$-\varepsilon^*$	$-\varepsilon$	−1	ε^*	ε	−1	ε^*	ε	a	
$^1E_{1/2,g}$	1	$-i\varepsilon^*$	$-\varepsilon$	i	$-\varepsilon^*$	$i\varepsilon$	1	$-i\varepsilon^*$	$-\varepsilon$	i	$-\varepsilon^*$	$i\varepsilon$	b	
$^2E_{1/2,g}$	1	$i\varepsilon$	$-\varepsilon^*$	−i	$-\varepsilon$	$-i\varepsilon^*$	1	$i\varepsilon$	$-\varepsilon^*$	−i	$-\varepsilon$	$-i\varepsilon^*$	b	
$^1E_{3/2,g}$	1	−i	−1	i	−1	i	1	−i	−1	−i	−1	i	b	
$^2E_{3/2,g}$	1	i	−1	−i	−1	−i	1	i	−1	i	−1	−i	b	
$^1E_{5/2,g}$	1	$-i\varepsilon^*$	$-\varepsilon^*$	i	$-\varepsilon$	$i\varepsilon^*$	1	$-i\varepsilon$	$-\varepsilon^*$	i	$-\varepsilon$	$i\varepsilon^*$	b	
$^2E_{5/2,g}$	1	$i\varepsilon^*$	$-\varepsilon$	−i	$-\varepsilon^*$	$-i\varepsilon$	1	$i\varepsilon^*$	$-\varepsilon$	−i	$-\varepsilon^*$	$-i\varepsilon$	b	
$^1E_{1/2,u}$	1	$-i\varepsilon^*$	$-\varepsilon$	i	$-\varepsilon^*$	$i\varepsilon$	−1	$i\varepsilon^*$	ε	−i	ε^*	$-i\varepsilon$	b	
$^2E_{1/2,u}$	1	$i\varepsilon$	$-\varepsilon^*$	−i	$-\varepsilon$	$-i\varepsilon^*$	−1	$-i\varepsilon$	ε^*	i	ε	$i\varepsilon^*$	b	
$^1E_{3/2,u}$	1	−i	−1	i	−1	i	−1	i	1	−i	1	−i	b	
$^2E_{3/2,u}$	1	i	−1	−i	−1	−i	−1	−i	1	i	1	i	b	
$^1E_{5/2,u}$	1	$-i\varepsilon$	$-\varepsilon^*$	i	$-\varepsilon$	$i\varepsilon^*$	−1	$i\varepsilon$	ε^*	−i	ε	$-i\varepsilon^*$	b	
$^2E_{5/2,u}$	1	$i\varepsilon^*$	$-\varepsilon$	−i	$-\varepsilon^*$	$-i\varepsilon$	−1	$-i\varepsilon^*$	ε	i	ε^*	$i\varepsilon$	b	

$\varepsilon = \exp(-i\pi/3)$.

A3.5 The C_{nv} groups

$2mm$ C_{2v}

C_{2v}	E	C_{2z}	σ_x	σ_y	TR	Bases
A_1	1	1	1	1	a	z, x^2, y^2, z^2
A_2	1	1	−1	−1	a	R_z, xy
B_1	1	−1	−1	1	a	x, R_y, zx
B_2	1	−1	1	−1	a	y, R_x, yz
$E_{1/2}$	2	0	0	0	c	

$3m$ C_{3v}

C_{3v}	E	$2C_3$	$3\sigma_v$	TR	Bases
A_1	1	1	1	a	z, x^2+y^2, z^2
A_2	1	1	−1	a	R_z
E	2	−1	0	a	$(x, y), (R_x, R_y), (xy, x^2-y^2), (yz, zx)$

Table $3m$ (*cont.*)

C_{3v}	E	$2C_3$	$3\sigma_v$	TR	Bases
$E_{1/2}$	2	1	0	c	
${}^1E_{3/2}$	1	−1	i	b	
${}^2E_{3/2}$	1	−1	−i	b	

For $\mathbf{n} = \mathbf{z}$, $3\sigma_v$ are $\sigma_\mathbf{d}$, $\sigma_\mathbf{e}$, $\sigma_\mathbf{f}$ in Figure 12.10.

$4mm$ C_{4v}

C_{4v}	E	$2C_4$	C_2	$2\sigma_v$	$2\sigma_d$	TR	Bases
A_1	1	1	1	1	1	a	z, x^2+y^2, z^2
A_2	1	1	1	−1	−1	a	R_z
B_1	1	−1	1	1	−1	a	x^2-y^2
B_2	1	−1	1	−1	1	a	xy
E	2	0	−2	0	0	a	$(x, y), (R_x, R_y), (yz, zx)$
$E_{1/2}$	2	$\sqrt{2}$	0	0	0	c	
$E_{3/2}$	2	$-\sqrt{2}$	0	0	0	c	

$5m$ C_{5v}

C_{5v}	E	$2C_5$	$2C_5^2$	$5\sigma_v$	TR	Bases
A_1	1	1	1	1	a	z, x^2+y^2, z^2
A_2	1	1	1	−1	a	R_z
E_1	2	$2c_5^2$	$2c_5^4$	0	a	$(x, y), (R_x, R_y), (yz, zx)$
E_2	2	$2c_5^4$	$2c_5^2$	0	a	$xy,\ x^2-y^2$
$E_{1/2}$	2	$-2c_5^4$	$2c_5^2$	0	c	
$E_{3/2}$	2	$-2c_5^2$	$2c_5^4$	0	c	
${}^1E_{5/2}$	1	−1	1	i	b	
${}^2E_{5/2}$	1	−1	1	−i	b	

$c_n^m = \cos(m\pi/n)$.

$6mm$ C_{6v}

C_{6v}	E	$2C_6$	$2C_3$	C_2	$3\sigma_d$	$3\sigma_v$	TR	Bases
A_1	1	1	1	1	1	1	a	z, x^2+y^2, z^2
A_2	1	1	1	1	−1	−1	a	R_z
B_1	1	−1	1	−1	−1	1	a	
B_2	1	−1	1	−1	1	−1	a	
E_1	2	1	−1	−2	0	0	a	$(x, y), (R_x, R_y), (yz, zx)$
E_2	2	−1	−1	2	0	0	a	(xy, x^2-y^2)
$E_{1/2}$	2	$\sqrt{3}$	1	0	0	0	c	
$E_{3/2}$	2	0	−2	0	0	0	c	
$E_{5/2}$	2	$-\sqrt{3}$	1	0	0	0	c	

∞m $C_{\infty v}$

$C_{\infty v}$	E	$2C_{\infty}(\phi)$	C_2	$\infty\sigma_v$	TR	Bases
$A_1(\Sigma^+)$	1	1	1	1	a	z, x^2+y^2, z^2
$A_2(\Sigma^-)$	1	1	1	−1	a	R_z
$E_1(\Pi)$	2	$2\cos\phi$	−2	0	a	$(x, y), (R_x, R_y), (yz, zx)$
$E_2(\Delta)$	2	$2\cos 2\phi$	2	0	a	(xy, x^2-y^2)
$E_3(\Phi)$	2	$2\cos 3\phi$	−2	0	a	
E_n	2	$2\cos n\phi$	$2(-1)^n$	0	a	
$E_{1/2}$	2	$2\cos \frac{1}{2}\phi$	0	0	c	
$E_{(2n+1)/2}$	2	$2\cos\left(\frac{2n+1}{2}\right)\phi$	0	0	c	

$n = 1, 2, 3, \ldots$

A3.6 The D_{nh} groups

mmm D_{2h}

D_{2h}	E	$C_{2\mathbf{z}}$	$C_{2\mathbf{x}}$	$C_{2\mathbf{y}}$	I	$\sigma_{\mathbf{z}}$	$\sigma_{\mathbf{x}}$	$\sigma_{\mathbf{y}}$	TR	Bases
A_g	1	1	1	1	1	1	1	1	a	x^2, y^2, z^2
B_{1g}	1	1	−1	−1	1	1	−1	−1	a	R_z, xy
B_{2g}	1	−1	−1	1	1	−1	−1	1	a	R_y, zx
B_{3g}	1	−1	1	−1	1	−1	1	−1	a	R_x, yz
A_u	1	1	1	1	−1	−1	−1	−1	a	
B_{1u}	1	1	−1	−1	−1	−1	1	1	a	z
B_{2u}	1	−1	−1	1	−1	1	1	−1	a	y
B_{3u}	1	−1	1	−1	−1	1	−1	1	a	x
$E_{1/2,\,g}$	2	0	0	0	2	0	0	0	c	
$E_{1/2,\,u}$	2	0	0	0	−2	0	0	0	c	

$\bar{6}m2$ D_{3h}

D_{3h}	E	$2C_3$	$3C_2'$	σ_h	$2S_3$	$3\sigma_v$	TR	Bases
A_1'	1	1	1	1	1	1	a	x^2+y^2, z^2
A_2'	1	1	−1	1	1	−1	a	R_z
E'	2	−1	0	2	−1	0	a	$(x, y), (xy, x^2-y^2)$
A_1''	1	1	1	−1	−1	−1	a	
A_2''	1	1	−1	−1	−1	1	a	z
E''	2	−1	0	−2	1	0	a	$(R_x, R_y), (yz, zx)$
$E_{1/2}$	2	1	0	0	$\sqrt{3}$	0	c	
$E_{3/2}$	2	−2	0	0	0	0	c	
$E_{5/2}$	2	1	0	0	$-\sqrt{3}$	0	c	

$4/mmm$ D_{4h}

D_{4h}	E	$2C_4$	C_2	$2C_2'$	$2C_2''$	I	$2S_4$	σ_h	$2\sigma_v$	$2\sigma_d$	TR	Bases
A_{1g}	1	1	1	1	1	1	1	1	1	1	a	x^2+y^2, z^2
A_{2g}	1	1	1	−1	−1	1	1	1	−1	−1	a	R_z
B_{1g}	1	−1	1	1	−1	1	−1	1	1	−1	a	x^2-y^2
B_{2g}	1	−1	1	−1	1	1	−1	1	−1	1	a	xy
E_g	2	0	−2	0	0	2	0	−2	0	0	a	$(R_x, R_y), (yz, zx)$
A_{1u}	1	1	1	1	1	−1	−1	−1	−1	−1	a	
A_{2u}	1	1	1	−1	−1	−1	−1	−1	1	1	a	z
B_{1u}	1	−1	1	1	−1	−1	1	−1	−1	1	a	
B_{2u}	1	−1	1	−1	1	−1	1	−1	1	−1	a	
E_u	2	0	−2	0	0	−2	0	2	0	0	a	(x, y)
$E_{1/2,g}$	2	$\sqrt{2}$	0	0	0	2	$\sqrt{2}$	0	0	0	c	
$E_{3/2,g}$	2	$-\sqrt{2}$	0	0	0	2	$-\sqrt{2}$	0	0	0	c	
$E_{1/2,u}$	2	$\sqrt{2}$	0	0	0	−2	$-\sqrt{2}$	0	0	0	c	
$E_{3/2,u}$	2	$-\sqrt{2}$	0	0	0	−2	$\sqrt{2}$	0	0	0	c	

$\overline{10}m2$ D_{5h}

D_{5h}	E	$2C_5$	$2C_5^2$	$5C_2'$	σ_h	$2S_5$	$2S_5^2$	$5\sigma_v$	TR	Bases
A_1'	1	1	1	1	1	1	1	1	a	x^2+y^2, z^2
A_2'	1	1	1	−1	1	1	1	−1	a	R_z
E_1'	2	$2c_5^2$	$2c_5^4$	0	2	$2c_5^2$	$2c_5^4$	0	a	(x, y)
E_2'	2	$2c_5^4$	$2c_5^2$	0	2	$2c_5^4$	$2c_5^2$	0	a	(xy, x^2-y^2)
A_1''	1	1	1	1	−1	−1	−1	−1	a	
A_2''	1	1	1	−1	−1	−1	−1	1	a	z
E_1''	2	$2c_5^2$	$2c_5^4$	0	−2	$-2c_5^2$	$-2c_5^4$	0	a	$(R_x, R_y), (yz, zx)$
E_2''	2	$2c_5^4$	$2c_5^2$	0	−2	$-2c_5^4$	$-2c_5^2$	0	a	
$E_{1/2}$	2	$-2c_5^4$	$2c_5^2$	0	0	$2c_{10}^1$	$2c_{10}^3$	0	c	
$E_{3/2}$	2	$-2c_5^2$	$2c_5^4$	0	0	$-2c_{10}^3$	$2c_{10}^1$	0	c	
$E_{5/2}$	2	−2	2	0	0	0	0	0	c	
$E_{7/2}$	2	$-2c_5^2$	$2c_5^4$	0	0	$2c_{10}^3$	$-2c_{10}^1$	0	c	
$E_{9/2}$	2	$-2c_5^4$	$2c_5^2$	0	0	$-2c_{10}^1$	$-2c_{10}^3$	0	c	

$c_n^m = \cos(m\pi/n)$.

$6/mmm$ D_{6h}

D_{6h}	E	$2C_6$	$2C_3$	C_2	$3C_2'$	$3C_2''$	I	$2S_3$	$2S_6$	σ_h	$3\sigma_d$	$3\sigma_v$	TR	Bases
A_{1g}	1	1	1	1	1	1	1	1	1	1	1	1	a	x^2+y^2, z^2
A_{2g}	1	1	1	1	−1	−1	1	1	1	1	−1	−1	a	R_z
B_{1g}	1	−1	1	−1	1	−1	1	−1	1	−1	1	−1	a	
B_{2g}	1	−1	1	−1	−1	1	1	−1	1	−1	−1	1	a	
E_{1g}	2	1	−1	−2	0	0	2	1	−1	−2	0	0	a	$(R_x, R_y), (yz, zx)$
E_{2g}	2	−1	−1	2	0	0	2	−1	−1	2	0	0	a	(xy, x^2-y^2)
A_{1u}	1	1	1	1	1	1	−1	−1	−1	−1	−1	−1	a	
A_{2u}	1	1	1	1	−1	−1	−1	−1	−1	−1	1	1	a	z

Table 6/*mmm* (*cont.*)

D_{6h}	E	$2C_6$	$2C_3$	C_2	$3C_2'$	$3C_2''$	I	$2S_3$	$2S_6$	σ_h	$3\sigma_d$	$3\sigma_v$	TR	Bases
B_{1u}	1	−1	1	−1	1	−1	−1	1	−1	1	−1	1	*a*	
B_{2u}	1	−1	1	−1	−1	1	−1	1	−1	1	1	−1	*a*	
E_{1u}	2	1	−1	−2	0	0	−2	−1	1	2	0	0	*a*	(x, y)
E_{2u}	2	−1	−1	2	0	0	−2	1	1	−2	0	0	*a*	
$E_{1/2,g}$	2	$\sqrt{3}$	1	0	0	0	2	$\sqrt{3}$	1	0	0	0	*c*	
$E_{3/2,g}$	2	0	−2	0	0	0	2	0	−2	0	0	0	*c*	
$E_{5/2,g}$	2	$-\sqrt{3}$	1	0	0	0	2	$-\sqrt{3}$	1	0	0	0	*c*	
$E_{1/2,u}$	2	$\sqrt{3}$	1	0	0	0	−2	$-\sqrt{3}$	−1	0	0	0	*c*	
$E_{3/2,u}$	2	0	−2	0	0	0	−2	0	2	0	0	0	*c*	
$E_{5/2,u}$	2	$-\sqrt{3}$	1	0	0	0	−2	$\sqrt{3}$	−1	0	0	0	*c*	

∞/mm $D_{\infty h}$

$D_{\infty h}$	E	$2C_\infty(\phi)$	C_2	$\infty\sigma_v$	σ_h	$2S_\infty(\phi)$	I	$\infty C_2'$	TR	Bases
$A_{1g}(\Sigma_g^+)$	1	1	1	1	1	1	1	1	*a*	x^2+y^2, z^2
$A_{2g}(\Sigma_g^-)$	1	1	1	−1	1	1	1	−1	*a*	R_z
$E_{1g}(\Pi_g)$	2	$2\cos\phi$	−2	0	−2	$-2\cos\phi$	2	0	*a*	(R_x, R_y), (yz, zx)
$E_{2g}(\Delta_g)$	2	$2\cos 2\phi$	2	0	2	$2\cos 2\phi$	2	0	*a*	(xy, x^2-y^2)
$E_{3g}(\Phi_g)$	2	$2\cos 3\phi$	−2	0	−2	$-2\cos 3\phi$	2	0	*a*	
$E_{n,\,g}$	2	$2\cos n\phi$	$2(-1)^n$	0	$2(-1)^n$	$2(-1)^n\cos n\phi$	2	0	*a*	
$A_{1u}(\Sigma_u^+)$	1	1	1	1	−1	−1	−1	−1	*a*	z
$A_{2u}(\Sigma_u^-)$	1	1	1	−1	−1	−1	−1	1	*a*	
$E_{1u}(\Pi_u)$	2	$2\cos\phi$	−2	0	2	$2\cos\phi$	−2	0	*a*	(x, y)
$E_{2u}(\Delta_u)$	2	$2\cos 2\phi$	2	0	−2	$-2\cos 2\phi$	−2	0	*a*	
$E_{3u}(\Phi_u)$	2	$2\cos 3\phi$	−2	0	2	$2\cos 3\phi$	−2	0	*a*	
$E_{n,\,u}$	2	$2\cos n\phi$	$2(-1)^n$	0	$-2(-1)^n$	$-2(-1)^n\cos n\phi$	−2	0	*a*	
$E_{1/2,\,g}$	2	$2\cos\,\frac{1}{2}\phi$	0	0	0	$2\sin\frac{1}{2}\phi$	2	0	*c*	
$E_{(2n+1)/2,\,g}$	2	$2\cos\,(\frac{2n+1}{2})\phi$	0	0	0	$2\sin(\frac{2n+1}{2})\phi$	2	0	*c*	
$E_{1/2,\,u}$	2	$2\cos\,\frac{1}{2}\phi$	0	0	0	$-2\sin\frac{1}{2}\phi$	−2	0	*c*	
$E_{(2n+1)/2,\,u}$	2	$2\cos\,(\frac{2n+1}{2})\phi$	0	0	0	$-2\sin(\frac{2n+1}{2})\phi$	−2	0	*c*	

$n = 1, 2, 3, \ldots$

A3.7 The D_{nd} groups

$\bar{4}2m$ D_{2d}

D_{2d}	E	$2S_4$	C_2	$2C_2'$	$2\sigma_d$	TR	Bases
A_1	1	1	1	1	1	*a*	x^2+y^2, z^2
A_2	1	1	1	−1	−1	*a*	R_z
B_1	1	−1	1	1	−1	*a*	x^2-y^2
B_2	1	−1	1	−1	1	*a*	z, xy
E	2	0	−2	0	0	*a*	$(x, y), (R_x, R_y), (yz, zx)$
$E_{1/2}$	2	$\sqrt{2}$	0	0	0	*c*	
$E_{3/2}$	2	$-\sqrt{2}$	0	0	0	*c*	

$\bar{3}m$ D_{3d}

D_{3d}	E	$2C_3$	$3C_2'$	I	$2S_6$	$3\sigma_d$	TR	Bases
A_{1g}	1	1	1	1	1	1	a	x^2+y^2, z^2
A_{2g}	1	1	−1	1	1	−1	a	R_z
E_g	2	−1	0	2	−1	0	a	$(R_x, R_y), (xy, x^2-y^2), (yz, zx)$
A_{1u}	1	1	1	−1	−1	−1	a	
A_{2u}	1	1	−1	−1	−1	1	a	z
E_u	2	−1	0	−2	1	0	a	(x, y)
$E_{1/2, g}$	2	1	0	2	1	0	c	
${}^1E_{3/2, g}$	1	−1	i	1	−1	i	b	
${}^2E_{3/2, g}$	1	−1	−i	1	−1	−i	b	
$E_{1/2, u}$	2	1	0	−2	−1	0	c	
${}^1E_{3/2, u}$	1	−1	i	−1	1	−i	b	
${}^2E_{3/2, u}$	1	−1	−i	−1	1	i	b	

$\bar{8}2m$ D_{4d}

D_{4d}	E	$2C_4$	C_2	$4C_2'$	$2S_8^3$	$2S_8$	$4\sigma_d$	TR	Bases
A_1	1	1	1	1	1	1	1	a	x^2+y^2, z^2
A_2	1	1	1	−1	1	1	−1	a	R_z
B_1	1	1	1	1	−1	−1	−1	a	
B_2	1	1	1	−1	−1	−1	1	a	z
E_1	2	0	−2	0	$-\sqrt{2}$	$\sqrt{2}$	0	a	(x, y)
E_2	2	−2	2	0	0	0	0	a	(xy, x^2-y^2)
E_3	2	0	−2	0	$\sqrt{2}$	$-\sqrt{2}$	0	a	$(R_x, R_y), (yz, zx)$
$E_{1/2}$	2	$\sqrt{2}$	0	0	$2c_8^1$	$2c_8^3$	0	c	
$E_{3/2}$	2	$-\sqrt{2}$	0	0	$2c_8^3$	$-2c_8^1$	0	c	
$E_{5/2}$	2	$-\sqrt{2}$	0	0	$-2c_8^3$	$2c_8^1$	0	c	
$E_{7/2}$	2	$\sqrt{2}$	0	0	$-2c_8^1$	$-2c_8^3$	0	c	

$c_n^m = \cos(m\pi/n)$.

$\bar{5}m$ D_{5d}

D_{5d}	E	$2C_5$	$2C_5^2$	$5C_2'$	I	$2S_{10}^3$	$2S_{10}$	$5\sigma_d$	TR	Bases
A_{1g}	1	1	1	1	1	1	1	1	a	x^2+y^2, z^2
A_{2g}	1	1	1	−1	1	1	1	−1	a	R_z
E_{1g}	2	$2c_5^2$	$2c_5^4$	0	2	$2c_5^2$	$2c_5^4$	0	a	$R_x, R_y, (yz, zx)$
E_{2g}	2	$2c_5^4$	$2c_5^2$	0	2	$2c_5^4$	$2c_5^2$	0	a	xy, x^2-y^2
A_{1u}	1	1	1	1	−1	−1	−1	−1	a	
A_{2u}	1	1	1	−1	−1	−1	−1	1	a	z
E_{1u}	2	$2c_5^2$	$2c_5^4$	0	−2	$-2c_5^2$	$-2c_5^4$	0	a	(x, y)
E_{2u}	2	$2c_5^4$	$2c_5^2$	0	−2	$-2c_5^4$	$-2c_5^2$	0	a	
$E_{1/2,g}$	2	$-2c_5^4$	$2c_5^2$	0	2	$-2c_5^4$	$2c_5^2$	0	c	
$E_{3/2,g}$	2	$-2c_5^2$	$2c_5^4$	0	2	$-2c_5^2$	$2c_5^4$	0	c	

Table $\bar{5}m$ (*cont.*)

D_{5d}	E	$2C_5$	$2C_5^2$	$5C_2'$	I	$2S_{10}^3$	$2S_{10}$	$5\sigma_d$	TR	Bases
$^1E_{5/2,g}$	1	-1	1	i	1	-1	1	i	b	
$^2E_{5/2,g}$	1	-1	1	$-$i	1	-1	1	$-$i	b	
$E_{1/2,u}$	2	$-2c_5^4$	$2c_5^2$	0	-2	$2c_5^4$	$-2c_5^2$	0	c	
$E_{3/2,u}$	2	$-2c_5^2$	$2c_5^4$	0	-2	$2c_5^2$	$-2c_5^4$	0	c	
$^1E_{5/2,u}$	1	-1	1	i	-1	1	-1	$-$i	b	
$^2E_{5/2,u}$	1	-1	1	$-$i	-1	1	-1	i	b	

$c_n^m = \cos(m\pi/n)$.

$\overline{12}2m$ D_{6d}

D_{6d}	E	$2C_6$	$2C_3$	C_2	$6C_2'$	$2S_{12}^5$	$2S_4$	$2S_{12}$	$6\sigma_d$	TR	Bases
A_1	1	1	1	1	1	1	1	1	1	a	x^2+y^2, z^2
A_2	1	1	1	1	-1	1	1	1	-1	a	R_z
B_1	1	1	1	1	1	-1	-1	-1	-1	a	
B_2	1	1	1	1	-1	-1	-1	-1	1	a	z
E_1	2	1	-1	-2	0	$-\sqrt{3}$	0	$\sqrt{3}$	0	a	(x, y)
E_2	2	-1	-1	2	0	1	-2	1	0	a	(xy, x^2-y^2)
E_3	2	-2	2	-2	0	0	0	0	0	a	
E_4	2	-1	-1	2	0	-1	2	-1	0	a	
E_5	2	1	-1	-2	0	$\sqrt{3}$	0	$-\sqrt{3}$	0	a	$(R_x, R_y), (yz, zx)$
$E_{1/2}$	2	$\sqrt{3}$	1	0	0	$2c_{12}^1$	$\sqrt{2}$	$2c_{12}^5$	0	c	
$E_{3/2}$	2	0	-2	0	0	$\sqrt{2}$	$-\sqrt{2}$	$-\sqrt{2}$	0	c	
$E_{5/2}$	2	$-\sqrt{3}$	1	0	0	$2c_{12}^5$	$-\sqrt{2}$	$2c_{12}^1$	0	c	
$E_{7/2}$	2	$\sqrt{3}$	1	0	0	$-2c_{12}^5$	$\sqrt{2}$	$-2c_{12}^1$	0	c	
$E_{9/2}$	2	0	-2	0	0	$-\sqrt{2}$	$\sqrt{2}$	$\sqrt{2}$	0	c	
$E_{11/2}$	2	$\sqrt{3}$	1	0	0	$-2c_{12}^1$	$-\sqrt{2}$	$-2c_{12}^5$	0	c	

$c_n^m = \cos(m\pi/n)$.

A3.8 The cubic groups T, T$_h$, T$_d$, O, O$_h$

23 T

T	E	$3C_2$	$4C_3^+$	$4C_3^-$	TR	Bases
A	1	1	1	1	a	$x^2+y^2+z^2$
^1E	1	1	ε^*	ε	b	$\}(x^2-y^2, 3z^2-r^2)$
^2E	1	1	ε	ε^*	b	
T	3	-1	0	0	a	$(x, y, z), (R_x, R_y, R_z), (xy, yz, zx)$
$E_{1/2}$	2	0	1	1	c	
$^1F_{3/2}$	2	0	ε^*	ε	b	
$^2F_{3/2}$	2	0	ε	ε^*	b	

$\varepsilon = \exp(-i2\pi/3)$.

$m3$ $\mathrm{T_h}$

$\mathrm{T_h}$	E	$3C_2$	$4C_3^+$	$4C_3^-$	I	3σ	$4S_6^-$	$4S_6^+$	TR	Bases
A_g	1	1	1	1	1	1	1	1	a	$x^2+y^2+z^2$
$^1\mathrm{E}_g$	1	1	ε^*	ε	1	1	ε^*	ε	b	$\left.\right\}$ $(x^2-y^2, 3z^2-r^2)$
$^2\mathrm{E}_g$	1	1	ε	ε^*	1	1	ε	ε^*	b	
T_g	3	−1	0	0	3	−1	0	0	a	$(R_x, R_y, R_z), (xy, yz, zx)$
A_u	1	1	1	1	−1	−1	−1	−1	a	xyz
$^1\mathrm{E}_u$	1	1	ε^*	ε	−1	−1	$-\varepsilon^*$	$-\varepsilon$	b	
$^2\mathrm{E}_u$	1	1	ε	ε^*	−1	−1	$-\varepsilon$	$-\varepsilon^*$	b	
T_u	3	−1	0	0	−3	1	0	0	a	(x, y, z)
$\mathrm{E}_{1/2,\,g}$	2	0	1	1	2	0	1	1	c	
$^1\mathrm{F}_{3/2,\,g}$	2	0	ε^*	ε	2	0	ε^*	ε	b	
$^2\mathrm{F}_{3/2,\,g}$	2	0	ε	ε^*	2	0	ε	ε^*	b	
$\mathrm{E}_{1/2,\,u}$	2	0	1	1	−2	0	−1	−1	c	
$^1\mathrm{F}_{3/2,\,u}$	2	0	ε^*	ε	−2	0	$-\varepsilon^*$	$-\varepsilon$	b	
$^2\mathrm{F}_{3/2,\,u}$	2	0	ε	ε^*	−2	0	$-\varepsilon$	$-\varepsilon^*$	b	

$\varepsilon = \exp(-\mathrm{i}2\pi/3)$.

$\bar{4}3m$ $\mathrm{T_d}$

$\mathrm{T_d}$	E	$3C_2$	$8C_3$	$6S_4$	$6\sigma_\mathrm{d}$	TR	Bases
A_1	1	1	1	1	1	a	$x^2+y^2+z^2$
A_2	1	1	1	−1	−1	a	
E	2	2	−1	0	0	a	$(x^2-y^2, 3z^2-r^2)$
T_1	3	−1	0	1	−1	a	(R_x, R_y, R_z)
T_2	3	−1	0	−1	1	a	$(x, y, z), (xy, yz, zx)$
$\mathrm{E}_{1/2}$	2	0	1	$\sqrt{2}$	0	c	
$\mathrm{E}_{5/2}$	2	0	1	$-\sqrt{2}$	0	c	
$\mathrm{F}_{3/2}$	4	0	−1	0	0	c	

432 O

O	E	$3C_2$	$8C_3$	$6C_4$	$6C_2'$	TR	Bases
A_1	1	1	1	1	1	a	$x^2+y^2+y^2$
A_2	1	1	1	−1	−1	a	
E	2	2	−1	0	0	a	$(x^2-y^2, 3z^2-r^2)$
T_1	3	−1	0	1	−1	a	$(x, y, z), (R_x, R_y, R_z)$
T_2	3	−1	0	−1	1	a	(xy, yz, zx)
$\mathrm{E}_{1/2}$	2	0	1	$\sqrt{2}$	0	c	
$\mathrm{E}_{5/2}$	2	0	1	$-\sqrt{2}$	0	c	
$\mathrm{F}_{3/2}$	4	0	−1	0	0	c	

432 O_h

O_h	E	$3C_2$	$8C_3$	$6C_4$	$6C_2'$	I	$3\sigma_h$	$8S_6$	$6S_4$	$6\sigma_d$	TR	Bases
A_{1g}	1	1	1	1	1	1	1	1	1	1	a	$x^2+y^2+z^2$
A_{2g}	1	1	1	−1	−1	1	1	1	−1	−1	a	
E_g	2	2	−1	0	0	2	2	−1	0	0	a	$(x^2-y^2, 3z^2-r^2)$
T_{1g}	3	−1	0	1	−1	3	−1	0	1	−1	a	(R_x, R_y, R_z)
T_{2g}	3	−1	0	−1	1	3	−1	0	−1	1	a	(xy, yz, zx)
A_{1u}	1	1	1	1	1	−1	−1	−1	−1	−1	a	
A_{2u}	1	1	1	−1	−1	−1	−1	−1	1	1	a	
E_u	2	2	−1	0	0	−2	−2	1	0	0	a	
T_{1u}	3	−1	0	1	−1	−3	1	0	−1	1	a	(x, y, z)
T_{2u}	3	−1	0	−1	1	−3	1	0	1	−1	a	
$E_{1/2,\,g}$	2	0	1	$\sqrt{2}$	0	2	0	1	$\sqrt{2}$	0	c	
$E_{5/2,\,g}$	2	0	1	$-\sqrt{2}$	0	2	0	1	$-\sqrt{2}$	0	c	
$F_{3/2,\,g}$	4	0	−1	0	0	4	0	−1	0	0	c	
$E_{1/2,\,u}$	2	0	1	$\sqrt{2}$	0	−2	0	−1	$-\sqrt{2}$	0	c	
$E_{5/2,\,u}$	2	0	1	$-\sqrt{2}$	0	−2	0	−1	$\sqrt{2}$	0	c	
$F_{3/2,\,u}$	4	0	−1	0	0	−4	0	1	0	0	c	

A3.9 The icosahedral groups Y, Y_h

53 Y

Y	E	$12C_5$	$12C_5^2$	$20C_3$	$15C_2$	TR	Bases
A	1	1	1	1	1	a	$x^2+y^2+z^2$
T_1	3	$2c_5^1$	$2c_5^3$	0	−1	a	$(x, y, z), (R_x, R_y, R_z)$
T_2	3	$2c_5^3$	$2c_5^1$	0	−1	a	
F	4	−1	−1	1	0	a	
H	5	0	0	−1	1	a	$(x^2-y^2, 3z^2-r^2, xy, yz, zx)$
$E_{1/2}$	2	$2c_5^1$	$2c_5^2$	1	0	c	
$E_{7/2}$	2	$2c_5^3$	$2c_5^4$	1	0	c	
$F_{3/2}$	4	1	−1	−1	0	c	
$I_{5/2}$	6	−1	1	0	0	c	

$c_n^m = \cos(m\pi/n)$.

$53m$ Y_h

Y_h	E	$12C_5$	$12C_5^2$	$20C_3$	$15C_2$	I	$12S_{10}^3$	$12S_{10}$	$20S_6$	15σ	TR	Bases
A_g	1	1	1	1	1	1	1	1	1	1	a	$x^2+y^2+z^2$
T_{1g}	3	$2c_5^1$	$2c_5^3$	0	−1	3	$2c_5^1$	$2c_5^3$	0	−1	a	(R_x, R_y, R_z)
T_{2g}	3	$2c_5^3$	$2c_5^1$	0	−1	3	$2c_5^3$	$2c_5^1$	0	−1	a	
F_g	4	−1	−1	1	0	4	−1	−1	1	0	a	
H_g	5	0	0	−1	1	5	0	0	−1	1	a	five d orbitals[a]
A_u	1	1	1	1	1	−1	−1	−1	−1	−1	a	
T_{1u}	3	$2c_5^1$	$2c_5^3$	0	−1	−3	$-2c_5^1$	$-2c_5^3$	0	1	a	(x, y, z)

Table 53m (*cont.*)

Y_h	E	$12C_5$	$12C_5^2$	$20C_3$	$15C_2$	I	$12S_{10}^3$	$12S_{10}$	$20S_6$	15σ	TR	Bases
T_{2u}	3	$2c_5^3$	$2c_5^1$	0	−1	−3	$-2c_5^3$	$-2c_5^1$	0	1	*a*	
F_u	4	−1	−1	1	0	−4	1	1	−1	0	*a*	
H_u	5	0	0	−1	1	−5	0	0	1	−1	*a*	
$E_{1/2,\,g}$	2	$2c_5^1$	$2c_5^2$	1	0	2	$2c_5^1$	$2c_5^2$	1	0	*c*	
$E_{7/2,\,g}$	2	$2c_5^3$	$2c_5^4$	1	0	2	$2c_5^3$	$2c_5^4$	1	0	*c*	
$F_{3/2,\,g}$	4	1	−1	−1	0	4	1	−1	−1	0	*c*	
$I_{5/2,\,g}$	6	−1	1	0	0	6	−1	1	0	0	*c*	
$E_{1/2,\,u}$	2	$2c_5^1$	$2c_5^2$	1	0	−2	$-2c_5^1$	$-2c_5^2$	−1	0	*c*	
$E_{7/2,\,u}$	2	$2c_5^3$	$2c_5^4$	1	0	−2	$-2c_5^3$	$-2c_5^4$	−1	0	*c*	
$F_{3/2,\,u}$	4	1	−1	−1	0	−4	−1	1	1	0	*c*	
$I_{5/2,\,u}$	6	−1	1	0	0	−6	1	−1	0	0	*c*	

[a] Five d orbitals $= (x^2 - y^2,\ 3z^2 - r^2,\ xy,\ yz,\ zx)$. $c_n^m = \cos(m\pi/n)$.

A4 Correlation tables

The following tables show how irreducible vector representations of point groups are re-labeled or reduced when the symmetry of the point group is lowered. The tables are in the reverse order to that given at the beginning of Appendix A3. For groups with pairs of complex conjugate representations, E means the direct sum ${}^1E \oplus {}^2E$, and similarly for E_g and E_u.

O_h	O	T_d	T_h	D_{4h}	D_{3d}
A_{1g}	A_1	A_1	A_g	A_{1g}	A_{1g}
A_{2g}	A_2	A_2	A_g	B_{1g}	A_{2g}
E_g	E	E	E_g	$A_{1g} \oplus B_{1g}$	E_g
T_{1g}	T_1	T_1	T_g	$A_{2g} \oplus E_g$	$A_{2g} \oplus E_g$
T_{2g}	T_2	T_2	T_g	$B_{2g} \oplus E_g$	$A_{1g} \oplus E_g$
A_{1u}	A_1	A_2	A_u	A_{1u}	A_{1u}
A_{2u}	A_2	A_1	A_u	B_{1u}	A_{2u}
E_u	E	E	E_u	$A_{1u} \oplus B_{1u}$	E_u
T_{1u}	T_1	T_2	T_u	$A_{2u} \oplus E_u$	$A_{2u} \oplus E_u$
T_{2u}	T_2	T_1	T_u	$B_{2u} \oplus E_u$	$A_{1u} \oplus E_u$

O	T	D_4	D_3
A_1	A	A_1	A_1
A_2	A	B_1	A_2
E	E	$A_1 \oplus B_1$	E
T_1	T	$A_2 \oplus E$	$A_2 \oplus E$
T_2	T	$B_2 \oplus E$	$A_1 \oplus E$

T_d	T	D_{2d}	C_{3v}	S_4
A_1	A	A_1	A_1	A
A_2	A	B_1	A_2	B
E	E	$A_1 \oplus B_1$	E	$A \oplus B$
T_1	T	$A_2 \oplus E$	$A_2 \oplus E$	$A \oplus E$
T_2	T	$B_2 \oplus E$	$A_1 \oplus E$	$B \oplus E$

T_h	T	D_{2h}	S_6
A_g	A	A_g	A_g
E_g	E	$2A_g$	E_g
T_g	T	$B_{1g} \oplus B_{2g} \oplus B_{3g}$	$A_g \oplus E_g$
A_u	A	A_u	A_u
E_u	E	$2A_u$	E_u
T_u	T	$B_{1u} \oplus B_{2u} \oplus B_{3u}$	$A_u \oplus E_u$

T	D_2	C_3
A	A	A
E	$2A$	E
T	$B_1 \oplus B_2 \oplus B_3$	$A \oplus E$

D_{6d}	D_6	C_{6v}	D_{2d}
A_1	A_1	A_1	A_1
A_2	A_2	A_2	A_2
B_1	A_1	A_2	B_1
B_2	A_2	A_1	B_2
E_1	E_1	E_1	E
E_2	E_2	E_2	$B_1 \oplus B_2$
E_3	$B_1 \oplus B_2$	$B_1 \oplus B_2$	E
E_4	E_2	E_2	$A_1 \oplus A_2$
E_5	E_1	E_1	E

D_{5d}	D_5	C_{5v}
A_{1g}	A_1	A_1
A_{2g}	A_2	A_2
E_{1g}	E_1	E_1
E_{2g}	E_2	E_2
A_{1u}	A_1	A_2
A_{2u}	A_2	A_1
E_{1u}	E_1	E_1
E_{2u}	E_2	E_2

D_{4d}	D_4	C_{4v}	S_8
A_1	A_1	A_1	A
A_2	A_2	A_2	A
B_1	A_1	A_2	B
B_2	A_2	A_1	B
E_1	E	E	E_1
E_2	$B_1 \oplus B_2$	$B_1 \oplus B_2$	E_2
E_3	E	E	E_3

D_{3d}	D_3	C_{3v}	S_6	C_3	C_{2h}
A_{1g}	A_1	A_1	A_g	A	A_g
A_{2g}	A_2	A_2	A_g	A	B_g
E_g	E	E	E_g	E	$A_g \oplus B_g$
A_{1u}	A_1	A_2	A_u	A	A_u
A_{2u}	A_2	A_1	A_u	A	B_u
E_u	E	E	E_u	E	$A_u \oplus B_u$

D_{2d}	S_4	D_2	C_{2v}
A_1	A	A	A_1
A_2	A	B_1	A_2
B_1	B	A	A_2
B_2	B	B_1	A_1
E	E	$B_2 \oplus B_3$	$B_1 \oplus B_2$

D_{6h}	D_6	$D_{3h}{}^a$	$D_{3h}{}^b$	C_{6h}	C_{6v}	$D_{3d}{}^b$	$D_{3d}{}^a$	D_{2h}	$C_{2v}{}^c$
A_{1g}	A_1	A_1'	A_1'	A_g	A_1	A_{1g}	A_{1g}	A_g	A_1
A_{2g}	A_2	A_2'	A_2'	A_g	A_2	A_{2g}	A_{2g}	B_{1g}	A_2
B_{1g}	B_1	A_1''	A_2''	B_g	B_2	A_{2g}	A_{1g}	B_{3g}	B_1
B_{2g}	B_2	A_2''	A_1''	B_g	B_1	A_{1g}	A_{2g}	B_{2g}	B_2
E_{1g}	E_1	E''	E''	E_{1g}	E_1	E_g	E_g	$B_{2g} \oplus B_{3g}$	$B_1 \oplus B_2$
E_{2g}	E_2	E'	E'	E_{2g}	E_2	E_g	E_g	$A_g \oplus B_{1g}$	$A_1 \oplus A_2$
A_{1u}	A_1	A_1''	A_1''	A_u	A_2	A_{1u}	A_{1u}	A_u	A_2
A_{2u}	A_2	A_2''	A_2''	A_u	A_1	A_{2u}	A_{2u}	B_{1u}	A_1
B_{1u}	B_1	A_1'	A_2'	B_u	B_1	A_{2u}	A_{1u}	B_{3u}	B_2
B_{2u}	B_2	A_2'	A_1'	B_u	B_2	A_{1u}	A_{2u}	B_{2u}	B_1
E_{1u}	E_1	E'	E'	E_{1u}	E_1	E_u	E_u	$B_{2u} \oplus B_{3u}$	$B_1 \oplus B_2$
E_{2u}	E_2	E''	E''	E_{2u}	E_2	E_u	E_u	$A_u \oplus B_{1u}$	$A_1 \oplus A_2$

[a] C_2'; [b] C_2''; [c] $\sigma_v \rightarrow \sigma_x$.

D_{5h}	D_5	C_{5v}	C_{5h}	C_5	$C_{2v}{}^a$
A_1'	A_1	A_1	A'	A	A_1
A_2'	A_2	A_2	A'	A	B_1
E_1'	E_1	E_1	E_1'	E_1	$A_1 \oplus B_1$
E_2'	E_2	E_2	E_2'	E_2	$A_1 \oplus B_1$
A_1''	A_1	A_2	A''	A	A_2
A_2''	A_2	A_1	A''	A	B_2
E_1''	E_1	E_1	E_1''	E_1	$A_2 \oplus B_2$
E_2''	E_2	E_2	E_2''	E_2	$A_2 \oplus B_2$

[a] $\sigma_h \rightarrow \sigma_y$.

D_{4h}	C_{4h}	$D_{2h}{}^{b}$	$D_{2h}{}^{c}$	$D_{2d}{}^{b}$	$D_{2d}{}^{c}$	$C_{2h}{}^{a}$	$C_{2h}{}^{b}$	$C_{2h}{}^{c}$	C_{4v}
A_{1g}	A_g	A_g	A_g	A_1	A_1	A_g	A_g	A_g	A_1
A_{2g}	A_g	B_{1g}	B_{1g}	A_2	A_2	A_g	B_g	B_g	A_2
B_{1g}	B_g	A_g	B_{1g}	B_1	B_2	A_g	A_g	B_g	B_1
B_{2g}	B_g	A_g	B_{1g}	B_2	B_1	A_g	B_g	A_g	B_2
E_g	${}^1E_g \oplus {}^2E_g$	$B_{2g} \oplus B_{3g}$	$B_{2g} \oplus B_{3g}$	E	E	$2B_g$	$A_g \oplus B_g$	$A_g \oplus B_g$	E
A_{1u}	A_u	A_u	A_u	B_1	B_1	A_u	A_u	A_u	A_2
A_{2u}	A_u	B_{1u}	B_{1u}	B_2	B_2	A_u	B_u	B_u	A_1
B_{1u}	B_u	A_u	B_{1u}	A_1	A_2	A_u	A_u	B_u	B_2
B_{2u}	B_u	B_{1u}	A_{1u}	A_1	A_2	A_u	B_u	A_u	B_1
E_u	${}^1E_u \oplus {}^2E_u$	$B_{2u} \oplus B_{3u}$	$B_{2u} \oplus B_{3u}$	E	E	$2B_u$	$A_u \oplus B_u$	$A_u \oplus B_u$	E

${}^a C_2$; ${}^b C_2'$; ${}^c C_2''$.

D_{3h}	D_3	C_{3h}	C_{3v}	C_{2v}	$C_s{}^{a}$	$C_s{}^{b}$	C_3
A_1'	A_1	A'	A_1	A_1	A'	A'	A
A_2'	A_2	A'	A_2	B_1	A'	A''	A
E'	E	${}^1E' \oplus {}^2E'$	E	$A_1 \oplus B_1$	$2A'$	$A' \oplus A''$	${}^1E \oplus {}^2E$
A_1''	A_1	A''	A_2	A_2	A''	A''	A
A_2''	A_2	A''	A_1	B_2	A''	A'	A
E''	E	${}^1E'' \oplus {}^2E''$	E	$A_2 \oplus B_2$	$2A''$	$A' \oplus A''$	${}^1E \oplus {}^2E$

${}^a \sigma_h$; ${}^b \sigma_v$.

D_{2h}	D_2	$C_{2h}{}^{a}$	$C_{2h}{}^{b}$	$C_{2h}{}^{c}$	$C_{2v}{}^{a}$	$C_{2v}{}^{b}$	$C_{2v}{}^{c}$
A_g	A	A_g	A_g	A_g	A_1	A_1	A_1
B_{1g}	B_1	A_g	B_g	B_g	A_2	B_1	B_2
B_{2g}	B_2	B_g	B_g	A_g	B_1	B_2	A_2
B_{3g}	B_3	B_g	A_g	B_g	B_2	A_2	B_1
A_u	A	A_u	A_u	A_u	A_2	A_2	A_2
B_{1u}	B_1	A_u	B_u	B_u	A_1	B_2	B_1
B_{2u}	B_2	B_u	B_u	A_u	B_2	B_1	A_1
B_{3u}	B_3	B_u	A_u	B_u	B_1	A_1	B_2

${}^a C_{2\mathbf{z}}$; ${}^b C_{2\mathbf{x}}$; ${}^c C_{2\mathbf{y}}$.

C_{6v}	C_6	$C_{3v}{}^{a}$	$C_{3v}{}^{b}$	$C_{2v}{}^{c}$
A_1	A	A_1	A_1	A_1
A_2	A	A_2	A_2	A_2
B_1	B	A_1	A_2	B_1
B_2	B	A_2	A_1	B_2
E_1	E_1	E	E	$B_1 \oplus B_2$
E_2	E_2	E	E	$A_1 \oplus A_2$

${}^a \sigma_v$; ${}^b \sigma_d$; ${}^c \sigma_v \rightarrow \sigma_{\mathbf{y}}$.

C_{5v}	C_5	C_s
A_1	A	A'
A_2	A	A''
E_1	E_1	$A' \oplus A''$
E_2	E_2	$A' \oplus A''$

C_{4v}	C_4	C_{2v} [a]	C_{2v} [b]
A_1	A	A_1	A_1
A_2	A	A_2	A_2
B_1	B	A_1	A_2
B_2	B	A_2	A_1
E	E	$B_1 \oplus B_2$	$B_1 \oplus B_2$

[a] σ_v; [b] σ_d.

C_{3v}	C_3	C_s
A_1	A	A'
A_2	A	A''
E	E	$A' \oplus A''$

C_{2v}	C_2	C_s [a]	C_s [b]
A_1	A	A'	A'
A_2	A	A''	A''
B_1	B	A'	A''
B_2	B	A''	A'

[a] $\sigma_{\mathbf{y}}$; [b] $\sigma_{\mathbf{x}}$.

C_{6h}	C_6	C_{3h}	S_6	C_{2h}
A_g	A	A'	A_g	A_g
B_g	B	A''	A_g	B_g
E_{1g}	E_1	E''	E_g	$2B_g$
E_{2g}	E_2	E'	E_g	$2A_g$
A_u	A	A''	A_u	A_u
B_u	B	A'	A_u	B_u
E_{1u}	E_1	E'	E_u	$2B_u$
E_{2u}	E_2	E''	E_u	$2A_u$

C_{5h}	C_5	C_s
A'	A	A'
E'_1	E_1	$2A'$
E'_2	E_2	$2A'$
A''	A	A''
E''_1	E_1	$2A''$
E''_2	E_2	$2A''$

C_{4h}	C_4	S_4	C_{2h}
A_g	A	A	A_g
B_g	B	B	A_g
E_g	E	E	$2B_g$
A_u	A	B	A_u
B_u	B	A	A_u
E_u	E	E	$2B_u$

C_{3h}	C_3	C_s
A'	A	A'
E'	E	$2A'$
A''	A	A''
E''	E	$2A''$

C_{2h}	C_2	C_s	C_i
A_g	A	A'	A_g
B_g	B	A''	A_g
A_u	A	A''	A_u
B_u	B	A'	A_u

D_6	D_3[a]	D_3[b]	D_2
A_1	A_1	A_1	A
A_2	A_2	A_2	B_1
B_1	A_1	A_2	B_3
B_2	A_2	A_1	B_2
E_1	E	E	$B_2 \oplus B_3$
E_2	E	E	$A \oplus B_1$

[a] C'_2; [b] C''_2.

D_4	C_4	D_2[a]	D_2[b]	C_2[c]	C_2[d]	C_2[e]
A_1	A	A	A	A	A	A
A_2	A	B_1	B_1	A	B	B
B_1	B	A	B_1	A	A	B
B_2	B	B_1	A	A	B	A
E	${}^1E \oplus {}^2E$	$B_2 \oplus B_3$	$B_2 \oplus B_3$	2B	$A \oplus B$	$A \oplus B$

[a] C_2'; [b] C_2''; [c] C_2; [d] C_2'; [e] C_2''.

D_3	C_3	C_2
A_1	A	A
A_2	A	B
E	E	$A \oplus B$

D_2	C_2[a]	C_2[b]	C_2[c]
A	A	A	A
B_1	A	B	B
B_2	B	B	A
B_3	B	A	B

[a] $C_{2\mathbf{z}}$; [b] $C_{2\mathbf{x}}$; [c] $C_{2\mathbf{y}}$.

S_8	C_4
A	A
B	A
E_1	E
E_2	2B
E_3	E

S_6	C_3	C_i
A_g	A	A_g
E_g	E	$2A_g$
A_u	A	A_u
E_u	E	$2A_u$

S_4	C_2
A	A
B	A
E	2B

C_6	C_3	C_2
A	A	A
B	A	B
E_1	E	2B
E_2	E	2A

C_4	C_2
A	A
B	A
E	2B

References

Allnatt, A. R. and Lidiard, A. B. (1993) *Atomic Transport in Solids*. Cambridge: Cambridge University Press.

Altmann, S. L. (1977) *Induced Representations in Crystals and Molecules*. London: Academic Press.

(1979) Double groups and projective representations. I General theory. *Mol. Phys.* **38**, 489–511.

(1986) *Rotations, Quaternions and Double Groups*. Oxford: Clarendon Press.

(1991) *Band Theory of Solids: An Introduction from the Point of View of Symmetry*. Oxford: Oxford University Press.

Altmann, S. L. and Herzig, P. (1982) Double groups and projective representations. III Improper groups. *Mol. Phys.* **45**, 585–604.

(1994) *Point Group Theory Tables*. Oxford: Clarendon Press.

Atkins, P. W. (1983) *Molecular Quantum Mechanics*, 2nd edn. Oxford: Oxford University Press.

Atkins, P. W., Child, M. S., and Phillips, C. S. G. (1970) *Tables for Group Theory*. Oxford: Oxford University Press.

Ballhausen, C. J. and Moffitt, W. (1956) On the dichroism of certain Co(III) complexes, *J. Inorg. Nucl. Chem.* **3**, 178–81.

Bethe, H. A. (1929) Term splitting in crystals. *Ann. Phys.* **3**, 133–208.

(1964) *Intermediate Quantum Mechanics*. New York: Benjamin.

Bhagavantam, S. (1966) *Crystal Symmetry and Physical Properties*. London: Academic Press.

Biedenharn, L. C. and Louck, J. D. (1981) *Angular Momentum in Quantum Physics. Theory and Application*. Reading, MA: Addison-Wesley.

Bilz, H. and Kress, W. (1979) *Phonon Dispersion Relations in Insulators*. Berlin: Springer.

Birman, J. L. (1962) Space group selection rules, diamond and zinc blende. *Phys. Rev.* **127**, 1093–106.

(1963) Theory of infra-red and Raman processes in crystals: selection rules in diamond and zinc blende. *Phys. Rev.* **131**, 1489–96.

Born, M. and Huang, K. (1954) *Dynamical Theory of Crystal Lattices*. Oxford: Clarendon Press.

Bouckaert, L. P., Smoluchowski, R. and Wigner E. (1936) Theory of Brillouin zones and symmetry properties of wave functions in crystals. *Phys. Rev.* **50**, 58–67.

Bradley, C. J. (1966) Space groups and selection rules. *J. Math. Phys.* **7**, 1145–52.

Bradley, C. J. and Cracknell, A. P. (1972) *The Mathematical Theory of Symmetry in Solids*. Oxford: Oxford University Press.

Bradley, C. J. and Davies, B. L. (1968) Magnetic groups and their co-representations. *Rev. Mod. Phys.* **40**, 359–79.

Burns, G. and Glazer, A. M. (1978) *Space Groups for Solid State Scientists*. New York: Academic Press.

Callen, H. B. (1960) *Thermodynamics*. New York: John Wiley and Sons.

Casimir, H. B. G. (1945) On the Onsager principle of microscopic reversibility. *Rev. Mod. Phys.* **17**, 343–50.

Catlow, C. R. A., Diller K. M., and Norgett, M. J. (1977) Interionic potentials for alkali halides. *J. Phys. C* **10**, 1395–411.

Cochran, W. (1965) In *Proceedings of International Conference on Lattice Dynamics, Copenhagen 1963*, pp. 75–84. Oxford: Pergamon.

Condon, E. U. and Shortley, G. H. (1967) *The Theory of Atomic Spectra*. Cambridge: Cambridge University Press.

Cornwell, J. F. (1984) *Group Theory in Physics*, vols I, II. London: Academic Press.

Cracknell, A. P. (1968) Crystal field theory and the Shubnikov point groups. *Adv. Phys.* **17**, 367–420.

(1975) *Group Theory in Solid State Physics*. London: Taylor and Francis.

De Groot, S. R. and Mazur, P. (1962) *Non-equilibrium Thermodynamics*. Amsterdam: North Holland.

Dick, B. G. and Overhauser, A. W. (1958) Theory of the dielectric constant of alkali halide crystals. *Phys. Rev.* **112**, 90–118.

Elliot, R. J. (1954) Spin-orbit coupling in band theory – character tables for some "double" space groups. *Phys. Rev.* **96**, 280–7.

Evarestov, R. A. and Smirnov, V. P. (1997) *Site Symmetry in Crystals*, 2nd. edn. Berlin: Springer.

Fano, U. and Racah, G. (1959) *Irreducible Tensorial Sets*. New York: Academic Press.

Flurry, R. L. (1980) *Symmetry Groups*. Englewood Cliffs, NJ: Prentice-Hall.

Flygare, W. H. (1978) *Molecular Structure and Dynamics*. Englewood Cliffs, NJ: Prentice-Hall.

Gale, J. D. (1997) GULP: A computer program for the symmetry-adapted simulation of solids. *J. Chem. Soc., Faraday Trans.* **93**, 629–37.

Golding, R. M. and Carrington, A. (1962) Electonic structure and spectra of octacyanide complexes. *Mol. Phys.* **5**, 377–85.

Griffith, J. S. (1964) *The Theory of Transition-Metal Ions*. Cambridge: Cambridge University Press.

Grimes, R. W., Harker, A. H., and Lidiard, A. B. (eds.) (1996) Interatomic potentials. *Phil. Mag.* B **73**, 3–19.

Hahn, T. (1983) *International Tables for Crystallography*. Dordrecht: Reidel.

Hall, M. (1959) *The Theory of Groups*. New York: Macmillan.

Hammermesh, M. (1962) *Group Theory and Its Applications to Physical Problems*. Reading, MA: Addison-Wesley.

Harris, D. C. and Bertolucci, M. D. (1978) *Symmetry and Spectroscopy*. New York: Oxford University Press.

Herring, C. (1942) Character tables for two space groups. *J. Franklin Inst.* **233**, 525–43.

Hurley, A. C. (1966) Ray representations of point groups and the irreducible representations of space groups and double space groups. *Phil. Trans. Roy. Soc. (London)* **A260**, 1–36.

Jackson, R. A., Meenan P., Price G. D., Roberts, K. J., Telfer, G. B., and Wilde, P. J. (1995) Potential fitting to molecular ionic materials. *Mineralogical Mag.* **59**, 617–22.

Jansen, L. and Boon, M. (1967) *Theory of Finite Groups. Applications in Physics*. Amsterdam: North-Holland.

Johnston, D. F. (1960) Group theory in solid state physics. *Rep. Prog. Phys.* **23**, 55–153.

Jones, H. (1962) *The Theory of Brillouin Zones and Electronic States in Crystals*. Amsterdam: North-Holland.

(1975) *The Theory of Brillouin Zones and Electronic States in Crystals*, 2nd edn. Reading, MA: Addison-Wesley.

Jones, H. F. (1990) *Groups Representations and Physics*. Bristol: Adam Hilger. p. 394

Karavaev, G. F. (1965) Selection rules for indirect transitions in crystals. *Sov. Phys. Solid-State* **6**, 2943–8.

Kim, S. K. (1999) *Group Theoretical Methods*. Cambridge: Cambridge University Press.

Koster, G. F., Dimmock, J. O., Wheeler, R. G., and Statz, H. (1963) *Properties of the Thirty-two Point Groups*. Cambridge, MA: M.I.T. Press.

Kovalev, O. V. (1993) *Representations of the Crystallographic Space Groups: Irreducible Representations, Induced Representations and Corepresentations*, 2nd edn. Philadelphia, PA: Gordon and Breach.

Kramers, H. A. (1930) General theory of paramagnetic rotation in crystals. *Koninkl. Ned. Akad. Wetenschap, Proc.* **33**, 959–72.

Landsberg, P. T. (ed.) (1969) *Solid State Theory*. London: Wiley Interscience.

Lax, M. (1962) Influence of time reversal on selection rules connecting different points in the Brillouin zone. *Proc. Intl Conf. on Physics of Semicondutors*. London: Institute of Physics and Physical Society.

(1965) Sub-group techniques in crystal and molecular physics. *Phys. Rev.* **138**, A793–802.

(1974) *Symmetry Principles in Solid State and Molecular Physics*. New York: John Wiley and Sons.

Liebfried, G. (1955). In *Handbuch der Physik* vol. VII/I, pp. 104–324. Berlin: Springer.

Littlewood, D. E. (1958) *A University Algebra*. London: Heinemann.

Lyubarskii, G. Ya. (1960) *The Application of Group Theory in Physics*. Oxford: Pergamon Press.

Maradudin, A. A. and Vosko, S. H. (1968) Symmetry properties of normal vibrations in a crystal. *Rev. Mod. Phys.* **40**, 1–37.

Maradudin, A. A., Montroll, E. W., Weiss, G. H., and Ipatova, I. P. (1971) *Theory of Lattice Dynamics in the Harmonic Approximation*, 2nd edn. New York: Academic Press.

Margenau, H. and Murphy, G. M. (1943) *The Mathematics of Physics and Chemistry*. New York: D.Van Nostrand.

McWeeny, R. (1963) *Symmetry: An Introduction to Group Theory and Its Applications*. Oxford: Pergamon Press.

Miller, S. C. and Love, W. F. (1967) *Tables of Irreducible Representations of Space Groups and Co-representations of Magnetic Space Groups*. Boulder, CO: Pruett.

Morgan, D. J. (1969) Group theory and electronic states in perfect crystals. In Landsberg, P. T. (ed.) *Solid State Theory*, chaps. IV and V. London: Wiley Interscience.

Nowick, A. S. (1995) *Crystal Properties via Group Theory*. Cambridge: Cambridge University Press.

Nowick, A. S. and Heller, W. R. (1965) Dielectric and anelastic relaxation of crystals containing point defects. *Adv. Phys.* **14**, 101–66.

Nye, J. F. (1957) *Physical Properties of Crystals*. Oxford: Oxford University Press.

Onodera, Y., and Okazaki, M. (1966) Tables of basis functions for double groups. *J. Phys. Soc. Japan* **21**, 2400–8.

Onsager, L. (1931a) Reciprocal relations in irreversible processes Part I. *Phys. Rev.* **37**, 405–26.
(1931b) Reciprocal relations in irreversible processes Part II. *Phys. Rev.* **38**, 2265–79.

Purcell, K. F. and Kotz, J. C. (1980) *Inorganic Chemistry*. Saunders, Philadelphia, PA: Saunders.

Rose, M. E. (1957) *Elementary Theory of Angular Momentum* (Wiley, New York).

Sangster, M. J. L. and Attwood, R. M. (1978) Interionic potentials for alkali halides. *J. Phys. C* **11**, 1541–54.

Shockley, W. (1937) The empty lattice test of the cellular method in solids. *Phys. Rev.* **52**, 866–72.

Shubnikov, A. V. and Belov, N. V. (1964) *Colored Symmetry*. Oxford: Pergamon Press.

Stokes, H. T. and Hatch, D. M. (1988) *Isotropy Subgroups of the 230 Crystallographic Space Groups*. Singapore: World Scientific.

Sugano, S., Tanabe, Y., and Kamimura, H. (1970) *Multiplets of Transition-Metal Ions in Crystals*. London: Academic Press.

Tanabe, Y. and Sugano, S. (1954) The absorption spectra of complex ions I. *J. Phys. Soc. Japan* **9**, 753–766.

Tinkham, M. (1964) *Group Theory and Quantum Mechanics*. New York: McGraw-Hill.

Venkataraman, G., Feldkamp, L. A. and Sahni, V. C. (1975) *Dynamics of Perfect Crystals*. Cambridge, MA: MIT Press.

Weber, W. (1977) Adiabatic bond charge model for the phonons in diamond. *Phys. Rev. B* **15**, 4789–803.

Whittaker, E. T. and Watson, G. N. (1927) *A Course of Modern Analysis*. Cambridge: Cambridge University Press.

Wigner, E. P. (1959) *Group Theory and Its Application to the Quantum Mechanics of Atomic Spectra*. New York: Academic Press.

Wooster, W. A. (1973) *Tensors and Group Theory for the Physical Properties of Crystals*. Oxford: Clarendon Press.

Worlton, T. G. and Warren, J. L. (1972) Group theoretical analysis of lattice vibrations. *Comp. Phys. Commun.* **3**, 88–117.

Yamada, S. and Tsuchida R. (1952) Spectrochemical study of microscopic crystals. II. Dichroism of praseocobaltic salts. *Bull. Chem. Soc. Japan* **25**, 127–30.

Yamada, S., Nakahara, A., Shimura, Y., and Tsuchida, R. (1955) Spectrochemical study of microscopic crystals. VII. Absorption spectra of *trans*-dihalobis(ethylenediamine) cobalt(III) complexes. *Bull. Chem. Soc. Japan* **28**, 222–7.

Zak, J. (ed.) (1969) *The Irreducible Representations of Space Groups*. Elmsford, NY: Benjamin.

Index